The Living Soil

The Living Soil

Fundamentals of Soil Science and Soil Biology

Jean-Michel Gobat
Professor
University of Neuchâtel
Switzerland

Michel Aragno
Professor
University of Neuchâtel
Switzerland

Willy Matthey
Professor
University of Neuchâtel
Switzerland

Translated from French by
V.A.K. Sarma

Science Publishers, Inc.
Enfield (NH), USA   Plymouth, UK

CIP data will be provided on request.

SCIENCE PUBLISHERS, INC.
Post Office Box 699
Enfield, New Hampshire 03748
United States of America

Internet site: *http://www.scipub.net*

sales@scipub.net (marketing department)
editor@scipub.net (editorial department)
info@scipub.net (for all other enquiries)

ISBN 1-57808-212-9 (Hard cover)
ISBN 1-57808-210-2 (Papar back)

Published by arrangement with Presses polytechniques et universitaires romandes, Lausanne, Switzerland.

Translation of: ***Le sol vivant***, *Bases de pédologie Biologie des sols* (Second Edition), Presses polytechniques et universitaires romandes, Lausanne, Switzerland, 2003.
French edition: © Presses polytechniques et universitaires romandes, Lausanne, 2003.
ISBN 2-88074-501-2

Published by Science Publishers Inc., NH, USA
Printed in India.

To the reader—student or specialist—who, while accompanying the authors on their interdisciplinary journey, will learn to share their enthusiasm, in his/her turn contributing to better understanding of and, thereby, protection of a living soil!

FOREWORD

It is accepted that soil science was born in 1883, with the thesis of V.V. Dokouchaev devoted to the Russian Chernozem. Just prior to this, in 1881, Charles Darwin had published his last work *The Formation of Vegetable Mould through the Action of Worms*. These pioneers deserve the credit for showing that living animals play a considerable role in the formation, functioning and development of soils. Of course, the role of physicochemical factors such as climate or nature of the parent material should not be forgotten.

The soil, perhaps more than the canopy of tropical forests, represents one of the last great unknowns, so far as its functioning as well as the role and diversity of organisms living in it is concerned. Realization of this fact explains the large number of researches done nowadays in soil biology. Not only do findings relative to fundamental soil biology often lead to practical applications, in agronomy for example, but also to many other areas such as soil conservation or the fight against pollution.

Too often books that cover soil science are by and large restricted to one aspect of soil studies. They favour the physicochemical and more or less static study of the subject by describing horizons, specifying their structure and chemical composition, or they are even devoted solely to the study of the life in soils, particularly microflora and microfauna.

The advantage of this book is that it brings together these two aspects of the study of soils without introducing imbalance in favour of either point of view. Study of a milieu by restricting oneself to just one of its constituents is a method perhaps no longer used nowadays. A holistic, ecological point of view is necessary. This is what the three authors of this book have achieved by bringing together their abilities in botany, in microbiology and in zoology. The soil is considered an 'ecological system' with numerous functions. It is a support for living organisms, a reservoir of organic and mineral matter, the regulator of fluxes and exchanges in the biogeochemical cycles, the site of transformation of organic matter, and a purifying system. It is also the site for agricultural and forest production necessary for human life.

I shall mention here only some of the most interesting subjects appearing in this book. The characteristics and classification of soils are highlighted by particular attention to the special features important to life in the soil. In-depth study of the various humus forms enables demonstration that they can serve as indicators of soil development. Also to be found is a complete analysis of the relations between soil and vegetation, equally at a very fine scale, which is that of interactions between roots and microflora at the level of the rhizosphere, as at a much larger scale, which may be that of the phytocoenose or even that of the landscape. The great systematic diversity of soil animals and their ecological niches are emphasized as is the complexity of their roles, resulting in a detailed study of the food webs, particularly complex in the soil. The variety and role of symbioses form the subject of one chapter, given the importance of these structures in, for example, the nitrogen cycle.

The study of soils has made great progress due in part to the introduction of new methods, of which the authors give numerous examples. This is evidenced by the example of the chapter devoted to soil enzymes, whose fundamental role is often forgotten.

The various specific case studies, some of which are unpublished and attributable to the researches of the authors, render more tangible the theories outlined previously. Applied soil science is not forgotten either. Its importance is highlighted, for example, in the practice of composting, and the fight against pollution or erosion of soils. One will also appreciate the numerous reminders that could be useful to the reader approaching certain subjects for the first time.

The authors of this book have succeeded in presenting clearly and attractively the different aspects of soil biology, including the most recent and most complex. But they also remind us that much effort should be taken to fill in the lacunae in our knowledge of subjects such as the role, still poorly understood, of the soil in the carbon cycle or the study of the systematics and biology of soil animals, which represent a large reservoir of biodiversity.

Soil biology, fundamental and applied, is a discipline of the future. This book, constituting a remarkable synthesis, will be an excellent tool for teaching and a source of literature for teachers, and will certainly stimulate talents.

Roger Dajoz
Professor, National Museum of Natural
History, Paris

PREFACE

'One of the last great frontiers in biological and ecological research is the soil' (Coleman and Crossley, 1996).

Conception and Key to Reading This Book

Scientific works are broadly grouped into two categories, well portrayed by de Rosnay in his famous *Le Macroscope* published in 1975:

A 'linear' book...

- The 'linear' book, in which each chapter logically follows the preceding one. This type of book demands an ordered reading from the first page to the last. While revealing knowledge progressively it also confines the reader to the direction chosen by the author. This linear approach is as much the consequence of physical constraints due to bound and numbered pages as it is a certain philosophy of teaching and thought.

... or a 'network' book?

- The 'network' book, accessible through any of the chapters, none of them having precedence over the others. Here the reader chooses his/her own path, giving preference to areas considered the easiest or in which a special interest is felt. This systematic organization is that of computer media for transmission of information, CD-ROMs, the Internet and other relational databases. Starting from the first hub, one explores in all directions without much prerequisite needed and often by means of a simple click.

A linear' book *and* a 'network' book!

- This book is like the soil, with which it is concerned (see figure):

General conception of the book. Part One, linearly organized, provides the fundamentals of general soil science; Part Two, modular in nature, gives various themes of soil biology, linked together by the evolution of organic matter.

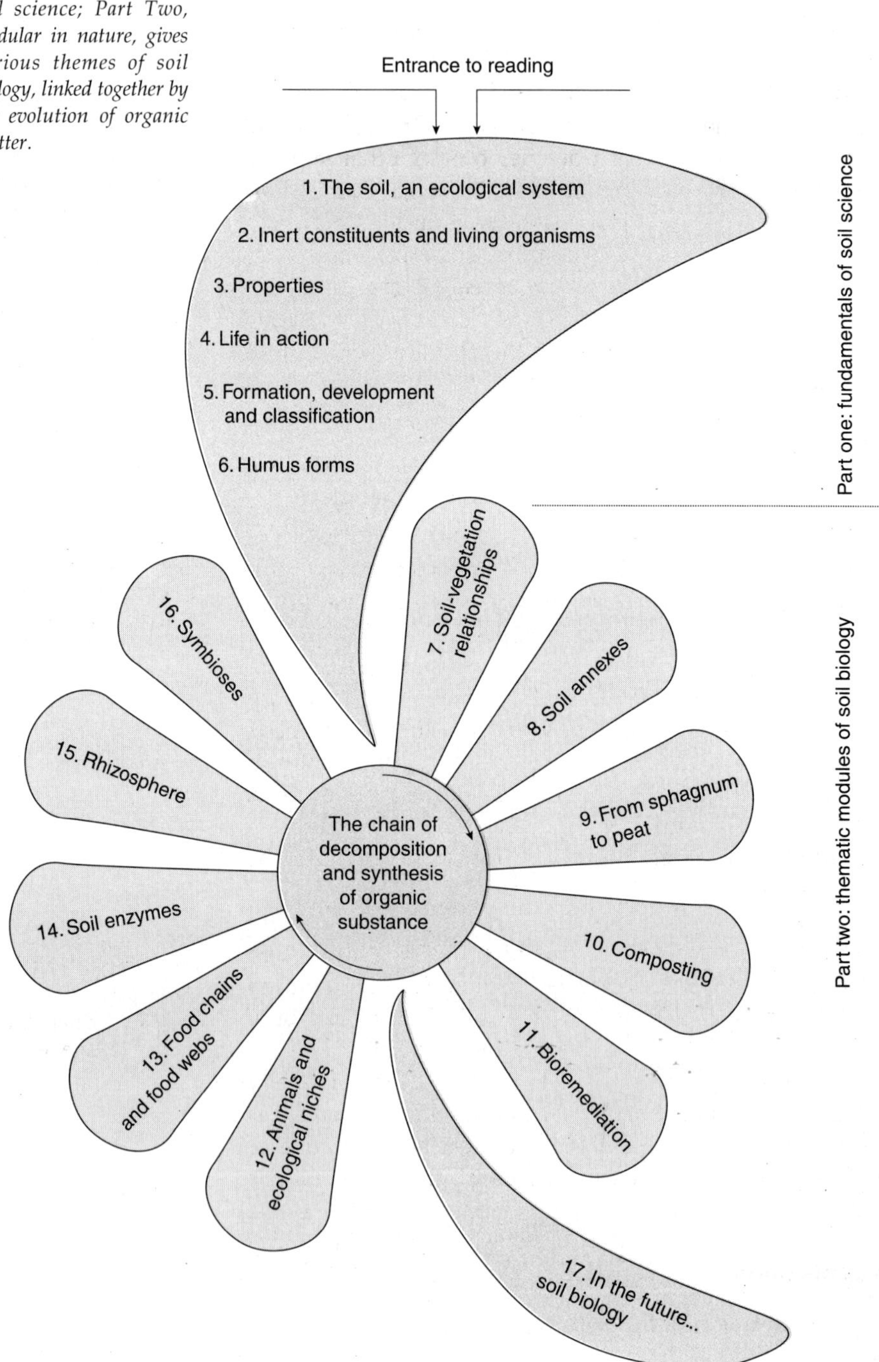

• linear in its first part, like the vertical sequence of horizons observed in the profile,
• systematic in its second part, like the interacting functions of soil constituents,
• open in its entirety, just as the soil is subject to climatic or human interference.

This option enables us to approach the soil and its organisms in the classic manner in Part One (Chaps 1-6) or, on the contrary, in a more personal manner starting from particular themes that form the objective of Part Two (Chaps 7-17). Generally speaking, understanding of the first part is necessary for comprehension of the second, where basic information is not repeated, except for in-depth analysis. Numerous cross-references between chapters, sections and paragraphs are given throughout the text to enable systemic correlation.

'Soil biology is a theme which runs through many of the major areas of modern science; the subject encompasses global issues concerning the environment, conservation and food production and the tools used for its study range from molecular biology to the common spade' (Wood, 1995).

Materials Covered

Unlike the many excellent works available, this book aims at a concise and coherent clarification of the ensemble of edaphic organisms and their activities, and the consequences of the latter on the functioning of the soil, providing the necessary fundamentals of general soil science, soil zoology and microbiology.

The Living Soil: fundamentals of soil science and, at the same time, those of soil biology, often separated or weighted differently in existing books.

The three authors, respectively botanist, microbiologist and zoologist originally, give priority to the living organisms of the soil, emphasizing their ecological, physiological and systematic aspects. Fundamental cellular processes, such as those that concern evolution, population dynamics or genetics for example, are presented only when necessary for understanding the other aspects.

Despite their importance, some areas have been deliberately left out for reasons of space or ability; such is the case of relations between fertilizers and soil fauna or of the role of viruses in the soil. Others are only outlined, for example, soil parasites, the role of living organisms in agricultural soils or the effect of heavy metals on the microflora. For more details the reader is referred to several specialized publications. Because of the fundamental interdisciplinary and pedagogic nature of the book, special attention has been devoted to the choice or establishment of definitions of very many concepts (more than a thousand terms covered!).

'The "courage to omit" seems to us to be as important as the research of exhaustiveness' (Lüttge et al., 1996).

Organization of the Book

In *Part One*, importance is given to *fundamentals of general soil science*; indeed understanding the role of living organisms (Chap. 4) demands that they be placed in their physical, chemical and pedogenetic context. After recognition and definition of the soil (Chap. 1), the constituents and properties are presented (Chaps 2 and 3), such as the processes of its development and the major principles of its classification (Chap. 5). In these areas, the role of living organisms is given prominence compared to physicochemical mechanisms, which can be found described in greater detail in other works. The importance of soil organisms is highlighted in Chapter 6, which discusses humus forms.

From vegetation and its wastes in Chapters 7 and 8 to soil bioremediation in Chapter 11, all the processes involved are ensured by edaphic organisms, the veritable stars of the ***Living Soil***.

In *Part Two*, which could be titled 'Selected Chapters on Soil Biology', eleven themes are developed from ecological niches (Chap. 12), to enzymology (Chap. 14) covering on the way soil annexes (Chap. 8) or food webs (Chap. 13). By virtue of the selections made, the reader can approach soil biology through his/her preferred field. For example, the forester or the geographer will be attracted by the relations between soil and vegetation (Chap. 7) and the agronomist by the symbioses in soil (Chap. 16). Chapter 10 on composting and Chapter 11 on soil bioremediation will be equally useful to waste managers and cultivators, and Chapter 15 on the rhizosphere to the specialist in plant nutrition. Lastly, the manager of nature protection will perhaps start off with Chapter 9 on peats and bogs, endangered ecosystems. But all will be concerned with Chapter 17, the last, devoted to the importance of biological knowledge and its foreseeable development for soil science in its entirety.

'One of the stimulating developments in soil biology in recent years has been the general recognition that soil cannot be studied from a solely chemical, microbiological, botanical or zoological standpoint. The ever-changing patterns of processes going on in the soil is made up of components from many disciplines, inextricably interwoven, which must be synthesized, not studied in isolation' (Benckiser, 1997).

Each of the eleven thematic chapters of Part Two could be read independently. The eleven chapters are, however, arranged according to a certain logic, that of the sequence of evolution of organic substances. This will reassure those not much accustomed to reading in the 'CD-ROM' manner and, furthermore, demonstrate the uniqueness of the functioning of the soil system.

Specific Remarks

Some points may be useful to note before reading the book:

Bibliographic references

- The books and scientific articles consulted are not cited for each piece of information used unless they are specific.

Most of the publications that served as the basis for this work are nonetheless listed in Bibliography. Citation of authors in collected works (X.X. in N.N.) always relates back to the *original articles* present in the collection, even if their original title is not listed in the Bibliography for reasons of space.

Methodological references

- With rare exceptions, especially when they are recent or specifically adapted to soil biology, methods are neither detailed nor discussed. Specialized books are referred to: Aubert (1978), Baize (1988), Carter (1993), SSSA (1994), Baize and Jabiol (1995), Schlichting et al. (1995), Schinner et al. (1996), Dunger and Fiedler (1997), Sala et al. (2000).

Definitions

- Definitions are taken from existing books or from the authors' reflections. An Index refers back to the major terms used, which appear in ***bold italic*** at the place of their definition—in the text or in the margin.

Emphasis

- Titles of books and some important points appear in *light italic*.

Nomenclature of soils

- One of the great constraints for the universality of soil science is that the nomenclature of soils and horizons has still not been standardized. In this book, for which discussion of the concepts of soil classification is not a primary objective, the choice fell on the typology of the French Soil Reference Base (AFES, 1995, 1998). However, other classifications or reference bases are presented in Chapter 5, § 5.6. According to the rule currently in force (AFES, 1995, 1998), the designations of the sola described in the taxonomy of the French Reference Base are printed in SMALL CAPITALS (for example, BRUNISOL SATURÉ, ORGANOSOL CALCIQUE); the **qualifiers** characterizing the ***Type*** category are printed in lower case (for example, RENDOSOL HUMIFÈRE, HISTOSOL MÉSIQUE with fibric horizon). Sometimes, for example in the case of reproduction of figures or quotations from other authors, the original nomenclature has been retained.

Biological systematics

- The systematic classifications used are, in botany, Tutin et al. (1964ss), Augier (1966), Isoviita (1967), Aeschimann and Burdet (1994) and, in microbiology, Krieg and Holt (1984ss). In zoology, as no recent and detailed book covers the ensemble of soil animals, reference is made to the organisms closest to those considered. Names of organisms appear in upper case if they relate to a precise taxonomic reference; in other cases lower case is preferred, with complete freedom. Many organisms are illustrated in the margin; these figures are lettered rather than numbered.

Concrete examples

• As a didactic tool, concrete examples are presented as often as possible. Many are original, often drawn from the work of the authors; however, these sometimes go back to certain 'great classics' of soil science, in particular in fields remote from personal researches or concerning specialities not much studied.

Particular themes

• In order to awaken the critical capabilities of the reader, certain eye-catching or ambiguous themes are discussed in a more subjective fashion: soil and the greenhouse effect, the notion of biological activity of the soil or even adaptive strategies. They are indicated by a boxed insert. So too are details of specific topics (for example Stokes law, types of nutrition, fundamentals of molecular biology).

Use of margins

• The margins contain:
  - flags serving as a connecting thread, set off by shading,
  - definitions,
  - citations, biographical references, certain supplementary illustrations,
  - legends to the figures

Units of measure

• The physical, chemical and biological symbols and units used are listed at the end of the book.

Acknowledgements

The authors very sincerely thank the following persons and institutions whose help was decisive in the preparation of this publication:

• Illustrations: Pierre-Olivier Aragno (computer graphics), Sylvette Gobat (cover page), Cécile Matthey (line drawings).

• Preliminary reading and correction of certain chapters: Jean-Marc Bollag, Alexandre Buttler, Jean-Daniel Gallandat, François Gillet, Elena Havlicek, Daniel Job, Johanna Lott Fischer, Laurent Marilley, Francine Matthey, Enrico Martinoia, Jean-Marc Neuhaus.

• Providing source material (photographs, original drawings, etc.): Yves Baer, Trello Beffa, Daniel Borcard, Yves Borcard, Rudolf Brändle, William Broughton, Martin Burkhard, François Felber, Karl Föllmi, Willy Geiger, Dittmar Hahn, Jean-Pierre Hertzeisen, Jean Keller, Philippe Küpfer, Pierre-François Lyon, Edward Mitchell, Pierre-Alain Mouchet, David Read, Sally Smith, Catherine Strehler, Eric Verrecchia, Verena Wiemken, Jürg Zettel.

• Various assistance: Elizabeth Boss, Georges Boss, Ernest Fortis, Gabriel Gobat, Nicole Jeanneret, Marie-Claude Santschi, Roberta Ventura.

• Financial support: Swiss Agency for Environment, Forests and Landscape, Berne; Swiss Academy of Natural Sciences, Berne; Swiss National Science Foundation, whose many research grants to the authors served as the indirect basis of this book; Cantonal Department of Public Instruction and Cultural Affairs, Neuchâtel; University of Neuchâtel.

We give very particular thanks to Professor Roger Dajoz of the National Museum of Natural History in Paris, who has done us the great honour of writing the Foreword to this book.

We also want to recall the constant, efficient help of the publishers of the original book, the Presses Polytechniques et Universitaires Romandes in Lausanne, Switzerland; our gratitude goes in particular to the late Mrs Claire-Lise Delacrausaz and to Messrs Olivier Babel, Christophe Borlat, Jean-Philippe Galley, Lucien Yves Maystre and Patrick Rième. We also thank the experts—anonymous to us—called upon by the publishers, for their suggestions during review of the manuscript prior to publication.

The authors are happy to share with the reader the pleasure they had in discovering and in introducing the soil, *the living soil* ... through its organisms and its organization. They will be grateful for any suggestions or criticism concerning this first edition.

Jean-Michel Gobat, Michel Aragno, Willy Matthey
Neuchâtel, Switzerland, autumn 2003

A NOTE FOR THE READERS OF THE ENGLISH TRANSLATION

Partly because of its youthfulness and also due to its occasionally different fundamental perceptions, soil science today still has not achieved conceptual uniformity world-wide. Many schools or approaches subsist in the world, reflecting different priorities accorded to soil processes or a more or less marked emphasis on fundamental or applied soil science, or even different ways of approaching the functioning of the soil.

But the most prominent differences pertain to nomenclature and soil classification, in spite of the recent fruitful attempts at harmonization (see Chapter 5, § 5.6). In this difficult situation, the present English translation maintains the choice made in the original French version of opting for the nomenclature of the *Référentiel Pédologique* (AFES, 1995) since an English version of that work has appeared (AFES, 1998). Not being specialists in the taxonomy of soils, the authors did not want to enter in the very hazardous and difficult game of equivalents.

Spaargaren (in Sumner, 2000) proposed nomenclatural correspondence among the different world classifications.

Otherwise, physical units have been maintained in the metric system even though the American system is still used in the United States. This is the case, for example, of the size of holes in sieves, expressed in number of meshes per square inch, whereas it is given in micrometres or millimetres in Europe.

On the other hand, quotations from the literature in French or German have generally been translated. The Bibliography is the same as in the French version, completed with some general references specific to this edition.

CONTENTS

Chapter 1

THE SOIL, AN ECOLOGICAL SYSTEM

How else to begin a book on soil except by speaking ... of the soil?

And if we begin with the soil?

Seemingly a truism, this question is not without foundation. Too often the scientific books that speak of complex subjects such as the organism, the ecosystem, a country, a mountain massif, the ocean, the soil, begin with an extremely reductionist approach. The organism is entered through the catalogue of molecules that constitute it or—a little better—through the functioning of its cells. Mountains are approached through the listing of constituent minerals of the rocks. We plunge into the ocean through the detailed explanation of variations in pressure and salinity!

Our idea is not to neglect the reductionist approach; it is indispensable even in the systems sciences. But why start through it, why not confront the complexity straightaway? Complexity is there, omnipresent in the organism, the ecosystem, the ocean or the soil! There will always be time for cutting the object into little pieces and analysing them after understanding, although in a simple manner, it in its totality.

No one seeking to better understand a painting by Leonardo da Vinci or Vincent van Gogh would first demand information on the type of pigments used or the thickness of the brush!

'The true means of attaining a good understanding of an object, even in its finer details, is to begin by visualizing it in its entirety' (Lamarck, *Philosophie zoologique*, 1809).

This chapter is devoted in the first stage to the apprenticeship step; brushing a quick portrait of the soil as such, without much consideration of its interior but without hiding the complexity of its organization and of its relations with the rest of the ecosystem. Imperceptibly we will thus approach its scientific definition. The chapters that follow, from the very analytical Chapter 2 to the more synthetic Chapter 6 will succeed one another in the progressive reconstruction of the soil system.

'Wisdom leads to combination of the two approaches—reductionist and holistic: replacing the processes in a global context and later analysing them at the scale of the mechanisms that explain them' (Lavelle, 1987).

Two possible approaches

The ***reductionist*** (or analytical, Cartesian) ***approach*** consists of reducing a system or complex phenomena into simpler components and of considering the latter as more basic than the complex whole (Schwarz, 1997). The reductionist approach—this epithet does not by itself carry a pejorative connotation (Garfinkel, in Boyd et al., 1992), contrary to some accepted ideas—is opposite to the ***holistic*** (or systemic, synthetic) ***approach***. This approach is an attitude that consists of considering a complex system as an entity possessing emergent characteristics related to its totality, properties that are not reducible to a simple summation of those of its elements (after Schwarz, 1997). Also see Chapter 17, § 17.1.3.

1.1 AS MANY SOILS AS PERSONS CONCERNED WITH SOIL

1.1.1 At the limits of the scientific approach, a universal soil

'How, in future, would they (the scientific disciplines) be able to claim to exhaust reality and render obsolete every other approach in the world?' (Hubert Reeves, *Malicorne*, 1990).

The theme of soil and the life it shelters concerns the ensemble of humanity in all its philosophical, cultural, economic, aesthetic and scientific diversity. Even if the actual natural soil is incontestably at the root of the various perceptions humans have of it, differing perceptions do exist, with distinctive features, nuances and evolutions. The scientific approach is but one of many, all equally respectable and necessary to understand as soon as we leave the academic confines of research for its application.

1.1.2 And why not some literature?

'Science is perhaps simply the laborious rediscovery of evidences...' (Chauvin, 1967).

Every practitioner of ecology, particularly soil science, feels this need to remain in touch with other views of soil existing in human society: that of the farmer who tills the field, that of the property speculator who buys or sells in square metres, that of the stressed city-dweller who relaxes in the forest or that of the poets who express their attachment to the land. The soil the scientist studies with difficulty, using instruments, has long been understood by others from different standpoints.

'The soil is the stomach of plants, which receive nutrition from it in a form ready for digestion. It has an immense quantity of forces for nourishing plants. The fertility and infertility of a soil, as also the geographical distribution of plants, depend (...) on the moisture necessary for plants in a given soil. The characteristics of the soil vary widely from one place to another.'

A student of Hippocrates, *c.* 400 BC, quoted by Boulaine (1989).

"Il ne faut pas toujours le bon champ labourer:
Il faut que reposer quelquefois on le laisse,
Car quand chôme longtemps et que bien on l'engraisse,
On en peut pris après double fruit retirer."

Oliver de Magny, *c.* 1550.

"Dès le commencement, il (Messer Gaster) inventa l'art fabrile et agriculture pour cultiver la terre, tendant à fin qu'elle luy produisit grain."

François Rabelais, *Le Quart Livre*, 1552.

"Je te salue, ô terre porte-grains,
Porte-or, porte-santé, porte-habits, porte-humains,
Porte-fruits, porte-tours, calme, belle, immobile,
Patiente, diverse, odorante, fertile,
Vêtue d'un manteau tout damassé de fleurs,
Passementé de flots, bigarré de couleurs,
Je te salue, ô cœur, racine, base ronde,
Pied du grand animal qu'on appele le Monde."

Guillaume de Salluste du Bartas, *c.* 1580.

'And the earth alone stays immortal, the mother from whom we come and to whom we return, that which we love even up to crime, which continually renews life for our ignorant purposes, even with our abominations and our miseries.'

Emile Zola, *La Terre*, 1887.

'Eternal is the blue of the sky and the land will last long and bloom anew in spring. But you, man, how much time do you aim for?'

Hans Bethge, in Gustav Mahler, *Das Lied von der Erde*, 1908.

'This earth! This earth that extends wide in all directions, rich, heavy with its load of trees and water, its rivers, its streams, its forests, its mountains and its hills (...), as though it were a living creature, a body?'

Jean Giono, *Colline*, 1929.

'To the red country and to part of the gray country of Oklahoma, the last rains came gently, and they did not cut the scarred earth... The surface of the earth crusted, a thin hard crust, and as the sky became pale, so the earth became pale, pink in the red country and white in the gray country.'

John Steinbeck, *The Grapes of Wrath*, 1947.

'It smells strong and hot, it smells of the earth that has smoked under the sun, the dry plant, thyme and mint, because he walks on it, and it is soft under his feet.'

Charles-Ferdinand Ramuz, *Derborence*, 1954.

"L'humidité de la terre monte à mes narines: odeur de champignons et de vanille et d'oranger... on croirait qu'un invisible gardénia, fiévreux et blanc, écarte dans l'obscurite ses pétales, c'est l'arôme même de cette nuit ruisselante de rosée (...). A moi le dessous gras de la terre, la demeure profonde du ver, le corridor sinueux de la taupe, à moi, encore plus bas, le roc que n'a jamais vu la lumière; à moi, si je veux, l'eau prisonnière et noire, enfouie à cent pieds."

Colette, *La Retraite sentimentale*, 1957.

'One sees well, while traversing these plateaus, that they are not devoid of resources, unlike this ungrateful soil. "In this region," a shepherd I met in front of the hamlet of Chaldas told me in the morning, "the crows fly upside down so that they do not see the misery of the land".'

Jacques Lacarrière, *Chemin faisant*, 1977.

'Soil is remarkably absent from those things clarified by philosophy in our western tradition (...). As philosophers, we emphasize the duty to speak about the soil. We offer resistance to those ecological experts who preach respect for science, but foster neglect for historical tradition, local flair and the earthy virtue, self-restraint.'

Ivan Illich and friends, *Declaration on Soil*, University of Kassel, 1991.

1.1.3 From the literary soil to the natural

The 'external' complexity of soil.

It is striking to observe in these quotations that the soil is linked to a cultural, landscape, economic or ecological environment: the 'ungrateful' soil of Lacarrière, the 'mother from whom we come' of Zola, the soil that 'varies widely from one place to another' of the student of Hippocrates, the 'pied du grand animal qu'on appelle le Monde' of de Salluste. The soil is part of a much larger all-inclusive totality (Fig. 1.1).

The internal complexity of soil, its constituents, its functional processes, its organization.

But it is also seen that soil is not a simple entity, homogeneous and static. Steinbeck points to 'scars, 'crusts' and 'crusting'; in Ramuz's view, on the contrary, 'it is soft under the feet.' The student of Hippocrates talks of an 'immense quantity of forces', de Salluste describes a land that is 'diverse, odorant, fertile', but Colette says 'the moisture of the earth rises to the (my) nostrils'.

The soil, its organisms and functions.

Lastly, the earth 'continually renews life' (Zola), it will 'last long and bloom anew in spring' (Bethge), but for this "il faut que reposer quelquefois on le laisse" (de Magny). The earth harbours life, it is 'a demeure profonde du ver' and exudes 'l'odeur de champignons' (Colette). Perhaps it is even animate: 'as though it were a living creature, a body?' suggests Giono. At the bottom of this question lies the perpetuation of life by degradation of dead organic material and its being kept available for those who follow.

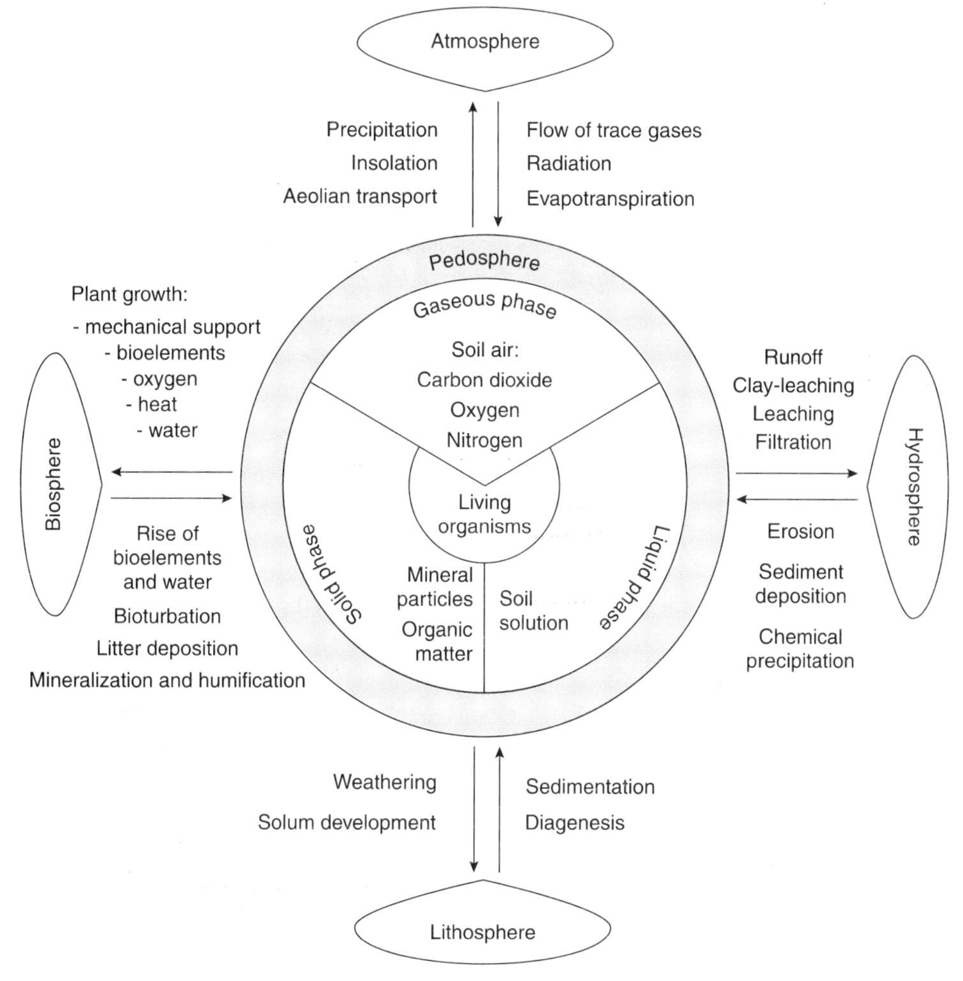

Fig. 1.1 *The external and internal complexity of soil.*

1.2 AND THE SOIL OF THE SCIENTIST?

Three major features of the soil lead us to its definition.

Among the preceding examples, there are some very close to the soil as understood by the scientist. Several characteristics appear: its complexity, its capacity to nourish plants and to promote life, its subjection to environmental agents, its temporal and spatial variability, its colour—a reflection of its mineralogical composition, its fertility. These varied characteristics are related to three features:

- the soil is a multifunctional crossroads,
- it exhibits a systemic internal organization,
- it has an earthly 'exclusivity'.

The soil is one of the essential compartments of the ecosystem, working as controller and pointer to numerous ecological processes by its physical, chemical and biological processes, in the short term and in the long: 'Soils should be the best overall reflection of ecosystem processes' (Paul, in Grubb and Whittaker, 1989).

The soil, skin of the Earth: indispensable, irreplaceable, but terribly fragile... (Chap. 17, § 17.2).

'We stand on soil, not on earth.' (Illich et al., 1991).

1.2.1 A multifunctional crossroads

Every global scientific approach to soil must first be functional. In the ecosystem, before containing 12% clay, having a macroporosity of 26% or harbouring billions of protozoans in a square metre, the soil is first of all a crossroads with multiple roles (Fig. 1.2).

According to its natural functions, soil is

- a support for living creatures,
- a reservoir of organic and mineral substances,
- a regulator of exchanges and flows in the ecosystem,
- a site for transformations of organic matter,
- a purifying system for toxic substances.

In direct relation to humans, the soil is:

- one of the essential bases of human life,
- the site of agricultural and forest production,
- a place for storing primary substances and wastes,
- a constituent element of the landscape,
- a mirror of the history of civilizations and cultures.

At the junction of all these functions and subject to their constraints, the soil plays an irreplaceable role on the surface of the Earth. It is, nevertheless, only a very

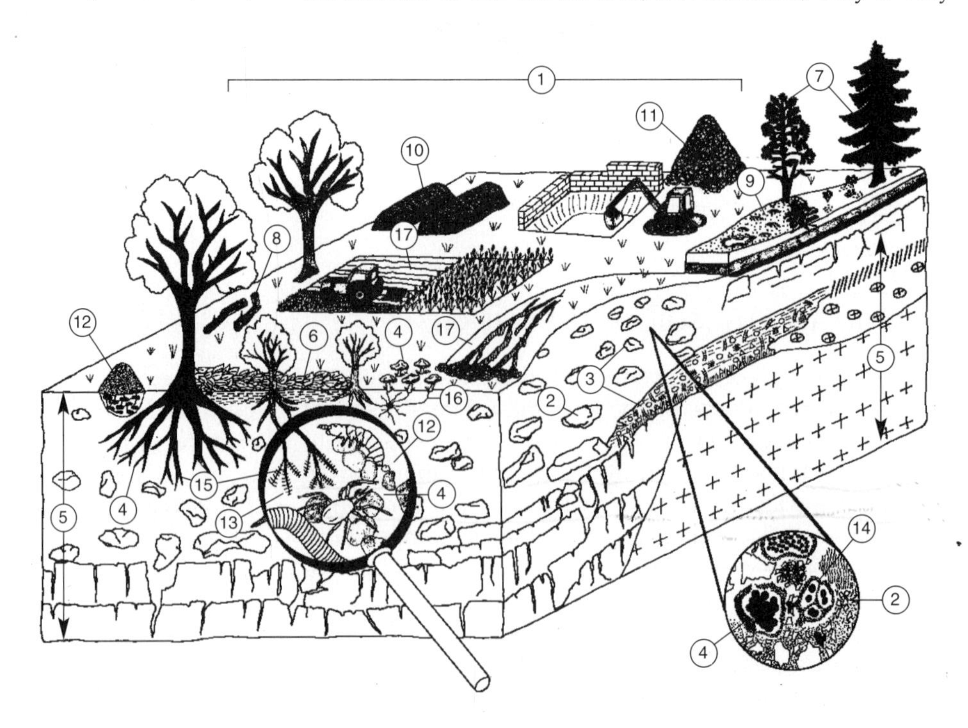

Fig. 1.2 The soil, a multifunctional crossroads. The numerals indicate corresponding chapters.

thin film at the interface of the lithosphere and the atmosphere (Fig. 1.3). Barely one to two metres thick, the soil is more fertile in its upper part, the humiferous episolum (Chap. 6, § 6.1.1).

1.2.2 A systemic internal arrangement

The soil manifests itself at all levels of spatial organization from chemical structure of clays to remote sensing by satellite. Its manifold functions well illustrate its 'external' complexity. Its internal organization is revealed by its features, expressed sometimes at a scale smaller than a millimetre, sometimes at a scale of hundreds of metres. The soil is thus organized in a multiscale manner (Fig. 1.4).

Its elements are of various sizes, often nested one in the other, related more or less closely. The structure of

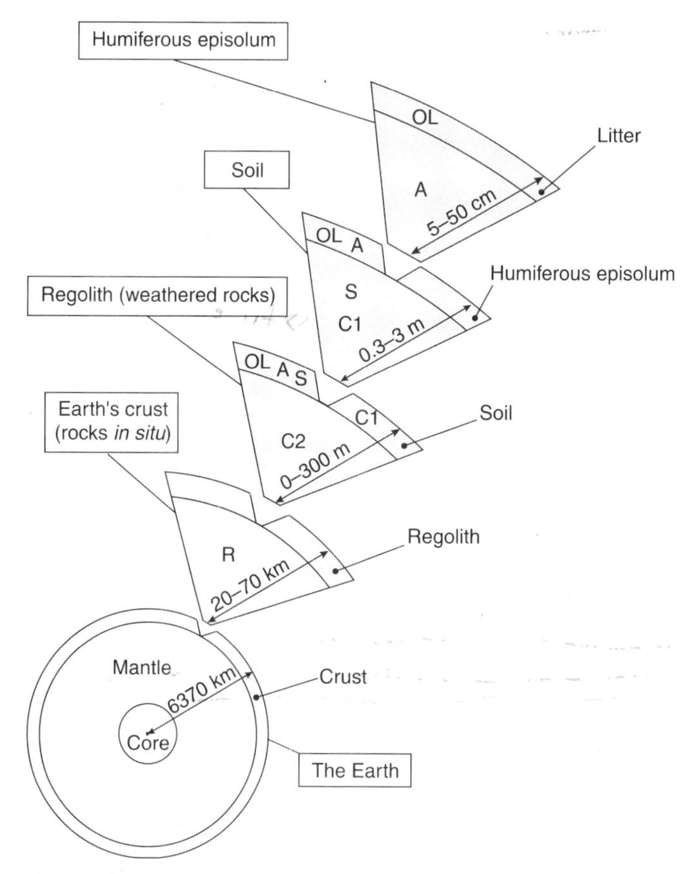

Fig. 1.3 The thickness of the soil and the humiferous episolum compared to the Earth (for horizon nomenclature, see Chap. 5, Table 5.3).

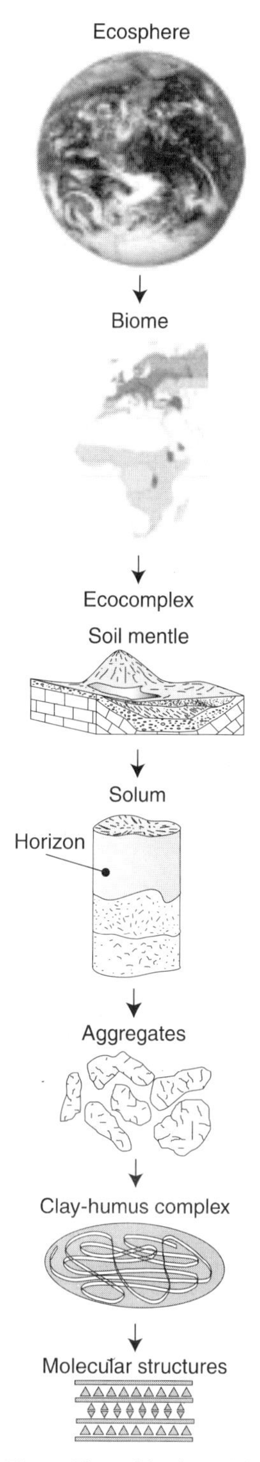

Fig. 1.4 The multiscale spatial organization of the soil.

the soil (Chap. 3, § 3.2) is a good example: large clods contain macroaggregates, themselves formed of microaggregates, the functional elementary units of soil (Lavelle, 1987). At the 'immediate' scale of the observer with natural senses, its more or less parallel layers, the horizons (Chap. 5, § 5.4), are the expression of the organization of the soil. But the soil itself is a functional element of other large systems such as the biogeocoenosis, ecocomplex or ecosphere (Chap. 7).

A soil which speaks.

Discovery of the soil on the terrain, a sensory approach!

On the terrain, the discovery of a soil is primarily a sensory approach! The eye distinguishes boundaries, colours, gradations, shapes; the sense of touch reveals sticky clay, silky silt or 'grating' sand; the nose senses hydrogen sulphide or mushrooms; the ear detects a slight effervescence with hydrochloric acid or the dull sound of a compact layer being dug. At times—but this is strongly advised against by dentists—we 'taste' the mineral particles....

Behind the spatial arrangement of the soil is hidden a temporal organization.

'The possibility for soil scientists to cover in their research the entire range of observation levels is, in our view, a highly scientific character' (Boulaine, 1989).

In the soil, some almost instantaneous processes, such as the capture of an iron atom by an organic molecule, coexist with much slower movements such as the progressive accumulation of clays by clay-leaching. Intermediate between these two, the dynamics of bacterial populations are counted in days and changes in overall pH in decades. However, from the nanosecond to the millennium, all temporal levels influence soil development 'simultaneously'.

Wagenet et al. (in Bryant and Arnold, 1994) delineate eleven degrees of spatio-temporal organization. In a simpler manner, Fournier and Cheverry (in Auger et al., 1992) recognize five, listed below:

'The scale of resolution chosen by ecologists is perhaps the most important decision in the research program, because it largely predetermines the questions, the procedures, the observations, and the results.' (Dayton and Tegener, in Schneider, 1994).

- The point scale, where the assemblage of elementary solid particles of the soil is attained (Chap. 2, §§ 2.1, 2.2). It corresponds to the microaggregates (Chap. 3, § 3.1.2), ionic exchange (Chap. 3, § 3.7.2) or the rhizosphere (Chap. 4, § 4.1.3; Chap. 15). The spatial dimension is of the order of mm^3 or cm^3; temporally, the processes are instantaneous to short-duration.

'Spatial and temporal variability causes problems for all environmental scientists; however these problems are exacerbated for soil biologists, who are faced with the inherent variability of biological

- The scale of the site (Chap. 7, § 7.7.1), which is that of the horizon or the solum (Chap. 5, § 5.4.2). The general principles of functioning of the soil are manifest here: hydric regime (Chap. 3, § 3.4), pedogenesis (Chap. 5), soil-plant relations (Chaps 4 and 7), etc. The dimensions are of the order of dm^2, m^2, dm^3 or m^3 spatially; for time, the season, year, decade or a few centuries.

• The scale of the slope, often approached through catenas or toposequences (Chap. 7, § 7.3), which best reflect the influence of relief on pedogenesis (Chap. 5, § 5.5.4) or erosion (Chap. 17, § 17.2.1), as also the linked modifications of soils and phytocoenoses (Chap. 7, § 7.4). The spatial scale is between the hectometre and the kilometre; the temporal scale is between the year and millennia.

• The scale of the drainage basin, well suited to the integration of the determinations made at the preceding scales, for example in the estimation of transport of substances. It reveals human actions relative to land utilization (Chap. 17, § 17.4) as well as mesoclimatic influences, observable for example by means of Geographical Information Systems (Collet, 1992; Legros, 1996). The drainage basin is depicted in km^2 and the processes take place over a year to a millennium or more.

• The scale of biogeographical zones, which is that of the great equilibria among macroclimate, the soil and vegetation (Chap. 5, § 5.5.1), leading to the formation of biomes (Chap. 7, § 7.2). The soil is here considered in relation to biogeochemical processes touching the ecosphere, such as the carbon cycle (Chap. 5, § 5.2.3; Chap. 9, § 9.2.1). The spatial dimension is of the order of millions of km^2, the time-scale of the order of centuries to hundreds of millennia.

systems, together with the fact that the processes of interest often take place in localised favourable areas of the soil, which may be anything from a few micrometres to a few millimetres in size' (Wood, 1995).

'Lying at the interface of the geosphere, hydrosphere, biosphere and atmosphere, soils represent the end-product of a complex set of interacting processes, operating over a vast range of time-scales' (Ellis and Mellor, 1995).

According to its organization and functioning, the soil is an ecological ***system*** to which the customary properties of systems are applicable, particularly those related to living organisms (Odum, 1971, 1996; Lavelle, 1987; Delcourt and Delcourt, 1988; Schwarz, 1988; Frontier and Pichod-Viale, 1991; Brady and Weil, 2002):
• respect for the principles of thermodynamics (Chap. 5, § 5.1.1),
• the necessity for defining the boundaries (Chap. 5, § 5.6.1),
• a spatio-temporal hierarchical organization, in which each level includes the attributes of the level below it (Chap. 5, § 5.4),
• an internal evolution determined by the combination of energy flux and the cycle of matter (Chap. 5, § 5.1.1),
• emergence of new properties, according to the rule that the whole is more than the sum of the parts (Chaps 3 and 5).

The soil, an ecological system *par excellence*.

System: ensemble of interdependent phenomena and events that are drawn from the external world by an arbitrary intellectual process, with a view to considering this ensemble as a whole (*Encyclopedia universalis*, 1985); ensemble of parts interconnected by functional links (Frontier and Pichod-Viale, 1991).

• functioning resulting from interactions between the constituents, a functioning of which the mechanisms are to be investigated at the lower level but the constraints at the higher (Chaps 2 and 3),
• feedback of the whole to the parts (Chaps 5 and 6),
• an opening to other systems, ecological or economic (Chaps 10, 15 and 16).

Four sources of heterogeneity: energy, space, time, function.

The complexity of the soil system results from the interference of four great sources of heterogeneity (Lavelle, 1987):
• energetic, with three types of energy being dissipated in the soil: physical (gravity, capillary), chemical (oxidation, reduction) and biological (production, bioturbation);
• spatial, from microaggregates to the soil mantle;
• temporal, from instantaneous chemical reactions to multimillennial development;
• functional, with mineral or organic development.

1.2.3 A terrestrial 'exclusivity'

Show me your complexes; from them I assume exclusivity!

Soil science: Soil science (...) involves the study of soil formation, classification and mapping, the physical, chemical, biological and mineralogical properties of soil from microscopic to macroscopic scales of resolution as well as the processes and behavior of soil systems and their use and management. It is an integrative science that interlinks knowledge of the atmosphere, biosphere, lithosphere and hydro-sphere (Sumner and Wilding in Sumner, 2000).

Allusions to the history of soil science are made throughout the book.

But the soil is not only a 'beautiful systemic organization'! Numerous other parts of the ecosphere share this characteristic: a meteorological disturbance, an ocean current, a river, a plant, an animal, an enzyme (Aragno, in Schwarz, 1988). It differs from them, however, in one unique property: its capacity to bond mineral and organic substances intimately in the clay-humus complex, at the molecular level.

With time, the weathering of rock minerals and degradation of organic matter from living creatures (Chap. 5) make them capable of binding one to the other in a new combination, the clay-humus complex (Chap. 3, § 3.6). This original entity, this emergent property, neither geological nor biological, but both, crystallizes the originality of the soil and the science that studies it, ***soil science***. In this science it becomes, according to the famous concept of Dokouchaev (1883), an 'independent natural body' studied by a new science. Soil science is first and foremost at the intersection of geology and biology. But it also integrates climatology, chemistry, physics and mathematics, which provide fundamental tools for understanding it.

1.3 EVOLVING DEFINITIONS

Everything is now in place for defining this complex object, the soil. Let it be pointed out rightaway that a straightforward reference is impossible because that would be excessively simplistic. For this reason, definitions of soil have succeeded one another in the history of soil science—and are still doing so! Each is the reflection of a particular school of thought corresponding to a specific objective or is addressed to a different user. Three definitions are proposed here, the third being the most appropriate in the context of this book:

- By ***soil*** is meant the external horizons of rocks naturally modified by the mutual influence of water, air and living and dead organisms; it is an independent, variable natural body.' (Dokouchaev, 1883). In this definition, 'independent body' would today be translated as 'functional entity'. Blume et al. (1996) speak of 'Boden als Naturkörper'.

- 'The ***soil*** is the product of weathering, reworking and arrangement of the upper layers of the earth's crust under the action of life, the atmosphere and exchanges of energy that are manifested in it' (Aubert and Boulaine, 1980, in Lozet and Mathieu, 1997).

- 'The ***soil*** is the outermost layer of the earth's crust, marked by living creatures. It is the site of intensive exchange of matter and energy among air, water and rocks. The soil, as a part of the ecosystem, occupies a key position in the global cycles of substances' (Swiss Society of Soil Science, 1997).

By virtue of its location in the ecosystem at the interface between the mineral world and the organic world, the soil is a veritable ***ecotone***. Like all zones in contact between two systems it concomitantly possesses the constituents and properties of both (here the aerial biocoenosis and the subjacent rock) and of others typically transitional, that is, itself (for example, humified organic matter, Chap. 2, § 2.2.4, or exchangeable cations, Chap. 3, § 3.8.1). The soil assumes the role of the interface at all its organizational levels, from the clay-humus complex smaller than 50 μm (Chap. 3, § 3.6.1) to the soil mantle under an extensive plant formation (Chap. 7, § 7.2).

In all cases the soil is a dynamic ecological system. In the succeeding chapters we shall 'open the box' by observing the contents in detail to begin with (Chap. 2),

Pedology has a narrower definition: science that quantifies the factors and processes of soil formation (Wilding in Sumner, 2000).

Edaphic: related to the soil; in interaction with the soil.

Pedological: concerning the soil.

The term 'soil', general and conceptual, must not be confused with 'solum', which has a methodological and descriptive purpose (Chap. 5, § 5.4.2).

Vassili Vassiliévitch Dokouchaev (1846-1903) is unanimously considered the 'father' of soil science, since his thesis of 1883, *The Russian Chernozem*. He was the first to learn to see in the soil a new object with its own laws and properties, forming by the action of numerous interacting ecological factors.

'A definition is neither true nor false; it is useful or useless' (H. Saner, student of K. Jaspers, German philosopher).

Ecotone: transitional zone between two adjacent ecologic systems, possessing an ensemble of spatio-temporally dependent characteristics defined by the strength of interaction between the two systems (Holland, in Holland et al., 1991). The ecotone is characterized by a diversity and a specific wealth greater that those of each of the communities it separates, because in it are found constituents and biocoenoses situated on both sides (...) and others that are typical of the particular biotope representing the ecotone (Ramade, 1993; Lachavanne, in Lachavanne and Juge, 1997).

then progressively in an integrated manner (Chaps 3 and 4). We shall progressively close it again for characterizing it in its entirety (Chap. 5), then in its external relationships (Chap. 6).

CHAPTER 2

BUILDING BLOCKS OF THE SOIL SYSTEM: INERT CONSTITUENTS AND LIVING ORGANISMS

The aim of this chapter is to present the principal features and functions of the fundamental elements of the 'soil system'. Mineral and organic constituents form its skeleton; the soil solution is one of the favoured vectors of substances whereas the soil atmosphere represents a gaseous interface between the interior of the soil and the external environment (Tab. 2.1).

Soil constituents: mineral and organic, solid, liquid and gaseous, living and inert.

2.1 MINERAL CONSTITUENTS

2.1.1 Origin and kinds of mineral constituents

Products of weathering of rocks, inherited or neoformed.

Mineral constituents of the soil are primary—directly inherited from the parent material—and secondary—resulting from the chemical transformation of the former and then recombined in the ***weathering complex.*** The former comprise salts (e.g. calcium and magnesium carbonates) or ***silicates*** (e.g. micas and clay minerals); the latter are ***colloids*** such as iron and aluminium hydroxides and other secondary minerals. If weathering is 'complete', it releases isolated ions or macromolecules.

Silicates: minerals based on the elementary design of a tetrahedron whose centre is occupied by an Si^{4+} ion and the apices by O^{2-} ions.

Colloid: (from Gk. *kolla*, glue, and *eidos*, kind of): substance formed of particles of very small size, the micelles, capable of being flocculated (gels and aggregates) or dispersed in

The mineralogical composition of the weathering complex differs from that of the original rock because of chemical transformations, with grave effects on the

Table 2.1 Principal constituents of soil (after Soltner, 1996a)

	Solid constituents		Liquid constituents (soil solution)	Gaseous constituents (soil atmosphere)
	Mineral	Organic		
Origin	Physical disintegration and biochemical weathering of rocks	Decomposition of living organisms	Precipitation, groundwater, runoff	Air outside the soil, decomposing matter, respiration
Grouping criteria	Size (particle-size distribution), Nature (mineralogy)	State (living, dead) Chemical nature (original, transformed)	Origin (meteoric, phreatic) Physical state (hydric potential) Chemical nature	Origin (air, organisms) Chemical nature
Categories	By particle-size distribution • skeleton (>2 mm) • fine earth (<2 mm) By mineralogy • quartz • silicate minerals • carbonate minerals, etc.	• living organisms • dead organisms • inherited organic matter: cellulose, lignin, resins • humified organic matter: fulvic and humic acids, humins	• water • dissolved substances: carbohydrates, alcohols, organic and mineral acids, cations and anions	• gases of the air: N_2, O_2, CO_2 • gases emanating from respiration and decomposition of organisms: CO_2, H_2, CH_4, NH_3

a liquid (colloidal suspension) (See Goldberg et al. in Sumner, 2000).

direction of pedogenesis (Chap. 5, § 5.5.2). Two processes govern evolution of rocks—disintegration and alteration.

2.1.2 Physical disintegration and biogeochemical alteration

Rocks that are fragmented and weathered at rates depending on temperature and properties of the rock.

During ***physical disintegration***, climatic agents such as wind, frost and water break the rock down into smaller and smaller pieces, the original composition being preserved. The rate of transformation is particularly high in climates with contrast. Demolon and Bastisse (in Soltner, 1995) calculated that a block of gneiss 20 cm in diameter, transported by a torrent down a 2% slope, is broken down into 2-cm grains at the end of just 6 km, and into 2-mm sand after 12 km.

From hydration to hydrolysis, five increasingly intense weathering processes.

Biogeochemical weathering of rocks, which involves water, with or without oxygen, carbon dioxide or organic acids, follows five paths: hydration, dissolution, oxidation, reduction and hydrolysis:

• ***Hydration***, which weathers ferruginous rocks primarily, slightly modifies their mineralogical nature by addition of water molecules, thereby rendering them fragile. For example, haematite, Fe_2O_3, is hydrated to goethite, FeOOH.

• ***Dissolution***, whereby water and the substances contained in it weather the rock in several ways: dissolution in alkaline medium, ***chelation***, acid dissolution. The last is illustrated by solubilization of calcite by carbonic acid, H_2CO_3, in a reaction that is basic to carbonate leaching in the soil (Chap. 5, § 5.3.2):

$$\underset{\text{(insoluble)}}{CaCO_3} + H_2O + CO_2 \rightarrow CaCO_3 + H_2CO_3 \rightarrow \underset{\text{(soluble)}}{Ca(HCO_3)_2}$$

Chelation: dissolution of metal complexes by anions of biological origin or colloids.

• ***Oxidation*** enables the release in ferric form Fe(III) of the iron present as Fe(II) in the crystal lattice of some silicates, thereby destabilizing them.
• ***Reduction*** solubilizes Fe(II) from the oxides and hydroxides of Fe(III) present in the cements of some sandstones, for example. Reduction pertains to poorly aerated environments.
• ***Hydrolysis*** results in considerable rearrangement of the crystal lattices. It depends on climatic conditions, high temperature and humidity being favourable. It affects not only simple minerals such as the ***nesosilicates*** (silicates with isolated tetrahedra, for example olivine) but also more complex ones like the chain silicates (***inosilicates***, for example pyroxenes and amphiboles), sheet silicates (***phyllosilicates***, for example clay minerals and micas) or framework silicates (***tectosilicates***, for example feldspars and quartz). Hydrolysis, often aided by microbial activity (Chap. 4, § 4.4.5), may be acid ***(acidolysis)*** or alkaline ***(alkalinolysis)***.

Physical disintegration and biochemical weathering give a mixture of constituents of varied sizes and mineralogical properties, which can be grouped according to two criteria—granulometric and mineralogical.

Table 2.2 Grouping and nomenclature of coarse fragments (after Baize and Jabiol, 1995).

Category	Limits, cm
Blocks	>20
Stones	5-20
Pebbles	2-5
Gravel	0.2-2

2.1.3 Granulometric analysis

Particle-size distribution separates the large particles from the small by sieving and sedimentation.

Particle size distribution is obtained by ***granulometric analysis***, which distributes the mineral constituents in size classes. A prior separation at 2 mm separates the ***coarse fraction***, also called ***skeleton grains*** (Tab. 2.2) from the ***fine earth***.

Two techniques are applied in sequence to the latter. Sieving separates the coarse, medium and fine ***sands*** on finer and finer sieves, down to the 50-μm limit. Below this the ***silts*** and ***clays*** are sedimented for a time defined by Stokes law. Finally the proportion of each fraction is expressed as a percentage of the dry weight of the original

Coarse fraction: by international convention, this comprises the individual mineral constituents of size greater than 2 mm and this differentiates it from the ***fine earth***. Most pedological analyses pertain to the fine earth.

Sands: mineral particles from 50 to 2000 μm; ***silts,*** from 2 to 50 μm; ***clays,*** less than 2 μm.

sample; it may be reported in the texture triangle (Chap. 3, Fig. 3.2).

Stokes law

Stokes law states that a spherical particle in laminar fall in a fluid is subject to two opposing forces. The first, resulting from the attraction of the Earth is expressed by $F_1 = 4/3\ \pi r^3(d_1 - d_2)\mathbf{g}$, where r = radius of the particle in cm, d_1 = density of the particle, d_2 = density of the liquid (for water, 1 g cm^{-3}) and $\mathbf{g}$ = acceleration of gravity = 9.81 m s^{-2}. The second is an upward force, $F_2 = 6\pi\eta rv$, where r = radius of the particle, v = velocity of fall in cm s^{-1}, and η = viscosity of the liquid in decapoises, which depends on the temperature.

Granulometric analysis is primarily concerned with the mineral constituents, but it is equally applicable with methodological modifications to organomineral or organic constituents (Tab. 2.3).

Table 2.3 Comparison of mineral, organo-mineral and organic granulometric analyses

Type of particle-size distribution	**Mineral**	**Organo-mineral**	**Organic**
Purpose	Proportion by weight of elementary mineral particles	Fractionation of organic matter Separation of structured matter and organo-mineral aggregates	Proportion of elementary organic particles
Materials studied	Mineral horizons and organo-mineral horizons	Organo-mineral horizons	Holorganic horizons, composts, peats
Type of sieving	Wet, with a prior sieving at 2 mm	Wet, without prior 2-mm sieving	Closed-circuit wet, without prior 2-mm sieving
Preliminary sample treatments	Dissolution of carbonate cements and organic matter	Dissolution of carbonate cements; destruction of lumps by crushing in water with agate balls	Separation of particles by agitation in water without agate balls
Limits for sieving and sedimentation (in μm)	Coarse sand 200–2000 Fine sand 50–200 Coarse silt 20–50 Fine silt 2–20 Clay <2 (after Lozet and Mathieu, 1997)	Free organic matter, structured >100 Organo-mineral macroaggregates 50–100 Organo-mineral microaggregates <50 (after Bruckert in Bonneau and Souchier, 1994)	Coarse fibres >2000 Medium fibres 500–2000 Fine fibres 200–500 Non-fibrous material (organic aggregates) <200 (after Bascomb et al., 1977)
Further results	Mineral texture (Chap. 3, § 3.1.2)	Degree of aggregation (Chap. 3, § 3.2.1)	Organic texture, fibre content (Chap. 3, § 3.1.2, Chap. 9, § 9.1.4)

2.1.4 Mineralogical nature

Mineralogical nature specifies the origin and defines the role of the particle-size fractions of soil.

Sands are usually siliceous, formed of very resistant quartz grains that survive a long time, even with intense acidification of soil (Fig. 2.1). Carbonate sands, rapidly weathered, are rare and are restricted to certain immature soils.

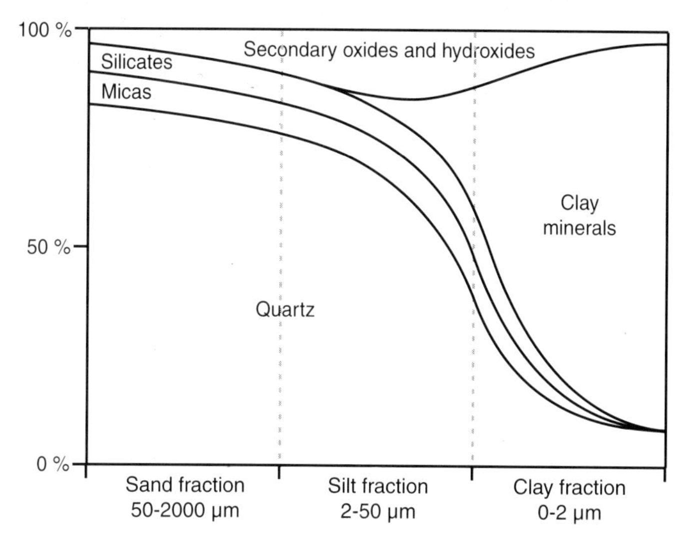

Fig. 2.1 *Mineralogical composition of particle-size fractions (after Schroeder, 1978). Carbonate constituents do not appear in the diagram because they are destroyed prior to granulometric analysis.*

Silts, like sands, originate from physical disintegration of rocks. They contain:

- negligibly weatherable quartz;
- other silicates, slowly weatherable and constituting a long-term nutrient reserve, such as pyroxenes, amphiboles, micas and feldspars;
- carbonate minerals, rapidly weathered by water containing CO_2 and supplying calcium and magnesium to the exchange complex.

Lastly, the ***clay fraction***, or the fraction less than 2 µm in size, is principally composed of clay minerals, defined below, but may also contain metal oxides and colloidal gels. Contrarily, clay mineral particles larger than 2 µm are found in the silt fraction.

Be careful of the two definitions of clay!

2.1.5 Clays, a pivotal role in soil

Structure of clay minerals

Clay minerals: microlayers of silicon-oxygen (silica) tetrahedra and aluminium-oxygen (alumina) octahedra.

Three types of ***clay minerals*** are distinguished according to the number of sheets constituting the layers (Fig. 2.2):

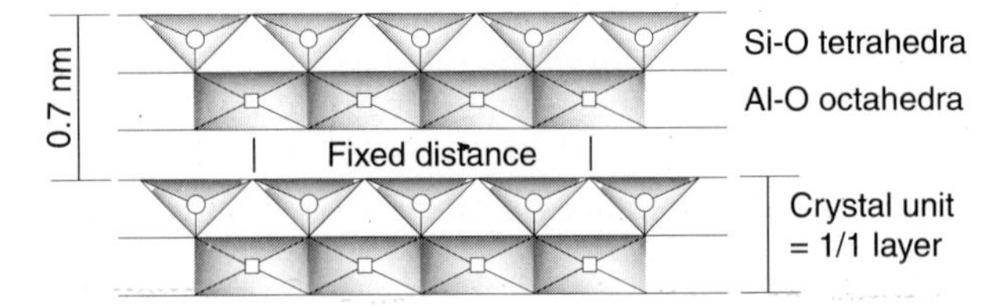

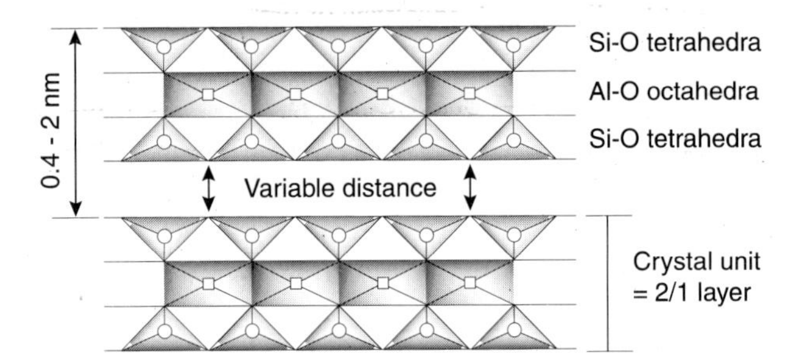

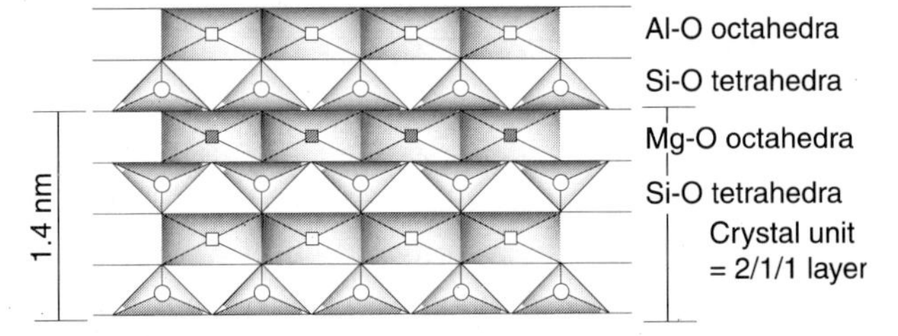

Fig. 2.2 *Crystal structure of clay minerals (explanations in the text).*

Clay minerals: minerals composed of layers with sheets of silica and alumina; they belong to the phyllosilicates and are similar to micas, but the size is much smaller.

- The 1/1 type is the simplest, with a layer composed of two sheets, one of silica tetrahedra and the other of alumina octahedra, with fixed thickness of 0.7 nm, including the interlayer space. ***Kaolinite*** forms part of this group.

- The 2/1 type is composed of a layer with three sheets—two silica sheets with an alumina sheet in-between. This type is close to the micaceous clays such as the ***illites*** (constant 1-nm thickness), the ***vermiculites*** (thickness variable from 1 to 1.5 nm) and the ***swelling clay minerals,*** which are the ***smectites-montmorillonites,*** in which the thickness of the sheets can reach 2 nm through hydration.

- The 2/1/1 type has a layer of three sheets, completed by a supplementary octahedral sheet based on magnesium. These clay minerals with low cation exchange capacity, the ***chlorites***, do not swell, the layer thickness remaining constant.

Origin of clay minerals

Clay minerals originally result from the weathering of rocks by hydrolysis of silicate minerals. The wide variety of minerals and biogeochemial conditions results in several types of weathering of silicates (Chamayou and Legros, 1989; Righi and Meunier in Velde, 1995; Churchman in Sumner, 2000). By way of example, the formation of kaolinite by hydrolysis of orthoclase, a potash feldspar, is as follows:

$$\underset{\text{orthoclase}}{2KAlSi_3O_8} + \underset{\text{water}}{3H_2O} \rightarrow \underset{\text{kaolinite}}{Si_2Al_2O_5(OH)_4} + \underset{\text{potash}}{2KOH} + \underset{\text{silica}}{4SiO_2}$$

In soils, the clay minerals often originate from the disintegration and dissolution of ***detrital rocks*** that had trapped them during submarine sedimentation and ***diagenesis***, and in which they may constitute a considerable portion. Thus soils derived from carbonate rocks rich in clay ***(marls)***, rapidly decarbonated in cool, humid climate, are found to be under the predominant pedogenetic influence of clay minerals and iron oxides, with calcite playing no part (Fig. 2.3).

Detrital rock: sedimentary rock formed by accumulation of rock debris, as distinct from ***organogenic rock,*** which contains animal and plant debris.

Diagenesis: all the processes that progressively transform a sedimentary deposit into a solid rock.

Marl: composite carbonatic sedimentary clay rock, soft and plastic (after Lozet and Mathieu, 1997). Marl, a rock, should not be confused with clay, which is a mineral... even if some marly geological layers are called clays (for example, flint clay, chert clay)!

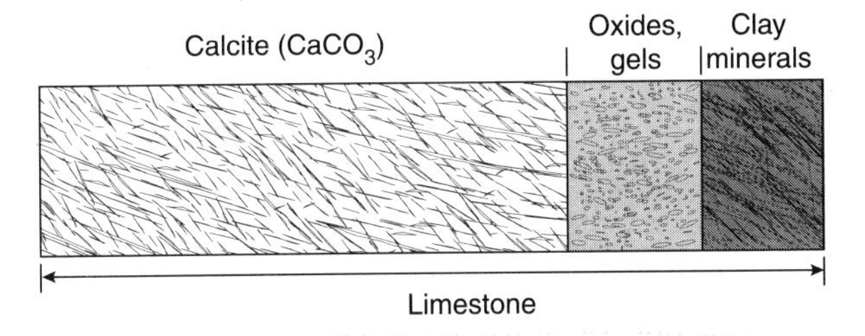

Fig. 2.3 General composition of a carbonate rock. The proportions range from pure limestone with high calcite content to marls, in which clay minerals dominate.

The situation is different under hot climates, in which the soil clays often originate from recombination of alumina and silica. This ***neoformation*** results in clay minerals more or less rich in silica, such as montmorillonite formed from albite:

$$\underset{\text{albite}}{8NaAlSi_3O_8} + \underset{\text{proton}}{6H^+} + \underset{\text{water}}{28H_2O} \rightarrow$$

$$\underset{\text{montmorillonite}}{3Na_{0.66}Al_{0.66}Si_{0.33}O_{10}Al_2(OH)_2} + \underset{\text{silicon hydroxide}}{14Si(OH)_4} + \underset{\text{sodium}}{6Na^+}$$

Properties of clay minerals

All clay minerals play a central role in the soil (Righi and Meunier, in Velde, 1995), influencing its structure, porosity and exchange capacity. They owe this role to

Three properties that clay minerals share with humified organic matter confer on them a major function in the soil.

three particularly important properties: electronegativity, hydrophilicity and the ability to ***disperse*** and ***flocculate***. These properties vary according to the structure of the minerals and their respective specific surface areas: 7–30 $m^2\ g^{-1}$ for kaolinite, 25–150 $m^2\ g^{-1}$ for chlorite and 600–800 $m^2\ g^{-1}$ for montmorillonite.

Clay minerals are ***electronegative*** because of the presence of unsatisfied negative valences at the edges of the layers, following substitutions in the tetrahedral sheets (Si^{4+} by Al^{3+}) or octahedral sheets (Al^{3+} by Mg^{2+}). These sites retain cations and anions, the latter through a bridging cation. The release of plant nutrients and soil fertility depend on this property (Chap. 3, § 3.11; Chap. 4, § 4.2).

Vertic movement: mechanical mixing of soil due to seasonal changes in volume of swelling clays with hydration.

Dispersed state: homogeneous mixture of colloidal particles in water.

Flocculated state: state in which the colloids are joined to form small, fluffy aggregates.

Clay minerals are also ***hydrophilic***, especially swelling minerals of the 2/1 type in which the layers are assembled in a veritable network. Water enters the spaces thus created and augments the hydric reserve of the soil, with proportional changes in volume. During the dry period, on the other hand, shrinkage cracks appear, which express ***vertic movements*** (VERTISOLS; Chap. 5, Tab. 5.2).

Lastly, like all colloids, the clay minerals can appear in ***dispersed*** or ***flocculated*** form, in accordance with forces of repulsion and attraction acting on the particles.

2.2 ORGANIC CONSTITUENTS

From litter to humified organic matter: on the death of living organisms, their wastes and secretions supply the soil with its organic matter, termed 'fresh' or 'particulate' up to its transformation into humus.

2.2.1 Particulate organic matter: litter

Litter, an ensemble of still intact organic matter, with the occasional exception, scarcely transformed.

The first category of organic matter, ***litter***, broadly speaking, is composed of all the organic materials of biological origin at different stages of decomposition, representing a potential energy source for the species that consume them. It comprises the organisms and their parts that have just died and sloughed, and also animal excrement and various compounds directly released into the environment. In a narrower but more usual sense, the term litter refers only to plant debris fallen on the soil—leaves, fruits, twigs and needles—forming the OL horizon.

Some quantitative data

The quantity of above-ground litter is variable according to the plant formation, which in turn depends on the climate (Tab. 2.4).

Quantitatively, litter reaching the soil is dominated by plant debris, although that of animal origin may be locally abundant.

Table 2.4 Annual fall and total weight of above-ground litter of some plant formations (from various sources)

Vegetation	Annual fall of above-ground litter, t ha^{-1} y^{-1}	Total weight on the Earth, 10^9 t
Arctic and alpine tundra	1.0–4.0	8.0
Shrub tundra, thickets	2.5–5.0	5.1
Boreal spruce forest	3.5–7.5	48.0
Temperate deciduous forest	11.0	14.0
Savanna	9.5	3.0
Semi-desert	0.6–1.1	0.4
Equatorial and tropical forests	20.0–40.0	7.2
Temperate grassland	7.5	3.6
Cultivated lands, agricultural ecosystems	0.3–2.0	1.4
Marshy zones	5.0–35.0	5.0

In certain cases though, litter of animal origin may be important as the three following examples show. In an oak grove, the droppings of caterpillars of mixed bombyx reach 400–1000 kg ha^{-1} y^{-1} when they proliferate (Duvigneaud, 1980). A 500-kg cow deposits daily about 25–30 kg dung on the soil of its pasture, covering nearly 1 m^2 (§ 8.3.2)! It has also been determined that the elephants of Tsavo National Park in Kenya deposit 550 kg (dry weight) of excrement per hectare per year (Mason, 1976).

Above-ground litter indeed supplies energy to the soil, but also to streams that traverse ecosystems. This supply is essential to the functioning of the aquatic food chains even far downstream (Mason, 1976).

In general, published data pertain to the narrow definition of litter, forgetting that other very large additions, such as the below-ground litter resulting from the death of roots. However, Fogel (in Fitter, 1985) states that root production can represent 85% of the total net primary production and the annual loss in fine roots, 92% of their biomass. Also, the rate of renewal of the biomass of rootlets is high, especially in soils rich in bioelements (Aber and Melillo, 1991). Killham (1994)

Not to be forgotten is the below-ground litter, often major in the additions of organic matter to soils, even greater than above-ground litter (Chap. 4, § 4.1.6)!

reported 30–90% annual turnover in forest; similarly all the rootlets less than 1 mm in diameter can be replaced in 16 weeks under Sitka spruce, *Picea sitchensis*.

Another 'litter' often forgotten in the organic matter balance: liquid secretions, exudates and urine.

Liquid plant and animal secretions can be very rich in carbon and play a major role in the balance sheets (Chap. 15, § 15.2.3; Kuzyakov and Domanski, 2000; Gaudinski et al., 2001). We just have to think of the slurry coloured by organic acids or of root exudates. With root exfoliation, the carbon secreted by rootlets in the rhizodeposition can represent 20–50% of the organic carbon supplied to the soil, or more (Coleman and Crossley, 1996). Heal and Dighton (in Fitter, 1985) opine that more than 80% of the organic substances metabolized by plants returns to the soil in this form under Alaskan maritime pine, *Pinus contorta*. In bulk these secretions attain 66.5 kg ha^{-1} y^{-1} in boreal mixed forest (Fogel, in Fitter, 1985).

'Unfortunately, although there is a wealth of information on surface-litter input from plants, we are still remarkably ignorant of the quality and quantity of below-ground inputs.' (Heal and Dighton, in Fitter, 1996).

Many volatile industrial compounds in liquid litter.

Many volatile industrial compounds occur in this liquid litter, brought down to the soil by rain, such as

Concepts of production and of biomass; units of mass and energy

Net primary production: portion of the gross primary production serving to increase biomass. It is expressed in units of dry weight per unit area (or volume) and time: kg m^{-2} y^{-1}, t ha^{-1} y^{-1}, etc.

Gross primary production: amount of energy utilized, not only for the generation of organic matter (growth of biomass and other biopolymers) but also for the maintenance of the plant. It is expressed in energy units or mass units per unit area (or volume) and time: kJ m^{-2} y^{-1}; kg m^{-2} y^{-1}, etc.

Biomass: strictly speaking, this is the total mass of living cells of a given site; it is usually reported per unit area or volume. Biomass can be subdivided according to the taxonomic category considered. It is different from ***necromass***, which represents the mass of dead cells still attached to the living organism (for example, dead branches on a tree, dried leaves of the previous year in a tuft of grass). But in reality, a tree for example is often considered biomass in its entirety—except possibly dead branches—even if more than 97% of its cells are no longer living (Lovelock, 1992)!

In the calculation of production or biomass, the energy expressed in joules or calories is equivalent to the mass expressed in kg, because each mass of organic matter contains chemical energy. On average, 1 g protein supplies 4 kcal, 1 g carbohydrate 3.8 kcal and 1 g lipid 9 kcal (Moore and Bellamy, 1974).

In physics, the calorie does not belong to the international system, SI, which defines the joule (J) as the unit of energy, expressed in kg m^2 s^{-2}. It is related to the calorie by the relation 1 cal = 4.19 J. In biology, the orders of magnitude necessitate common use of the kilocalorie (kcal); unfortunately, this is often called calorie or large calorie for convenience.

'Until recently, leaf litter was considered the main source of carbon entering the soil ecosystem. Studies of soil cores have changed this picture to one in which fine roots and mycorrhizae dominate the supply of organic matter (...); root exudates and rhizodeposition of sloughed cells, root caps, etc., might also contribute a significant source of energy for soil processes' (Fogel, in Edwards et al., 1988).

polychlorinated hydrocarbons (Benckiser, 1997). Some are degraded by soil organisms, but others can persist for a very long time (Chap. 14, § 14.4.4, Chap. 17, § 17.2.2).

A better accounting of the liquid litter could well modify the mass and energy balances established otherwise. Some models of evolution of organic matter try to integrate it by including it in the active organic fraction of soil (for example, the CENTURY model of Parton, in Bryant and Arnold, 1994, and in Powlson et al., 1996; Chap. 5, § 5.2.3).

Qualitative aspects

Plant litter presents two opposing characters:

A litter that ameliorates, another that acidifies ... and all possible intermediates!

- It could be ***ameliorating***, that is, rich in nitrogen and cellulose, but relatively poor in lignin; it activates the bacterial processes in soil, especially the availability of nitrogen to plants. Ameliorating litter, rich in easily available energy (Chap. 6, § 6.1.3; Heal and Dighton, in Fitter, 1985), is supplied by broad-leaved trees such as ash, maple, willow, alder, elm and lime, as well as most herbaceous plants.

- It could be ***acidifying***, on the contrary, poor in nitrogen but rich in lignin: it therefore inhibits bacterial activity in soil. This is because of ***tannins***, toxic ***phenols*** and organic acids, which are released directly by the litter, such as salicylic acid, or released by transformation of lignin. Lignin of this type, of low energy level, comes from spruces, pines, heathers, myrtles and rhododendrons.

Tannins: complex group of water-soluble polyphenolic compounds of relative molecular mass from 500 to 3000, with certain properties in common, such as the ability of coagulating proteins or combining with other polymers like cellulose and pectins (after Ribéreau-Gayon, 1968).

Phenol: molecule composed of at least one 6-carbon ring with conjugate double bonds and an alcoholic functional group (hydroxyl).

Between the two, litter from chestnut, oak, beech and white fir has its ameliorating or acidifying characters strongly or weakly expressed according to the general pedochemical situation (Toutain, 1974).

Lavelle et al. (in Dindal, 1980) fed young tropical earthworms *(Millsonia anomala)* with leaves or roots of grasses. Their growth was less in the latter case because of the smaller energy supplied by the substances contained in root wastes.

The nature of the below-ground litter, different from the above-ground litter, influences the soil fauna.

Animal litter, which includes among others, dead bodies (Chap. 8, § 8.3.1), excretions (Chap. 8, § 8.3.2), fur and feathers, is composed of biochemical compounds in proportions different from those in plant litter. Proteins and lipids are better represented, whereas cellulose and lignin are in low quantity and present only in excrements. These latter also contain large amounts of vitamins, minerals and growth factors (Mason, 1976). ***Chitin*** is characteristic of this kind of litter.

Feathers, fur, carcasses and excrements ...

Chitin: nitrogenous polysaccharide present particularly in the cell walls of fungi and the exoskeletons of arthropods.

C/N ratio: ratio by weight of the total organic carbon to total nitrogen in a soil. Total nitrogen comprises organic and mineral forms, but the latter scarcely exceeds a few per cent.

UV or IR spectrometry: measurement of the absorption by a pigment as function of the wavelength, in these cases, in the ultraviolet or infrared.

Gel chromatography: method for qualitative and quantitative separation of components of a mixture by their retention on a support—in this case a gel—which sorts them according to their molecular size.

High-performance liquid chromatography, HPLC: liquid-chromatography system with high performance, working under high pressure and controlled temperature.

Nuclear magnetic resonance, NMR: analysis of the behaviour of electrons in a molecule subjected to an intense magnetic field, enabling determination of functional chemical groups: alcohol, acid, amino, keto, etc.

Aromatic: said of a molecule comprising one or more rings of 6 carbon atoms linked by conjugate double bonds. The simplest is that of benzene, C_6H_6 (Fig. 2.4).

Polycondensation: reaction of macromolecule formation involving many different monomers.

Livestock add large quantities of elements through their excrement, modifying the reserve in soil: a milch cow and a horse respectively provide per year 105 and 65 kg nitrogen, 15 and 12 kg phosphorus, 149 and 91 kg potassium, 12 and 8 kg magnesium and 37 and 25 kg calcium.

A good index used for litter quality is the ***C/N ratio***. High values of this ratio, above 25 to 30, indicate poorly degradable litter, resistant to organisms and biochemical attack. Low values, nevertheless higher than 6 or 7, point to nitrogen-rich organic materials easily accessible to decomposing agents. The C/N ratio is also applied to other categories of soil organic matter such as the products resulting from humification, or to entire horizons provided they contain organic matter.

2.2.2 Inherited or humified macromolecules

If litter, at least in its restricted sense, is easily visible and definable, the same is not true of other categories of soil organic matter. Their delimitation was at first empirical, based on physical separations or chemical extractions by various reagents: water, alcohol, acetone, acids, bases, etc. According to their affinity for these extractants, macromolecules were grouped into more or less homogeneous categories. Then, gradually, precise analytical techniques such as ***UV*** or ***IR spectrometry***, ***gel chromatography***, ***high performance liquid chromatography (HPLC)*** and also ***nuclear magnetic resonance (NMR)*** enabled their characterization, in particular their functional groups (Stevenson, 1982; Tate, 1987; Baldock and Nelson in Sumner, 2000). But despite the many methods available, a large fraction of the soil organic matter, very strongly bound to minerals, still defies identification (Paul, in Grubb and Whittaker, 1989).

Although certain macromolecules are directly inherited from organic debris, most of them are synthesized in the soil, following complex chemical and biochemical processes that constitute humification (Chap. 5, § 5.2.3). Among the former, cellulose, lignin, proteins and lipids are predominant. In the latter are found ***aromatic*** compounds with varying degrees of ***polycondensation***, such as very stable macromolecules forming part of humin. But the actual boundary between the two categories is not easy to locate, especially in the humins, which can be inherited from plant material or neoformed.

Their respective proportions in soils are very variable, according to the horizon considered, the physico-chemical environment and even the type of vegetation. Data from different authors should be compared with care, as methods of separation are not always identical (Tab. 2.5).

Table 2.5 Average proportions of broad categories of organic substances in plant and soil (after Foth, 1990).

Category of organic substance	Proportion in plants (% of category)	Proportion in soil (% of category)
Cellulose	20 – 50	2 – 10
Hemicelluloses and proteins	10 – 30	0 – 2
Lignin	10 – 30	35 – 50
Proteins	1 – 15	28 – 35
Lipids, waxes, others	1 – 8	1 – 8

More than 8000 different structures of phenolic compounds have been enumerated in living organisms, mostly of plant origin. About one-half belong to the ***flavonoid*** group (Harborne, in Crawley, 1997).

2.2.3 Inherited organic substances

Dry plant material is composed, to the extent of 99%, of eleven major elements: C, H, O, N, P, S, Ca, Mg, K, Cl and Na (Callot et al., 1982). The molecules forming it are of many types and their proportions vary with category of litter (Tab. 2.6; Fig. 2.5). These are ***carbohydrates***, lignins, lipids and nitrogenous compounds (proteins, amino acids, nucleic acids, nucleotides (Fig. 2.6). In the soil, these molecules are degraded by specific enzymes (Chap. 14, § 14.3.1).

Among the carbohydrates, ***cellulose*** is the major constituent of plant cell walls. Formed of chains containing 1400 to 10,000 glucose units, it is rapidly decomposed in the soil, except when it is impregnated with lignin. The ***hemicelluloses*** are degraded still more rapidly. These ***polysaccharides*** can constitute more than 15% of the soil organic matter; they are essential for structure formation because of their aggregation properties. ***Monosaccharides*** or simple sugars (glucose, fructose) are present only in small amounts because they are rapidly consumed by microorganisms.

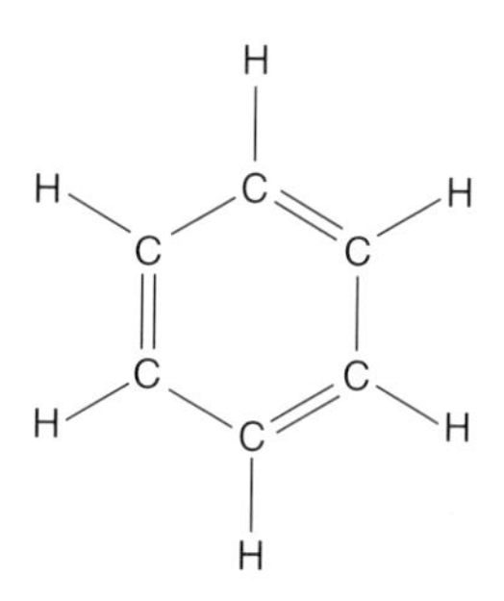

Fig. 2.4 *Benzene, the simplest of aromatic molecules.*

Flavonoid: phenolic compound containing a C_6–C_3–C_6 backbone, the C_6 being a benzene ring and the C_3 variable in different compounds, often containing oxygen (Chap. 16, § 16.2.1). Flavones, flavonols and anthocyanins are flavonoids (Ribéreau-Gayon, 1968).

Four great families of inherited substances: carbohydrates, lignins, lipids and proteins.

Carbohydrate (or sugar): component of living matter of general formula $C_n(H_2O)_p$.

Hemicelluloses: all the polysaccharides of plant cell walls other than cellulose. They are chiefly composed of xylans, arabans and pectins.

Polysaccharide: sugar macromolecule resulting from the condensation of a few hundred to several thousand mono saccharides.

Monosaccharide: basic unit of sugar with a molecule containing 3 to 8 atoms of carbon, most often 5 or 6.

Table 2.6 Major types of biomolecules inherited by soil (various sources)

Type	Example	Formula	Extractant (choice)	Origin
Carbohydrates				
Monosaccharides and derivatives	D-glucose	$C_6H_{12}O_6$	Water, sulphuric acid	Cytoplasm
	D-xylose Galacturonic acid	$C_5H_{10}O_5$		
Polysaccharides	Cellulose	$(C_6H_{12}O_6)_n$	Hot water, formic acid, caustic soda	Cell walls
	Hemicelluloses			Cell walls
Lignins				
Lignins	Based on the phenylpropane structure; 3 types:	$RR'OHC_6C_3$		Cell walls
	p-coumaryl alcohol	R=H, R'=H		
	coniferyl alcohol	R=H; R'=OCH_3		
	synapyl alcohol	R=OCH_3; R'=OCH_3		
Lipids and associated compounds				
Fatty acids	Lignoceric acid	$CH_3(CH_2)_{22}COOH$	Ether	
Waxes	Esters of fatty acids	$CH_3(CH_2)_xCOO-(CH_2)_y(CH_3)$	Chloroform, ether, benzene, alcohol	Cuticles, residues of inhibited biological activity
Resins				Wood of conifers
Hydrocarbons, alkanes, terpenes	Pyrene, toluene			Neoformed or atmospheric deposition
Carotenoids	Alpha-carotene			Plant pigments
Porphyrins	Chlorophyll derivatives		Acetone	Plant pigments
Nitrogenous compounds				
Proteins	Collagen		Hydrochloric acid, caustic soda	Living cells, skin, bone
Amino acids	Leucine	$(CH_3)_2CH_2NH_2-CHCOOH$		Root exudates
	Beta-alanine	$NH_2(CH_2)_2COOH$		Antibiotics

Lipids: hydrophobic and largely aliphatic compounds, a major part of which are fats and steroids. ***Aliphatic*** denotes a molecule in which the carbon atoms form a linear chain.

Lignins have a more complex chemical structure than cellulose, with very varied arrangements starting from a structure based on phenylpropane (Stevenson, 1982). More resistant to decomposition because of their aromatic nuclei, except in acid soils where certain ligninolytic fungi are very active (Chap. 8, § 8.6; Hammel, in Cadisch and Giller, 1997), they can represent up to 30% of wood. After cellulose, lignin is the most abundant organic substance on the Earth (Lüttge et al., 1996).

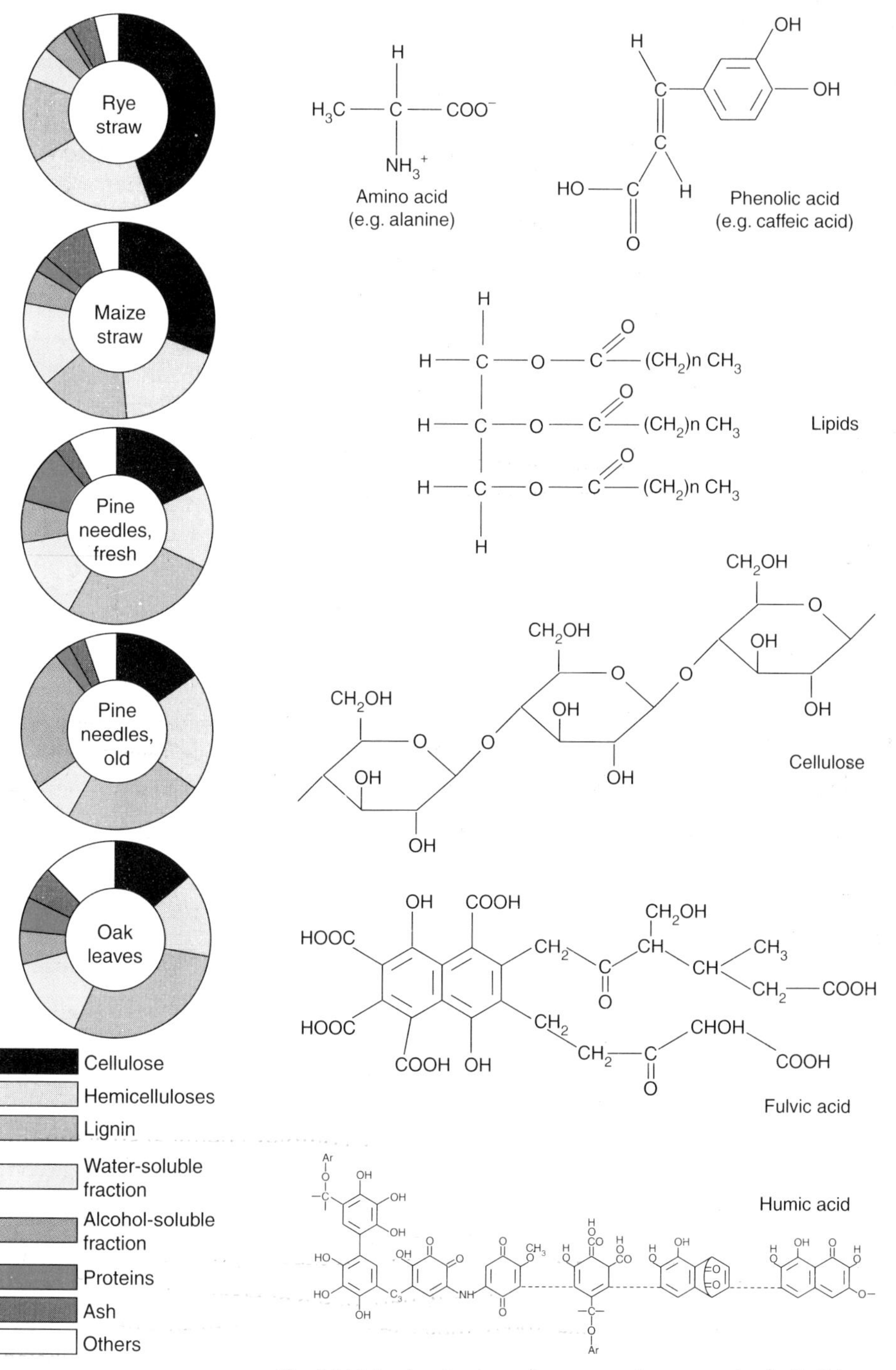

Fig. 2.5 Composition of some types of litter (after Ross, in Ellis and Mellor, 1995).

Fig. 2.6 Molecular structure of some organic compounds of soil. Other model structures of fulvic and humic acids are presented and described in detail by Stevenson (1976), Ziechmann (in Blume et al., 1996), Schulten and Schnitzer (1997) and Van Breemen and Buurman (1998).

Murein: complex macromolecule, characteristic of the bacterial wall. It is composed of linear polysaccharide chains with short peptide side chains made of four amino-acids. Peptides of different chains in murein are bonded to each other, thus giving the bacterial wall the structure of a rigid network, an enormous macromolecule ***(saccule)*** totally surrounding the cell.

Urea: soluble organic nitrogenous waste excreted by mammals, most adult amphibians and numerous fishes.

Uric acid: scarcely soluble organic nitrogenous waste excreted by snails, insects, birds and terrestrial reptiles.

Amine: organic compound bearing an amino group; general formula $R–NH_2$.

Protein: macromolecule (molecule mass > 10 kDa) synthesized in a living cell at ribosome level. Highly organized structure; it is composed of one or many linear amino-acid chains.

Soil ***lipids*** and associated compounds, fatty acids, waxes and resins can be as high as 1% (grasslands) to 20% (PODZOSOLS and acid HISTOSOLS) of the total organic matter. Some fatty acids—numbering about fifteen in soils—are capable of bonding to humic and fulvic acids. Waxes and resins come directly from plant tissue, the former from the cuticles, often thick, that protect leaves and needles. Porphyrins, derived mostly from chlorophyll, are in small amounts, excepts in soils poor in oxygen.

Nitrogenous compounds, which can constitute more than a third of the soil organic matter and can contain 95% of the total nitrogen, form a wide variety: nucleic acids (DNA, RNA and their bases—see the definitions in Chap. 15, § 15.5.5), chitin, ***murein***, ***urea*** and ***uric acid***, ***amines***, ***proteins*** and their constituents, the free amino acids. The last-mentioned can be leached, leading to loss of nitrogen. They can also be fixed in humified compounds.

2.2.4 Humified organic substances

Organic substances resulting from humification are broadly grouped according to their molecular mass, which also reflects their behaviour *vis-à-vis* extraction procedures (Tab. 2.7).

Table 2.7 Characteristics of macromolecules resulting from humification

Type	Molecular mass, Da	Solubility			
		Water	Alcohol	Alkali	Acid
Crenic acids	100 – 500	Yes	Yes	Yes	Yes
Hymatomelanic acids	500 – 900	No	Yes	Yes	No
Fulvic acids	$900 - 2{\cdot}10^3$	Yes	Yes	Yes	Yes
Humic acids:		No	No	Yes	No
• grey	$2{\cdot}10^3 - 5{\cdot}10^4$	• insoluble in salt solution			
• brown	$5{\cdot}10^4 - 10^5$	• soluble in salt solution			
Humins	10^5–$5{\cdot}10^5$	No	No	No	No

Crenic acids, short-lived precursors in the first stages of humification (Chap. 5, § 5.2.3); hymatomelanic acids, often inherited.

Water-soluble ***crenic acids*** are formed by polycondensation of two or three aromatic nuclei. Recognized as early as 1806 by the Swedish chemist Berzelius—who also later identified humic acids and humin (Boulaine, 1989)—they are now often included in the very light fulvic acids. A little larger are

hymatomelanic acids, composed of polycondensates of crenic acids, and also of largely aliphatic inherited molecules, such as ***bitumens***, common in raw humus and peats.

Bitumens: ensemble of largely aliphatic compounds, comprising various lipids, resins and waxes.

The yellow-coloured ***fulvic acids*** have long aliphatic or peptide side chains and a small aromatic nucleus. They are very reactive because of dissociated —COOH groups that confer on them, on account of their electronegativity, a strong ability to bond with divalent or trivalent cations. The ratio of ***absorbances*** at 465 and 665 nm (Fig. 2.7), ***Q4/6***, is 6—8.5 against 2.2—5 for ***humic acids***. The latter differ from the former in having shorter side chains attached to a larger aromatic nucleus (Figs. 2.6, 2.8; Chap 14, § 14.4.3), and are generally derived by polycondensation.

Fulvic acids to humic acids, variations on a single basic structure.

Absorbance (E): in spectrophotometric measurement, the logarithm of the ratio between the intensity of the light beam before (I_0) and after (I_t) passing through the sample. The term ***optical density***, long used as a synonym for absorbance, is presently restricted to photometry of turbid media such as bacterial suspensions (Widmer and Beffa, 1997).

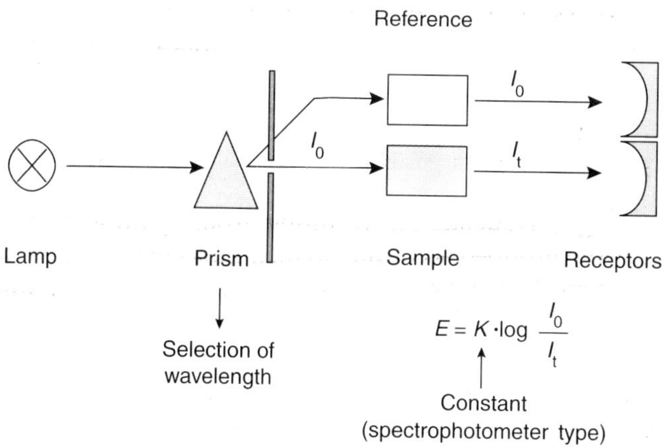

Fig. 2.7 *Principle of spectrophotometric measurement. Symbols are explained in the text.*

With a very large nucleus and short chains, humin is extremely stable and is firmly fixed on clay minerals or colloidal gels, ensuring structural permanence (Chap. 3, §§ 3.2, 3.6). It results from polycondensation of fulvic and humic acids (***insolubilized humin***, Chap. 14, § 14.4.3) or bacterial neosynthesis (***microbial humin***, Chap. 4, § 4.4.1) or even inheritance of substances already present in litter (***residual*** or ***inherited humin***, Chap. 5, § 5.2.3).

The largest humic molecules in soil: humin.

In the biochemical sense, the term ***humus*** encompasses all the organic compounds of soil resulting from humification. However, Lozet and Mathieu (1997) also considered under the term 'young humus'—as against 'stable humus'—the transitory organic substances directly

Just as there are clays and clays, there are also different kinds of humus: the first is biochemical . . .

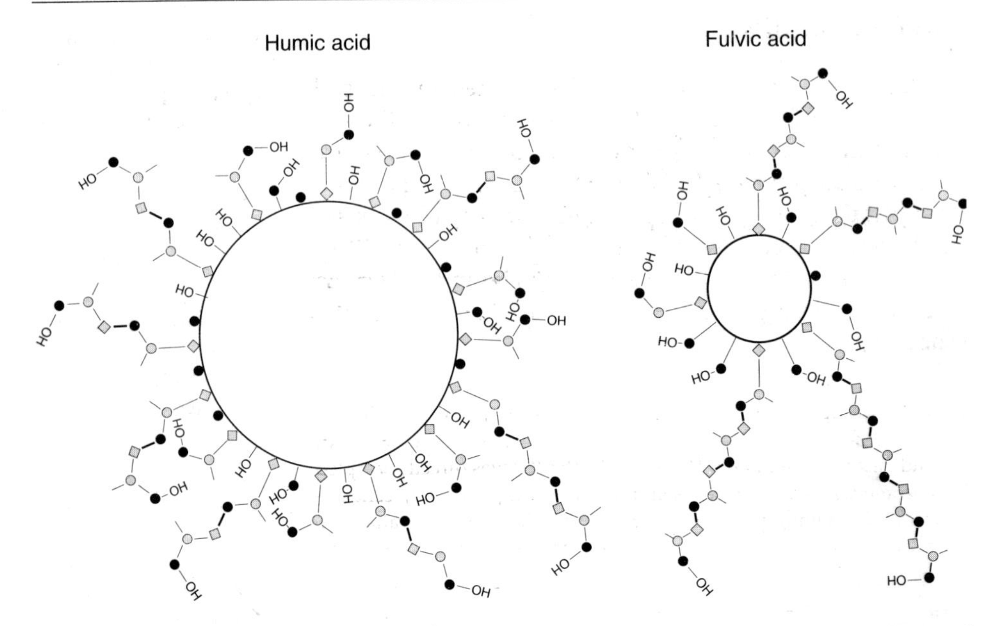

Fig. 2.8 *General structure of fulvic acid and humic acid (after Andreux and Munier-Lamy, in Bonneau and Souchier, 1994). Also see Fig. 2.6 and Chap. 14, § 14.4.3 for further details.*

inherited from fresh organic matter, hence not yet humified. Wallerius was the first, in 1761, to define soil humus as 'decomposed organic matter' (Morel, 1996).

. . . and the second, macromorphological—but it would be better to forget it!

On another scale, this term is sometimes applied to the sequence of upper horizons of the soil, which contain organic matter. Here the macroscopic effect of biochemical humus is reflected. To better differentiate the two cases the term 'humus form' is preferred for describing the morphology of the upper part of the soil, the humiferous episolum (Chap. 6, § 6.1.1).

2.3 THE SOIL SOLUTION

2.3.1 Definition and functions

The soil solution, water enriched in mineral and organic ions and molecules.

Very mobile, the liquid fraction of soil or ***soil solution*** is an important functional crossroads in the soil because of its ability to transport substances between, for example, the solid fraction and plant rootlets (Chap. 4, § 4.2.1). It ensures leaching of cations during pedogenesis (Chap. 5, § 5.3.2) and is the site of numerous dissolution and insolubilization processes. Compared to solid constituents, which often pertain to long-term evolution, it mirrors the present functions.

Soil solution: the entire soil water and the substances dissolved in it.

2.3.2 Determination

Water appears in three principal states in the soil (Chap. 3, § 3.4.2) and it is not always easy to find the soil solution in it. The total water content of a soil (moisture content) is subject to rapid fluctuations in accordance with precipitation, evapotranspiration and capillary rise. Furthermore, free groundwater has chemical qualities different from those of more strongly bound water, as seen in the example of the swamp of Lake Neuchâtel (Tab. 2.8).

Seeking the solution ... in the soil!

Table 2.8 Comparative chemical qualities of groundwater and the soil solution in four phytocoenoses of the banks of Lake Neuchâtel, Switzerland (after Cornali, 1992). The groundwater (G) was directly sampled in piezometers, while the soil solution (S) was expressed by centrifugation to the wilting point after the soil had drained. The values are the means of several summer samplings, scarcely variable at one point.

Item	Type of water	Pine forest *(Pinus sylvestris)*	Grassland *(Molinia coerulea)*	Grassland *(Schoenus nigricans)*	Grassland *(Cladium mariscus)*
Soil type		RENDISOL, redoxic	RÉDUC-TISOL with mull	RÉDUC-TISOL with hydromull	RÉDUC-TISOL with anmoor
Conductivity ($\mu S\ cm^{-1}$)	G S	424 440	235 398	746 557	881 408
Organic C (abs. at 270 nm–1 cm)	G S	417 1647	58 680	85 697	308 738
K^+ ($mg\ L^{-1}$)	G S	1.7 6.1	1.2 5.4	2.6 6.8	3.1 3.0
NO_3^- ($mg\ L^{-1}$)	G S	0.3 10.3	0.2 0.9	0.1 0.5	0.1 1.7

Which then is the true soil solution from which plants draw their nutrition? No absolute reply can be given, the plant adapting its suction to the immediate conditions, at times saturated, at times unsaturated.

On another scale, the soil solution is also the integrator of phenomena of the entire drainage basin, which reflect the interplay of geochemical alterations and precipitation as well as biological mechanisms of absorption and excretion. The soil solution progressively changes in quality from the top to the bottom of a slope. This is

From the drainage basin to the soil solution, or 'the multiscale funnel'.

particularly true during passage through agricultural lands in which artificial additions and drainage often accelerate the processes. But the permanence of the soil solution along a drainage basin depends considerably on the permeability of the underlying rock: on impermeable gneisses, the same water can successively pass through several soils, while on a fissured limestone, it immediately disappears underground!

2.4 THE SOIL ATMOSPHERE

2.4.1 Determination

The soil atmosphere: free or dissolved gases difficult to determine (Scanlon et al., in Sumner 2000).

If the soil solution is at times difficult to determine, what can be said of the soil atmosphere, when the simple step of creating an opening for withdrawal of a sample modifies its composition? Technical difficulties in its study result in knowledge of the soil gases being much less advanced than that of other constituents, in spite of the essential role of these gases in the regulation of exchanges within the soil and with the outer air.

2.4.2 Location and composition

Considerably more carbon dioxide and a little less oxygen.

Gaseous exchanges in soil: very difficult problems of fluid mechanics!

Air in the soil occupies the pores left empty of water when the water withdraws from them, the largest first, then the finer. Its quantity thus depends on a combination of texture, structure and moisture content (Chap. 3). But it is also in contact with the outer atmosphere with different relative concentrations of free gases (Tab. 2.9). The composition of soil air shows seasonal fluctuations related to biological activity: root respiration, aerobic microflora and the fauna consume oxygen and release carbon dioxide. Bacterial nitrogen fixation and denitrification (Chap. 4, § 4.4.3) modify the nitrogen concentration while methane can be consumed or produced (Chap. 4, § 4.4.4, Chap. 9, § 9.2.1).

The average production of carbon dioxide in the soil is estimated to be 15 t ha^{-1} y^{-1}, two-thirds of which is by microbial activity. If the structure is aerated, this production does not accumulate because the air is renewed by diffusion with the outside, regulated by differences in concentration. Specific quality of air in microsites is essential for microbial activity and can vary greatly over distances shorter than a millimetre (Chap. 3, § 3.6.1, Chap. 15, § 15.5.1). As for the soil fauna, it is more resistant to CO_2 than that living in free air.

Table 2.9 Composition of soil air and the outer atmosphere

Constituent	Soil air, %	Outer atmosphere, %
Oxygen	8–20.5 in well-aerated soil 10 after a rain 2 in compact structure 0 in reduced horizons	21
Nitrogen	78.5–80	78
Carbon dioxide	0.2–3.5 5–10 in rhizosphere	0.03
Water vapour	Generally saturated	Variable
Various gases	Traces of H_2, N_2O, Ar Under anoxic conditions: NH_3, H_2S, CH_4	1 (largely Ar, others in traces)

Gaseous exchanges in the soil are distinguished from liquid exchanges by the presence of zones of production and consumption *in situ*, whereas soil water is essentially of external origin, neglecting the additions from decomposing cells of edaphic organisms.

2.5 LIVING ORGANISMS: THE MICROFLORA

The living organisms of the soil are bacteria, fungi, algae, below-ground parts of plants and also very varied animals, protozoa to mammals. All participate in one way or another in formation and evolution of soil, especially its organic fraction. Their number and biomass in the soil are often beyond imagination (Tab. 2.10).

'Are living organisms part of the soil? We would include the phrase 'with its living organisms' in the general definition of soil. Thus, from our viewpoint soil is alive and is composed of living and nonliving components, having many interactions' (Coleman and Crossley, 1996).

Table 2.10 Abundance of living organisms in the soil. These estimates pertain to all the continents (various sources; n.d. = not determinable)

Organisms	Approximate number		Average biomass	
	per g dry soil	per m^2	kg ha^{-1} in 20 cm depth	% (without roots)
Bacteria	10^6–10^9	10^{11}–10^{14}	1500	25
Fungi	n.d.	n.d.	3500	59
Algae	1000–10^5	10^8–10^9	10–1000	traces
Protozoa	10^4–10^6	10^9–10^{11}	250	4
Soil fauna (except protozoa)	0.1–1000 (according to groups)	10–$5 \cdot 10^6$ (according to groups)	1–5000 (according to groups)	12
Roots	n.d.	n.d.	6000	—
Total	n.d.	n.d.	~12,000	100

Living organisms belong to three fundamental domains: *Bacteria, Archaea* and *Eukarya*.

Phylogeny: evolutionary history of a species or a group of related species.
Domain: highest hierarchical level in the classification of living organisms.
Methanogenic: able to anaerobically produce methane, CH_4.
Cytoplasm: zone between the nuclear envelope and the membrane enclosing the cell.
Plastid: typical plant cell organelle, whose basic type is the chloroplast, site of photosynthesis.
Mitochondrion: organelle constituting, in the *Eukarya*, the site of cell respiration.

Eukaryotic: said of a cell with nucleus enclosed by a double membrane, whose cytoplasm contains internal membrane systems and organelles (mitochondria, chloroplasts) surrounded by a double membrane.
Prokaryotic: said of a cell with poorly differentiated cytoplasm, without true nucleus or internal organelles, plastids or mitochondria; its respiratory and/or photosynthetic functions are ensured by the cell membrane or membranes derived from it.

2.5.1 Three broad categories of organisms

The ***phylogeny*** of living organisms shows that two of the three fundamental ***domains*** of their classification have a prokaryotic cellular structure: the *Bacteria*, which include most of the common bacteria, and the ***Archaea***. The latter comprise, on the one hand, bacteria living in conditions of extreme temperature or salinity, as also the ***methanogenic bacteria***, characteristic of highly anaerobic environments such as submerged sediments or soils (Chap. 4, § 4.4). The third domain is the *Eukarya*, with eukaryotic cellular organization; it comprises animals, plants, fungi and numerous other forms (among other different categories of protozoans).

The term microflora, like 'microorganism', is very elastic. Two of the principal characteristics typical of the microflora being small size and often lack of morphological expression, one of the best delimitations possible is methodological: ***microflora*** comprises the organisms whose direct observation provides little or hardly any information on identity and function. To observe and identify, and to characterize their functions, these organisms require cultivation or molecular methods (Chap. 15, § 15.5.5). Four groups have been recognized in soils: bacteria, fungi, microscopic algae and protozoa. In soil science, the last one is conventionally included in the microfauna (Chap. 12, § 12.4.1).

2.5.2 Bacteria with multiple functions

Bacterial organization and reproduction

Broadly speaking, the term bacteria groups the unicellular organisms with ***prokaryotic*** structure (*Bacteria* and *Archaea*), in contrast to protozoans and yeasts, which are ***eukaryotic*** (Fig. 2.9).

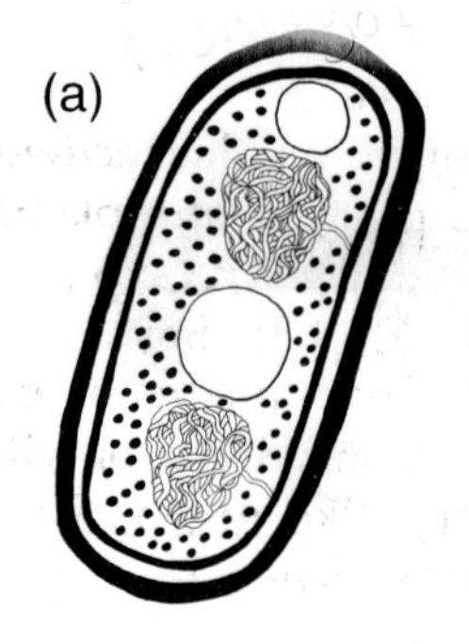

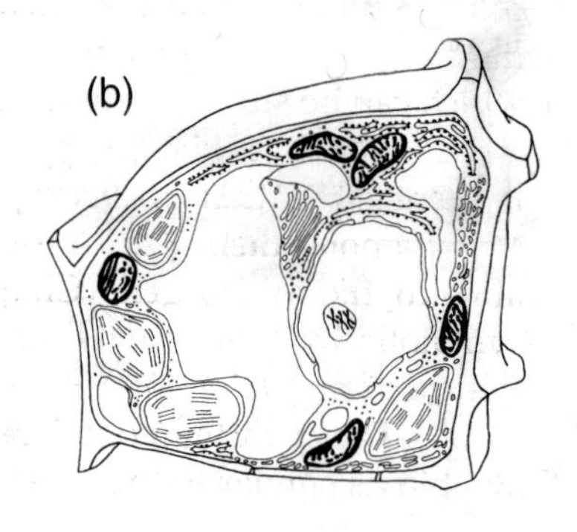

Fig. 2.9 *Diagram of (a) prokaryotic cell and (b) plant eukaryotic cell. Explanations given in the definitions.*

With size close to a micrometre, the bacterial cell is on average 10 times (in linear dimension) or 1000 times (by volume) smaller than that of an eukaryote. On the other hand, its active cellular surface area per unit volume is much higher, leading to an almost unimaginable potential metabolic activity: in a rich medium, *Escherichia coli* doubles its biomass every twenty minutes. Thus a single cell weighing $5 \cdot 10^{-13}$ g will give, if there are no environmental constraints, 2360 tons of cells after 24 hours, a mass equal to that of the Earth after 44 hours 20 minutes, while after four days the compacted mass of bacteria will occupy the volume of our Galaxy! In nature, fortunately, bacteria are subject to strict control of their activity by ecological factors such as limited supply of nutrients or predation by other microorganisms, thereby maintaining their population relatively constant.

Small size but great variety and intense activity.

Potential growth of *Escherichia coli:* from the micrometre to the Galaxy in four days!

Quantitatively, bacterial biomass is estimated at about 10^9 living germs per gram soil, that is, an always modest biomass of less than 500 μg g^{-1}; but the active surface area is still 50 cm^2 g^{-1}. Soil microorganisms under cows in a pasture thus have a potential metabolic activity several times higher than that of the animal!

Where bacteria work better than the cow!

Morphology

Bacteria are very modest in their morphological expression; most have the shape of rods or small spheres, some among them are curved or spiral, others branched (Plate VII-1) to the extent of forming an actual mycelium, as in the group of Actinomycetes (Chap. 16, § 16.2.2).

Poorly expressed morphology.

Some bacteria move themselves by means of ***flagella***, which function as ships screws set in motion by a rotary motor. Others, the **myxobacteria**, common in soils and litter, move by gliding like minute slugs on solid bodies. Also they have a 'social' behaviour; they move, grouped in 'swarms' in search of prey, generally other bacterial cells or fungi. When the environment is impoverished, they assemble to form true fruit-bodies, the shapes of which can be surprisingly differentiated.

Flagellum: specialized organelle for locomotion.

The **cyanobacteria** are photosynthetic and their activity is identical to that of chloroplasts of algae and plants. Many among them (e.g. *Nostoc* spp., *Anabaena* spp.) are able to fix dinitrogen (Chap. 4, § 4.4.3). Some live symbiotically with plants, like *Anabaena azollae* in the tissues of small floating ferns of the genus *Azolla*, or with fungi. These latter constitute, just like green algae, symbioses similar to the lichens (cyanolichens).

Because of their photosynthetic activity, the cyanobacteria have long been considered algae (blue-green algae, Cyanophyceae).

Bacteria and soil structure

Bacterial populations mirror soil structure, principal factor in the diversity of edaphic organisms.

At the scale of bacteria, a soil presents a mosaic of very distinct ecological niches. Thus, bacteria with mutually exclusive living conditions, such as obligate aerobes and obligate anaerobes, live together at times at a distance of a fraction of a millimetre. Furthermore, the conditions can rapidly change at such a scale. One can imagine what a sequence of events at the submillimetric scale accompanies the decomposition of a small arthropod!

In turn, bacterial populations act on soil structure.

Bacteria synthesize compounds, in particular polysaccharides, very resistant to enzymatic degradation. By their long life these compounds constitute a large fraction of the humified organic matter—microbial humin—participating in the formation of microaggregates. They also condition the mode of development of bacteria in their environment, ensuring the cohesion of microcolonies (Plate VII-1).

Bacterial functions and physicochemical environment of soil

Bacteria are the essential regulators of gaseous equilibria in soil and biogeochemical cycles.

The effects of the microflora on physicochemical characteristics of soil are primarily related to bacterial functions. For example, aerobic respiratory activity, which consumes oxygen, can lead to anoxia; this is pertinent not only to hydromorphic soils in which the diffusion of air is restricted, but also to the centre of large particles in aerated soils (Chap. 3, § 3.10.1). When excess of carbon-rich substrate is present, bacteria hoard the available nitrogen. On the other hand, under conditions of shortage of nitrogen, others are able to fix elemental nitrogen, N_2.

Directly or indirectly, bacteria act on the entire biocoenose.

By synthesis of growth factors, (vitamins) as well as antibiotics, certain bacteria exert control, positive or negative, on other organisms.

Their biogeochemical functions are most important.

But it is above all by their biogeochemical functions, such as the mineralization of organic matter, oxidation of reduced inorganic compounds, anaerobic reduction of oxidized inorganic compounds, dissolution or precipitation of minerals, not to forget the transformation of certain organic compounds to humin, that bacteria play an essential role in soil formation and development (Chap. 4, § 4.4; Chap. 5, § 5.2).

2.5.3 Filamentous fungi, a circulatory system in soil

Taxonomy and metabolism

Compared to bacteria, fungi present a relative uniformity in metabolism: they are all aerobic ***heterotrophic*** eukaryotes with extracellular digestion. They belong to very different taxonomic groups, though.

It is not only boletus or morels. There are moulds and lower fungi too!

The most spectacular are the ***Macromycetes***, which form macroscopic fruit-bodies, the carpophores: everybody knows Basidiomycetes such as boletus, chanterelles, amanita and russula or **Ascomycetes** such as morels or truffles. However, culturing of soil suspensions on appropriate media shows that most of the colonies obtained are those of moulds, microscopic Ascomycetes or lower fungi (Zygomycetes and Oomycetes). Their ***spores***, particularly abundant but most often dormant in soils, germinate easily on most of the usual culture media (Davet and Rouxel, 1997). On the other hand, when samples are diluted, the vegetative mycelia of the Macromycetes are most often broken up and destroyed.

Heterotrophic: said of an organism in which most of the cellular carbon is derived from an organic food (as in animals), as opposed to ***autotrophic***, denoting an organism that uses inorganic carbon (carbon dioxide, carbonic acid, bicarbonate) as the sole source of cellular carbon (as in plants).

It is therefore very difficult to approach the reality of mycoflora by conventional culturing methods (Chap. 15, § 15.5). Other approaches, such as introduction in the soil of 'fungus traps' or pieces of nylon gauze, allow observation *in situ* of the growth of mycelial filaments, which often present the characteristics of the vegetative mycelium of **Basidiomycetes**.

Sometimes fungi are trapped!

Abundance and distribution

Most soil fungi have a vegetative apparatus made up of branched filaments ***(hyphae)***, the ***mycelium***. These are septate or aseptate (Fig. 2.10).

Colonizers with wide area of action.

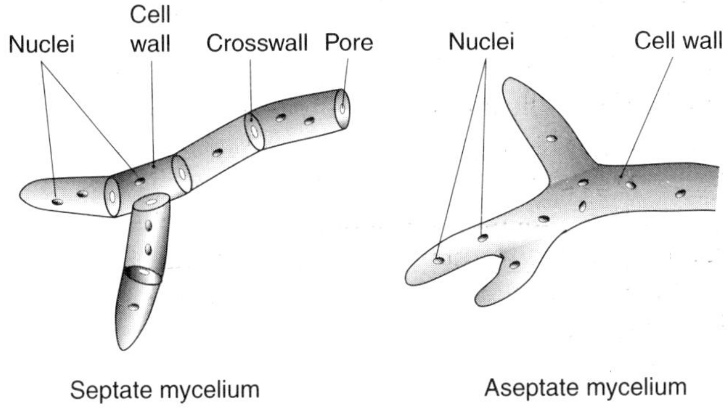

Fig. 2.10 *Two types of mycelium, septate and aseptate. In the former case, pores traverse the crosswalls and allow circulation of the cytoplasm.*

Spore: specialized cell, often resistant to external factors (drying, freezing, etc.) serving to multiply, disperse and preserve the organisms that form them (fungi, bacteria, some protozoa and some algae).

...about ecology of fungi.

Contrary to that of bacteria, the ecology of fungi is mostly conditioned by the overall characteristics of the soil, less so by those of microenvironments. The length of the mycelium of an 'individual' fungus is often considerable, up to several metres. Smith et al. (1992) have also noted a case of a single individual *Armillaria bulbosa* (Basidiomycete) spreading over more than 15 hectares! Its size is not at all similar to that of a bacterial microcolony, which is restricted to a fraction of a millimetre. In one square metre of fertile soil the network of mycelial filaments can attain a total length of 10,000 km.

Functions in the soil

Translocation: active movement within an organism (fungal mycelium, plant) of water and dissolved substances, organic and inorganic, or even particles or vesicles.

By virtue of its size and structure, a mycelium can actively transport large quantities of water and substances from one zone of the soil to another. ***Translocation*** of organic nutrients enables the formation of fruiting bodies in a day or two, a large portion of the stored materials accumulated in a mycelium thus being transported to the growing carpophores.

Soil fungi hide a wide range of functions behind their morphological uniformity.

Translocation of mineral salts assumes significance in the ***mycorrhizae***, symbiotic associations between a fungus and the roots of a plant (Chap. 16, § 16.1). The fungus in this case is a collector of mineral salts that it transfers to the plant or holds in reserve during the non-growing season.

The mycelia of mycorrhizal fungi: 'super-absorbing hairs', veritable nutrient touts favouring the root.

Because of their ramified structure, mycelia increase the cohesion of particles in the surface layers of soil. The effect of adhesion of the mycelium can be seen by merely lifting certain fungal fruit-bodies growing on leaf litter, clearly shown by the morphology of the OF horizons (Chap. 6, Tab. 6.3).

Fungi participate in the degradation of litter and its progressive transformation to humus.

Some fungi are specialized in the utilization of plant polysaccharides or lignin, and can accumulate melanized precursors of humic substances in their mycelia or spores. Others are used to living with plants, directly taking the organic nutrients they need from the living webs. This adaptation is often mutualistic, as in the case of mycorrhizae already mentioned, or parasitic (Chap. 16, § 16.1).

Figure 2.11 summarizes the principal functions of fungi in the soil.

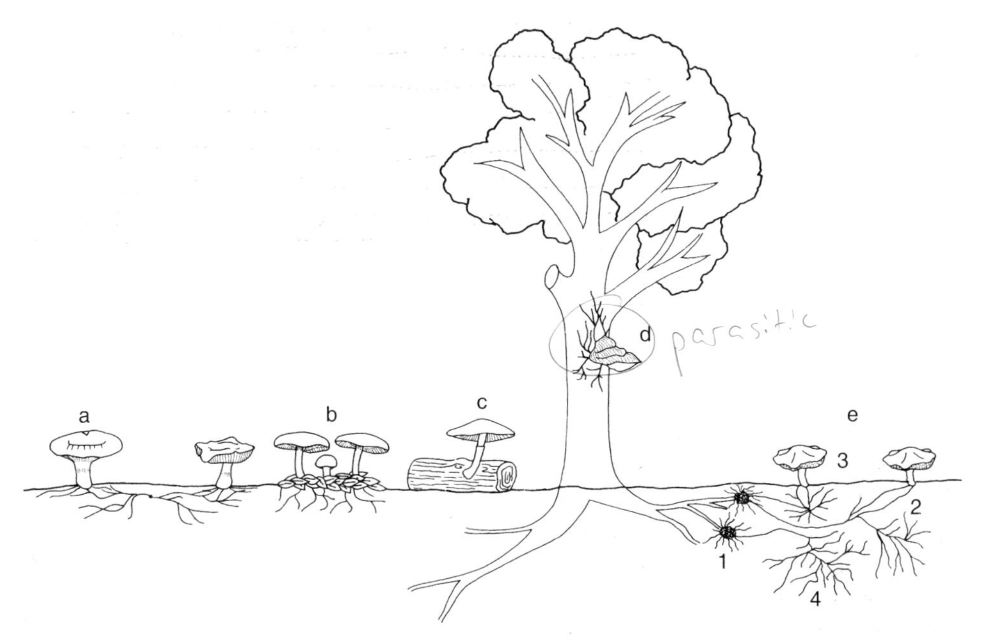

Fig. 2.11 *Principal functions of fungi in the soil.*
(a) humus-inhabiting fungi (e.g. formation of fairy rings),
(b) saprophytic fungi of litter,
(c) wood-inhabiting saprophytic fungi (destruction of dead wood),
(d) parasitic fungus,
(e) mycorrhizal fungus: 1. mycorrhiza (exchange with the plant); 2. translocating hyphal cords; 3. fruiting bodies (carpophores); 4. diffuse mycelium (absorption of ions and water from the soil).

2.5.4 Algae with underestimated role in soil

Microscopic algae, unicellular or in filamentous colonies, are often abundant in soil, but remain localized at its surface or in large cracks.

Three eukaryotic taxonomic groups are represented: green algae (Chlorophyceae: *Chlamydomonas, Chlorella, Pleurococcus*), yellow-green algae (Xanthophyceae: *Heterococcus, Vaucheria*)—these two groups dominating acid soils—and diatoms (Bacillariophyceae: *Achnanthes, Navicula, Pinnularia*), mostly in neutral or alkaline soils. The 'blue-green algae', the Cyanophyceae, are actually bacteria (§ 2.5.2).

Blue-green 'algae', green algae, yellow-green algae ... and diatoms.

According to the soil, the algal biomass, cyanobacteria included, comprises between 10 and 1000 kg dry matter per hectare, with a maximum of 24 tons for the only cyanobacteria of rice paddies. Davet (1996) counted 100 to 10^9 individuals per gram of soil.

On average, a few thousand algae per gram of soil, but highly variable!

Algae take part in pedogenesis.

Because of their photosynthetic activity, algae rapidly colonize rough mineral surfaces, the weathering of which is speeded up by solubilizing substances. They also produce extracellular polysaccharides, which aggregate solid particles and strengthen their cohesion. In an aquatic environment, some form chalk by precipitating calcite and thus participate in the evolution of submerged soils (Chap. 6, § 6.2.1).

2.6 LIVING ORGANISMS: THE FAUNA

'The diversity of animal life in soil and litter is far greater than might generally be believed, and many groups frequently encountered are unfamiliar even to the trained zoologist. This is particularly so in the case of immature forms' (Kevan, 1962).

'The other last biotic frontier.' Thus did André et al. (1994) describe the soil fauna while deploring the lack of knowledge about soil animals, despite appearances to the contrary. The recent application of new techniques for extraction of the soil fauna shows that the systematic picture we still have today of certain groups is very incomplete, to say nothing of their functional aspects!

However, by their various capacities, animals are essential role-players in all soils of the world: macroarthropods of the litters of temperate countries, termites of tropical soils, microarthropods of peaty soils, earthworms of all climates, each group intervening in transfer of matter and energy in the soil (Chap. 13, § 13.6.3). Relative ignorance about them is chiefly due to lack of specialists in systematics: only correct identification of organisms can enable us to proceed to the next step, that of their ecology and the role each plays in the soil.

In this chapter, only general information has been given; a more detailed description of the pedofauna and its functions, inseparable from aspects of systematics, is the subject of Chapter 4, § 4.5 and of Chapter 12.

'Soil is one of the most diverse habitats on the Earth and contains one of the most diverse assemblages of living organisms' (Giller et al., 1997).

2.6.1 Taxonomic diversity of the pedofauna

In spite of fluctuations in systematic nomenclature we shall continue to use the term Protozoa, firmly established in usage.

With the exception of Porifera, Cnidarians and Echinodermata, all the great phyla are represented in the soil fauna:

- Protozoa: Flagellata, Sarcodina (naked and testate amoebae) and Ciliata or Ciliophora,
- Plathelminthes: class Turbellaria (terrestrial planarians),
- Nematoda,
- Rotifera,
- Tardigrada

- Annelida: class Oligochaeta (earthworms and enchytraeids),
- Mollusca: class Gastropoda (snails and slugs),
- Arthropoda: classes Arachnida (e.g. spiders, mites), Crustacea (e.g. woodlice), Insecta, Myriapoda (this is today divided into four new classes—Pauropoda, Symphyla, Chilopoda and Diplopoda),
- Chordata: subphylum Vertebrata, in particular class Mammalia.

The term phylum

In zoology, the term ***phylum*** is situated at a precise hierarchical level in classification. It can also denote, as in microbiology and botany, an evolutionary line of organisms without reference to a precise level in systematics.

2.6.2 Abundance of the pedofauna

Estimates of the abundance found in literature vary widely according to the soils studied (Tab. 2.11).

How many would it take to weigh 1 gram? $16 \cdot 10^6$ protozoa, 10^6 nematodes, 100,000 collembola, 100,000 woodlice, 2 crane-fly larvae, 0.25 snail or 0.1 blackish slug!

Table 2.11 Abundance of the pedofauna in temperate regions (various sources)

Group	Individuals m^{-2}	Biomass g m^{-2}	Refer to
Protozoa	10^9–10^{11}	6–30+	Chap. 12, § 12.4.1
Nematoda	10^6–$30 \cdot 10^6$	1–30	Chap. 12, § 12.4.2
Earthworms	50–400	20–400	Chap. 12, § 12.4.3
Acari	$20 \cdot 10^3$–$40 \cdot 10^4$	0.2–4	Chap. 12, § 12.4.8
Collembola	$20 \cdot 10^3$–$40 \cdot 10^4$	0.2–4	Chap. 12, § 12.4.9
Insect larvae	up to 500	4.5	Chap. 12, § 12.4.9
Myriapoda			
• Diplopoda	20–700	0.5–12.5	Chap. 12, § 12.4.7
• Chilopoda	100–400	1–10	Chap. 12, § 12.4.7
Isopoda	up to 1800	up to 4	Chap. 12, § 12.4.6

There are, on average, 150 g of animals in 1 m^2 of grassland soil, representing, on average, 260 million individuals. This means that the greater part of the fauna is very small in size. But these estimates are enormously variable in space and time. Thus, during proliferation of crane flies (Diptera) in grasslands of Normandy, Ricou (1967) counted, on average, 500 to 1000 larvae per square metre, which destroy part of the plant cover. In normal times fewer than 10 are found per m^2. In the transition zones on the banks of peat bogs, Vaucher-von Balmoos (1997) counted 1500 hatchings of Diptera per square metre per year.

Under the sole of a stroller may be found as many invertebrates as there are inhabitants in Switzerland, or nearly 7 million!

Comparison of the weights of vertebrates and soil invertebrates in the same ecosystem proves the abundance of the latter. In a pasture supporting 2 to 3 heads of cattle per hectare, the biomass of earthworms (1000 to 1500 kg ha^{-1}) is almost comparable to that of the livestock (around 1800 kg). In a beech forest, Lemée (in Lamotte and Bourlière, 1978) found that the bird biomass was 0.240 kg ha^{-1} while that of soil oribates was 25–35 kg ha^{-1}! Other instances also underscore the abundance of soil animals:

'In spite of its apparent harsh aspect for growth of life, soil accommodates most of the living biomass of the Earth' (Bourguignon, 1996).

- the average animal biomass in soil is estimated at 2.5 t ha^{-1};
- the population of earthworms ranges from 100 m^{-2} in poorly organic soils to more than 1000 in richer soils (by weight, 500 to 5000 kg ha^{-1}); the smaller enchytraeid worms number 1400 to 143,000 m^{-2} (Didden et al., in Benckiser, 1997) or even 290,000 m^{-2} according to the type of soil;
- Borcard (1991) counted between 140,000 and 250,000 oribatids per square metre in a HISTOSOL FIBRIQUE, to a depth of 5 cm, whereas an alpine greensward of Graubünden (Switzerland) contained between 100,000 and 200,000 of these Acari (Matthey et al., 1981);
- the testate amoebae of moist organic soils renew their generations every one to eleven days according to climatic and edaphic conditions; thus they supply 3 to 26 g of annual biomass per square metre in the top 5 mm of soil (Schönborn, 1982);
- the biomass of protozoans in an agricultural soil can be almost equal to that of the earthworms (Wood, 1995);
- lastly, in extreme cases, nearly 30 million roundworms are found in one square metre!

2.6.3 Variety and size of organisms

From microfauna to megafauna, a range of abundance!

Generally, writers divide the soil fauna into three or four categories according to size and role of the organisms. In this book, the option of four categories is retained (Fig. 2.12; Chap. 12).

The ***microfauna*** is composed of animals less than 0.2 mm in length (diameter < 0.1 mm), mainly the protozoans, unicellular organisms: Amoebae (naked and testate amoebae), Flagellata (Euglenophyta), Ciliata (paramecia), Rotifera as well as Nematoda.

The ***mesofauna***, with length ranging from 0.2 to 4 mm (diameter 0.1–2 mm), comprises mainly the Nematoda,

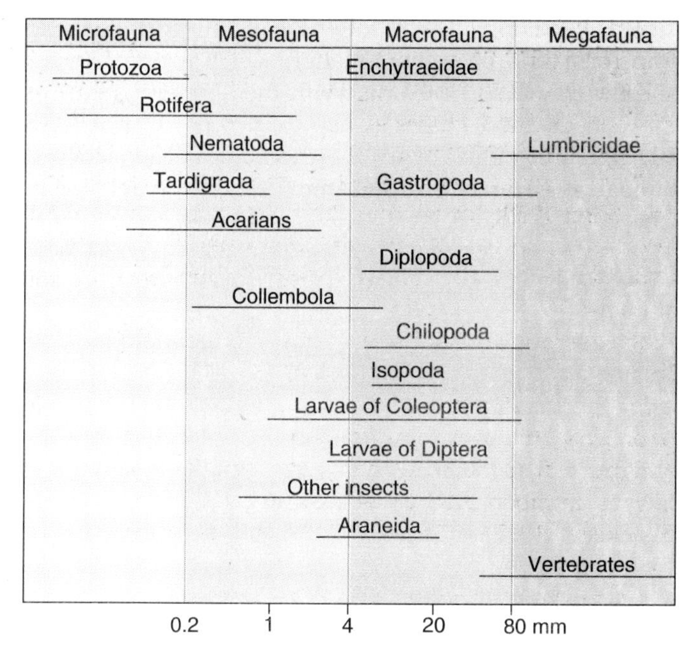

Fig. 2.12 Respective sizes of animals of the microfauna, mesofauna, macrofauna and megafauna.

the Acari (prostigmatic, mesostigmatic and oribatid mites) and the Apterygota (springtails, diplurans; Plate I–1 to 3). The arthropods belonging to the mesofauna are termed ***microarthropods***.

The ***macrofauna*** pertains to animals about 4 to 80 mm in length (diameter 2 to 20 mm). Their principal representatives are

- Oligochaeta such as enchytraeids and earthworms;
- Gastropoda such as slugs and snails;
- Arthropoda other than insects: Isopoda (woodlice), Diplopoda (millipedes), Chilopoda (centipedes), Arachnida such as scorpions, harvestmen, spiders;
- Insecta: Isoptera (termites), Orthoptera (crickets, mole-crickets), Coleoptera (ground beetles, staphylinid beetles), Diptera (scuttleflies, non-biting midges, crane flies), Hymenoptera (ants).

The arthropods of the macrofauna are ***macro-arthropods***.

The ***megafauna***, exceeding 80 mm in length, comprises vertebrates that work in the soil through their burrows: reptiles, burrowing mammals such as voles, moles, kangaroo rats, prairie dogs and marmots. Also included are large earthworms, of which examples 75 cm long have been discovered in the Basque provinces and 3 m long in Australia!

Chapter 12, § 12.5.2 defines the ecological niche of these four categories.

Functions complementary to those of the microflora.

The fundamental role of the soil fauna pertains to transformation of organic matter, which it prepares for soil fungi and soil bacteria. With them, it forms part of the detrital food chains (Chap. 5, § 5.2.4, Chap. 13, § 13.3.5). Three modes of action—mechanical, chemical and biological—characterize soil animals; they are detailed in Chapter 4, § 4.5.

CHAPTER 3

SOIL PROPERTIES

The constituents of soil interact among themselves to confer on it its properties. Their proportions, their spatiotemporal variations, and the flow rates that link them influence the functioning of the system. This chapter presents eleven essential properties, physical and physicochemical (Fig. 3.1). In accordance with the general orientation of the book, the biological properties form the subject of another chapter (Chap. 4).

Physical, chemical and biological, the properties of soil enable its functioning. They are particularly important factors controlling soil organisms.

3.1 TEXTURE, AT THE ROOT OF (ALMOST) EVERYTHING

3.1.1 Definitions

Texture refers to the respective proportions of the constituents sorted by size (Chap. 2, § 2.1.3). We distinguish ***mineral texture***, the proportion of sand, silt and clay determined by granulometric analysis, from ***organic texture*** that reflects the proportions of fibres and fine microaggregated matter in holorganic materials.

Texture, basic to (almost) all the other properties!

Texture: soil property defining in a global way the granulometric composition of the fine earth.

3.1.2 Determination and types

Mineral texture

This expresses by a simple (e.g. sand, clay) or compound (e.g. sandy loam, silty clay) term, the texture located in a mineral ***texture triangle***, in which the

Mineral texture, a matter of a triangle … or touch!

The division into texture classes dates to 1927, based on proposals by the Swedish chemist Atterberg (1846–1916) (Boulaine, 1989).

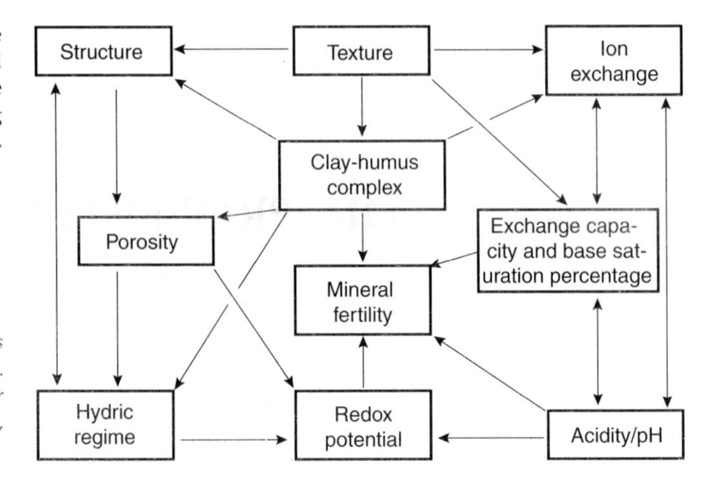

Fig. 3.1 *Essential relations among ten soil properties. Temperature is a factor underlying all the others, influencing them indirectly.*

categories are demarcated. Figure 3.2 presents the standard triangle (USDA, 1975; Singer and Munns, 1996), but others are also in use.

Mineral texture is also estimated in the field; an experienced observer can identify the thirteen textural divisions in the triangle.

Sand, visible, is rough to the touch. Silt, silky to the touch, does not stick to the fingers but often leaves glistening flakes of mica in wrinkles on the skin; the finest allow small brittle rods to be rolled between the fingers. Clay sticks to the skin, retains fingerprints and forms very flexible rods that do not break when bent. But in this evaluation, it is necessary to take into account the moisture content of the soil, organic matter content and the presence of specific phyllosilicates (Baize and Jabiol, 1995).

Organic texture: property of certain holorganic materials such as peats and composts, reflecting the proportions of fibrous and non-fibrous organic materials (e.g. microaggregates formed in the casts of enchytraeids).

Organic texture

Organic texture is also determined in a triangle, which enables assigning the sample to the fibric, mesic or sapric domains, basis of the classification of peats (Fig. 3.3; Chap. 9, § 9.1.4; Gobat et al., 1991). In addition to granulometric indications, it gives—this is a difference from mineral texture—information on the microstructure of the material.

Organic texture enables classification of peats.

3.1.3 Functions

Texture, a stable property of the soil.

Mineral or organic texture directly controls soil structure and therefore porosity and the hydric regime. In particular, the proportion of clay influences formation of clay-humus complexes, exchange capacity, fertility and rooting depth. Texture is a stable property, changing only with long-term soil development, of which it is a useful index (Chap. 5, § 5.3.2).

In view of its interpretative importance, texture should be determined in all cases, even promptly in the field!

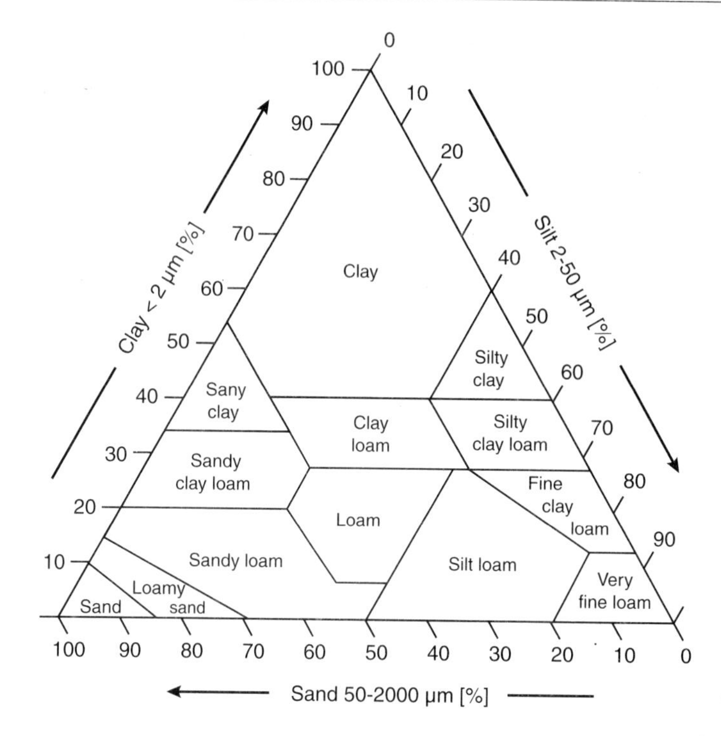

Fig. 3.2 Mineral-texture triangle (after USDA, 1975): each sample is located on it based on its gravimetric content of sand, silt and clay, the three adding up to 100%. For each percentage of clay, silt and sand, draw a straight line parallel to the corresponding axis. The intersection of the three lines denotes the texture of the sample.

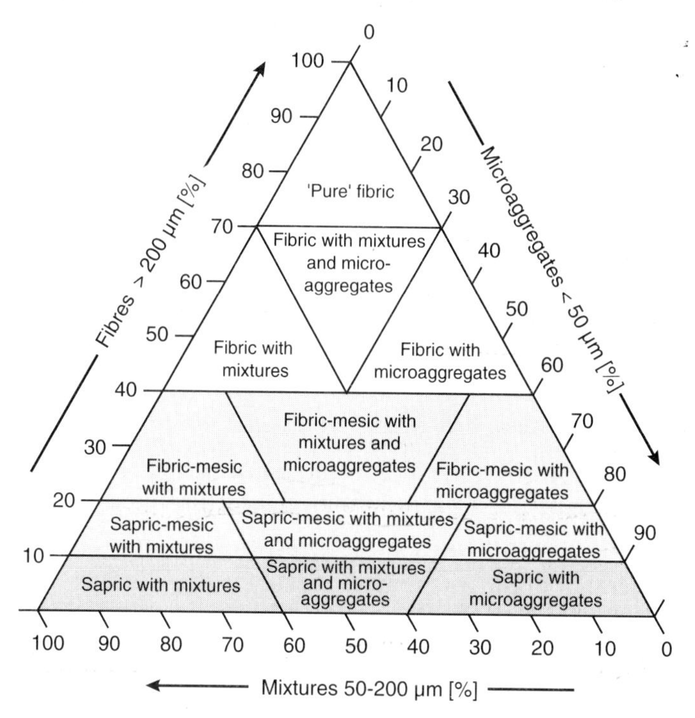

Fig. 3.3 Organic texture triangle (after Gobat et al., 1991). Identical in use to the mineral-texture triangle. Mixtures are composed of residues of plant tissue, often disintegrated and also sometimes of macro-aggregates. Unshaded: peats of the fibric domain; light grey: peats of mesic domain; dark grey: peats of sapric domain.

3.2 STRUCTURE, A CHANGING PROPERTY

3.2.1 Definition

To a great extent resulting from texture, soil structure is an integrative property essential for fertility.

Structure: mode of organization of solid constituents of soil, mineral and/or organic. They may be aggregated (***pedal*** structure) or not (***apedal*** structure).

Structure is a soil condition, variable in the short term, according to season for example. It directly depends on the texture (the reverse is not true!) and also on the state of the colloids, the moisture content and organic matter and, to a large extent, faunal activity (bioturbation, Chap. 5, § 5.3.3). It is observed at macroscopic scale—structure proper—or microscopic, in which case it is called ***microstructure***. Strength of the structure, its resistance to destructive agents, is evaluated by tests of ***structural stability***.

Microaggregates and macroaggregates.

Aggregated structures are divided into four types according to size, reflecting different physical or chemical properties (Wilson, 1991; Elliott et al., in Powlson et al., 1996):

- ***microaggregates*** from 2 to 20 μm, very stable, formed of highly aromatic organic substances bound to clays and fine silt, and bacterial polysaccharides;
- microaggregates from 20 to 250 μm, containing coarse silt and sands, aggregated by bacterial polysaccharides;
- ***macroaggregates*** from 250 to 2000 μm, formed from the preceding and coarse sand bound by polysaccharides, bacterial cells and mycelium;
- macroaggregates larger than 2000 μm, composed of the above, associated with particles of free organic matter, roots and mycelium, the last consolidating the assemblage.

3.2.2 Determination and types

First of all, visual appreciation.

While texture can be precisely measured, determination of structure is more empirical; structure should be appreciated visually. In spite of the great variety of possible arrangements of the particles, soil structure can be subdivided into five broad categories (Baize and Jabiol, 1995, simplified; Fig. 3.4):

- Absence of aggregates, structure inherited from the parent material: lithic or lithologic structure.

- Absence of aggregates, structures generally seen in mineral or organomineral materials:
  - — Coherent material: massive structure, continuous, compact (Fig. 3.4a). The elements are embedded in a mass of dispersed clay,

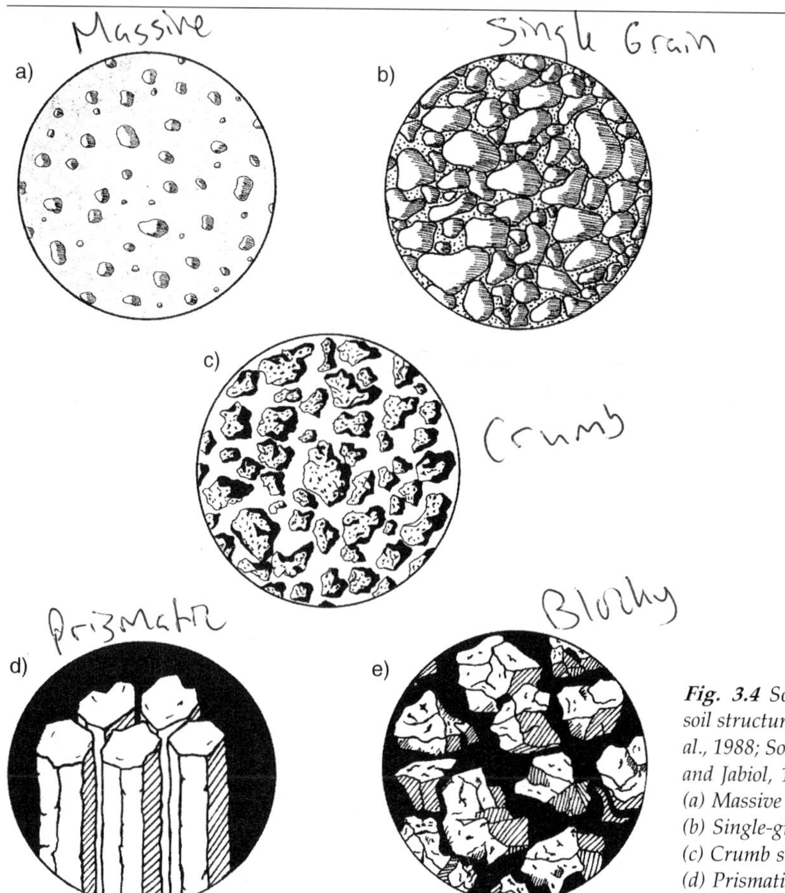

Fig. 3.4 Some examples of soil structure (after Kuntze et al., 1988; Soltner, 1995; Baize and Jabiol, 1995):
(a) Massive structure
(b) Single-grain structure
(c) Crumb structure
(d) Prismatic structure
(e) Blocky structure.

forming a homogeneous block. The soil is asphyxiating, unfavourable for biological activity: poor root penetration (Polomski and Kuhn, 1998), difficulty in burrowing, etc.

— Material formed of individual free particles: single-grain structure (Fig. 3.4b). The elements are touching or packed together, without colloidal binding. The soil is incoherent, free-draining and dries out rapidly. If the texture is fine silt, there is risk of the surface becoming impermeable.

• Presence of rounded aggregates. These aggregates are conducive to soil fertility, particularly the crumb structure, in which the organic and mineral elements are bound in the clay-humus complex. It retains water well, at the same time permitting its circulation and soil aeration; it leaves voids that can be colonized by animals or roots:

— aggregates more or less spherical: granular structure,
— irregular aggregates, more or less agglomerated: crumb structure (Fig. 3.4c),
— crumbs smaller than a millimetre: micro-crumb or fluffy structure.

- Presence of aggregates with angular vertices. Structures of this type often result from physical processes affecting clay minerals like the swelling and shrinking of smectites. Blocky structure is common in the S horizons of weathering, whereas prismatic and columnar structures mostly characterize alkali soils:
  — horizontal orientation of aggregates: lamellar or platy structure,
  — vertical orientation, elongated aggregates: prismatic structure (Fig. 3.4d),
  — same as above, but with rounded caps: columnar structure,
  — plane faces with oblique orientation, grooved: sphenoid structure,
  — few distinct plane faces, edges of same size, cubic structure;
  — many flat faces, very sharp unequal edges: angular blocky structure (Fig. 3.4e),
  — same as above, with rounded corners: subangular blocky structure.

- Structure formed from mostly plant material:
  — coarse organic residues, fibrous (mosses, peat, compost): fibrous structure,
  — organic residues from leaves or needles, arranged horizontally: laminated or rod-like structure,
  — millimetre-size spherical masses (excrements), individualized: coprogenous or granular structure.

3.2.3 Functions

Structure primarily determines porosity, that is, the volume and arrangement of voids in soil.

Changes in soil structure modify the circulation of water, which is very rapid in coarse single-grain structures, medium in crumb structure and almost zero in strongly compacted soils. In practice, structure is an essential property that the farmer should understand: frequency of working, type of ploughing, machines used, germination of seeds are as much elements influencing structure as are influenced by it! Crumb structure, for example, is an unstable condition of cropped soil, rapidly destroyed by a lack of organic matter, too much of certain fertilizers or extreme compaction by excessively heavy machines.

Beware, one structure can hide another in itself!

Agronomists often talk of 'destruction of structure', which is almost nonsensical to soil scientists. Structure, which is fundamentally the mode of assemblage of soil particles, cannot be destroyed. It can only be modified, for example from the aggregated state to the single-grain state. When talking of structure, the agronomist implies 'built-up structure', especially crumb. In this more restricted sense, it is clear that structure can be destroyed and replaced by another.

Maintenance of a fertile structure in agricultural soils is primordial (Chap. 17, § 17.2.1).

3.2.4 Microstructure or micromorphology

Techniques of study

Study of soil with the binocular hand lens, optical microscope or electron microscope using techniques such as ***thin sections***, ***polarized light*** and ***scanning*** reveal the fine organization of its structure. It shows up the coarse fractions (twigs, wood, coal), which can be described and precisely quantified. It also details the modes of assemblage of the fine fraction, for example distribution of clayey zones within a silty matrix. Soil voids are also interesting for micromorphological study: are they isolated or joined in a network, round or thread-like, of large or small diameter (§ 3.3.3)?

Thin section: undisturbed preparation of a soil sample 20–25 μm thick embedded in a resin for study under a polarizing microscope.

Polarized light: light that has its vibrations produced in a defined plane, unlike ordinary light in which they are uniformly produced in all planes.

Scanning: electron microscopic technique in which the surface of a sample, previously coated (or not in ESEM microscopes) with gold, carbon or platinum, is 'scanned' by an electron beam that produces a three-dimensional image on a cathode-ray screen.

Pedofeatures

But the major interest in a study at this scale is analysis of ***pedofeatures***, which often result from biological activity and can indicate the effect of such or such organism on soil functioning. Lavelle and Spain (2001) provided a complete table of them while Bullock et al. (1985) described six types (Plates II–1 to II–6):

- textural features, formed by the localized predominance of one granulometric fraction (e.g. ***argillans***);
- crystalline features, resulting from the neoformation of various minerals in the soil; the most frequent are calcitic features, $CaCO_3$ accumulations of physicochemical as well as biological origin: shells, calcitized root cells, calcium oxalate coating on mycelial filaments of certain basidiomycetes (Keller, 1985);

From rust-coloured mottles to prismatic droppings; the fine details are important!

Pedofeature: small unit pedologically different from the adjacent material and formed by soil functional processes.

Argillan: coating of clay accumulation in the BT horizons.

- amorphous and pseudocrystalline features, accumulations of organic matter and ferruginous materials (accumulation of ferric iron around rootlets in the Gr horizon, rust-coloured mottles of the Go horizons, etc.);
- depletion features such as white mottles of incipient eluviation resulting from loss of a constituent;
- assembled features, often created by the fauna (Chap. 4, § 4.5.1): packing of particles around a burrow, material buried by earthworms, etc.;
- excremental features, droppings of the fauna: faecal pellets of enchytraeids, earthworm casts, prismatic excrement of woodlice.

Micromorphology details the fine organization of the humiferous episolum.

Micromorphology is especially used for understanding the functioning of the humiferous episolum, which presents a multitude of pedofeatures indicating the general or local physicochemical conditions and concomitantly the activity of living organisms.

3.3 POROSITY, OR SOIL VOIDS

3.3.1 Definition and types

Where the voids, even filled, are significant.

Porosity: soil property reflecting the volume of soil voids, expressed as a percentage of the total volume.

Depending on moisture content, the soil voids are mostly occupied by water or air. Their total represents the ***porosity***. Porosity gives a good idea of the structural condition with, certainly usefully, the possibility of comparative measurements. According to pore size it is subdivided into ***macroporosity*** (voids > 50 μm, which can be filled with gravitational water and often colonized by medium roots), ***mesoporosity*** or ***capillary porosity***, composed of voids 0.2 to 50 μm in diameter that retain the water useful to plants, and ***microporosity*** with voids smaller than 0.2 μm, retaining unusable water (§ 3.4.2). Mesoporosity depends greatly on texture, macroporosity primarily on structure.

3.3.2 Determination

Total porosity, P.

Total porosity = macroporosity + mesoporosity + microporosity. The limits of these categories vary according to author. For example, Callot et al. (1982) placed microporosity

Porosity is measured in a soil sample of known volume, drawn without modifying its structure and weighed after drying at 105°C. The ***total porosity*** P is:

$$P = \frac{(d - \rho A)}{d} \cdot 100 \qquad [\%],$$

where *d* represents the ***specific gravity*** of the solid constituents of the soil; it is estimated to be 2.65 for an average soil, 2.4 for a highly calcareous soil, 2.0 for a humiferous soil and 1.5 for a peaty soil. ***Bulk density***, ρA (dimensionless in the formula above) is determined by the quotient *m/V*, *m* being the weight (in grams) of the soil after drying and *V* the volume of the soil core, in cm^3. Porosity ranges from 30% in very fine textured soils to 80% in peats.

Mesoporosity and microporosity are estimated by the proportion of water, expressed in volume, retained by a given volume of soil after it has drained; macroporosity is calculated from the total porosity by difference.

between 0.2 and 6 μm and mentioned matric porosity below 0.2 μm. The diameter 6 μm is considered the lower limit of pores accessible by rootlets.

Specific gravity: ratio of the density (e.g. bulk density) of the solid constituents of soil to the density of water at 4°C and standard pressure (Skopp in Sumner, 2000).

3.3.3 Functions

Porosity gives information on the water and air retention properties of soil, in volume or flow rate. In the latter case, however, just indication of porosity is not enough, because circulation of water (and of air) depends as well on the relations among soil voids and their arrangement (Fig. 3.5).

At the same porosity, the arrangement of voids makes the difference: for soil water the shortest path is not always the best!

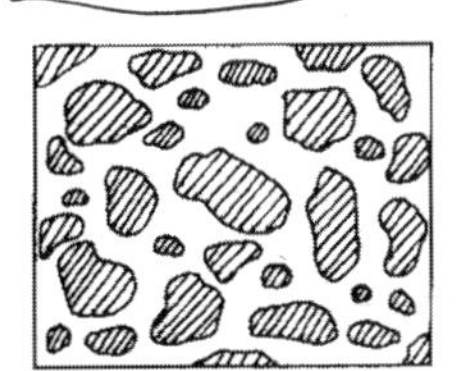
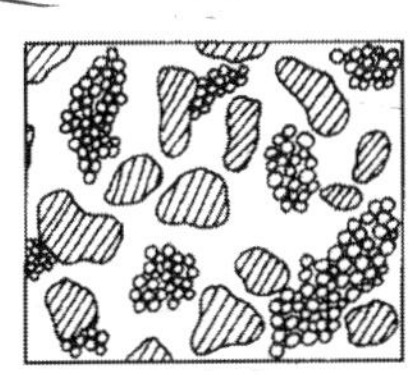

Fig. 3.5 *Three possible arrangements of soil voids. At the same total porosity the three soils have very different hydric regimes: the first generally retains water poorly and hence dries rapidly; the second does not retain gravitational water and holds capillary water very well, but the latter is locked up in anoxic microsites; the third retains gravitational water well by ensuring smooth flow through the soil and long-term retention.*

This was also shown by Gisi et al. (1997), who mentioned the combined effect of ***tortuosity*** and ***connectivity*** of the pores. Musy and Soutter (1991) use the term ***residual porosity*** for the total of closed pores unconnected to the other voids, the latter forming the ***effective porosity***.

Very commonly, the general potential of water circulation in soil is revealed by its ***hydraulic conductivity***. This property reflects, better than porosity, the capacity to transmit water because it integrates structure, porosity, tortuosity and connectivity.

Tortuosity: average ratio of the actual path length between two points and the linear distance separating them (Musy and Soutter, 1991; Singer and Munns, 1996). Tortuosity indicates the straightness or the twists of the connections between soil voids.

Connectivity: degree of relation among pores, well or poorly connected with one another.

Hydraulic conductivity (or ***permeability):*** speed of percolation of gravitational water, expressed in centimetres per hour (the coefficient *K* in Darcy's law—Musy and Soutter, 1991). *K* ranges from less than 0.4 cm h^{-1} in clayey soils to more than 20 cm h^{-1} in sands and gravel.

Diffusion: flow of one particular compound compared to overall flow of the mixture in which it exists. Diffusion results from individual displacement of molecules under the effect of Brownian movement (also see Chap. 4, § 4.2.1).

Hydric regime: result of variations in soil-water content over the year.

In peats, the water content is often reported on moist weight basis.

The total water content of soil, or moisture content, teaches us almost nothing ...

When water is all worked up!!

As for circulation of air, especially of oxygen, it is decisive for root growth and activity of the microflora, very sensitive to the degree of anoxia. Gaseous exchanges take place by equilibration of the three categories, outside air, soil air and gases dissolved in the soil solution. Availability of oxygen is generally assured by exchange with the outer air, except if the microporosity is discontinuous (second case in Fig. 3.5). The oxygen of the soil solution then functions as reserve for microorganisms. In this case, however, its ***diffusion*** is about 10,000 times slower than in air (Gisi et al., 1997).

3.4 THE HYDRIC REGIME, SOIL WATER

3.4.1 Generalities and water content

The ***hydric regime*** of soil depends directly on the three preceding properties:

- texture determines the strength of water retention,
- structure influences water circulation,
- porosity defines the volume of the hydric reservoir in soil.

The 'total' quantity of water retained by a soil is the difference in the weights of a sample before (moist weight) and after (dry weight) drying at 105°C. Reported on dry sample, it enables us to calculate the ***gravimetric water content***, *θ*. This does not, however, include the chemically bound water, for example on clay minerals, which can be driven out only at 500°C.

Easily determined, water content is actually a poor ecological index because of its extreme sensitivity to recent precipitation, drainage and plant cover. Much more useful is information on the distribution of water according to retention capacities of soil: how much water is remaining in the microporosity? Is this water accessible to plants? How does this reserve vary in soils of different textures? Answers to these questions depend on precise knowledge of the states of water in the soil, themselves dependent on retentive forces.

3.4.2 States of water in the soil

Three states of water are distinguished in the soil according to the strength with which the soil retains it and according to its availability to plants: gravitational water, plant-available water and plant-unavailable water. Their proportions in the water content θ depends to a large extent on texture and to a lesser degree on the organic matter content. The exposed area of the particles is also determinant for the processes of surface adsorption. For example, the exposed area of the solid constituents of one square metre of loamy soil to 150-cm depth approaches 10 km^2!

Gravitational water (or ***free water):*** water filling the macroporosity and draining by gravity to the point of drying.

The most mobile, ***gravitational water,*** remains in the soil only a few hours or days following a rain, or when a permanent water table exists. When the force of gravity is balanced by the retentive force of the soil, the ***point of drying*** is attained; the remaining water is retained in the soil and constitutes the ***field capacity***.

A fugitive water, gravitational water.

More strongly retained than gravitational water, the ***plant-available water*** fills pores of diameter between 0.2 and 50 μm or forms films 5 to 10 nm thick on the surface of particles. Roots absorb it to the temporary wilting point, which is reversible, then to the ***permanent wilting point,*** which is attained when the retentive force of the soil for water equals the maximum suction force exerted by the plant. Numerous representatives of the soil microfauna, such as protozoa and small nematodes, reside within these water films, like true aquatic animals (Chap. 13, § 13.7.4).

Plant-available water (or ***capillary water):*** water occupying the meso-porosity or adsorbed in the form of relatively thick films on the surface of solid particles.

Up to the wilting point, a water available to plants.

Below the permanent wilting point is found the ***plant-unavailable water.*** Only intense evaporation enables its elimination. But even at high temperatures a thin water film remains in the soil, around certain minerals whose hydration is ensured (Chap. 2, § 2.1.2).

Trapped in the finest pores or bound to minerals, the plant-unavailable water

Plant-unavailable water: water contained in the finest pores or strongly held in the form of films thinner than 5 nm on particle surfaces.

3.4.3 Forces, pF, soil water and organisms

Forces exerted on soil water

Three forces act on soil water, demarcating the above categories: the force of gravity resulting from the Earth's attractive force ***P,*** the retentive force of the solids ***F*** and, lastly, the suction force of plants ***S*** (Fig. 3.6). These forces, sometimes compared with pressures, are expressed in megapascals (MPa), bars, atmospheres, centimetres of water or of mercury!

1 MPa = 10 bars = 9.87 atm = 10,197 cm H_2O = 750 cm Hg.

Definition of pF

The exponential increase in ***F*** makes its transformation to logarithmic form and use of the symbol pF analogous to pH more convenient:

$$\text{pF} = \log_{10} \mid F\text{(bars)} \mid$$

Which is the winner, ***P***, ***F*** or ***S***?

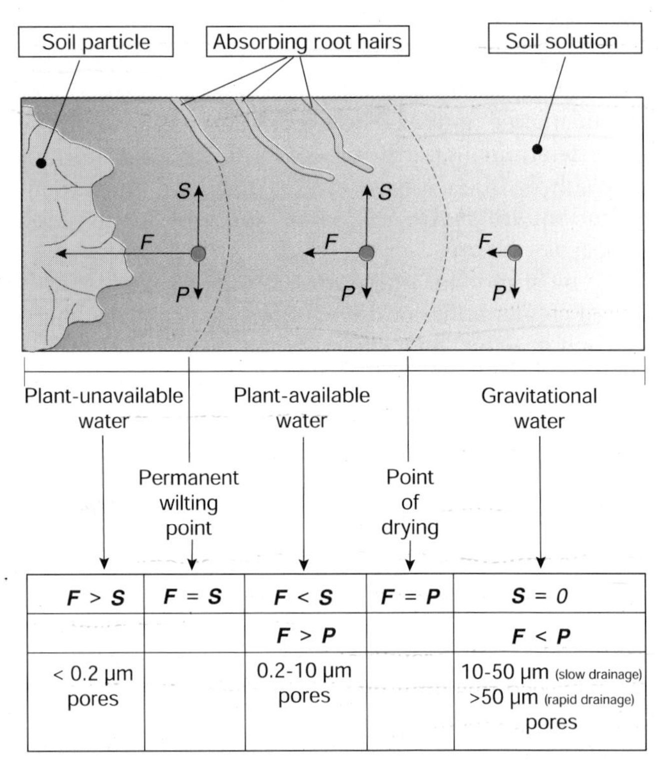

F > S	F = S	F < S	F = P	S = 0
		F > P		F < P
< 0.2 μm pores		0.2-10 μm pores		10-50 μm (slow drainage) >50 μm (rapid drainage) pores

__Fig. 3.6__ Forces exerted on soil water. The categories of water are demarcated in relation to gravity __(P)__, soil retention __(F)__ and plant suction __(S)__ (after Soltner, 1996a). Explanations in text.

The pF corresponds to the ***matric potential*** or ***capillary potential*** ψ_m, defined as the energy resulting from the pressure of the water due to effects of binding around solid particles and to capillary effects in the pores. The matric potential does not distinguish the adsorbed water from capillary water; it reflects the total affinity of water for the entire solid matrix of the soil (Musy and Soutter, 1991; Or and Wraith in Sumner, 2000). In practice, pF is less and less used; metres of water or, better still, kPa or MPa are preferred to it. Table 3.1 gives some important pF values.

Soil water and organisms

A more or less empirical permanent wilting point.

By studies on mesophilic plants, the permanent wilting point was fixed at –1.6 MPa or pF = 4.2. But it was soon found that numerous organisms succeed in living or surviving well beyond this limit. For example, yeasts and fungi (especially all *Aspergillus* spp. and *Penicillium* spp.; Wood, 1995) survive in the form of spores to –20 MPa. One of the most resistant fungi is *Xeromyces bisporus*, which reaches –69 MPa (Chap. 8, § 8.6.2; Davet, 1996).

Table 3.1 Some values of the matric potential, expressed as the retentive force ***F*** and pF.

State of water or limit	Force *F* (MPa)	Force *F* (bars)	Force *F* (g cm^{-2})	pF
Maximum saturation	−0.001	−0.01	−10	1.0
Field capacity	−0.006	−0.06	−63	1.8
Point of drying	−0.05	−0.5	−500	2.7
Rupture of capillary bond	−0.25	−2.5	−2,500	3.4
Temporary wilting point	−1	−10	−10,000	4.0
Permanent wilting point	−1.6	−16	−16,000	4.2
Air-dry soil	−100	−1000	−1,000,000	6.0

Lastly, in a very specific 'soil', the moulds that develop on jams do so by virtue of a suction reaching −10 MPa, the sole means of counterbalancing the extremely high water retention of a medium containing almost 50% sugar!

Osmotic adjustment in plants of dry regions

The value −1.6 MPa is a limit for plants that scarcely undergo hydric stress: crop plants, grassland plants and plants of the undergrowth of temperate regions. However, others exceed this limit by virtue of ***osmotic adjustment***, which enables them to colonize very dry areas (Tab. 3.2). The record belongs to the wormwood *Artemisia herba-alba* of the Negev desert, which develops a suction force of −16.3 MPa (Etherington, 1982)! Against this, aquatic plants wilt below −1.6 MPa, not tolerating any hydric stress.

Numerous species succeed in counterbalancing the very high retentive forces *F*, even beyond −1.6 MPa.

Osmotic adjustment: rapid adaptation of internal hydric potential by increase in concentration of cellular solutes, in particular carbohydrates and potassium.

Soil-plant-atmosphere hydric relations

The osmotic adjustment that plants are capable of in order to absorb water better is not, however, the primary cause of the hydric flux through the soil-plant-atmosphere system (Carlier, in Bonneau and Souchier, 1994; Campbell and Mathieu, 1995; Lüttge et al., 1996; Evett in Sumner, 2000). The circulation of water is actually imposed on the plant by external conditions; the plant's only role—but very important!—is control of flow by regulating it at certain stages along its progress.

Circulation of water in the plant: a flow the plant is subject to.

Transfer of water in the soil-plant-atmosphere system has been likened to an electric current passing through three resistances (Van den Honert's model; Fig. 3.7):

Three successive increasing resistances.

- resistance to the passage of water from the soil to the xylem of the roots (R1), where the osmotic potential of the plant opposes the matric potential of the soil; regulation here is done by osmotic adjustment;

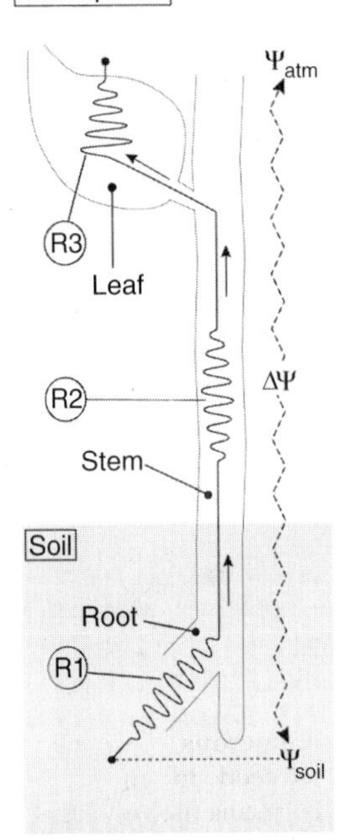

Fig. 3.7 Van den Honert's model of circulation of water in the soil-plant-atmosphere system (after Carlier, in Bonneau and Souchier, 1994). Symbols explained in the text.

Vacuole: cellular organelle with a single membrane, the tonoplast, reserved for storing various water-soluble compounds and playing an essential part in the regulation of exchange of water with the outside (Widmer and Beffa, 1997).

General climatic conditions determine hydric flows through plants. The energy necessary for these flows is of solely physical origin.

Hydrogen bond: weak chemical bond formed when a hydrogen atom already covalently bonded to an electronegative atom is subject to the attraction of another electronegative atom.

Table 3.2 Permanent wilting point of some species

Ecological group	Species	Permanent wilting point (MPa)
Aquatic species	*Nymphaea alba* *Nuphar luteum*	–0.7 –1.1
Mesophilic species	*Anemone nemorosa* *Achillea millefolium* *Lolium perenne*	–1.5 –1.5 –1.9
Mesoxerophilic species	*Vitis vinifera* *Picea abies* *Prunella grandiflora*	–1.9 –2.2 –2.3
Xerophilic species	*Buxus sempervirens* *Quercus coccifera* *Hippocrepis comosa* *Potentilla arenaria* *Aster linosyris* *Artemisia herba-alba*	–3.4 –4.4 –5.6 –8.1 –10.2 –16.3
Fungi	Moulds of preserves	–10.0

- resistance to circulation from the roots to the leaves within the plant (R2); regulation is ensured by variations in pressure of water in certain cells of the xylem, with possibility of evaporation in the ***vacuoles***;
- resistance to passage of water from the leaves to the atmosphere (R3); regulation is done by activity of the stomata, which act on transpiration (Chap. 4, § 4.2.3).

Certain plants, such as fir and pine, promote control at the end, closing the stomata as soon as hydric stress appears in the soil. Others, such as ash, work mainly at the entry point by osmotic adjustment and maintenance of their transpiratory activity despite rise in the matric potential.

The column of water between the soil and atmosphere is continuous because of ***hydrogen bonds*** that ensure cohesion among its molecules (Fig. 3.8). Therefore movement of the entire water column is a consequence of the difference in potential $\Delta\psi$ between the two ends, otherwise termed that between the hydric potential of the atmosphere ψ_{atm} and that of the soil ψ_{soil}. The plant uses none of its metabolic energy to circulate water unless it is for the necessary regulations indicated above. In reality, however, there are some deviations from this general model (Carlier, in Bonneau and Souchier, 1994):

- the acceleration of gravity ***g*** creates a slight depression in tall plants;

- all the roots of a plant are not subjected to the same soil-water potential, given the heterogeneity of the latter;
- a lag appears between absorption and transpiration in higher plants because of entry into the xylem of water earlier contained in the cell reserves of the plant;
- adhesion of water to cell walls and hydraulic conduction between the soil and leaf are not constant.

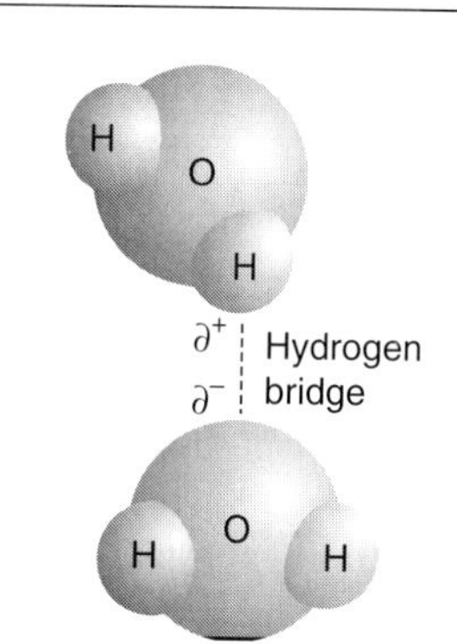

Fig. 3.8 Hydrogen bond between water molecules.

3.4.4 Water content and matric suction

Influence of soil texture

The value of matric potential ψ_m at the point of drying (pF = 2.7) and wilting point (pF = 4.2) are the same for all soils. On the other hand, the water content θ corresponding to these values varies with texture. A mesophilic plant definitely wilts with about 4% water in a sandy soil, 15% in a clay loam soil or even 50% in certain peats, which trap water very efficiently (Fig. 3.9; Chap. 9, § 9.6.1).

A variable water content for constant point of drying and wilting point.

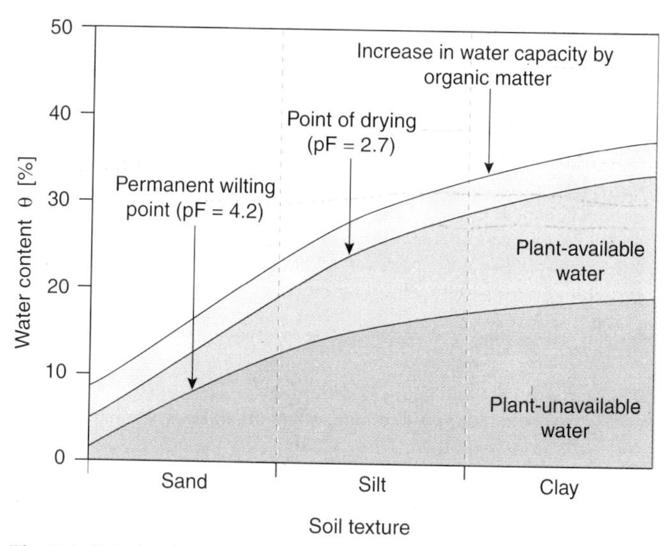

Fig. 3.9 Relation between the water content and pF at different textures. Note the increase in water capacity when organic matter is present (after Soltner, 1996a).

The desert of peat bogs . . .

The high water-retention capacity of peat stimulates paradoxical morphological adaptations in vascular plants of raised bogs, particularly in those of hummocks with sphagnum (Fig. 3.10; Chap. 9, § 9.2.3). Apparently living in water, they actually face conditions of extreme physiological drought when the water has diminished. Their main problem then is to avoid excessive water loss by

A paradoxical adaptation: peat-bog plants.

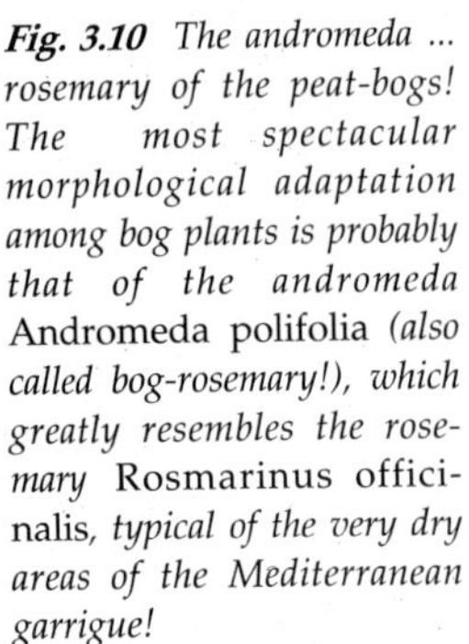

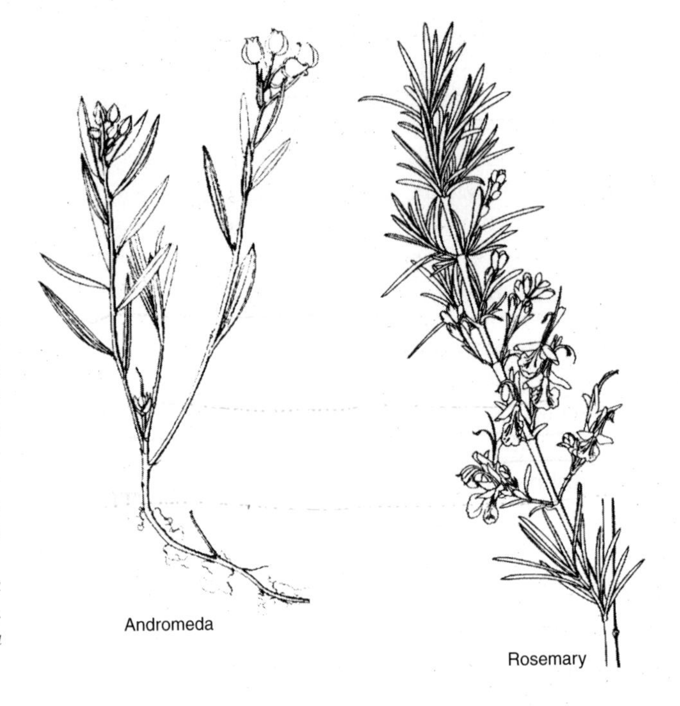

Fig. 3.10 *The andromeda ... rosemary of the peat-bogs! The most spectacular morphological adaptation among bog plants is probably that of the andromeda* Andromeda polifolia *(also called bog-rosemary!), which greatly resembles the rosemary* Rosmarinus officinalis, *typical of the very dry areas of the Mediterranean garrigue!*

With permission from Birkhaüser Verlag, Basle, Switzerland. Extracted from Hess et al. (1967ff).

transpiration, which they achieve because of their tough leaves with thick cuticle, like the cranberry *(Vaccinium vitis-idaea)*, or by reduced evaporation like heather *(Calluna vulgaris)*. Lowering of transpiration slows down the hydric flow near roots and diffusion of ions (Chap. 4, § 4.2.1). This is particularly true for dissolved iron, which could become toxic without this barrier to entry (Chap. 4, § 4.3.5). According to Jones (1972), here would lie the true physiological reason for the xeric morphology of certain raised-bog plants.

The paradox of seeing plants of 'xerophilic' appearance in water-rich environments, which occurs in peat-bogs or salterns, has long puzzled physiologists.

There is drought and there is drought!

As early as 1898, the botanist A.R.W. Schimper wrote: 'A very wet substrate is quite dry for a plant if it cannot absorb water, while a soil that appears quite dry to us can supply water in adequate quantity to many plants with low demand. (Similarly) a soil that is rich in soluble salts is totally dry for a plant even if it is completely flooded.' Schimper knocked down the old dogma of the nineteenth century that hydrophytes live on physically wet lands and xerophytes on physically dry lands (Acot, 1988).

3.4.5 Total soil-water potential

Four potentials to make just one.

Actually the matric potential ψ_m is only one of the four components of the ***total soil-water potential*** ψ_t of the soil (Hillel, 1980; Singer and Munns, 1996; Or and Wraith in Sumner, 2000), which also includes the ***pressure***

potential ψ_p, the ***osmotic potential*** ψ_s and the ***gravitational potential*** ψ_z:

$$\psi_t = \psi_p + \psi_s + \psi_m + \psi_z$$

where ψ_p represents the pressure exerted by a column of water moving through the soil; its value is zero except in the case of percolation after a downpour or during movements of a water table. The potential ψ_s reflects the forces caused by differences in chemical concentration of dissolved substances; it is negligible in the soil except in highly saline soil (Chap. 4, § 4.2.3). Lastly, ψ_z corresponds to the water removed by gravity to the point of dryness. In a 'normal' aerated soil, just ψ_m can be taken as the single true determinant of water availability.

Soil-water potential: amount of work required per unit quantity of pure water to transport reversibly and isothermally an infinitesimal quantity of water from a pool of pure water at a specific elevation and atmospheric pressure to the soil water at the point considered (Musy and Soutter, 1991).

3.5 TEMPERATURE AND PEDOCLIMATE

In a soil, because of its heterogeneity and thickness, many different temperatures coexist simultaneously, reflecting as many exact energy balances. Structure, water content, colour and content of coarse fragments influence transmission of heat; however, one source of heat energy is truly important: the sun. It acts directly on the soil by heating it, and indirectly through photosynthesis and food chains (Chap. 13). Against the solar energy, the heat flux originating from the centre of the Earth (360 W m^{-2}) is negligible in the total heat budget.

From the sun into your soil!

Each temperature indicates a balance of heat energy in the measured area.

3.5.1 Direct or diffuse: solar energy

From the solar constant to net radiation.

The initial flux of solar energy is the ***solar constant***, which contains all the wavelengths between 200 and 4000 nm (Fig. 3.11). The thermal infrared, between 700 and 1000 nm, affects the soil the most, while the visible, between 400 and 700 nm, is vital for organisms.

The fraction of the solar constant that actually reaches the soil, direct (I) or diffuse (D) is the ***solar irradiance*** R_{si} (Evett in Sumner, 2000). The latter depends on the angle h between the solar rays and the soil (latitude, slope), a right-angle surface receiving the most energy. It is one of the components of ***net radiation*** R_n, which itself is distributed among four forms in the ecosystem:

- L·E, energy consumed for evaporation of water or released by condensation of water;
- q_a, convected heat, diffusion in liquid medium;
- q_s, conducted heat, diffusion in solid medium;

Solar constant: initial flux of solar energy in the upper atmosphere, by definition measured on a surface perpendicular to it. Its value is 2 cal cm^{-2} min^{-1} or 1.4 kW m^{-2}.

Net radiation R_n***:*** energy balance between the solar irradiance R_{si} (reduced by the albedo a), the ***atmospheric thermal radiation*** R_a (0.28 kW m^{-2}) and ***terrestrial thermal radiation*** R_t (0.35 kW m^{-2}). ***Albedo:*** fraction of the net radiation reflected by the soil. The albedo is maximum in light-coloured dry

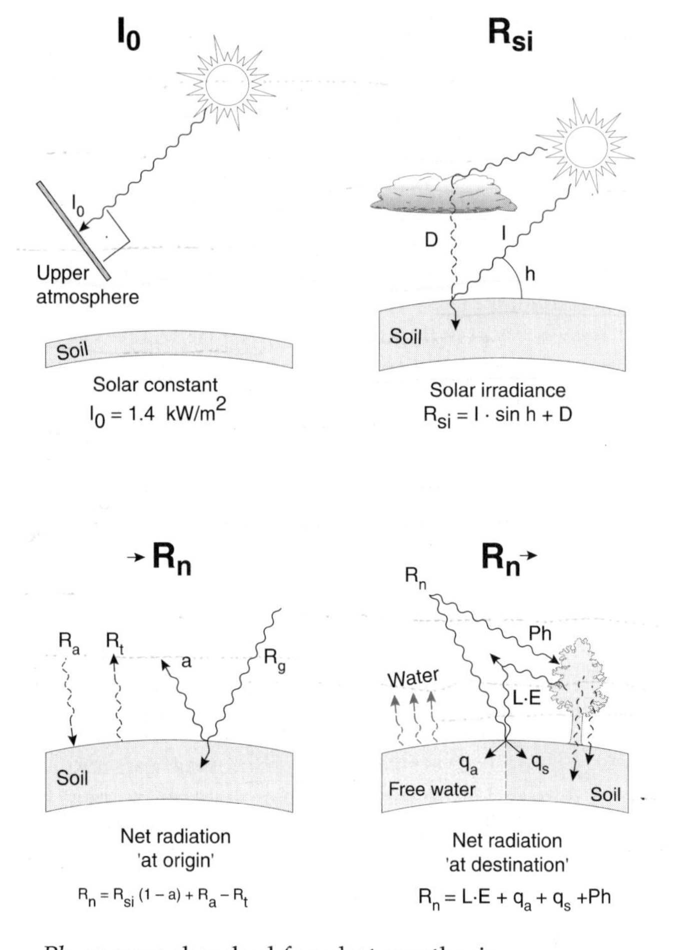

Fig. 3.11 The stages of transformation of solar energy from the upper atmosphere to the soil. Explanations in the text.

soils (e.g. ARÉNOSOLS of coastal dunes) and minimal in dark moist soils (e.g. HISTOSOLS, ORGANOSOLS).

- *Ph*, energy absorbed for photosynthesis.

Soil temperature integrates the first three forms, chiefly q_s, *Ph* being negligible in a quantitative heat balance.

3.5.2 Heating and cooling of the soil

Up to 70°C in dark peats!

Transmission of heat q_s in the soil depends in the first place on its colour, dark-coloured soils heating up faster. Grosvernier et al. (in Wheeler et al., 1995) measured a temperature of 55.9°C in the first centimetre of a dark peat on the way to mineralization. Schmeidl (1978) reported a value of even 71.1°C (Chap. 9, § 9.3.2)! That is at the denaturation limit of certain proteins, which does not fail to influence biological activity! The water content of the soil too plays an important part because of the ***heat capacity*** of water, nearly five times higher than that of air or solids. This explains the slow spring growth of

Heat capacity (or ***specific heat):*** amount of energy required to raise by one degree the temperature of one gram of any material. The heat capacity of water is 1 cal g^{-1}, of sand 0.20, of clay 0.22, of air 0.24, of peat

plants of swamps, a delay accentuated by, among others, the loss of many calories consumed for evaporation of water.

0.40 cal g^{-1}. A soil with water table thus heats up much more slowly than another formed from limestone gravel.

At a certain interval of time, for example a year, measurement of soil temperature at various depths enables establishment of a ***thermal profile***, reflecting the diffusion of heat (Fig. 3.12; Bachmann, in Blume et al., 1996).

3.5.3 Temperature, organisms and pedoclimate

Soil temperature and its variations modify many pedogenetic processes related to rock weathering, clay neoformation or modification of structure. But they also strongly influence life in the soil and on its surface. Bacteria are often very sensitive and have well-defined ranges of optimum activity, for most species between 21 and 38°C. Plant growth is dependent on temperature from germination to building of adult tissue. The mesofauna of a thick litter, for example, is divided according to the very gradual temperature gradients into different debris layers, while the fauna of sunny crags is able to resist fluctuations of more than 50°C in a day.

Bacteria, plants and animals of the soil are very sensitive to soil temperature.

Winter survival of soil organisms is ameliorated by good snow cover, which insulates them against very low temperatures. Some invertebrates, however, resist cold well: *Boreus* (Mecoptera), *Chionea* (Diptera) and Collembola. Among them can be mentioned *Isotoma saltans* or glacier flea, which lives on the snow or ice in high mountains. Another species, *Isotoma hiemalis*, makes a sugar-based antifreeze present in the haemolymph, enabling these insects to withstand temperatures as low as –15°C (Zettel and von Allmen, 1982).

'Les glaciers ont des puces à eux, tout comme les cuisinières et les chiens barbets (...). Elles sont grosses comme les nôtres à peu près, et velues, pour avoir chaud apparemment' (Toepffer, 1846).

The internal climate of the soil, or ***pedoclimate***, is linked to the general climate by heat flows depending on the seasons and by precipitation but, because of soil constituents and properties, it attenuates or amplifies the features of the general climate. For example, the large thermal buffering capacity of a light-coloured, water-saturated fibrous peat retains throughout the year a constant temperature below 5–10 cm depth, underneath a surface layer that may be very warm sometimes, very cold at others. Another example: a soil on the footslope receives continuously the water of upper zones by runoff; even if rainfall has ceased, the effect of the rain is amplified here and wetness prolonged accordingly.

Separated from, but dependent on its environment, the pedoclimate 'follows' with some liberty the general climate.

Pedoclimate: internal climate of the soil, integrating the combined effects of its temperature, water content and aeration.

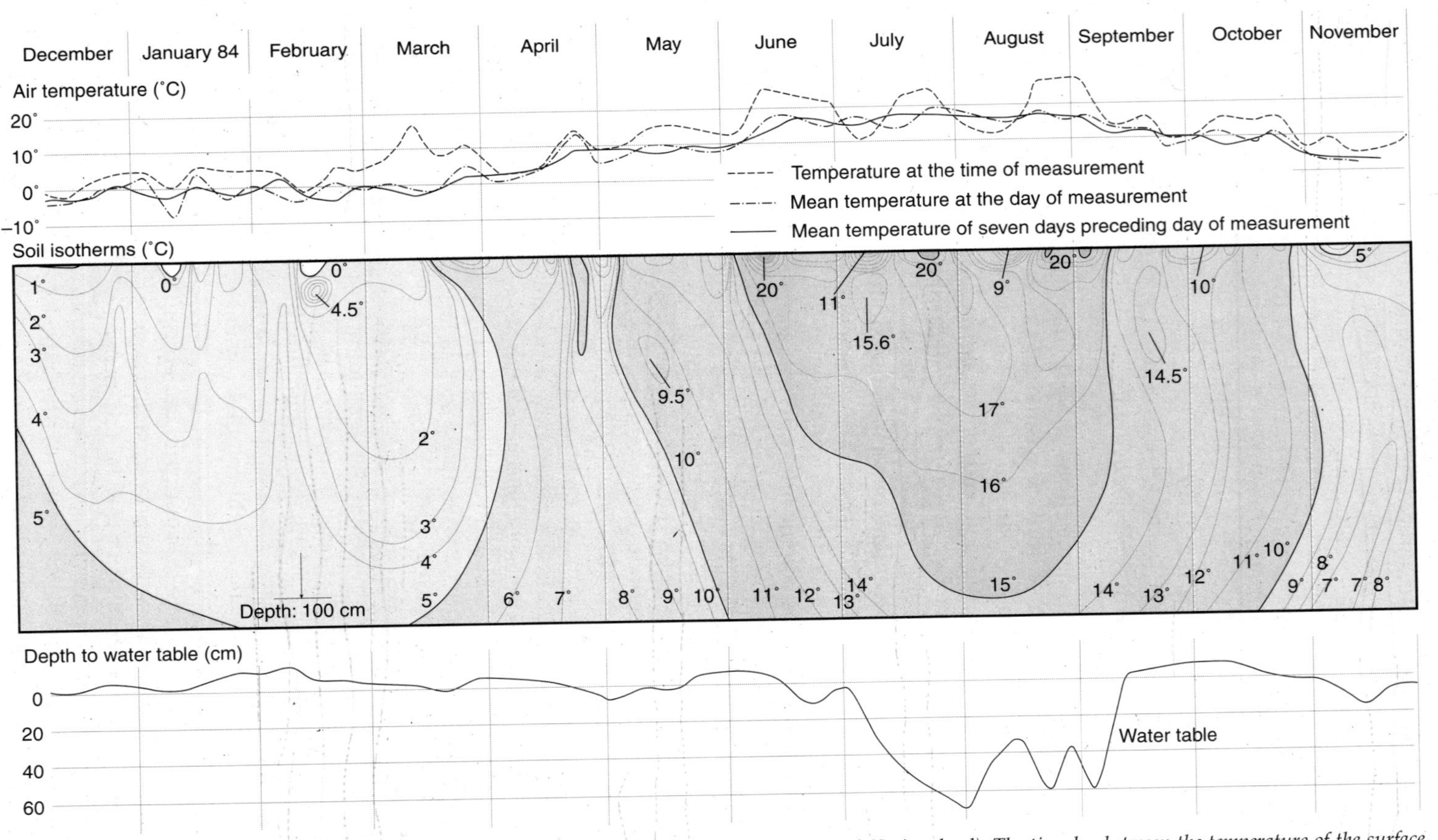

Fig. 3.12 Thermal profile of a RÉDUCTISOL of a small-sedges grassland on the banks of Lake Neuchâtel (Switzerland). The time-lag between the temperature of the surface and deep layers is seen, particularly during the spring warming, such that short-term variations reach only shallow depths. On the other hand, the summer lowering of the water table allows a very obvious warming to a depth more than one metre, as the heat capacity of water no longer opposes it (from Buttler, 1990).

At the scale of organisms, pedoclimate is composed of many ***microclimates***, sometimes reflecting very localized conditions that may vary over a few centimetres or millimetres (Chap. 8, § 8.2). Soil depth, colour, structure and water content are factors creating strong microclimatic heterogeneity.

Pedoclimate: a rather holistic concept.

Microclimate: ensemble of climatic conditions measured at the scale of the organism.

3.6 THE CLAY-HUMUS COMPLEX, EXCLUSIVE PROPERTY OF THE SOIL

3.6.1 Levels of structural organization of the soil

In the ecosystem, the soil is the preferred meeting place of the mineral and organic worlds (Chap. 1, § 1.2.3). At macroscopic scale, 'earth' is its expression through the various structures described earlier. Among them, crumb structure best accomplishes this union, in structural macroaggregates (Fig. 3.13).

The clay-humus complex, close encounter of two worlds ...

In reality every macroaggregate is itself composed of smaller units joined to each other, about 50 μm in size—the microaggregates—in which are distinguished silt particles and bacterial colonies held together by polysaccharides (Chap. 2, § 2.5.2; Chap. 4, § 4.4.1). General cohesion of the microaggregates is ensured by an often brownish matrix, the ***clay-humus complex***. The molecular scale reveals in it a close chemical bond between the clay mineral layers and large molecules of humus. Here appear the true organo-mineral bonds of soil, which give to 'earth' its specificity.

Clay-humus complex (or ***exchange complex***; Chap. 3, § 3.7.1): ensemble of soil materials constituted by the association of humified organic molecules and clay minerals (Jabiol et al., 1995).

3.6.2 Structure of the clay-humus complex

A cation, generally calcium or iron, ensures the bonding of clay minerals to organic polymers, by forming a bridge between the two. Calcium gives strong bonds, very stable, which impede very rapid mineralization of humified organic matter and oppose dispersion of the clays. The humus-calcium-clay complex confers on the soil a dark colour, often observed in carbonate-rich soils (RENDOSOLS, RENDISOLS). Iron replaces a little or much calcium in decalcified soils (BRUNISOLS, NÉOLUVISOLS) or in iron-rich calcic soils (CALCISOLS, CALCOSOLS). The bonding in them is weaker because of the water of hydration around the cations. The humus-iron-clay mineral complex colours the soil brown.

Calcium or iron binds together the two components basic to the clay-humus complex, clay mineral and humus, which are both electronegative (Chap. 2, §§ 2.1.5, 2.2.4).

'When finely divided humus is placed in intimate contact with moist clay, the two substances adhere to one another so tenaciously that a small quantity of humus is combined with each clay mineral particle, resulting in a dark mixture that looks like a fine-grained earthy substance when dried slowly...' (Müller, 1889).

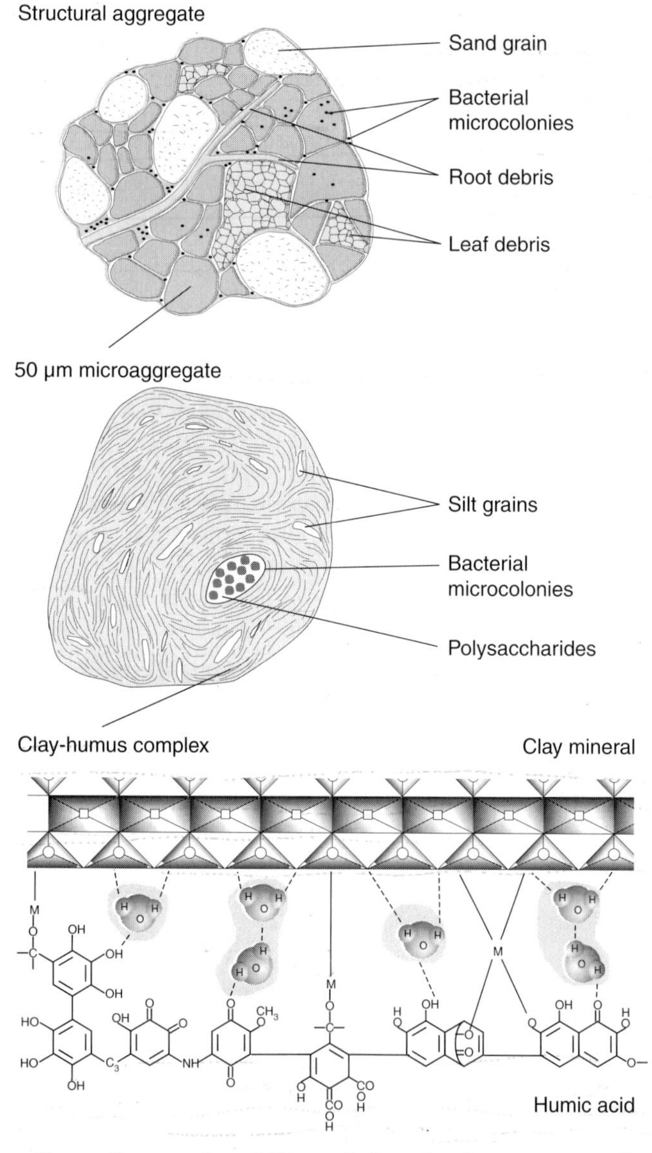

Fig. 3.13 *Levels of structural organization of soil, from macroaggregates to molecules (after Bruckert, in Bonneau and Souchier, 1994).*

The digestive tract of earthworms, an excellent factory of the clay-humus complex (Chap. 4, § 4.5.2).

The clay-humus complex, turntable of the soil's functioning between rock, plant and soil.

Formation and stability of the clay-humus complex thus depends on the quantity and quality of the organic matter, presence of certain clay minerals and bonding cations; they are promoted by the fauna and the microflora.

3.6.3 Effects of the clay-humus complex

A stable clay-humus complex obtains for the soil new properties, all promoting its fertility:

- flocculation of clay and humic colloids is favourable for an aerated structure and adequate water storage;
- the clay mineral-humus bond slows down mineralization of humified organic matter,
- by being bound with the clay, humus impedes its dispersion, preventing clogging and compaction of the soil;
- integration of the clay and humus in one compound enhances the capacity of the soil to retain biolements indispensable to plants (Fig. 3.14).

3.7 IONIC EXCHANGES IN THE SOIL

3.7.1 Discovery

In 1850, J. Thomas Way poured liquid manure, with characteristic odour and colour (!) on sand and on a clayey soil. The manure emerged intact, coloured and odoriferous, from the sand, while the clayey soil released a colourless, odourless liquid. The clay-humus complex of the second substrate retained the compounds of the liquid manure, especially ammonia and coloured pigments: being able to fix ions, it has an ***absorbing power.*** In another experiment, a KCl solution was poured on the same substrates. The emerging percolates were enriched in calcium and impoverished in potassium, especially in the clayey soil! The clay-humus complex also has an ***ion-exchanging power,*** potassium replacing calcium.

A game of antagonists, in which liquid manure becomes... water!

_Ab_sorption complex or _ad_sorption complex?

J. Thomas Way, a British chemist, conducted many experiments on different exchanging substrates and exchanged substances, following the demand of a Yorkshire farmer who had published his practical experience (Boulaine, 1989). He attributed *ab*sorbing power to the complex (whence *ab*sorption complex), which is correct, looking from the outside; but the actual process is rather an *ad*sorption on the electronegative charges on the surface of the clay-mineral layers or on the periphery of organic macromolecules (whence *ad*sorption complex).

3.7.2 Mechanisms and laws of ion exchange

All ion exchanges need an intermediary, the soil solution (Fig. 3.15). For example, addition of KCl increases the concentration of potassium in the solution. In accordance with the law of equilibrium, numerous K^+ ions are retained on the complex, taking the place of ions

The exchanges can pertain to all the ions retained on the complex, whether positive or negative; in the latter case, they are held through a cation bridge.

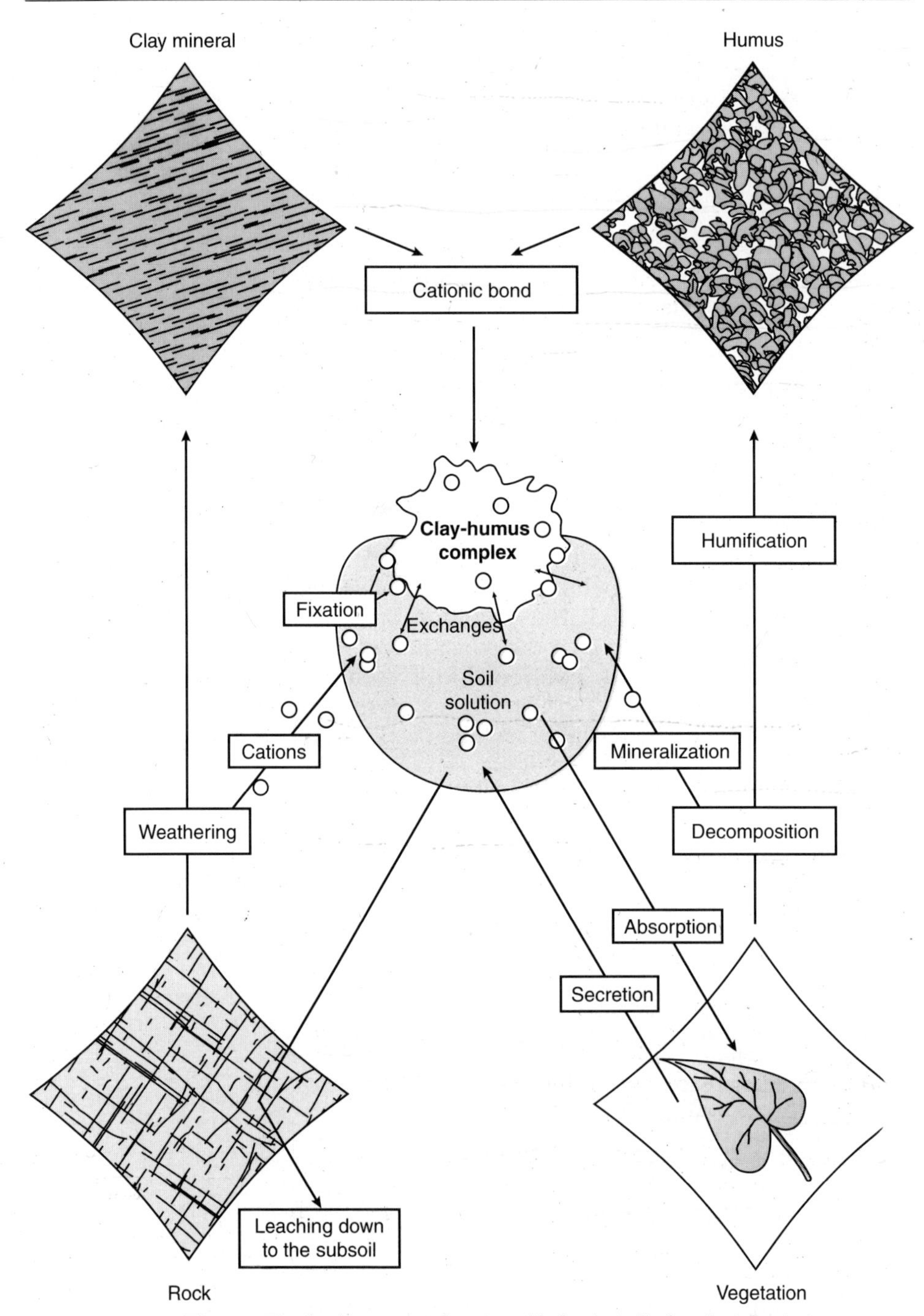

Fig. 3.14 *The clay-humus complex, turntable for the soil's functioning.*

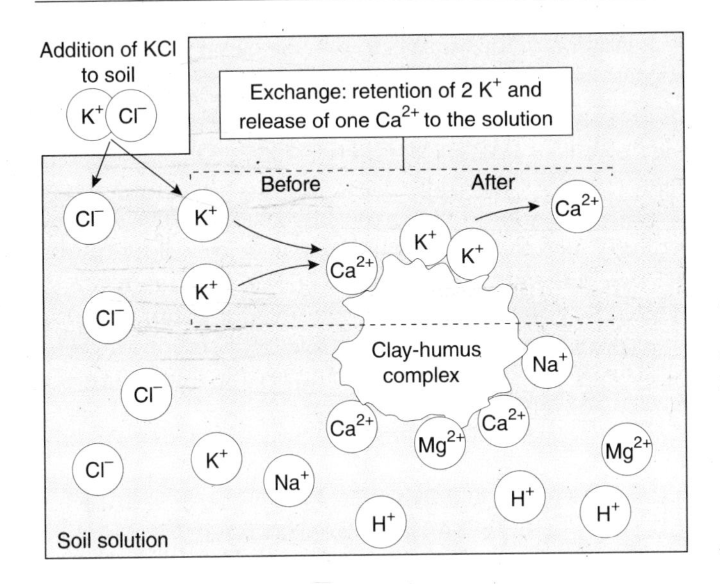

Fig. 3.15 Ion exchanges in the soil, illustrated by the exchange of Ca^{2+} by K^+ when potassium chloride is added. By exchange the potassium chloride decalcifies the soil leading to grave problems if this kind of fertilizer is applied in excess to certain fragile soils.

that are found there, Ca^{2+} in particular. The general equation for the ***ion exchange*** is the following:

$$\text{Exchanger}^-A^+ + B^+ + X^- \Leftrightarrow \text{Exchanger}^-B^+ + A^+ + X^-.$$

A second type of ion exchange exists, in which a monovalent cation expels a divalent bonding cation and releases an anion. For example, potassium can displace calcium holding nitrate on the clay-humus complex. The equation is:

$$\text{Exchanger}^-A^{2+}X^- + B^+ \rightleftharpoons \text{Exchanger}^-B^+ + A^{2+} + X^-.$$

Ion exchange obeys three rules:

- For a given ion, equilibrium, but not equality, should be maintained between the quantities of ions released to the soil solution and those replacing them on the complex. This ***law of ionic equilibrium*** results in displacement of ions from the soil solution to the complex if external addition takes place, and in the reverse direction in case of uptake, for example by roots.
- Ion exchange obeys a relation of equivalence. Thus, one ***gram-equivalent*** of K^+ is exchanged for one gram-equivalent of Ca^{2+}. For quantitative purposes what is often used in soil science is the milligram-equivalent, abbreviated meq, expressed in meq/100 g dry soil; however this unit has to be abandoned in favour of the official unit to which it corresponds, the centimole$^+$ expressed per kg dry soil [cmol (+) kg^{-1}]. This cationic equivalent relation is obviously not respected in the second type of exchange presented below, since the positive charges are unequal.

Three key words of ion exchange: equilibrium, equivalence, intensity.

Gram-equivalent: atomic mass of an element divided by its electronic valence. For example, one meq of K^+ will be 39 mg/1 or 39 mg, while one meq of Ca^{2+} becomes 40 mg/2 or 20 mg. In exchange therefore 39 mg potassium replace 20 mg calcium. The gram-equivalent can also be presented as a unit 'quantity of charge', based on the charge carried by 1 g H^+; thus, 40 g calcium carry twice the charge as one g H^+ (Jabiol, pers. commun.).

Let us discard the old unit of gram-equivalent! Yield to the units based on molarity!

- Retention of ions is selective, some being retained more strongly on the complex than others. The retentivity of cations increases in the order $Na^+ < NH_4^+ < K^+ < H^+ < Mg^{2+} < Ca^{2+} < Al^{3+} < [Mn^{2+}, Hg^{2+}, Cd^{2+}, Fe^{2+}, Cu^{2+}, Zn^{2+}$, etc.]. This force depends on the atomic radius, the valence (for similar atomic radii, the monovalent ions are less strongly held than the divalent ions), degree of hydration of ions and soil pH. Knowledge of this intensity of retention is important in agriculture for estimating the chances of retention of an inorganic fertilizer for example, as also in the study of soil pollution for evaluating the hazards of release of heavy metals.

3.8 CATION EXCHANGE CAPACITY AND BASE SATURATION PERCENTAGE

Cation exchange capacity and base saturation percentage, two useful indices of fertility and degree of soil development.

Not all exchange complexes have the same capacity to retain ions since the type of clay mineral or biopolymer constituting them differs. The proportions of different cations also vary according to the general physicochemical conditions. Nonetheless it is very useful to compare different soils with respect to the amount of basic cations retained versus acid cations; this in turn is a useful index of the mineral fertility and also of the general degree of soil development. But this procedure requires definition and estimation of four values, namely: cation exchange capacity CEC, sum of exchangeable basic cations S, exchange acidity EA and base saturation V.

3.8.1 Total cation exchange capacity

CEC = all the cations held on the clay-humus complex and also, as function of the degree of hydration of the soil, the cations dissolved in the soil solution.

The ***total cation exchange capacity*** is determined by exchange of soil ions with a salt solution, at buffered pH or at the pH of the soil (effective CEC) and expressed in cmol (+) kg^{-1} dry soil. It represents the sum of S and EA. Its value ranges from 1 to 1400 cmol (+) kg^{-1} according to the constituents (Tab. 3.3). The CEC is relatively constant in a soil since it depends on the texture as well as on the amount and quality of organic matter.

Total cation exchange capacity (CEC or T)***:*** maximum quantity of cationic charges that a defined mass of soil can retain and exchange.

Compared to that for cations, the capacity for anion exchange is smaller, of the order of 5%. Anions are only weakly exchanged and generally stay bound to the cations. For example, exchangeable phosphorus represents just 1% of the total phosphorus (Callot et al., 1982).

Table 3.3 Cation-retaining power of constituents and horizons of soil (various sources)

Constituent or horizon type	**CEC, cmol (+) kg^{-1}**
Minerals	
1/1 clay mineral—kaolinite	2–15
2/1 clay mineral—illite	10–50
2/1 clay mineral—montmorillonite	80–150
2/1 clay mineral—vermiculite	100–150
2/1/1 clay mineral—chlorite	5–40
Allophanes (oxyhydroxides)	5–350
Organic materials	
Scarcely humified organic matter, peat	100
All soil organic matter *in situ*	60–280
Pure humified organic matter	200–500
Pure humic acids	485–870
Pure fulvic acids	→ 1400
Pedological horizons (selected)	
Sandy soils, C horizons	1–5
PODZOSOL on sand, E horizon	12
ALOCRISOL on weathered silt, Sal horizon	18
RENDISOL on loess, Aca horizon	28
Humic-clayey soils, A horizons	60–80

3.8.2 Sum of exchangeable basic cations

The ***sum of exchangeable basic cations*** is determined by selective measurement of each cation after exchange in a salt solution, usually KCl or NH_4Cl. These cations are termed 'basic' because they alkalinize the soil and neutralize the acids. In practice, we determine only the four most important, Ca^{2+}, Mg^{2+}, K^+ and Na^+, which together often constitute 99% of the total! S changes more rapidly than exchange capacity as function of the seasons, plant nutrition or degree of hydromorphy.

S = the basic cations Ca^{2+}, Mg^{2+}, K^+, Na^+, etc.

Sum of exchangeable basic cations (S): sum of exchangeable basic cations retained on the complex, expressed in cmol (+) kg^{-1}. By misuse of language, we often speak of the sum of exchangeable 'bases', but this should be avoided!

3.8.3 Exchange acidity

Exchange acidity is often calculated by difference between the exchange capacity CEC and the sum of exchangeable basic cations S; it is also directly determined by potentiometric titration. It comprises the proton H^+, the only true acid cation, and aluminium Al^{3+}, which acidifies the soil through its hydroxyl ions OH^- releasing free protons by dissociation of a water molecule (§ 3.9.1).

EA = the acid cations H^+ and Al^{3+}.

Exchange acidity (EA): sum of acid cations held on the complex.

3.8.4 Basic cation saturation percentage

$V = S/T$ (%)

Basic cation saturation percentage is calculated as follows:

$$V = \frac{S\,[\text{cmol}\,(+)\,\text{kg}^{-1}] \cdot 100}{T\,[\text{cmol}\,(+)\,\text{kg}^{-1}]} \qquad [\%]$$

Basic cation saturation percentage (V): ratio between the sum of exchangeable basic cations S and the cation exchange capacity CEC or T, expressed as a percentage.

The base saturation percentage of some horizons is presented in Table 3.4. It can be seen that the least developed soils on limestone (RENDOSOLS, CALCOSOLS) have a high base saturation percentage while the most developed soils (NÉOLUVISOLS, LUVISOLS, PODZOSOLS), irrespective of the parent material, have the lowest values as a result of the leaching of basic cations. Furthermore, a certain correlation exists between pH and base saturation percentage.

Table 3.4 Basic cation saturation percentage (S/T) and pH of some soil types (various sources)

Soil type	Horizon	S/T (%)	pH
SODISOL	A	100	9.0–9.5
RENDOSOL	Acah	100	7.5–8.0
CALCOSOL	Aca	100	7.0–7.5
CALCISOL	Aci	80–100	6.0–7.0
NÉOLUVISOL	BT	50–60	4.5–6.0
BRUNISOL OLIGOSATURÉ	A	40–60	4.5–5.0
LUVISOL TYPIQUE	A	40–50	4.0–4.5
ALOCRISOL	A	10–30	4.0–5.0
ORGANOSOL INSATURÉ	OH	15–25	4.0–4.5
RANKOSOL	Ah	5–15	3.5–4.5
PODZOSOL MEUBLE	Ah, E	5–15	3.0–4.5

3.9 SOIL pH, TWO-SIDED

The pH, a two-sided physicochemical index.

3.9.1 Exchange acidity and actual acidity

pH_{KCl}: pH measured in a salt solution of potassium chloride, designed to dislodge the retained H^+ by exchange with excess K^+. Some measure ***pH_{CaCl_2}***, using calcium chloride as the exchanging salt.

The exchange acidity is quantified or estimated by measuring ***pH_{KCl}***. But it is also possible to determine ***pH_{water}*** by placing a sample in distilled water. In the latter case the electrode measures only the protons in the soil solution since no exchange has been effected; this is termed ***actual*** or ***active acidity***. The pH_{water} is thus always a little higher than pH_{KCl}, by 0.2 to 1.5 units depending on the soil.

In soils, pH between 2 and 10.

The pH_{KCl} is the 'true' pH of the soil since it integrates in one index the physicochemical properties of the soil solids, hence the long-term processes (Chap. 5, § 5.5). The pH_{water} is more useful in the study of soil-plant relations or for understanding the short-term functional processes, such as leaching of very mobile cations. The pH_{water} ranges from 2 in acid polders with high concentration of H_2SO_4 (THIOSOLS) to nearly 10 in some alkali soils (SODISOLS SOLODISÉS). Most soils of temperate regions have a pH between 4 and 7.5.

3.9.2 Buffering power of soils

When the soil resists acidification.

Buffering power: capacity of a soil to resist change in pH when acids or bases are added to it.

Each soil has a ***buffering power,*** which acts in five successive phases resisting acidification:

- buffering by carbonates, pH 8.6 to 6.2,
- buffering by silicates, pH 6.2 to 5.0,
- buffering by clay minerals, pH 5.0 to 4.2,
- buffering by aluminium hydroxides, pH 4.2 to 2.8,
- buffering by iron hydroxides, pH < 3.2.

Knowledge of the currently active buffering phase enables evaluation of the degree of development of a soil (Chap. 5, § 5.5.5) as well as its potential resistance to atmospheric pollutants.

The buffering power of soils may also concern other ecological factors, for example temperature (§ 3.5.3). On the whole it thus guarantees maintenance of a stable physicochemical system beneficial to plants and the microflora.

3.10 REDOX POTENTIAL

3.10.1 Redox potential and soil organisms

Soil air contains, on average, 18–20% oxygen, which is sufficient for aerobic organisms. But in specific cases the oxygen concentration may be considerably lowered, to reducing conditions. At that stage, higher organisms get asphyxiated, yielding place to groups of microorganisms adapted to anoxia. Anaerobiosis is facultative or strict according to the functions or groups affected (Chap. 4, Fig. 4.19). But bacteria, by feedback to their environment, can in turn accentuate reducing conditions.

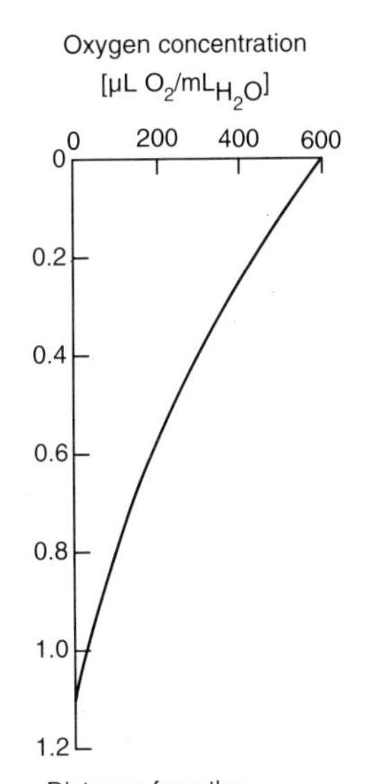

Fig. 3.16 *Oxygen profile from the surface of a spherical microaggregate of silty clay soil (after Greenwood and Goodman, 1967). Different bacteria colonize the 'anoxic interior of the microaggregates and their aerated surface.*

Redox heterogeneity of a microaggregate

The critical diameter of an aggregate faced with anoxia is given by $a^2 = 6C \cdot D/R$ (with a = diameter of aggregate, cm; C = difference in O_2 concentration between the centre of the aggregate and its surface, ml O_2/ml; D = O_2– diffusion coefficient in the aqueous phase, $cm^2\ s^{-1}$; R = level of respiration in the interior of the aggregate, ml $O_2\ cm^{-3}$) (Fig. 3.16; Killham, 1994).

From microsites to the entire horizon.

Even if at the bacterial scale, the soil is made up of microsites with very different conditions such as in the rhizosphere (Chap. 15, § 15.6.1), it is justified for a more general purpose to consider the redox potential of an entire horizon. For example, general physicochemical conditions of the horizon can be related to the movements of the water table.

3.10.2 Determination and values

In soils the redox potential ranges from 800 to –300 mV.

Oxidation-reduction potential (or ***redox potential,*** Eh): value reflecting the quantitative transfer of electrons from donors to acceptors; measured by means of a combined platinum electrode; expressed in millivolts (mV).

The ***redox potential*** of soil is divided into four domains:

- 800 to 450 mV, oxygen is present and nitrification active; organic matter is more or less rapidly decomposed;
- 450 to 0 mV, the milieu is becoming poor in oxygen, which retards decomposition of organic matter; if the pH is low, reduction processes start earlier;
- 0 to –200 mV, the milieu is anoxic; organic matter is decomposed by anaerobic fermentation, while ferrous salts accumulate, colouring the material greenish grey (Gr horizons);
- –200 to –300 mV, reduction is complete; organic matter decomposes releasing hydrogen sulphide or methane, as in peat-bogs (Chap. 9, § 9.2.1); iron sulphides form in mangrove soils (Chap. 4, § 4.4.4).

3.11 FROM MINERAL FERTILITY TO OVERALL FERTILITY

Which fertility for soil? Numerous definitions for different objectives!

Base saturation percentage and exchange capacity are good indices for the mineral fertility of soil. The plant draws useful cations from the soil, their availability being directly a function of the lining of the complex. But the ***overall mineral fertility*** also requires a good supply of anions, which are not accounted for in the base saturation percentage. If potassium is essential for plant growth, nitrate and phosphates are equally so. This overall mineral fertility, provided by cations and anions, is however only one part of the ***general natural fertility*** of soil, in which all the properties described in this chapter take part.

A totally anthropocentric view of soil fertility!

Some definitions of fertility have been very narrow, like that of Mulder et al. (in UNESCO, 1969): 'Fertility is the ability of the soil to support growth of crop plants.' Even in agronomy this one-sided view of fertility is now outdated!

In agronomy, the notion of ***acquired fertility*** adds to natural fertility the human effort directed towards augmentation of soil productivity through ploughing, addition of fertilizers and use of various pesticides and herbicides (Amberger, 1983; Morel, 1996).

When the problems of conservation of agricultural soils became crucial at the end of the twentieth century (Chap. 17, § 17.2), a new view of ***fertility*** emerged. This approach integrates the nutritional needs of the human species and those of all living organisms. It broadens the notion of 'producer of crops' of the preceding definition to a duration that guarantees the edaphic conditions of the entire biocoenosis (principle of sustainability). It does full justice to the living organisms of soil, whose actions and considerable effects have too often been underestimated in the past.

In Switzerland, for example, ***fertility*** has been defined officially in an Ordinance on Soil Pollutants (1998) in a way that well recalls the ideas stated above: 'The soil is considered fertile when:

- it has a diversified and biologically active biocoenosis, structure typical for the site and intact degradation capacity;
- it enables plants and plant associations, natural or cultivated, to grow and develop normally, without being injurious to its properties;
- the fodder and plant products it provides are of good quality and do not threaten the health of man and animals;
- its ingestion or inhalation do not endanger the health of man and animals.

And the fertility of soilless cultures?

The fertility of soilless cultures can be evaluated only by the notion of acquired fertility, the others being inapplicable to this totally artificial milieu. Also, this type of agricultural technique is not generally considered by the provisions of law applied to soil. It is at least difficult to term 'soil' 'a portion of air in which we disperse nutrient solutions' ... even if plant roots live well in it!

Acquired fertility: Aptitude of a soil to produce under its climate. It is measured from the abundance of the yield it bears when most suitable agricultural techniques are applied and from the quality and long-term persistence of this aptitude to produce (Lozet and Mathieu, 1997).

Fertility: ability of a soil to produce the entire food chain going from micro organisms to humans, passing through plant and animal, and this for generations (Soltner, 1996a, who used the term fecundity).

'For tomorrow's agriculturist to fertilize his fields correctly demands that he be concomitantly pedologist, soil microbiologist and plant and animal physiologist (...). While a simple chemical approach to fertility indicates our cultivated soils are fertile, a physical (internal surface area of soil, exchange capacity and quality of the exchange complex) or biological (microbial activity) approach shows that they are being impoverished' (Bourguignon, 1996).

To learn more about soilless cultures, refer to Morard (1997).

CHAPTER 4

LIFE IN ACTION

Many essential functions of the soil and some of its physicochemical properties depend on the action of living organisms, which up to now has only been outlined (Chap. 2, §§ 2.5, 2.6). The objective of this chapter, the largest in the book, is to emphasize the fundamental role of living organisms: plant-soil relationship and root functioning (§ 4.1), plant nutrition (§ 4.2), ***bioelements*** and their utilization by plants (§ 4.3), principal functions of microorganisms (§ 4.4) and of the fauna (§ 4.5), and biological indicators of pollution (§ 4.6).

No soil can be formed without living organisms: no plant, animal or human nutrition, no biological regulation of the hydrologic cycle, no structure formation or energy flow!

Bioelement: chemical element entering the composition of living matter or utilized in the metabolism of a living organism. For convenience of language—which is what we have done in this book—small molecules such as NO_3^- and SO_4^{2-} are included in this term.

4.1 PLANT AND SOIL: AN INTIMATE AND 'TOTAL' RELATION

4.1.1 Edaphic peculiarities of plants

If the higher plants growing on the soil have not been the subject of any particular presentation, unlike bacteria, fungi, algae or invertebrates, it is because it is impossible to define a 'soil plant' as for fungi or bacteria. Acting at the same time at depth through their roots and above the ground by their aerial organs, plants influence the soil as much by the active processes of their living parts as by the passive effects of their necromass and litter. They also permanently exchange water and dissolved, absorbed or evacuated substances by ***secretion*** and ***excretion***.

There are soil bacteria, soil fungi and soil invertebrates, but there are no 'soil plants'!

Secretion: evacuation towards another organ or outside the organism, of substances elaborated by cells.

Excretion: process ensuring the rejection of metabolic wastes outside the organism.

A single plant, for example a tree, is active at all the spatial scales described till now, from the micrometre of bacteria to the metre of the mycelium, passing through the millimetre of microarthropods. Also, the complex and multiscale spatial organization of plant communities vertically into strata or horizontally into mosaics further complicates the relations between soil and plant, modifying them compared to isolated individuals (Chap. 7, § 7.1). For all these reasons, the edaphic functions of plants cannot be described in a 'split' manner, as functions of exact pedologic processes. Figure 4.1 nevertheless attempts to summarize these manifold activities.

The plant is autotrophic, surely, but heterotrophic as well!

Is the plant truly autotrophic? And is the soil an ecosystem?

The plant, which occupies space above and below the ground, is sometimes considered a dual organism from the trophic viewpoint: the above-ground portion contains the autotrophic organs (chlorophyll-bearing tissues) while the below-ground portion is entirely heterotrophic, depending for life on metabolites elaborated by the former. To this spatial segregation of functions is added a temporal change, because the plant is autotrophic only during the day, by virtue of sunlight, but is heterotrophic by night, continuing to respire! At the higher level of organization (Chap. 8, § 8.1) the aerial portion of the phytocoenosis is the autotrophic subsystem of the ecosystem, whereas the soil, including the roots, represents its heterotrophic portion.

The soil is not an ecosystem, but an ecological system.

Therefore, soil should not itself be considered an ecosystem. It is actually incapable of capturing and transforming solar energy by itself and, thus, of ensuring the function of the primary producer (Chap. 2, § 2.2.1; Swiss Soil Science Society, 1997), a task nevertheless inseparable from the global trophic-dynamic function of any ecosystem (Lindeman, 1942). On the other hand—and this is the point of view adopted in this book—nothing prevents us from considering the soil as an ecological system (Chap. 1, § 1.2.2).

4.1.2 The root and its relation to soil

The root or, better, the root system is the preferred site for interaction of the soil and the vascular plant.

The root system, illustrated here by that of beech *Fagus sylvatica* (Fig. 4.2), consists of different parts characterized by smaller and smaller diameter of its ramifications. The stump at the base of the trunk forms the link between the above-ground and below-ground parts; the principal roots grow from here in preferred directions, more or less deeply into the soil. They ensure mechanical anchorage of the tree; they extend by medium roots that penetrate large soil pores (Chap. 3, § 3.3.1), then by fine roots. These latter, the ***rootlets***, ensure nutrition of the plant with water and bioelements by means of their absorbing hairs and/or mycorrhizal fungi associated with them

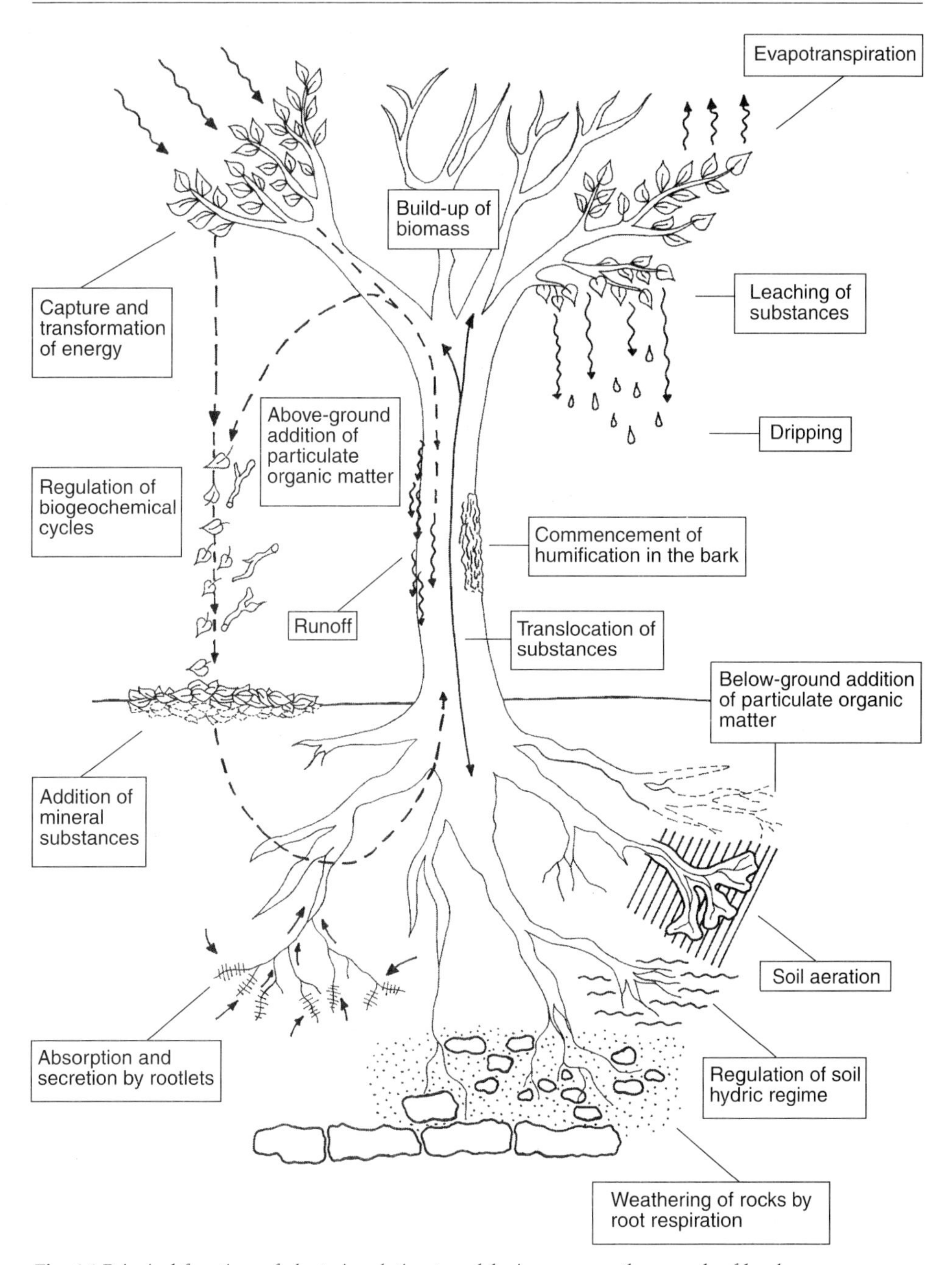

Fig. 4.1 *Principal functions of plants in relation to pedologic processes: the example of beech.*

Fig. 4.2 *Anatomy of the root system. Example of beech,* Fagus sylvatica, *at the edge of a gravel-pit (photo J.-M. Gobat).*

'Roots account for between 40 and 85% of net primary production in a wide range of ecosystems from grassland to forest (...). It seems axiomatic, therefore, that an understanding of the functioning of plants within natural communities must demand an equal understanding of the behaviour of roots and root systems' (Fitter, 1987).

'It is very important for this plant (beech) that soil particles be compactly bound around the roots to better enable their transformation into a small bag of earth that encompasses the roots' (Müller, 1889).

Some organisms come into direct contact with the root to form associations with one-sided (parasitic) or mutual (symbiotic) benefit (Chap. 15).

Rhizosphere: region of the soil under the direct influence of the root. This term was introduced in 1904 to denote the zone in

(Chap. 16, § 16.1). The functional interface between plant and soil, the rhizosphere, is found in the vicinity of the rootlets. In some cases adventitious roots appear from above-ground stems and act as props.

The differences in ecological behaviour between two species are better explained by the functioning and morphology of the root system in its entirety than by the anatomy of roots taken individually, very similar to each other (Fitter, 1987; Polomski and Kuhn, 1998). Among the differentiating criteria can be mentioned the relative proportions of primary and secondary roots, the degree of ramification or even the capacity to put out lateral roots.

The relation between the soil and the roots cannot be described as a simple juxtaposition of the properties of the root (such as observed for example in a sterile physiologic solution) and those of the soil (such as at a good distance from the roots). By its growth, activity and products the root considerably modifies the physicochemical properties of the soil it comes into contact with, and therefore the biocoenosis of that soil. By feedback, organisms also modify the conditions of the root environment.

4.1.3 The rhizosphere

The fine roots in the soil determine in their vicinity a transition zone, an interface, termed the ***rhizosphere*** (Chap. 15). From here begins the ***bulk soil*** which, in fact, represents the earth easily detached from the roots by simply shaking the clods. Because of the extent of the mycorrhizosphere and the root density in most soils, it is in reality difficult to identify a soil fraction that agrees with the definition.

Root tissues too, however, can serve as a habitat for microorganisms. Apart from the well-known and visible mycorrhizae (Chap. 16, § 16.1) and nitrogen-fixing nodules (Chap. 16, § 16.2), certain bacteria live in direct contact with the root, or even penetrate the rhizodermal and cortical tissues without being parasitic or predatory. These organisms remain attached to the root even after thorough washing. The concept of rhizosphere is often extended to these habitats, designated ***endorhizosphere***. This underscores the fact that the interface between the root and the microflora extends into the root. The term endorhizosphere, however, is very general. The ***rhizoplane***, which is the surface of the root tissues, the ***histosphere***, the region in which microorganisms inhabit the interior of the tissues but outside the cells, and the ***cytosphere***, where microorganisms inhabit the interior of the cells, are more precisely recognized.

The external boundary of the rhizosphere is indistinct and depends on the factor considered. The length of the mycelium of a fungus forming mycorrhizal symbioses (Chap. 16, § 16.1) may at times attain several metres; this zone, under the indirect influence of the root, is termed the ***mycorrhizosphere***.

which the root of Leguminosae attracts soil Rhizobia prior to establishment of symbiosis (Chap. 16, § 16.2.1); it has now gained a more general meaning.

Bulk soil: fraction of the soil at the level of the root system not under the direct influence of the root.

Endorhizosphere: intraroot habitat of some bacteria, without inducing formation of specific morphologically recognizable organs such as nodules or rhizothamnia (Chap. 15, § 15.3.3; Chap. 16, § 16.2). The term endorhizosphere pertains to a region with poorly defined boundaries but 'has the merit of evoking gradual passage from the soil to the interior of the root' (Davet, 1996). The true functional boundary between the root and the soil is the endodermis (§ 4.2.1).

4.1.4 Structure of the root environment

The root and its immediate environment: two mutually perpendicular spatial gradients and one temporal gradient.

The first gradient, radial, expresses the change in conditions from the surface of the root to the bulk soil. The second, longitudinal, corresponds to development of the root from its apex, the ***tip***, towards its base; furthermore, the extremity of roots is generally more acid than their base (Davet, 1996) but, depending on soil pH or N concentration in the soil solution, the contrary situation is sometimes observed (Marschner, 1995). Differentiation of the root thus corresponds to an important structural and functional evolution. One could also consider a third gradient, temporal. The root at a point in soil in contact with it manifests *in time* the same growth observed at a given instant along the longitudinal gradient. Figure 4.3 illustrates this spatio-temporal gradient.

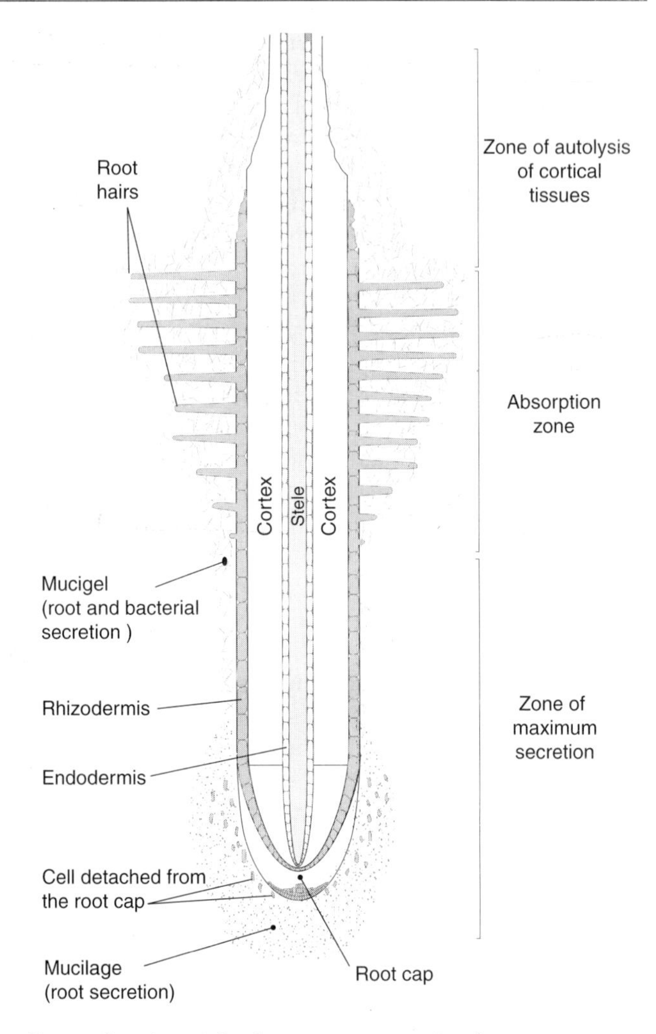

Fig. 4.3 *Diagram of a rootlet and its environment. Explanations in the text.*

Stele: central part of the root containing in particular the xylem and the phloem. ***Xylem:*** conductive tissue of plants that transports water and minerals (xylem sap) from the roots to the leaves. Its conductive cells are dead. ***Phloem:*** conductive tissue of plants transporting the phloem sap, rich in carbohydrates, from the leaves to the roots. Its conductive cells are living.

From the tip to the base are recognized:

- The ***root cap***, a mass of proliferating cells that constitute a sort of protective shield for the apex. When in contact with the soil the outer cells of the root cap are easily shed and are continually renewed.
- The ***apical meristem***, zone of cell multiplication where the primary growth of the root takes place.
- The zone of principal cell elongation and of differentiation of the ***stele***, where the secondary growth takes place.
- The cells of the ***rhizodermis***, or ***root epidermis***, and their extensions, the ***root hairs***. These extensions not only have the effect of augmenting the absorbing surface area of the rootlet, but also anchor it in the soil, serving as a point of support of apical growth. Some trees count many

millions of root hairs, totalling 10 to 50 km in length. Their life is very short, between 2 and 3 days, but they are renewed equally rapidly (Finck, 1976; Davet, 1996).

- A zone of lysis of rhizodermal cells and of the subjacent ***cortex***. It is generally accepted, at least for grasses, that this lysis occurs under genetic control of the plant and is not the effect of an attack by rhizosphere microorganisms.

Further back on the root, in the case of plants with persistent roots, formation of a ***lateral meristem*** responsible for growth in girth of the root is often observed.

Cortex (or ***bark***): zone of the root between its epidermis and its innermost portion, the stele.

Lateral meristem: zone of secondary cell growth, which extends peripherally from the root or stem, resulting in their thickening.

4.1.5 Root products in the soil

The ***rhizodeposition*** represents an important part of the total photosynthesized matter, generally 20–50%, with a maximum of 80%. It exceeds root tissue in annual carbon production (Davet, 1996). Gross production is roughly subdivided into three equal fluxes: net primary production, cellular respiration and rhizodeposition (Fig. 4.4). The root thus creates in the rhizosphere a considerable flux of carbon and energy, to the benefit of heterotrophic organisms capable of assimilating this production.

Together with dead roots the rhizodeposition constitutes a veritable below-ground litter (Chap. 2, § 2.2.1).

Rhizodeposition: ensemble of organic substances of various types elaborated by the root and rhizosphere organisms associated with it (Chap. 15, § 15.2.3). Its precise determination is difficult, depending not only on the plant species considered, but also its age and the conditions around it.

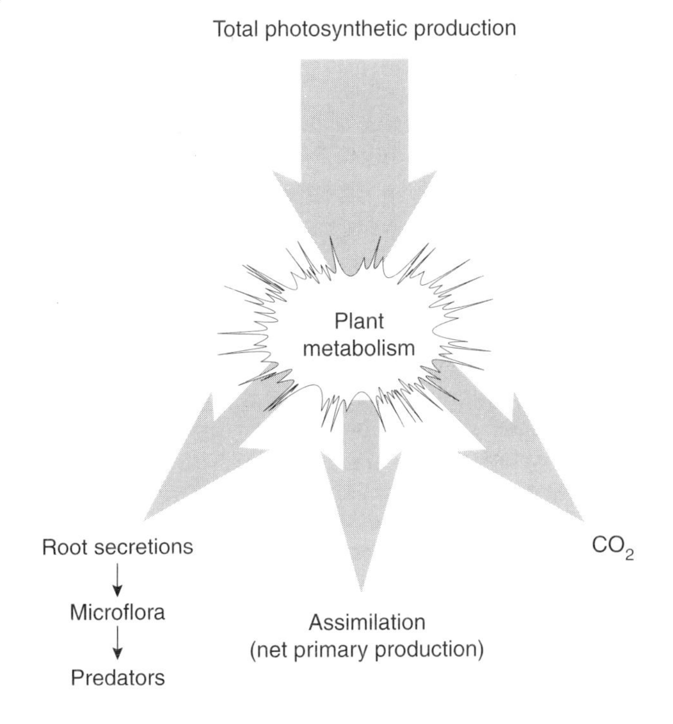

Fig. 4.4 *Distribution of gross photosynthetic production by plant metabolism.*

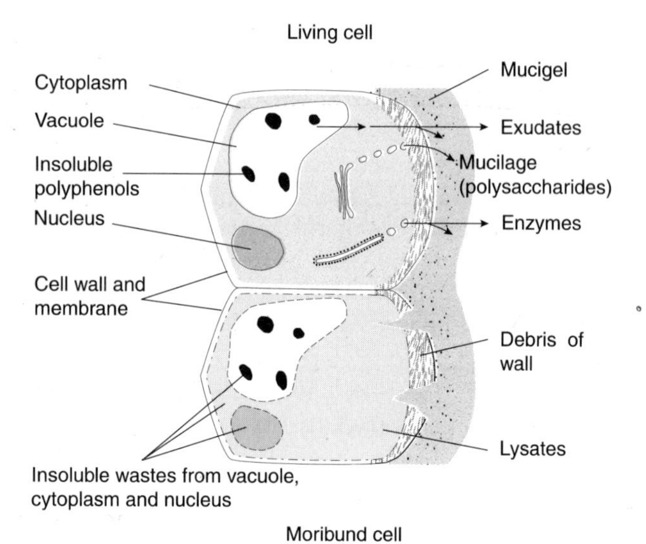

Fig. 4.5 *Constituents of the rhizodeposition (after Foster and Martin, in Paul and Ladd, 1981).*

By its growth and activity, the living root releases into the soil products of various kinds, commonly grouped under the term rhizodeposition.

The rhizodeposition comprises (Fig. 4.5; see also Marschner, 1995):

• Detached cells, mainly from the root cap. Production from the root cap is large, 400–600 cells per tip per day for bean to 21,000 for maize. These cells do not divide further but survive for some time in the soil and can secrete enzymes and proteins (Davet, 1996). Left behind during rootlet growth, they are finally invaded and lysed by bacteria.

Lysate: liquid substance released by destruction of tissues and cells.

Mucilage: gelatinous compound formed from polysaccharides produced by the cells of the root cap and rhizodermis. Not to be confused with mucus (§ 4.5.2)!

• ***Lysates*** resulting from autolysis of rhizodermal and cortical tissues, which benefit saprophytic bacteria.

• ***Mucilages*** secreted by many zones of the root, especially the cap. They remain behind the tip during root growth and are important for aggregation of microorganisms and soil particles. The zone of root hairs also forms a mucilage of pectic nature (polygalacturonic acid). Plant mucilages are quite rapidly metabolized by rhizospheric microflora. In turn, the latter, because of large additions of substrates rich in carbon, secrete other polysaccharides which are often much more resistant to enzymatic decomposition. The root environment is thus structured in the form of a ***mucigel*** of mixed plant and microbial origin.

Mucigel: sheath of mucilages forming a gel surrounding the root.

Exudate: soluble organic compound released by the root, particularly from the apical region.

• ***Exudates*** form the major part of the rhizodeposition and, at the same time, the one most rapidly metabolized by microorganisms. They comprise sugars, amino acids, organic acids, ***growth factors*** and hormones. Some of these compounds result in a specific response in some microorganisms present in the rhizosphere environment, at times as a prelude to closer interaction. This is true for volatile compounds carried by higher plants to 'their' symbiotic fungi, inducing them to grow towards the host plant (Chap. 16, § 16.1.2). For example, *Sclerotium cepivorum* is sensitive to these signals up to more than a centimetre from the plant (Davet, 1996). This is also true of flavonoids released by the roots of Leguminosae prior to nodulation (Chap. 16, § 16.2.1).

Growth factor: precursor of a cell constituent that a particular organism is incapable of synthesizing. It should therefore be found ready-made in the organism's food (example, vitamins—precursors of enzyme cofactors).

4.1.6 Root growth in the soil

Elongation of the root takes place mainly between the apical meristem and the zone of root hairs. These hairs, as already mentioned, enable progression of the apex. The cap, given its intense proliferation and detachment of cells, represents a constantly renewed 'shield against wear and tear' protecting the very delicate structure of the apical meristem. Here polysaccharides act as an essential lubricant, facilitating root penetration in the soil. Also, by their ***gel*** structure, they protect the tissues from desiccation, by forming a water interface between them and the soil atmosphere. Thus they facilitate movement of water towards the root.

The environment of the root plays an important part in its growth and functioning.

Gel: mixture of a colloidal substance and a liquid formed spontaneously by flocculation and coagulation.

Root production can be very large, reaching 10 t ha^{-1} y^{-1} in forests, or 40 to 70% of the net primary production NP_1, while not exceeding 5 t ha^{-1} y^{-1} in meadows. In the latter case, however, its proportion can attain 85% of the NP_1! Among rootlets with diameter less than 5 mm, 99% of the production is from the finest, smaller than 2 mm (Fogel, in Fitter, 1985). Some species are capable of large production in a very short time: a strawberry plantation at optimum growth can produce nearly 129 kg ha^{-1} of new roots in a week (Atkinson, in Fitter, 1985).

The proportion of root production in the net primary production of the phytocoenosis ranges from 40 to 85%.

From 15 to 25% in forest, the root biomass can be as high as 75–98% in grasslands or tundra, or more than 100 t ha^{-1}! For example, in tall-sedge meadows *(Caricetum elatae)* of Lake Neuchâtel (Chap. 6, § 6.4.2; Chap. 7, § 7.6), the root biomass reaches 85 t ha^{-1} in the driest zones and 157 t ha^{-1} in very moist regions where peat is formed. For the entire swamp the below-ground biomass is between 79 and 94% of the total biomass (Buttler, 1987). Below-ground biomass is determined by drying and weighing the roots separated from a known volume of

As a result of production, the root biomass forms a variable portion of the total plant biomass, difficult to estimate for methodological reasons.

soil and washed. Unfortunately this method never gives the total biomass because the very fine rootlets remain attached to the soil or are broken off by washing.

Longitudinal growth of a major root ranges from a few millimetres to a few centimetres per day (Davet, 1996).

Root length is determined as much by specific genetic factors as by edaphic constraints (Polomski and Kuhn, 1998). Among the latter, hydric stress compels plants of arid areas to extend roots growth very far in search of water sources. For example, the roots of tamarisk *(Tamarix aphylla)* extend laterally 30 metres from the trunk, while roots of *Prosopis*, a thorny shrub of the dry regions of America and Africa, reportedly reached a depth of 53 m (Etherington, 1982)! Generally, plants lengthen their roots in soils poor in bioelements and water.

4.1.7 Death of roots

Root death and consequent loss of mass are very difficult to characterize, even more so than production or biomass.

Even if browning is a good criterion, dead roots do not always differ morphologically from living ones, making their quantification very tricky. Fogel (in Fitter, 1985) estimated that 30–90% of rootlets, measured by length, volume or number, disappeared every year from the root system of trees, evidently compensated for by production of new ones. In the turnover of root biomass this represents 40% for poplar, 66% for pine and a remarkable 80–92% for beech. The below-ground litter from roots and mycorrhizae related to them is then two to five times greater than that from the above-ground parts (Chap. 2, § 2.2.1).

At death, roots release organic and mineral substances to the soil.

'What is missing from all this speculation (on the dynamics of the root) is any extensive information on root longevity' (Fitter, 1987).

Dead roots nourish the food chains or the exchange complex via the soil solution. In temperate regions, death and decomposition of roots are highest in September and October, while their elongation is greatest in August and September. On the other hand, large woody roots increase in thickness in spring (Atkinson, in Fitter, 1985). To the 'natural' death of roots may be added losses (up to 10%) due to phytoparasitic nematodes and rhizophagous larvae of Coleoptera.

4.2 PLANT NUTRITION

4.2.1 General theory

A unifying general theory of plant nutrition.

By greatly simplifying chiefly the chronology of events it is rather easy to derive a general theory of mineral nutrition of plants (Heller, 1989; Campbell and Mathieu,

1995; Marschner, 1995; Lüttge et al., 1996; Soltner, 1996b; Campbell et al., 1999). But it should not be forgotten that two of the essential elements, carbon and oxygen, cannot be covered under this theory since they first enter through the leaf!

Transport of substances from the soil to the plant is assured by ***mass flow*** (entrained in water) or by ***diffusion***; it is ***passive*** in both cases (Fig. 4.6). It pertains to circulation in the bulk soil as well as in the intercellular spaces of the outer layers of the rootlets (***apoplastic route*** in the root cortex). ***Active transport*** concerns the elements *within* the cells ***(symplastic route)*** at the level of root hairs, or at that of the ***endodermis*** at the boundary of the stele. Here, passage through the walls is blocked by a barrier (the Casparian strip) formed of hydrophobic substances such as ***suberin***; all the elements are then

A passive transport towards the cells ...

Diffusion: transport of a substance from a compartment in which the electrochemical potential of the substance is higher, to a compartment in which the potential is lower (also see Chap. 3, § 3.3.3; Leij and van Genuchten in Sumner, 2000). This transport requires no expenditure of energy. If the substance is bound to a protein during transport, the term used is ***facilitated diffusion.*** Diffusion, unlike facilitated diffusion, allows no regulation by the plant of the transport.

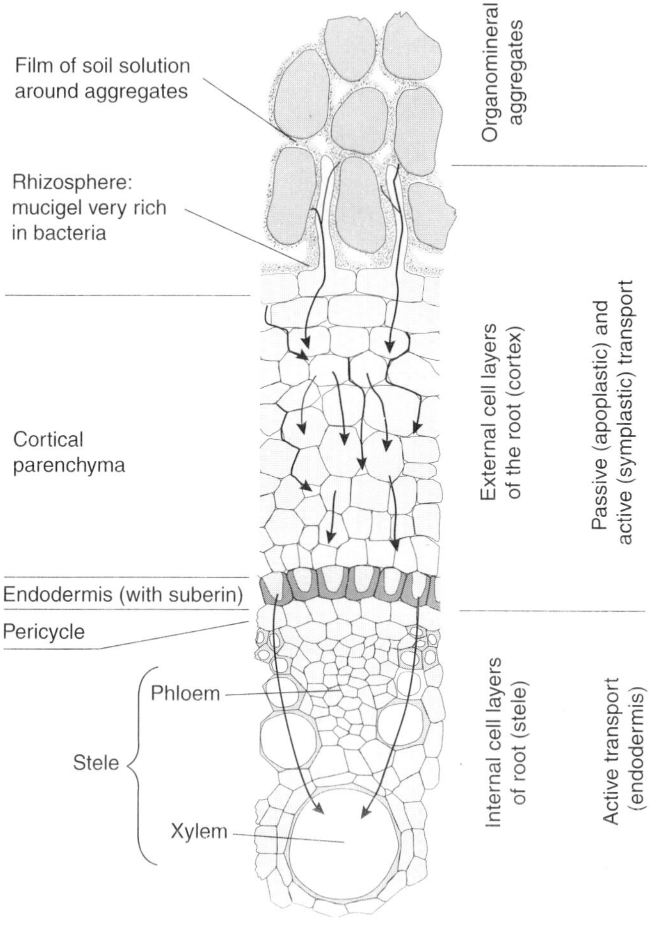

Active transport: transport of a substance from a compartment, in which the electrochemical potential of the substance is lower, to a compartment in which the potential is higher (Callot et al., 1982). This transport requires expenditure of energy but enables concentration, sometimes several hundredfold, of a solute in the cell. On the contrary, ***passive transport*** requires no expenditure of energy.

Fig. 4.6 *Transport of bioelements from the soil solution to the plant (cross-section of a rootlet at the level of root hairs). Explanations in the text.*

Endodermis: innermost cell layer of the cortex forming a physiological barrier to the entry of water into the stele of the root. The endodermis 'forces' water and mineral salts to take the symplastic route from this zone.

... then an active transport to penetrate them.

Suberin: biopolymer very resistant to decomposition composed of equal parts of carbohydrates and phenols.

compelled to penetrate the cytosol and use the symplastic route (Amberger, 1983; Marschner, 1995; Lüttge et al., 1996; Campbell et al., 1999).

Active transport is then required to overcome the adverse osmotic gradient and to lead the ions into the cells with higher electrolyte concentration. This active transport is indispensable up to the xylem. By means of an expenditure of energy the ions can cross the cell membranes using three mechanisms in succession (Fig. 4.7):

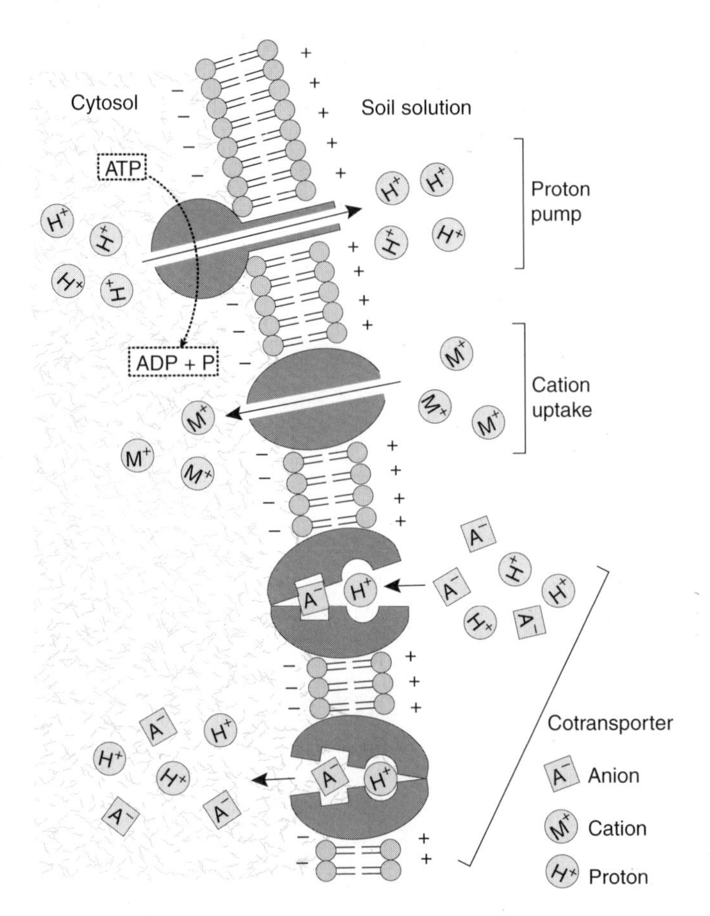

Fig. 4.7 *Model of active transport of bioelements between soil solution and cytosol through the plasma membrane (after Campbell and Mathieu, 1995; Campbell et al., 1999). Explanations in the text.*

- a 'proton pump' generates a difference in membrane potential and in pH by expelling H^+ ions;
- following this difference in potential, cations penetrate the cytoplasm, which has an electrochemical potential lower than the exterior; this entry is of course passive and possible solely through the preceding expenditure of energy of the proton pump;

• anions enter against the adverse electrochemical gradient by means of a transport protein, the ***cotransporter***, which concomitantly sends protons back to the interior of the cell.

Cotransporter (or ***symporter***) protein responsible for transfer of substances across the plasma membrane (channel). The cotransporter uses the electrochemical gradient of a substance to transport another with opposite gradient. In root absorption such a system is frequent between a proton and an anion, at the level of the plasma membrane.

4.2.2 Physicochemical factors influencing ion absorption

Distribution of ions in the short-term (soil solution), medium-term (exchange complex) or long-term (mineral reserves) largely depends on variations in their solubility for which pH and redox potential are the principal controllers (Fig. 4.8).

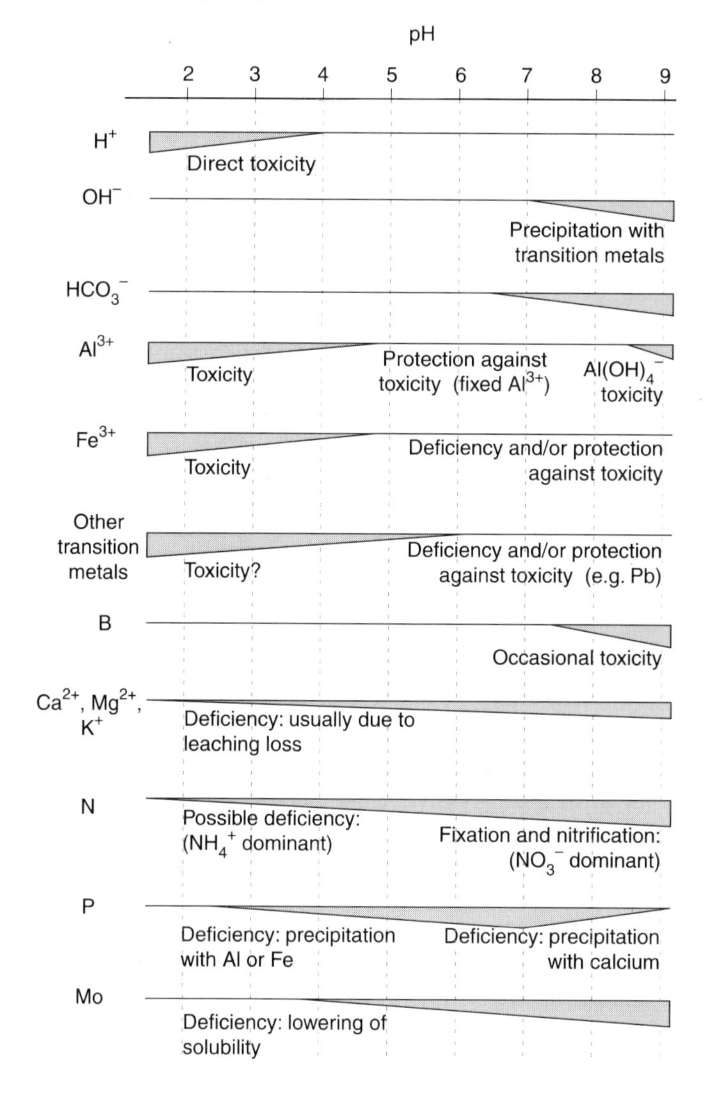

Fig. 4.8 *Solubility of ions as function of pH. Lowering or augmentation of pH, especially to extreme values, solubilizes many ions, notably metals, which can then become quite toxic to plants (§ 4.3.5) (after Etherington, 1982).*

The speed of water fluxes near the root influences nutrition directly.

The general composition of the medium also acts, with possibilities of dependence as between nitrogen and phosphorus, or of antagonism as seen between calcium and potassium (§ 4.3.3). Also, because of the law of ionic equilibrium between the exchange complex and the soil solution (Chap. 3, § 3.7.2) the latter is regularly replenished when absorption by the plant takes place. Other physical properties of soil also influence ion absorption (Callot et al., 1982; Marschner, 1995; BassiriRad, 2000):

- Texture promotes or restricts spatial development of the root system.

- Structure modifies morphology of the roots according to the obstacles they encounter during colonization of the soil. Generally straight, few in number and thread-like in massive-structured soils, roots are often sinuous, branched and rich in root hairs in substrates with crumb or blocky structure.

- Porosity affects the mean root length, which respectively attains 10.9 and 6.5 cm for porosities of 50% and 35% in the case of cat's-tail grass, *Phleum* sp. (Callot et al., 1982).

- Permeability influences soil aeration, important for root activity, especially root growth, If low, it opposes root penetration. This is the case of ***plough pans***, which sometimes prevent crop plants from accessing deeper water reserves.

Plough pan (or ***plough sole***): plane smoothed out by a ploughshare, rendering the soil impermeable at that level and, in some cases, arresting the downward growth of roots. Earthworms often pierce and 'repermeabilize' the plough pan (§ 4.5.1).

4.2.3 Biological factors influencing and regulating nutrition by ions

Anatomical adaptations

With a reserve of identical quantities of bioelements not all plant species behave similarly with respect to absorption. Differences are also seen between individuals of the same species, for example of different ages, or even between rootlets of the same plant according to their location in specific microsites. The type of rooting, its architecture or capacity for regeneration are also fundamental in this regard, enabling the plant to explore a larger or smaller volume of soil (Fig. 4.9; Polomski and Kuhn, 1998).

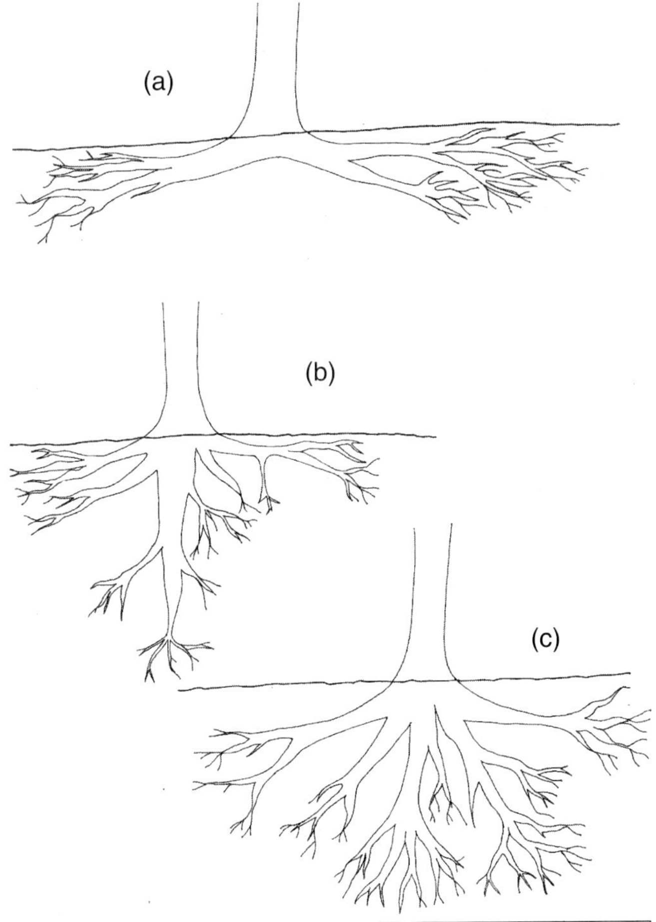

Fig. 4.9 *Three types of root systems (greatly simplified): (a) running, (b) tap root, (c) multidirectional.*

Anatomical and physiological adaptations of plants growing in anoxic conditions

A possible adaptation of plants to hydromorphic situations is the rapid production of an internal air-containing tissue, the ***aerenchyma***. Examples are rice *(Oryza sativa)*, sticky alder *(Alnus glutinosa)*, horsetail (*Equisetum* spp.) or reed *(Phragmites australis)*, in which the roots are often enveloped by a thin red sheath contrasting sharply with the greenish grey of the bulk soil with negative redox potential. The iron present in the oxic sheath is, however, not solely due to oxidation of native reduced iron in the soil; it also results from active evacuation that reduces its concentration in the root cells.

Other adaptations are the constitution of adventitious roots, tolerance to toxins in soil or to internal metabolites originating from anaerobic processes, or even rapid adaptation to temporary flooding (Etherington, 1982). Some plants survive for a more or less long time by means of alcoholic fermentation that substitutes for respiration in providing energy to the plant. On the contrary, others such as maize produce lactic acid by fermentation. They are thus rapidly inhibited or killed by anoxia resulting from temporary flooding (Fig. 4.10).

The aerenchyma, a tissue useful for the plant ... and for insects.

Aerenchyma: plant tissue with cells capable of transporting air from the atmosphere to depth, to zones temporarily or permanently deprived of it (Chap. 15, § 15.4). The aerenchyma of bog plants supplies oxygen to buried larvae of long-horn leaf beetle (Coleoptera, Chrysomelida), which pump it by means of a siphon crossing the walls of plant cells.

Fig. 4.10 *Difference in behaviour of maize (left) and barley (right) after flooding of the soil (photo: R. Brändle). Explanations in the text.*

Plants: great gourmets who do not swallow just anything! In nature, myrtle prefers ammonium to nitrate, while the reverse is true of nettle!

Regulation of entry of bioelements

Physiologically speaking, each species has a specific requirement of ions related to its actual metabolism, and specific resistance to toxins. For example, it is known that cereals are 'potassium-loving' plants, that tomato demands manganese and that the volume of sugar-beet can be increased by supplying sodium chloride. Each species thus regulates absorption of ions needed by it, selecting them by different physicochemical processes. This selectivity is proved by measuring the concentrations of elements in a nutrient solution before and after nutrition (Fig. 4.11).

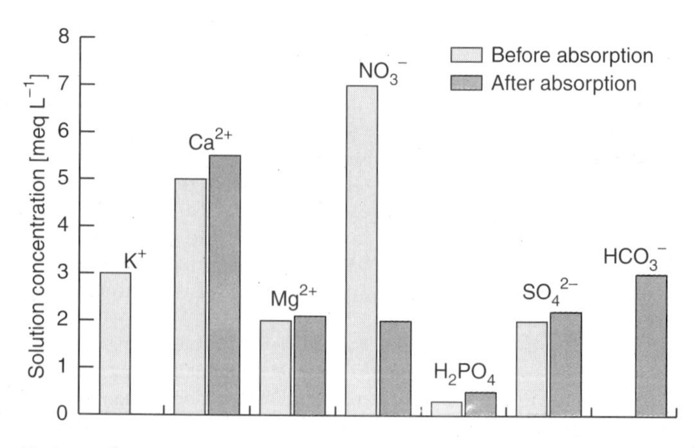

Fig. 4.11 *Nutrient selection of certain bioelements by the alga* Nitella *(after Hoagland, 1948, in Amberger, 1983).*

Internal storage

On soils rich in a specific mobile ion (e.g. sodium in coastal soils), the plant and its associated organisms often

Halophilic: literally 'salt-loving'; qualifies organisms that live in soils very rich

cannot prevent its entry in large quantities. So the plant conducts it to particular tissues or cell vacuoles, and concentrates and neutralizes it there, thus precluding interference in metabolism. Malate and citrate, which chelate heavy metals, ensure their non-toxic translocation across the cytosol to the vacuole where they are stored (Mathys, 1977; Martinoia et al., 1985). Wood (1995) cites the case of tea, *Camellia sinensis*, with old leaves very rich in aluminium. Other examples are myrtle, which accumulates manganese up to 3500 ppm (mean = 20–200 ppm), or milk vetch, *Astragalus ramosus*, which tolerates 5600 ppm selenium against 3 ppm for *Astragalus missouriensis*. The detoxifying role of the vacuole with regard to herbicides has also been suggested (Gaillard et al., 1994).

The peculiar regime of halophilic plants
Unlike other plants, the ***halophiles*** of saline lands live in a soil solution with ionic concentration often higher than that in their cytosol (Lüttge et al., 1996). Two challenges should be taken up: preclusion of toxicity of ions they are unable to prevent from entering, and ensuring water supply against an adverse electrochemical gradient, the osmotic potential ψ_s being very high (Chap. 3, § 3.4.5). They take up the former by accumulating salts in inactive tissues and the latter by reducing evapotranspiration by anatomical (e.g. reduction of leaves to small scales) or physiological (e.g. regulation of water flux through the ***stomata***, Fig. 4.12) adaptation. Some halophilic plants accumulate so much salt that they can be used to add zest to hors-d'oeuvre.

in salts, especially NaCl. Examples: *Salicornia, Salsola, Suaeda, Arthrocnemum.*

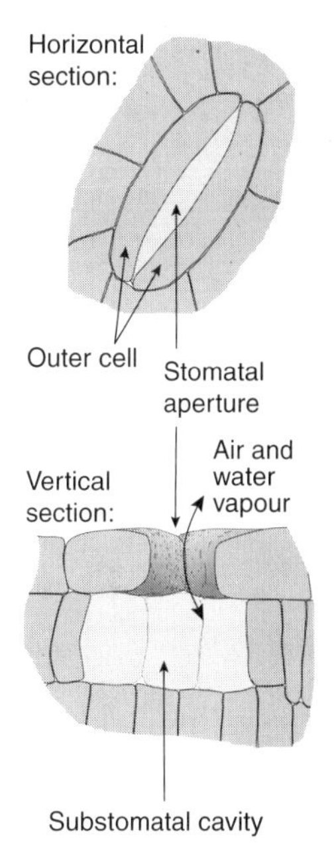

Fig. 4.12 *Anatomy of the stoma. Regulation of air and water flow between the substomatal cavity and the exterior is ensured by variations in volume of the elongated outer cells, the guard cells.*

Many acid-tolerant species (myrtle, rhododendron, lycopodium) are also aluminium-tolerant, an adaptation that enables them to live on soils with very low pH, in which this metal is solubilized (Fig. 4.8). Lastly, the resistance of certain species to very high metal concentrations makes them bio-indicators of metal-bearing strata, for example *Buchnera cupricola* for copper in the Congo and *Viola calaminaria* for calamine (zinc carbonate and silicate) in Europe.

From resistance to bio-indication.

Stoma: anatomical structure in the epidermis of leaves and stems, consisting of a controllable microscopic pore surrounded by cells.

Accelerated evacuation

In a second mechanism of internal defence against excessive additions, the plant speeds up the passage of the ion and evacuates it through salt glands. Thus tamarisk, a bush of seacoasts, eliminates large quantities of salt through its very thin leaves with large surface area. This has the effect of desalinizing soil near the

If internal storage capacity is insufficient, salt may be evacuated.

rhizosphere (Etherington, 1982), thus promoting entry of water into the plant (Chap. 3, § 3.4.3).

The plant can also adapt itself in a very short time to changes in nutritional conditions.

Storage and evacuation, two lasting physiological defences, are sometimes complemented by other more immediate mechanisms. For example, a plant deficient in iron can 'command' a change in root activity at the rhizosphere level such that this metal is better solubilized and made assimilable in ***chelate*** form (Chap. 15, § 15.3.5). It can also expel protons, thereby enabling better assimilation of zinc, phosphorus and manganese (Rorison, 1987).

Chelate: metal-organic complex in which an organic molecule traps a metal ion by means of chemical bonds laid out in the shape of 'pincers'.

4.2.4 'Love' for bioelements

Potential and actual ecologic ranges

'Love' or tolerance?

We have seen that plants regulate the entry , storage or discharge of bioelements according to their requirements or toxicity hazards. These physiological properties influence the distribution of plants in ecosystems, for example specific plants growing only on saline, acid or calcareous soils. They have been very logically termed halophilic, acidophilic or calciphilic, in order to indicate their affinity, or their 'love' (*philein* = to love!) for salt, protons or calcium. Originally these terms reflect the reality of the land, where it is effectively seen that saltwort *Salicornia perennis* grows only on saline soils, spikenard *Nardus stricta* only on acid soils and blue sesleria *Sesleria coerulea* only on calcic soils with high pH.

Autoecology (or ***autecology, ecophysiology)***: study of the action of ecological factors on individuals or species considered singly. ***Ecological factor:*** physical, chemical or biological agent of the environment able to have a direct physiological effect on a living organism. Not to be confused with ***ecological descriptor:*** element or condition characterizing the environment necessary not having direct physiological effect on living organisms (e.g. slope, altitude).

A more precise study of their physiology reveals that in fact these species can live perfectly well in less extreme environments, when they are grown alone, not mixed with others as in natural communities (Gigon, 1971; Lebreton, 1978). Their ***autecology*** allows a wider growth range than that realized in nature, where competing ***synecological*** processes intervene. Figure 4.13 shows the difference between ***physiological range***, or potential range, of a species referred to an ecological factor and its realized ***ecological range***, influenced by interspecific competition.

... but some plants have phobias.

Often the 'love' of species for salt, acidity or calcium is just a *tolerance* with respect to these factors, which they endure better than other plants. For example, the pine of peat-bogs, *Pinus mugo* ssp. *uncinata*, characteristic of acid, very wet and oligotrophic environments—a good example of stress-tolerant species S (Chap. 12, § 12.6.2)—would ask for no more than to grow in a good Brown soil with

'For certain ultra-specializeds plants, acidophily is only an indication of a veritable phobia with respect to any alkali and alkaline-earth ion.' (Lebreton, 1978).

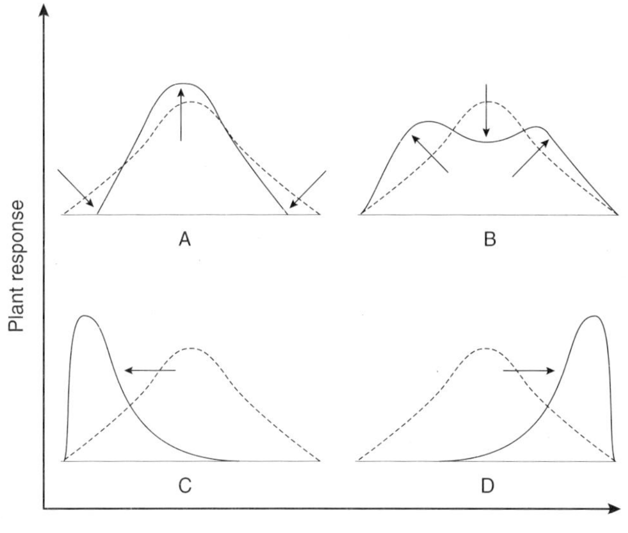

Fig. 4.13 *Physiological (in pure culture, without interspecific competition; broken line) and ecological (natural, with interspecific competition; solid line) ranges compared: A. Increase of the optimum and reduction of range; B. Lowering of original optimum and appearance of two shifted optima; C. Possible survival exclusively near the minimum; D. possible survival exclusively near the maximum.*

medium pH, but that space has already been taken by beech and fir! All the same, to identify better this proclivity of organisms for particular conditions, between 'love' and tolerance, the suffixes *-cline* and *-tolerant* may be used: acidocline, halotolerant, etc.

Synecology: study of communities and biocoenoses within the ecosystem. ***Community:*** collection of living organisms assembled at a given moment in an area under precise conditions. The community, in this sense, is a fraction of the biocoenosis. ***Biocoenosis:*** integrated assemblage of all the greatly inter-dependent synusia of producers, consumers and decomposers sharing the same biotope during a given period (Chap. 7, § 7.1.4; Gillet et al., 1991; Gillet and Gallandat, 1996); structured community of organisms inhabiting a particular biotope. Compared to the biocoenosis and the community, the ***population*** is a local assemblage of individuals *of the same species* at a given instant. The study of populations is ***demo-ecology***, to which are attributed the dynamics and genetics of populations.

Do not confuse nitrophilic and nitratophilic!

Nitratophilic plants are often termed ***nitrophilic***, a misused simplification! Actually two forms of mineral nitrogen nourish plants: nitrate and ammonium. But nitratophilic plants, which occur for example on the borders of farms, are not the same as those preferentially nourished by ammonium, often located on acid soils. And ammonium is not the total nitrogen in the soil, which is 95% organic, for example in the form of proteins!

Nitratophilic: nitrate-loving.
Nitrophilic: nitrogen-loving.

Elimination of the suffix *-philic* from these ecological attributes and replacement by the suffixes *-tolerant* or *-cline* must therefore be encouraged, except when there is clear physiological proof of an actual affinity for such or such ecological factor (§ 4.4.6). For example, nettle *Urtica dioica* and the Bon-Henri spinach *Chenopodium bonus-henricus* that actually require much nitric nitrogen are

Let us replace wherever possible the suffix -philic by *-tolerant* or *-cline*!

Halobic: organism closely and obligatorily dependent on high concentrations of salt. The most extreme halobic species are crustaceans like *Parartemia zietzana* and *Artemia salina*, which only live in water fully saturated with salt (35.5%)!

Calcifuges are often not just . . . calcifuges!

Jules Thurman of Porrentruy, naturalist of the Swiss Jura, was a pioneer in studies on the affinities of plants to their substrate. In his book *Essai de phytostatique appliquée à la chaîne du Jura* (Thurmann, 1849), he coined the terms xerophilic and hydrophilic (Acot, 1988; Deléage, 1991).

The effect of an ecological factor on a living organism depends on the state of other factors.

true nitratophiles. So too are *Archaea* of the genus *Halobacterium* (Chap. 2, § 2.5.1), typical halophiles of salt marshes, which require at least 12% NaCl in their environment (or four times the salinity of sea-water). Among animals staphylinid beetles of the genus *Bledius* are also true halophiles depending on saline clays containing up to 20% salt; certain ***halobic*** weevils feed exclusively on saline plants such as *Statice* or *Cakile*.

On the other hand, the suffix *-fuge* actually expresses the 'flight' of an organism faced with an element noxious to it. A ***calcifuge*** species such as myrtle does not endure high concentrations of calcium that rapidly become toxic for it, while the earthworm is an acidofuge, incompatible with soils of pH lower than 4.4. Calcifuge species are generally also very frugal with phosphorus but tolerant of aluminium, which they precipitate in their cell walls, thus avoiding any damage to the cytoplasm (Clarkson in Rorison, 1969).

By analogy with bioelements, adjectives such as ***hydrophilic*** (water-loving), ***mesophilic*** (loving moderate conditions), ***orophilic*** (mountain-loving), ***psychrophilic*** (cold-loving), ***thermophilic*** (heat-loving), ***xerophilic*** (dryness-loving) and many others have been coined. The same discussion applies to them too. With respect to temperature, the sequence of decreasing ecological affinity is expressed as: thermo*bic*, thermo*philic*, thermo*clinic*, thermo*tolerant* and thermo*fuge*. But some organisms are thermo-indifferent!

Multifactor influences

While seeking the living range of an organism, it should be borne in mind that no ecological factor acts alone. Calcifugy is thus more or less pronounced according to the average hydric regime of the soil on which the plants grow. A climate with heavy rainfall widens the range of calcifuges towards higher pH; they can then access the needed iron, which is usually blocked in the trivalent form if the pH exceeds 5. The added effect of excess water and bacteria render the centre of microaggregates anoxic, thereby releasing iron in the Fe^{2+} form, available to plants (Chap. 3, § 3.10.1; Etherington, 1982).

Calcifugy is also reduced by the antagonism between two cations. Thus, the 'acidophilic' chestnut tree grows on carbonate-rich dolomitic scree in southern France near Saint-Guilhem-le-Désert (Fig. 4.14). The abundant magnesium from dissolution of dolomite $CaMg(CO_3)_2$

probably reduces the toxic action of calcium and widens the tolerance range of this calcifuge species.

The form in which bioelements exist in the soil is thus primordial: are they available to plants or on the contrary not attainable by rootlets? Section 4.3 gives some elements of response, while § 4.4 explains how microorganisms are supplied with these bioelements and transform them.

Fig. 4.14 Chestnut trees of Saint-Guilhem-le-Désert on carbonate-rich dolomitic scree (photo: J.-M. Gobat).

4.3 AT THE JUNCTION OF SOIL, PLANTS AND MICROORGANISMS: BIOELEMENTS

4.3.1 The concept of 'available ionic reserve', a difficult problem

From the clay-humus complex to the plant: not so simple!

Along with photoautotrophic 'nutrition' of the chlorophyll-bearing plant by photosynthesis and maintenance of cell ***turgor***, absorption of nutrients is the *sine qua non* of plant life. But where is the nutrient reserve of the soil? What are the relations between elements, what are their effects on the plant? How does the plant utilize them? These questions underscore the difficulty of the problem!

On the menu for plants: cations, anions, chelates and small organic molecules.

The mineral bioelements essential for the plant are cations (Ca^{2+}, Mg^{2+}, K^+, Cu^{2+}, Mn^{2+}, NH_4^+, etc.) and anions (NO_3^-, HPO_4^{2-}, $H_2PO_4^-$, SO_4^{2-}, etc.). Plants also absorb small organic molecules like monosaccharides, and chelates such as siderophores (Chap. 15, § 15.3.5).

In soil, the mineral elements are found in very varied forms.

Mineral elements are originally found in the crystal lattice of intact rocks (e.g. calcium in limestones) or weathered rocks (e.g. silicon of clays); after their release they are held on the clay-humus complex or, by exchange, are released into the soil solution. Can we then, under these categories, define a plant-available ionic reserve, similar to that for water (Chap. 3, § 3.4.2)? The answer is more difficult in this case for the following reasons (see also Or and Wraith in Sumner, 2000; Duchaufour, 2001)):

- seasonal changes in microstructure alter the accessibility of certain sites to rootlets and to mycelia of extraradicular mycorrhizae;
- water content, pH and redox potential influence the solubility of ions;
- ionic concentrations can change very rapidly according to the law of ionic equilibrium;
- not all plants have the same absorbing power or identical requirements for nutrient elements (§ 4.2.3);

Turgor: condition of a cell caused by a lower osmotic concentration in the solution outside its plasmic membrane than in its cytosol. This causes entry of water in the cell, which exerts pressure on the cell wall.

No method is able to correctly define the reserve of available nutrients!

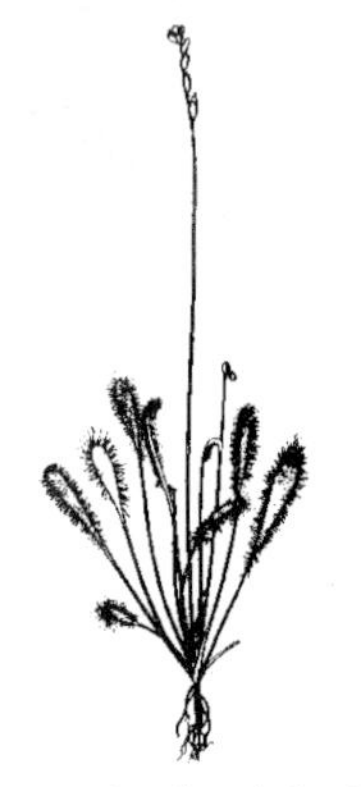

Fig. 4-A *Sundew (after Hess et al., 1967ff). With permission of Birkhäuser Verlag, Basle.*

Fig. 4-B *Butterwort (after Hess et al., 1967ff). With permission of Birkhäuser Verlag, Basle.*

- antagonism between cations modifies their assimilation;
- symbiosis between plants and microorganisms influences the nutritional conditions of the plant, with manifold locations and processes (Chap. 16).

Every soil at every moment is a specific case, with changing conditions for availability of ions. For comparing soils, which is essential particularly from the agronomic viewpoint, different analyses have nevertheless been proposed, every one of which lends its partial solution to the general problem, but none is really satisfactory.

The peculiar nutrition of carnivorous plants

Although most plants follow the general nutritive principles, carnivorous plants partially form an exception. Taking few nutrients from a substrate that is generally poor, particularly in combined nitrogen, they at times supplement them by absorption of organic substances of animal origin, extracted by extracellular digestion of prey they have captured (Lüttge et al., 1996). The best known carnivorous species are *Utricularia* in fresh water, catchfly *(Dionaea)*, sundew *(Drosera)* (Fig. 4-A) and *Sarracenia* of peat-bogs, and butterwort *(Pinguicula)* (Fig. 4-B) of fens. Darwin (1877) had long ago proved the production of proteolytic enzymes and formic acid by digestive glands in *Dionaea*!

4.3.2 Some methods of analysis of bioelements

Determination of total elements in the soil, for example by acid hydrolysis, is of no use here because it mixes up all the categories, extracting for example cations of the crystal lattices, inaccessible to plants. This is why two other milder analyses are usually done:

Two mild extractants: distilled water and electrolyte solution.

- Soil ions can be extracted by distilled water to give a good indication of the status of the soil solution. This method is not immune to criticism since it can extract elements *ad infinitum*, each passage of water extracting ions of the exchange complex in accordance with the law of equilibrium! The duration and volume of extractant must then be properly standardized. Besides, the solubilizing power of distilled water is not that of the soil solution, often with different pH and Eh; thus natural conditions are not respected. This method, commonly used in agronomy, gives the actual concentration of ions, similar to actual acidity (Chap. 3, § 3.9.1).

Extraction by distilled water gives information on the ionic reserve immediately available to the plant.

Jean-Baptiste Boussingault published his *Économie agricole* in 1843. He showed in it that total chemical analysis of soil is of little practical use so far as plant nutrition is concerned (Boulaine, 1989).

- Ions are also extracted by an electrolyte solution. The most commonly used salts are KCl and NH_4Cl. This method sums the ions in solution and those held on the exchange complex, all without attacking the ions constituting rocks. Unfortunately it does not furnish the quantities actually available at the moment of analysis, because there is an equilibrium but no equality between the ions held on the complex and those of the soil solution (Chap. 3, § 3.7.2).

Extraction with an electrolyte solution estimates the reserve available in the short or medium term.

But these two methods are too crude for revealing what takes place in the rhizosphere, in the area of absorption. Other methods approach nutrition from the side of the plant and determine, for example, the absorption of certain ions from nutrient solutions of known concentration (§ 4.2.3). Here, the quantities actually extracted are determined by difference between the concentrations before and after absorption. But these experiments can be conducted only under controlled conditions, without soil, far from the reality on the land.

Mild techniques are still too aggressive for the very fine gradients in the rhizosphere (§ 4.1.3; Chap. 15, § 15.5.1).

To approach this reality, the cations and anions contained in the plant biomass can be analysed, this being termed the ***mineralomass***. It is assumed that the concentrations within the plant have integrated all the ecophysiological constraints in transport between soil and plant. This is true in most cases but not all, because plants select the ions they need, excreting others or storing them in non-vital organs (§ 4.2.3; Marschner, 1995). Here too we are far from defining the available reserve in the soil!

The mineral content of plants can also be determined to give information directly on their status (foliar diagnosis; Morel, 1996).

In short, there is no method for measuring in a simple and reliable way the ionic reserve actually available to plants! The only way to proceed is to apply many methods, array the results and compare them with those of other analyses (texture, pH, exchange capacity, plant production, etc.).

4.3.3 Relationships between nutrient elements: dependence and antagonism

Experiments on selective absorption often call for simple situations in which one ion is tested at a time. But the reality is totally different . . .

Bioelements coexist in soil and are exchanged for one another (Chap. 3, § 3.7.2). The capacity of a plant to absorb an ion is thus also subject to presence and concentration of other elements in the vicinity of the absorption site, in addition to internal factors, as was seen for the chestnut (§ 4.2.4). Thus there are ionic interactions of dependence or antagonism with respect to possibilities of absorption.

Dependence and antagonism, two possible interactions between ions.

In ***dependence*** two ions favour each other, absorption of one facilitating that of the other. The best known in agronomy is that of nitrogen and potassium: a high concentration of nitrogen in rootlets speeds up entry and translocation of potassium to young or growing organs such as leaves, flowers or fruits. Phosphates improve magnesium nutrition.

On the other hand, in ***antagonism*** an ion can expel another from the exchange complex and thereby reduce its medium-term reserve; this is termed exchange antagonism. But antagonism can also reflect a competition for the same absorption site on a protein (facilitated-diffusion protein or cotransporter, § 4.2.1), and thus slow down or prevent entry of one of the two ions into the plant. These phenomena are important in practical agriculture: a wrong combination of fertilizers can lead to a deficiency, increased susceptibility to parasites or to depletion of the soil. This is true particularly for iron in $CaCO_3$-rich soils (Lüttge et al., 1996).

Some hostile brothers: potassium and calcium; potassium and rubidium; magnesium and ammonium; iron and copper.

A well-known antagonism is that which sets calcium against potassium, readily exchanged in the soil. This pedological opposition is biological as well, with very different, often competing roles in the plant. Potassium, the most important cation in plant production, is concentrated in the young and growing organs; very mobile, it controls the osmotic potential and enables entry of water into the plant. On the contrary calcium, which is scarcely mobile, serves as a building element of tissues, to which it imparts rigidity. It also promotes export of water and reduces the permeability of membranes.

The mineralomass often ignores resistance to toxicity, luxury uptake or the effect of deficiencies.

4.3.4 State and role of elements in the plant

From deficiency to toxicity

Although used to understand plant nutrition, the mineral composition does not truly reflect the physiological requirements of plants because some species accumulate toxic elements by isolating them (§ 4.2.3). Others develop ***luxury uptake***. Against this, because of antagonisms or real weakness of the exchange complex, some deficiencies can develop: for example, chlorosis (deficiency of chlorophyll) can indicate a deficiency of iron, prevented from entering by high content of HCO_3^- in calcareous soils (Marschner, 1995). Trace elements, often with important qualitative function in metabolism, can also be deficit.

'Only the dose makes the poison.' This remark of Paracelsus, famous physician of the sixteenth century, shows that deficiency and toxicity are two faces of the same element.

Actually, deficiency, optimum and luxury uptake, and toxicity are four physiological expressions of increasing contents of an element (Fig. 4.15).

Luxury uptake: absorption of ions in excess by a plant without their serving toward growth, yet not being toxic. Luxury uptake

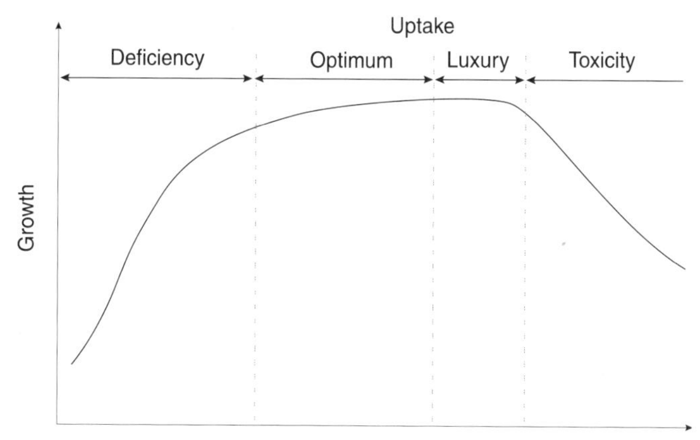

*Fig. 4.15 From deficiency to toxicity. At low content in soil solution, rhizosphere or mineralomass as the case may be, an element may be deficit for the plant, which then is **deficient**. Rise in concentration (or supply) enables it to be properly nourished (optimum), eventually leading to luxury uptake. Later the element becomes toxic.*

is infrequent because the cost:benefit ratio is very high: the plant expends energy for absorption and translocation of an ion that is of no use to it, at least in current conditions. On the other hand, luxury uptake can certainly be a profitable long-term investment because it enables formation of reserves easily available in the event of sudden necessity.

Macroelement (or ***macronutrient***)**:** major bioelement composing living matter: C, H, O, N, P, S and the ions K^+, Mg^{2+}, Ca^{2+}, sometimes Na^+, Cl^-, SiO_4^{4-}.

Trace element (or ***microelement, micronutrient***)**:** indispensable bioelement, albeit required in very small quantities in the cells of living organisms. Generally it enters the structure and function of enzymatic cofactors (e.g. Fe, Mn, Cu, Mo, Co, W, Zn, Ni).

Role of nutrient elements in plants

Nutrient elements are divided into ***macroelements*** (or macronutrients) and ***trace elements*** (or micronutrients) (Marschner, 1995; Lavelle and Spain, 2001). Tables 4.1 and 4.2 give, without comment, information synthesized from numerous sources on their states, roles and contents in soil and plants.

4.3.5 Negative effect of elements: toxicity

How is toxicity to be defined?

Current harmlessness or potential toxicity?

While some consider the soil—at least under certain conditions—as a filter that retains dangerous substances and prevents them from contaminating the food-chains (Fournier and Cheverry, in Auger et al., 1992; Duchaufour, 1997), others estimate that this filter is soon saturated (Häberli et al., 1991): the elements will then be solubilized and their toxicity will not yet be neutralized. This choice well shows the difficulty in correctly defining toxicity of an element: is it actual or only potential?

We have seen that ***toxicity*** is the expression of too high a concentration of an element compared to the requirements and resistance of a species, another under identical conditions not being harmed. It integrates the concept of the element, the environment and the species. This is in conformity with observations made on the behaviour of plants in which, for example, increase in

Toxicity: feature shown by a nutrient element when its concentration exceeds a proportion suitable for growth of an organism (adapted from Carles, 1967). Toxicity of an element depends on its interaction with others, for example in a culture medium or in soil, and the species considered.

Table 4.1 Major bioelements in soil and plant (various sources)

Element	Form in the soil	Concentration in soil (S) and in plant (P) (‰ of dry matter)	Principal roles in the plant, deficiency, toxicity
N	• organic: more than 95% of total • NH_4^+ : transitory form, retained on complex • NO_3^- : principal source of nitrogen for plants; easily lost by leaching	S: 0.3 to 3 P: 5 to 50	• constituent of amino acids, proteins, nucleic acids and lipids • promotes multiplication of cells and chloroplasts • promotes synthesis of carbohydrates • forms reserves in seeds • constituent of hormones
P	• organic: in debris of litter • mineral: unavailable constituent of some minerals (e.g. apatite) • adsorbed PO_4^{3-}; scarcely available • free HPO_4^{2-} or $H_2PO_4^-$	S: 0.1 to 1 P: 1 to 5	• major constituent of phospho-proteins (e.g. lecithins) • constituent of DNA, RNA and phospholipids • role in metabolism of carbohydrates and in fruit-setting • transport of energy in the cell (ADP, ATP) • moves at end of growing season to storage organs
S	• mineral of gypseous rocks, pyrite • oxidized to sulphate by bacteria, starting from sulphides or elemental sulphur • H_2S originating from decomposed organic matter or sulphate-reducing bacteria	S: 0.1 to 1 P: 0.5 to 5	• constituent of sulphoamino acids (methionine, cysteine) • constituent of certain enzyme cofactors • toxic in excess, with resistant species by accumulation up to 7% of dry matter or by restriction of absorption
K	• constituent of silicates (micas, feldspars): 95–98% of total soil K • fixed within clay minerals • retained (loosely) on exchange complex • free in soil solution	S: 2 to 30 P: 5 to 50	• found in soluble form, very mobile • chief regulator of osmotic pressure, therefore of water movement and passive phase of absorption • activator of enzymes • promotes synthesis and storage of carbohydrates • very easily leached out of leaves by rain
Ca	• constituent of carbonate rocks • 'active' as fine powder of $CaCO_3$ • exchangeable on exchange complex • free in soil solution	S: 2 to 15 P: 0.5 to 50	• constituent of cell walls making them rigid and resistant • activator of enzymes • promotes fruit maturation • neutralizes organic acids formed by metabolism • accumulates in old organs (bark, wood)
Mg	• constituent of dolomites • exchangeable on exchange complex • free in soil solution	S: 1 to 10 P: 1 to 10	• constituent of chlorophyll • activator of enzymes • prevents chlorosis (absence of chlorophyll formation) • selects species if its concentration in soil is high

Table 4.2 Trace elements in soil and plant (various sources)

Element	Form in the soil	Concentration in soil (S) and in plant (P) (ppm of dry matter)	Principal roles in the plant, deficiency, toxicity
Fe	• more than 20 mineral compounds: haematite, goethite, lepidocrocite, hydroxides, etc. • constituent of clay-humus complex • chelated in organic matter • ionic form in solution (Fe^{2+} in anoxic soils)	S: up to 40,000 P: 50 to 1000	• prevents chlorosis • conductor of oxidation-reduction processes • constituent and activator of enzymes • chelated by siderophores, cellular carriers • regulator of nitrite reduction and nitrogen fixation
Mn	• similar to those of iron	S: 200 to 4000 P: 20 to 200	• promotes growth and prevents chlorosis • possible deficiency in alkaline soils and toxicity in acid soils • constituent and activator of enzymes • role in the oxidation of water during photosynthesis
Cu	• constituent of minerals (e.g. chalcopyrite $CuFeS_2$) • chelated in organic matter	S: 5 to 100 P: 2 to 200 (cuprophytes up to 1600)	• regulator of oxidation-reduction processes • constituent of enzymes ensuring lignin synthesis • stimulates growth • toxic at high concentrations, except in adapted plants
Zn	• constituent of ferromagnesian silicates • very little Zn^{2+} in solution	S: 10 to 300 P: 10 to 100	• constituent of oxidation enzymes (oxidases) • synthesis and protection of growth hormones • help in synthesis of chlorophyll
Mo	• constituent of minerals • MoO_4^{2-} or $HMoO_4^-$ held on exchange complex or free in soil solution	S: 0.5 to 5 P: 0.2 to 10	• required for nitrogen metabolism (constituent of the enzyme nitrate-reductase and also of the nitrogenase complex)
B	• constituent of silicates (e.g. 3–4% in tourmaline)	S: 5 to 100 P: 2 to 100	• constituent of enzymes • help in synthesis of chlorophyll • frequent deficiencies, leading for example to heart-rot of sugar-beet or bark lesions on apple trees
Al	• fundamental constituent of minerals, along with Si (8% of the lithosphere) • numerous forms according to soil acidity; among them $Al(OH)_4^-$, $Al(OH)_2^+$ and Al^{3+} (free ion)	S: 50 to 200 P: 2 to 3	• in very low concentrations, promotes yield of cultivated species • very rapidly toxic when soil pH is lower than 5.5 • action of species selection on acid soil: only aluminium-tolerant species (e.g. Ericaceae) are resistant

Limiting factor: among an assemblage of ecological factors, the one closest to the critical physiological minimum or maximum for the organism.

aluminium concentration is the only ***limiting factor***, in other cases it is rather the general level of concentration of salts of all sorts that causes toxicity.

Principal toxic mineral elements in soil

Thirty-eight weighty poisoners ... at times indispensable for plants!

It is estimated that there are thirty-eight heavy metals (density >5 g cm^{-3}) potentially toxic to plants (Etherington, 1982). Many are very widespread in soils and, in small quantities, necessary for metabolism. This is true of iron, manganese, copper, zinc or molybdenum, which form many compounds with inorganic or organic molecules, especially enzymes. Others (e.g. cadmium, mercury, lead) are major pollutants in soils even if all of them have not yet revealed their toxicity in the food chains leading to humans (see Chap. 11).

Resistance to toxicity

Thanks to fungi!

To resist heavy metals, plants have developed two physiological strategies (Baker, 1987):

- internal adaptations that enable evacuation of toxic elements or their storage in inactive form (§ 4.2.3);
- external mechanisms slowing down or preventing entry of toxic cations; such protective agents can be mycorrhizal fungi, which filter the ion flows before entry into the plant.

But there are also genetic resistances, by selection of metallo-resistant mutants, as seen on freshly contaminated soils (e.g. waste dumps at mining sites). Some species present before the deposition are adapted to them by favourable ***mutations***. Thus, *Agrostis tenuis*, a grass, has established populations resistant to zinc in less than thirty years, a short span of time for such an evolution (Etherington, 1982). In these instances of rapid genetic adaptation, the physiological processes are rather of the first type mentioned above, namely, the plant allows entry of the toxic metal and develops mechanisms to inactivate it.

Mutation: modification of genes present on DNA, entering in the lineage of the organism and introducing changes in the characteristics of the product of the gene (enzyme or other).

Plants and microorganisms are more and more being used to detoxify soils contaminated by heavy metals and organic pollutants. This topic is discussed in Chapter 11.

4.4 MICROORGANISMS: THE SOIL 'PROLETARIAT'

'Proletarian: person carrying out a manual profession and having for livelihood only the wages, generally low, allowed him by those to whom he sells his labour.' (*Petit Larousse*, 1995).

Microbes are not well thought of: they are usually considered purveyors of disease, which actually is true only for a small minority among them. In fact, most of the species accomplish in nature and in our environment tasks as important as they are unrecognized. Furthermore, they maintain favourable or unfavourable relations with other soil organisms. Their roles pertain to:

- transformation of plant and animal wastes;
- oxidation, reduction, precipitation and solubilization of mineral ions;
- fixation of molecular dinitrogen;
- control of bioelement cycles, in particular those of carbon, oxygen, nitrogen, sulphur and iron;
- transformation of the parent material.

4.4.1 Transformation of plant and animal wastes

The major part of dead matter, plant and animal, is composed of macromolecules (Chap. 2, § 2.2.3). Among them can be mentioned:

- Polysaccharides of plant (e.g. cellulose, pectin, starch), animal and fungal (e.g. chitin) origin;
- phenolic polymers of plant and fungal origin (e.g. lignin);
- proteins;
- nucleic acids;
- lipids.

Biochemical decomposition of dead organic matter is essentially the task of microorganisms—bacteria and fungi (Chap. 5, § 5.2.4). Yet, neither is capable of ingesting particles or macromolecules. The 'digestion', more precisely hydrolysis, of these macromolecules into their monomeric or dimeric constituents is done by extracellular enzymes (Chap. 14, § 14.3.1).

Extracellular enzymes: a prerequisite for the digestion of biopolymers (Chap. 14, § 14.2.1).

Sometimes this digestion takes place directly at the contact of the bacteria and the substrate, as with *Cytophaga*. Their cellulases are located on the surface of the cell wall; during the attack on cellulose, no by-product of hydrolysis appears in the medium. But in most cases, the enzymes are released into the surrounding medium and their activity serves not just the organisms that have synthesized them. Commensals capable of assimilating the products also benefit from them. The monomers and dimers resulting from enzymatic hydrolysis are absorbed by bacteria and fungi, which use for this purpose the very efficient systems of active transport they are generally provided with (§ 4.2.1; Chap. 16, § 16.1.1).

Cytophaga, the most active bacteria in cellulolysis (Chap. 14, § 14.3.1), profit selfishly from their digestive enzymes. But this is an exception!

Mineralization: return to the sources.

Aerobe (noun), aerobic (adj.): an organism that uses oxygen as the acceptor of respiratory electrons. Life under these conditions is ***aerobiosis.*** There are strict aerobes and also facultative aerobes that can equally live in anoxic conditions through another metabolic path (fermentation, anaerobic respiration).

Metabolism: ensemble of biochemical processes of transformation of matter and energy in the cell or organism, which results in the formation of its constituents (anabolism) and in release of the energy necessary for its functioning (catabolism).

Anabolism: ensemble of metabolic reactions leading to synthesis of cell material from nutrients drawn from the external environment.

Catabolism: ensemble of metabolic reactions that result in production of energy utilizable by the cell (ATP, trans-membrane gradients) starting from an organic substrate. The reactions of catabolism, as the case may be, culminate in complete mineralization of the substrate or excretion of organic compounds (e.g. fermentation products, methane).

Metabolite: product of metabolism.

Under ***aerobic*** conditions nearly one-half the absorbed organic carbon is assimilated (transformed to cellular material) by reactions of ***anabolism***, the remainder most often being completely oxidized (mineralized) by reactions of ***catabolism*** (Fig. 4.16). Under conditions of permanent anaerobiosis (Eh < –200 mV; Chap. 3, § 3.10.2) about 90% of the carbon is transformed to biogas (CH_4 + CO_2) while only 10% is assimilated. Under conditions of temporary or less marked anaerobiosis (Eh between 0 and –200 mV), catabolism of organic substances involves fermentation reactions accompanied by excretion and accumulation of ***metabolites*** (alcohols, organic acids), molecular hydrogen and CO_2.

Other bioelements are assimilated by microorganisms in proportion to their concentration in the cellular material. The surplus, if any, is often excreted into the environment, mainly as inorganic compounds. Thus,

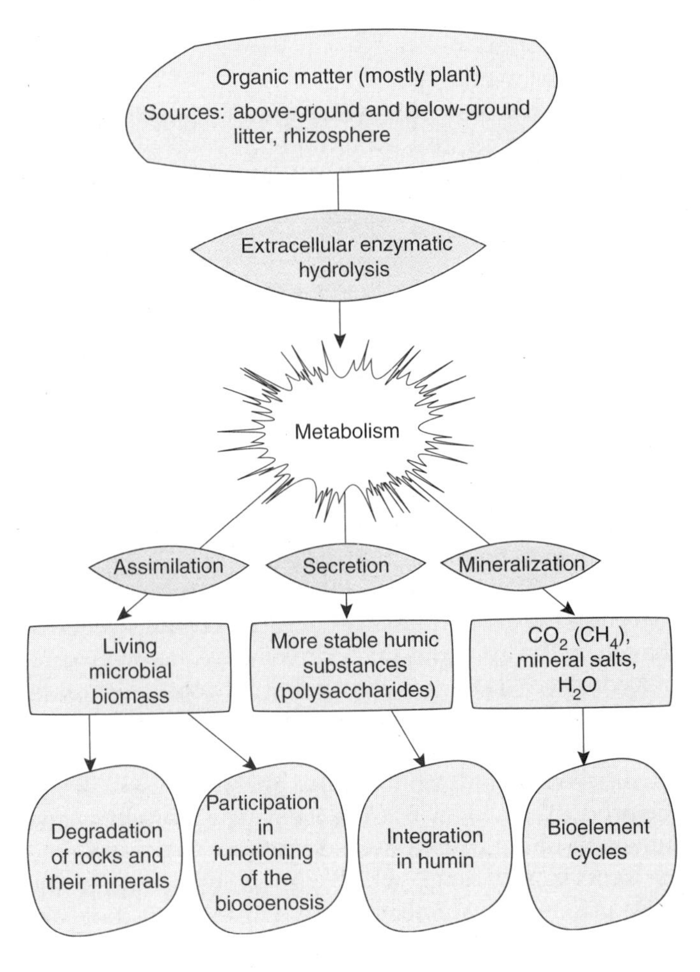

Fig. 4.16 Functional aspects of the soil microflora.

excess nitrogen is evacuated in the form of NH_4^+, sulphur as H_2S and phosphorus as orthophosphates. In aerobic conditions, iron is released as insoluble ferric oxides or chelated iron; in ***anaerobic*** conditions as Fe^{2+}, generally soluble but liable to be precipitated by some anions, in particular sulphide.

In addition to their activity of destroying organic matter, some bacteria often secrete large quantities of biopolymers, chiefly polysaccharides (Fig. 4.16). These macromolecules, mostly very resistant to enzymatic degradation, accumulate in soils where they form an important part of the humified organic fraction, microbial humin (Chap. 2, § 2.2.4; Chap. 5, § 5.2.3; Chap. 14, § 14.4.2). They most often occur as mucilages and therefore participate in aggregation of soil particles, especially in the rhizosphere (Chap. 15, § 15.3.1) and in the first phases of aggregate formation (Davet, 1996; Lavelle and Spain, 2001).

Relatively stable building blocks of soils: bacterial polysaccharides.

Anaerobic (adj.), anaerobe (noun): qualifies an organism that can live in absence of oxygen. Life under these conditions is ***anaerobiosis.*** Energy metabolism of various types enable anaerobiosis: fermentation, respiration with an acceptor other than oxygen (§ 4.4.2) and photosynthesis. There are strict anaerobes and facultative anaerobes. Facultative anaerobe = facultative aerobe, this is a question of viewpoint!

4.4.2 Oxidation and reduction of inorganic substances: a quasi-exclusivity of bacteria

Although reduction of sulphates and nitrates occurs during assimilation of these anions by plants and a weak heterotrophic nitrifying activity has been observed in some fungi, the magnitude of these reactions in natural situations is far below those effected by chemolithoautotrophic bacteria and those with anaerobic respiratory metabolism. Detailed explanation of their physiology in these reactions is beyond the purview of this book; the reader is referred to books on general microbiology (Lengeler et al., 1999; Madigan et al., 2002).

Bacteria surpass-by far!-plants and fungi in redox phenomena.

Chemolithoautotrophy involves utilization of a reduced inorganic compound (nitrogen, sulphur, iron) as a source of energy as well as of respiratory electrons: this enables bacteria to use carbon dioxide as the sole source of cellular carbon (Fig. 4.17). Specifically mentioned here are:

Chemolithoautotrophy: a mode of respiration based on inorganic molecules.

- ***nitrifying*** bacteria, comprising ***nitrous*** bacteria that oxidize ammonium to nitrite and ***nitric*** bacteria that in turn oxidize nitrite to nitrate;
- ***sulphur-oxidizing*** bacteria that oxidize reduced compounds of sulphur (sulphides, hydrogen sulphide and elemental sulphur) to sulphate;
- ***iron-oxidizing*** bacteria that oxidize divalent ferrous iron to trivalent ferric (hydr)oxides;

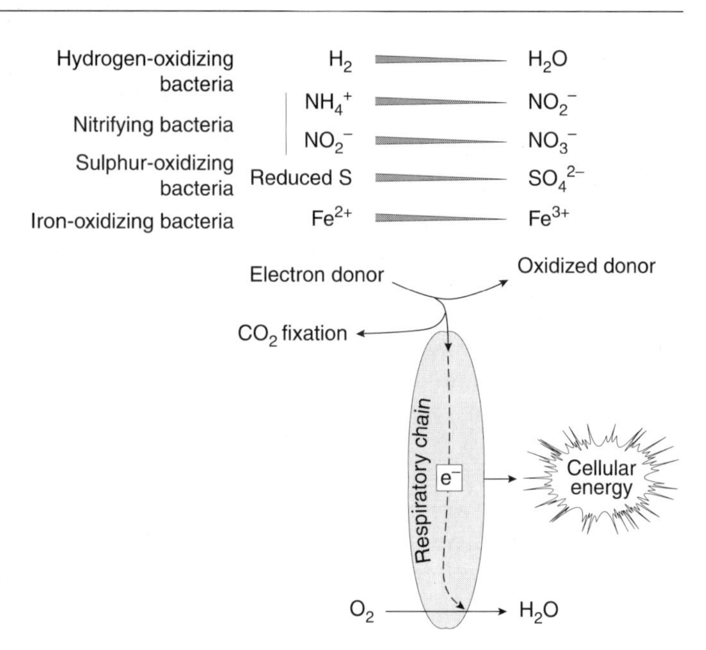

Fig. 4.17 Simplified diagram of aerobic chemolitho-autotrophy. Explanations in the text.

- ***hydrogen-oxidizing*** bacteria that oxidize molecular hydrogen to water.

An energy source.

An electron source.

Eight ways for organisms to be nourished

In a functional approach, organisms can be grouped according to the manner in which they obtain the energy, electron and carbon necessary for ensuring synthesis of their cell constituents and expenditure of energy for maintaining the cell (Fig. 4.18).

The energy source may be:

- light, electromagnetic radiation (Chap. 3, § 3.5.1) captured by an 'antenna' of photoreceptive pigments and transformed to energy utilizable by the cell; this is ***phototrophy***;
- a chemical substance whose exergonic transformation provides energy utilizable by the cell (***fermentation***, respiration); this is ***chemotrophy***.

It can be seen that the term ***photosynthesis***, widely used, is rather ambiguous because it denotes concomitantly utilization of light as energy source (phototrophy) and utilization of CO_2 as carbon source (autotrophy). Actually, it applies to two distinct phenomena independently manifested by photoorganoheterotrophic bacteria and chemolithoautotrophic bacteria.

The electron source may be:

- an inorganic substance (NH_4^+, NO_2^-, H_2S, S^0, H_2, Fe(II), H_2O); this is ***lithotrophy***;
- an organic substance; this is ***organotrophy***.

The processes of conversion of energy such as phototrophy and respiration involve electrons whose flux through a redox chain induces in the form of transmembrane protonic or electronic gradients a potential energy utilizable by the cell.

The source of cellular carbon may be:

- an inorganic substance (CO_2, HCO_3^-); this is ***autotrophy***;
- an organic substance; this is ***heterotrophy***.

Practically, four of these eight possible combinations are really important. In the photolithoautotrophs (some specialized bacteria and almost all plants), the sources of energy, electrons and carbon are distinct. In the photoorganoheterotrophs (exclusively some bacteria, e.g. *Rhodospirillum rubrum*), an organic substrate serves altogether as carbon and electron source. In chemolithoautotrophs (also an exclusivity of some bacteria) one and the same inorganic compound may serve as source of electrons and energy. Lastly, in the chemoorganoheterotrophs (many bacteria, all fungi and animals), the source of carbon, energy and electrons can be the same organic substrate.

A carbon source.

Eight theoretically possible combinations of sources of energy, electrons and carbon corresponding to eight different trophic types.

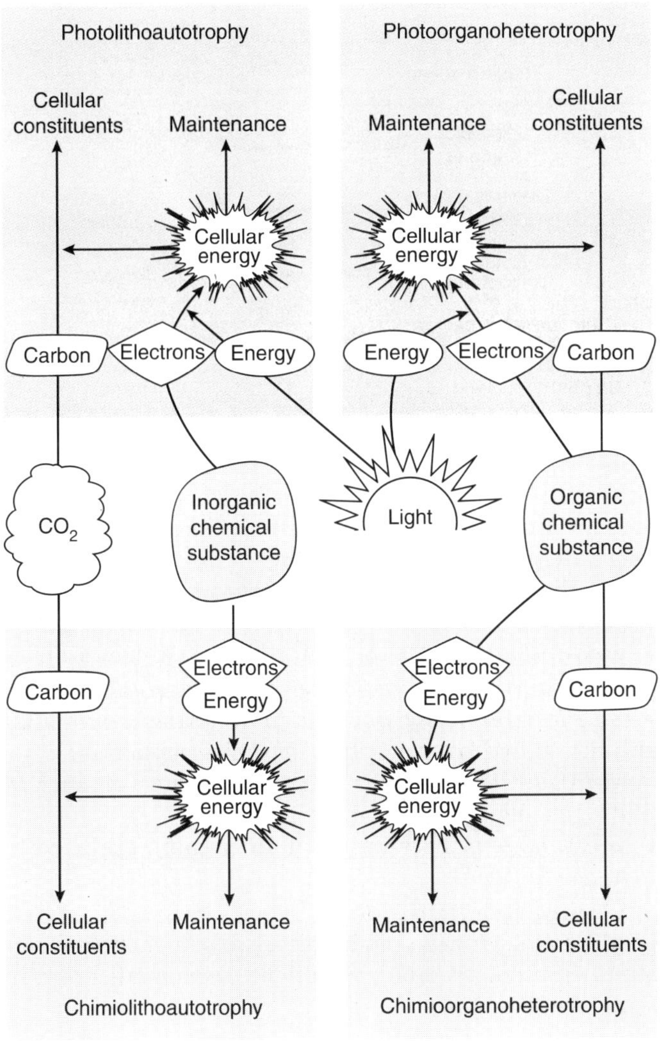

'Without the intervention of microorganisms the progress of all metabolism in the biosphere would be slowed down, and finally halted' (Lebreton, 1978).

Fig. 4.18 *The four principal trophic types. Explanations in the text.*

Anaerobic respiration, or how to do it without oxygen.

In the absence of oxygen, bacteria with anaerobic respiratory metabolism utilize oxidized inorganic compounds as final acceptors of respiratory electrons. They thus effect almost exactly the reverse reactions of the preceding (Fig. 4.19).

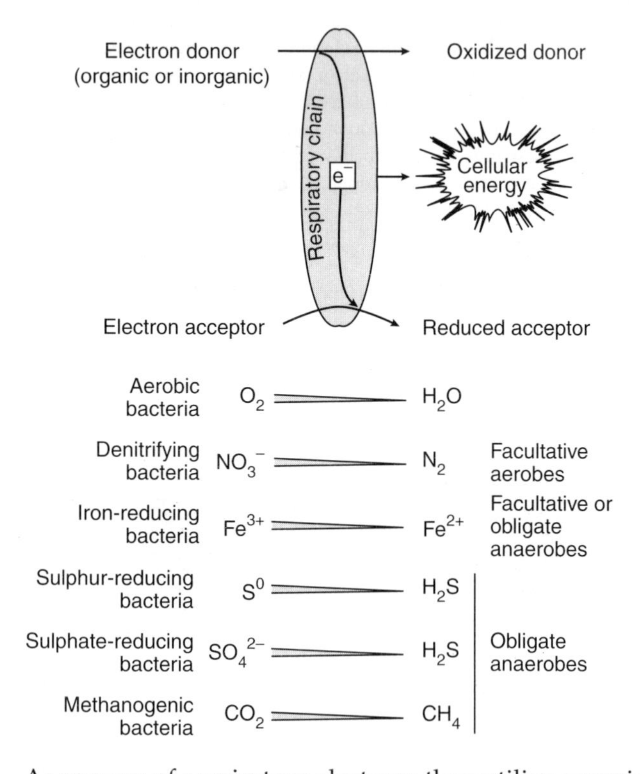

Fig. 4.19 *Simplified diagram of aerobic and anaerobic respiration.*

As sources of respiratory electrons they utilize organic substrates or hydrogen. They are:

- ***denitrifying*** bacteria that reduce nitrate to gaseous nitrogen, molecular nitrogen N_2 mainly ***(denitrification)***;
- ***iron-reducing*** bacteria that reduce trivalent iron to divalent iron ***(iron reduction*** or ***'iron respiration')***;
- ***sulphur-reducing*** bacteria that reduce elemental sulphur to hydrogen sulphide ***(sulphur reduction)***
- ***sulphate-reducing*** bacteria that reduce sulphate to hydrogen sulphide ***(sulphate reduction)***;
- ***methanogenic*** bacteria that reduce carbon dioxide to methane.

Oxic: qualifies an environment, a milieu containing molecular oxygen in gaseous or dissolved form.

Anoxic: qualifies an environment, a milieu without oxygen. Absence of oxygen is ***anoxia.***

Different reductions require different redox potentials.

During the transition between ***oxic*** and ***anoxic*** conditions, such as occurs in soils with a fluctuating water table, different anaerobic metabolic phenomena occur in sequence (Fig. 4.20).

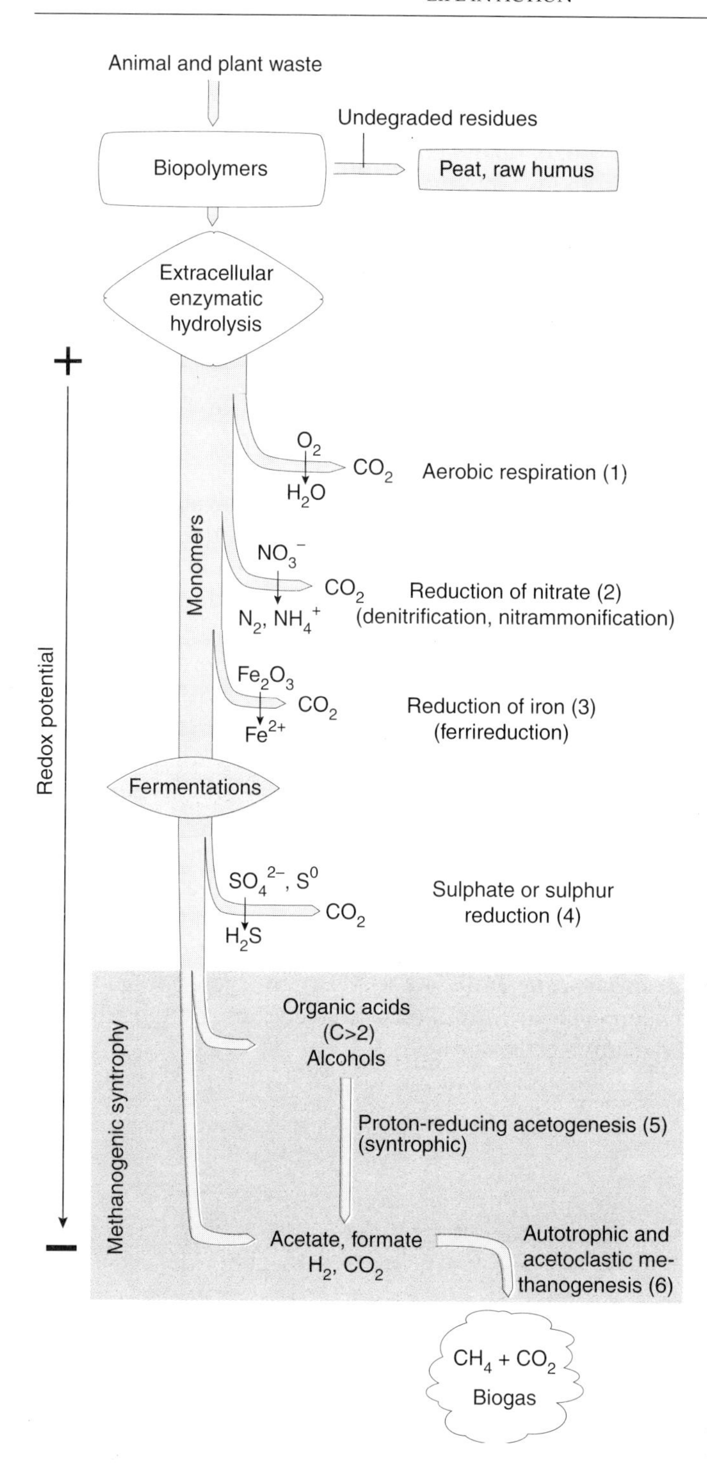

Fig. 4.20 *Successive reductions during the aerobic-anaerobic transition. After aerobic respiration (1) has exhausted available oxygen, facultatively anaerobic denitrifying bacteria (2) and iron-reducing bacteria (3) reduce the nitrate and iron, with a concomitant lowering of redox potential of the medium. Fermentations can then take place, giving rise to metabolites (alcohols, organic acids) that can be utilized by obligate anaerobic bacteria, the sulphate- and sulphur-reducing ones occurring first (4). The 'climax' stage of methanogenic syntrophy (5 and 6) produces marsh gas or biogas, a mixture of methane and carbon dioxide. It occurs only at the end of the process, if oxidizable substrates are left.*

4.4.3 Fixation of molecular dinitrogen, an essential link in the nitrogen cycle

The rhizosphere environment and root symbioses provide fixing organisms with a particularly favourable environment for the induction of dinitrogen fixation (Chap. 15, § 15.3.3; Chap. 16, § 16.2). In the ecosphere, this fixation is the major pathway and the only biological one for drawing from the pool of atmospheric nitrogen. It complies with three requirements that condition its ecology.

A strong inhibition by oxygen

The enzyme complex responsible for fixation, ***nitrogenase***, requires very low redox potentials. It is therefore particularly sensitive to oxygen, which acts as a powerful inhibitor and repressor. By means of efficient protection mechanisms, free bacteria in the soil, such as *Azotobacter* are still able to fix dinitrogen in the presence of air. Others, even though aerobic, can do it only under very low oxygen concentration. This is true of numerous dinitrogen-fixing bacteria in the rhizosphere (Chap. 15, § 15.3.3) or in root symbioses (Chap. 16, § 16.2.1). Among the other N_2 fixers can be mentioned strictly anaerobic heterotrophs (e.g. *Clostridium pasteurianum*), which obviously do not have to deal with oxygen, and cyanobacteria. Cyanobacteria produce oxygen by photosynthesis; some show differentiation of their cells into two types—vegetative, non-fixing but equipped with the complete photosynthetic apparatus, and ***heterocysts***, fixing cells not provided with the oxygen-evolving ***photosystem II*** (§ 4.4.4, Oxygen cycle) (Fig. 4.21). Heterocysts are supplied with electrons by the adjacent vegetative cells.

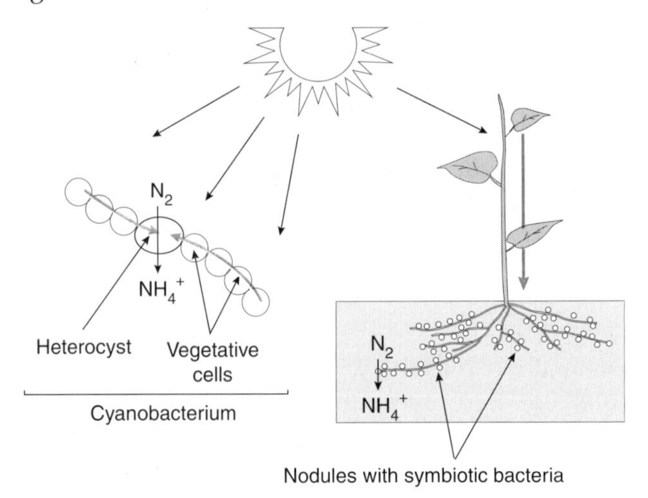

Fig. 4.21 *Biological fixation of dinitrogen using light energy. Left, direct fixation by a cyanobacterium; right, indirect fixation by symbiotic bacteria of roots.*

Nitrogen fixation requires much energy: to activate and assimilate one molecule of dinitrogen at least 20 molecules of ATP are used. Photosynthetic bacteria and those living in association with plants are particularly favoured in this because they utilize light energy directly or indirectly.

High energy requirements.

Considering its high energy requirement, dinitrogen fixation is actuated only when really necessary, that is, for removing growth limitation caused by a deficiency of combined forms of this element. If the nitrate or ammonium concentration is adequate, fixation and its associated reactions are inhibited, and the enzymes responsible are not formed (repression; Chap. 14, § 14.1).

Repression by combined forms of nitrogen.

4.4.4 Five important biogeochemical cycles of oxidation and reduction of bioelements

The different reactions associated with metabolic functions mentioned above enter the cycles of oxidation and reduction of the concerned elements, in particular carbon, oxygen, nitrogen, sulphur and iron.

The carbon cycle

To present a more or less complete carbon cycle is a challenge, considering the large number of compounds, almost all organic, it involves and the still larger number of transformations that occur in it. The carbon cycle presented here is greatly simplified (Fig. 4.22). The cell is considered here with two compartments: metabolites and biomass. We have directly represented the end forms of biological carbon: the most oxidized (CO_2) and the most reduced (CH_4). The innumerable 'organic substances' occupy all intermediate oxidation states (see Chap. 9, Fig. 9.11).

A very complex cycle, but with two simple end products, carbon dioxide and methane.

In the presence of oxygen, almost all the carbon is finally transformed to CO_2 by aerobic heterotrophic organisms. In stable and deeply anoxic conditions, the reductive power of organic matter cannot be transmitted to oxygen. After exhaustion of other potential acceptors (Fig. 4.20), it is found in the form of methane.

Carbon dioxide is in complex equilibrium with the forms dissolved in water and with carbonate rocks (Chap. 2, § 2.1.2).

Some organic substances have a very short life because they are easily absorbed and transformed by heterotrophic organisms, chiefly bacteria and fungi. Others, such as lignins and melanins, are much more resistant to microbial attack. They then accumulate in soils in the form of inherited humic compounds (Chap. 5, § 5.2.3). A very small portion of the organic matter is fossilized.

Resistant organic compounds and other labile compounds.

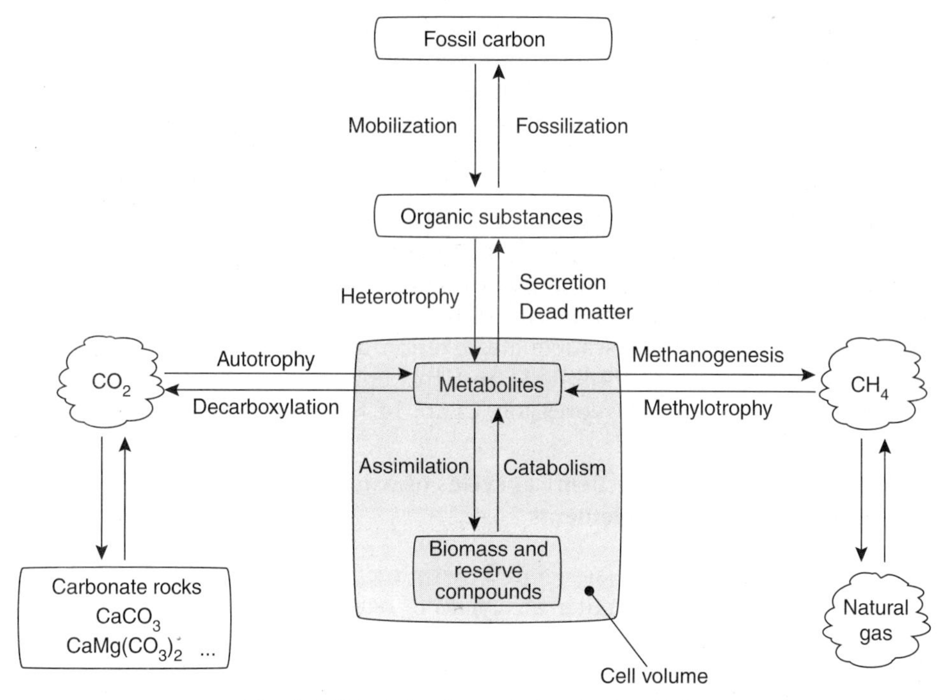

Fig. 4.22 *Biological carbon cycle (greatly simplified).*

No accumulation of oxygen in the air without fossilization of organic matter!

Fossilization of carbon and atmospheric oxygen

Photosynthesis by plants produces sugars and oxygen from carbon dioxide and water. The aerobic mineralization of sugars by heterotrophic organisms (respiration) is exactly the reverse reaction. There will therefore be no net accumulation of oxygen in the atmosphere (all the oxygen produced will be reutilized) if there is no concomitant accumulation of organic matter, essentially in fossil form. Only fossilization has enabled the attainment of the present oxygen concentration in the atmosphere, all by forming deposits of coal, petroleum, bitumen, peat and natural gas! These originate from ancient decomposition of organic materials trapped in anoxic sediments.

The oxygen cycle

In today's biosphere, most organisms (including almost all eukaryotes) depend on the presence of molecular oxygen (O_2) in their environment. In most cases, oxygen is used as electron acceptor in aerobic respiration (Fig. 4.19). But this was not always so. It is generally estimated that life evolved without molecular oxygen during the first half of its history. Thus the biological oxygen cycle (Fig. 4.23) appeared later, about 2 billion years ago, when oxygen began to accumulate in the atmosphere due to the "invention" of ***oxygenic photosynthesis*** and because of fossilization of part of the produced biomass (see box).

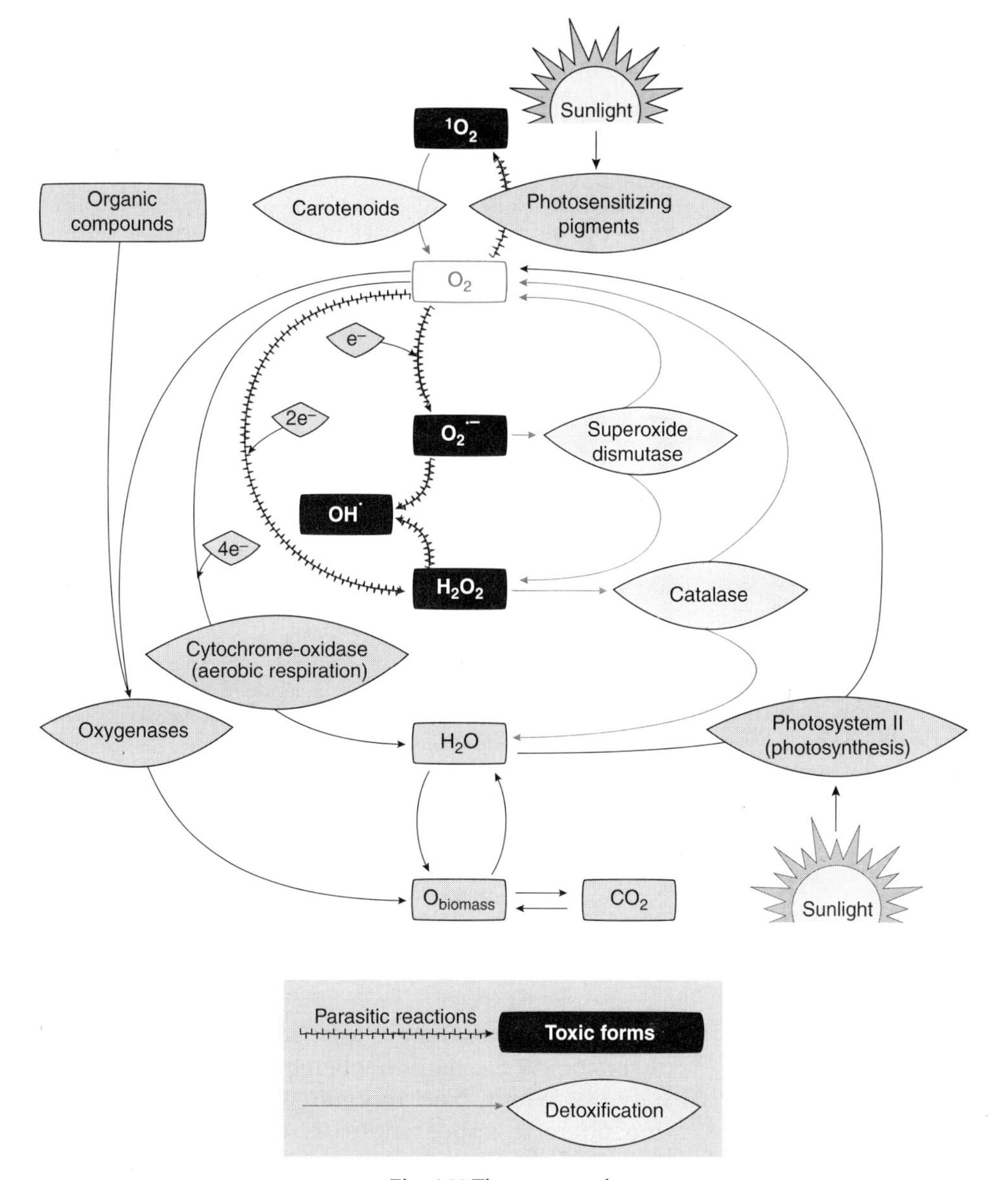

Fig. 4.23 *The oxygen cycle*

Anoxygenic and oxygenic photosyntheses

During the evolution of life, the "invention" of photosynthesis played a major role. Indeed, it enabled living things to become independent of terrestrial energy sources that are of chemical nature and available in limited amounts, to make use of the abundant solar energy always present at the time-scale of life evolution. The general principle of the transformation of light energy to cellular energy is shown in Fig. 4.24a. Light is absorbed by an "antenna" of pigments (e.g. chlorophylls, carotenoids) that are excited by the absorption of a photon. This excitation is transmitted to ***photosystem 1*** (PS1), which is a protein-reduced chlorophyll complex. Thus excited, PS1 becomes able to transmit electrons to a low redox potential compound (X), which will further transmit these electrons. In this way, PS1 is oxidized and reverts to the unexcited state.

Some of the electrons will feed an electron transport chain similar in principle to a respiratory chain: this chain will generate energy (as trans-membrane gradients and ATP) which can be used by the cell. At the end of the chain, the electrons will be transmitted to PS1, which gets reduced and ready for a new light-induced excitation. These electrons will thus be recycled.

Another part of the electrons from X is used for reductive purposes, in particular for the synthesis of cell material from CO_2 (autotrophic metabolism). These electrons are not recycled, therefore the system needs to be again fed with electrons from an external donor. In the most primitive types of chlorophyll-dependent photosynthesis, the donors are reduced compounds (H_2, H_2S, S°, Fe^{2+}) able to transmit electrons at a potential compatible with the transport chain (around 0 V). The oxidation products of these donors (H_2O, SO_4^{2-}, Fe(III)) are *not* molecular oxygen, therefore such metabolism is termed ***anoxygenic photosynthesis.*** The availability of such donors was very limited, so the next step in evolution, ***oxygenic photosynthesis,*** was of immense importance: the ability to use water, the most widespread molecule in the biosphere, as electron donor with concomitant evolution of molecular oxygen (Fig. 4.24b). The redox potential of water oxidation:

$$H_2O \rightarrow 2e^- + 2H^+ + \frac{1}{2}O_2$$

is however much too high to allow these electrons to be injected directly in the transport chain. A second photosystem (***photosystem 2,*** PS2) is therefore necessary to bring electrons from water to a potential compatible with the transport chain (about 0 V). Molecular oxygen, the oxidation product of water, is then a metabolic waste of this type of photosynthesis.

Oxygenic photosynthesis (plants, Cyanobacteria) is the only biological mechanism whereby molecular oxygen is evolved. When free oxygen accumulated in the atmosphere it appeared toxic, through compounds resulting from its excitation or partial reduction rather than by itself. Living organisms therefore had to invent protective mechanisms:

- in the presence of cellular pigments activated by light, the oxygen molecule becomes excited to give the highly reactive and toxic singlet oxygen. Other cell pigments, carotenoids, are however able to catalyse the return of oxygen molecule to its basic (triplet) form.
- in the presence of certain electron donors (flavoproteins, quinones, iron-sulphur proteins), oxygen

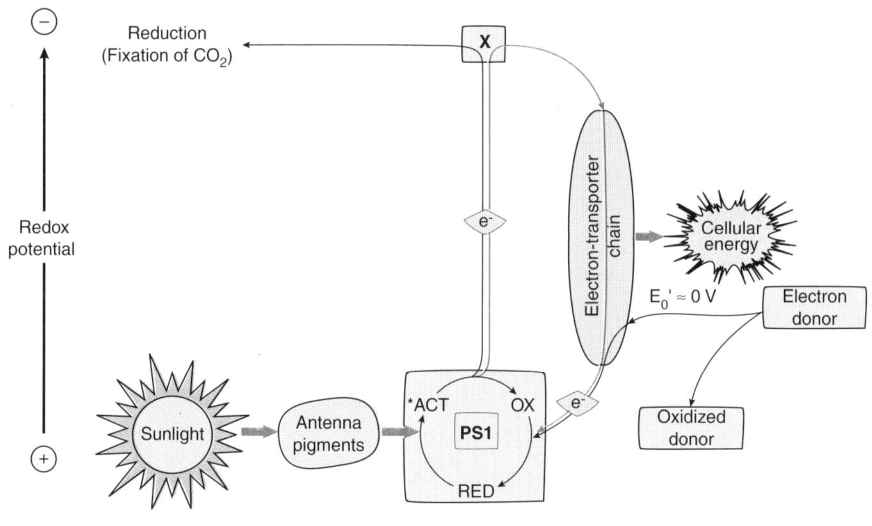

a) Anoxygenic photosynthesis

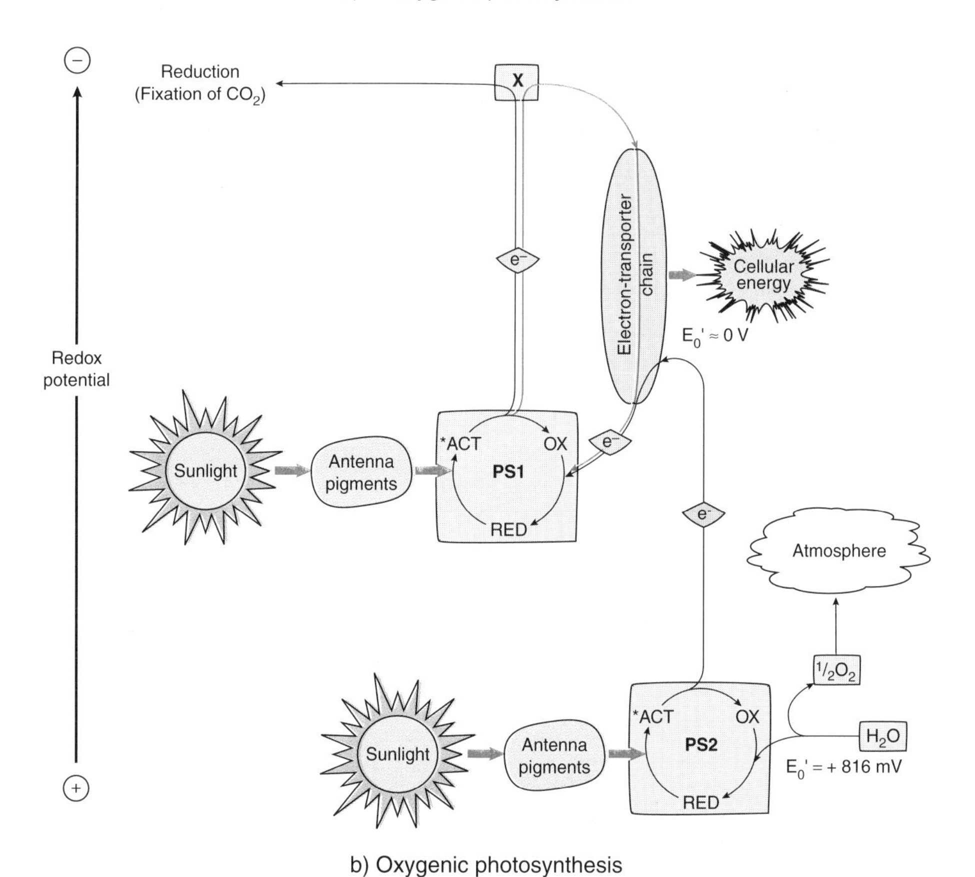

b) Oxygenic photosynthesis

Fig. 4.24 *The two main types of photosynthesis. Explanation in text box.*

is partially reduced to the superoxide radical (one-electron reduction) or hydrogen peroxide (two-electron reduction). Both compounds are very reactive and toxic, causing fatal damage to the cell. Superoxide and peroxide may then react to produce hydroxyl radicals, still more toxic. Therefore, the organisms living in the presence of molecular oxygen had to develop enzymatic protective mechanisms: superoxide-dismutase transforms superoxide into molecular oxygen and hydrogen peroxide, whereas catalase transforms peroxide to molecular oxygen and water. Thus both enzymes work together to protect the cell.

Once protected against the pernicious effects of oxygen, the organisms were able to learn to use it, in two ways:

- as a respiratory electron acceptor: because of its high redox potential ($E_0' = +816$ mV), complete reduction of oxygen (4 electrons, with the production of two water molecules) yields much more energy than the electron acceptors of anaerobic respirations;

- as a reactant capable of attacking recalcitrant organic compounds, such as aliphatic and aromatic hydrocarbons, by means of ***oxygenases.*** These enzymes introduce one or two oxygen atoms from O_2 into an organic molecule, allowing it to be further utilized in metabolic reactions for assimilation and energy production.

In the biomass, oxygen is an important component of most organic molecules. It originates mainly from CO_2 and water, to which forms it returns when the organic compounds are mineralized.

The nitrogen cycle

The nitrogen cycle (Fig. 4.25) illustrates the integration of the three types of functions mentioned earlier.

Biological fixation.

The major pool is represented by atmospheric dinitrogen; its fixation (reduction to ammonia) by microorganisms is the main way of integration in the cycle. However, a small quantity of combined nitrogen reaches the biosphere in the form of nitrogen oxides, following cosmic irradiation, storms and human actions.

Nitrification.

Ammonium is the chief source of nitrogen for microorganisms and fungi; most plants assimilate it poorly. They require nitrate, which they utilize by the mechanism of ***assimilative reduction***. The nitrate is supplied to them by ***nitrifying*** bacteria. In the absence of plants, production of high concentration of nitrate results

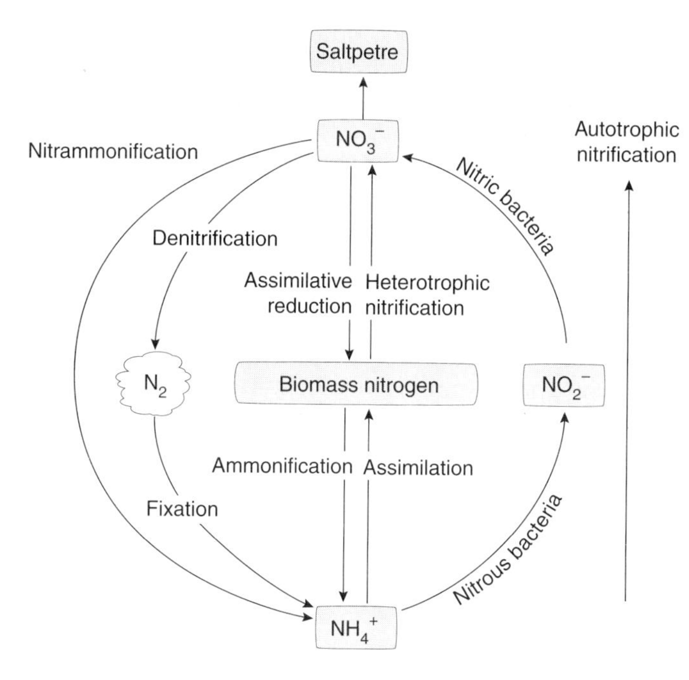

Fig. 4.25 *Biological nitrogen cycle.*

in solubilization of cations (§ 4.3) and thus provokes leaching of soils. Nitrate is then found in groundwater, and sometimes makes it unsuitable for consumption.

Denitrification.

If oxygen has become exhausted, denitrifying bacteria reduce the nitrate to molecular nitrogen N_2. For an ecosystem, denitrification results in a loss of combined nitrogen. But overall this is the means of replenishing the atmospheric pool, enabling at the same through fixation to make this element available at all levels of the food-chain. It takes place primarily in systems where oxic and anoxic conditions alternate, as in Ag or Go horizons.

Nitrammonification.

Nitrammonification, on the other hand, is performed by some otherwise anaerobic bacteria, strict or facultative. It takes place mainly in permanently anoxic environments. As there is no loss of combined nitrogen, these milieus better conserve this element than those subject to alternation of oxic and anoxic conditions.

The sulphur cycle

The sulphur cycle closely resembles the nitrogen cycle, but the pools are located elsewhere.

In organisms, sulphur is found in two amino acids, cysteine and methionine, in some growth factors such as thiamine (vitamin B_1) and also in iron-sulphur proteins. Beyond this role as constituent element of living matter, it also acts as the chief redox component in numerous anoxic aquatic ecosystems, which are therefore termed '***sulphuretums***'.

In sedimentary rocks, sulphates, in crystalline rocks, sulphides.

In sedimentary rocks, sulphur exists in sulphate minerals, in particular gypsum. In crystalline rocks, it occurs in the form of metal sulphides, such as pyrite. Sulphate is the most oxidized form; it serves as the source of sulphur for plants and numerous microorganisms. In anaerobic conditions, it is the electron acceptor for sulphate-reducing bacteria.

From sulphides to elemental sulphur and sulphate.

Hydrogen sulphide is oxidized by two groups of bacteria that transform sulphides to elemental sulphur, then to sulphate (Fig. 4.26). One group, strictly anaerobic, is photosynthetic: abundant in some lakes, these bacteria are of no special significance in soils. The other group, strictly aerobic, comprises sulphur-oxidizing chemolitho-autotrophs, such as those found in the U horizon of SULFATOSOLS (Chap. 5, Tab. 5.13). Elemental sulphur is directly assimilated by some bacteria.

'Sulphuretum': anoxic ecosystem, generally aquatic, in which the sulphur cycle is basic to oxidation and reduction reactions involving, among others, sulphate-reducing bacteria and sulphide-oxidizing, anoxygenic photosynthetic bacteria. This term has been adopted by analogy with those of phytosociology (Chap. 7, § 7.1.4).

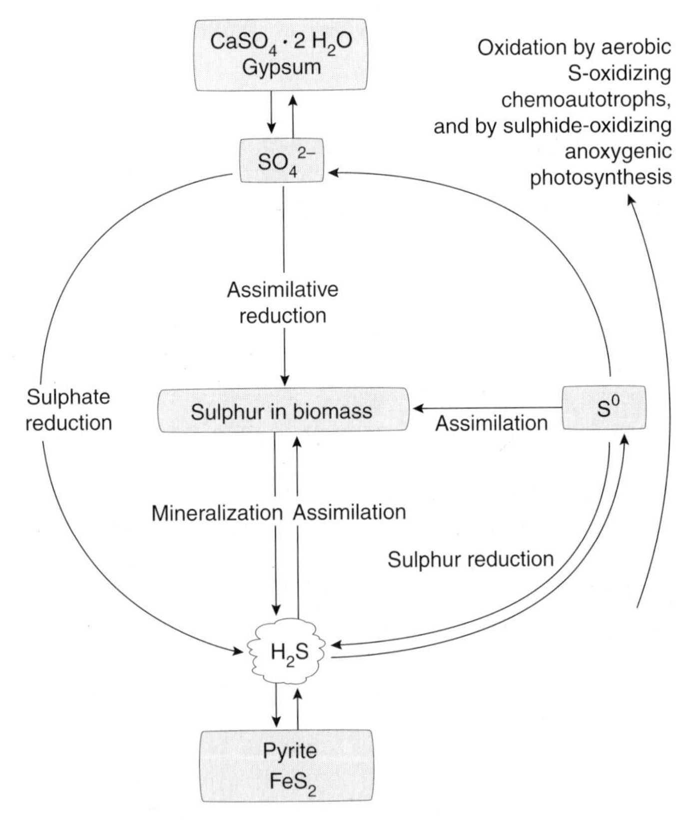

Fig. 4.26 *Biological sulphur cycle.*

Cofactor: non-protein molecular structure that directly participates in the reaction catalysed by an enzyme. If this structure is permanently bound to the enzyme, it is a ***prosthetic group***. A coenzyme, on the other hand, can leave the enzyme after the reaction and participate in a reaction catalysed by another enzyme.

Iron-sulphur proteins: class of redox proteins participating in electron transfer at the level of iron atoms (e.g. ferredoxins, thioredoxins, nitrogenase, hydrogenases). Iron is coordinated to S atoms (cysteine or inorganic S).

Haemoproteins: class of redox proteins taking part in electron transfer and oxygen transport. Iron atoms are coordinated within a prosthetic group, haeme (e.g. catalases, peroxidases, haemoglobin, myoglobin).

The iron cycle

Iron is the redox element *par excellence* in enzymes and in the electron-transporting ***cofactors***: $Fe^{2+} \leftrightarrows Fe^{3+} + e^-$. It is found mostly in two types of molecules: the ***iron-sulphur proteins*** and the ***haemoproteins***. The major pools are represented by (Fig. 4.27):

- sedimentary iron, chiefly in the form of ferric hydroxides and oxides (e.g. haematite),
- iron of crystalline rocks, chiefly in the form of divalent iron (e.g. pyrite).

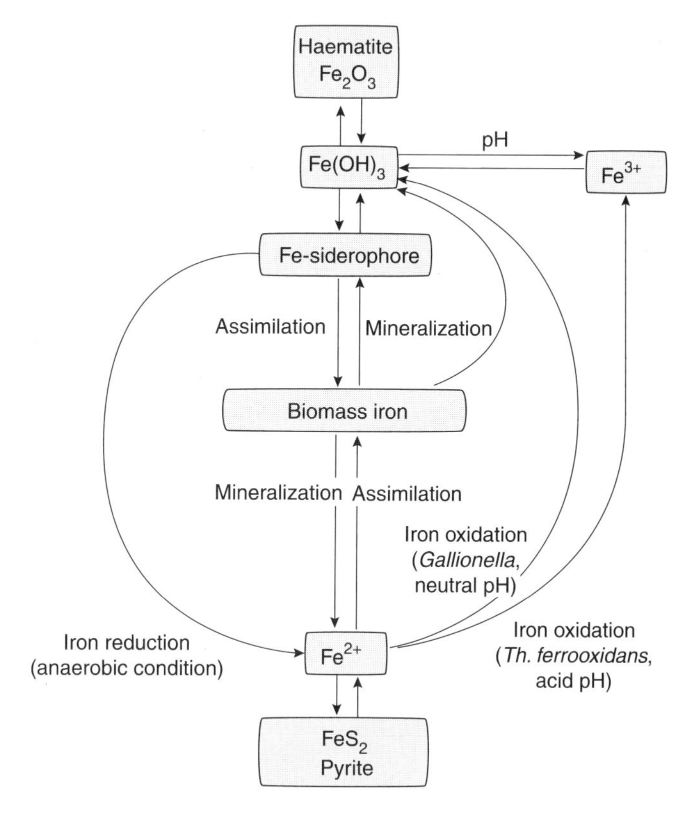

Fig. 4.27 *Biological iron cycle*

The ecology of iron-reducing bacteria is still in its early stages (Nealson and Saffarini, 1994)!

It was long thought that anoxic reduction of Fe(III) and its solubilization as Fe^{2+} ions was an abiotic process. Münch and Ottow (1977) showed that the presence of microorganisms is almost indispensable for solubilization of iron under anaerobic conditions (Fig. 4.28). This reduction is done either by specialized obligatorily anaerobic iron-reducing bacteria comprising the genera *Geobacter* and *Geothrix* or by facultative anaerobes, including *Shewanella*. These organisms play an essential part in the reduction of iron and oxidation of organic substrates, particularly in the elimination of organic contaminants from anoxic aquifers.

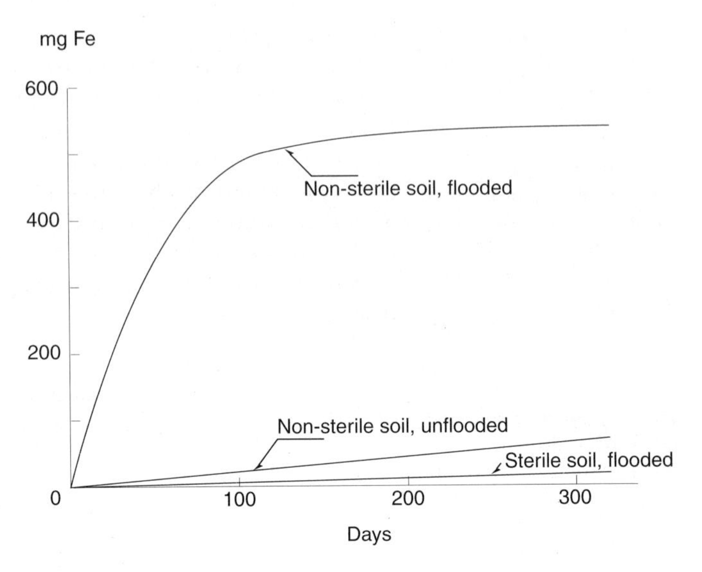

Fig. 4.28 *Anaerobic bacterial solubilization of iron. In a non-sterile flooded soil, bacterial activity consumes all the available oxygen: the resulting anoxia then permits anaerobic solubilization of iron (also see Fig. 4.20).*

In aerobic conditions, iron is solubilized as free ferric ions only in strongly acidic media. Near neutrality, its solubilization requires the intervention of ***siderophores,*** chelates that complex it and allow uptake by the cells (Chap. 15, § 15.3.5). Oxidation of ferrous iron can occur spontaneously, without participation of living organisms. However presence of iron-oxidizing bacteria (e.g. *Thiobacillus ferrooxidans* below pH 4, *Gallionella ferruginea* near neutrality) speeds up the process considerably.

The cycles are joined

A close coupling of carbon, oxygen, nitrogen, sulphur and iron cycles.

The major energy flux is due to organic carbon. It conditions numerous phenomena requiring energy such as nitrogen fixation, lowering of oxygen concentration (consumed by aerobic respiratory activity), and also sulphate reduction and iron reduction, which depend on establishment of anaerobic conditions and on excretion of alcohols and acids, the products of fermentation of sugars.

The presence or absence of oxygen is an important parameter; through these cycles it conditions the predominance of certain forms of the principal bioelements:

- reduced (e.g. ammonia, hydrogen sulphide, etc.) in anoxic conditions;
- oxidized (e.g. nitrate, sulphate) in oxic conditions.

Significant concentrations of intermediate forms (nitrite, thiosulphate and polythionates) can be an indicator of transition phenomena or disequilibrium between two phases.

The presence of oxygen is also a prerequisite for the assimilation and/or mineralization of many organic compounds, e.g. aromatic derivatives and hydrocarbons.

4.4.5 Relations with the geosphere: do microbes 'eat' stones?

By their metabolism, certain microorganisms degrade minerals directly. They can also provoke their solubilization by absorbing ions that they immobilize in their biomass, causing a new solubilization by shift of equilibrium (Chap. 3, § 3.7.2).

Directly or not, many bacterial activities are involved in rock weathering (Chap. 2, § 2.1.2), jointly with purely chemical or physical reactions.

Anaerobic respiratory reduction of iron, as already seen, results in its solubilization. Oxidation of metal sulphides supplies certain chemolithoautotrophic bacteria the respiratory electrons they require. Here the mineral is the energy source. These bacteria, like *Thiobacillus ferrooxidans*, oxidize sulphides and iron together. Strongly acidophilic (pH 0.5 to 4), they have little pedological significance except in soils of polders and mangroves (THIOSOLS or SULFATOSOLS with TH or U horizons; Chap. 5, Tab. 5.3).

Organisms take part indirectly in corrosion of rocks by modifying the environment, rendering it more aggressive. Such changes pertain to pH, oxygen concentration and redox potential, as well as production of organic compounds capable of complexing mineral ions in the form of chelates (Chap. 5, § 5.3.2). Acid attack of minerals or acidolysis (Chap. 2, § 2.1.2) is indicated by solubilization of minerals into the ionized form. In aerobic conditions, mainly nitrifying and sulphide-oxidizing chemolithoautotrophic bacteria cause this solubilization by producing sulphuric and nitric acids. Thus nitrification results in dissolution of calcium (Fig. 4.29).

Chelation, acidolysis and alkalinolysis.

Production of organic acids by microorganisms and roots is also an important factor in weathering of certain minerals. Here acidolysis must be distinguished from the effect of chelating organic substances. The former is primarily the work of the monocarboxylic formic, acetic, butyric and lactic acids, and solubilization affects chiefly calcium and potassium. Consumption of organic anions by microorganisms leads to alkalinization of the environment by release of the corresponding bases and, occasionally, to solubilization of silica (alkalinolysis).

A number of divalent and trivalent mineral elements (Fe, Al, Cu, Zn, Ni, Mn, Ca, Mg) are liable to form complexes with di- or trianionic organic compounds.

Chelators are powerful potential agents of rock weathering directly linked to degradation of organic matter.

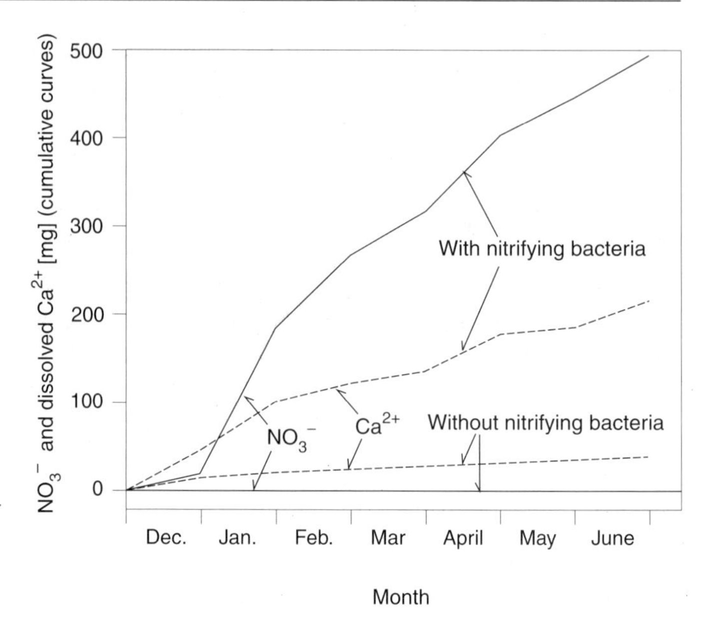

Fig. 4.29 Influence of nitrification on solubilization of calcium in a granitic sand (after Berthelin, in Krumbein, 1983).

These complexes are represented in the form of cyclic compounds incorporating the cations, stabilizing them in soluble form, through a wide range of pH and redox potential. Such compounds are often secreted by microorganisms, in particular oxalic, citric, 2-ketogluconic and tartaric acids as well as polyphenols (salicylic and dihydroxybenzoic acids). Crenic and fulvic acids (Chap. 2, § 2.2.4), are particularly active as iron chelators in PODZOSOLS. Specific siderophores are involved in iron nutrition of plants and microorganisms (Chap. 15, § 15.3.5).

An interesting relation between metabolism and weathering by secreted organic compounds.

If its growth is limited by elements other than carbon, a bacterium continues to absorb carbon-containing substrates available to it. It transforms them into metabolites subsequently secreted into the environment, making it more aggressive. The resulting solubilization of minerals can in turn remove the limitation.

4.4.6 Abundance or famine: different strategies

The autochthonous and zymogenous bacteria of Winogradsky

Sergei Nikolaevitch Winogradsky (1856–1946) is considered the father of soil microbiology. A Russian biologist from a family of bankers, he wandered many years through European laboratories before settling in 1922 in the

The portion of soil remote from both roots and litter is subject to very weak energy flows (Chap. 6, § 6.2.1), a principal limiting factor for microbial activity. Substances transported from the litter by the fauna, dead bodies of animals, soluble substances leached by percolating waters and slowly degrading humic substances represent the bulk of the energy added. Such bulk soil mainly accommodates organisms adapted to living parsimoniously, thus to be nourished by substrates with very low concentration. Winogradsky (1949) empirically termed them ***autochthonous***. These populations, by virtue of the high affinity they show for their nutritive substrates, have developed a competitive strategy under limiting nutrient conditions. Autochthonous populations scarcely fluctuate, even with sudden addition of energy-rich sources. The concept of autochthony corresponds very well to the K strategy developed in ecology (Chap. 12, § 12.6.2), one application of which to soil protozoans has been presented by Bamforth (in Benckiser, 1997).

When he added to a soil sample a solution rich in nutrients, for example glucose and/or yeast extract, Winogradsky observed during the following hours an intense growth of bacteria, so far not at all abundant in the bulk soil, which eventually largely dominated the bacterial community. He called them zymogenous, literally 'resulting from yeast extract'. These populations were very ephemeral, however, and after a few days or

Pasteur Institute in Paris. He demonstrated the autotrophic nature of sulphide bacteria and isolated numerous organisms responsible for the nitrogen cycle. His treatise of 800 pages on soil microbiology was published posthumously in 1949 (Boulaine, 1989).

Peaceful autochthonous bacteria and expansive but elusive zymogenous bacteria!

Adaptive strategies: an attempt at comparison

Many types of adaptive strategies have been independently defined by ecologists studying different groups of living organisms (animals, plants, protists, bacteria) and at different levels of integration (populations, biocoenoses). Without being absolutely equivalent, as they are based on different models and concepts, these types nonetheless present striking similarities (Chap. 12, § 12.6.2). In the final analysis it is highly probable that the strategies of different life forms resemble each other more than the models used to describe them! Table 4.3 attempts to summarize these similarities by comparing in particular the approach of microbiologists to those of other biologists.

A model for bacterial growth

An interesting parallel can be drawn between the empirical statements of Winogradsky and the very simple model proposed by Monod (1950) for bacterial growth, a model that establishes the relation between the growth-limiting substrate concentration *(s)* and the growth rate *(μ)*:

$$\mu = \mu_{max} \cdot s/(k_s + s),$$

where μ_{max} is the maximum growth rate in absence of any limitation and k_s the ***affinity constant*** (the smaller the k_s the greater the ***affinity*** for the substrate). An autochthonous bacterium (K strategist) will thus be characterized by small values of μ_{max} and k_s, a zymogenous one (r strategist) by high values of both these parameters (Fig. 4.31).

Affinity: in enzymology and microbial-growth physiology, affinity denotes the ability of an enzyme or bacterium to utilize a substrate with low concentration. The ***affinity constant*** k_s represents the concentration of substrate enabling bacterial growth to take place at half its maximum speed. In enzymology, the equivalent constant is the Michaelis-Menten constant, k_m.

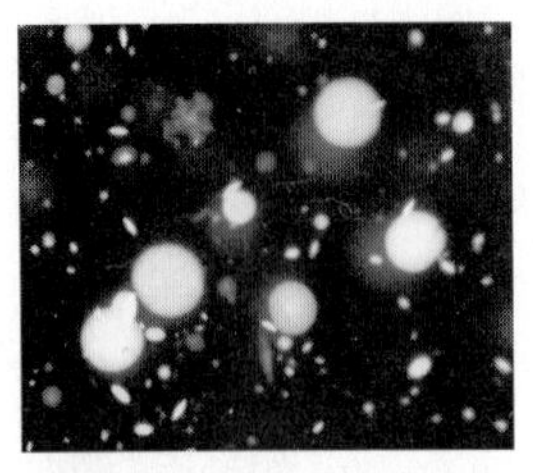

Fig. 4.30 *Bacterial colonies formed in a culture on nutrient agar in a Petri dish, inoculated with a diluted soil suspension and incubated one week at 20°C. Some large colonies (zymogenous bacteria) and much more numerous minute colonies (autochthonous bacteria) are distinguishable (photo: N. Jeanneret and M. Aragno).*

Not all soil bacteria are culturable. Only molecular methods can prove the existence of some autochthonous populations (Chap. 15, § 15.6.5).

weeks without fresh addition of the nutrient solution, the autochthonous bacteria dominated again. Contrary to autochthonous bacteria, ***zymogenous*** bacteria rapidly form large colonies in culture on nutrient broth agar in Petri dishes (Fig. 4.30). The strategy of abundance of zymogenous bacteria corresponds well to the concept of the r strategy in ecology.

Culturable bacteria . . . and others

When a soil sample is cultured on a nutrient agar, the number of units that form colonies on it is much smaller than the number of living bacterial cells observed microscopically in the sample. This number is also very low compared to the biomass estimated by the ATP content of the soil (Chap. 15, § 15.5.2). The ratio is often only 1:100 in agricultural soils or 1:10,000 in environments more limited in energy, such as sands from deep aquifers. The reasons for this are many: a ***colony-forming unit*** (cfu) may be composed of many agglomerated cells; some populations (e.g. strict anaerobes, autotrophs...) are inhibited by culturing conditions or by not finding their required nutrients in the medium. But the major cause probably stems from the dominating presence of very

Table 4.3 Comparison of adaptive strategies

	Juvenile environments, temporarily unstable; nutrients abundant, high food/energy flow	**Mature environments, permanent, stable; nutrients limited, low food/energy flow**	**Often extreme environments, selective**
Demographic approach, level of integration: the population (Chap. 6, § 6.1.3; Chap. 9, § 9.2.2; Chap. 12, § 12.6.2)	r strategists	K strategists	—
Coenotic approach, level of integration: the biocoenosis (§ 4.2.4)	i strategies	s strategies	—
Grime's classification (plants) (Chap. 9, § 9.2.2)	R (ruderal plants)	C (competitive plants)	S (stress-tolerant plants)
Winogradsky's classification (soil bacteria) (§ 4.4.6)	Zymogenous	Autochthonous	—
Classification according to Monod's model (§ 4.4.6)	High μ_{max}, high k_s	Low μ_{max}, low k_s	—
Other concepts (bacteriology)	Opportunistic	Competitive, VBNC (extreme)	Extremophilic (extremotolerant)

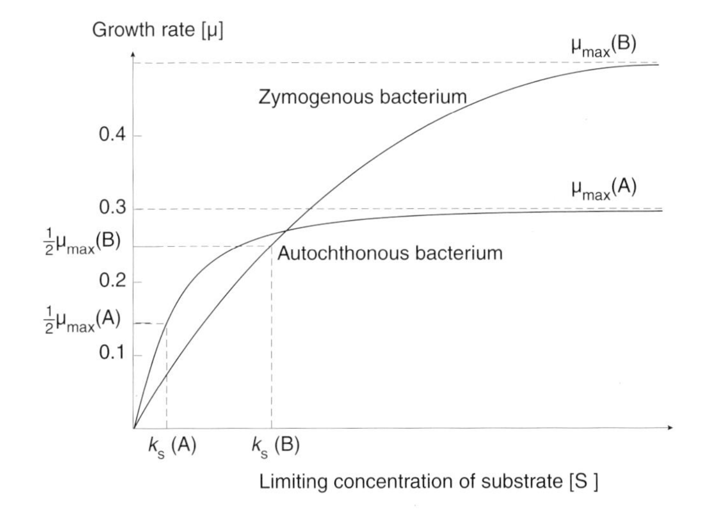

Fig. 4.31 *Competition between an autochthonous bacterium (K strategist) and a zymogenous bacterium (r strategist) according to Monod's model of bacterial growth. It can be seen that the autochthonous bacterium is favoured (faster growth) by low concentrations of the limiting substrate, while the zymogenous bacterium dominates at higher concentrations (symbols explained in the text).*

slow growing autochthonous organisms that often are even incapable of forming colonies on a rich medium.

When they are maintained for a certain time in a medium very limited in nutrients, cells of common and usually easily culturable species can enter a viable but not culturable state (viable but not culturable cells, VBNC). These cells preserve their genome and part of their activities. In some cases, reversion (sometimes termed 'resurrection'!) of the VBNC state to the earlier culturable state has been obtained.

Where bacteria come back to life!

4.4.7 Microorganisms and biocoenosis of the soil

Other than the strategies mentioned, microorganisms act in a more targeted fashion in direct and indirect interactions among themselves or with other soil organisms.

Beneficial or harmful, accelerators or brakes for biological activity, symbiotic, parasitic or predatory, the micro-organisms are, in the final analysis, even prey—starting points of secondary food-chains (Chap. 13, § 13.6.2).

Some populations exert through their secretions positive or negative regulatory effects on other members of the biocoenosis or even on themselves (see box below). Many living organisms require growth factors, that is, organic molecules (amino acids, vitamins, etc.) they are incapable of synthesizing and must therefore find in their environment. They are termed ***auxotrophs***. Some organisms are capable not only of synthesizing these substances, but also of secreting them in excess and thus making them available to other organisms in their vicinity (Fig. 4.32).

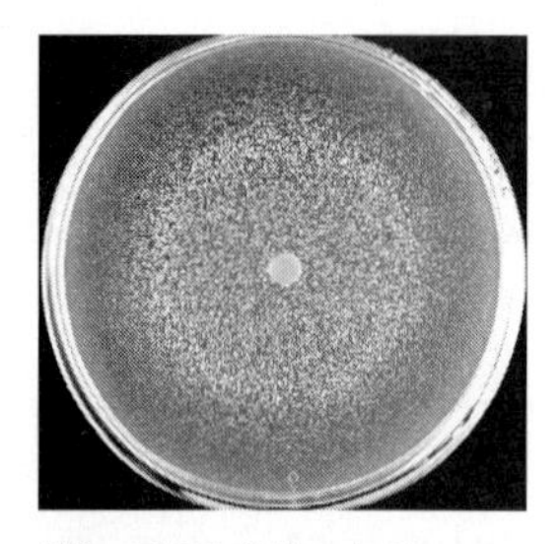

Fig. 4.32 Stimulation on vitamin-free agar medium of the growth of colonies of Xanthobacter flavus *by biotin, a vitamin, impregnating a filter paper disc (at centre of image) (photo: M. Aragno and N. Jeanneret).*

Bacterial populations attain quorum

In a number of cases, bacterial functions are expressed only when the density of the related population is high enough. Bacterial secretions, particularly homoserine lactones in Gram-negative organisms, are responsible for this phenomenon. Homoserine lactone concentration informs the cells of the density of their related population. This auto-inductive system is referred to as 'quorum sensing'. Therefore, the population as a whole reacts collectively, rather than the cells individually.

Conversely, many microorganisms secrete substances with antagonistic effect into their environment. Some act broadly, such as acids that lower the pH and render the environment unsuitable for growth of neutrophilic organisms. Others have a more specific effect, and act at much lower concentrations: these are antibiotics. Information on their effect in soils is very fragmentary; they undoubtedly play a part in the regulation of populations, enabling organisms producing them to limit development of competitors (Fig. 4.33).

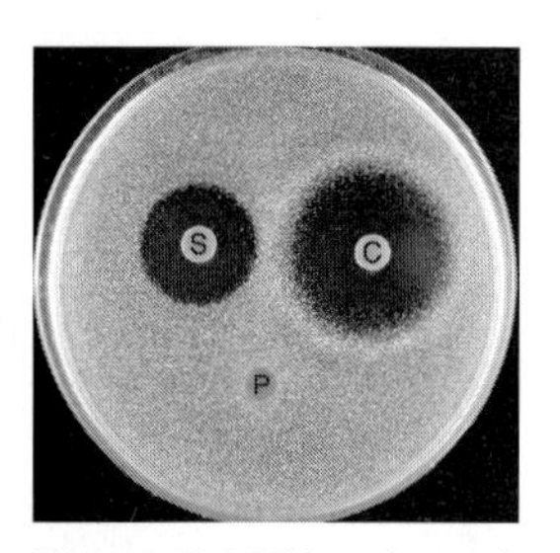

Fig. 4.33 Inhibition of growth of a bacterium by three antibiotics: streptomycin (S), chloramphenicol (C) and penicillin (P). The strain used is resistant to penicillin (photo: N. Jeanneret and M. Aragno).

An analogy in higher plants: allelopathy

Higher plants also inhibit growth of their congeners by means of substances elaborated into the medium: this is ***allelopathy*** (Lavelle and Spain, 2001). The inhibitory effect of the walnut tree *Juglans regia* on certain crop plants that grow directly below it has been known since Roman times through Pliny. The toxic substance, juglone, is formed in the leaves and washed down by rain. But allelopathy can be more direct, by root secretions as proved, for example, between grasses and legumes cultivated together (Wood, 1995). Lastly, volatile substances (such as camphor) elaborated by the leaves can inhibit root growth when the leaves fall to the ground. Collinson (1988) cites the case of sage *Salvia leucophylla* which is thus protected against invasion by annual grasses.

Bacteria and fungi maintain more intimate relations with other soil organisms. Some relations (commensalism, parasitism) are for the benefit of only one of the partners. Others (syntrophy, mutualism) benefit both. Chapter 16, devoted mainly to soil symbioses, presents in the introductory section an overview of all these interactions.

The devourers and the devoured: predation.

Some bacteria and fungi can be described as predatory microbes in soils (Chap. 13, § 13.6.3):

- Myxobacteria (Chap. 2, § 2.5.2) produce powerful extracellular lytic enzymes that destroy the cell walls of living fungi and bacteria; they then feed on the released substances.

• Certain specialized filamentous fungi make traps designed to ensnare soil nematodes (Chap. 12, § 12.4.2). At rest these traps appear as rings or small mycelial networks. When a nematode passes through one by chance in its peregrinations, it is trapped. In the case of *Arthrobotrys dactyloides* (Fig. 4.34) the rings contract violently a few seconds after contact with the worm. Activation of the trap is caused by interactions between the fungus and the bacterial flora in the alimentary canal of the nematode (Davet, 1996). Following this, the fungus sends out feeding filaments from the ring that penetrate the animal and digest the contents. Promotion of growth of predatory fungi is a means of biological warfare against plant-parasitic nematodes (Valloton, 1983; Zunke and Perry, in Benckiser, 1997).

Bacteria and fungi that eat ...

Fig. 4.34 *Nematode captured by a nematophagous fungus.*

In their turn, bacteria and fungi serve as food for other organisms, protozoans and invertebrates. They can thus be considered secondary producers, at the bottom of important food-chains (Chap. 13, § 13.3). This predation is important in that it augments overall mineralization. The turnover of microflora is also speeded up, the predators 'making room' and thus ensuring faster growth of bacteria and fungi (Chap. 15, § 15.3.2).

. . . bacteria and fungi that are eaten!

When organic wastes are consumed by saprophytic microorganisms, part is assimilated (production of biomass) and the rest is mineralized (Fig. 4.35). When the microbial biomass is consumed in turn by predators, only a part is found in their biomass, the rest again being mineralized. And so on, when predators in turn are eaten.

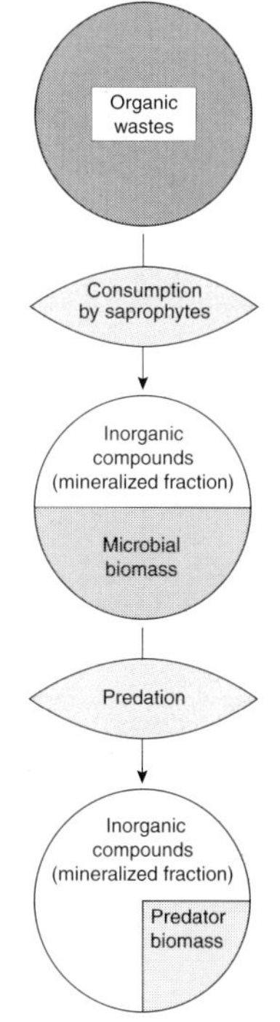

Fig. 4.35 *Augmentation of mineralization of organic wastes following the action of predators on microbial biomass. This simplified scheme does not take into account the residual (unutilized) organic substances.*

4.4.6 Conclusion

Microorganisms, soil 'proletariat'? The epigraph as definition of this section suits them very well, it must be admitted. Small in size, often passing unseen, they are however what 'run the machine' of the soil: they process animal and plant wastes, manufacture structure-forming molecules, provide plants with mineral elements in a suitable form and attack the rock for forming soil. They behave collectively through quorum sensing. If some rebel at times, parasites, predators, inhibitors of Lord Animal or His Majesty the Plant, even Man, the capitalist who thinks himself all-powerful, is it not also their right? Anyway, at the end of the accounting they are eaten... .

4.5 THE ESSENTIAL ROLE OF THE FAUNA

The soil fauna comprises very many species (Chap. 2, § 2.6; Chap. 12, § 12.4). They live on the surface, in soil annexes (Chap. 8) and abound in the humiferous episolum, especially in zones of preferential rooting. Soil animals have direct or indirect impact on their habitat by moving through it, drawing nourishment from it, and excreting and dying in it.

The soil shelters a varied and abundant fauna. The latter's vital activities are essential to functioning of ecosystems and soil formation.

In science, errors are sometimes fruitful!

Some roles of the microfauna were discovered following errors in handling, for example in the laboratories of the great British soil scientist Sir John Russell (1872–1965). He tested changes in redox potential of soils sterilized beforehand and then inoculated with microorganisms. An accidental partial sterilization enabled him to observe a much higher mineralizing activity, which was attributed for good reason to edaphic protozoans (from Boulaine, 1989).

4.5.1 Physical functions of the soil fauna

Five principal mechanical effects, the first two at least relevant to bioturbation (Chap. 5, § 5.3.3).

Like all processes dependent on the soil fauna, mixing, burrowing and fragmentation progress at many scales, simultaneously or successively according to the size, feeding habits and behaviour of the organisms involved. Five major mechanical effects of the soil fauna have been proved:

- macro-mixing,
- micro-mixing,
- burrowing,
- fragmentation,
- aggregate formation.

Macro-mixing

Ants, termites, earthworms, beetles and certain mammals: particularly effective diggers.

Ants, termites, earthworms, beetles and certain mammals (moles, voles, prairie dogs, etc.) dig up large quantities of soil, bringing to the surface horizons rich in mineral substances and burying surface organic horizons, litter and manure. It is however necessary to modify this statement slightly by mentioning that *the* earthworms, *the* termites, etc. do not refer to the entirety of these taxa, but to a certain number of species in each. Not all termites build cathedral-like termitaria and not all earthworms are anecic species (Chap. 12, § 12.4.3).

One to three tons of soil per hectare are brought up to the surface every year by yellow pasture ants.

For example, a large nest of fungus-growing ants *Atta* houses many million individuals. In the soil, it forms a cavity with many chambers from which the earth is evacuated to the surface nearby. For digging one of these nests, 22.7 m^3 earth weighing nearly 40 tons was dug up (Hölldobler and Wilson, 1996)! In the soil of pastures of the Jura mountains, the yellow ant *Lasius flavus* also digs underground nests from which the material is thrown out on the surface, forming domes of fine soil covered by a particular vegetation. Up to 1000 colonies have been counted per hectare, their domes covering about 1% of the area.

Termites play an essential role in the tropical zone (Chap. 12, § 12.4.9; Chap. 13, § 13.6.4; Lavelle and Spain, 2001). The drawing of fine material from depth occasionally results in porous zones below the nests where temporary water tables accumulate. In the case of *Bellicositermes rex*, the termitarium can measure 5–6 metres in height and 30 metres in diameter (Grassé, 1982). The volume of ten such termitaria in a hectare will correspond to a covering 10 centimetres thick if erosion has spread them out in a continuous layer, thus burying ancient stony or gravelly horizons.

It has been estimated that termites bring up 1000 t soil km^{-2} y^{-1} in northern Cameroon.

In the temperate zone, and also in the tropical zone, the activity of macro-mixing by earthworms is of capital importance for soils. It is measured by the quantity of worm casts seen on the surface (Fig. 4.36; Tab. 4.4; Binet and Le Bayon, 1999). Worms produce 40–250 t ha^{-1} y^{-1} of them in the temperate zone; the entire soil of a grassland thus passes through their digestive tract in 10 years. The quantities are still larger in the tropics, attaining 500 t ha^{-1} y^{-1} in the savanna and exceeding 2500 t ha^{-1} y^{-1} in the very fertile fields of the Nile valley!

Fig. 4.36 *Earthworm cast in a pasture of the Jura mountains (photo: J.-M. Gobat).*

Table 4.4 Quantity of worm casts in different environments (various sources)

Country/region	Plant formation	Worm casts (t ha^{-1} y^{-1})
Switzerland	Cultivated meadow	18 to 81
Germany	Cultivated meadow	91
England	Old grassland	19 to 40
Germany	Beech	7
Ghana	?	50
Ivory Coast	Savanna	507
Cameroon	?	2100
Nile valley	Cultivated area	2500
Kansas	Tallgrass prairie	13 to 75

Charles Darwin (1809–1882) was a brilliant naturalist, most famous for his theories of evolution, summarized in *The Evolution of Species*, his best-known work. But he also published in 1881 a work on earthworms that has remained classic, *The Formation of Vegetable Mould through the Action of Worms, with Observations of their Habits.*

The Scarabaeinae or dung beetles (Chap. 8, § 8.3.2; Chap. 12, § 12.4.9) comprise many species very efficient

Scarabaeids are irreplaceable for burying all sorts of excrement.

at hiding excrement of the surface soil. For example, a single pair of *Heliocopris dilloni*, a large African species (Fig. 8-c), is capable of burying one cowpat in one night (Waterhouse, 1974)! Often, these coleopterans specialize in one type of excrement. *Aphodius elevatus* prefers human excreta, while *Onthophagus drescheri* is particularly attracted by tiger droppings. The latter species slowly disappears along with its food supplies (Paulian, 1988).

Micro-mixing

The micro-mixers, veritable 'pellet factories' to borrow the expression of Travé et al. (1996).

Other macro-invertebrates and also the mesofauna influence soil structure in a manner admittedly less spectacular than macro-mixing, but equally necessary. Their weaker penetrating power confines them to the humiferous episolum where they conduct another activity indispensable for incorporation of organic matter: production of enormous quantities of small faecal pellets (Chap. 5, § 5.2). These phytosaprophagous organisms (mites, springtails, larvae of dipterans and enchytraeids) incorporate organic matter in this form in the soil. On the other hand, they hardly bring up any mineral material.

Pellets constitute a pedofeature very useful for determination of humus forms (Chap. 3, § 3.2.4; Chap. 6, § 6.2.1).

These minuscule pellets are washed down by percolating water-leaching and can accumulate to 60-cm depth, in masses or thin layers, quite visible to the trained eye of the field pedologist (Kevan, 1962).

Burrowing (Chap. 12, § 12.5.4)

The third mechanical activity described here, formation of burrows or galleries, is important for soil aeration (Chap. 2, § 2.4; Chap. 3, § 3.10) and the hydric regime (Chap. 3, § 3.4). Earthworms, termites and rodents dig a permanent network, sometimes of great length, augmenting macroporosity by 20 to 100% (Edwards and Bohlen, 1996).

His Majesty the Earthworm, King of Digging.

It is estimated that earthworms dig 400 to 500 metres of tunnels below one square metre of grassland, or an air volume of 5 to 9 dm^3. In the top 40 centimetres, where they are densest, these tunnels represent up to 3% of the total volume. Under these conditions, the water capacity can rise by 80% and penetration of water can be four to ten times faster. Bouché recounts that he once poured more than 100 litres of water into the hole of a large worm in the Basque provinces without flooding the network (pers. comm.). The work of worms has great importance in cultivation of strongly compacted soils with a more or less impermeable plough pan (§ 4.2.2). These

'Development of animal life in the earth's crust, because of the tendency of animals to move in all directions and to dig the soil, will necessarily result in a mixture of organic residues with the mineral soil (...). According to my observations in the beech forest, there must be intervention of an active

two constraints diminish entry of water and increase surface runoff, therefore erosion. Worms perforate the plough pan, thereby improving infiltration of water and offering new routes for root penetration (Fig. 4.37).

element in order for the mixture to become a veritable compost, and it is the earthworm that appears to be called upon for this role' (Müller, 1889).

Fig. 4.37 Maize root having benefited from a worm tunnel for passing through the plough pan (source unknown).

Although quantification is not precise, the excavation work carried out by termites is also very important, comparable in tropical soils to that of earthworms in the temperate zone (Gullan and Cranston, 1994).

Termite galleries sometimes descend 55 metres to reach deep groundwater (Bachelier, 1978).

At their scale, enchytraeids too construct tunnels located close to the surface and to 40-cm depth.

The action of enchytraeids complements that of macro-invertebrates and improves permeability of the soil and its exchanges with the atmospheric air.

Lastly, mammals are not at rest among the miners. For example, a family of ground voles digs networks of galleries 30 to 80 metres long in the top 30 centimetres of soil, occasionally including a vertical tunnel that goes very deep (Meylan, 1977).

Of varying diameter, at the scale of different actors, the burrows constitute a system of drains that collect rain-water and facilitate its flow. Also, water carries fine uncompacted material through these tunnels, which then become the routes for preferential penetration of roots or eluviated clay (Chap. 5, § 5.3.2). Being penetration paths for epigeal invertebrates, the burrows facilitate access to roots and enable deposition of excrements at depth.

All types mixed up, the burrows and their walls are zones of transport and air-soil contact, with an area of nearly 5 m^2 per square metre of soil for a network of total length 380 m.

Fragmentation

Along with bioturbation, fragmentation of litter (Chap. 5, § 5.2.4), dead wood (Chap. 8, § 8.4) and dead bodies (Chap. 8, § 8.3.1) is one of the most important consequences of the activity of the soil fauna. It greatly

Grinders, cutters and fragmenters augment surface area for bacterial and fungal attack.

influences the evolution of soil organic matter and largely conditions the size of bacterial, fungal and microfaunal populations. This effect is due to numerous phytosaprophages, the action of which is specially discussed in Chapter 13, and is present even in 'extreme soils', such as CRYOSOLS for example.

Aggregate formation

Worm casts contain a larger quantity of microaggregates than the surrounding soil.

Once the litter is fragmented, the organic matter is suitable for humification, then for aggregate formation. Earthworms and macroarthropods (chiefly termites, but also diplopods, isopods and chilopods) that ingest soil particles along with their food contribute to this formation by mixing organic matter and mineral materials in their digestive tract (Plate II-3). According to Edwards and Bohlen (1996), soil invertebrates and microflora stabilize aggregates by:

Fragments of excrements of macroarthropods (3 to 5 μm) are stabilized by cohesive forces, which thereby ensure permanence (Webb, in Mattson, 1977).

- intestinal secretions of invertebrates and bacterial colloids from the alimentary canal, which act as cement;
- the network of fungal hyphae and plant fibres originating from the leaves consumed;
- fragmentation, because then formation of organomineral bonds is possible, according to the size and nature of organic particles.

The faecal pellets of microarthropods and enchytraeids are very stable, except that they cannot rightly be called organomineral structures. They are mostly 'micropellets', which differ in nature from the original food (litter or macroarthropod faeces) in pH, water content and microflora. They will be poor in fungi and will have only low bacterial activity. They are found almost intact in the faeces of macroinvertebrates that have absorbed them, thus entering the constitution of aggregates.

4.5.2 Direct and indirect chemical effects

Direct chemical effects

In the case of chemical effects of the soil fauna, it is often difficult to distinguish the specific role of soil invertebrates from that of their associated intestinal microflora (Chap. 5, § 5.2.4). The most prominent chemical effect is modification of quality of the food during its passage through the food-chain (Chap. 13), in particular mineralization of organic matter and consequent release of nutrient ions.

Decomposition of dead bodies restores to the soil the bioelements stored during the life of the animals (Chap. 8, § 8.3.1). In this the addition by the microfauna and mesofauna is quite small. On the contrary, that by earthworms is large. For example, decomposition of one ton of earthworms releases 36 to 60 kg nitrogen; this is equivalent to the nitrogen dose recommended by agronomists for sparsely to semi-intensively grazed pastures, or to 14% of the nitrogen utilized by trees in spring in a tulip-tree forest (McBrayer, in Mattson, 1977). The dead bodies of ground voles (adult weight between 65 and 130 g) also represent a large quantity of organic matter for recycling. Winter mortality may cause the density of ground voles to drop from 1000 individuals per hectare to a very low level in a single season (Meylan and Saucy in Hausser, 1995).

Mucus: Viscous substance comprising mucoproteins. In earthworms and gastropods it is secreted by unicellular cutaneous glands and functions by keeping the tegument moist and facilitating locomotion. It is also produced in abundance by mucous membranes, e.g. intestinal. Not to be confused with mucilage, which is based on polysaccharides (§ 4.1.5)!

The fauna also affects the chemical composition of soil by its excreta which, in six months, will produce a nitrogen flux equivalent to that exported through haymaking: for example, the casts of about 1 t ha^{-1} of earthworms corresponds to 18–50 kg N ha^{-1} y^{-1}. The quantity of ***mucus*** secreted is more difficult to estimate: it will represent daily nearly 0.2% by weight of the total nitrogen in a worm (Edwards and Bohlen, 1996).

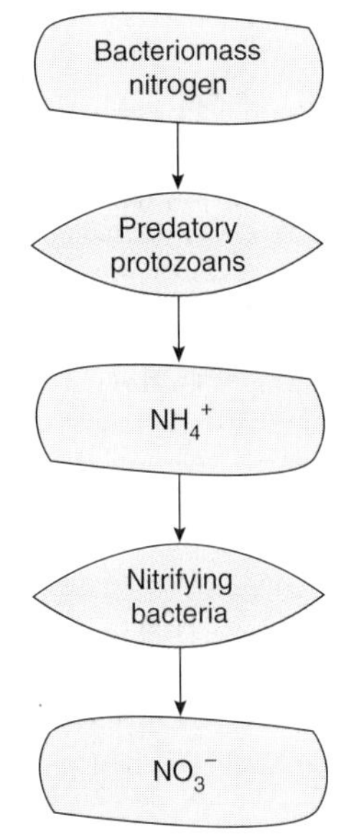

Fig. 4.38 *Effect of predatory protozoans on activation of nitrification. Protozoans are nourished by the entire bacteriomass whereby the nitrogen in it is mineralized to ammonium ions. These ions then serve as a source of energy and electrons for chemolithoautotrophic nitrifying bacteria (§ 4.4.4).*

Indirect chemical effects

Among indirect chemical effects of the soil fauna, that of protozoans is important. They are capable of mineralizing nitrogen, phosphorus and sulphur from their food, that is bacteria that they consume in large quantities (Fig. 4.38). Predatory amoebae thus augment the quantity of nitrate directly utilizable by the plant from the rhizosphere (Chap. 15, § 15.3.2; Benckiser, 1997).

Mites and springtails function in the nitrogen cycle through the fungi they consume and are important bioaccumulators. They select fungal communities striking out old colonies and advantaging the most dynamic (§ 4.4.7). In this manner they intervene in the equilibrium between bacteria and fungi, and indirectly influence decomposition processes (Chap. 13, § 13.6.2).

Enchytraeids exert the same indirect effects as other organisms of the mesofauna. The enzymes secreted by the digestive tract or by the intestinal microflora impregnate the faecal pellets; they continue to function after the pellets have been deposited in the soil (extracellular enzymes; Chap. 14, § 14.2.1).

Enchytraeids will be responsible for 2% of the total mineralization of organic matter in soils in which they are abundant.

Larvae of Sciaridae (Diptera), abundant in moders, speed up the decomposition of oak leaves, which are rich in tannins.

Lastly, some animals are able to metabolize the tannins of leaves rich in them (e.g. those of oak). This is contrary to the general rule that aromatic compounds are inhibitors of bacterial activity and digestion in many arthropods (acidifying litter; Chap. 2, § 2.2.1). The function that normally devolves on bacteria is here passed on to animals.

4.5.3 Biological effects, or living organisms among themselves

Predation: trophic relations between heterotrophic organisms. This definition envisages predation in its broadest sense; others limit it, for example, to the action of animals. Some predator-prey pairs frequent in soils can be mentioned (*predators* in italics): *protozoans*-bacteria, *nematodes*–bacteria; *nematodes*–fungi, *fungi*–nematodes, *springtails* and *mites*–fungi, *ground beetles*–larvae of dipterans, *moles*–earthworms, etc.

Competition: biological mechanism of regulation of populations, tending to maintain the actuals in equilibrium with the capacity of the environment, that is, with the quantity of available resources (food, shelter, room for reproduction, etc.). Intraspecific competition limits the number of individuals; interspecific competition controls the number of species in the biocoenosis.

Intraspecific competition adjusts the numbers to the capacity of the habitat; it depends on population density.

The soil is governed by a complex and dynamic equilibrium among the different compartments of the food-web. This equilibrium is governed partly by the physicochemical conditions of the environment (the 'life frame' or biotope) and partly by biotic factors (interactions between living organisms). Two of the latter are preponderant: predation and competition.

Predation

The effects of ***predation*** on the population of prey are important because they place the latter more or less rapidly in equilibrium with available resources such as food, shelter, etc. (Chap. 13; Chap. 15; § 15.3.2).

Soil predators are very varied, exploiting in all imaginable ways the abundant populations of phytophages, saprophages... and predators. They reduce competition between individuals of the same species and protect in some way the future and quality of action of their prey. For example, in cultivation of wheat without fertilizer on a sterilized and reinoculated soil, the quantity of nitrogen absorbed by the plants, when protozoans and bacteria were present, was at least three times that in control plots where only bacteria were present (Chap. 15, § 15.3.2; Ingham et al., 1985).

Competition

Competition is exercised between individuals of the same species (***intraspecific*** competition) or between those of different species (***interspecific*** competition). In both cases, the individual or species fights to be assured of adequate access to resources of the environment.

Intraspecific competition functions by the play of density-dependent factors of mortality: malnutrition and its consequences, juvenile mortality, cannibalism. For example, the larger a population of woodlice, the faster the fragmentation of litter (positive effect with regard to recycling of bioelements), and the more intense the competition between individuals for food and favourable shelter (negative effects on individuals who die in large numbers). In the medium term, intraspecific competition thus has positive effects on the population of woodlice, which remains in equilibrium with the resources of the environment. In another case, it was observed that the number of sites for common cockchafer larvae was limited in the rooting zone. Intraspecific competition was then manifested between the development stages of the insect: when the number was excessive, old larvae, stronger, eliminated the youngest by scissoring them with their mandibles.

In interspecific competition, two species enter into competition when they fight, directly or otherwise, for the same ecological niche. According to the rule, only one may occupy it at a time, which limits the number of potential species in a given soil (principle of competitive exclusion; Chap. 12, § 12.5.1). Travé et al. (1996) reported an experiment in which two species of mites were raised separately or together. Alone in the environment, each occupied the OL and OF horizons. Together, they entered into competition and both survived by virtue of displacement to their spatial niches, one upwards to the OL and the other down to the OF.

Interspecific competition regulates the occupancy of ecological niches.

It is also known that in the case of African scarabaeids imported into Australia, indirect interspecific competition led to a considerable reduction in populations of biting horn flies (*Haematobia* spp.). Actually the larvae of the latter, maggots, grow in the dung that the scarabaeids quickly deplete by consuming or burying it (Waterhouse, 1974).

If different species occupy neighbouring ecological niches, a competition can occur only for one specific resource: for example, when earthworms monopolize a large part of the litter at the cost of diplopods and woodlice.

Disappearance of carcasses

Necrophages and coprophages are the cleaners of the soil surface. They ensure sanitation by activating degradation of carcasses and, in pastures, by burying the eggs of parasites or pathogenic agents, thus preventing their transmission to cattle. However, Pasteur showed that earthworms could bring up spores of the anthrax bacillus 'eaten' in the buried carcasses of cattle dead of this grave disease. Infection is thus possible during grazing.

A very anthropocentric view: 'cleaning' of the soil.

Dissemination agents

Spores of fungi are activated by passage through the digestive tract of arthropods, enabling their later germination, somewhat like the mistle-thrush dispersing mistletoe.

All organisms of the meso- and macrofauna disseminate spores and bacteria. Earthworms determine their vertical distribution while microarthropods and enchytraeids sow them in small holes. Propagation is effected through casts and pellets dispersed through the soil or by transport on animal bodies. In peat-bogs, some animals disseminate, apart from spores and bacteria, humic compounds adsorbed on the cuticle (Chap. 9, § 9.1.3).

4.6 BIOINDICATION

Besides having truly pedological effects, reviewed in § 4.5, the soil fauna also brings out specific environmental conditions (pollution, microclimatic changes) or significant ecological processes (bioaccumulation). This concept is fundamental to the methods of detection that come under ***bioindication***.

Bioindication: ability of certain species to reflect, by their presence, absence and/or demographic behaviour (change in density), the characteristics and evolution of a milieu (Blandin, 1986). Bioindication can be an efficient diagnostic method in soil biology, and appears to promise great development (Schubert, 1991).

Bioindication is not the opposite of physicochemical analyses, it complements them. It must be based on thorough knowledge of the organisms used and certainly on the systematics and physiology as well as ecology and behaviour. In soil zoology, only a small number of organisms answer to these requirements, which means that bioindication by the pedofauna is still in its infancy.

Most living things can qualify as bioindicators for such or such descriptor. By choice, this section will mostly concern the animal kingdom. The reader will find elsewhere in this book numerous examples of plant bioindication, particularly in Chapter 3, § 3.4.4 (hydric regime), § 4.2.4 (tolerance and ecological range), Chapter 6, § 6.4.2 (humus forms) as well as in the whole of Chapter 7 (soil-vegetation relationships). Some examples of microbial bioindication, less developed than the above for obvious reasons of methodology and fundamental knowledge, are nonetheless provided in Chapter 8, § 8.6.3 (wood fungi) and Chapter 16, § 16.1.6 (mycorrhizae).

4.6.1 An approach based on knowledge of organisms and soil

The ecological niche (Chap. 12, § 12.6) is a concept summarizing the demands of organisms *vis-à-vis* physicochemical and biological descriptors of the environment. It places living things in the ecosystem and defines their habitat, actual or potential. Therefore each organism is the reflection of the habitat it occupies; it thus becomes an indicator species or ***bioindicator***. The same species are found in nature every time living conditions are similar; the 'health' of populations, sensitive to changes in the environment, becomes indicative of natural transformations, pollution and degradation suffered by the milieu they occupy.

Qualities that make certain species sentinels of the environment.

In practice, only specialized species with narrow ecological range or with well-defined strategy have high indicator value. The terms associated with them such as acidophilic, calcicole, hygrophilic, saxicole, xerophilic, etc. are themselves very descriptive, taking into account the exact significance of these terms in ecology (see discussion in § 4.2.4). But generally an assemblage of species better reflects the conditions of the milieu that a single indicator species...this is the basis of phytosociology (Chapter 7, § 7.1.4). However, in the practice of ecological monitoring, most often a restricted number of key species is selected, these species being considered representative of the entire community and having at least one of the following qualities:

- they are easy to recognize, abundant and are clearly positioned in taxonomy (Chapter 17, §§ 17.1.1, 17.1.2);
- they are characteristic of their milieu and are present in several homologous ecosystems;
- their biology and ecology are known;
- they lend themselves to regular monitoring because quantitative sampling methods are available for estimating changes in their populations;
- they prove to be sensitive to pollution and to changes in the their environment by reacting progressively to the growing impact of a disturbance.

'The health budget of ecosystems is an integrated approach that emphasizes the "quality" of ecological diversity and provides the lines of conduct for restoration of damaged ecosystems (...). A healthy system is defined as one capable of maintaining its structure and functional autonomy over time (...). the *biological integrity of ecosystems* may be defined as the capacity of the milieu to shelter and to maintain a balanced and adapted community of organisms having a diversity and functional organization comparable to those of the natural habitats of the region (or at least of the least disturbed habitats)' (Levêque and Mounolou, 2001; also see Chap. 7, § 7.1.4).

In hydrobiology, the idea of using sensitive organisms to characterize waterways and to detect changes in the milieu, particularly those due to the human activities, is at least 150 years old. This path of research is well developed and today there are more than a hundred methods for qualifying continental waters (zonations, biotic indices, saprobic systems, etc.).

Bioindication: born in water ...

In the terrestrial ecosystem, description of the natural habitats rests primarily on the vegetation (Chap. 7, § 7.1.4), and that of their soils generally makes use of physicochemical methods. The fauna, less understood in its totality, is a descriptor more difficult to handle. However, certain zoological groups such as birds,

... and slowly developed to the terrestrial milieu.

butterflies, ground beetles, orthopterans and molluscs are more and more being used in this direction (Baur et al., 1996; Gonseth and Mulhauser, 1996; Delarze et al., 1998; Lugon et al., 2001).

Three forms of bioindication can be thought of:

- determination of composition of reference communities inhabiting milieus 'in good health'. From this it is possible to estimate their natural evolution in the medium and long term (***ecological bioindication***, § 4.6.2);
- evaluation, through changes that have taken place in the composition of communities, of the magnitude of anthropic impacts (pollution, compaction, drying out, etc.) and their consequences for the ecosystem in the short term and medium term (***perturbation bioindication***, § 4.6.3);
- use of reference species that reveal contaminants present in the soil or used as test organisms in the laboratory (***toxicological bioindication***, § 4.6.4).

4.6.2 Ecological bioindicators in soil

Compared to hydrobiology, soil biology shows some tardiness in the domain of bioindication, chiefly because no general method as precise as that of biotic indices for water is at the moment applicable. But, considering the importance of its practical fall-out, this direction of research should be given priority. Lack of knowledge of the pedofauna is the principal cause and explains why few attempts have been made in this direction (see, for example, Ducommun, 1991).

For want of such a well-structured method, we mostly find in the literature reference to specific cases. We have chosen a few designed to illustrate the utility and also the complexity of ecological bioindication in soils:

Plants and figures ...

Ellenberg (1979) proposed a 10-points system including some bryophytes.

- Landolt (1977) assigned an index value to pteridophytes and phanerogams of Switzerland based on ten descriptors: water content and acidity of soil, content of nutrients (nitrogen most of all) and humus, soil aeration in relation to particle-size distribution, salinity, light intensity, mean soil temperature during the growing period, continentality (evaluated from temperature and humidity ranges) and biological forms. A scale of values from 1 to 5 is used to quantify each descriptor. Thus, each species can be characterized by a formula such as 3–1–2–5–4—2–3–3–z for *Vaccinium myrtillus*. In concrete terms, this signifies that this species is a ***subfrutescent chamaephyte*** (z) growing on non-saline, moderately dry, acid, nutrient-poor, humus-rich soils low in coarse fragments. It avoids highly continental climatic conditions, its sites of occurrence are shady and located in the mountain zone. Thus, bilberry characterizes caducifoliate forests on acid soil (beech and chestnut groves), spruce plantations and peat-bog forests. This approach should be compared with that proposed for animals in Chapter 12, § 12.6.1.

Chamaephyte: bushy plant with buds located below 25–50 cm height and therefore protected from frost by the snow layer during winter. Chamaephytes are described as ***subfrutescent*** when their stems are entirely lignified.

- Baur et al. (1996) compared the floral and invertebrate populations of three dry grasslands on calcareous soil, on the southern foothills of the Swiss Jura mountains. Baur's aim was to demonstrate the homology of the flora and fauna at three edaphically and climatically close sites separated from each other by about ten km. The fidelity of biocoenoses to this type of biotope overall is variable, but certain groups have very good bioindicative capacity [vascular plants (54% of the species living at the three sites), gastropods (59%) and other lesser ones (mites, 32%; ground beetles, 18%; diplopods, 12%)].

Attention should be drawn to the monumental work, very useful in bioindication, which assembles knowledge of the coleopterans of Central Europe in thirty or so volumes (Freude et al., 1965*ff*). Nine volumes (nearly 3000 pages) cover current knowledge of the ecology of coleopterans, including coleopterans of the pedofauna, and of the associations of species by milieu.

Vascular plants, snails and slugs, good indicators of natural grasslands.

- Also located on the southern slope of the Jura mountains are six well-defined forest phytocoenoses forming a vertical zonation (Chap. 5, § 5.5.1) from the hill stage to the upper mountain stage. A total of 36 species of ground beetles and 26 diplopod species were captured from that toposequence by means of pittfall traps (Chap. 12, § 12.2). Each phytocoenosis comprises a community composed of a mixture of ***ubiquitous*** and characteristic species, the bioindication manifesting itself mainly by differences in species abundances between the milieus.

The composition of these macroinvertebrate communities is clearly related to altitude, which influences the temperature and water content in the soil, and, indirectly, the structure of the litter (Borcard, 1981, 1982; Pedroli-Christen, 1981).

Ubiquitous: said of a species undemanding with respect to environmental conditions, therefore found in widely different habitats.

- Other investigations on cultivated soils have confirmed that ground beetles, diplopods and staphylinid beetles are good indicators of edaphic microclimate (Chap. 3, § 3.5.3) and that the composition of their populations is influenced by altitude, soil moisture content and plant cover, in that order (Matthey et al., 1990). Thus *Bembibion obtusum* is a small carabid beetle of pasture lands with dense vegetation, whereas *Bembibion andreae* characterizes cereal fields where the stems are sparser.

Earthworms, champions of bioindication!

• Earthworm populations vary in quality and quantity according to nature of soil. Delhaye and Ponge (1993) showed that in the climactic beech grove of Tillaie reserve in the Fontainebleau forest, only epigeal earthworms are found in PODZOSOLS, while NÉOLUVISOLS contain only endogeal earthworms. In the same beech grove, Arpin et al. (2000) compared the stages in recolonization of clearings and the composition of worm populations. In recent treeless clearings, epigeal worms diminished considerably following reduction in additions of litter. Recolonization of clearings was accompanied by a depression in anecic, mostly endogeal worms, while epigeal worms returned in force. This evolution was attributed to exhaustion of mineral reserves in soil by growing trees and to return of litter. By integrating several groups of the pedofauna (nematodes, earthworms, microarthropods), these authors demonstrated a parallel evolution of the natural forest cycle and an edaphic cycle associating dynamics of humus and of animal populations. They underline the essentially indicative role of the soil fauna in the framework of state of the milieu or its dynamics.

In the grasslands of the Swiss Plateau, *Lumbricus terrestris* is practically the sole representative of anecic *Lumbricus* spp. Its biomass serves as an index of water-holding capacity, which could be related to resistance of the soil to erosion and compaction (Cuendet et al., 1997). Moreover, earthworm density tends to diminish with increase in altitude (Table 4.5).

Table 4.5 Altitudinal changes in earthworm populations in eleven permanent grasslands in western Switzerland (Matthey et al., 1990)

Altitude (m)	Density (individuals m^{-2})	Biomass (g m^{-2})
445	345	272
730	95	140
1140	29	34

A collective memory of springtails and earthworms?

• Microarthropods are good indicators of soil structure. According to Dunger (1982), communities of endogeal springtails characteristic of viticulture soils can persist for at least two centuries after use of the land in which they live had changed and the vegetation transformed. Cuendet (1985) pointed out the same 'memory effect' in earthworm communities of the meadows of Lower Engadin (Switzerland), where species associated with human activities (*Lumbricus terrestris, L. castaneus*, etc.) survived long after the situation had changed. Although this long inertia can help in reconstituting the history of a soil, it upsets the clarity of bioindication by provoking a faunal mixture, for example in grasslands under rotation.

4.6.3 Bioindicators of disturbance

These bioindicators are particularly useful in permanent programmes of environmental monitoring. Ramade (1993) divided them into two incompletely watertight categories.

Bioindicators of anthropization

Starting from zero time, ***bioindicators of anthropization*** reveal, among others, the effects on pedofloral and pedofaunal activity of the upkeep of ski runs on Alpine and mountain soils, of reparcelling and multiplication of concreted byways, of putting grasslands to grazing or cultivation and also of compaction and erosion of cultivated soils. Thus:

Bioindicators of anthropization: organisms responding to physical changes caused by man (Ramade, 1993).

- The biomass and density of earthworm populations are greatly reduced by deep ploughing (25 individuals m^{-2}, weighing 15 g) in comparison to direct sowing (200 individuals m^{-2}, weighing 152 g) (Matthey et al., 1990).

- Grazing intensity favours certain anecic earthworms such as *Nicodrilus* spp. This is in all probability due to the fact that crushed and fragmented grass mixed into the surface soil by the hooves of cattle forms a food source more beneficial to these species than to other anecic species such as *Lumbricus terrestris*. Besides, these *Nicodrilus* would have better aptitude for a soil compacted by trampling (Cuendet et al., 1997).

Pollution bioindicators

Among known examples of the use of ***pollution bioindicators*** can be mentioned that of communities of urban lichens growing on trunks and walls (soil annexes, Chap. 8, §§ 8.2, 8.5), whose modification or disappearance is indicative of the intensity of atmospheric pollution by SO_2, known since the nineteenth century (Ramade, 1992). Similar observations have been made in forest in spread of industrial fumes loaded with sulphur dioxide. Impact on the pedofauna has also been the object of many investigations, for example:

Pollution bioindicators: organisms that generally react negatively to application of pesticides and excessive spreading of inorganic fertilizers or manures, or to air pollution (Ramade, 1993).

• The spread of fluoridated fumes in Fricktal (northern Switzerland) affected the pedofauna of the beech groves of neighbouring regions. In order to locate the boundaries of the impact, soil arthropods were counted at different distances from the source (Bader, 1974). Microarthropods and insect larvae, that is the most sedentary organisms, were the most clearly affected by the fumes (Table 4.6).

Table 4.6 Impact of fluoridated fumes on density of pedofaunal organisms (Bader, 1974)

Distance from source	0.3 km	3 km	6 km
	individuals m^{-2}		
Myriapods	29	40	22
Spiders	6	10	18
Oribatid mites	257	493	3936
Other Acari	108	318	852
Springtails	547	891	1168
Insects (imagos)	79	27	136
Insects (larvae)	13	29	72

• Indicator species can quantify the degree of pollution of a milieu by their demographic behaviour (change in density or abundance). From this viewpoint, Freuler et al. (2001) proposed consideration of the abundance of *Bembibion quadrimaculatum* as reliable indication of the impact of springtime surface treatment with broad-spectrum insecticides in market-garden cultivation.

The evolution of populations of springtails and mesostigmatic mites, respectively prey and predator, following insecticide treatment is characteristic of this kind of situation (Fig. 4.39).

Bioindication, bioaccumulation, bioconcentration and bioamplification . . . here too bio is in fashion!

4.6.4 Ecotoxicological bioindicators

Ecotoxicological bioindication rests on a very important ecological process: ***bioaccumulation*** or ***bioconcentration*** (we consider these two terms synonymous, though all authors do not). It was demonstrated for the first time at Clear Lake, California, in a case that has made a mark in ecology. From 1949 to 1957 many applications of DDD were done to water to fight a dipterous insect with aquatic larvae, non-stinging but considered insupportable by tourists because of its rapid multiplication. This resulted in high mortality of some 3000 grebes with just about thirty sterile pairs surviving. To understand this disaster,

Bioaccumulation (or ***bioconcentration***)***:*** process of increasing concentration of xenobiotic substances in the tissues of organisms along a food chain. The principle is applied particularly to potential pollu-

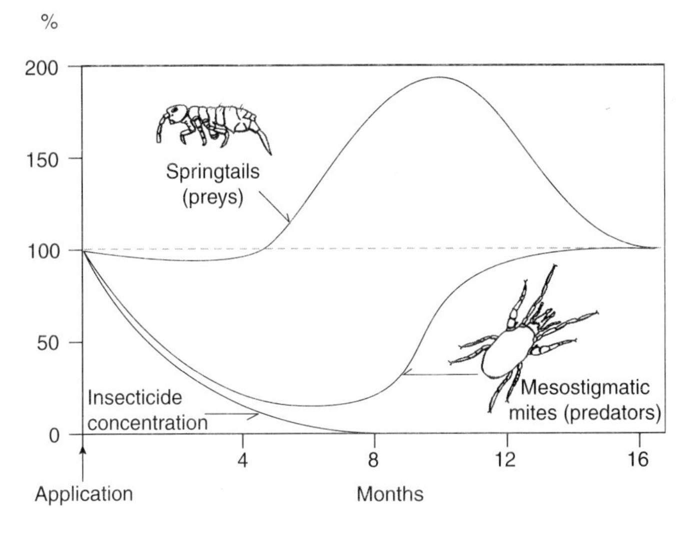

Fig. 4.39 Characteristic demographic curves of mesostigmatic mites and springtails after insecticide application (from Massoud, in Pesson, 1971). 100% = populations at time of application.

the concentration of DDD at different levels in the food chain *phytoplankton* → *zooplankton* → *microphagous fishes* → *carnivorous fishes* → *grebes* was measured. Published in 1960, the figures are eloquent: the concentration went from 0.014 mg kg^{-1} in the water to 2500 mg kg^{-1} in the fat tissue of grebes, or a 178,500-fold concentration.

From many such examples a pyramid of concentration of ***pesticides*** can be established, the reverse of that of numbers or biomasses (Fig. 4.40) and the process observed at Clear Lake can be generalized: in the food chains and food webs, an amplification of concentration of toxic or radioactive substances persisting in the biomasses is observed from the base to the top of the ecological pyramid (§ 4.3.5; Chap. 13, §§ 13.1.1, 13.3.1).

tants (organochlorines, heavy metals, fluorine). As these bioconcentrating organisms accumulate these substances in concentration sometimes much higher than those in the surrounding environment, we can also say ***bioamplification***. A broad review of the problems linked to bioaccumulation has been given by Arndt et al. (1987).

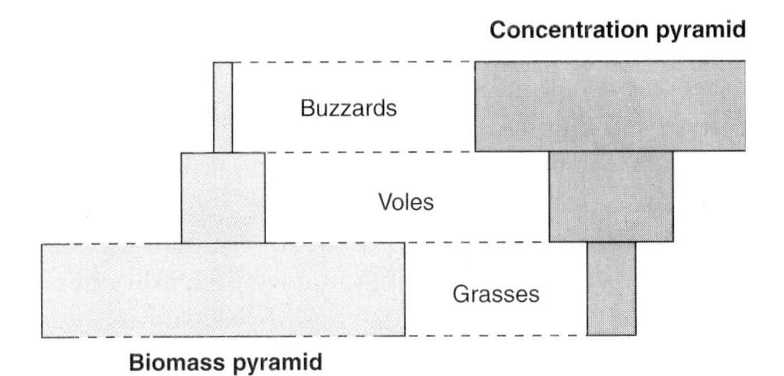

Fig. 4.40 *During passage from one trophic level to another, more than nine-tenths of the biomass is lost. It is spent in respiration, in excrement or is not eaten (rosulate leaves, bones, skin). Loss of organochlorines is less; actually they are not eliminated by respiration, hardly through excrement and are concentrated in the fat tissues eaten first (diagram not to scale).*

Xenobiotics, molecules foreign to natural milieus, often dangerous and always difficultly degradable (see also Chap. 11, § 11.3).

Pesticides, which wish us well, but ... !

Pesticides (from L *pestis*, pest; *occidere*, kill): toxic substances used by man to fight pests, a general term designating organisms harmful to man (insects, fungi, nematodes, etc.) directly—medical domain—or indirectly—agricultural and veterinary domains. According to the organisms they work against, pesticides are called acaricides (acari), bactericides (bacteria), herbicides (weeds), insecticides (insects), nematicides (nematodes) or rodenticides (rodents). These compounds are deliberately spread in the environment unlike heavy metals, PCBs or fluorine, that are industrial wastes, or radioactive substances accidentally or militarily spread in the environment. Fertilizers are not pesticides, being used to directly augment crop yields. Pesticides give protection to crops and thereby indirectly increase crop yield.

Bioconcentration is better known in aquatic systems than in terrestrial milieus.

Insects ... a healthy diet?

The most numerous quantified examples pertain to the chains of predators born in aquatic milieu, where it is easier to measure initial concentrations, sometimes also easier to determine the structure of food webs. In soil, the predation food chains, often branching into decomposition food chains, may be as long as in lacustrine or marine milieus, as in the example *litter* → *decomposers* (e.g. *woodlice*) → *predators* (e.g. *woodmice, shrews*) → *superpredators* (e.g. *long-eared and tawny owls, stoats*); however, more than two successive concentration levels have rarely been quantified.

Three major themes guide research in the domain of ecotoxicological bioindication:

- effects of toxic substances, chiefly organochlorines and heavy metals, on population dynamics (bioindicators of ecological effects);
- tolerance of organisms and determination of lethal doses at species level.

Edwards and Bohlen (1996), for example, gave a table of toxicities to earthworms (bioindicators of toxicological effects) of chemical substances used in agriculture.

- the capacity of species to concentrate toxic substances (accumulation bioindicators).

Organochlorine compounds in soil

A very large number of studies are available on the effect of organochlorines on food chains, all of which begin in the soil. Six examples illustrate this important and worrying problem, while an introduction to these substances is given in a box.

Soil *arthropods* are more sensitive than earthworms and molluscs to pesticides; evolution of their strength is thus a more convincing indication of pollution by these substances than the amounts stored in their tissues. Generally *holometabolous insects* are not regarded as good bioaccumulators (Mayeux and Savanne, 1996), even if they are able to accumulate high concentrations neurotoxic organochlorine compounds in their fat-bodies. Consequently, if we accept Hauser's (1995) statement that the daily intake of a common shrew is the equivalent of 2000 coleopterans 5-mm long, some accumulation through the food chain can be presumed!

Acronyms and compounds.

Organochlorine compounds

Among the pesticides dreaded for their long-term biocidal effects, ***organochlorine compounds*** are very often mentioned. They are synthetic organic compounds obtained by chlorination of unsaturated hydrocarbons. Long-persisting because non- or only partially biodegradable, they may persist in the soil longer than 20 years. Some of these compounds with complicated names are designated by their acronyms such as

- DDT for dichloro-diphenyl-trichloroethane
- DDD for dichloro-diphenyl-dichloroethane
- DDE for dichloro-diphenyl-dichloroethene
- BHC for benzene hexachloride
- HCH for hexachlorocyclohexane.

Others have names such as aldrin, dieldrin, chlordane heptachlor, lindane, etc.

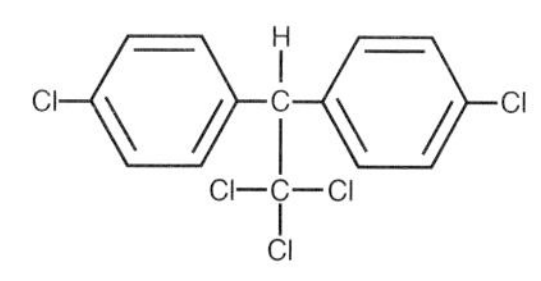

Fig. 4C. *Chemical formula of DDT.*

The first synthetic insecticide was DDT, discovered in 1932 by the Swiss chemist P.H. Müller, who received the Nobel prize for this discovery. It works as a contact and ingested poison, showing great affinity for fat tissues and easily passing through the cuticle of arthropods. It paralyses the nervous system. This property makes it a dread weapon against many disease vectors and crop pests. Inexpensive moreover, it was used the world over for decades, even after problems of ***resistance*** in invertebrates were discovered and those linked with bioconcentration in vertebrates including man. DDT and most other similar insecticides were gradually banned in North America and Europe. However, by 1990, exports to developing countries still had not reduced (Goldsmith and Hildyard, 1990).

Resistance: capacity of certain individuals in a genetically heterogeneous population to survive normally lethal doses of biocides. It leads to selection of resistant lines among invertebrate vectors or pests. Resistance to insecticides may be acquired by a population in about 15 generations. Houseflies, sometimes used as bioindicators, today tolerate organochlorine doses 1000 times the initial lethal dose. This resistance is shown to most commercially available insecticides.

The polychlorobiphenyls (PCBs) are chlorine compounds widely used in industry (paints, insulators, etc.), but their manufacture has been banned for about twenty years. However, they are still introduced into the environment through poorly monitored wastes, but mostly by destruction of used mineral oils and by discarding after use transformers, condensers and hydraulic systems (2–6 t y^{-1} in Switzerland, according to Müller, 1964; 600 t y^{-1} in the United Kingdom according to Goldsmith and Hildyard, 1990). The PCBs are very persistent and accumulate in soils and lake and sea bottoms, then are concentrated in food webs like other organochlorines, with the same consequences: concentration coefficients of 80,000,000 have been measured in marine milieu.

DDT and PCB are examples of ***ubiquitous*** compounds, spread all over the earth in low concentrations to be later accumulated in food webs.

Bioaccumulation in terrestrial systems often starts with earthworms.

Bioaccumulation of organochlorines has mostly been studied in *earthworms*. These concentrate from five to ten times the levels measured in soil, while the latter levels were 0.001–0.01 mg kg^{-1} (Edwards and Bohlen, 1996). The concentrations vary with earthworm group, the highest having been measured in an epiendogeal species *Allobophora chlorotica* (Chap. 12, § 12.4.3). They also vary with compound. Old determinations show in the case of earthworms a DDT concentration level 150 times that in the surrounding environment. The effects on food webs outside the soil are large, because earthworms are at the base of several chains in which the predator and superpredator levels are occupied by birds or mammals.

From soil to the hawk and little owl.

The case of diurnal and nocturnal *birds of prey* has been cited in all books on ecology, that of the peregrine being the best known. The effect of organochlorines is manifested mostly by great mortality at the egg stage, followed by disturbance of functioning of the shell glands. Although the organochlorine level in eggs of predatory birds has been determined in numerous studies, the origin of the poisons is often difficult to locate (Juillard et al., 1978). Analysis of 254 unhatched eggs of little owl (Juillard, 1984) showed that all contained on average 1 mg kg^{-1} DDE (in Germany up to 58 mg kg^{-1} of BHC), and that 93% of them contained traces of PCB. As the little owl feeds 65% on earthworms (in Switzerland), the earthworm-owl link in bioconcentration is more than probable.

To illustrate the problem of bioconcentration in terrestrial mammals, we must take recourse to an aquatic mammal!

Data regarding the otter are the best for illustrating the effects of bioconcentration of PCBs in *mammals*. Otter populations in Europe are in great decline. It can be said that wherever PCB in the fat of the otter reaches a level of 0.5 mg kg^{-1}, reproduction is upset and the populations decline. In Switzerland in 1990, reintroduction of this carnivore was abandoned as the PCB levels in the environment did not show any sign of decreasing (Weber, 1990), Terrestrial mammals seem to be less studied but the food regime of many of them—the European mole, common shrew, Eurasian badger—includes a large proportion of earthworms. We may then presume that they absorb some quantity of pesticides with their food. A mole, for example, an essentially drilophagous animal, consumes about 50 g earthworms daily. According to the milieu, it surely concentrates large quantities of pesticides; astonishingly few data on the physiological consequences of this poisoning are found.

Lettuce, slugs and hedgehogs.

Among *gastropods*, slugs are the best accumulators of organochlorines, showing concentrations four times higher than earthworms in the same milieus. Almost all gastropods consume their own faeces, thereby constantly enhancing the pesticide content in the hepatopancreas and gonads particularly. Under laboratory conditions, concentrations up to 648 mg kg^{-1} have been found in slugs fed on lettuce treated with DDT. In crop fields in the USA, slugs contain up to 200 mg kg^{-1} organochlorines against 50–100 mg kg^{-1} in earthworms (Godan, 1983). Hedgehogs being great consumers of slugs, it is not surprising that they accumulate lethal doses of organochlorines in the vicinity of urbanized zones where levels in the soil are particularly high (Berthoud, 1982). For this species, poisoning is one of the two major causes of mortality, the other being highway traffic.

Because of effects on animals and also possible risks for *man*, most of the industrialized countries have regulated the use of organochlorines on their territory since 1972. The situation has improved because there is less poisoning of wild species by these compounds. However, we should remember that their great persistence (more than 20 years for DDT) means that at present soils of the world still retain large quantities, some 300,000 t according to Bliefert and Perraud (2001). But other compounds have taken over with the same effects.

And ... man?

Bioaccumulation of heavy metals and fluorine

Although some of the problems caused by heavy metals are becoming less acute, for example those related to lead near highways in countries that have adopted unleaded petrol, pollution by these elements remains on the whole a broad topic.

In spite of slight improvements, pollution by heavy metals remains a highly topical issue.

Three groups of bioaccumulating invertebrates forming part of the pedofauna are specially studied in this problem: earthworms, woodlice and diplopods, microarthropods (springtails and mites among others). They have a common factor in feeding on litter and having a low assimilation ratio for food; this compels them to consume a lot and to produce abundant faeces, whence their importance as fragmenters (§ 4.5.1). They are the prey of such varied predators as moles, foxes, shrews, hedgehogs, blackbirds, woodpeckers, little owls, gulls, ground beetles, spiders and even mesostigmatic mites.

The theme of heavy metals in soil is also discussed in Chapter 11, § 11.2, devoted to phytoremediation of contaminated soils, while § 4.3.4 presents the role of elements including some heavy metals in physiology of the plant.

According to van Straalen et al. (2001), the best bioaccumulators of heavy metals are the invertebrates that store these substances taken from soil water, particularly earthworms and certain small springtails. But isopods and diplopods are equally efficient in this domain.

The ability of earthworms to concentrate pesticides has already been mentioned. It is found again *vis-à-vis* heavy metals present in soil and litter. For example, near highways their tissues accumulate lead at levels several

Where earthworms are found once more in the front row!

orders of magnitude higher than in soil. Edwards and Bohlen (1996) proved this for *Lumbricus rubellus*, an epigeal species (Chap. 12, § 12.4.3) in which the bioaccumulation factor (ratio of concentration in mg kg^{-1} in worms to that in soil) is 9 to 188 for cadmium, 2.8 to 8.3 for zinc and 0.08 to 0.18 for lead. However the resistance of earthworms has limits:

- soils containing 80 mg kg^{-1} copper (from fungicides) do not contain earthworms;
- regular application of pig slurry containing more than 1000 mg kg^{-1} copper leads to large decline in earthworm populations;
- earthworm populations are affected by the following concentrations (in mg kg^{-1} soil): Cd>33, Cu>287, Pb>4800. According to other sources: Cu>78, Pb>36, Zn>171;
- in a soil polluted by brass (alloy of copper and zinc), the LD_{50} is 190 mg kg^{-1};
- reproduction of *Eisenia fetida* is inhibited at the following concentrations (in mg kg^{-1} soil): Cd 2000, Cu 50, Ni 400, Zn 5000;
- growth of *Eisenia fetida* is inhibited at the following concentrations (in mg kg^{-1} soil): Cd 1800, Cu 1100, Ni 1200, Zn 1300.

The 'common' woodlouse, an unequalled accumulator of fluorine!

Woodlice and *diplopods* have physiological mechanisms for regulating the level of heavy metals and fluorine whereby they can survive high levels in litter and soil: storage in fat body, in cuticle impregnated with calcium carbonate or in the hepatopancreas. Among decomposer arthropods, the woodlouse *Porcellio scaber* (Isopoda) is a common species often used as reference animal (see box) and has been studied near metallurgical works. We may cite here two examples in which this crustacean has been used as bioindicator:

- Pollution of soil by gases rich in fluorine carried to the environment from the aluminium-smelting works at Steg (in canton Valais, Switzerland) has been analysed by Contat et al. (1998). In 1994, this establishment emitted 7680 kg of gaseous and particulate fluorine compounds. The major portion was deposited in the predominant wind direction 2 km from the factory, where a very sharp accumulation of fluorine in decomposers was noted compared to that in the soil (Table 4.7). Among the non-selective predators of woodlice, only spiders showed relatively higher levels, as high as 537 mg kg^{-1}. Vertebrates were not studied.

Table 4.7 Fluorine contents measured 425 m from the works, the site of maximum deposition.

	Total fluorine	Soluble fluorine
	mg kg^{-1}	
Soil (A horizon)	833	129
Litter (OL horizon)	1329	190
Isopods (*Porcellio*)	2234	n.d.
Diplopods (Iulidae)	3172	n.d.

Toxicological bioindicators

Biological accumulation takes place in all species, but this property is particularly developed in some of them. Affected invertebrates are used on land as pollution indicators; in laboratories these are 'guinea pigs' that enable testing the toxicity of compounds. Their ability to concentrate must be well understood, even their systematic position (species, groups of species, ecotypes) and their biology (growth rate at given temperatures, number of eggs, food regime, physiology of nutrition and excretion, ecological range in natural conditions *vis-à-vis* temperature, humidity and light). Also, these bioindicators should support standardized study conditions in the laboratory, consequently only a small number of species, belonging to different phyla, satisfy these requirements.

In soil, we love squealers ... in the laboratory too!

In the tests, two parameters are generally studied:

- Mortality in the population. The ***Lethal Dose 50*** or ***LD_{50}*** is calculated. This, expressed in mg kg^{-1} body weight, corresponds to the quantity of toxic substance that leads to the death of 50% of the individuals tested in a given time.
- Sublethal effects, which do not directly lead to death of individuals. This term primarily designates upsets in reproduction and growth caused by ingestion of toxic compounds or by contact with them. The concentration at which disturbance of these vital processes appears in 50% of the individuals tested is determined. The ***Efficacious Concentration 50*** or ***EC_{50}*** expresses this limit.

Lethal dose or efficacious concentration?

Seven reference species are more particularly used in tests of toxicity (Mayeux and Savanne, 1996):

Species	Observed descriptor
Cognettia sphagnetorum (Enchytraeidae)	Growth, reproduction
Eisenia fetida (Lombricidae)	Mortality, reproduction
Helix aspersa with two subspecies *H. a. aspersa* and *H. a. maxima* (Gastropoda)	Growth
Folsomia candida (Collembola) (Fig. 4D)	Reproduction
Platynothrus peltiger (Oribatida, Acari) (Fig. 4D)	Growth, reproduction
Hypoaspis aculeifer (Gamasida, Acari)	Growth, reproduction
Porcellio scaber (Isopoda)	Growth, reproduction

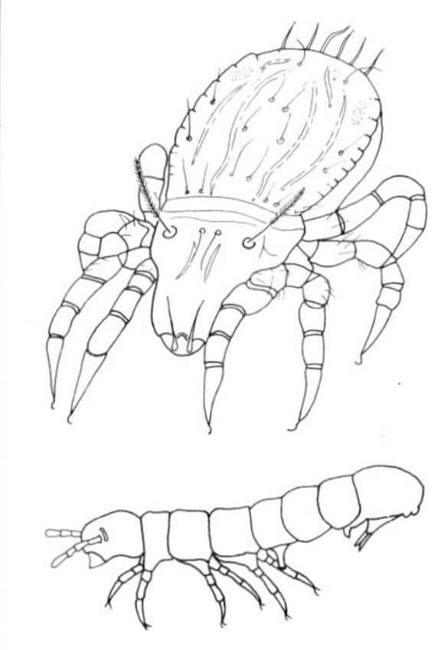

Fig. 4-D Platynothrus peltiger *and* Folsomia candida, *a mite and a springtail highly valued in ecotoxicology.*

In the laboratory, the following characteristics should be standardized (from Mayeux and Savanne, 1996):

- species and stage of development,
- mode of contamination and the method of study,
- duration of the experiment.

For example, the recommended conditions for laboratory testing of the influence of a xenobiotic on growth of *Porcellio scaber* are:

- development stage: 10–16 weeks (weight 20–40 mg);
- food: mixture of oak litter, rabbit food and potato powder;
- mode of administration of tested compounds: added to the food;
- conditions of test: temperature 20°C, in glass petri dishes, on a plaster substrate saturated with water (water content >90%);
- duration of test: 8 weeks.

• In the Netherlands, isopods collected from a site contaminated by industrial fumes bore in their tissues concentrations of zinc three times those in a reference population sampled from an uncontaminated milieu (Donker et al., 1996). When the food contains high levels of zinc, its consumption does not decrease but its assimilation is reduced in isopods of the contaminated zone without their growth being affected. The authors saw in this an adaptation for survival in the severely polluted sites.

Oribatid mites, microarthropods very useful at small scale.

According to Lebrun and van Straalen (1995) and van Straalen (2001), *mites* can be indicators of localized accumulations of heavy metals in soils. By virtue of their high population density, therefore of their important part in the second compartment of the detritus food chain (Chap. 13, § 13.7.3), and also of their limited ability to move, they enable study of soils at rather small scale. Zaitsev and van Straalen (2001) showed this in a zone polluted by deposition from the metallurgical works of Kosogorsky in Russia. As for cadmium and manganese, *Platynothrus peltiger* (see box) seems to be one of the best concentrators among the microarthropods. Its lethal dose, very high compared to cadmium goes up to 234 mg g^{-1}. Mycophagous oribatids are the best indicators of zinc, this metal being concentrated by fungi in the first stage. Some other studies, not cited here, pertain to springtails, *Folsomia candida* in particular (see box).

Vertebrates at the end of the chain.

Birds of prey at the top of the food pyramids often contain in their eggs high levels of heavy metals, though no reliable correlation can be established between the content and mortality of eggs. The species mentioned in Table 4.8 maintain a trophic relationship with soil through rodents, insectivores and birds that feed on the ground.

The high mercury contents measured in the case of the peregrine falcon and the sparrow-hawk actually shows that both often hunt birds that feed in the agricultural zone on cereal seeds coated with mercury salts (protection from rodents).

A parallel should be drawn between the behaviour of organochlorines and heavy metals in the grazing food chains.

Table 4.8 Average heavy metal contents measured in eggs of birds of prey in Western Europe (Juillard et al., 1978).

	Mercury	Lead	Cadmium
	mg kg^{-1}		
Peregrine falcon	0.216	0.163	0.014
Kestrel	0.026	0.173	0.002
Common buzzard	0.056	0.162	0.003
Red kite	0.040	n.d.	0.001
Sparrow-hawk	0.930	0.230	0.006
Long-eared owl	0.035	0.155	0.001

Bioconcentration and its effects on predatory mammals are still debated, particularly in the case of ***homothermal*** animals that appear to excrete toxic compounds more efficiently than ***heterothermal*** (Laskowski, 1991). However, other authors (van Straalen and Ernst, 1991) opine that an animal with very specialized food regime such as the mole can accumulate heavy metals from contaminated prey to levels above the pathological tolerance limits.

Homothermal: said of an animal that maintains its temperature constant. ***Heterothermal*** or ***poikilothermal:*** said of an animal whose temperature changes according to the external temperature.

The food webs studied in forest ecosystems by Grodzinski et al. (1984) however do not allow us (yet?) to adopt without restrictions the idea of bioaccumulation of heavy metals in terrestrial webs, in contrast to measurements made of mercury and other toxic heavy metals in aquatic milieu (Fig. 4.41).

4.7 CONCLUSION

This chapter presented the living actors of the soil system. Plants, bacteria, fungi and soil animals play a role that no physicochemical process can replace: to be the principal internal motor for soil formation. Admittedly, climatic and mineralogical factors among others fix the frame within which living organisms evolve. But only the latter are able to actuate, directly or indirectly, processes as essential to pedogenesis as nitrogen fixation, digging of burrows or secretion of

Life in action? The principal internal motor for soil formation!

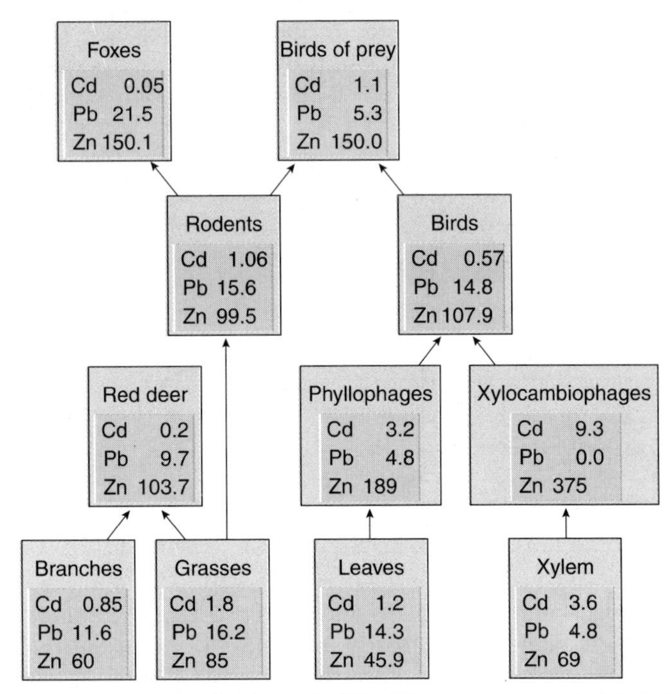

Fig. 4.41 *Heavy metal levels (mg kg^{-1}) in different compartments of the food web in the Niepolomice forest (Grodzinski et al., 1984).*

enzymes. They become adapted to changing environment and thus become testimonies of natural evolution as well as anthropic modifications of their habitats.

In common with abiotic agents, enable the meeting of mineral and organic matter during different phases of soil formation and evolution. This is what we shall see in the next chapter.

CHAPTER 5

FORMATION, DEVELOPMENT AND CLASSIFICATION OF SOILS

This chapter presents soil formation and development—***pedogenesis***—with its constraints, general features and also its exceptions. Special emphasis is placed on incorporation of organic matter and the role of organisms, two fundamental aspects for understanding clearly the thematic modules of the second part of the book. A glimpse of soil classification concludes this chapter.

Physical and chemical constituents, organisms and properties: a soil in its totality and its dynamics!

Entropy: thermodynamic parameter that enables evaluation of the degradation of energy and characterization of degree of disorder in a system. The greater the disorder, the higher the entropy.

5.1 BASIC PRINCIPLES AND PHASES OF PEDOGENESIS

5.1.1 The soil thermodynamic system

Entropy and soil development

Ecological systems, for example the soil (Chap. 1, § 1.2.2), are open thermodynamic systems, exchanging matter and energy with the outside (Odum, 1971; Runge, 1973; Schwarz, 1988; Hoosbeck and Bryant, 1992). The entry of energy in a noble form capable of doing work enables the soil system to oppose increase in ***entropy***, which is the rule in closed systems, and to organize itself.

In a closed system, entropy actually can only increase because the energy available to do work—free energy—is degraded to residual energy of lower quality, namely, heat. On the contrary, in an open system, as are all

'By analysing the manner of external agents, we have always seen them, on their own, tending towards a state of relative equilibrium in which, if the external circumstances remain constant, their mechanical power will be at least considerably reduced, if not annihilated. In regions where this equilibrium has been established, the play of external dynamics is not suspended , but takes on a new shape characterized by the intervention of living organisms.' (De Lapparent, 1911).

'A well-conducted study of biology should have as a starting point the theme of energy and its transformations' (Lehninger, in Lebreton, 1978). The latter, in his book, makes an excellent presentation of the application of the principles of thermodynamics to the ecosystem.

'Little is to be gained by considering the soil without the plant' (Addiscot in Bryant and Arnold, 1994): in the ecosystem, the build-up of biomass is the biological activity *par excellence* that enables the soil subsystem to lower its entropy.

The Earth's attraction, with energy such as that of the wind, contributes to increase of entropy in specific soil compartments or for limited periods.

biological systems from the cell to the ecosystem, free energy is continually renewed by external addition. In ecosystems the sun is the principal source of energy, supplemented by addition of the primary production of chemolithoautotrophic bacteria (Chap. 4, § 4.4.2).

With time, through the action of microorganisms, the ecosystem follows an 'ordered process of rationally directed development leading to a steady state' (Odum, 1971). Work done by the free energy of the soil system enables it to organize itself by means of the flow established between the reservoir with large work capacity, the sun, and that which receives the residual heat of respiration, namely space (Reeves, 1990; Addiscot in Bryant and Arnold, 1994).

In soil, mainly two types of energy enable organization:

- The kinetic energy of a particle moved by gravity. For example, water transports substances downwards from above and contributes to the ordered creation of horizons (Gerrard, 1992). Here the flow of water decreases entropy but works in reverse when it erodes the soil. The balance between these opposing thermodynamic actions of water depends on the soil type, the area affected by rain and the amount of rain (Addiscot, in Bryant and Arnold, 1994).

- Solar energy, fixed by photosynthesis in the biomass of organisms or utilized in their metabolism, which indirectly enables bioturbation (§ 5.3.3) and the organization of food-webs (Chap. 13). It also actuates water flow among soil, plant and atmosphere (Chap. 3, § 3.4.3), thus raising the soil bioelements towards the above-ground organs.

Erosion and mineralization are two processes that increase entropy similar to respiration and senescence of living organisms or even destructuration of soil (Addiscot in Bryant and Arnold, 1994). For example, the covering of a mature soil by colluviated material annuls at a stroke an earlier organization and reinitiates pedogenesis just as erosion does. Similarly, in the OF and OH horizons, organic matter is degraded by mineralization of built-up structures (e.g. tissues, cells) to simple forms (e.g. cations and anions), with dissipation of energy as heat (Chap. 10, § 10.3.1).

Entropy and humification

In humification, formation of microbial humin H3 undoubtedly decreases entropy: building of complex, structured and stable molecules, namely polysaccharides, necessitates a large injection of energy during bacterial metabolic processes (Chap. 4, § 4.4.1; Chap. 14, § 14.4.2). On the contrary, polycondensation, which results in insolubilization humin H2, creates entropy. Chemical bonds are formed at random from collisions between molecules and free radicals, giving rise to an infinite variety of forms in very great disorder (Chap. 14, § 14.4.3). Stability is also high, but only because enzymes cannot easily identify clear biochemical sites for their degrading action.

While mineralization increases entropy, humification is more ambiguous. Nevertheless, Addiscot (in Bryant and Arnold, 1994) groups it among processes that decrease entropy.

The stability characteristic of microbial humin H3 or insolubilization humin H2 thus corresponds to two different thermodynamic causes:

- H3 owes its stability to the decrease in entropy accompanying an ordered formation of new molecules;
- H2 finds its stability in increase in entropy related to disordered making of new molecules.

But why do enzymes not rapidly degrade microbial humin, composed as it is of easily identifiable polysaccharides?

One mystery remains..

As for residual or inherited humin H1, the varied composition of which depends on the parent plant material, it seems to owe its stability to two simultaneous thermodynamic paths: some of its macromolecules originate from ordered processes, such as cellulose preserved in peat (Chap. 9, § 9.1), whereas others such as lignin seem to be highly entropic.

In the intermediate position, inherited humin.

In the final accounting, soil formation is energy-dissipative, with acquisition of order and decrease in entropy. Locally and/or temporarily a retrograde development can take place by which entropy is increased.

Overall, pedogenesis reduces entropy.

5.1.2 Factors of soil formation

Soil formation is a multidimensional phenomenon controlled by a combination of five ecological factors (Ellis and Mellor, 1995; Van Breemen and Buurman, 1998):

- climate,
- mineral parent material,
- living organisms and their organic matter,
- relief,
- time.

One principle, three phases, five factors, seventeen processes ... and an infinite number of soils!

While climate is determinant at the global scale, the other agents act powerfully at regional and local levels.

The five factors and their variations combine to give a nearly infinite range of soils on the Earth's surface (§ 5.4.1; Fig. 5.1). However, one single common principle of evolution unites the latter into three phases, concomitantly successive and simultaneous, while seventeen fundamental processes suffice to explain all the pedogeneses.

The principle: Environment → Processes → Characters.

This common principle means that environmental factors determine development processes that impart certain characters to the soil. Or, more precisely: the physical, chemical and biological environment in the soil and outside it causes alterations, constructions, reactions and transfers; in turn these processes modify the colour or morphology of the soil and form horizons and their boundaries.

Defining a causal principle enables us to (Legros, 1996):

- rationally study the soil mantle, the distribution of which is not random;

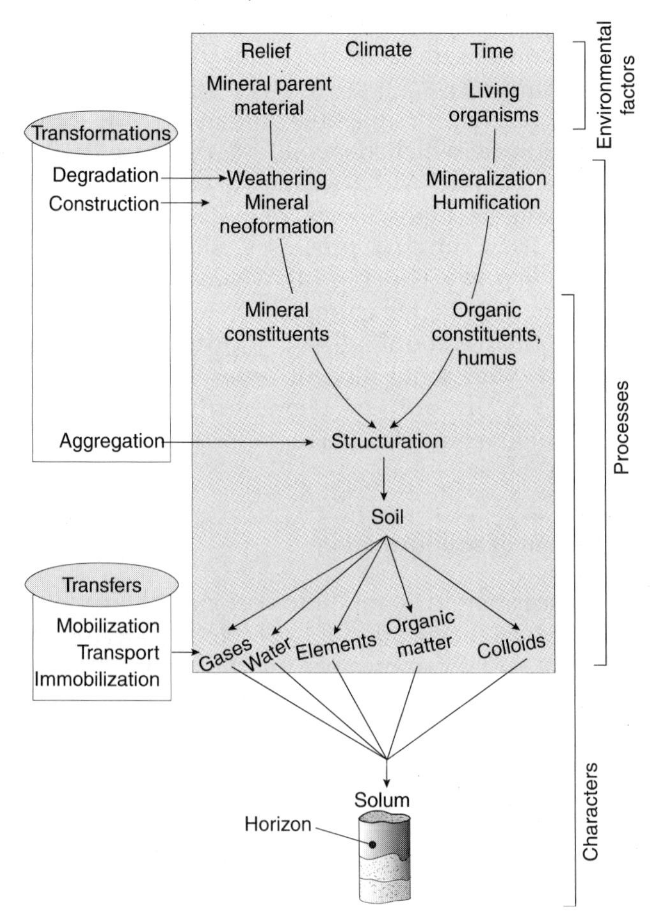

Fig. 5.1 General theory of pedogenesis (after Schroeder, in Gisi et al., 1997).

• use a deductive and predictive step for determining the soil types of comparable landscapes, for example during soil mapping;
• to formulate an explanatory rationale if the general law is not obeyed: is the environment homogeneous, is sampling correct, is the scale used appropriate?

5.1.3 The three phases of soil formation

Soil formation, from fresh rock to a system at equilibrium, is divided into three phases (Fig. 5.2).

The three phases: rock weathering, incorporation of organic matter and transport of matter.

The first phase results in small mineral particles—sand, silt and clay—with or without mineralogical modification; this is weathering. Physical and chemical processes of

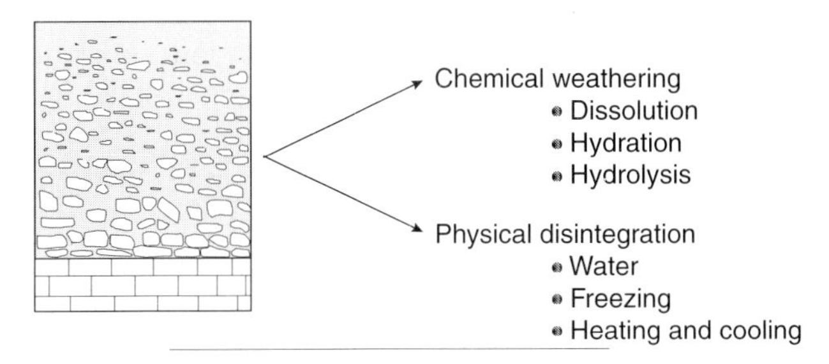

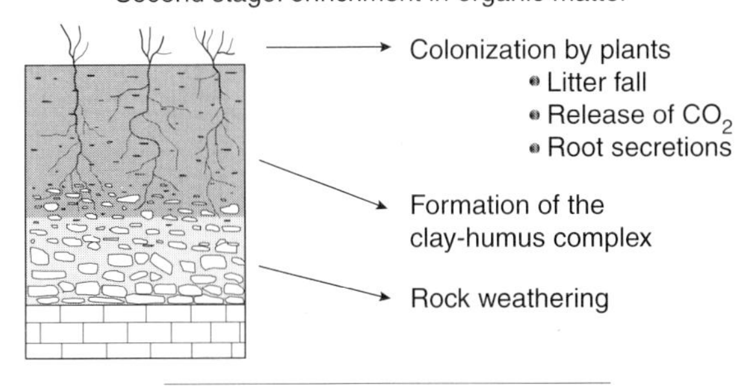

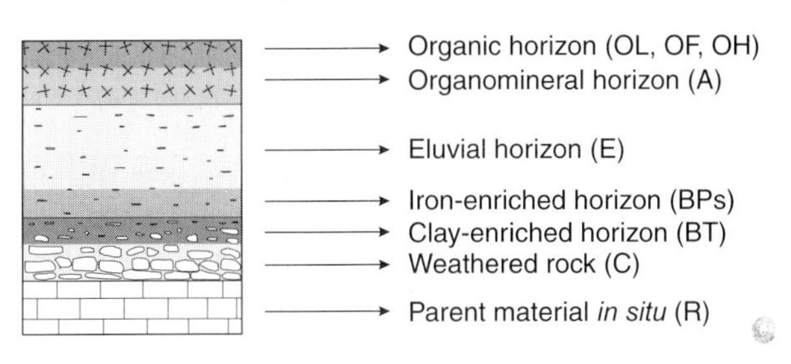

Fig. 5.2 *The three phases of soil formation: rock weathering, incorporation of organic matter and transport of matter (after Soltner, 1996a).*

weathering have been presented in Chapter 2, § 2.1.2, those of biological nature in Chapter 4, § 4.4.5; they are not again discussed here. Their product consists of mineral substances suitable for incorporating, during the second phase, with the organic substances in a new entity to be created, 'earth'.

5.2 INCORPORATION OF ORGANIC SUBSTANCES

5.2.1 General process

Because of organic additions, the mineral substrate is transformed little by little to 'soil'; we pass, so to speak, from geology to soil science.

In the second phase of soil development, the edaphic organisms intervene first; their effect on the processes of transformation of organic substances is summarized in Fig. 5.3. Somewhat separated in the schematic diagram for didactic reasons, the latter are often simultaneous in nature. Furthermore, these transformations that certainly begin in the second phase of soil formation continue during the entire soil development (third phase) as long as new additions take place.

Besides being mineralized and humified, part of the organic matter serves as food for microorganisms.

When organic matter falls on the mineral substrate (above-ground litter) or forms within it (below-ground litter), it undergoes three types of transformations:

- ***mineralization***, a physical, chemical and biological process leading to transformation of organic constituents to mineral constituents;
- ***humification***, biochemical process of neosynthesis of organic substances by increase in size of certain molecules;
- assimilation by microorganisms, the ultimate consumers at the end of the debris food chains (Chap. 13, § 13.6).

All the litter is not transformed on (or in) the soil!

In dense lawns, dry leaves can be trapped among living leaves and thus constitute a suspended litter, transformed above the soil. This is also the case of leaves and twigs that remain held in the forks of large branches or in cavities in the trunks of trees. Veritable 'hanging soils' can then form, sheltering a diversified fauna (Chap. 8, § 8.8.1).

5.2.2 Mineralization

Rapid primary mineralization.

Primary mineralization M1 degrades particulate organic matter, especially fragile compounds such as carbohydrates, proteins and amino acids, and also lipids and nucleic acids. If degradation is total, the products

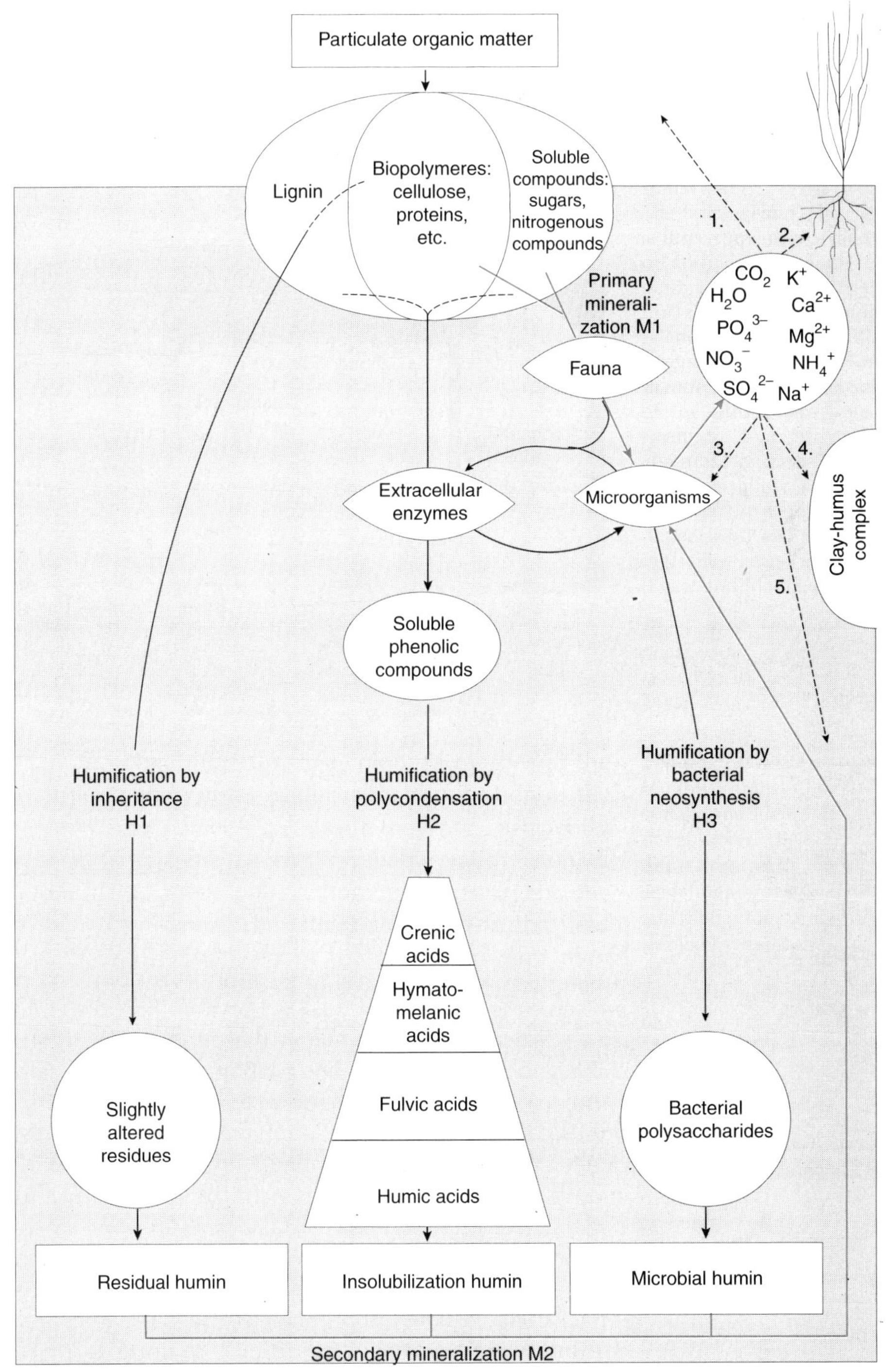

Fig. 5.3 *General principles of evolution of organic substances of soil (modified from Soltner, 1996a). Explanations in the text.*

are cations, anions and simple molecules such as water and CO_2. These substances, soluble in the soil solution, face five possible fates:

1. evacuation to the atmosphere by gaseous exchange (e.g. CO_2, H_2O, NH_3, N_2, H_2S);
2. absorption by plants (e.g. cations, anions, H_2O);
3. absorption by microorganisms (e.g. CO_2, NH_4^+, NO_3^-, SO_4^{2-}, PO_4^{3-});
4. retention on the exchange complex (e.g. Ca^{2+}, K^+, NH_4^+, H^+);
5. removal by leaching (e.g. K^+, Na^+, Ca^{2+}, NO_3^-).

Primary mineralization is rapid, of the average order of one to five years for an ameliorating litter. Table 5.1 gives more precise data for some types of litter.

Table 5.1 Speed of degradation of leaves of different species (after Mason, 1976)

Species	Scientific name	Organ	Humus form	Time for loss of 50% of the mass (days)
White birch	*Betula pendula*	Leaves	Mull	110
Small-leaved lime-tree	*Tilia cordata*	Leaves	Mull	165
Chestnut-tree	*Castanea sativa*	Leaves	Moder	220
Tall fescue	*Festuca arundinacea*	Roots	Mull	240
White birch	*Betula pendula*	Leaves	Moder	350
Durmast oak	*Quercus petraea*	Leaves	Mull	351
Beech	*Fagus sylvatica*	Leaves	Moder	700
Beech	*Fagus sylvatica*	Twigs	Mull	1000
Scots pine	*Pinus sylvestris*	Needles	Mor	2500

A slow secondary mineralization.

Secondary mineralization **M2** destroys more slowly (1 to 3% of the humified matter per year) but with the same final result as primary mineralization, organic molecules previously synthesized by humification. Unlike most particulate organic matter these molecules are very stable and strongly resist degradation. Without this safety valve of secondary mineralization, accumulation of humified organic matter would have suffocated all forms of life. Actually all the particulate organic matter will end up at one time or another stabilized by humification. But this 'fine-control' tap can also remain open in soils that no longer receive adequate organic matter, like some cultivated soils to which application of manure is no longer done. A slow but regular decline in organic matter content follows, very prejudicial to long-term soil conservation (Davet, 1996).

A welcome safety valve!

5.2.3 Humification

Under the general term humification lie hidden three pathways for synthesis of stabilized organic matter, forming humus in the biochemical sense (Chap. 2, § 2.2.4; Chap. 14, § 14.4):

Three paths for humus making: inheritance, polycondensation and bacterial neosynthesis.

- humification by inheritance H1, which gives residual or inherited humin;
- humification by polycondensation H2, which provides insolubilization humin;
- humification by bacterial synthesis H3, which produces microbial humin.

In ***humification by inheritance*** H1, the most resistant compounds released during fragmentation of the litter (lignins, resins, phenolic acids) are directly incorporated in the clay-humus complex without much change. They constitute the residual humin, which is not truly neosynthesized *in situ*. This humification is favoured by low pH or high content of ***active lime***.

Directly from plants: H1!

Active lime: the finest fraction, often powdery, of the total lime.

In ***humification by polycondensation*** H2, simple phenolic compounds and polysaccharide and polypeptide chains from the first stages of mineralization or inherited from the litter, are polycondensed into larger and larger molecules, from crenic acids to humic acids, ending with insolubilized humin. This veritable neosynthesis *in situ* is favoured by moderate edaphic conditions, neither too acid nor too basic.

From crenic acids to humic acids, H2!

Lastly, in ***humification by bacterial synthesis*** H3, certain soluble organic molecules resulting from enzymatic degradation or from root secretions are absorbed by microorganisms that transform them and secrete them in the form of extremely stable polysaccharides. These compounds give coherence to bacterial microcolonies and structure the soil. Like the preceding, bacterial humification is favoured by a 'moderate' physicochemical environment.

Products of bacteria, H3!

Soil organic matter and the greenhouse effect: black box and missing carbon ...

The increase now proved (Wood, 1995) of the level of mean temperature of the globe by accentuation of the greenhouse effect forces investigators to a global approach to the carbon cycle (Chap. 4, § 4.4.4; Chap. 9, § 9.2.1). Numerous models have been proposed with the aim of simulating the flows and storage of this element in the compartments of the ecosphere: carbonate rocks, oceans, atmosphere, biomass, organic horizons of soil, etc. (Bryant and Arnold, 1994; Powlson et al., 1996).

In modelling the carbon cycle, the soil is not just a 'black box'!

'Organic C in litter and humus plays a key role (to understand the response of the C-cycle under climatic change). The number of compartments and the pathways of C-flows influence both the transient phase and equilibrium of the system, (...) but has not been investigated systematically for any of the models. Hence, the multitude of aggregation levels used to represent detritus and the variety of decomposition formulations used in the models may result in inconsistencies of simulation results' (Perruchoud and Fischlin, 1994).

Will the soil of forest ecosystems be the site of accumulation of the 'missing carbon'?

Too long has the soil been considered in these models as a simple 'black box', without distinguishing the categories of organic matter. The latter nevertheless present very different biochemical behaviours, regulating their stability, their degradation and the energy they provide to the food-chains (Chap. 13). A detailed accounting of soil organic matter is thus indispensable for establishing more sensitive models, which alone will be able, for example, to quantify and qualify the famous 'missing carbon' of global budgets. Actually the carbon introduced to the ecosphere following human activity is estimated at around 7 Gt y^{-1}. Of this, the atmosphere accumulates 3.4 Gt and the oceans 2 Gt. The destination of the remainder, evaluated at 1.6 Gt, remains unknown (Beeby and Brennan, 1997). Like others, Gifford (1994) feels that it is hidden somewhere in the soils of terrestrial ecosystems: 'Atmospheric carbon budgets that ignore the possibility of terrestrial ecosystem responses to global atmospheric change do not balance; there is a "missing sink" of about 0.4–4 Gt C per year.'

5.2.4 Principal mechanisms of integration

Successive and simultaneous processes

The mechanisms of integration of organic matter with the soil are greatly varied, considering the multitude of organisms involved and the diversity of physicochemical conditions.

Many phases succeed one another from the death of a leaf, some starting even in the plant. Caused by decomposing organisms, these transformations are physical, chemical and biochemical, but it is not possible to assign a precise chronological order to these phenomena. While it is obvious that a freshly fallen leaf is immediately cut up by certain macroarthropods, it is possible that at the same time or even before this bacteria or fungi adhering to the limb dissolve some compounds from it by means of their extracellular enzymes. It is also easier to attack a pine needle from within, taking advantage of a stoma that passes through its thick cuticle. Every sequence is thus only indicative, the processes being able to succeed one another and to coexist as well, or be reversed in order according to the nature of the litter (Fig. 5.4). Lastly, interactions are ubiquitous among the groups of decomposers involved, each species being subject to the influence of other organisms that can modify its activity.

Mechanisms linked to the soil fauna

Specialists and ... earthworms, handymen!

Figure 5.5 shows schematically a chronological model of the degradation of a dead leaf by animals and the associated microorganisms. The following are the stages:

- physical destruction of superficial tissues and opening up of the epidermis by large springtails;
- digging out of larger windows by small dipteran larvae;

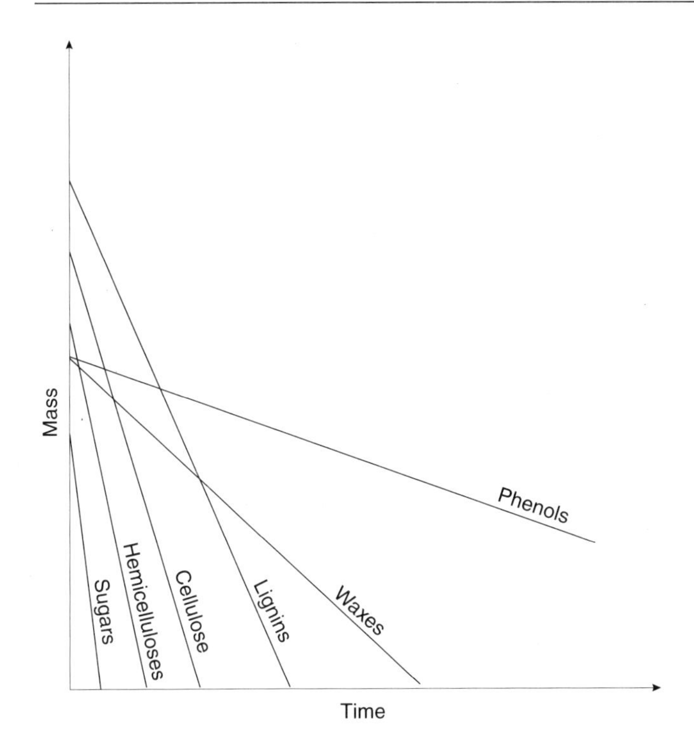

Fig. 5.4 *Speeds of degradation of principal constituent compounds of the litter (after Minderman, in Ellis and Mellor, 1955).*

- cutting up of leaves by macroarthropods, first the limb, then large ribs;
- reduction in size of debris and pellets by springtails, mites and enchytraeids;
- burying of pieces of leaves and pellets by the soil fauna and by leaching;
- humification of the organic matter in the microorganism-rich alimentary canal of earthworms and incorporation of mineral matter in the clay-humus complex;
- dissemination and control of bacteria and fungi by nematodes, mites and springtails;
- intervention of earthworms at each stage of the sequence.

According to Nef (1957) the surface area of a pine needle 'worked' by arthropods could go from 180 mm^2 to 1.8 m^2, or 10,000 times larger!

Towards a finer modelling of the evolution of organic matter

A better accounting of the evolution of soil organic matter—even if it is not yet perfect!—is gradually appearing, taking into account the integration mechanisms (Paustian et al. in Cadisch and Giller, 1997; Davidson et al., 2000; Lal et al., 1998, 2000). One of the best attempts is the CENTURY model developed by Parton et al. (in Bryant and Arnold, 1994 and in Powlson et al., 1996; Paul and Clark, 1996). It defines three pools of organic matter (rapidly, moderately and slowly decomposing), two pools of litter (above-

Many models presented by Christensen (in Powlson et al., 1996) subdivide the microbial biomass according to its activity, specifically based on its autochthonous and zymogenous characters (Chap. 4, § 4.4.6).

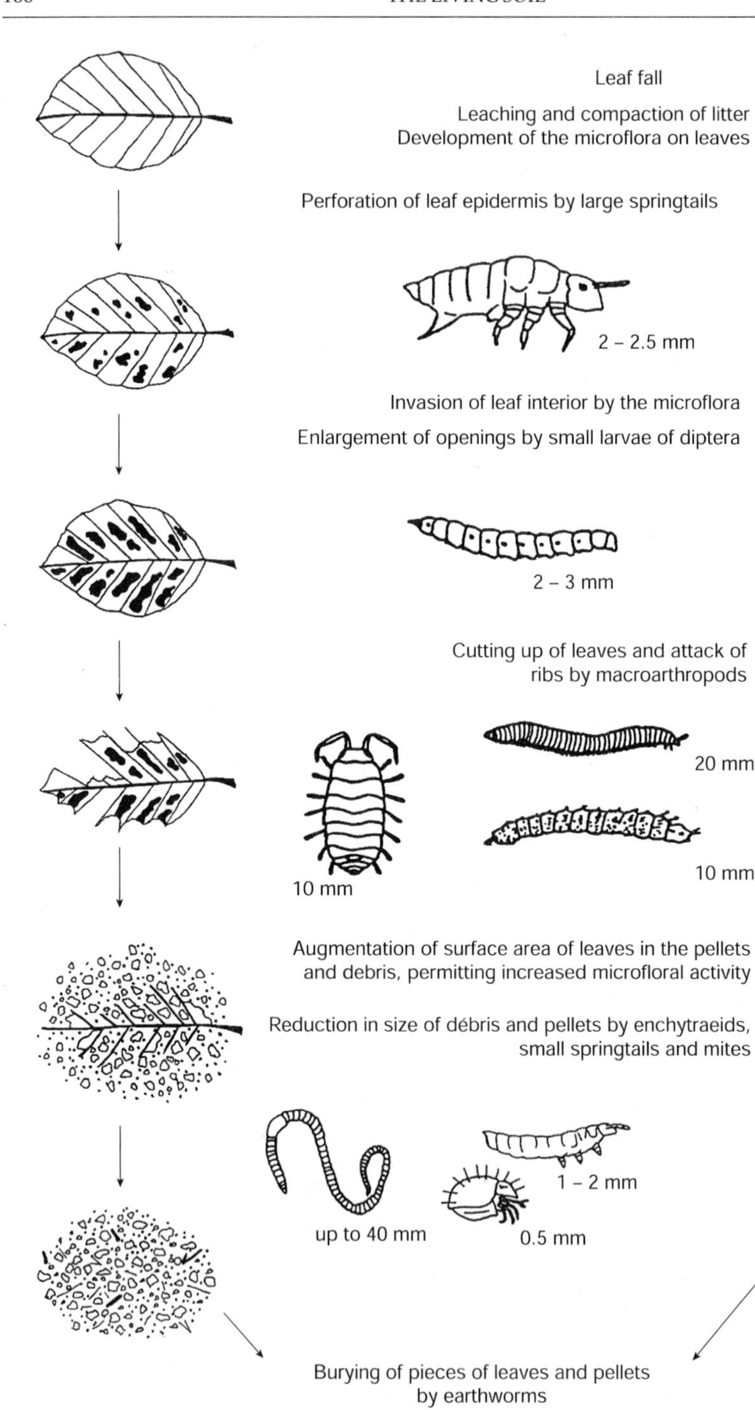

Fig. 5.5 *Sequence of transformation of a beech leaf by decomposing organisms.*

ground and below-ground), and one pool of microbial biomass. Animal wastes are also taken into consideration. The CENTURY model, developed for grassland soils, is equally valid for croplands, forests and savannas.

But despite the performance of recent models, numerous aspects still make the correct modelling of edaphic carbon difficult:

- comparison of results obtained by different methods;
- lack of experimental data to validate the models;
- shift of scale between a particular soil and the entire soil mantle (Slater et al. in Bryant and Arnold, 1994; Wilding in Sumner, 2000);
- multiscale organization of structure, with organic matter sequestered in the aggregates (Chap. 3, § 3.2; Elliott in Powlson et al., 1996);
- evolution of vegetation and of its relation with the soil (Chap. 7);
- quality of the litter (Jolivet et al., 2001; Chap. 2, § 2.2.1);
- stability and ages of different categories of humus (§ 5.5.5);
- heterogeneity of physical, chemical and climatic conditions that influence the organic matter (Chap. 6, § 6.3), favouring formation of CO_2 or of CH_4, according to aeration conditions (Chap. 4, § 4.4.4);
- importance of liquid additions in the balance (Chap. 2, § 2.2.1);
- influence of nitrogen on carbon (Coûteaux et al., 1995; Aerts and de Caluwe, 1999);
- activity of the pedofauna, which is difficult to quantify. We can cite here the discharge of methane by termites, estimated by different authors at 15 to 150 t y^{-1} (Chap. 12, § 12.4.9; Mooney et al., 1987).

'Yet to be plausible, any explanation of how the carbon stock evolves in the soil would have to include some detailed understanding of the dynamics of organic material, and in particular the role that nitrogen content plays in the litter, and of the conditions for releasing complexing acids' (Gobat et al. in Cebon et al., 1998).

'Simulation models are probably the only way to predict climate-derived changes on soil organic matter storage' (Christensen in Powlson et al. 1996).

Mechanisms linked to bacteria and fungi

The chief microbial processes linked to the evolution of organic matter are the following:

- synthesis of extracellular enzymes linked to the phenomena of enzymatic hydrolysis of organic polymers, of enzymes responsible for activation of phenolic compounds as prelude to their polycondensation and, lastly, of those concerned with mineralization of organic compounds (Chap. 14, §§ 14.2, 14.3);
- mineralization of monomers and oligomers resulting from enzymatic degradation;
- production and secretion of stable polysaccharides (bacterial humin);
- biological fixation of nitrogen;
- translocation of organic substances by fungi;
- production of growth factors or antibiotics, formation of symbioses, predation and parasitism;
- formation of soil structure;
- participation in the soil biocoenosis.

Microorganisms participate in the degradation of organic matter, its transformation, integration and mineralization (Chap. 4, § 4.4).

Chemical and biochemical mechanisms

In the soil, many transformations of organic matter take place with no direct intervention of living organisms.

It is often difficult to distinguish abiotic reactions from enzymatic reactions and those catalysed by cells (Chap. 14, § 14.2).

Here the reactions catalysed by soil enzymes of biotic origin must be distinguished from purely abiotic reactions. These latter reactions are sometimes controlled by previous enzymatic reactions, such as the examples of redox reactions between phenolic compounds and manganese oxides: these are formed again by an enzymatic reaction (manganese-peroxidase; Chap. 14, § 14.3.3). But others are totally abiotic, such as the digestion of some humic compounds under the action of light (Aguer and Richard, 1996).

Numerous lytic reactions concern the decomposition process, with different rates according to the substances to be degraded (Chap. 14, § 14.3). On their part, others favour polycondensation of molecules, chiefly aromatic, to which are added peptide and carbohydrate oligomers (Chap. 14, § 14.4). Oxidative polymerization gives humic compounds that become darker and darker (***melanization***) by accumulation of oxidized structures such as quinones, which are ***chromatophores*** (Fig. 5.6).

Chromatophore: molecular structure that absorbs certain spectral bands of visible light and therefore appears coloured.

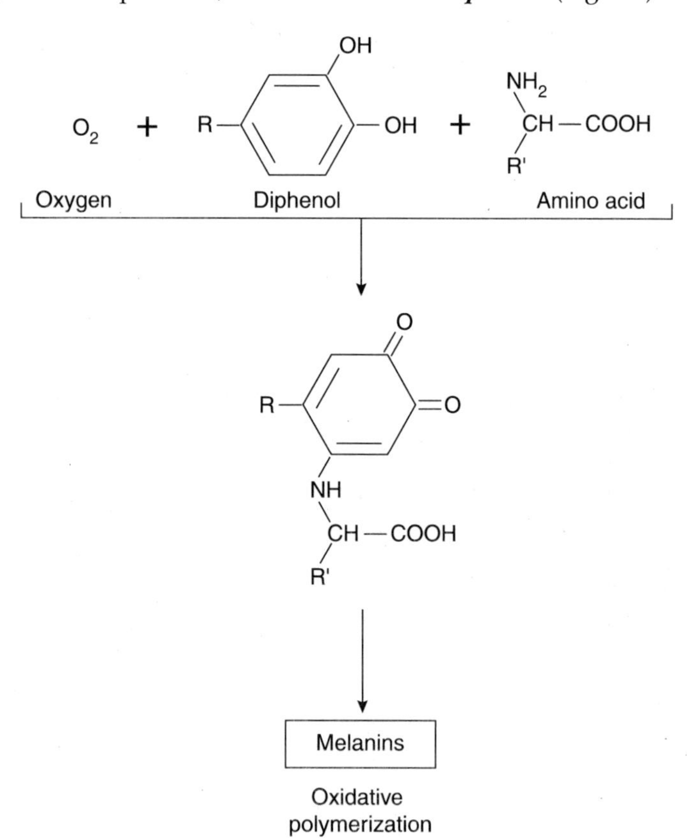

Fig. 5.6 *Formation of melanins by polymerization of phenols (after Andreux and Munier-Lamy in Bonneau and Souchier, 1994).*

5.3 TRANSPORT OF SUBSTANCES

After the second phase of pedogenesis, humified organic substances are bound to fine mineral particles in the clay-humus complex. A real soil is then formed but is still immature and often comprises, except for organic layers, only an organomineral A horizon and the parent material C. Gradually other horizons will be constituted, resulting from transport of various substances.

In the third phase—while the first and second phases are continuing—the soil is progressively organized into well-differentiated layers.

5.3.1 Transporting agents

Water is the chief carrier of substances, capable of transport in all directions. In temperate climate, where precipitation exceeds evaporation, downward movements predominate. In dry tropical climate the reverse holds true. As for lateral transport, it depends on local conditions, in particular the slope and permeability of the underlying rock. Other transporters are soil animals, especially anecic earthworms, ants and termites, capable of strong mixing (bioturbation processes).

The agents of transport of substances in soil are relatively few: water, the soil fauna and gravity. Fungi and plants, which also participate, do it by internal circulation of water and dissolved salts (translocation).

Water, mineral salts and organic compounds can also be translocated by fungal mycelia, the very dense network of which is extremely efficient (Chap. 2, § 2.5.3; Chap. 16, § 16.1), as well as vascular plants in the biogeochemical cycle. Lastly, gravity, sometimes in combination with flow, is an important agent of incorporation of organic matter to depth in VERTISOLS, in which wide shrinkage cracks open up during dry periods.

Humphreys (in Ringrose-Voase and Humphreys, 1994) and Paton et al. (1995) attribute very great importance to biological processes of transport of matter, compared to the physicochemical paths that until now have been overestimated.

5.3.2 Mechanisms of transport by water

Downward movements

Three downward movements are due to water (Fig. 5.7):

- leaching (lixiviation),
- clay-leaching,
- cheluviation.

The last two are forms of ***eluviation***, phenomena of depletion of superficial soil horizons to the benefit of deeper horizons that are enriched by ***illuviation***.

Three downward movements: leaching (lixiviation), clay-leaching and cheluviation.

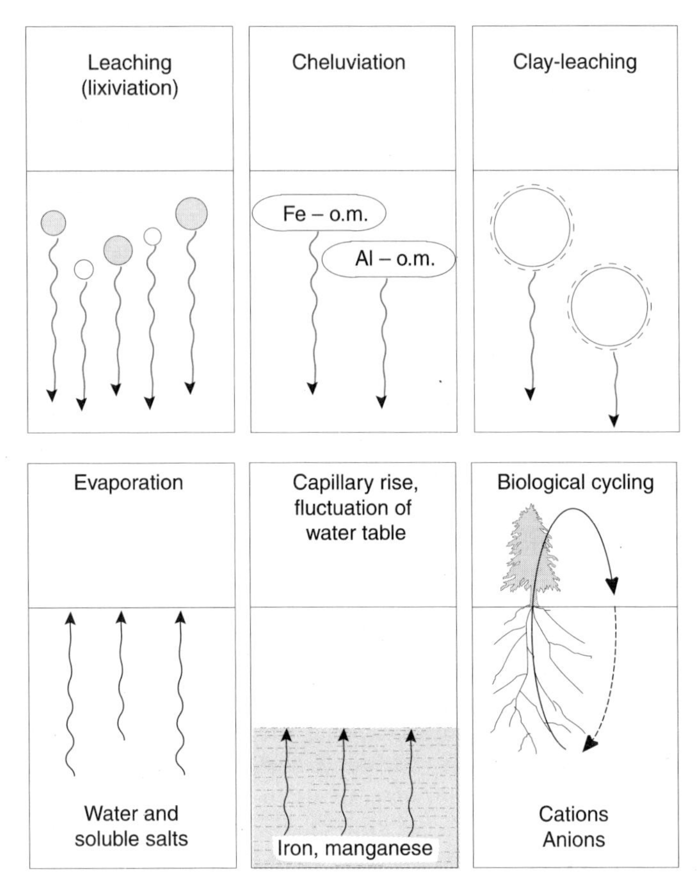

Fig. 5.7 *Principal movements of substances due to water in a soil (after Soltner, 1996a).*

Leaching (or ***lixiviation):*** migration, at the expense of the solum or some of its parts, of cations, anions or small soluble molecules.

Clay-leaching: mechanical transport of the fine mineral particles of the soil, the clays.

Decalcification: leaching of calcium, often replaced by protons on the exchange complex.

Leaching has three effects on soil development:

- in a non-calcareous environment, it desaturates the complex and, as a corollary, acidifies it by exchange mechanisms; monoculture of conifers accelerates the phenomenon;
- in calcareous medium, it transports carbonates after dissolving calcium carbonate (Chap. 2, § 2.1.2); the calcium retained on the complex resists better but it too is finally evacuated;
- in humid equatorial climate, silicon can be eliminated, leaving only aluminium and iron (Ferrallitic soils).

In soils rich in active lime and calcium, the clay is flocculated, which prevents all ***clay-leaching*** (e.g. RENDOSOLS, CALCOSOLS). But as soon as ***decalcification*** appears, the clay-humus bond is weakened, provoking entrainment of very fine clay particles, even without distinct pedofeatures in the soil (e.g. some CALCISOLS and

BRUNISOLS). If decalcification continues, accompanied by great reduction in base saturation percentage and pH, clay-leaching is intensified and is manifested by formation of eluvial E and illuvial BT horizons (NÉOLUVISOLS, LUVISOLS) (Plate II-2).

In ***cheluviation***, the clay mineral layers are broken up, releasing aluminium and iron, which are then bound to complexing organic molecules—crenic or fulvic acids (***acidocomplexolysis***). The chelates thus formed are entrained to depth (Plate II-6). Morphologically, cheluviation is indicated by the eluvial E horizon and illuvial BPh (precipitation of organic molecules after breakup of the chelates) and BPs (precipitation of ***sesquioxides*** of iron and aluminium) horizons (Plates II-4, II-5). Cheluviation is the fundamental process of formation of PODZOSOLS, under a vegetation with acidifying litter (see details in Van Breemen and Buurman, 1998, for example).

Cheluviation: entrainment of metal-organic complexes, the metal chelates, under generally reducing and acid conditions.

Sesquioxide: metal oxide of general formula M_2O_3; *sesqui-* signifies 'one and a half', referring to the proportion of oxygen and metal atoms.

Upward movements

Three types of upward movements have been recognized in which water is the chief carrier (Fig. 5.7):

- ascent by evaporation,
- ascent by capillary rise and/or fluctuation of water table,
- ascent by biological cycling.

Three upward movements: evaporation, capillary rise and water table fluctuation, biological cycling.

In hot dry climate, at least seasonally so, water ascends by capillary rise, then evaporates on contact with air. It caries along dissolved elements, which precipitate in the upper layers of the soil or on its surface when the water evaporates. This is true for iron and aluminium in tropical regions, the former colouring the soil dark red by its oxides (FS horizon). Sodium can form white efflorescences on the surface of a soil such as solonchak.

In hydromorphic conditions, iron is generally in the form of barely soluble salts of the $FeCO_3$ type in the part of the soil permanently saturated with water (Gr horizon). In the first stage, atmospheric precipitation reaching the soil passes through the rooting zone, and accumulate CO_2; on contact with the water table, it solubilizes iron in the form of $Fe(HCO_3)_2$. Following this first stage, capillary rise takes the groundwater with dissolved iron close to the surface, where it mixes with rain-water richer in oxygen. The resultant increase in redox potential provokes oxidation of the iron chemically, or biologically by bacteria of the *Gallionella ferruginea* type (ferrooxidation), and its

In reduced soils, fluctuation of the water table leads to mixing and transformation of the iron according to variations in redox potential.

«L'eau ferrugineuse, comme son nom indique, contient du fer (...). Et pourquoi y a-t-il du fer l'eau ferrugineuse ? C'est parce que l'eau a passé et repassé sur le fer, et le fer ... a dissous. Et le fer à dix sous, c'est pas cher ... », as Bourvil, a famous french humorist, says in a well-known sketch!

Biological cycling conserves the bioelements in the soil.

precipitation as $Fe(OH)_3$ (Chap. 4, § 4.4.4). If drying is intense, precipitation takes place as Fe_2O_3. Among hydromorphic soils, HISTOSOLS are veritable iron-solubilizing machines because of the abundant water-soluble complexing organic matter (Chap. 9, § 9.1.3; Bouyer and Pochon, 1980; Bouyer, 1999).

Of the agents that ensure upward movements, vascular plants, especially trees, are often capable of counterbalancing the downward movements, thus maintaining the soil at a specific stage of development (§ 5.5.6). They nourish themselves by pumping the bioelements by their roots and taking them to the leaves; these bioelements later return to the soil surface by stemflow, throughfall rain leaching or litter fall, thus closing the biological part of the biogeochemical cycle (Fig. 5.8).

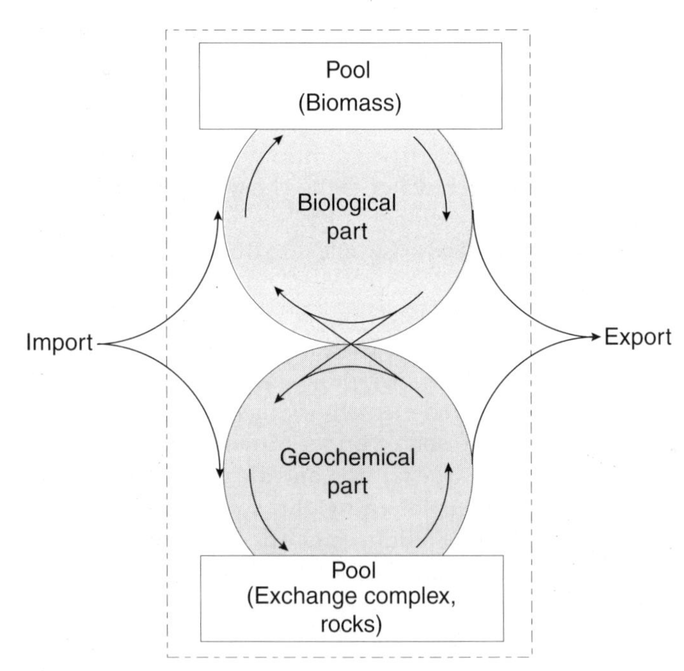

Fig. 5.8 Simplified representation of the ***biogeochemical cycle****. In the soil the biological part of the cycle represents the main pathway for upward movements since it exists in all soils covered by vegetation.*

5.3.3 Pedoturbation

The great bioturbators: earthworms, ants, termites as well as man ... and birds!

Bioturbation: displacement of matter caused by movements of living organisms. It has five principal effects on soil:

Pedoturbation groups the processes of local cyclic movements of soil materials not due to liquid water or gravity. ***Bioturbation*** and ***cryoturbation*** are distinguished in it.

Earthworms are the champions of bioturbation in temperate regions (Chap. 4, § 4.5.1). In tropical climate, ants and termites, which bring up matter from more than

10-m depth, surpass them in efficiency (Lavelle and Spain, 2001). Conversely, erosion of termitaria replenishes the soil around with, on average, 0.2 millimetre of new material every year. Curiously enough, birds can also be soil bioturbators. In Australia, for example, 104 species rework the land in one way or another, by their search for food, construction of underground nests or mounds (Paton et al., 1995). Lastly, it should not be forgotten that man is a great 'bioturbator' by ploughing cultivated fields and reworking the soil in construction sites.

mechanical loosening, oxygenation of deep parts, redistribution of organic matter, bringing buried materials to the surface and neutralizing soil pH.

Cryoturbation: displacement of matter due to mixing of particles by alternate freezing and thawing.

The uprooted tree, agent of bioturbation and soil development

Certain trees with superficial roots, like the spruces, very locally 'bring up' large quantities of soil when they are toppled by wind; their root system is brought to the vertical, temporarily carrying lumps of soil to several metres above the ground (Fig. 5.9). This is indispensable for maintaining soil fertility and regeneration of certain forests of the American Pacific Coast (Spaltenstein, pers. comm.). This uprooting process has been well described by Paton et al. (1995).

Fig. 5.9 *Spruce uprooted by a sudden gale; border of a bog in the Swiss Jura (photo: J.-M. Gobat).*

By cryoturbation, periglacial soils are mixed and their constituents sorted according to their physical resistance to displacement by alternate freezing and thawing. ***Polygonal soils*** form on the surface, as seen not only in the circumpolar zones but also at altitude in high mountain massifs. Cryoturbation incorporates organic matter more or less deeply in the mineral material.

Polygonal soil: rocky soil of periglacial regions, whose surface is characterized by juxtaposition of polygons with sides formed of material texturally different from that of the centre.

5.4 THE HORIZON: PRODUCT OF SOIL DEVELOPMENT

5.4.1 General processes of pedogenesis

The fundamental pedogenetic processes are presented without comment in Table 5.2. They refer to mechanisms described in earlier sections and are summarized in Fig. 5.10. Books on general pedology should be consulted for their detailed analysis and fields of application under different conditions of vegetation, rock and climate.

Physical, chemical and biological activities in soil are integrated in seventeen fundamental processes of development.

5.4.2 Definition and nomenclature of horizons

Transformation and transport processes mark the soil with morphological or analytical characters: crumb structure of a developed clay-humus complex, bleached

Table 5.2 Principal processes of pedogenesis (various sources)

PROCESSES	DESCRIPTION	PEDOLOGICAL CHARACTERS	PRINCIPAL RÉFÉRENCES
I. Processes linked to humification (cold and temperate climates)			
Carbonate leaching	Dissolution of carbonates by water enriched with carbon dioxide.	No effervescence (or slight effervescence) in the fine earth with HCl. Release of organic matter blocked by calcareous cements.	CALCOSOL, CALCISOL, MAGNÉSISOL, DOLOMITOSOL, FLUVIOSOL BRUNIFIÉ, ORGANOSOL CALCIQUE, RENDISOL
Decalcification	Leaching of the calcium ion.	Diminution of base saturation percentage, pH and stability of the clay-humus complex.	BRUNISOL MÉSOSATURÉ, BRUNISOL OLIGOSATURÉ, CALCISOL, MAGNÉSISOL, ORGANOSOL CALCIQUE
Brunification	Augmentation of the role of iron in pedogenesis, particularly in organo-mineral bonding.	Formation of the S horizon. Brown coloration of soil. Diminution of organic matter reserve. Activation of the biogeochemical cycle.	BRUNISOLS, FLUVIOSOL BRUNIFIÉ
Clay-leaching	Mechanical leaching of fine clay from the upper horizons to the horizons below.	Formation of E and BT horizons. Acidification. Diminution of biological activity.	LUVISOLS, NÉOLUVISOL
Podzolization	Destruction of clay minerals, migration of metal-organic chelates and precipitation of organic matter and iron-aluminium complexes at depth.	Formation of an ash-coloured E horizon and the spodic BPh and BPs horizons. Horizons with generally clear boundaries and bright contrasting colours.	PODZOSOLS
Andosolization	Rapid precipitation of humic precursors by active alumina.	Formation of very dark, stable aggregates. Process on volcanic rocks.	ANDOSOLS
II. Processes conditioned by strong seasonal contrasts (continental climates)			
Melanization	Deep incorporation of organic matter by bioturbation. Simultaneous rapid mineralization and strong stabilization of organic matter. Moderate neoformation of swelling clay minerals.	Thick, black Ach horizon. Formation of a K horizon at depth.	CHERNOSOLS, KASTANOSOLS, PHAEOSOLS

(Contd.)

Table 5.2 *(contd.)*

PROCESSES	DESCRIPTION	PEDOLOGICAL CHARACTERS	PRINCIPAL RÉFÉRENCES
Calcification	Formation of calcitic crusts by precipitation of dissolved calcium.	Formation of a K horizon at depth, soft (Kc) or indurated (Km).	KASTANOSOLS
Vertisolization	Neoformation of swelling clay minerals. Deep incorporation of organic matter by vertic movements (alternate shrinking and swelling of clay).	Structure with shrinkage cracks in the dry period. Formation of SV and V horizons.	VERTISOLS
III. Processes due to prolonged geochemical weathering (Mediterranean, tropical, equatorial climates)			
Fersiallitization	Strong neoformation of clay minerals. Considerable rapid crystallization of iron oxides released by weathering (rubefaction).	Predominance of 2/1 clay minerals. Soil coloured red by iron oxides (FS horizon). Possible presence of a BT horizon formed by clay-leaching in the rainy season.	FERSIALSOLS
Ferrugination	Strong weathering of primary minerals. Great loss of bases. More or less strong desilication.	Predominance of neoformed 1/1 clay minerals. Reduction of base saturation percentage. Very little clay-leaching.	FERRISOLS (Ferruginous soils)
Ferrallitization	Complete weathering of primary minerals except quartz. Almost complete loss of silica. Complete neoformation of clay minerals (1:1 type).	No 2/1 clay minerals. Very deep soil, reaching several metres. Crystallization of aluminium (gibbsite) = allitization. Very stable kaolinite-haematite aggregates. No clay-leaching.	FERRALSOLS (Ferrallitic soils)
IV. Processes linked to physicochemical conditions at the site (all climates)			
a) Intervention of reductive water			
Hydromorphy	Reduction and local segregation of iron by permanent or temporary saturation of the pores by reductive water. Temporary hydromorphy in the surface layer (g horizon) or permanent at depth (G horizon).	Formation of An, g and G horizons. Iron-oxidation mottles in the temporarily aerated zones (Go horizons). Greenish-grey colour in reduced zones (Gr horizon).	HISTOSOLS, RÉDOXISOLS, RÉDUCTISOLS

(Contd.)

Table 5.2 *(contd.)*

PROCESSES	DESCRIPTION	PEDOLOGICAL CHARACTERS	PRINCIPAL RÉFÉRENCES
b) Intervention of the sodium ion			
Salinization	In presence of a saline water table with comparable levels of salts of sodium and alkaline earths, increase in proportion of Na^+ on the clay-humus complex.	Upward movement of salt by capillary rise and precipitation on the surface as white efflorescence. pH < 8.7. Aerated crumb structure. Exchangeable sodium < 15% (Sa horizon).	SALISOLS
Sodization	Great increase in saturation of exchange complex by sodium.	Sodium saturation > 15%, often reaching 30% (Na horizon). Structure becoming powdery.	SODISOLS *p.p.*
Alkalization	In presence of fresh water (rain, lowering of saline water table), clay-leaching and hydrolysis of sodic clays, releasing Na^+ to the soil solution.	Formation of a natric BT horizon, with pH > 9. Sodium saturation reaching 50% or more.	SODISOLS *p.p.*
Sulphate reduction	Process with participation of sulphur, varying in state according to Eh (sulphides or sulphates).	Massive structure. Greenish-grey colour speckled with black (iron sulphides, TH horizon) in the most reduced zones, or yellow and rust-coloured mottles in temporarily aerated zones (iron sulphates, U horizon).	SULFATOSOLS, THIOSOLS

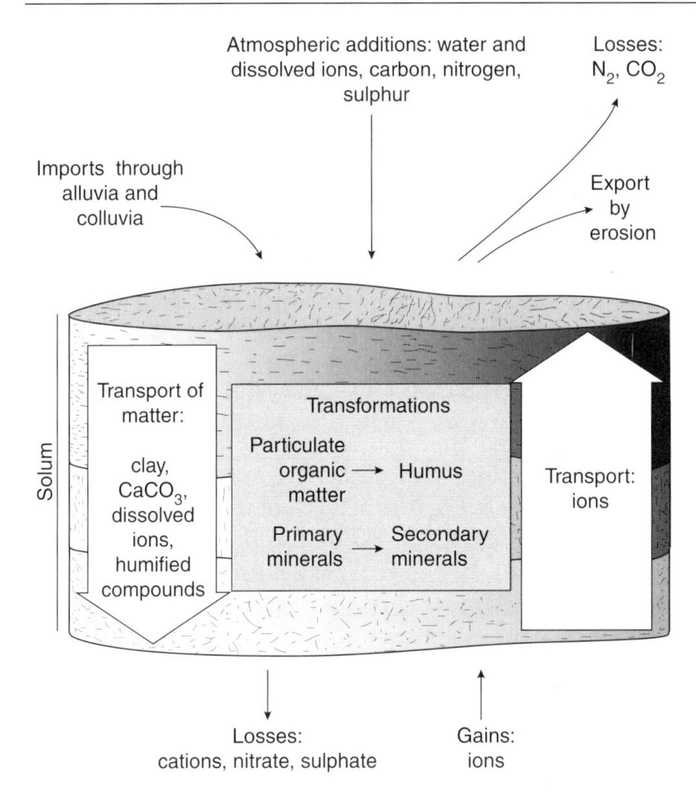

Solum: vertical section of the soil or, better still, of the soil mantle observable in a pit or cut. It is described by a ***profile,*** the vertical ordered sequence of all data concerning the solum (AFES, 1995, 1998). The term ***pedon*** concerns a three-dimensional body of soil large enough to permit complete study of horizons, shapes and relations (Wysocki et al. in Sumner, 2000). ***Soil mantle:*** superficial portion of the lithosphere transformed by physical, chemical and biological processes into a complex, organized structure, three-dimensional and developing, which bears vegetation (Lozet and Mathieu, 1997).

Fig. 5.10 *Mechanisms fundamental to pedogenesis (after Coleman and Crossley, 1996).*

layer of an eluvial horizon, rust-coloured mottles of the zone of fluctuation of the water table, diminution of the base saturation percentage, etc. If they are sufficiently well expressed, these characters enable subdivision of the ***solum*** into more or less homogeneous layers, the ***pedological horizons***. The latter, described in the field and occasionally analysed in the laboratory, are then compared with the ***reference horizons*** (AFES, 1995, 1998) and assigned to those with the best-matched criteria. Each reference horizon mirrors one or a few fundamental processes.

In pedology, the variety of approaches and schools is wide (§ 5.6.2). While the development processes are known everywhere, their translation into a single nomenclature has not been achieved because of different weightage given to such or such transformation. A revised list of horizons of universal utility was recently published (AFES, 1995, 1998); it has been adopted here. Its correspondence with the other systems can be found in books on general soil science cited in the Bibliography.

Pedological horizon: layer roughly parallel to the soil surface and identifiable by the observer. The horizons differ from each other in constituents, organization and behaviour; they are formed by transformations undergone by the material since the beginning of its development (Lozet and Mathieu, 1997).

Reference horizon (or according to the authors, ***diagnostic horizon***): interpretative horizon having a quantitatively defined ensemble of properties (Lozet and Mathieu, 1997).

There is no uniform international system as yet for defining reference horizons.

Basic principle: one capital letter per principal horizon. For example, 'O' denotes all the ***holorganic*** horizons formed under aerated conditions, 'OL' represents the litter and 'OLn' recent litter, freshly deposited on the soil and still barely weathered.

5.4.3 Principal reference horizons

Without exception, the horizons are denoted by a capital letter reflecting their essential characteristic. They can be subdivided into subhorizons by a second or third capital or lower case letter, or by a numeral, indicating more subtle features (Tab. 5.3). The vertical sequence of reference horizons defines the model sola, the Références (§ 5.6.3; AFES, 1995, 1998).

5.5 FACTORS INFLUENCING PEDOGENESIS

Development under manifold influences.

Holorganic: describes a horizon that contains more than 30 g of organic carbon per 100 g, or a loss on ignition at 600 °C > 50 g/100 g (AFES, 1995, 1998).

We have seen that soil formation and development are subject to five ecological factors: climate, mineral parent material, living organisms and their organic matter, relief and time (§ 5.1.2). While the first is the principal determinant at the global scale, the others became essential in certain locations and can mask the effect of general climate. Some examples from the multitude of cases described in the literature are given below.

5.5.1 Climate (macroclimate)

Climate determines the broad directions of pedogenesis on the Earth, according to bioclimatic zones corresponding to biomes. This latitudinal zonation is also found in high mountain massifs, according to altitude (Fig. 5.11; Chap. 7, § 7.2.3).

Organic matter 'commands' if the climate is cold, mineral matter if it is hot.

Generally, pedological development dominated by organic matter corresponds to cold temperate climates. Climate works directly through precipitation and transport of substances caused by it, or indirectly by influencing vegetation and biological activity of the soil. It regulates the balance between the additions of organic matter to the soil and its decomposition, thus its accumulation (Fig. 5.12; Chap. 9, § 9.2.1).

Under hot climates, the temperature is favourable for mineralogical modifications, as in neoformation of clay minerals. The geochemistry of oxides of iron, aluminium and silicon dominates pedological development, compared to organic matter, which hardly accumulates.

The macroclimatic distribution of soils (***zonality*** of soils) is basic to the Russian classification (§ 5.6.2). ***Zonal soils*** (e.g. CRYOSOLS, PODZOSOLS, CHERNOSOLS), strongly influenced by the macroclimate, correspond to the biomes

Table 5.3 Principal pedological horizons and subhorizons (simplified, after AFES, 1995, 1998)

Horizon	Sub-horizon	Sub-division	Characteristics	Origin of symbol	Examples of RÉFÉRENCES containing the horizon
ORGANIC HORIZONS					
H	**Histic horizon**, peaty material			*h*istic	HISTOSOLS, RÉDUCTISOLS
	Ha		highly decomposed drained material	*h*istic-cleansed (in French: *a*ssaini)	ditto
	Hf		rich in fibre	*h*istic-*f*ibric	ditto
	Hm		moderately rich in fibre	*h*istic –*m*esic (hemic in Soil Taxonomy)	ditto
	Hs		poor in fibre	*h*istic -*s*apric	ditto
O	**Organic horizon**, non-peaty material			*o*rganic	*ALMOST ALL RÉFÉRENCES*
	OL		litter, particulate matter composed of debris identifiable by the naked eye; no humified fine material	*o*rganic-*l*itter	*ALMOST ALL RÉFÉRENCES*
		OLn	debris without distinct transformation	recent, *n*ew litter	ditto
		OLt	distinctly fragmented debris	*t*ransitional litter	ditto
		OLv	chemically altered debris	*v*erwittert (Ger. for weathered) litter	ditto
	OF		more or less fragmented identifiable debris, < 70% humified fine material by visual examination	*o*rganic-*f*ragmentation *o*rganic-*f*ermentation	ORGANOSOLS, LUVISOLS, PODZOSOLS, RANKOSOLS, etc.
		OFr	< 30% by volume fine material (faecal pellets)	foliar *r*esidues	ditto
		OFm	30–70% by volume fine material (faecal pellets)	fine organic *m*atter	ditto
		OF.c	numerous mycelial filaments	*c*hampignons (Fr. for fungi)	ditto

(*Contd.*)

Table 5.3 *(contd.)*

Horizon	Sub-horizon	Sub-division	Characteristics	Origin of symbol	Examples of RÉFÉRENCES containing the horizon
ORGANIC HORIZONS					
	OH		> 70% by volume fine humified material	*o*rganic-*h*umification	ORGANOSOLS, PODZOSOLS, etc.
		OHr	70–90% by volume fine organic matter	foliar *r*esidues	ditto
		OHf	> 90% by volume fine organic matter	*f*ine organic matter	ditto
		OHc	numerous mycelial filaments	*c*hampignons (Fr. for fungi)	ditto
		OHta	'fat' material, staining the fingers, calcic	*ta*ngel	ditto
ORGANOMINERAL HORIZONS					
A	**Organomineral surface** horizon, structured, site of the clay-humus complex			First tier from the surface	BRUNISOLS, FERSIALSOLS, NÉOLUVISOLS, LUVISOLS
	Aca		effervescence with cold HCl	cal*ca*reous A	RENDOSOLS, CALCOSOLS, CALCARISOLS, FERSIALSOLS
	Ach		rich in highly developed organic matter	*ch*ernic A	CHERNOSOLS
	Aci		non-calcareous, saturated or subsaturated	cal*ci*c A	RENDISOLS, CALCOSOLS, CALCISOLS, FERSIALSOLS
	Ado		effervescence with hot HCl	*do*lomitic A	DOLOMITOSOLS
	Ahs		dark, very rich in organic matter highly developed *in situ*; unsaturated (sombric horizon)	*h*umus-rich, *s*ombre (Fr. for dark) A	GRISOLS, PHAEOSOLS
	An		black, plastic horizon, in zones with slight fluctuation of water table	a*n*moor A	RÉDUCTISOLS
	And		predominance of amorphous minerals (allophanes)	a*nd*ic A	SILANDOSOLS
	Alu		aluminium generally complexed by organic acids; little allophane	A with complexed a*lu*minium	ALUANDOSOLS

(Contd.)

Table 5.3 *(contd.)*

Horizon	Sub-horizon	Sub-division	Characteristics	Origin of symbol	Examples of RÉFÉRENCES containing the horizon
ORGANOMINERAL HORIZONS					
	Avi		material rich in volcanic glass	*vi*tric A	ANDOSOLS, VITROSOLS
		A..h	rich in humified organic matter	*h*umus-rich A	ALOCRISOLS, ORGANOSOLS, PODZOSOLS, RANKOSOLS, VERACRISOLS, etc.
J	**Slightly differentiated or structured organomineral horizon**, little organic matter			*j*eune (Fr. for young)	COLLUVIOSOLS, CRYOSOLS, FLUVIOSOLS, THALASSOSOLS
	Js		located on the surface	*s*urface J	ditto
	Jp		located at depth	J at *p*rofondeur (Fr. for depth)	COLLUVIOSOLS, CRYOSOLS, FLUVIOSOLS, RÉDOXISOLS, RÉDUCTISOLS, THALASSOSOLS
L	**Ploughed organomineral horizon**, cultivated			*l*abouré (Fr. for ploughed)	*NUMEROUS RÉFÉRENCES*
MINERAL HORIZONS					
B	**Illuvial horizon** or accumulation horizon			B: located between surface A and weathered rock C	*see subdivisions*
	BT		clay illuviation	*T*on (Ger. for clay)	NÉOLUVISOLS, LUVISOLS
		BTβ	layer located between a BT and a non-clayey carbonate rock	Greek letter β	ditto
		BTd	BT horizon with tonguing of the E horizon	'*d*egraded' BT	LUVISOLS DÉGRADÉS, VERACRISOLS
	BP		podzolic illuviation resulting from cheluviation (podzolization, spodic horizons)	*p*odzolized B	PODZOSOLS
		BPh	accumulation of humified organic mater	BP with *h*umified matter	ditto
		BPs	accumulation of metal oxides	BP with *s*esquioxides	ditto

(*Contd.*)

Table 5.3 *(contd.)*

Horizon	Sub-horizon	Sub-division	Characteristics	Origin of symbol	Examples of RÉFÉRENCES containing the horizon
MINERAL HORIZONS					
C	**Mineral horizon without pedological structuration** but with fragmented and/or weathered mass			C: situated at base of solum	ARÉNOSOLS, COLLUVIOSOLS, AND ALMOST ALL RÉFÉRENCES
E	**Eluvial horizon** or depleted horizon			*e*luvial	NÉOLUVISOLS, LUVISOLS, PODZOSOLS
		Ea	highly weathered, disappearance of clay minerals and iron oxides; often only quartz remains	*a*lbic (= white) E	DERNIC LUVISOLS
		Eg	with rust-coloured mottles or iron-manganese redox nodules	E with pseudo*g*ley	LUVISOLS DÉGRADÉS, PLANOSOLS, VERACRISOLS
		Eh	with organic materials	E with *h*umified matter	GRISOLS, PODZOSOLS
FS	**Fersiallitic horizon**, with moderate geochemical weathering of silicates			*fersiallitic*	FERSIALSOLS
G	**Reduced horizon**, depending on permanent groundwater			*g*ley	RÉDUCTISOLS
	Go		zone of fluctuation of permanent ground water	*o*xidized G	ditto
	Gr		permanently saturated zone	*r*educed G	ditto
g	**Redoxic horizon**, depending on temporary water tables			pseudo*g*ley	FLUVIOSOLS, NÉOLUVISOLS, RÉDOXISOLS
K	**Calcaric horizon**, with *discontinuous* accumulation of calcium carbonate (= typical K)			*K*alk (Ger. for limestone)	CALCARISOLS
	Kc		with soft *continuous* accumulation	*c*ontinuous K	ditto
	Km		with hard *continuous* accumulation	*m*assive K	ditto
Na	**Sodic horizon**, with high exchangeable sodium percentage			sodium = *Na*	SODISOLS
S	**Structural horizon** or horizon of weathering			*s*tructure	BRUNISOLS, etc.

(Contd.)

Table 5.3 *(contd.)*

Horizon	Sub-horizon	Sub-division	Characteristics	Origin of symbol	Examples of RÉFÉRENCES containing the horizon
MINERAL HORIZONS					
	Sal		soil solution dominated by aluminium compounds; blocky *and* microgranular, fluffy structure	*al*uminic S	ALOCRISOLS
	Snd		dominance of amorphous minerals (allophanes)	S with a*nd*ic properties	SILANDOSOLS
	Sca		effervescence with cold HCl	cal*ca*reous S	CALCOSOLS
	Sci		non-calcareous, saturated or subsaturated	cal*ci*c S	CALCISOLS
	Sdo		effervescence with hot HCl	*do*lomitic S	DOLOMITOSOLS
	Slu		aluminium usually complexed by organic acids; little allophane	S with complexed a*lu*minium	ALUANDOSOLS
	Sp		more than 45% clay, non-calcareous, with prismatic superstructure	*p*elosolic S	PÉLOSOLS
	SV		with vertic properties	*v*ertic S	VERTISOLS
Sa	**Salic horizon** with accumulation of salts more soluble than gypsum			with high *sa*linity	SALISOLS
Si	subsurface **silicic horizon**, cemented by silica			*si*lica-rich	SILANDOSOLS
		Sim	petrosilicic horizon; indurated (duripan)	*m*assive Si	ditto
TH	**Sulphidic** or **thionic horizon** containing at least 0.75% sulphur as sulphide (on dry weight basis)			*th*ionic (= sulphur-rich)	THIOSOLS
U	**Sulphuric horizon**, with sulphates of bacterial origin (sulpho-oxidation), with pH < 3.5			s*u*lphate-rich	SULFATOSOLS
V	Deep **vertic horizon**, containing > 40% clay in the fine earth, mostly swelling; blocky or oblique platy structure			with *v*ertic properties	VERTISOLS

(Contd.)

Table 5.3 *(contd.)*

Horizon	Sub-horizon	Sub-division	Characteristics	Origin of symbol	Examples of RÉFÉRENCES containing the horizon
MINERAL HORIZONS					
X	**Gravelly** or **stony horizon** with > 60% by weight of particles > 2 cm			rich in caillou*x* (Fr. for pebbles)	PEYROSOLS
	Xc		stone content < 40% of total soil	*c*ailloux-rich X	PEYROSOLS CAILLOUTIQUES
	Xp		stone content > 40% of total soil	X rich in *p*ierres (Fr. for stones)	PEYROSOLS PIERRIQUES
Y	**Gypsic horizon**, with gypsum crust			g*y*psum-rich	GYPSOSOLS
		Yp	located at depth	Y of *p*rofondeur (Fr. for depth)	ditto
		Ys	located on the surface	*s*urface Y	ditto
	Ym	petrogypsic horizon, with indurated crust		Y with *m*assive crust	GYPSOSOLS PÉTROGYPSIQUES
GEOLOGICAL SUBSTRATA					
D	**Displaced material**, fragmented, hard			*d*isplaced material, *d*ur (Fr. for hard)	COLLUVIOSOLS, FLUVIOSOLS, etc.
M	**Unconsolidated rock**, in place			*m*euble (Fr. for unconsolidated) rock	FLUVIOSOLS, RÉGOSOLS, etc.
R	**Hard rock** in place			hard *r*ock	LITHOSOLS, ORGANOSOLS, etc.
IID, IIIM...	**Material with** mineralogical **discontinuity**			II = second type of material, III = third, etc. (I is understood)	*ALMOST ALL RÉFÉRENCES*

Notes

- The nature of the D, M and R layers can be defined more precisely by adding a suffix (e.g. Rdo = *do*lomitic rock, Mcr = *cr*aie (Fr. for chalk), Dsi = *si*liceous pebbles).
- Many suffixes can be added to the horizons for specifying certain properties or peculiarities (see Bibliography).
- Combinations of horizons are possible (e.g. LH = ploughed peat, EBT = transitional horizon between E and BT).

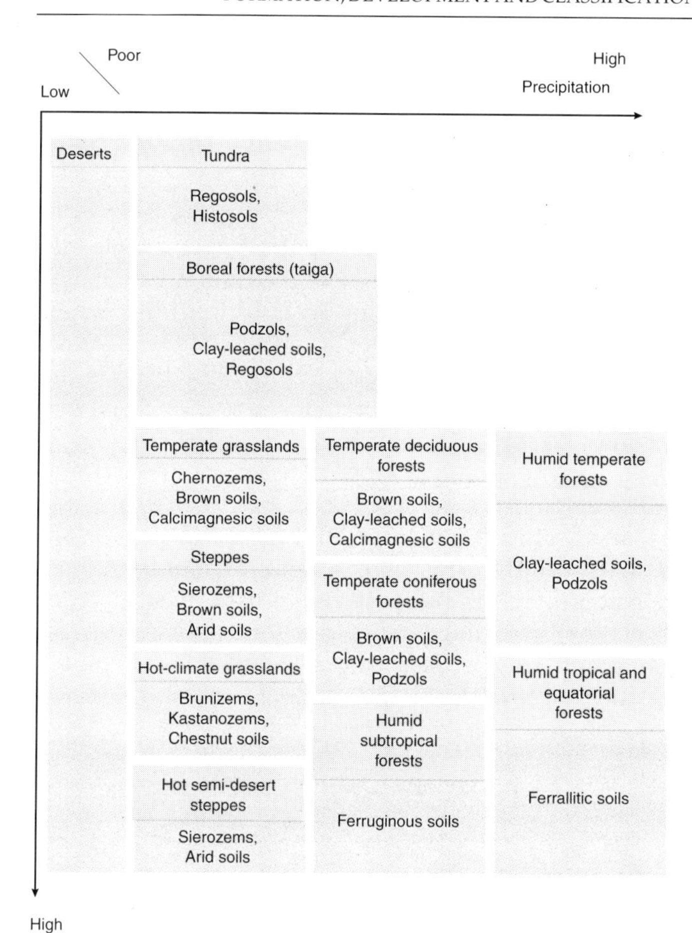

Fig. 5.11 Distribution of soils according to biomes (after Aber and Melillo, 1991; Ellis and Mellor, 1995).

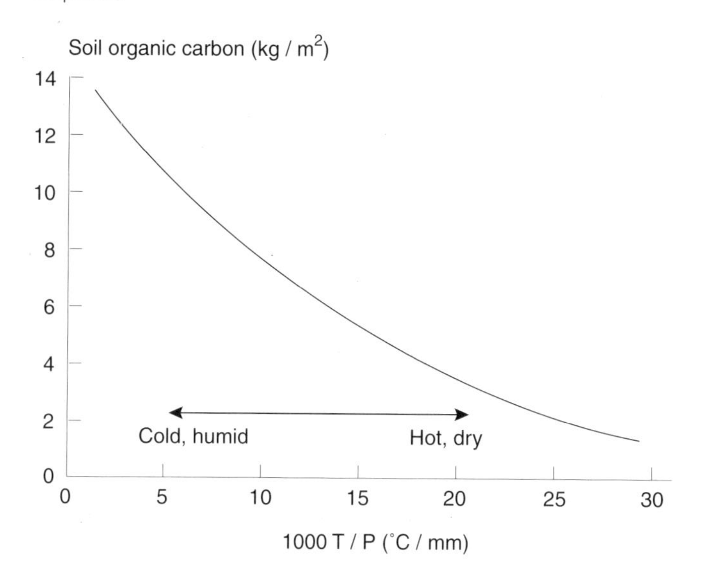

Fig. 5.12 Relation between organic carbon pool and climate in soils of New Zealand (after Tate in Ellis and Mellor, 1995).

(Strakhov in Ellis and Mellor, 1995); ***intrazonal soils*** (e.g. RÉDUCTISOLS, SODISOLS) are under the dominant influence of local factors such as hydromorphy or salinity, whereas ***azonal soils*** comprise all those that are immature (FLUVIOSOLS, RÉGOSOLS).

Permafrost (or ***pergelisol***): permanently frozen layer, mineral or organic, at a mean annual temperature < 2°C during at least two consecutive years. In summer, the permafrost thaws on the surface in its active layer (AFES, 1995, 1998).

At meso- or microclimatic scale, some soils develop differently from the general rule. For example, the ORGANOSOLS on hard limestone of the Jura mountains are mostly formed in extremely cold regions where degradation of organic matter is very slow. A well-known case (Lesquereux, 1844; Duchaufour, 1976, 1983) is that of the Creux du Van, Switzerland, where the subsoil does not thaw in summer (Fig. 5.13; Richard, 1961; Pancza, 1988). The ORGANOSOL INSATURÉ, consisting of a hydromor with sphagnum moss, rests on a ***permafrost***, and that too at an altitude of just 1100 metres (Delaloye and Reynard, 2001)! The parent mineral material (hard limestone), relief (foot of the northern slope) and vegetation (acidifying litter of sphagnum moss, ericaceous plants and spruce) also contribute to this exceptional situation for a 47° lat. N in Europe.

'At the bottom of the Creux du Vent (...) sphagnum (...) have extended their numerous stems on the calcareous debris fallen from the neighbouring rock. They form thin peat deposits, accidentally it is true, on the pure limestone (Lesquereux, 1844).

Fig. 5.13 *ORGANOSOL INSATURÉ with hydromor on permafrost (Creux du Van, Neuchâtel, Switzerland) (photo: J.-M. Gobat).*

A zonation does not always reflect a succession.

Not to be confused: zonation and succession; zonation and zonality!

Zonation indicates the distribution of biomes, ecosystems or plant communities in adjacent bands, often aligned perpendicularly to a major ecological gradient. A zonation, which is an 'instantaneous' concept, does not necessarily correspond to a ***succession***, in which the ecosystems are replaced more or less rapidly in a given area. Some zonations are stable, at the time-scale of the ecosystem, such as staggering of algal communities of the sea-coast according to the mean amplitude of the tides. Others in turn reveal rapid successions, for example when a shallow pond fills up in some decades by the centripetal advance of riverine vegetation.

Zonation of soils at the scale of biomes has led to the paradigm of zonality of soils (§ 5.6.2), one of the pioneers of which was, according to Boulaine (1989), the naturalist of Neuchâtel, Léo Lesquereux. In his book *Quelques recherches sur les marais tourbeux en général* published in 1844, he pointed out 'the curious connection' that can be made between the peat-bogs of Ireland and those of the Falklands, situated in two different hemispheres, but at the same latitude and the same mean temperature. The actual proponent of the concept of the zonality of soils was the Russian soil scientist Sibirtzev, pupil of Dokouchaev.

5.5.2 Mineral parent material

Mineralogical influences

Crystalline rocks (granites, gneisses, mica-schists, etc.) determine a pedogenesis by the acid path, in which the processes of ***brunification***, clay-leaching and cheluviation dominate. The major soils are RANKOSOLS, BRUNISOLS, LUVISOLS and PODZOSOLS. On limestones and dolomites, the predominant influence of carbonates and basic cations results in a pedogenesis in neutral or slightly alkaline medium, at times in acid medium after alkalinizing agents have been leached away. RENDOSOLS, RENDISOLS, CALCOSOLS and CALCISOLS are found here and, if the soil is acidified and the clay leached down, NÉOLUVISOLS. Volcanic materials, the recent ones in particular, induce other processes. Under hot humid climate and in weakly acid to slightly alkaline medium, hydrolysis of volcanic glass results in formation of amorphous minerals, the ***allophanes*** (SILANDOSOLS). If the material is acid or particularly rich in aluminium, in colder humid environment, this metal is chelated (ALUANDOSOLS).

Rocks define the evolutionary series.

Brunification: pedogenetic process of non-calcareous (or already carbonate-leached) environments with moderate acidity (...), characterized by the formation of an S horizon of weathering, coloured brown by iron oxides bound to clay. In the aggregates ferric iron acts as a bond between the humified organic matter and clay minerals (after Lozet and Mathieu, 1997).

Allophane: poorly crystallized aluminosilicate, seen as a gel.

The soil looks for its parents!

The actual origin of the parent material is not always easy to identify because of frequent overturning and reworking on the surface of the Earth by geological or geomorphological processes. For example, glaciers deposit moraines whose mineralogical composition depends on the nearby rocks and deposited till. This explains the presence of clay-leached soils (NÉOLUVISOLS, LUVISOLS) on the south side of the Jura Mountains, where crystalline or mixed deposits of Würm age cover the limestone with several metres of material.

Similarly, after the ice retreated, winds swept the periglacial plains still bare of vegetation, transporting silt

Loess: silty formation of aeolian origin, chiefly composed of quartz, micas, feldspars and calcite.

to occasionally distant areas. The ***loess*** thus deposited on calcareous substrata determined an acid pedogenesis, quite different from the usual calcic path. This process has been proved for example in Texas (Rabenhorst et al., 1984), northern France (Jamagne, 1973), the Swiss Alps (Spaltenstein, 1984) and even in the Jura (Pochon, 1978; Havlicek and Gobat, 1996).

The underlying hard rock is not always the 'parent' of the soil.

Lastly, accumulations of materials transported by streams (***alluviation***) or by gravity (***colluviation***) isolate the soil from the underlying rock and often redirect pedogenesis along a different path. The deposited material that comes from upstream erosion could have undergone a preliminary pedogenesis under other conditions. The new development thus depends not only on the intrinsic mineralogical characters of the material, but also on its earlier pedological development.

Physical influences

Certain physical properties of a rock correspond to its mineralogical characters.

A gneiss is much more resistant to fracturing than a carbonate rock. Similarly, the latter shows all the degrees of fissuring possible between very pure, very hard limestones (e.g. Portlandian and Kimmeridgian) and clayey limestones (e.g. Hauterivian or Oxfordian). The development, distribution and proportion of numerous types of soils depend on this gradient (Fig. 5.14).

On limestone, three developmental series of similar duration end in ORGANOSOL INSATURÉ, RENDISOL or CALCISOL.

The strongly acid ORGANOSOLS INSATURÉS form where the resistant limestone releases little calcium and clay, preventing incorporation of organic matter. They are found on ***lapiés*** or on large blocks. If the limestone is split into small pebbles or gravel ('crushed' limestone according to Bruckert and Gaiffe, 1985), incorporation of organic matter is promoted, and the large reserve of calcium prevents acidification and clay-leaching. The climactic soil is a RENDISOL that remains highly calcic. Lastly, if the limestone contains much iron oxide and/or clay, brunification and decalcification predominate; CALCISOL represents the end-member of this series. In certain conditions, development can continue by clay-leaching up to a NÉOLUVISOL.

Lapiés: superficial geological formation of hard limestone with dissected relief caused by more or less deep, often parallel, solution grooves and fissures. According to slope and altitude, lapiés are bare or covered by vegetation, for example, spruce forest or ericaceous heathlands.

Of these three development paths, two are acid, and this even in the absence of allochthonous deposition of the loess type for example. In the third, loss of carbonates and decalcification leave behind only iron oxides and/or clay minerals 'controlling' soil development. Thus, after just a few centuries, the soils of the humid calcareous massifs (the Jura, and also the Vercors and the Dolomites-Havlicek, 1999) were rapidly taken beyond the influence of carbonates and even of calcium!

But where have the carbonate soils of the Jura, the Vercors and the Dolomites gone?

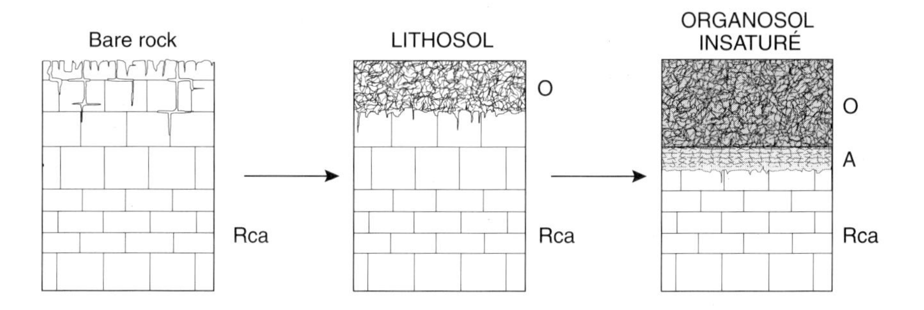

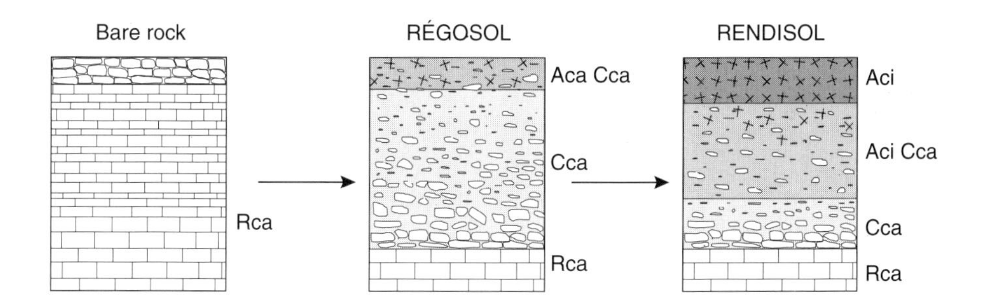

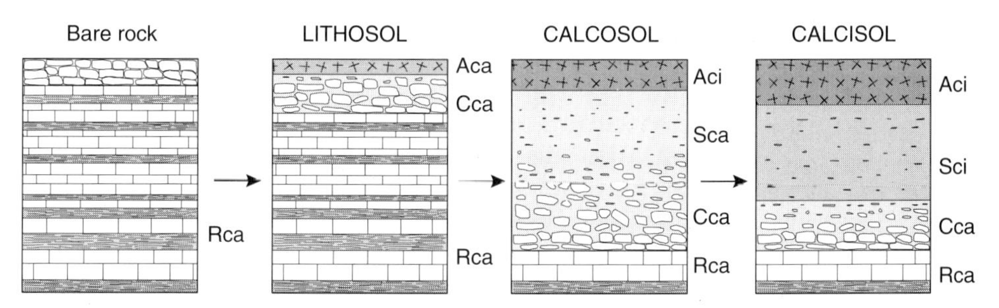

Fig. 5.14 *Comparison of development of soils on three types of limestones (after Havlicek, 1999). Explanations in the text.*

5.5.3 Living organisms and their organic matter

The major role of living organisms in the fundamental pedological mechanisms has been mentioned many times in Chapters 2 and 4.

In the general framework of pedogenesis, the most marked influence on living organisms is that of plant litter, particularly its quality. Ameliorating, mixed or acidifying, it modifies biological activity, selects the microorganisms and certain phytosaprophagous invertebrates, and releases compounds more or less resistant to decomposition (Chap. 2, § 2.2.1; Cadisch and Giller, 1997).

The high concentration of phenols and tannins in acid peats slows down mineralization.

HISTOSOLS and ORGANOSOLS are excellent examples of a direct influence of the quality of organic matter, variably attacked by decomposing agents. Litter of sedges or alder, plants of alkaline peat-bogs, provide a peat relatively poor in tannins and rich in ash, more quickly decomposed than that of sphagnum moss, ericaceous plants or pine, plants of acid peat-bogs (Chap. 9, § 9.1.4). Acidifying litter also allows cheluviation, active in PODZOSOLS for example.

Vegetation progressively 'isolates' the episolum from its original substratum, resulting in similar soils, though of different origin. This is what is called convergent soil development.

More generally, plants, especially trees, act through their litter on the entire biogeochemical cycle. On acid substratum they concentrate little by little the sparse basic cations released by the rock or added through precipitation, thus raising the pH. If this action continues long enough, the acidity can be totally neutralized and the pH can attain values similar to those observed in soils on limestone after carbonate-leaching (between 5.5 and 7.0). The CALCISOL with A-Sci-C horizons on limestone corresponds to the BRUNISOL SATURÉ with A-S-C horizons on acid rock.

Man, biological agent of all or nothing!

Man is the living organism that modifies pedological development the most, often drastically. Soils show all possible reactions to human activity, from their destruction pure and simple by tarring for example to their (re)creation by use of composts and activated sludge in the planting of road embankments or other raw mineral surfaces (Chap. 10, § 10.7.5). In certain less extreme cases, man accelerates soil development if he plants a monoculture of conifers on soils already acid and 'ripe' for podzolization. On the contrary, the farmer slows down leaching of potassium by replenishing the exchange complex by judicious application of potash fertilizer on a not excessively acid soil.

In a conifer plantation, the change in quality of the litter inhibits bacterial activity, often already quite low, and promotes activity of fungi and cheluviation.

5.5.4 Relief

Relief, often a strong restraint for soil development.

With similar climate, rock and biological activity, pedogenesis is modified by the topographic position of the soil, by debris, landslides, colluviation and erosion. Slope and relief combine to determine whether additions or losses will take place (Wysocki et al. in Sumner, 2000).

In the first case, the soil receives additional material, still mineral and unweathered as at the foot of a rock cliff, or already developed by an earlier pedogenesis.

Solifluction can occasionally turn the soil 'head over heels' and invert the horizons completely (Plates III-1 and III-2).

Solifluction: slow mass flow of a soil in the form of mud, particularly in cold climate with frozen subsoil (Lozet and Mathieu, 1997, modified).

In the second case, the soil is rejuvenated by erosion and removal of its superficial layers, especially if it is on a summit. A new development then begins from the underlying layers, as seen in the Mediterranean zone. Erosion of FERSIALSOLS following deforestation of oak forests and introduction of pasturage brought to light the FS horizons, more resistant because of their subangular blocky structure. The recent diminution of agricultural activity now sees garrigue, maquis or forest reclaiming their rights, but in a different bioclimatic situation that rather favours brunification (Godron, pers. comm.).

Relief influences hydromorphy too. An entire range of soils can be seen in a drainage basin, from soils of dry regions (e.g. CALCOSOL) to those of humid areas (RÉDUCTISOL) with numerous possible intermediates (CALCOSOL rédoxique, CALCOSOL réductique, CALCOSOL-RÉDUCTISOL, etc.). The oxidation-reduction mechanisms characteristic of G and g horizons replace carbonate leaching, brunification and clay-leaching.

Hydromorphy affects the development processes in aerated soils even to masking them completely, thus directing pedogenesis.

5.5.5 Time

Time: human scale or ecological interval?

Time is a peculiar ecological factor: it can be expressed as 'conventional' time, that of the clock, or as an 'ecological' duration, corresponding to the interval during which a development process has operated. In the former case, we talk of age of the soil, in the latter, its degree of development. What is the relation between these two approaches to pedological time? Bockheim (in Ellis and Mellor, 1995) proposed a function of the type $y = a + b \cdot \log t$, or $\log y = a + b \cdot \log t$, y being a soil property dependent on time t. An example is the rate of loss of carbonates in FLUVIOSOLS (Bureau, 1995).

Generally, if there is no major disturbance, the degree of development of a soil is a positive non-linear function of time.

'Time, this insidious factor that works on so many phenomena ...' (Bourguignon, 1996).

A soil develops in different phases, from its birth to a stage at which the macroclimatic, biological and geological conditions modify it no more. The climax is reached, a state of equilibrium expressed by biomes for example. Thus it can be assumed theoretically that all PODZOSOLS of the taiga attained their climax at the same age, after an identical duration of development, subsequent to retreat of the glaciers. Here, age and degree of development would signify the same thing.

All post-Würmian soils have almost the same age, with the exception of those of recent periglacial regions and of frequently reworked areas.

Soils of the same age may exhibit different degrees of development according to their own reaction to micro-climatic, mineralogical and topographic factors.

In reality, this is rare because of other pedogenetic factors that supplant the general climate and slow down or accelerate climatic development of the soil (§ 5.1.2). Without their action, all soils of the regions freed at the end of the Würm would have been identical, probably PODZOSOLS! In fact, the present extreme pedological diversity shows that soils of the same age in the regions affected by this last glaciation have very different degrees of development.

Age and development of soil, between positive and zero correlations.

Thus the age of a soil, expressed in years, should be clearly distinguished from its degree of development determined on the basis of morphological and analytical characters:

- ***age of a soil*** indicates the time that has elapsed until now from the beginning of pedogenesis, marked by the first organic depositions on a mineral substratum;
- its ***degree of development*** defines the state in which it is presently found, as a product of physical, chemical and biological processes that have operated on it.

What is the age of horizons and of soils?

The age of the horizons? Between one year and several thousand years!

Duchaufour (1983), Ellis and Mellor (1995) and Righi and Meunier (in Velde, 1995) give the time necessary for formation of certain horizons and for their equilibration with environmental conditions (Fig. 5.15):

- OL 1 year,
- OF and OH 3 to 10 years,
- A 600 to 1500 years,
- BPh and BPs 500 to 4000 years,

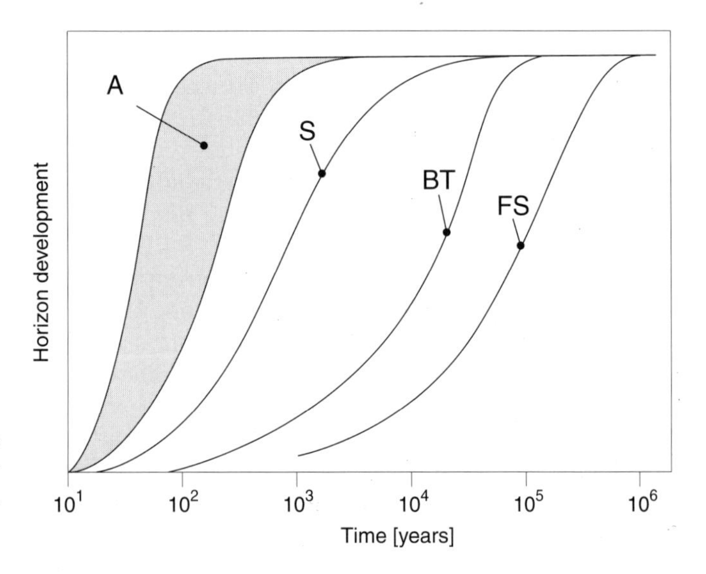

Fig. 5.15 *Development time for some horizons (after Mellor, in Ellis and Mellor, 1995).*

- Sci 3000 to 5000 years,
- S at least 5000 years,
- E and BT 6000 to 10,000 years.

For the whole soil, just one 'overall' estimate of its age can be made based on a general, slightly intuitive evaluation of the time required to end up in a 'steady-state' functioning of the entire solum. Fortunately, precise determination of certain soil constituents can be of help, such as composition of the weathering complex or the rate of loss of carbonates (Bureau et al., 1994); dating of the organic matter is also used, with ^{14}C for relatively recent times (Fig. 5.16) or with ^{13}C for longer periods.

How to determine the time necessary for formation of an entire soil, which is an assemblage of horizons with different degrees of development or ages?

Based on ^{14}C, Balesdent (1982) dated the organic matter of NÉOLUVISOLS of the Jura at 700 y in the E horizon, 1860 y in the BT1 and 4800 y in the BT2. These ages are the means of young and older organic matter mixed in each horizon by intense bioturbation by earthworms. In an American Prairie soil, the fulvic acids had an average age of 630 y, humic acids 1308 y and humin 1240 y (Aber and Melillo, 1991).

Organic matter categories of different ages.

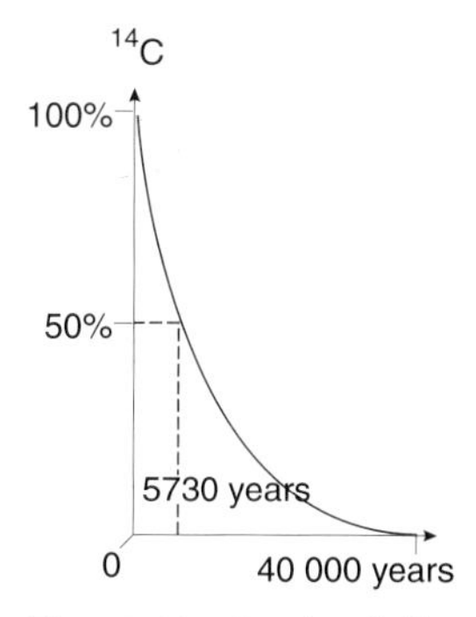

Fig. 5.16 Graph of disappearance of ^{14}C as function of time (after Roberts, in Ellis and Mellor, 1995).

Schwartz (1988), using ^{13}C, reconstructed the history of PODZOSOLS of the Congo from 30,000 years ago. Well adapted to the study of tropical soils, ^{13}C permits distinction of organic matter originating from grasses and from trees (Guillet et al., 2001). The relative dynamics of these two plant formations can thus be compared to that of the soils (Chap. 7, § 7.7.2). Again, using ^{13}C, Calderoni and Schnitzer (1984) found ages of 8000 to 31,000 years in some palaeosols of southern Italy, with an organic matter that has undergone practically no change during this period of time.

Another dating technique makes use of the property of certain amino acids of changing their molecular configuration with passage of time, after the death of organisms. In Australia for example, Milnes (in Martini and Chesworth, 1992) showed that the ratio between the ***dextrorotatory*** and ***laevorotatory*** forms of aspartic acid increases with age of the soil.

Dextrorotatory, laevorotatory are said of a molecule that rotates the plane of polarization of light clockwise (dextrorotatory) or anticlockwise (laevorotatory). Amino acids are generally laevorotatory, while natural glucose is dextrorotatory.

But it should not be forgotten that in a soil the three phases—solid, liquid and gas—and their interactions do not develop at the same pace. Thus, the solid phase must be considered with its dynamics and changes by the pedologist who is retracing the history of the soil, but it can be seen as undeformable and invariant by the soil physicist who studies hydric flows (Musy and Soutter, 1991). Also, at the very interior of this phase, physicochemical reactions affecting minerals for example, require periods between a microsecond and a millennium to achieve equilibrium (Fig. 5.17).

Two simplified age categories

Long and short cycles.

Because of difficulties and lack of certainty of the actual age of soils, pedologists have established rather broad age categories as a function of the dominant macroclimatic influences (Fig. 5.18; Duchaufour, 2001)).

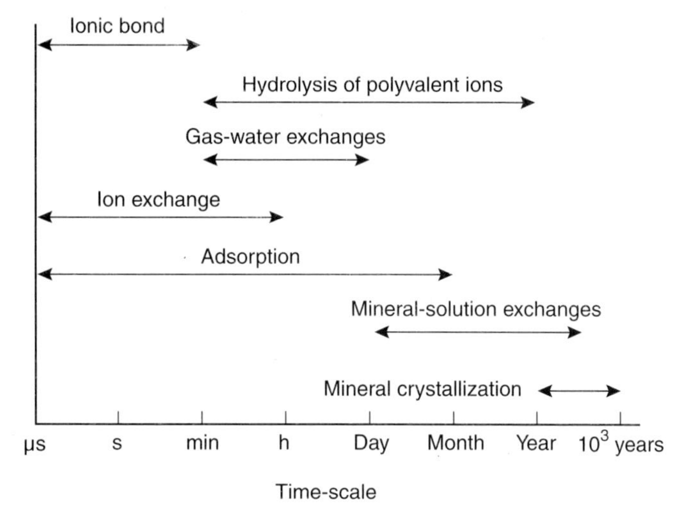

Fig. 5.17 *Time-scale of some physicochemical reactions of soil minerals (after Sparks, 1995).*

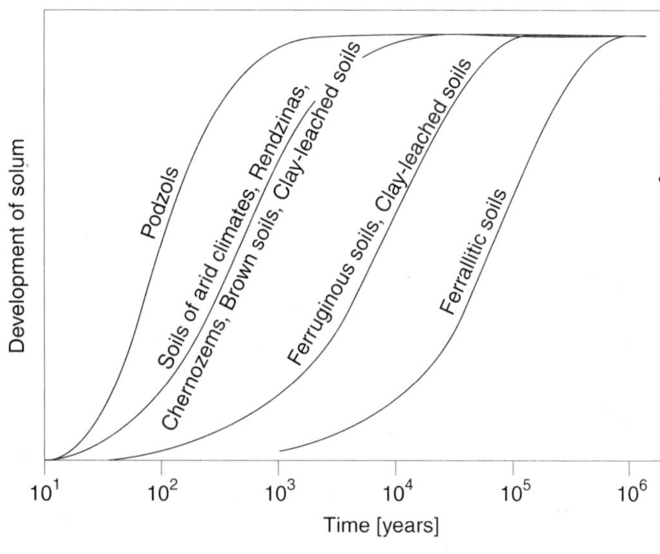

Fig. 5.18 *Time of development of some soils (after Birkeland, in Ellis and Mellor, 1995).*

When does pedogenetic development start?

The starting point, the date of birth of soils presently seen, is very difficult to fix! We can surely identify the precise moment of a recent alluvial deposition; but here it is a matter of geomorphology and not of soil science, because the soil is created when the organic and mineral worlds interact. In these conditions, at what time can the beginning of the 'real' soil be dated? Theoretically, from the arrival of organic matter. But can we really speak of soil when fungal spores, pollen grains or some dead leaves are deposited on a mineral substrate one or two days after a flood? Or must we wait a few months, with the first decomposing organisms acting, or several years, at the beginnings of clay-humus integration?

In reality, it is impossible to attribute a precise overall date of birth to the soil. It should not be forgotten that the soil is an ecological system structured in space and time (Chap. 1, § 1.2.2). Several temporal levels coexist in it, particularly at the scale of the horizons: an OL layer can be deposited suddenly after a flood, forming a solum of the OL-D type, perhaps only a few days old. After a few years, the solum will be of OL-Js-D type and one must wait several decades or centuries to see a true A horizon form. We conclude that fixing a starting point for development of a soil is arbitrary and totally determined by the property selected as 'initializing', and by the horizon corresponding to it.

'The point at which this material (weathered rock and mineral material) becomes soil is not clearly defined' (Wood, 1995).

The younger the soil, the more important the criteria linked to organic matter, and therefore to biological activity, are compared to mineral characters.

OL-D is not the oldest ...!

• Soils of hot regions, whose development depends primarily on that of oxides, are formed during ***long cycles***, between 100,000 years for certain Ferruginous soils and one million years for Ferrallitic soils. They can attain a thickness of several metres and are generally polycyclic.

• On the contrary, soils of cold temperate climates develop rapidly under the influence of precipitation and organic matter. Often monocyclic, they reach their climactic equilibrium in less than 10,000 years and rarely exceed one to ten metres in thickness. They are called ***short-cycle*** soils.

For example, Rehfuss (1981) attributed 2500 to 7000 years to PODZOSOLS on aeolian sands of northern Germany (short cycle), while Foth (1990) showed that clay-leaching in soils of California on granite has progressed without interruption for 140,000 years (long cycle).

This distinction into two broad development periods is not absolute, however, as pedogenesis can begin in a short cycle and continue in a long cycle, the processes themselves being of variable duration. Soil acidification and loss of iron observed on very old dunes of Australia began a few centuries after their formation and continued for 200,000 years. But the loss of carbonates was very

The distinction between short and long cycles is simplistic, but is very useful for comparing soil development to evolution of climates for example.

early, typical of the short cycle, and cheluviation late, starting only 50,000 years later (McArthur and Bettenay in Gerrard, 1992).

Anachronistic soils

When time plays a trick on the pedologist: polycyclic and related soils!

The example of dune soils of Australia shows that it is not always easy to ascertain whether a soil has undergone one or several cycles of development. Thus, some soils of temperate regions contain horizons surely formed in climatic conditions other than those of the present day. This holds true for the rubefied soils of the Jura in which iron-rich BT horizons reveal an ancient development under a climate permitting ***rubefaction*** (Guenat, 1987). There is another possibility, that of an ancient A horizon of a BRUNISOL buried under colluviated material, which acts as a BT horizon in a new pedogenesis, in a stable landscape.

Rubefaction: pedogenetic mechanism characterized by evolution of iron in FERSIALSOLS, with dehydration of iron oxyhydroxides released by fersiallitic weathering (Tab. 5.2) and bound to clay minerals (Lozet and Mathieu, 1997).

The variety and large number of soils not conforming to general monocyclic development have led to definition of the following terms (after Duchaufour, 1983):

Recent, monocyclic, ancient, 'palaeo-', fossil, polycyclic, composite, complex or polyphasic: the jungle of anachronistic soils!

- ***recent*** or ***monocyclic soils***: short-cycle soils formed in a unit of temporal, climatic and geological development, generally post-glacial;
- ***ancient soils***: short-cycle, particularly aged soils, scarcely or not at all touched by glaciations;
- ***palaeosols***: soils formed under conditions, especially climatic, different from those of the present; very general term;
- ***fossil soils***: palaeosols buried under thick recent alluvial deposits, which cannot be reached by plant roots and are not interfering with present-day pedogenesis;
- ***polycyclic soils***: soils subjected to at least two successive development cycles, corresponding to two different climatic periods; by extension, soils that have undergone great change in vegetation conditions with pedogenetic repercussions are also termed polycyclic;
- ***composite soils***: soils formed from two superposed geological beds but without present-day pedogenetic interference; the deeper one could be an ancient soil;
- ***complex soils***: similar to the preceding, they differ in having the deeper layer included in present-day pedogenesis;
- ***polyphasic soils***: monocyclic soils with buried horizons but in which the process of covering again forms part of the basic functioning, under the same macro- or mesoclimatic conditions (e.g. alluvial soils often covered

by new material); for Lozet and Mathieu (1997), polyphasic is synonymous with polycyclic.

5.5.6 General pattern of influences on soil development

Be careful of theoretical development series ...

The many possible combinations of factors influencing pedogenesis imply several development paths, in spite of the existence of the single principle 'milieu → processes → characters' (§ 5.1.2). Indeed, the literature presents broad dynamic schemes linking, for example, the RANKOSOL with the PODZOSOL, passing through the BRUNISOL and NÉOLUVISOL. If it is mentioned clearly that these development series are theoretical, established for chronologically ordering processes that can operate only in a specific sequence, the idea is admissible, even desirable!

... they must effectively reflect the reality of the land ...

... but cannot be applied to all actual cases!

On the other hand, their direct application to actual cases seen on the landscape will often be erroneous. For example, the change of a BRUNISOL to a PODZOSOL, theoretically possible, is very rare in nature, the respective durations needed for formation of S and E-BP horizons not often being comparable, at least in post-glacial development. Establishment of real, concrete development series must respect certain criteria of climatic, mineralogical and topographic uniformity, as Rehfuss (1981) and Kuntze et al. (1988) have for example done very well.

A theoretical explanatory model of soil development.

In view of the elaboration of development series which remains, despite everything, an essential aim of pedology, Figure 5.19 attempts to summarize the multifactor circumstances of soil formation and to demonstrate the constraints that must be taken into account.

Based on the theoretical model of development LITHOSOL – BRUNISOL – PODZOSOL, it is seen that pedogenesis is subject to factors, processes and events:

• accelerators (planting of conifers 'wrapping up' an already present but diffuse podzolization in an ALOCRISOL HUMIQUE; overfertilization with potash of an BRUNISOL OLIGOSATURÉ provoking the leaching of calcium and the start of clay-leaching);

• decelerators (control of acidification of a BRUNISOL MÉSOSATURÉ by application of basic cations; maintenance of a pool of basic cations by a deciduous forest on a soil with low base saturation percentage);

• blockers (regular addition of limestone pebbles at the foot of a cliff and persistence of a RENDOSOL; cold

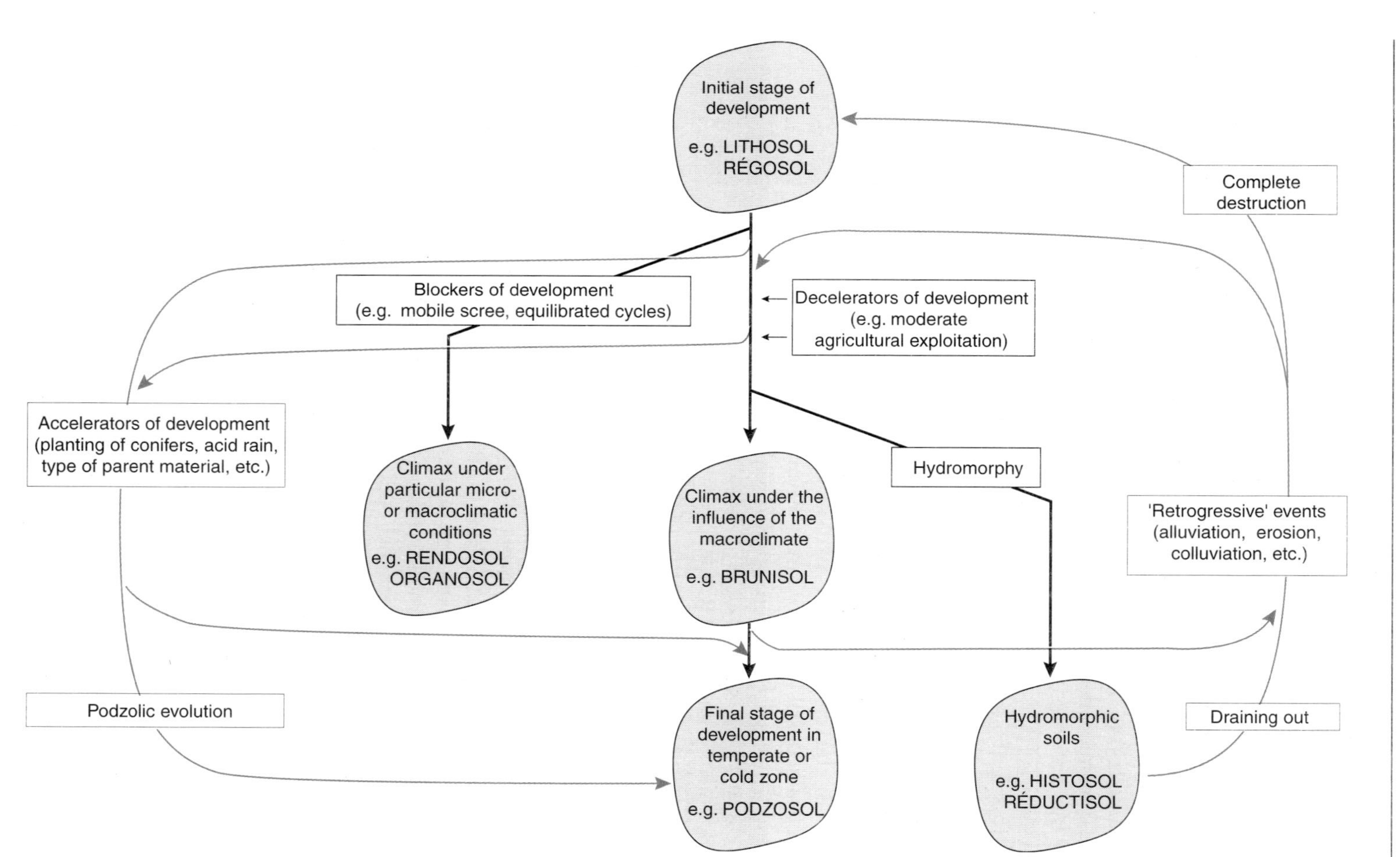

Fig. 5.19 *Theoretical explanatory model of the influences working on soil formation and development.*

microclimate preventing degradation of the organic matter of an ORGANOSOL);

- retrogressive factors (covering of a developed soil by a thin deposit of alluvial sands or of volcanic ash; erosion of a poorly structured organomineral horizon);
- destroyers (major landslide baring the underlying marl; thick deposit of construction-site materials without prior removal of the arable land cover);
- reorienters (change of aerated pedogenesis to hydromorphic development by rise of a water table; change in duration of frost in Pergelisol following climatic changes).

5.6 ORDERING THROUGH CLASSIFICATION AND NOMENCLATURE

In all sciences, man has to classify the objects studied. The soil, very varied because of the combined action of pedogenetic factors, is not an exception. This section presents the broad principles of soil classification, the theme of the book not requiring details indispensably. Two types of classification and nomenclature are presented. The reader will find more easily in pedological literature the definitions sought on such or such soil or the relations with other classifications, particularly Soil Taxonomy (USDA, 1999; Spaargaren, in Sumner, 2000) or the World Reference Base (ISSS, 1998). As for the humus forms, more relevant to the aim of this book, they form the subject of a special chapter (Chap. 6).

Classification, pedological or not, permits establishment of relational links between the objects studied and communication of the data.

Spaargaren (in Sumner, 2000) and Duchaufour (2001) have given a good review of the ensemble of classifications.

Three institutions, three abbreviations

For learning more about the classification and nomenclature of soils, consult CPCS (1967), USDA (1975), Duchaufour (1983, 2001), Mückenhausen (1985), Dudal (1990), Driessen and Dudal (1991), Spaargaren (1994), AFES (1995, 1998), Lozet and Mathieu (1997), Brunner et al. (2002).

CPCS = Commission de Pédologie et de Cartographie des Sols; USDA = United States Department of Agriculture; AFES = Association Française pour l'Étude du Sol.

Consult and decipher!

5.6.1 Difficulties in soil classification

While in biology agreement could be reached more or less on a single principle of classification (families, genera, species, binomial nomenclature, etc.), for want of agreement on its application (Chap. 12, § 12.3) this is not

The soil, though living, is not an organism!

It is always the soil scientist who decides where the soil 'begins' or 'ends'.

yet true for soil science. Indeed this is because of the relative youth of this science, but also because of an important peculiarity of the soil. The soil is a *continuous* natural object, the soil mantle, contrary to a daisy or a trout, of which at least the physical limits are evident. Thus it devolves on the observer to decide himself where the object of his study 'stops', as must be done for all ecological systems (Chap. 1, § 1.2.2).

Let us enter the third dimension! The horizon, as the fundamental functional unit of the soil, tends to be 'autonomous'.

Most soil descriptions have been made till now on vertical sections; but it is also possible to consider the horizon as an independent three-dimensional unit and to 'follow' it through the soil mantle (Fig. 5.20; Boulet et al., 1982; Girard, 1989; Ruellan et al., 1989). This three-dimensional approach will perhaps culminate in classifications of the entire soil mantle, at the catena level,

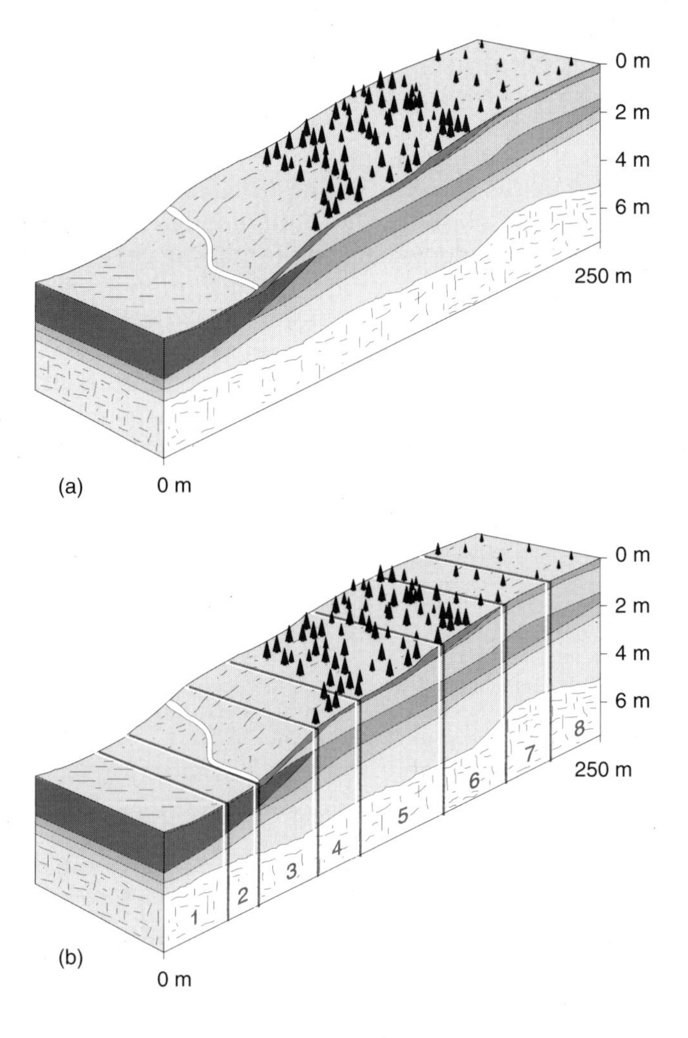

Fig. 5.20 *Two ways of approaching the soil mantle: (a) through horizons considered as homogenous volumes, (b) by the 'classic' vertical solum (after Brabant, in Legros, 1996).*

particularly through Geographic Information Systems (GIS) (King et al., 1994; Legros, 1996). However, for the time being, the great majority of classifications are based on study of sola described from vertical sequences of horizons.

The difficulties in fixing an objective descriptive limit to the soil, added to its complexity, mean that pedologists still do not agree on a single nomenclatural system. According to regions, climates, technical possibilities, socio-cultural environment, etc., soil scientists have established systems that often well suited their own way of looking at things, or their work conditions, but were not always useful otherwise. This apparent incompatibility of classifications has led scientists to avoid the perils of hierarchization of criteria and to rather define ***reference bases***, catalogues of model soils, theoretical or concrete, for comparing field observations against. These lists have the advantage of being very flexible in use, but their drawback is they do not always easily reveal the ecological factors determining such or such pedogenesis.

Soil classification: more a Tower of Babel than Esperanto... But we progress!

'By the diversity of its creations, Nature often escapes the classifications we establish to subject it to our impotence' (Lesquereux, 1844).

Scientific schools and socio-geographic context

In phytosociology (Chap. 7, § 7.1.4), the Scandinavian school has developed a highly landscape-based, physiognomic approach to vegetation (Lippmaa, 1939), while the alpine countries, with greater biodiversity, have favoured the floristic view (Braun-Blanquet, 1964). But these apparently opposing trends tend to come together in the new paradigm of the integrated synusial approach (Gillet et al., 1991).

In pedology, the Russians, Americans and Europeans study the soil with different priority criteria; the humus forms change according to whether the observer is a person 'of the hills' or more 'of the mountain' (Chap. 6, § 6.2.2).

In numerous areas of natural sciences there is a close link between scientific schools and their geographic or socio-cultural context.

5.6.2 The hierarchical classifications

Three great schools

Historically and by simplification, three great schools of classification developed, each with its own philosophy but yet in some aspects converging with the others:

- the Russian classification,
- the American classification *(Soil Taxonomy)*,
- the European classifications.

In the Russian classification of Dokouchaev and Sibirtzev, the broad categories of soils are delimited by the influence of macroclimate. The lithological and hydric aspects intervene at the second level and the development

The Russians rely much on climate; their classification is zonal; it respects the changes seen between the biomes.

processes (incorporation of organic matter, clay-leaching, podzolization, etc.) only at the lower levels of the hierarchy. The Russian classification inspired the first American classification, that of Baldwin in 1938, which was also climate-based. This paradigm of zonality of soils is occasionally criticized, notably by Paton et al. (1995), who do not accept an exclusively zonal view based on the study of vertical profiles alone, and who insist on supplementing it by an approach that gives more importance to the processes of lateral transport and pedoturbation.

Some great names of world soil science (after Boulaine, 1989, 1997)

Vassili Vassilievitch Dokouchaev (1846–1903) is considered the father of soil science in general and of the Russian school in particular, the other chief scientists of which were N.M. Sibirtzev (1860–1899, zonality of soils), K.D. Glinka (1867–1927, general concepts of soil science), V.I. Vernadsky (1863–1945, geochemistry of the ecosphere), A.A. Rode (1896–1979, moisture regime of soils) and also N.A. Dimo (1873–1959, soil systematics).

In the United States, Eugen Woldemar Hilgard (1833–1916) truly 'launched' soil science. Highly analytical with his many profile descriptions in the field, he is the inventor of the C/N ratio. His overall view of soil did however not neglect the importance of climate in pedogenesis. Among the American soil scientists can also be mentioned F.H. King (1848–1911, soil chemist and hydrologist) and M. Whitney (1860–1927), soil mapper, opposed to the ideas of Hilgard and of King.

European soil science, more than the Russian and American, underwent historic vicissitudes in the respective countries and progressed more chaotically. Among the numerous investigators may be mentioned A. Demolon (1881–1954), who completely reorganized French soil science after the First World War, a role taken by A.D. Hall (1864–1942) in Great Britain and by E. Blanck (1877–1953) in Germany.

Other soil scientists are mentioned elsewhere in this book, where their direct contributions are covered.

Americans analyse; they place emphasis on the precise physico-chemical characters of soil.

The American classification is highly hierarchized, from the upper levels down. The results of physicochemical analyses (organic matter content, base saturation percentage, texture, etc.) allow connection of real horizons described in the field and laboratory to diagnostic horizons defined by an assemblage of characters (§ 5.4.2). The diagnostic horizons differentiate orders, suborders or types based on a very logical nomenclature of Greek and Latin prefixes and suffixes bringing to mind the essential properties. The American classification approaches the European classifications in its morphogenetic character. It directly inspired the *FAO Soil Legend*, established in 1975 and revised in 1989, which serves as basis for the Soil Map of the World.

Lastly, in Central and Western Europe, soil scientists best took into account the basic principle 'milieu → processes → characters' (§ 5.1.2), without however reaching unity of viewpoint. Fortunately, the conceptual distance between the schools is relatively small since all or nearly all combine the internal analytical characters preferred by the Americans and the environmental criteria given priority by the Russians. These systems are called ***morphogenetic classifications***: morphology defines the solum, it is analytical; pedogenesis explains, it is synthetic.

Europeans try to synthesize the best aspects of the Russian and American classifications ...

... but they establish as many classifications as there are countries, even more!

An example of hierarchical classification, *Soil Taxonomy*

Eight major diagnostic horizons are defined in ***Soil Taxonomy***, the first four on the surface of the soil, the others at depth (USDA, 1999); the horizons are:

- ochric: light-coloured, poor in organic matter,
- mollic: dark, > 1% organic matter, base saturation percentage > 50%, C/N ratio < 17,
- histic: peaty material,
- umbric: same as mollic, except base saturation percentage < 50% and C/N > 17, acid,
- cambic: horizon of incomplete weathering of primary minerals, CEC < 16 cmol (+) kg^{-1} clay,
- oxic: horizon of complete weathering of primary minerals, CEC > 16 cmol (+) kg^{-1} clay, greatly coloured by iron,
- argillic: accumulation of illuviated clay,
- spodic: podzolic accumulation.

Soil Taxonomy is one of the fruits of the American school of soil science, very active at the beginning of the twentieth century. C.F. Marbut (1863–1935) synthesized numerous ideas on soil whereas C.E. Kellogg (1902–1980), through his position as Director of the Bureau of Soils *(Soil Survey)* from 1934 to 1971, created the grounds for a new classification. The drafting of *Soil Taxonomy* was directed by G.D. Smith (1907–1981) and took 23 years.

Secondary diagnostic horizons define particular processes: albic, calcic, gypsic, natric, etc. horizons.

The diagnostic horizons enable assigning the solum to one of the following eleven orders: Entisols, Vertisols, Inceptisols, Aridisols, Mollisols, Spodosols, Alfisols, Ultisols, Oxisols, Histosols and Andisols. These orders are subdivided into suborders, often delimited by the soil climate. For example, the order Mollisols, which roughly corresponds to Calcimagnesic soils (CPCS, 1967), is subdivided into *Rend*olls (calcareous Mollisols), *Aqu*olls (hydromorphic Mollisols), *Xer*olls (Mollisols of dry regions), etc., by adding a prefix. Each suborder contains great groups, defined by a second prefix (e.g. *Calci*xerolls, *Argi*xerolls).

By way of example, a CALCOSOL (Calcareous Brown soil) of temperate regions according to AFES (1995, 1998) corresponds to a *Mollic Eutrochrept* in *Soil Taxonomy*. It belongs to the order Inc*ept*isols (soils with cambic horizon, here Sca horizon) and to the suborder Ochr*epts* (Temperate Brown soils).

5.6.3 The reference bases

Types of reference bases

Let us avoid the errors, at times due to lack of control, of systematic botanists, microbiologists and zoologists.

While *Soil Taxonomy* and other hierarchical classifications require classification priorities in the choice of criteria, this is not needed by reference bases, which are meant to be simple to get to and flexible in use. A certain number of model soils are defined based on vertical sequences of horizons. They are (i) theoretical and synthetic reference bases such as the *Référentiel pédologique* (AFES, 1995, 1998), or (ii) actual profiles, as in the CPCS/Duchaufour reference base (Duchaufour, 2001). In spite of this difference, reference bases resemble each other more than hierarchical classifications, the danger of priorities generally being avoided. Here the risk is that of multiplying the References, unless strict control is exercised by systematic soil scientists.

'Most attempts at hierarchical classification have, it must be realized, ended in failure, because it is impossible to select a criterion (or a group of criteria) that would be fundamental enough to efficiently define two or three great divisions at the highest level' (Duchaufour, 1983).

In French-speaking countries, the most popular reference bases are the *Référentiel pédologique* (AFES, 1995, 1998) and the CPCS/Duchaufour reference base (Duchaufour, 2001); at world level, the WRB (*World Reference Base for Soil Resources)* (ISSS, 1998) gradually becomes the common denominator. A synthesis among the reference bases and classifications has been proposed under the authority of the FAO and the UNESCO (Driessen and Dudal, 1991).

The first example, the *Référentiel pédologique*

The ***Référentiel pédologique:*** 102 Références, subdivided into numerous ***Types*** by adding qualifiers.

The *Référentiel pédologique* (AFES, 1995, 1998) now presents 102 ***Références***, which might expand to 150 for all the soils of the world. The typology of the Références first takes into account the morphology of the sola, prefixing the major pedological characters that mirror the functioning of the soils (Fig. 5.10). The properties of reaction to human intervention (agronomic, silvicultural or geotechnical) and those of natural functioning, like the hydric regime, are also important. Lastly, the pedogenetic processes intervene in the general interpretation of the sola, thereby respecting the morphogenetic principle of the European schools.

Références are defined by the vertical sequences of horizons, occasionally by environmental criteria; they are listed in Table 5.4. Equivalents in the CPCS/Duchaufour reference base are also given; they are of variable precision because the respective conceptual bases may be different.

Table 5.4 Références according to the *Référentiel pédologique* (AFES, 1995, 1998) and possible equivalents in the CPCS/Duchaufour reference base (Duchaufour, 2001)

Référence	Equivalent in CPCS/Duchaufour	Référence	Equivalent in CPCS/Duchaufour
ALOCRISOLS (2)	Acid Brown soils, Ochrous Brown soils, Alpine Rankers	PÉLOSOLS (2)	Pelosols
ALUANDOSOLS (3)	Andosols	PEYROSOLS (2)	?
ANTHROPOSOLS	?	PHAEOSOLS (2)	Brunizems, Phaeozems
ARÉNOSOL (1)	?	PLANOSOLS (3)	Planosols
BRUNISOLS (4)	Acid Brown soils, Eutrophic Brown soils	PODZOSOLS (7)	Podzols, Cryptopodzolic Rankers, Podzolic soils, Ochrous Podzolic soils
CALCARISOL (1)	?	POSTPODZOSOL (1)	?
CALCISOL (1)	Calcic Brown soils	PSEUDOLUVISOL (1)	?
CALCOSOL (1)	Calcareous Brown soils, Brunified Rendzinas	QUASILUVISOL (1)	?
CHERNOSOLS (3)	Chernozems	RANKOSOL (1)	Alpine Rankers
COLLUVIOSOL (1)	Colluvial soils	RÉDOXISOL (1)	Pseudogleys
CRYOSOLS (2)	Cryosols, Arctic Polygonal soils	RÉDUCTISOLS (3)	Gleys, Stagnogleys, Amphigleys
DOLOMITOSOL (1)	Dolomitic Rendzinas	RÉGOSOL (1)	Regosols
FERSIALSOLS (4)	Fersiallitic soils, Desaturated Fersiallitic soils	RENDISOL (1)	Rendzinas, Humo-calcic soils
FLUVIOSOLS (3)	Alluvial soils	RENDOSOL (1)	Rendzinas, Humo-calcareous soils
GRISOLS (3)	Forest Grey soils	SALISODISOL (1)	Saline soils
GYPSOSOLS (2)	Subdesertic Grey soils	SALISOLS (2)	Saline soils, Solontchaks
HISTOSOLS (7)	Peats	SILANDOSOLS (4)	Andosols
LEPTISMECTISOL (1)	Vertisols	SODISALISOL (1)	Alkaline soils
LITHOSOL (1)	Lithosols	SODISOLS (3)	Alkaline soils, solonetz, soloths
LITHOVERTISOL (1)	Vertic soils, Vertisols	SULFATOSOL (1)	Acid Sulphate soils
LUVISOLS (4)	Clay-leached soils, Dernopodzolic soils	THALASSOSOL (1)	?
MAGNÉSISOL (1)	Dolomitic Rendzinas	THIOSOL (1)	Thiosols
NÉOLUVISOL (1)	Clay-leached Brown soils	TOPOVERTISOL (1)	Vertic soils, Vertisols
ORGANOSOLS (4)	Humo-calcareous soils, Humo-calcic soils, Lithocalcic soils	VERACRISOL (1)	?
PARAVERTISOLS (2)	Vertic soils, Vertisols	VITROSOL (1)	Andosols, Vitr isols

Note: The Références are given by their first term, their total number being specified in parentheses.

A three-step procedure analogous to that of phytosociologists (Chap. 7, § 7.1.4).

To summarize, taxonomic study calls for three different 'soils', successively envisaged (AFES, 1995, 1998; Legros, 1996):

- the actual soil: the reality, part of the soil mantle,
- the 'picture' soil: data on the actual soil (= profile),
- the conceptual soil: the Reference or the taxon in a classification.

Plate IV presents nine relatively frequent Références in Western Europe.

The second example, the CPCS/Duchaufour reference base

The CPCS/Duchaufour reference base (Duchaufour, 2001), briefly presented here in a manner suitable for making comparisons, is based on the description of type profiles (or better, solum-types) that reflect, by their organization and properties, one or more of the fundamental processes of development that were presented in Table 5.2. The type profiles are grouped into 13 classes, based on the predominance of one basic process, while each class is divided into subclasses as a function of particular environmental factors. Table 5.5 presents the classes and subclasses of this system and some possible matches with the *Référentiel pédologique*.

Table 5.5 Soil types according to the CPCS/Duchaufour Reference Base (after Duchaufour, 2001, simplified) and dominant pedogenetic processes

Dominant pedological process	Class	Criteria delimiting the subclasses	Soil types (generally subclasses)
None (little development)	I. Immature soils	Additions, erosion, climatic processes	Lithosols, Regosols, Alluvial soils, Colluvial soils
Cryoturbation, calcification	II. Soils under extreme climatic conditions	Cryoturbation, calcification	Cryosols, Desert soils, Regs, Ergs, Sierozems
Cryptopodzolization or andosolization	III. Poorly differentiated soils	Formation of 'alpine' humus, cryptopodzolization, andosolization	Rankers, Andosols
Carbonate-leaching, decalcification	IV. Calcimagnesic soils	Calcic melanization, brunification	Rendzinas, Humo-calcareous soils, Humo-calcic soils, Calcareous Brown soils, Calcic Brown soils
Brunification, clay-leaching	V. Brunified soils	Saturation, acidification	Acid Brown soils, Eutrophic Brown soils, Clay-leached Brown soils, Clay-leached soils, Forest Grey soils
Podzolization	VI. Podzolized soils	Hydromorphy	Podzols

(Contd.)

Table 5.5 (*Contd.*)

Dominant pedological process	**Class**	**Criteria delimiting the subclasses**	**Soil types (generally subclasses)**
Climatic melanization	VII. Melanized soils	Pedoturbation, brunification, fersiallitization, calcification	Chernozems, Chestnut soils, Brunizems, Phaeozems, Kastanozems
Vertisolization	VIII. Vertisols	Pedoturbation, melanization	Vertisols, Vertic soils
Fersiallitization	IX. Fersiallitic soils	Rubefaction, brunification, depletion, hydromorphy	Fersiallitic Red soils
Ferrugination	X. Ferruginous soils	Acidification	Ferruginous soils, Ferrisols
Ferrallitization	XI. Ferrallitic soils	Allitization	Ferrallitic soils, Allitic soils, Ferric soils, cuirasses
Hydromorphy and oxidation-reduction	XII. Hydromorphic soils	Intensity and type of hydromorphy	Pseudogleys, Stagnogleys, Gleys, Peats, Pelosols, Planosols
Salinization, sulphate-reduction, sodization	XIII. Salsodic soils	Salinization, sulphate-reduction, sodization, alkalization	Saline soils, Solontchaks, Acid Sulphate soils, Thiosols, Alkali soils, Solonetzes, Soloths

Chapter 6

BETWEEN LIFE AND SOIL: THE HUMUS FORMS

This chapter presents the humus forms, considered here as an integration of the surface layers of soil and as the crossroads of interactions between living organisms and inert constituents. Emphasis is given to the functional principles of the humus forms and their roles in the ecosystem, in particular as indicators of its dynamics.

6.1 GENERAL PICTURE OF HUMUS FORMS

6.1.1 The reality: humiferous episolum or humiferous topsoil

Soil classification generally does not allow sufficient detail of the holorganic and organomineral horizons, which have their own characters and functional rhythms (Chap. 5, § 5.5.5). For that reason most pedological taxonomies give one name to the solum in its entirety (including the surface layer) and another to the ***humiferous episolum*** alone. The two names can otherwise be combined in a more global typology: LUVISOL with mull-moder, RÉDUCTISOL TYPIQUE with hydromull, etc. (§ 6.2.1). Almost alone among the European classifications, that of Brunner et al. (2002) sometimes uses the humus form as a diagnostic element of the entire solum.

Peculiar aims for the taxonomy of soils and for the taxonomy of humus forms, which complement rather than compete with each other.

Humiferous episolum: assemblage of the upper horizons of a solum that contain organic matter (O, H and A horizons), with organization essentially dependent on biological activity (AFES, 1995, 1998).

A junction between the living and mineral worlds.

The humiferous episolum is the principal junction between the living and mineral worlds, site of their intimate relation within the clay-humus complex. The litter is transformed in the episolum; the fauna and microflora of the soil are most active in it, and vascular plants often preferentially root in it. Located at the surface of the soil mantle, it is the part of it most subject to disturbance, especially in agricultural soils.

6.1.2 The morphofunctional description: humus form

From humus to the humus form... or: from the constituents to their effects.

Humus contains all organic constituents of the soil that result from humification (Chap. 2, § 2.2.4, Chap. 5, § 5.2.3). But this term has also been applied on another scale, that of the humiferous episolum. This has resulted in ambiguity because it confuses biochemical constituents with their macroscopic effects in the solum. To better differentiate these two domains, the expression ***humus form*** is preferred for the macroscopic aspect. The term 'form' is well chosen as it effectively describes the morphology of the holorganic and organomineral horizons, enabling deduction of the essential principles of their functioning.

Humus form: ensemble of morphological and macroscopic characters of the humiferous episolum depending on its mode of functioning (AFES, 1995, 1998). It reflects the functioning of an extremely active integrator compartment of the soil where the progress of the soil system is the fastest (Chap. 5, § 5.5.5). Even though the expression 'humus form' may have been forgotten with time, it is, in fact, quite old. For example, it is even found in Müller (1889), who described the humiferous episolum of beech forests.

Every humus form is characterized by a specific vertical sequence of O, H and/or A horizons and their subhorizons, each indicating one type of evolution of the organic matter. Just as the Références are valid for the entire solum, conceptual humus forms serve as taxonomic benchmarks for the sequences observed in the field and described by their reference horizons (Jabiol et al., 1995; Zanella et al., 2001). For example, a hydromoder is composed of the horizons termed OL (litter, particulate organic matter), OF (site of fragmentation of debris), OH (principal site of humification) and Ag (organomineral horizon with slight hydromorphy).

6.1.3 The thermodynamically active compartment of soil

Soil formation: a thermodynamic process in which the contribution of energy is highest at the surface.

In soil, the site of maximum utilization and circulation of energy is the humiferous episolum (Chap. 5, § 5.1.1). Solar energy deploys all its effects there, with release of the chemical energy contained in the organic matter, earlier fixed by photosynthesis in the autotrophic subsystem of the ecosystem (Chap. 4, § 4.1.1; Chap. 13, Fig. 13.4). Among the biomolecules, entropy is the lowest in nucleic acids, which are the most organized, then becomes progressively higher in proteins, polysaccharides and phenolic polymers. According to the type of vegetation, from 50 to 90% of the energy contained in the primary production enters the food-chains of the soil; only the remaining 50 to 10% nourishes the grazing-predatory chains outside the soil (Chap. 13, § 13.5.1).

Addition of energy at the level of the humiferous episolum enables functioning of the food-chains and food-webs of the soil and their many compartments.

This energy enables growth of organisms, organization of their mutual relations and building of new biomolecules (Chap. 2, § 2.2.4) by work that reduces entropy. But contrary to this, evolution of the organic matter increases entropy when the original biomolecules are destroyed (Chap. 2, § 2.2.3) and converted to water and CO_2, two compounds incapable of providing work of organization in the soil by themselves. Surely water can play such a role during pedogenesis. But for this it should be activated by the force of gravity, which enables it to move substances and initiate formation of horizons (Chap. 5, §§ 5.1.1, 5.3.2).

'Lastly, whatever its mode of utilization, entropized hydrogen returns to water, a molecule which, it will be acknowledged, constitutes the least polluting by-product and the least troublesome waste' (Lebreton, 1978).

The amount and quality of energy provided by the litter acts on the speed of thermodynamic transformations, themselves dependent on the variety of decomposing organisms (Heal and Dighton, in Fitter, 1985).

Strong or weak energy resources for different strategies.

In the first phase, called 'exploitative', ameliorating litter with high energy level is rapidly colonized by a microflora with high production and great power of dispersal (r strategy). Populations increase rapidly but also decline abruptly. Animals are small in size but often able to survive in resistant forms (Chap. 13, § 13.7.4). In the second phase, called 'interactive', interspecific competition is set up between the organisms responsible for subsequent degradation of the compounds. The K strategists have supplanted the r strategists.

In ameliorating litter, a succession of r and K strategists. Such a functioning is seen in mulls (§ 6.3.2).

In acidifying litter with low energy or with difficultly accessible energy resources, the microflora and microfauna are not very diverse and intraspecific competition is intense because of the large number of individuals characterizing the populations. The best response then to the low availability of resources is mutual help as symbioses, and reciprocal spatial isolation; the colonies of living organisms are often organized in micromosaics. Here we can hardly define a phase as having r strategy or K strategy, the organisms being rather of stress-tolerant S type (Chap. 12, § 12.6.2; Grime et al., 1988). This is the case of mors (§ 6.3.2) and HISTOSOLS (Chap. 9, § 9.1.3).

'Regulatory biological systems make the humiferous episolum the principal crossroads of transformation, regulation and redistribution of energy of the entire soil. Reflecting its morphology and its diversity, the humus forms integrate all the environmental factors that influence biological activity: light and also thermal energy, soil texture, hydric regime, acidity, redox potential, or nature of litter' (after Toutain, 1981, 1987).

The humiferous episolum, distribution centre of energy in the soil.

Overall, the humiferous episolum appears to be the site where processes that increase or decrease entropy operate simultaneously in a never-stabilized balancing game and at variable speeds. It comprises the three types of biological systems regulating the soil (Lavelle, 1987):

- the litter-surface root system where the microflora consists mostly of fungi and where regulation is ensured by saprophagous invertebrates (Chap. 4, § 4.5.2; § 6.3.2);
- the rhizosphere that associates deeper roots with a mainly bacterial microflora, regulated by the microfauna (Chap. 15, § 15.3.2);
- the drilosphere, controlled by earthworms (Chap. 13, § 13.5.2).

6.2 CLASSIFICATION OF HUMUS FORMS

6.2.1 General classification

Biological activity determines principal orientations

As in the taxonomy of soils, there is no unanimity regarding the humus forms.

A general principle of classification, based on the intensity of biological activity, is accepted by everyone but the definition of reference horizons indicates conceptual differences, making uncertain any direct correspondence.

All classifications make first reference to the degree of biological activity of mineralization and humification revealed by the morphology of the horizons described in the field. As early as 1889 Müller distinguished mulls with strong earthworm activity from mors (which he named *Torf*), in which organic matter accumulates by remaining poorly decomposed. He also discovered differential animal species within the same taxonomic group, such as the testate amoebae or the ciliates in the two forms mull and mor (Foissner, 1987).

The third humus form, moder, owes its name to Romell and Heiberg (1931).

Specific regions, the 'hot spots of activity', concentrate the greater part of biological activity (Chap. 13, § 13.5.2).

However, the total biological activity of the humiferous episolum should not conceal the heterogeneity of the latter. Certain zones, often small, such as the rhizosphere, microaggregates, litter or the drilosphere are the sites of maximum exchange of energy and maximum chemical exchange. These are the 'hot spots of activity' of Beare et al. (1995). For example, Ingham et al. (1985) showed that 5–7% of the total volume of the soil occupied by the rhizosphere shelters more than 70% of the bacteria and fungivorous nematodes. Similarly, Foster and Dormaar (1991) discovered unsuspected activity of amoebae in micropores of very small aggregates where they insert their pseudopodia to consume the bacteria found there.

Biological activity or biological efficiency of soil?

Classically, in soil science, *intense biological activity* signifies that transformations of organic matter, mineralization in particular, are rapid. This is true for mulls and hydromulls. The reverse indicates poor decomposition of the material that is more or less accumulated, as in mors or hydromors. This way of viewing things obscures the fact that in certain humiferous episola with 'weak' biological activity, that of particular organisms may be very intense. For example, enchytraeids are abundant in certain mors and mites multiply rapidly in acid peats (Chap. 9, § 9.1.3). It is therefore a rather rapid short-cut for qualifying certain moders or mors as humus with weak biological activity. It is more precise to say that their *biological transformation efficiency* is low.

Even if strong, biological activity is sometimes indicated by a poor final result of the work of organisms, which must devote, especially in soils with high ecological stress, a good part of their energy expenditure to their survival (gross production, chiefly respiratory). Indeed, here too they utilize the energy contained in organic biomass, which they convert to CO_2. But this mineralization is scarcely indicated by visible macromorphological characters in the humiferous horizons. With the above thoughts, we have nonetheless retained in this book the better known term 'biological activity'.

It seems that the authors were themselves a little too focused only on the biological activity of earthworms—often forgetting other organisms—to interpret it directly in the morphology of humus forms and to make it the red thread of classification.

Let us retain in future the term 'biological efficiency', more consistent with the general laws of ecology and the concept of biological efficiency in ecosystems (Lindeman, 1942; Frontier and Pichod-Viale, 1991; Odum, 1996)!

Complementary to general biological activity, a second differentiating criterion is often adopted, that is, the degree of aeration, opposing the aerated humus forms to the hydromorphic ones. The mean level and amplitude of the water table determine oxic and anoxic conditions (temporary or permanent) that influence growth and activity of organisms.

A second classification criterion: the degree of hydromorphy.

Humus forms: eight broad categories

Though old, the differentiation of eight principal humus forms (three aerated and five hydromorphic) distributed in two groups retains its explanatory value, because it well reflects the interdependent effects of general biological activity and degree of hydromorphy (Tab. 6.1; Fig. 6.1).

Humus forms and C/N ratio

A good criterion for characterizing the three aerated humus forms is the C/N ratio, which is 8–15 in mulls, 15–25 in moders and greater than 25 in mors. These values are indicative and the ratio may differ between two humiferous horizons in the same soil.

It should be noted that the German soil classification (Mückenhausen, 1985) proposes three groups of humus forms according to the degree of hydromorphy, whereas the French classification gives only two:

And the subaquatic humus forms?

- terrestrial forms: mull, moder and mor;

Table 6.1 Principal categories of humus forms of temperate regions (after Duchaufour, 1983; Jabiol et al., 1995)

Nomenclature			
Aerated medium	**Mull**	**Moder**	**Mor**
Moist medium; fringe of capillary rise of a water table	**Hydromull**	**Hydromoder**	**Hydromor**
Medium temporarily saturated with water	Does not exist	**Anmoor**	**Hydromor**
Medium permanently saturated with water	Does not exist	Does not exist	HISTOSOL **(Peat)**
Major characteristics			
General biological activity (transformation efficiency)	High	Medium	Low
Clay-humus complex	Mature, stable	Immature, often unstable	Very immature or absent
Type of A horizon (in aerated conditions)	Biomacrostructured (or issued from insolubilization) A; crumb to micro-crumb	A issued from juxta-position; massive or single-grain	A issued from juxta-position, very thin or absent
C/N ratio	8–15	15–25	>25

- semi-terrestrial forms: hydromull, hydromoder, hydromor, anmoor and peat of transitional swamps and raised bogs (partly HISTOSOLS);
- subaquatic forms: peats of fens (partly HISTOSOLS), ***sapropel*** (black humic layer of reduced submerged zones, poor in multicellular organisms), ***gyttja*** (grey to black humic layer of oxygenated submerged zones, rich in plant residues and organisms) and ***dy*** (brown peaty deposits of acid waters, poor in organisms, for example on the banks of raised bogs).

It is necessary to take into account subaquatic forms in the study of ecotones between terrestrial and aquatic milieus, where the transition to semi-terrestrial forms is often very gradual, according to fluctuations of the water table (also see Lachavanne and Juge, 1997).

6.2.2 Two more elaborate classifications

Concepts and criteria

Two classifications, each with its advantages and disadvantages.

Two classifications are presented here, without either being preferred: that of the *Référentiel pédologique* (AFES, 1995, 1998; Jabiol et al., 1995) and that of Green et al. (1993). Between the two, different diagnostic priorities

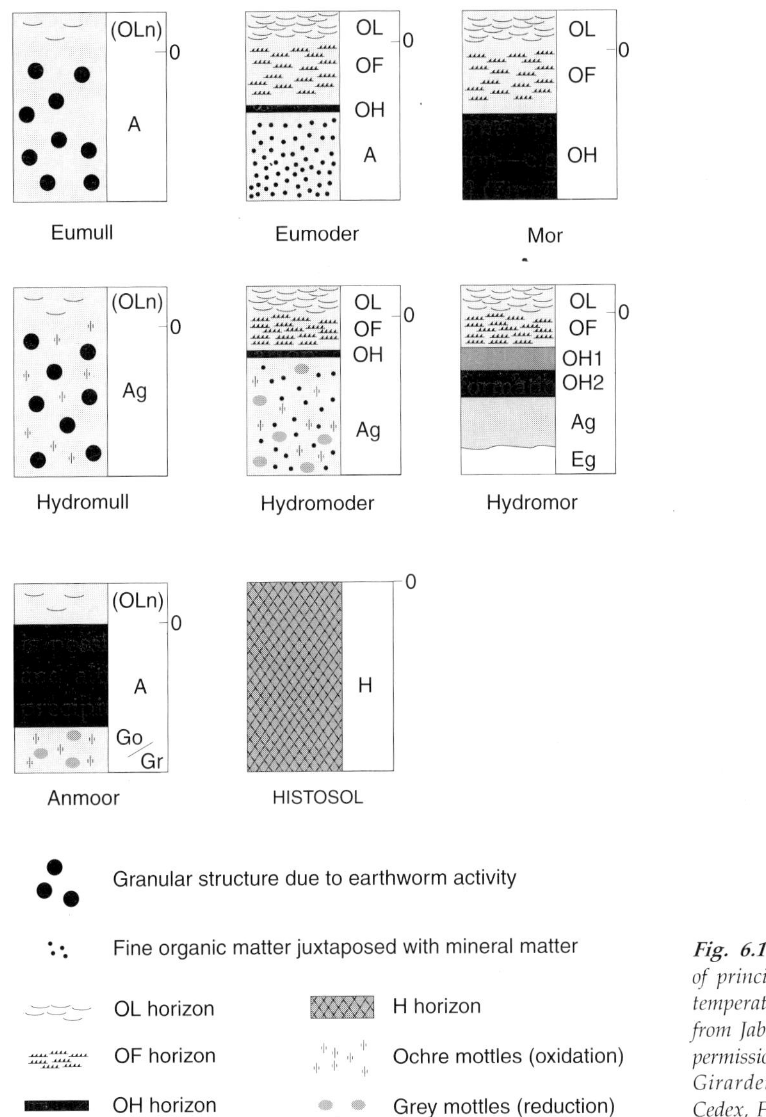

Fig. 6.1 Schematic diagram of principal humus forms of temperate regions (extracted from Jabiol et al., 1995; with permission of ENGREF*, 14 rue Girardet, F-54042, Nancy Cedex, France).*

are accorded to the A or O horizons, and to their morphology or functioning.

Different concepts of the Référentiel and of Green et al. ...

• The Référentiel first favours characters of the A horizon, then those of the holorganic horizons O and H. In the third stage, physicochemical or functional peculiarities are defined by the addition of qualifiers. Each horizon is defined on essentially morphological bases, while functioning appears mainly at the level of humus forms in their entirety.

• Green et al. (1993) adopted a reverse course of action: the typology relies first on criteria related to functioning of the holorganic O and H horizons; the A horizon is subsidiary. Also, the horizon approach is rather functional and that of humus form morphological.

... due to their different geographical origins.

According to the criteria of Green et al. (1993), the same humiferous episolum may end up, by the determinative keys, in a humus form less active than that according to AFES (1995, 1998). Overall, a shift is seen of mulls and moders of the Référentiel towards moders and mors according to Green et al. The humus forms with intermediate activity, for example at the join between a mull and a moder, can change in category from one classification to another. Thus, the eumoder of the Référentiel corresponds in certain cases to the humimor of Green et al. This is also due to the different geographical origins of the two schools: more montane and boreal in the case of Green et al. and more temperate and hilly in the case of Jabiol et al. (Chap. 5, § 5.6.1).

Classification according to the *Référentiel pédologique*

Table 6.2 summarizes the characteristics of fifteen humus forms described in the Référentiel (AFES, 1995, 1998) and detailed by Jabiol et al. (1995). The aerated forms are primarily forest forms, the authors not defining the intervening modifications, for example, in a grassland with dense hairy roots. Five hydromorphic humus forms are included.

The determinative key for humus forms of Jabiol et al. is presented with complements in Figure 6.2, while reference horizons have been described in Table 5.3.

Classification according to Green et al. (1993)

Table 6.3 presents the diagnostic horizons of this classification, differing sometimes from those of the Référentiel.

Sixteen humus forms are covered, six of them hydromorphic. Supplemented with anmoor, which was not considered by Green et al. in spite of its originality, these forms are determined according to the key presented in Figure 6.3.

Some correspondences between the humus forms in the two classifications are suggested in Table 6.4. They are only indicative and do not serve for direct translation because of the conceptual differences indicated earlier.

Table 6.2 Principal humus forms of temperate regions (extracted from Jabiol et al., 1995)

	A. Aerated humus forms				
Reference A horizons	Biomacrostructured. Granular structure of biological origin, resulting from crushing by earthworms.	Issued from insolubilization. Microgranular structure resulting from chemical precipitation of organic matter by iron and clay; low earthworm activity but high fungal activity.	Issued from juxtaposition. Massive or single-grain structure; no built-up structure of biological or chemical origin; inherited organic matter, juxtaposed with mineral materials.		No A horizon or only diffused organic matter in the subjacent mineral horizon (e.g. Eh horizon).
Clay-humus complex	Mature and stable		Immature or absent		
Transition between O and subjacent horizons	Clear discontinuity between O and A		Gradual transition between O and A		Clear discontinuity between O and subjacent mineral horizons
Reference O horizons ↓	**MULLS**		**MODERS**		**MORS**
(OLn)	**Eumull**	—	—		—
OLn + (OLv)	**Mesomull**	—	—		—
OLn + OLv + (OF)	**Oligomull**	**Mycogenic oligomull**	—		—
OLn + OLv + OF	**Dysmull**	—	**Hemimoder**		—
OL + OF + OH (or + (OH))	**Amphimull**	—	**Eumoder** (OH < 1 cm)	**Dysmoder** (OH ≥ 1 cm)	**Mor**
	B. Hydromorphic humus forms				
Slight hydromorphy Ag horizon	**Hydromull** OLn		**Hydromoder** OL + OF + thin OH		**Hydromor** OL + OF + thick OH
Strong hydromorphy An, H horizons	—	—	**Anmoor** OL (+ An)		**HISTOSOL** OL + OF (+ H)

Notes: The HISTOSOL—formed entirely of undecomposed plant debris (peat)—is simultaneously a humus form and a solum in itself. Besides, it is sometimes difficult to distinguish a thick OH hydromor horizon from an Hs HISTOSOL horizon, separated only by mode of saturation with water.

1	a	Well-drained or imperfectly drained sites; humiferous horizons not saturated with water for long periods. Aerated humus forms.	→ *see under* **2**
	b	Poorly drained or undrained sites; saturation for long periods. Hydromorphic humus forms	→ *see under* **12**
2	a	Presence of an OH (in addition to OL and OF) horizon	→ *see under* **3**
	b	No OH horizon	→ *see under* **6**
3	a	Distinctly granular A horizon	AMPHIMULL
	b	Non-granular or no A horizon	→ *see under* **4**
4	a	Abrupt discontinuity between OH (occasionally thin A) and the subjacent mineral horizon	MOR
	b	Gradual transition, with an A horizon issued from juxtaposition	→ *see under* **5**
5	a	OH horizon ≥ 1 cm thick	DYSMODER
	b	OH horizon < 1 cm thick, sometimes discontinuous	EUMODER
6	a	OF horizon present	→ *see under* **7**
	b	OF horizon absent	→ *see under* **9**
7	a	Continuous, more or less thick OF horizon	→ *see under* **8**
	b	Sporadic OF horizon	OLIGOMULL
8	a	Non-granular A horizon, massive (silts) or single-grain (sands)	HEMIMODER
	b	Granular or microgranular A horizon	DYSMULL
9	a	OLn and OLv horizons present	→ *see under* **10**
	b	Only OLn horizon present	→ *see under* **11**
10	a	Thick continuous OLv	OLIGOMULL
	b	Sporadic OLv	MESOMULL
11	a	Continuous OLn; A horizon with fine granular structure of low stability	MESOMULL
	b	Discontinuous OLn; A horizon with stable coarse granular structure	EUMULL
12	a	Predominant An horizon present	ANMOOR
	b	Predominant Ag horizon present	→ *see under* **13**
	c	Predominant H horizon present	→ *see under* **15**
13	a	Horizon sequence of OL/Ag type. No OF or OH horizon	HYDROMULL
	b	OF and/or OH horizon present	→ *see under* **14**
14	a	Horizon sequence of OL + OF + thin OH + Ag type	HYDROMODER
	b	Horizon sequence of OL + OF + thick OH + Ag type	HYDROMOR
15	a	Thickness of Hf > 50% of total of H horizons	FIBRIC HISTOSOL
	b	Thickness of Hm > 50% of total of H horizons	MESIC HISTOSOL
	c	Thickness of Hs > 50% of total of H horizons	SAPRIC HISTOSOL

Fig. 6.2 *Determinative key for humus forms (extracted from Jabiol et al., 1995, modified; with permission of* ENGREF*). The symbol + between two horizons indicates a diffuse boundary, the symbol / a clear boundary.*

6.3 WELL-DIFFERENTIATED FUNCTIONINGS: SOME EXAMPLES

6.3.1 A single basic functional scheme

Five processes are common to all humus forms: M1, M2, H1, H2 and H3.

All humus forms follow the same rough scheme of functioning in five modes of transformation of organic matter (Chap. 5, Fig. 5.3; § 5.2):

Table 6.3 Reference horizons for humus forms (after Green et al., 1993)

Horizon	Subhorizon	Characters
OL (litter)	**OLn** (n = *n*ew)	Freshly fallen plant material, essentially non-fragmented.
	OLv (v = *v*ieillie, Fr. for aged)	Material showing the beginnings of decomposition, but slightly fragmented. No fine organic material.
OF (fragmented)	**OFm** (m = *m*ycogenic)	Plant residues aggregated into a packed, sticky structure by abundant mycelia. Roots may be very numerous and can reinforce the packed structure. Very few or no faecal pellets.
	OFz (z = zoogenic)	Weakly aggregated plant residues, with friable, loose consistence. The structure reflects presence of active populations of soil mesofauna and microfauna. Numerous faecal pellets can be easily seen with a hand lens. Little or no mycelium.
	OFa (a = *a*mphi)	Plant residues aggregated into a more or less loose, intermixed structure. Poorly resistant structure; aggregates yielding to pressure. Alternation of masses of aggregated material with pockets of loose material. Mycelium and faecal pellets present, without predominance of either.
OH (humified)	**OHh** (h = *h*umic)	Fine materials predominate; few or no elements recognizable by the naked eye. Organic matter with plastic consistence when wet, with massive or angular blocky structure, black, staining the fingers.
	OHz (z = zoogenic)	Fine materials predominate; few or no elements recognizable by the naked eye. Horizon mostly composed of faecal pellets. Black organic matter with granular structure. Plenty of cylindrical or spherical faecal pellets, giving the appearance of fine, black 'sawdust'.
	OHr (r = *r*esidues)	Fine materials predominate, with plant fragments recognizable by the naked eye (fine roots, bark, wood). Organic matter with slightly plastic consistence, not staining the fingers. Dark reddish-brown colour, redder than the OHh horizon, if also present.
S layer		Layer of living Bryophyta 'rooted' in the litter (OLv); important in structuration of the humiferous episolum, especially in grassland.
An		Anmoor horizon (see Chap. 5, Tab. 5.3 ; Tab. 6.2)
Hf, Hm, Hs		Histic horizons (see Chap. 5, Tab. 5.3 ; Tab. 6.2)

- primary mineralization M1,
- secondary mineralization M2,
- humification by inheritance H1,
- humification by polycondensation H2,
- humification by bacterial neosynthesis H3.

The aspects that change from one to the other are:

- relative weightages of mineralization and humification;

No.		Criterion	Result
1	*a*	Well-drained or imperfectly drained sites; humiferous horizons not saturated with water for long periods. Aerated humus forms.	→ *see under* ***2***
	b	Poorly drained or undrained sites; saturation for long periods. Hydromorphic humus forms.	→ *see under* ***11***
2	*a*	Total thickness of OF and OH > 2 cm if thickness of A > 2 cm; or thickness of OF + OH ≤ 2 cm if thickness of A < 2 cm	→ *see under* ***3***
	b	Total thickness of OF + OH ≤ 2 cm if thickness of A > 2 cm (mull group)	→ *see under* ***10***
3	*a*	OF horizon of OFm type predominates (mor group)	→ *see under* ***4***
	b	OF horizon of OFz and/or OFa type predominates (weak OFm possible) (moder group)	→ *see under* ***7***
4	*a*	Degraded wood > 35% by volume of organic matter of OF + OH	LIGNOMOR
	b	Degraded wood ≤ 35% by volume of organic matter of OF + OH	→ *see under* ***5***
5	*a*	Thickness of OF > 50% of total thickness of OF + OH	HEMIMOR
	b	Thickness of OF < 50% of total thickness of OF + OH	→ *see under* ***6***
6	*a*	OH horizon of OHh type	HUMIMOR
	b	OH horizon of OHr type	RESIMOR
7	*a*	Degraded wood > 35% by volume of organic matter of OF + OH	LIGNOMODER
	b	Degraded wood ≤ 35% by volume of organic matter of OF + OH	→ *see under* ***8***
8	*a*	OFa horizon > 50% of thickness of OF horizons; or OFm present	MORMODER
	b	OFz horizon > 50% of thickness of OF horizons	→ *see under* ***9***
9	*a*	Total thickness of OF + OH ≥ thickness of A	LEPTOMODER
	b	Total thickness of OF + OH (OHh or OHz) < thickness of A	MULLMODER
10	*a*	Rhizogenic A horizon, formed by the decomposition of dense, fine roots	RHIZOMULL
	b	Zoogenic A horizon, formed by action of earthworms in abundance	VERMIMULL
11	*a*	Presence of a predominant An horizon	ANMOOR
	b	No An horizon or An horizon weakly expressed (e.g. at the base of a histic H horizon)	→ *see under* ***12***
12	*a*	Total thickness of OF + OH + H > 2 cm; or ≤ 2 cm if A < 2 cm	→ *see under* ***13***
	b	Total thickness of OF + OH + H ≤ 2 cm if A > 2 cm	HYDROMULL
13	*a*	Total thickness of OF + OH > thickness of H	→ *see under* ***14***
	b	Total thickness of OF + OH < thickness of H (typical humus of HISTOSOLS, but can surpass others)	→ *see under* ***15***
14	*a*	OF horizon of OFm type	HYDROMOR
	b	OF horizon of OFz and/or OFa type	HYDROMODER
15	*a*	Thickness of Hf ≥ 50% of total thickness of H horizons	FIBRIMOR
	b	Thickness of Hm ≥ 50% of total thickness of H horizons	MESIMOR
	c	Thickness of Hs ≥ 50% of total thickness of H horizons	SAPRIMODER

Fig. 6.3 *Key for determination of humus forms (after Green et al., 1993, supplemented)*

Table 6.4 Possible correspondences between humus forms according to Green et al. (1993) and according to the *Référentiel pédologique* (AFES, 1995, 1998) (not exhaustive)

Green et al. (1993)	AFES (1995, 1998)
Aerated humus forms	
Vermimull	Eumull
Rhizomull	Eumull, Mesomull
Mullmoder	Dysmull, Amphimull
Leptomoder	Dysmull, Eumoder, Dysmoder
Mormoder	Dysmoder, Hemimoder, Eumoder
Lignomoder	?
Hemimor	Oligomull, Hemimoder, Eumoder
Humimor	Eumoder, Dysmoder, Mor
Resimor	Mor
Lignomor	Mor
Hydromorphic humus forms	
Hydromull	Hydromull
Hydromoder	Hydromoder
Hydromor	Hydromor
?	Anmoor
Saprimoder	HISTOSOL SAPRIQUE
Mesimor	HISTOSOL MÉSIQUE
Fibrimor	HISTOSOL FIBRIQUE

- comparative importance of M1 and M2;
- relative intensities of H1, H2 and H3;
- speed of biochemical transformations and stability of the compounds (Chap. 14, § 14.4);
- mineralogical influence of the parent material (Chap. 2, § 2.1.4);
- influence of nature of the litter (Chap. 2, § 2.2.1);
- relative influence of various categories of organisms (proportions between bacteria, between fungi and the fauna, between the micro-, meso- and macrofauna, etc.).

6.3.2 From mull to mor, three benchmarks for aerated humus forms

Mull, moder, mor: the classic trilogy of aerated humus forms according to decreasing biological efficiency.

Aerated humus forms distinguished on the basis of general biological activity are distributed along a gradient from very active forms, the mulls, through intermediate forms, the moders, to forms with weak transformation of organic matter—the mors. A differentiated distribution of great groups of the pedofauna corresponds to this gradient (Fig. 6.4).

In ***mulls*** of forest environment, the boundary between the litter OL and the organomineral A horizon is abrupt.

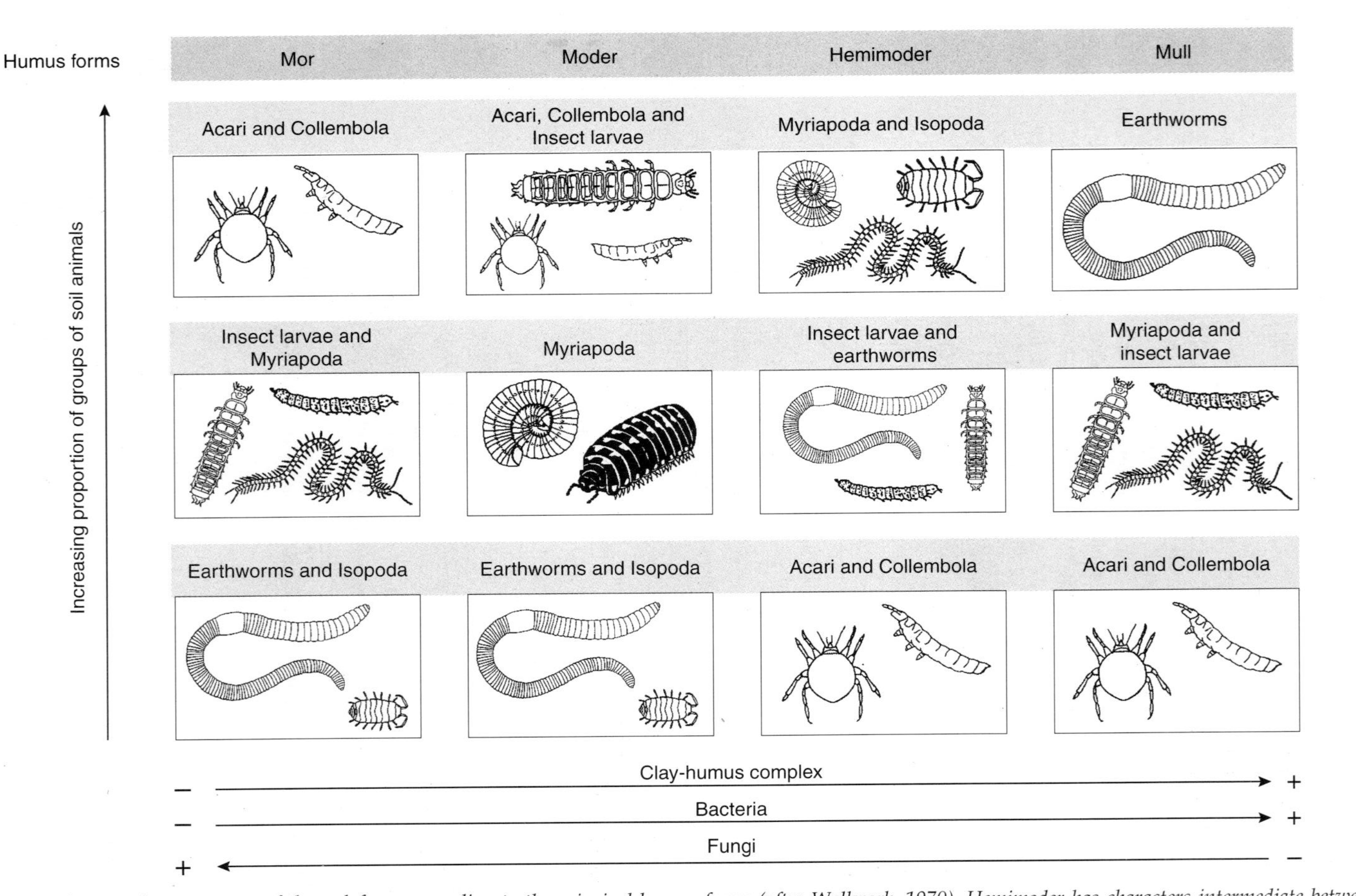

Fig. 6.4 *Distribution of great groups of the pedofauna according to the principal humus forms (after Wallwork, 1970). Hemimoder has characters intermediate between mull and moder.*

In grassland the transition is often less distinct, as in rhizomulls (Green et al., 1993). Living organisms rapidly transform all the foliar constituents, including the most resistant such as lignin and brown pigments. Mineralization and humification are intense. The A horizon is granular or microgranular because of biostructuration by termites or by anecic and endogeal worms or by chemical insolubilization of humified compounds (mycogenic oligomull in the latter case).

In ***moders*** the transition is gradual between whole leaves, fragmented leaves, animal droppings, holorganic then organomineral materials and, lastly, the organomineral horizon itself. Many compounds hard to decompose are preserved or concentrated. The A horizon often has massive or single-grain structure, with juxtaposition of silt or sand mineral particles and faecal pellets. OF and OH horizons (the latter with exception) are always present.

Acari, Collembola, insect larvae and Myriapoda are particularly abundant in moders, the reverse for earthworms.

Lastly, in ***mors,*** the boundary between the holorganic horizons (OL, OF and OH) and subjacent mineral horizons is clear, the A horizon being absent or thin. Mineralization is very weak, limited by physicochemical or climatic constraints (Chap. 5, § 5.5.1); addition of humin is large, but essentially through the inheritance pathway H1.

Acari and Collembola are the dominant animals in mor; earthworms and Isopoda are very poorly represented.

The more intense activity of fungi, springtails and mites in mors compared to mulls has been indicated by Ingham and Klein (1982) and Travé et al. (1996), for example in coniferous litter. Protozoans and testate amoebae are also more active in mors and in peats than in mulls; certain species form more that 50 generations a year in a mor against 28 in a moder or just 12 in a mull (Schönborn, 1982, 1986).

By way of example and based on Figure 5.3 we shall compare the functioning of three humus forms important in temperate environment: carbonate mull, saturated or mesosaturated mull, almost equivalent, and desaturated mor.

6.3.3 Carbonate-rich mull

Carbonate-rich mull is found on immature soils derived from a parent material that releases much carbonate in the fine earth (screes, calcareous moraines, heavily jointed strata cut at a high angle by topography, etc.). It is typical of RENDOSOLS, RENDISOLS and ORGANOSOLS CALCAIRES. Primary

Carbonate mull: active but quickly blocked.

The paradox of an intense primary mineralization and accumulation of scarcely transformed inherited organic matter.

The blocking of fulvic acids renders their maturation to humic acids or humin difficult or impossible.

Carboxylation: here, formation of carboxylic–COOH groups by various reactions such as oxidation of aldehyde groups or rupture of certain bonds.

The mineral fertility of carbonate-rich mull is rather low because of trapping of bioelements in the organic matter blocked by calcite. Nutrient deficiencies easily supervene (Chap. 4, § 4.3.4).

mineralization M1 predominates in it, with high biological activity of the macrofauna and its digestive microflora which lead to a speedy transformation of the organic matter (Fig. 6.5a; Plate V-1). The soil fauna is favoured by an aerated granular structure, which it helps establish, and which is ensured by plenty of Ca^{2+} ions.

However, fulvic acids formed at the beginning of humification H2 are rapidly immobilized by calcium, and are thereby removed from later biochemical processes of polycondensation and enzymatic degradation (secondary mineralization M2). Only inherited humin H1 is present in large quantity because it is directly released by the particulate organic matter and is not at all subject to chemical transformation, except to ***carboxylation*** that greatly increases its cation exchange capacity (Duchaufour, 1983). But with certain small-size plant debris, it undergoes ***sequestration*** by calcitic cements that fix it by covering it with a hard coat (Plate II-1). At their scale, bacteria find themselves facing an insurmountable wall of calcite. In these conditions of extreme protection of organic matter by calcium and calcitic cements, secondary mineralization M2 can only be very weak. A thick black Acah horizon is formed, very rich in organic matter.

In ***chernozemic mull*** with the A horizon calcareous only in its lower part, sequestration of the organic matter is less efficient. Formation of insolubilization humin H2 is possible thanks to continental climatic conditions with dry hot summers. Extremely stable, this humin is subject only to weak mineralization M2 and accumulates in the soil for centuries or millennia.

6.3.4 Saturated mull or mesosaturated mull

Saturated or mesosaturated mull: where animals and bacteria work the most!

Saturated mull and ***mesosaturated mull*** (eutrophic mull of old classifications) are typical of many widely distributed soils: CALCISOL, BRUNISOL SATURÉ, BRUNISOL MÉSOSATURÉ, FLUVIOSOL BRUNIFIÉ. The pH values are close to neutral or slightly acid—the latter is called mesosaturated mull—between about 5.5 and 7.0. Conditions are ideal at the same time for primary mineralization M1 (high faunal activity), for maturation of humus H2 (favourable pH and base saturation percentage) and for microbial humification H3 (Fig. 6.5b; Plate V-2).

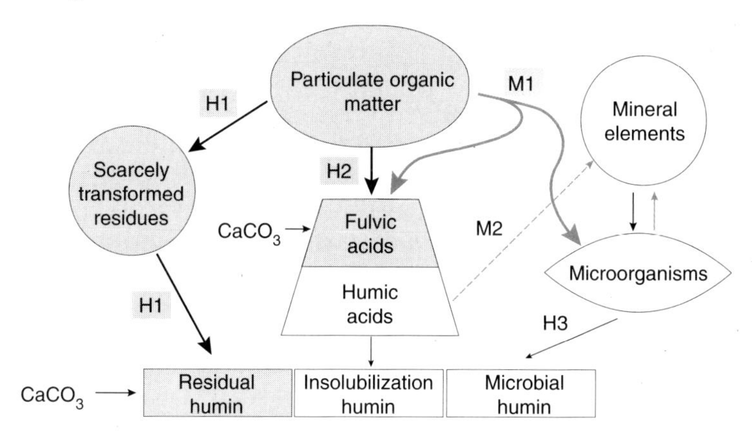

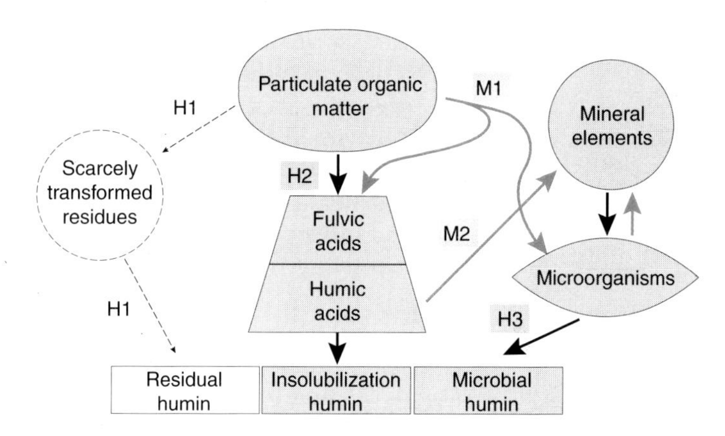

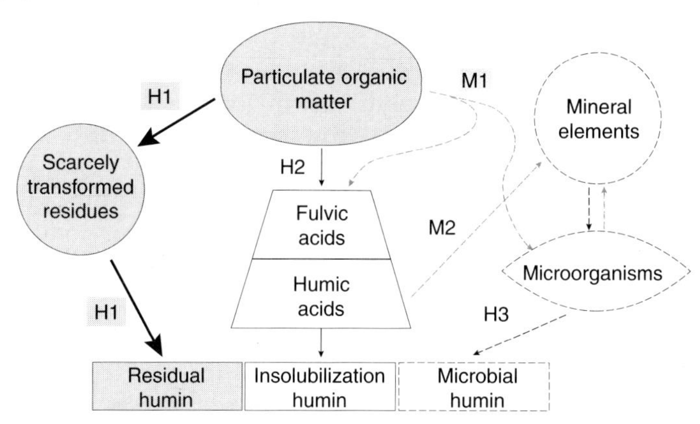

Fig. 6.5 *Comparison of functioning of (a) carbonate-rich mull, (b) saturated or mesosaturated mull and (c) unsaturated mor (inspired by Soltner, 1996a).*

A humus form productive for plants: rapid circulation of bioelements and well-aggregated structure.

Secondary mineralization M2 is more intense than in carbonate mull since the organic matter, even though stabilized, is accessible to decomposing agents. Long-term accumulation does not take place, so that the general hue is lighter. No very extreme limiting factor disturbs the high biological activity or the polycondensation of humic precursors, which remain transformable in the absence of blocking by calcite. Except for inherited humin H1, all categories of organic matter are represented, guaranteeing high resistance of the soil to changes in environmental factors (climate, vegetation type).

Three other forms of non-carbonate mull function a little differently from saturated and mesosaturated mulls.

In ***andic mull***, humus, in particular fulvic and humic acids, is blocked by amorphous minerals such as aluminium gels. Microbial degradation is rendered impossible and organic matter accumulates in thick Alu or And horizons, somewhat like in carbonate mull.

In ***vertic mull***, mechanical pedoturbation due to swelling and shrinking of 2/1 clay minerals (Chap. 2, § 2.1.5) incorporates the organic matter very deep in the soil. The organic matter is dominated by grey humic acids and insolubilization humin.

In ***ferrallitic mull*** of equatorial forests, the turnover of organic matter is very rapid, with abundant production of crenic and fulvic acids. The latter form microaggregates with haematite, goethite and/or kaolinite.

6.3.5 Unsaturated mor

In unsaturated mor, accumulation of scarcely transformed organic matter thickens the holorganic horizon.

As against saturated mull, ***unsaturated mor*** is typical of environments with strong climatic or physicochemical constraints that slow down biological activity. It is not, however, water in excess that inhibits organisms as in HISTOSOLS (Chap. 9, § 9.2.1), but rather cold, heat, dryness or acidity. This humus form is found on very varied soils, where conditions are too harsh for the edaphic microflora and fauna: LITHOSOLS, RÉGOSOLS, ORGANOSOLS, RANKOSOLS, LUVISOLS and PODZOSOLS.

Acid-tolerant plant species generally supply acidifying litter.

Generally, mors—desaturated mor in particular—are found under ericaceous plants or conifers, suppliers of leathery litter, poor in nitrogen but rich in phenolic acids and tannins. These compounds, toxic for most soil organisms, greatly limit their diversity. Selection of species in mors is also indirect, for example by the ability of the litter to solubilize toxic metals such as aluminium or to chelate useful bioelements for transporting them to depth.

With macrofauna nearly absent, primary mineralization M1 is almost nil. Only animals with particularly efficient and hard mouthparts (Chap. 13, Fig. 13.3) are capable of cutting leathery plant tissues (Fig. 6.5c; Plates V-3 and V-4). Moreover, the general environmental conditions, especially strongly acid pH, inhibit bacterial activity and humification H3, as well as processes of biochemical immobilization H2 and secondary mineralization M2. Room is left for residual humin H1 directly inherited from the original plant material. The only important transformation of organic matter is carried out by lignivorous fungi (Chap. 8, § 8.6), which are very resistant to low pH and heavy-metal toxicity; they act chemically or mechanically by piercing the walls of dead plant cells.

General decomposition of organic matter is slow compared to that in mulls or even in moders.

Residual humin gains all the way!

The macroscopic consequence of reduced biological activity is accumulation of poorly decomposed organic matter in the form of thick OF and OH horizons. This is particularly true on hard calcareous rocks, where an acidophilic vegetation is very superficially rooted with no contact with the carbonates below (ORGANOSOLS INSATURÉS with mor; Chap. 5, § 5.5.1).

Veritable carpets isolating the vegetation from the parent material hinder exchanges between the surface and deeper layers.

In addition to desaturated mors are ***saturated mors with tangel***, found at rather high altitude on fine calcareous debris (OHta horizon present); in the plains are found ***xeromors*** on calcareous flagstones exposed to the sun and high temperatures. In the former case, cold and moisture are the limiting factors, in the latter, heat and dryness.

Unsaturated mor, a good example of feedback

The functioning of unsaturated mor is a good example of a cyclic process, with an action of the plant on the soil (supplying acidifying litter) and a feedback of the soil on the plant (increasing isolation from the rock below and selection of acid-tolerant species). This feedback also affects the availability of bioelements, released in small quantities by an unproductive vegetation with diminishing growth following lower and lower reserves (Coleman and Crossley, 1996).

6.4 THE HUMIFEROUS EPISOLUM AS INDICATOR OF ECOSYSTEM DYNAMICS

The humiferous episolum is a perfect integrator of environmental conditions (§ 6.1; Zanella et al., 2001). Two examples drawn from the investigations of the Laboratory

The humiferous episolum better preserves marks of recent evolution of the environment that shelters it than, for example, the mineral fraction of soil with slower transformation (Bernier and Ponge, 1994; Bernier, 1997).

of Plant Ecology at the University of Neuchâtel illustrate this 'recent ecosystemic memory' the humiferous episolum comprises. In the first example, the humus forms concomitantly reveal past evolution of the soil and of the vegetation and their present-day general functioning; in the second they relate the first stages of pedogenesis in very young soils.

6.4.1 Humus forms and evolution of wooded pastures of the Swiss Jura

Wooded pasture: semi-natural sylvipastoral plant formation characterized by a mosaic of numerous arborescent, arbustive, herbaceous and moss synusia (Chap. 7, § 7.4.2; Gillet and Gallandat, in Étienne, 1996).

The soil of the ***wooded pasture*** studied, dominated by spruce, is a NÉOLUVISOL formed from loess (Fig. 6.6; Tab. 6.5). Three humus forms respectively located on an open acidocline meadow (hemimor according to Green et al., 1993), at the boundary of a spruce crown (leptomoder) or directly under the latter (mormoder) reveal many aspects of the present-day functioning of the ecosystem (Havlicek and Gobat, 1998).

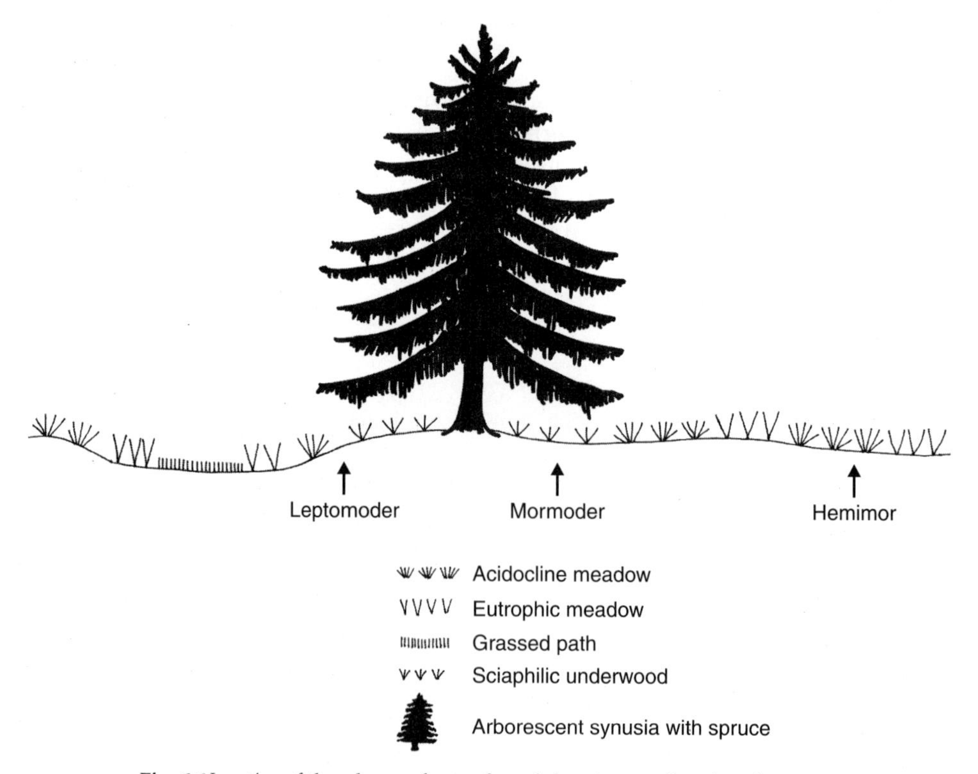

Fig. 6.6 *Location of three humus forms of wooded pastures as function of tree cover.*

Table 6.5 Comparison of three humus forms of wooded pastures of the Swiss Jura

Green et al. (1993)	**Hemimor**	**Leptomoder**	**Mormoder**
Jabiol et al. (1995)	Hemimoder	Dysmull	Dysmoder
Vegetation	Spikenard meadow	spruce: crown	under spruce
Soil moisture	high	medium	low
Cation content	low	high: dripping	high: sheltered
Brightness	high	medium	low
Biological activity	low	high	moderate

The hemimor of meadow indicates some surface hydromorphy due to trampling by cattle. Its upper part is dominated by activity of enchytraeids and fungi. Deeper down, anecic earthworms ensure effective integration of the organic matter with mineral matter.

Trampling by cattle is the primary ecological factor in the present-day context.

The greatest biological activity is seen at the outer limit of the crown of spruce, in leptomoder, in which the running roots (Chap. 4, Fig. 4.9) ensure oxygenation of the soil. This is also the zone where the bioelements leached from the needles preferentially fall, following the architectural 'umbrella' structure of spruce (Fig. 6.7). The humus form shows an ecotonal effect at the scale of a few decimetres.

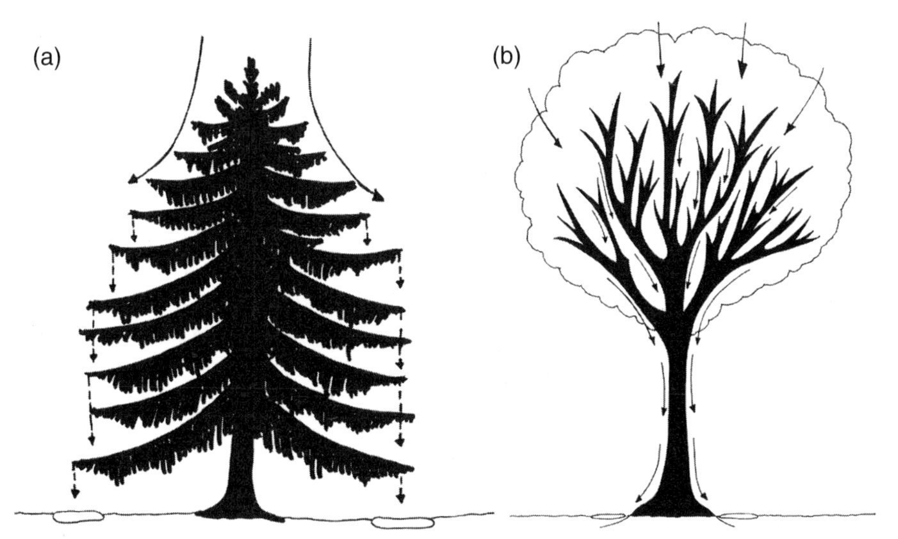

Fig. 6.7 *Comparative architectural structure of (a) spruce and (b) beech, and its influence on water dripping on soil. The architecture of spruce leads water drops and dissolved elements to the periphery, that of beech funnels them down the trunk. Whoever has found himself in forest during very heavy rain can observe the veritable sleeve of flowing water around the trunk of the beech, while that of the spruce remains almost dry!*

Biological activity under spruce is quite high in spite of the acidification brought about by the needles. This acidification is tempered by restoration of nitrogen by cattle, which use the place as shelter, and by ameliorating litter of herbaceous plants that grow there thanks to sufficient lateral light in these semi-open environments. The humus form is a mor-moder, while a closed spruce forest more often would shelter a resimor or a lignomor, with an exclusively acidifying litter.

But the humiferous episolum is also an indicator of the past dynamics of the wooded pasture:

Relict layers of bark or needles trace the history of the ecosystem.

- At depth, many layers of bark or twigs of spruce, preserved to date, trace a change in the spatial distribution of woody species at the scale of tens of metres at the site studied and over a few decades. These residues are found at variable depths, in typical grassland as well as under present-day trees.
- In mormoder under spruce, the formerly dominant mycogenic activity is today substituted by a more zoogenic decomposition, a consequence of the recent possibility of shelter offered to cattle by fully-grown spruce.

6.4.2 Hydromorphic humus forms and the first steps in pedogenesis

The end moraines left behind by glaciers, the freshly deposited alluvial terraces or volcanic materials: good situations for studying the beginning of soil formation. The humus forms here are valuable tools.

Soil scientists are often constrained to describe the first phases of pedogenesis, lacking examples in which the starting point of soil development is known precisely enough (Chap. 5, § 5.5.5). The banks of Lakes Bienne, Neuchâtel and Morat, in Switzerland, are of general interest for this subject. Actually, more than 2000 ha of land have been born through artificial lowering of the mean level of these lakes by close to three metres at the end of the nineteenth century (Buttler et al., 1985). From that period, the soils and vegetation have generally been left alone (Buttler and Gobat, 1991). This duration is certainly too short to allow significant mineral evolution, especially in hydromorphic environment with a permanent water table: all these soils are still calcareous, with pH often higher than 7.5 or 8.

On the other hand, dynamics of organic matter has differentiated many humus forms, according to the mean level and amplitude of the water table and the vegetation type (Fig. 6.8; Tab. 6.6). These forms indicate the first phases of pedogenesis:

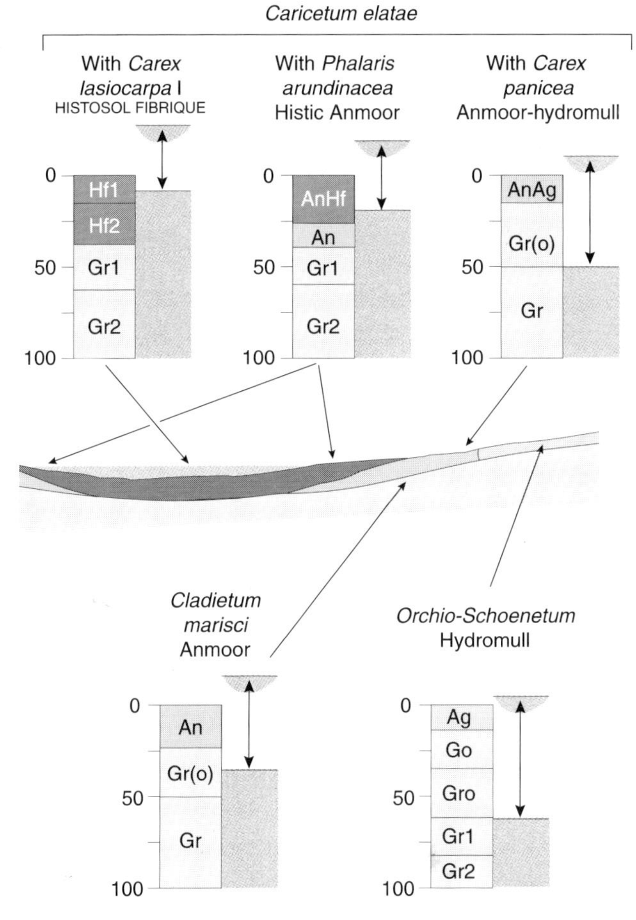

Fig. 6.8 Toposequence of soils and humus forms on the south shore of Lake Neuchâtel (Switzerland) after a century of soil development (after Buttler et al., 1985; Buttler and Gobat, 1991).

- after a century and under uniform general conditions (oceanic temperate climate, sandy loam soil, carbonate-rich water), their diversity is great as thirteen varieties of humus forms have developed, from HISTOSOL FIBRIQUE to saturated mor, passing through hydromull and eumull;
- The C/N ratio rises gradually from 14 (hydromull and carbonate eumull) to 41 (HISTOSOL FIBRIQUE), illustrating the variability of overall biological activity;
- ***extractability*** (EXT) of organic matter rises progressively from marsh (mean EXT = 7.6; scarcely humified organic matter) to broad-leaved forests (EXT = 12.3; more intense humification by polycondensation) and to pine forests (EXT = 16.6; humification by polycondensation and by inheritance);
- organic particle-size distribution distinguishes the fibric horizons from those that are richer in 0–50 µm aggregates.

Extractability of organic matter (EXT): ratio by weight of the carbon extracted by an alkaline reagent (NaOH, $Na_4P_2O_7$, $Na_2B_4O_7$) and the total organic carbon of a sample, expressed in %. Extractability is a good indicator of the degree of humification of organic matter: low values indicate scarcely transformed organic matter, while high values show an organic matter richer in humic acids and humin.

Table 6.6 Physicochemical characters of humus forms of the south shore of Lake Neuchâtel (Gobat, unpublished)

Vegetation type (phytocoenosis)	Humus form (AFES, 1995, 1998)	Horizon	pH	Org. C %	Total N %	C/N	EXT %	>200 µm fraction %	>50 µm fraction %	>5 µm fraction %	>0 µm fraction %
Marshes and forests on sand (RÉDUCTISOLS)											
Caricetum elatae with *C. lasiocarpa* I	HISTOSOL FIBRIQUE	Hf	7.1	48.9	1.54	32	6.0	85	6	6	3
Caricetum elatae with *C. lasiocarpa* II	HISTOSOL FIBRIQUE over anmoor	Hf	6.7	46.2	1.14	41	9.0	n.d.	n.d.	n.d.	n.d.
Caricetum elatae with *Phalaris*	Histic anmoor	AnHf	6.9	13.7	0.86	16	8.1	n.d.	n.d.	n.d.	n.d.
Cladietum marisci	Anmoor	An	7.6	12.6	0.80	16	7.8	22	24	44	10
Caricetum elatae with *C. panicea*	Anmoor-hydromull	AnAg	7.6	11.6	0.67	17	n.d.	n.d.	n.d.	n.d.	n.d.
Orchio-Schoenetum	Hydromull	Ag	7.8	13.1	0.92	14	7.2	83	9	6	2
Ash forest with white alder	Carbonate eumull	Aca	8.1	3.6	0.25	14	15.4	30	36	29	5
Ash forest with birch	Buried HISTOSOL FIBRIQUE	Hf	7.6	48.1	2.08	23	10.3	79	9	10	2
Ash forest with poplar	Calcic eumull	Aci	7.7	15.2	0.88	17	11.2	12	31	45	12
Forests on molasse (RENDISOLS rédoxiques (× 3), CALCISOL rédoxique with surface acidity*)											
Pine forest and molinia, with ash	Carbonate mesomull	Aca	7.2	5.8	0.30	19	13.2	n.d.	n.d.	n.d.	n.d.
Pine forest and molinia, with ash and oak	Calcic mesomull	Aci	8.2	7.9	0.38	21	14.1	56	18	20	6
Pine forest and molinia, with oak*	Hemimoder	A	5.9	6.4	0.23	28	17.2	n.d.	n.d.	n.d.	n.d.
Pine forest and molinia, with spruce	Saturated mor	OFAci	7.6	20.1	0.82	25	21.9	n.d.	n.d.	n.d.	n.d.

The humus forms clearly distinguish the vegetation types but not always at the same syntaxonomic level (Chap. 7, § 7.1.4). Sometimes, two humus forms correspond to two distinct phytocoenoses: such is the case of anmoor of meadow with *Cladium* (*Cladietum marisci*) and hydromull of meadow with *Schoenus* (*Orchio-Schoenetum*). But at a lower level, the humus forms differentiate variants in the same phytocoenosis; thus, meadow with tall sedge (*Caricetum elatae*) is subdivided into a dry variant with *Carex panicea* on anmoor-hydromull, a moderately moist variant with *Phalaris arundinacea* on histic anmoor and two moist variants with *Carex lasiocarpa* on HISTOSOL FIBRIQUE and on HISTOSOL FIBRIQUE with anmoor.

Correspondence between the pedological and phytosociological taxonomic levels is not always established at the same hierarchical level on both sides.

Lastly, the humus forms can reveal time gaps between the morphology of the solum and its functioning (Chap. 7, § 7.7.2). For example, ash forest with birch preserves old histic horizons below the surface carbonate eumull (Fig. 6.9). The current water-table fluctuation—more than 3 metres—does not allow peat development (Chap. 9, § 9.2). But peat, a relic of the time before artificial lowering of the lake level, still persists in the more aerated recent conditions. The humus form is polycyclic, composed of a superficial eumull conforming to the present behaviour and a subjacent HISTOSOL, memory of the wetter past of the site.

The humus form integrates two time-scales: that of recent decades at the surface and that of preceding centuries deeper down.

6.4.3 The humiferous episolum, an irreplaceable spatiotemporal pivot in the ecosystem

The humiferous episolum links the development time of the vegetation (season, decade or century according to the synusia) to that of the soil, slower overall (decade to millennium according to the horizons; Fig. 6.10), thus constituting an irreplaceable temporal and spatial pivot in the ecosystem.

The humiferous episolum and the humus form are seen to be very valuable for overall understanding of the dynamics of the ecosystem.

Created and reworked by bacteria, fungi and animals of the soil, the humiferous episolum is the original explanatory site of relations between soil and vegetation (Chap. 7, § 7.1.1). Explicitly or implicitly, it is at the centre of all the themes of the second part of this book, such as the site of biological activity, physical support, crossroads of energy, habitat of organisms, regulator of circulating substances and preferred interface between soil and vegetation.

Fig. 6.9 *Polycyclic humiferous episolum of riverain ash forest of Lake Neuchâtel (Switzerland). The surface eumull indicates the present functioning of the solum and the subjacent* HISTOSOL FIBRIQUE *the conditions that prevailed before lowering of the groundwater (photo: J.-M. Gobat).*

'Survey and interpretation of these surface horizons are greatly useful in pedological diagnosis and still more in site diagnosis in which correlation is sought between the soil characters and the floristic composition of the vegetation type the soil supports or the behaviour of forest species (AFES, 1995, 1998).

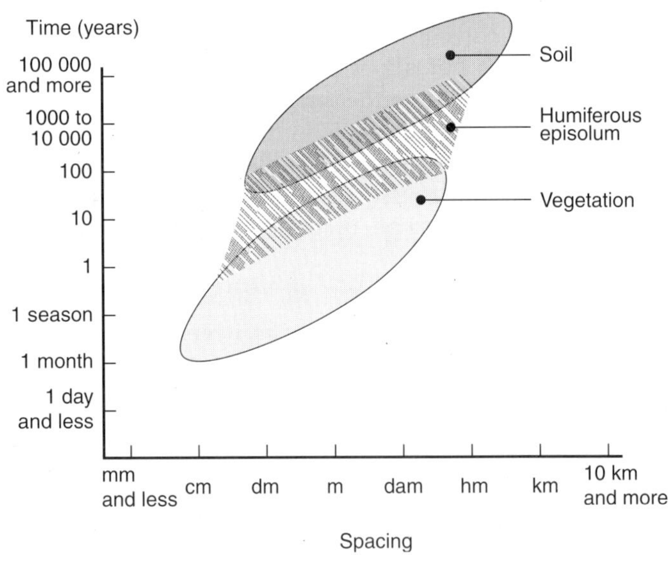

Fig. 6.10 *Spatiotemporal location of the humiferous episolum at the conjunction of whole soil and vegetation.*

Chapter 7

SOIL AND VEGETATION: RELATIONSHIPS AT MANY LEVELS

Relationships between the soil and the plant fall in autoecology; those established between soil and ***vegetation,*** discussed here, belong to synecology (Chap. 4, § 4.2.4). The closest relationship, at the scale of the individual, is located in the rhizosphere, while the broadest is that established at the level of biomes. Chapter 7 begins the second part of this book, dedicated to selected themes in soil biology, through discussion of the relationship between vegetation and soil in the ecosystem. It presents the main suppliers of organic materials—plants—in their community structure.

Autoecology and synecology; rhizosphere and biomes.

Vegetation: all the plant communities present in a given territory.

Reader beware: heed the definitions!

Indeed very many interactions are discussed in other chapters, but chiefly from a descriptive or functional viewpoint, giving importance to species and populations (Chaps 2, 4 and 5). Without ignoring these aspects, it is proposed here to discuss more generally the principles of the relationship established between the soil and vegetation. As this theme touches on many aspects of ***phytosociology,*** it is necessary to devote one section to it to define its principal concepts (§ 7.1.4).

Phytosociology: study of plant communities from the floristic, ecological, dynamic, chorologic and historic viewpoint (Guinochet, 1973). Phytosociology belongs to synecology, contrary to the study of species, which comes from autoecology. ***Chorologic:*** pertaining to ***chorology,*** discipline of biogeography with the objective of explaining the reasons for geographic distribution of living species (Ramade, 1993).

7.1 A THEORY, QUESTIONS, EXAMPLES... SOMETIMES ANSWERS!

7.1.1 Towards a global explanatory theory?

Voluminous literature, not cited here, presents

examples of relationships between a soil and the vegetation it supports:

- parallelism among climate, soil and vegetation,
- toposequences of soil types with corresponding vegetation types,
- comparative dynamics of vegetation and soils, for example on glacial margins, volcanic deposits or alluvial terraces,
- pedogenetic consequences of the colonization of a new sediment by plants,
- nutritional relationships between a substrate and production of the phytocoenosis,
- action of a specific pedological factor on a synusia, etc.

Paradigm: framework of reading, theoretical thread and hypotheses, or even totality of concepts on which all science rests and which governs the manner whereby the scientist thinks and interprets the results of experiments (Schwarz, 1997, modified).

But, in spite of the multitude of cases studied and discussed, there is hardly any universal explanation, a ***paradigm*** of the relationships between soil and vegetation. Simply put, this chapter sets some milestones useful for future establishment of such a theory. Two important aspects of systemic ecology will be specially approached—the spatiotemporal structure of the ecosystem and the change of scale (Bouché, 1990; Frontier and Pichod-Viale, 1991; Auger et al., 1992).

Good methodological questions to be set!

To highlight and then to understand relationships between soil and vegetation constitutes a very complex problem. Before approaching the concept through concrete examples, let us ask ourselves some questions on the methodology of approach to soil-vegetation relationships. From choice, these questions primarily concern the spatial aspect of these relationships. An identical concept could be applied to the temporal relations, such as the relative speeds of development of soils and of vegetation. Without going into detail, these aspects are nevertheless evoked in certain spatial examples; the concluding thoughts in the chapter also take them into account.

7.1.2 Good questions to be posed

How are soil-vegetation relationships to be highlighted?

Simultaneous or independent observations?

Is it appropriate to simultaneously describe the soil and the vegetation in the same area by means of a single protocol, aiming thereby to understand the relation at the site? If yes:

- At what scale and at which organizational level should we work?
- What is to be described in the vegetation?
- What is to be described in the soil (Frontier, 1983)?

On the contrary, should the vegetation and the soil be subjects of separate descriptions, for example at the regional level, without wanting to establish *a priori* relations between them? In that case, how are the soil and the vegetation to be compared later:

The Swiss soil scientist Hans Pallmann (1903–1965), close associate of the renowned phytosociologist Josias Braun-Blanquet, made the first serious attempt at correspondence between soils and vegetation in 1954 (Braun-Blanquet et al., 1954; Boulaine, 1989).

- By calculation: mutual information (Daget and Godron, 1982), regression, multivariate analysis (Legendre and Legendre, 1984; Wildi and Orloci, 1996)?
- By superposition of map layers in a Geographic Information System (Collet, 1992; Legros, 1996)?
- By intuitive approach?

Would it be satisfactory to study vegetation alone and use its bioindicative value (Landolt, 1977, see Chap. 4, § 4.6.2; Ellenberg, 1992) to know soil characters?

Lastly, the fourth suggestion, would an experimental study of key species, for example through crops on different soils, be useful? This approach is often used in agronomy and ecophysiology.

What must be observed on the land?

At what organizational level should the study of vegetation be approached: that of biomes, of phytocoenoses or of synusiae? Should it even go down to that of populations, of a particular individual or of a root-tip?

The level of the biome or of the rootlet?

And which is the pertinent level for approach to the soil:

- The soil mantle?
- A solum in a phytocoenosis? (But many soil types can exist in the same phytocoenosis!)
- The horizons explored by roots? (But those of all the plants or of just one species?)
- The humiferous episolum, under the direct influence of litter?
- A single edaphic factor known to be limiting?

How are the results to be interpreted?

If many soil types exist under the same phytocoenosis, does it mean:

Where the method influences the result!

- That the phytocoenosis is independent of the soil and is only influenced by the climate, for example?

Invariant: which does not vary, is constant (*Petit Larousse*). In the specific case, it applies to edaphic factors that have sufficiently uniform values or features from one area to another of the phytocoenosis to determine it overall, even if the soil identification based on vertical sequences of horizons leads to recognition of many References.

- That the 'functional separation' between the soils is so small that the phytocoenosis is determined by common, ***invariant*** pedological aspects?
- That the correspondence between many soils and one phytocoenosis is only due to a single determining edaphic factor?
- That the phytocoenosis is not as homogeneous as the phytosociological analysis reveals it to be?

In the reverse case, where many phytocoenoses colonize the same soil type, must one conclude:

- That the determining factor is not edaphic but climatic, historical or anthropic, for example?
- That we cannot be satisfied with a soil description made to the Référence or Type (Chap. 5, § 5.6.3) levels, which are global concepts? And that varied functions due to actual horizons of different thickness are hidden under the denomination obtained, homogeneous in its sequence of reference horizons?

Ecosphere: part of the globe in which living organisms are found and permanent life is possible. The ecosphere comprises lithosphere, hydrosphere, atmosphere (troposphere) and ***biosphere***. In the strict sense, the last term encompasses all the organisms living on the surface of the Earth; it was defined thus by the Viennese geologist Suess, who introduced the term in 1875. A more modern concept, including the functional processes was later proposed by Vernadsky (1929). The term biosphere is often used as a synonym for ecosphere.

Lastly, how can we explain the dominance on two identical soils separated by 20 metres, of one plant species in one case and another in the second, and this with apparently similar ecological factors (light, moisture, etc.)? Is this due:

- To random chance, that 'subtle dustbin' to which all that cannot be explained is consigned?
- To biotic factors relating to interspecific competition (Chap. 4, § 4.5.3), symbiotic relations (mycorrhizae, fixing nodules; Chap. 16), or to the physiological characteristics of the species, such as adaptive strategies (Chap. 4, § 4.4.6; Chap. 12, § 12.6.2)?

7.1.3 Six levels of spatial organization

Biome: living community found over large areas in continental environment, defined at the scale of broad bioclimatic zones (Ramade, 1993, modified). The biome corresponds to the biocoenosis specific to macroecosystems extending over large portions of land. Arctic tundra, deserts and tropical forests are biomes.

Some responses to these questions are provided by concrete examples of relationships between soil and vegetation, drawn from investigations of the Plant Ecology Laboratory at the University of Neuchâtel, Switzerland. They have been selected at specific levels of spatial organization of ecosystems and do not pretend to cover all. The six cases envisaged are:

- the ***ecosphere***, with correspondence between the ***biomes*** and the great pedogenetic processes (§ 7.2),
- the ecocomplex, with description of catenas (§ 7.3),

- the phytocoenosis (§ 7.4),
- the synusia (§ 7.4),
- some species in a community (§ 7.5),
- one species in a community (§ 7.6).

7.1.4 Some phytosociological concepts and definitions useful for soil science

Landscape ecology, biogeography and landscape phytosociology are concerned with the most general organizational levels.

Perception of the first levels of spatial organization pertains to landscape ecology (Forman and Godron, 1986; Lefeuvre and Barnaud, 1988; Leser, 1991; Turner and Gardner, 1991), biogeography (Elhai, 1968; Ozenda, 1982; Archibold, 1995; Bailey, 1996; Grabherr, 1997) or landscape phytosociology (by description of large assemblages of plant communities—geosigmassociation, sigmassociation; Tab. 7.1) on a floristic base (Géhu, 1974); they are not detailed here. In the lower organizational levels, phytosociology takes over with various concepts apparently close but often associated with different stages, scales and domains related to the study of vegetation. Many of these concepts are also used in soil science. Partly summarized in Table 7.1, phytosociological concepts are grouped under the following themes:

- methodology: from the concrete to concepts, through observation,
- organization of the ecosystem: levels of spatiotemporal observation,
- 'quality' of the vegetation: physiognomy or floristic composition,
- dynamics of the vegetation: primary succession or secondary succession,
- regulation: control of the functional processes.

From the concrete to concepts, through measurement and observation

The phytosociological diagnosis of vegetation (Braun-Blanquet, 1964; Guinochet, 1973; de Foucault, 1986; Gillet et al., 1991) is a procedure similar to that of soil survey (Chap. 5, § 5.6.3). This procedure pertains to three aspects:

- the concrete, the reality on land; ideas such as soil mantle or the solum in soil science, the synusia, phytocoenosis and plant community in phytosociology and the catena in both sciences belong to this ensemble;
- collection of data, measurement of the concrete; concerned here are the profile description of the solum

Table 7.1 Principal concepts of phytosociology applied at organizational levels

Organizational level	**Concrete objects**	**Observations**	**Abstract units of the classification**
At all levels	*Plant community*	*Phytosociological relevé*	*Vegetation unit*
Ecocomplex: • ensemble of successions with different climaxes • ensemble of successions converging to a single climax	• Catena • Tesela	• Relevé of catena (toposequence) • Relevé of tesela	• Geosigmassociation • Sigmassociation
Biogeocoenosis	Phytocoenosis	• Relevé of phytocoenosis (in integrated synusial approach) • Floristic relevé (in classic phytosociology)	• Coenassociation • Association
Elementary community	Synusia	Floristic relevé	Association (higher hierarchical taxonomic units: alliance, order, class; see Tab. 7.2)
One dominant species	Facies	—	—

Phytosociological relevé: exhaustive inventory of species, types of plant synusiae or of phytocoenoses (according to the chosen organizational level) present in a sample area representative of a plant community. Semi-quantitative coefficients are assigned to the object inventoried, taking into account its abundance, coverage and vigour (§ 7.5.2).

Catena (in soil science): chain of genetically linked soils, each of which has received from others (e.g. through erosion) or has given up to others (e.g. by oblique clay-leaching) some of its constituent elements (Lozet and Mathieu, 1997).

and physicochemical analyses in soil science, the ***phytosociological relevé*** in study of vegetation;

• typology, taxonomic units, classification; the diagnostic horizon or Référence in soil science, association or vegetation unit in phytosociology belong to it.

Through the organizational levels of the ecosystem

A landscape is composed of the overlapping of nested structures, through which we pass by changing the spatial scale (Fig. 7.1). Successively identified are:

• the catena, often located at the scale of the drainage basin (ecocomplex),
• the tesela, at the scale of the elementary geomorphological unit (e.g. a uniform slope with homogeneous geological substratum),
• the phytocoenosis, at the scale of the biogeocoenosis,
• the synusia, at the scale of the elementary community.

In soil science, a chain of soils beginning on a rocky cliff and ending in the coarsest scree on the footslope, but always nourished by the cliff, is an example of a catena. In phytosociology, the ***catena*** is a complex of teselas or phytocoenoses assembled in zones and/or mosaic within the same great geomorphological unit, which could be derived one from the other by primary

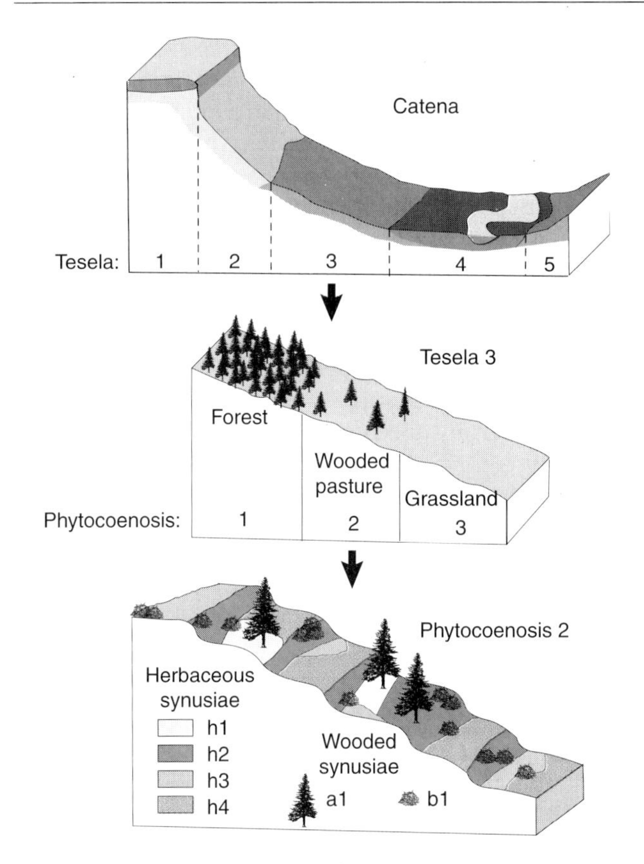

Fig. 7.1 *Organization of the landscape, exemplified by the wooded pasture of the Jura (after Gillet and Gallandat, in Étienne, 1996). Explanations in the text.*

successions (Gillet et al., 1991); this meaning is close to that of soil scientists.

Often at the same scale as the catena, but without apparent genetic link, a chain of soils or of phytocoenoses successively crossing an alluvial terrace, a morainic hill and a nearby limestone slope is a ***toposequence***. The toposequence is not strictly linked to the scale of the ecocomplex, but can also cross biomes for example (§ 7.2).

Toposequence: complex mapping unit of soils or connected plant communities whose geographical distribution is constantly found in a fixed order, ruled by the relief and topography, with no apparent genetic link between them (Lozet and Mathieu, 1997, modified).

Studies on catenas and toposequences are useful means of approaching the ***ecocomplex***, which is an integrated assemblage of interdependent biogeocoenoses (Gillet et al., 1991) or a system of **ecosystems** (Blandin and Lamotte, in Lefeuvre and Bernaud, 1988), this latter term being used here in the limited sense of biogeocoenosis! Examples are an alpine valley, the Dombes in France, the Creux du Van in the Swiss Jura (§ 7.3.1) or even a portion of the marine littoral.

At a slightly more precise scale, but with the idea of dynamics, a rich grassland, a cultivated land and a small thicket located on an old alluvial terrace belong to the same ***tesela***. When agriculture is abandoned, they will all evolve towards the same climatic climax, for example oak-hornbeam forest. In turn, the swamp that skirts them, though equally harvested, forms part of another tesela. Situated on peat soil, it will probably develop, if abandoned, into a black alder forest.

Tesela: complex of phytocoenoses assembled in zones and/or mosaic, derived from one another by secondary succession and corresponding to the same present-day potential climax (Gillet et al., 1991).

Still finer, the plants of a dry beech forest on a south slope, those of an alkaline fen or those of a relatively homogeneous portion of a wooded pasture constitute phytocoenoses. The ***phytocoenosis*** is a plant community formed from a complex of plant synusiae organized spatially, temporally and functionally in a biogeocoenosis, and showing strong relations of ecological, dynamic and genetic dependence (Gillet et al., 1991).

Synusia: ensemble of organisms sufficiently close in their living space, ecological behaviour and periodicity to share at a given moment the same milieu (Gillet et al., 1991, modified).

Within a phytocoenosis, all the mosses colonizing the vertical face of a stone block form a ***synusia***, like the tall trees of a hardwood forest or the living animals under a peel fallen on the soil.

Do not confuse biogeocoenosis and ecosystem!
The phytocoenosis represents the vegetation component of the biocoenosis (Chap. 4, § 4.2.4). The latter, delineated at the scale of phytocoenosis and supplemented by abiotic constituents (e.g. precipitation, soil clay minerals, water, etc.) constitutes a ***biogeocoenosis***. The concept of biogeocoenosis is differentiated from that of the ecosystem, in which it is included, because it pertains to *a specific spatial scale*, that of the phytocoenosis. In turn, the concept of ecosystem has no spatial content; rather it is defined by its functional aspects (Tansley, 1935; Lindeman, 1942; Odum, 1971, 1996; Auger et al., 1992). 'The systems thus formed are, from the viewpoint of the ecologist, the basic units of the nature of the surface of the Earth. These ecosystems, as we can call them, offer the largest diversity of type *and of size*' (Tansley, 1935). Thus, the ecosphere, a forest, a forest pool or a clump of sedge and its soil are nested ecosystems, but not the soil alone or an old stump (Chap. 4, § 4.1.1).

Biogeocoenosis: portion of the ecosphere in which, in a certain area, the biocoenosis and the lithosphere corresponding to it are uniform and, consequently, the interaction of all parts that form a unique complex also remain uniform (after Sukachev, 1954). The two terms biogeocoenosis and ecosystem are often confused for one another because the most 'obvious' ecosystems are located exactly at the scale of the biogeocoenosis: a forest, a grassland, etc.

At a scale lower than that of the synusia, we enter the domain of populations or individuals that are concepts belonging to the autoecology of species. The only exceptions are those of mono- or paurispecific synusiae occasionally found in environments with severe ecological stress (§ 7.6.1; Chap. 9, § 9.1.3) and that of ***facies***. In the alluvial zone, for example, tall invading species such as the Indian balsam *Impatiens glandulifera* or Canadian goldenrod *Solidago canadensis* forms facies in herbaceous synusiae of alder or ash forest.

Facies: in a plant community, zones with a particular physiognomy due to the local dominance of one species (Delpech et al., 1985).

From general physiognomy to floristic composition

In the nineteenth century, the first descriptions of vegetation preferred the *physiognomic* or *morphological aspect*, differentiating the hardwood forests from coniferous forests, greensward from heaths or even grasslands from pastures. This approach, always valid, is that of ***plant formations***; the floristic aspect is secondary. A peat-bog, a heath, an oligotrophic lawn, a wooded steppe, a fen, whatever their floristic composition, are plant formations.

Plant formation: vegetation unit defined by a relatively uniform general physiognomy because of the predominance of one or several biological form(s). The plant formation is defined at the organizational level of the phytocoenosis.

In a more subtle approach, greater importance was later accorded to the *floristic composition* of communities. This enabled the distinction, for example, of black bog-rush fen (*Orchio palustris–Schoenetum nigricantis*) from Davall sedge fen (*Primulo farinosae–Caricetum davallianae*). These two taxonomic units, corresponding to two different synusiae, are ***plant associations***. They belong physiognomically to the same plant formation, that of fens. The plant association is an abstract concept, a taxonomic unit justified and proved by statistical criteria, a 'nomenclatural label'.

Plant association: fundamental abstract unit of the hierarchic classification of plant synusiae (or phytocoenoses), based on floristic and statistical criteria.

The classification of plant associations pertains to ***synsystematics***. In this, the associations are grouped into ***alliances***, then into ***orders*** and finally into ***classes***. They can be subdivided into ***subassociations*** or ***variants*** (Tab. 7.2). If a statistically recognized taxonomic entity cannot be classified at a specific systematic level, it is called a ***vegetation unit*** (Ge: *Pflanzengesellschaft*; Fr: *groupement végétal*).

Synsystematics (or syntaxonomy): hierarchic phytosociological classification; it is so named to distinguish it from floristic systematics.

To which concrete organizational level does the plant association correspond?

In the integrated synusial approach (Gillet et al., 1991; Gillet and Gallandat, 1996), the plant association is, by principle, *always* situated at the level of the synusia. The taxonomic integration of plant associations into a ***coenassociation***, which is a taxonomic type of phytocoenosis, corresponds in the field to the actual integration of synusiae into a phytocoenosis.

On the other hand, in classic phytosociology (Braun-Blanquet, 1964; Guinochet, 1973), the plant association is *sometimes* situated at the level of the synusia (e.g. in grassland), *sometimes* at that of the phytocoenosis (e.g. in forest), thus not always respecting the spatiotemporal organization of the ecosystem and the correct scale for each organism.

Primary or secondary origin

Every plant community, just like the ecosystems in which it is set, tends to evolve more or less rapidly to a ***climatic climax*** or to a ***site climax***. The ***climax***, a concept

Climatic climax: climax essentially induced by the macroclimate.

***Site climax** (or **edaphic climax**):* Climax determined primarily by local factors: particular pedoclimate, micro- and mesotopography, floods, etc.

Table 7.2 Syntaxonomic levels and examples

Syntaxonomic level	**Suffix**	**Nomenclature (example taken from Gallandat, 1982)**	**Corresponding vegetation**
Class	-etea	*Molinio-Arrhenather*etea	Grasslands
Order	-etalia	*Molini*etalia *coeruleae*	Wet grasslands
Alliance	-ion	*Molin*ion *coeruleae*	Nutrient-poor wet grasslands
Association	-etum	*Trollio-Molini*etum	Wet grassland with globeflower and purple moor grass on soil with variable moisture
Subassociation	-etosum	*Trollio-Molinietum stachy*etosum	Wet grassland with globeflower and purple moor grass, subassociation with betony on relatively dry soils
Variant	—	*Trollio-Molinietum stachyetosum*, variant with *Galium verum*	Same as above, variant with Lady's bedstraw, on slightly acid soils

Exogenous environmental factors: ensemble of environmental elements that *existed prior to* the installation of a biocoenosis and contributed to conditioning its existence. It pertains to the original fraction of the biotope.

Primary succession: succession of biocoenoses starting from an environment determined by exogenous factors up to the climax, with creation and development of an environment determined by endogenous factors.

Endogenous environmental factors: ensemble of environmental elements *created or modified by living organisms* of a biocoenosis. It concerns the 'built-up' fraction of the biotope. The assemblage of exogenous and endogenous factors forms the ***biotope***.

introduced by the American F. Clements (Clements, 1916), represents the mature stage of the biogeocoenosis successions; it is constituted by a spatiotemporal complex of pioneer phases, transient and terminal, which ensures for them optimum autonomy, homoeostasy and resilience, and which indicates a dynamic equilibrium of exogenous and endogenous environmental factors (Gillet et al., 1991).

The evolution of an ecosystem towards the climax can start in two ways (after Gillet et al., 1991):

- in a totally abiotic milieu in which ***exogenous factors***—parent material, solar energy, mesoclimate, topography, etc.—dominate; we speak of ***primary succession*** such as the successive development of moss mats, clumps of grass, then a closed lawn on a large block of scree freshly detached from a rock cliff;
- in an environment already transformed by living organisms, in which ***endogenous factors***—soils, organic matter, microaggregates, microclimates, microtopography—have priority; we then have a ***secondary succession***, such as invasion of a pasture by herbaceous wasteland plants, then by thorny shrubs and finally by a pioneer forest, prelude to return of primitive forest.

It is essential to know the respective proportions of exogenous and endogenous factors in the study of plant successions established after a major disturbance such as a flood in the alluvial zone. A weak flood leaves the old soil in place (endogenous factors predominant) under a thin layer of freshly deposited sand; in this case, only the herbaceous stratum is modified and the trees *in situ* determine the future dynamics of the vegetation. On the contrary, a strong flood carries away the soil and the forest, leaving a field of bare pebbles (exogenous factors predominant); in this case, recolonization by plants starts again from zero, without any 'boost' from the old soil, lost in the waters. Lepart and Escarre (1983) and also Drouin (1994) made a very complete analysis of plant succession.

Secondary succession: succession of biocoenoses following the disappearance of the climax, starting from an environment of endogenous origin.

Homoeostasy: ability of systems (organisms, ecosystems, etc.) to maintain their functioning constant against changes in the external environment.

Resilience: ability of a system to survive changes and perturbations in its structure and/or its functioning, and to revert, after disappearance of the changes, to a state comparable to the initial situation (Ramade, 1993). It can be compared to healing of the ecosystem.

Controlled functional processes

The dynamics towards the climax and also the periodic internal fluctuations of the system at equilibrium are under the control of numerous processes relevant to ***homoeostasy***:

- negative feedbacks between soil and vegetation,
- functions of ***resilience***,
- interactions between diversity and stability.

7.2 ECOSPHERE, BIOMES AND PEDOGENETIC PROCESSES: GREAT LANDSCAPE ASSEMBLAGES

At the scale of biomes, the relationship between vegetation and soil is—very simply speaking!—causal in temperate and cold climatic zones, while it is often a simple coincidence without direct causality in tropical and equatorial zones.

7.2.1 From southern to northern Russia: the causality factor

The first example selected, soils of the Northern Hemisphere.

Russian soil scientists have excellently demonstrated the close relationship that binds soil types to the great biomes (Fig. 7.2; Chap. 5, § 5.6.2).

This toposequence reveals excellent concordance among macroclimate, plant formation and soil, within a single climatic climax, and this in each latitudinal section. This is hardly surprising, because most of these biomes, with the exception of the southernmost, are subject to temperate or cold climate. And yet, in these conditions, soil development largely depends on that of the organic matter, which acts through loss of carbonates, decalcification, brunification, clay-leaching or even cheluviation (Chap. 5, Tab. 5.2). The soils are generally

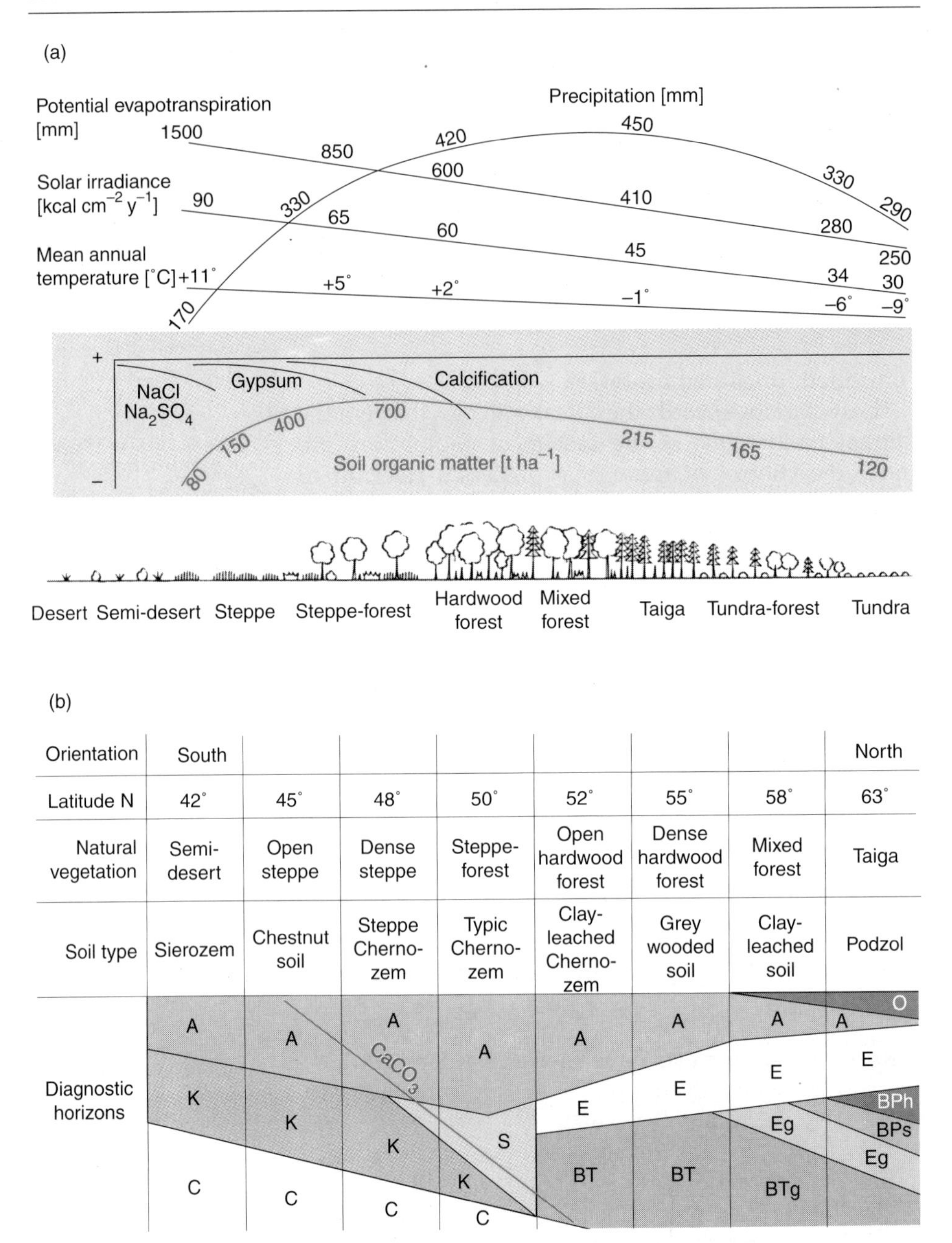

Fig. 7.2 *Toposequence across European Russia: (a) Evolution from south to north of some ecological descriptors as function of the biomes (plant formations) (after Duvigneaud, 1984, modified); (b) Principal soil types.*

of short-cycle type, post-glacial (Chap. 5, § 5.5.5). The vegetation, at the level of plant formation, actively influences pedogenesis, in an obviously causal relationship. Vegetation is in turn modified by the soil, by feedback for example in production of biomass.

What about soils of regions with a hot climate, whose development depends primarily on geochemical factors?

7.2.2 Africa: a non-causal correspondence

In very slowly developing long-cycle soils, the concept of a single climatic climax at the scale of a continent is hardly applicable, as the time-scales are so much different (Chap. 5, § 5.5.5; Paton et al., 1995). However, if a plant-formation map and a soil map of Africa are compared, concordance is very good between soil type and vegetation, as in the more temperate zones (Fig. 7.3).

Long-cycle soils, slow development...

... and good concordance between soil and vegetation.

In the tropical and equatorial zones, vegetation reacts rather quickly to climatic changes, sometimes caused by man, while the soils are formed by very slow processes, extending over hundreds of thousands of years. Medium-term climatic variations can thus greatly modify the vegetation pattern without being able to reorient pedogenesis. Evidence of this is the presence, under humid equatorial forests, of very ancient ferruginous or aluminous cuirasses (Ferrallitic soils), which could not have formed in the present-day climatic conditions (Trochain, 1980).

Vegetation and soil: very different speeds of development.

This exact superposition of plant formations and soil types cannot be explained by a causal relationship because, here, unlike in the previous example, the vegetation only very weakly influences pedogenesis. It is mostly a coincidence, juxtaposition of two compartments of the ecosystem that are relatively independent but together subject to the action of the macroclimate.

A relationship of coincidence dominated by a third partner, climate.

This principle of coincidence is valid for the entire African continent but is certainly expressed differently according to latitude (e.g. desert and RÉGOSOLS, savanna and Ferruginous soils). Also, abrupt ecological change at the pertinent time-scale (e.g. massive deforestation) enables a catastrophic process (e.g. accelerated erosion).

Correspondence between a soil type and a plant formation can reveal not only a causal relationship (e.g. a spruce forest on a PODZOSOL in the taiga), but also a simple coincidence, under the influence of a third agent (e.g. equatorial forest on Ferrallitic soil in Africa). In the

To summarize! Two possible cases: a causal mutual relationship or simple coincidence.

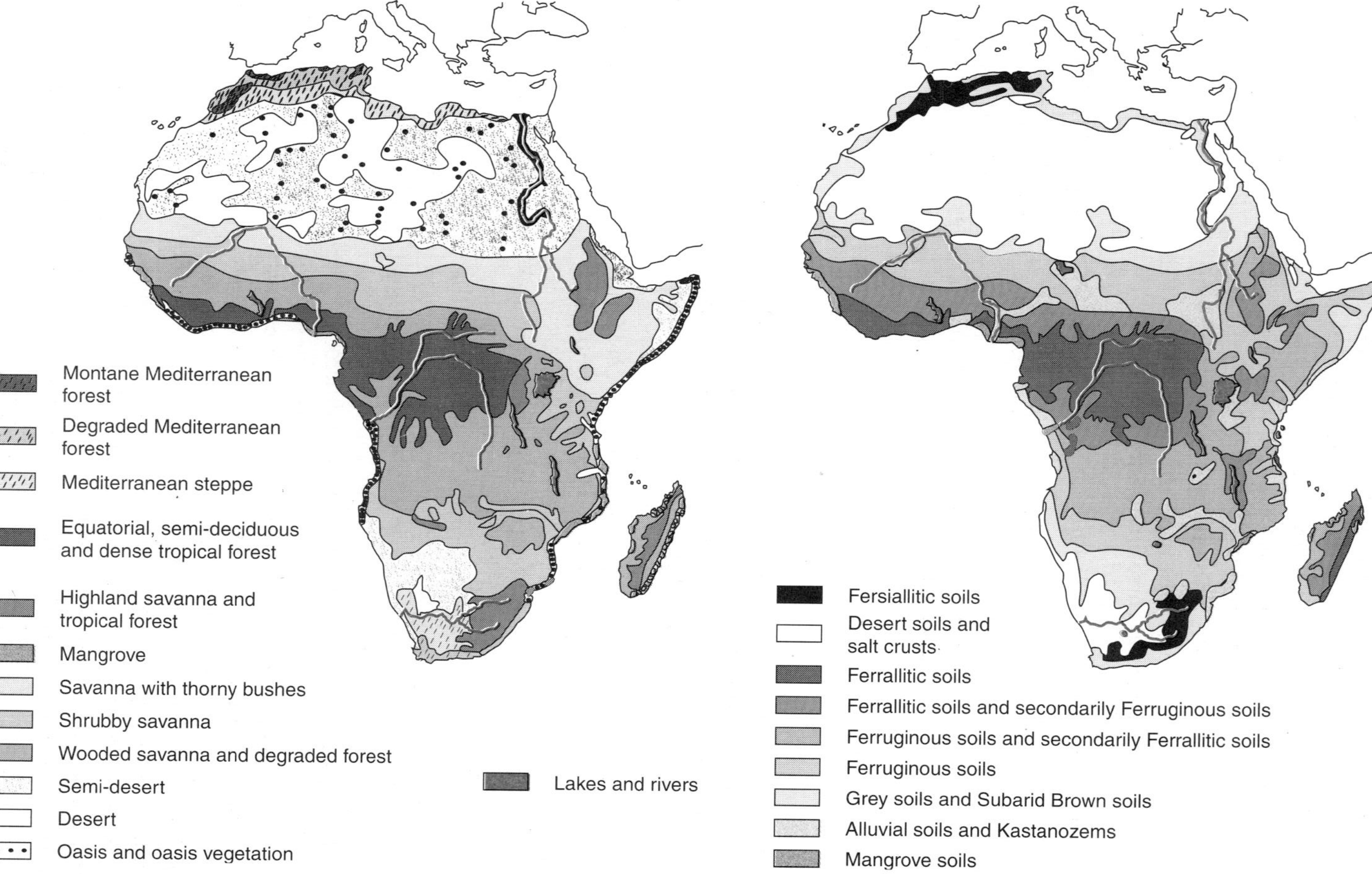

Fig. 7.3 *Comparison of biomes (plant formations) and soils in Africa (after Soltner, 1995).*

temperate zone, climate influences two elements in close causal interaction of the feedback type. In Africa, climate influences two rather poorly related compartments (maybe only at the level of the concerned biome, by a slight effect of the soil on the vegetation—e.g. selection of acid-tolerant or calcitolerant species) (Fig. 7.4).

7.2.3 From the ecosphere to the mountain massif: a relationship independent of spatial scale

Altitudinal zonation corresponds to latitudinal zonation: the passage from one plant formation and one soil to another plant formation and another soil takes place vertically over a few hectometres, against the thousands of kilometres in the first case (Fig. 7.5; also see Chapter 16, § 16.1.6, which illustrates application of this principle to mycorrhizae).

Correspondence between plant formations and soils, such as seen in Russia and Africa, is also established on a much more restrained scale.

Correspondence between soil type and plant formation is preserved in spite of great change in spatial scale. This is possible thanks to the similar behaviour, in latitude and altitude, of several ecological factors: temperature, precipitation, snow cover. Thus relational processes independent of the spatial scale exist between soil and vegetation. We shall see another example in § 7.5.3.

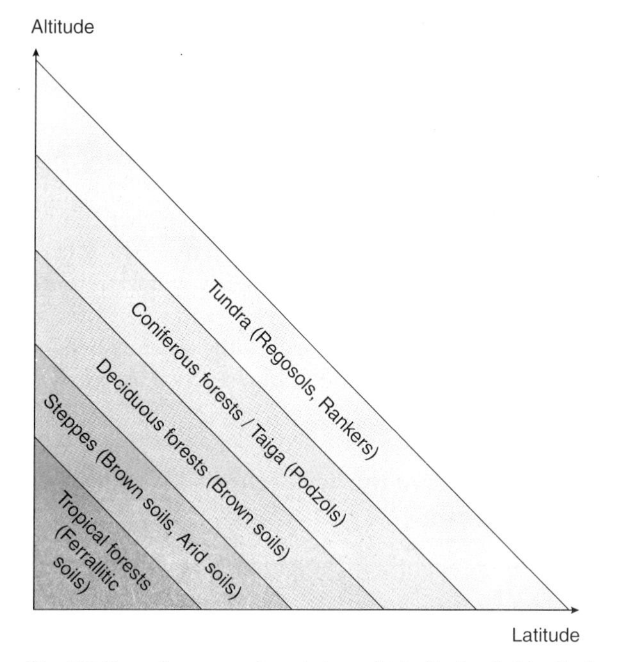

Fig. 7.5 General correspondence between latitudinal and altitudinal zonations of plant formations and soils.

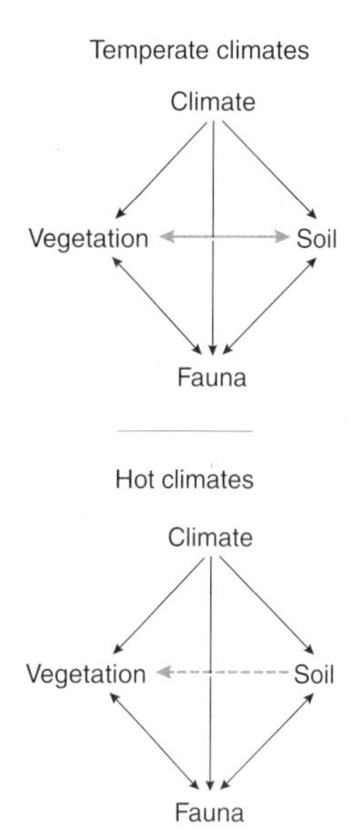

Fig. 7.4 Reciprocal influence among climate, soil, fauna and vegetation in temperate climates and hot climates.

7.3 SOILS OF AN ECOCOMPLEX: VERY TYPICAL OR LESS CLEAR-CUT

Organization of the ecocomplex... contribution of catenas and toposequences.

At the scale of the ecocomplex, soils are usefully described by means of surveying catenas or toposequences located along certain representative axes of the ecocomplex. Thus we can prefer altitudinal or climatic axes or those crossing various parent materials. At the same spatial scale, that of the kilometre for example, and at the same level of phytosociological description, here the phytocoenosis, the soils can be very 'contrasting' or, on the contrary, remain quasi-similar along the catena.

Two cases are presented here: one in which each phytocoenosis is characterized by a specific soil, and another in which many phytocoenoses are found on the same soil type. A third possible case of correspondence is that in which one phytocoenosis grows on different soil types. We return to this at the synusial scale, discussed in § 7.4.

7.3.1 Ecocomplex of the Creux du Van: to each soil its own phytocoenosis!

The rock cirque of the Creux du Van in the Swiss Jura presents a very varied topography: scoured plateaus with loess-filled depressions (§ 7.4.1), rock ledges, high limestone cliffs, stabilized fine talus, loose medium scree, large stable blocks, calcareous moraines (Fig. 7.6).

One phytocoenosis, one soil type, one climactic equilibrium. The Référence level is informative enough here to distinguish edaphic conditions of the phytocoenoses.

Geomorphological, climatic and plant diversity has resulted in formation of nine categories of soils at the Référence level (earlier at Type level). A different phytocoenosis corresponds to each Référence. The entire ecocomplex is characterized by stable soil-vegetation pairs in equilibrium with the meso- or microclimate. This is called a mosaic of climaxes, some climatic such as beech-fir forest, others site-controlled, like spruce or maple forest.

7.3.2 Alluvial terraces of the Sarine: one soil... but five phytocoenoses!

Only analysis of specific edaphic features separates the soils of different phytocoenoses. The Reference level is not sufficiently informative.

Compared to the Creux du Van, which is an old, very stable ecocomplex, the alluvial zone of the Sarine near Grandvillard (Fribourg canton, Switzerland) contains only one relatively equilibrated phytocoenosis among five: the

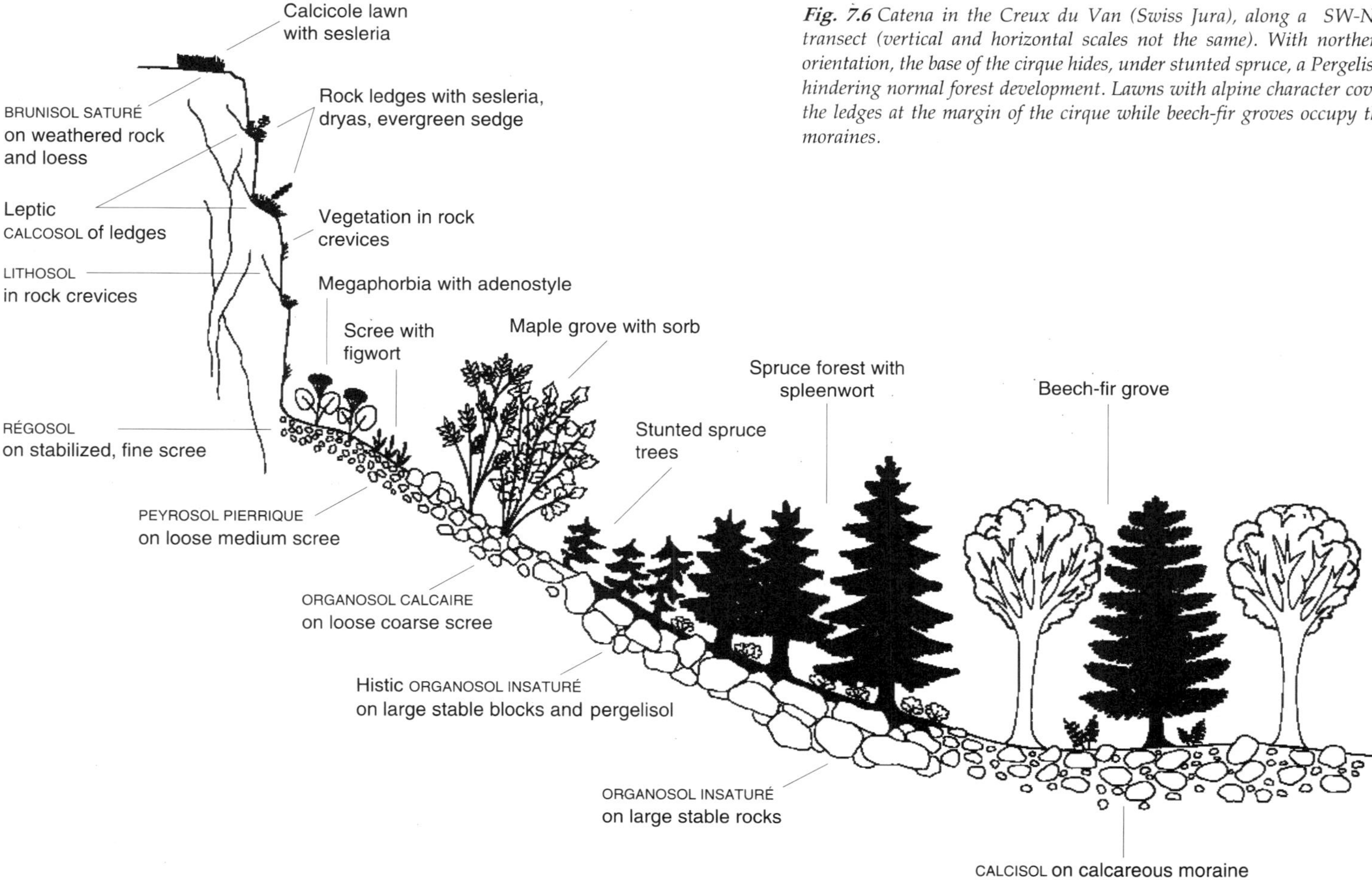

Fig. 7.6 *Catena in the Creux du Van (Swiss Jura), along a SW-NE transect (vertical and horizontal scales not the same). With northern orientation, the base of the cirque hides, under stunted spruce, a Pergelisol hindering normal forest development. Lawns with alpine character cover the ledges at the margin of the cirque while beech-fir groves occupy the moraines.*

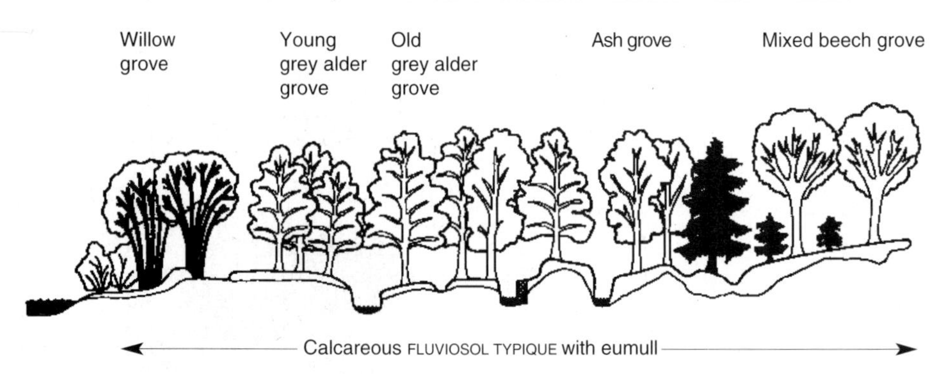

Fig. 7.7 *Toposequence through the alluvial zone of the Sarine near Grandvillard (Fribourg canton, Switzerland) (after Bureau, 1995, simplified).*

148-year-old beech forest. All the others represent development phases of more recent alluvial terraces, 18 to 38 years old (Fig. 7.7; Bureau et al., 1994; Bureau, 1995).

Subtle nuances in the case of FLUVIOSOLS!

In spite of their relatively young age, the five phytocoenoses are quite typical from the phytosociological viewpoint (Gallandat et al., 1993), which is not true for the soils. A single Type, *a fortiori* a single Référence, supports all the phytocoenoses here, including beech grove: the calcareous FLUVIOSOL TYPIQUE with eumull. The originality of the soils of each phytocoenosis has to be investigated at a finer level, that of specific pedological properties or processes such as the intensity of loss of carbonates (Bureau, 1995). The organic carbon and total nitrogen reserves in the organomineral horizon are also informative. The former ranges from 8.6 t ha^{-1} in the willow grove to 83.3 t ha^{-1} in the beech grove, the latter from 0.7 to 7.0 t ha^{-1}. In turn, the degree of evolution of organic matter, estimated by the C/N ratio, by the extractability of organic matter and by the FA/HA ratio scarcely brings out differences from one phytocoenosis to the other (Fierz et al., 1995).

7.4 PHYTOCOENOSES, SYNUSIAE AND SOIL TYPES: HOMOGENEITY AND HETEROGENEITY

Here too, correspondence between one level of plant organization and another of pedological order is not one-to-one.

The two examples chosen at this scale show that within a phytocoenosis the soil can be heterogeneous, reflecting synusial changes or, on the contrary, very homogeneous, identical in all the synusiae.

7.4.1 Lawn of the Haut-Jura: to each soil its own synusia!

Dip: angle made by the plane of the greatest slope of a geological bed with the horizontal (Fig. 7.8).

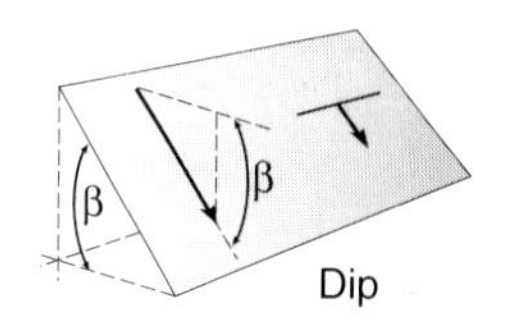

Fig. 7.8 Dip of a geological bed.

As in the Creux du Van, but over shorter distances, geomorphological diversity induces clear boundaries between soils and vegetation units. This is the case in karst zones with relatively small ***dip*** (Fig. 7.8), such as those found in the entire Jura massif, particularly in the Chasseron, from where the example below is taken (Fig. 7.9; Gobat et al., 1989; Havlicek et al., 1998).

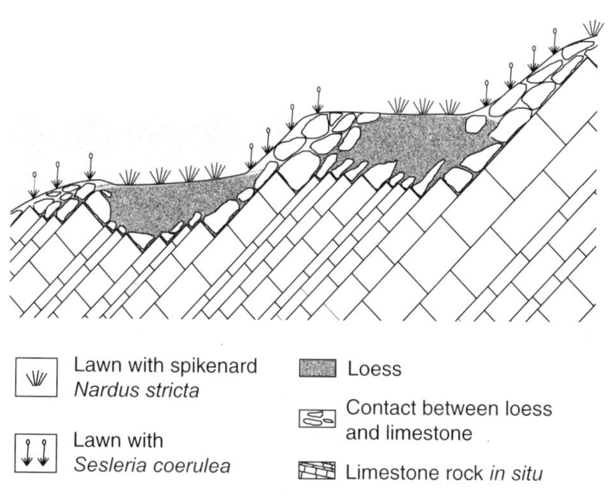

Fig. 7.9 Relationships between synusiae of the subalpine lawn and the soils in the Chasseron region (Jura, Switzerland).

On humps, the synusia with alpine lady's mantle and blue sesleria (*Alchemillo conjunctae–Seslerietum albicantis* association) colonizes a stone-rich ORGANOSOL CALCIQUE on the extremities of layers of massive Sequanian limestone. A few metres near-by and in the depressions an acid-tolerant synusia with spikenard, pill sedge and whortleberry (*Carici piluliferae–Nardetum strictae* association) grows on a strongly acid NÉOLUVISOL (pH_{KCl} = 4.2 in the A horizon), occasionally podzolized (Michalet and Bruckert, 1986). No dynamic link binds the synusiae of the phytocoenosis together.

The close correspondence between each synusia and 'its' soil type indicates the unrelated development of each pair thus formed with respect to the others.

7.4.2 Wooded pasture of Franches-Montagnes: one soil... but eight synusiae!

Unlike subalpine lawn, the wooded pasture of Franches-Montagnes (Swiss Jura) considered here colonizes a soil (NÉOLUVISOL) that remains uniform over many hectares (Fig. 7.10).

Geomorphological uniformity leads to evolution towards a single climax and imposes a single soil type.

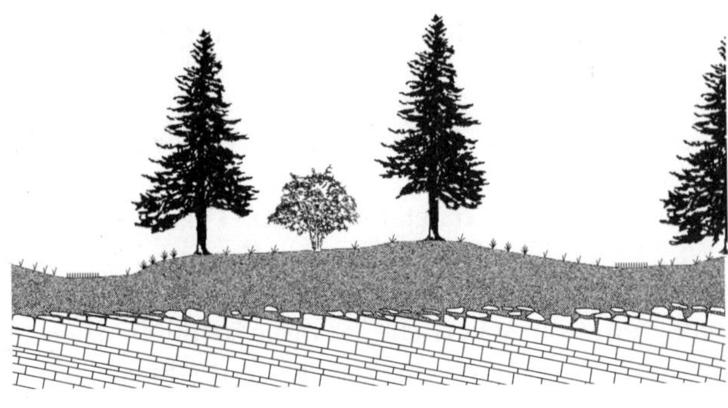

Fig. 7.10 *Wooded pasture of Franches-Montagnes; a phytocoenosis on a very uniform* NÉOLUVISOL *(Havlicek et al., 1998).*

The loess deposits that covered the French-Swiss arc of the Jura after retreat of the glaciers allowed formation of NÉOLUVISOLS, particularly on the shelves and in depressions (Pochon, 1978; Havlicek and Gobat, 1996). Because the soil is uniform, the synusiae constituting the phytocoenosis tend to evolve towards a single ***attractor*** (Gillet et al., 2002), determined by the uniform pedological and mesoclimatic conditions. Present-day synusial diversity is due to human and livestock action (exploitation of woods, trampling, grazing, dung dropping) and is not at all due to pedological variations at the Référence level. On the other hand, different conditions appear at the level of humus form, which 'sticks' better than the entire solum to the vegetation types (Chap. 6, § 6.4.1).

Attractor: in systems sciences, final equilibrium stage of convergent dynamics of processes starting under different conditions.

7.5 SPRUCE FOREST WITH BLECHNUM: A FEW SPECIES MAKE THE DIFFERENCE

Scale-independence, rare situations and bioindication: example of a spruce forest in the Jura.

Also of the phytocoenotic level but with population aspects, this example confirms the idea of scale-independence of certain soil-vegetation relationships and highlights the very informative aspect of rare ecological situations. Again selected in the Jura massif, veritable storehouse of pedological diversity behind its appearance of pure, hard limestone, the example of spruce forest with blechnum *Blechnum spicant* enables discussion of the bioindicative role of differential species of vegetation units.

7.5.1 A phytocoenosis with two feet, one peaty and the other mineral

Interested in paradoxical situations, Richard (1961) describes many 'acidophilic' vegetation units (Chap. 4, § 4.2.4) of the limestone range of the Jura. Among them, spruce forest with sphagnum (*Sphagno-Piceetum*) colonizes either the borders of raised bogs (Chap. 9) or certain small areas distributed sporadically on the entire range, on a mineral substrate with podzolized soils.

In addition to beech forests with wood-rush on crystalline moraines derived from the Alps, Richard (1961) highlights coniferous forests on highly acid substrata in the Jura limestone.

Phytosociological analysis shows that the spruce forest on peat constitutes a first 'subphytocoenosis' differentiated by some species of the peat bogs (*Betula pubescens, Pinus mugo* subsp. *uncinata, Carex nigra, C. echinata, Sphagnum* spp.). On the mineral substrate, a second 'subphytocoenosis' is distinguished by *Blechnum spicant, Prenanthes purpurea* and *Athyrium filix-femina*, or just three species out of a total of nearly 100 identified in the 21 relevés by Richard.

Through pedological analysis, Richard (1961) demonstrates a very close relationship between the presence of spruce forest with blechnum and that of podzolized soils. A recent more complete investigation (Vadi and Gobat, 1998a, 1998b) confirms this almost exclusive correspondence, valid for the entire French-Swiss Jura range: under every spruce forest with blechnum inventoried is hidden a soil with podzolization of varying intensity. In many cases, a true PODZOSOL has been identified on the basis of analysis of forms of iron and aluminium.

Against all expectations, veritable PODZOSOLS exist on the limestone rock of the Jura!

7.5.2 A very informative singularity

But for one exception, the areas of PODZOSOLS considered are small, a few ares or less. This peculiarity means that the relationship—indicated by just three differential species!—between vegetation and soil is very informative. It is a matter, in the case of spruce forest with blechnum, of a strong bond, totally specific, to bioindication of greatest value.

Even proved, presence of PODZOSOLS in the Jura remains a rare situation.

This confirms the possible use in ecology of mathematical information theory. This theory, initiated by Shannon in 1948 for the needs of telecommunication, has been successfully used in ecology, especially in situations wherein the usual multivariate analyses are poorly adapted, such as transects with high ecological

Amount of information: the amount of information furnished by a phenomenon is expressed by the logarithm to base 2 of the probability of its occurrence which in turn is calculated according to the formulae for calculation of probabilities (number of combinations realized divided by the number of possible combinations); its unit is the ***bit*** or ***shannon***. Thus calculated, the information is indicative of the entropy level of a system (Schwarz, 1988; Frontier and Pichod-Viale, 1991).

gradient (Daget and Godron, 1982). This theory informs us that the ***amount of information*** provided by a phenomenon is the greater when the probability of its occurrence is the smaller. As Frontier and Pichod-Viale (1991) put it: 'This concurs with the usual meaning: when a very probable event takes place, no one is astonished by it or feels that much information has been gained by verifying it. On the contrary, if it were hardly probable *a priori* but nonetheless happened, much information has been obtained.' This is exactly the case of PODZOSOLS of the Jura which provide the soil scientist, who did not expect to find them there, very specific information on, for example, the potential limits of soil development in a calcareous medium and the very great bioindicative value of some species 'lost' amidst a phytocoenosis.

Some quantitative properties of the species in the biocoenoses and of the biocoenoses

Many concepts, sometimes indicated by numerical indices, have been proposed for characterizing certain general properties of the species in the biocoenoses or of the biocoenoses themselves (Beeby and Brennan, 1997). Some of them are defined below:

Abundance

- ***Abundance*** corresponds to the number of individuals of a species counted in a census. The term ***relative abundance*** applies to a comparison of species.

Density

- ***Density*** is the abundance relative to a unit area or volume. The term abundance is occasionally used for density!

Relative frequency

- ***Relative frequency*** is the percentage of individuals of a species in relation to the total number of individuals counted in the same sample.

Constancy

- ***Constancy*** is the ratio (expressed in per cent) between number of relevés containing a given species and total number of relevés.

Dominance

- ***Dominance*** expresses the influence exercised by a species in a community. It often differs from abundance because a less abundant species (e.g. a few ruminants in a pasture) can have much greater influence on the environment than others (e.g. numerous phytophagous insects). In phytosociology, dominance is combined with abundance in the ***abundance-dominance index***, which integrates in a single estimate the number of individuals and spatial coverage of the species, synusia types and phytocoenosis types counted in a relevé. Dominance is often linked to coverage, biomass or the living space of a species.

Fidelity

- ***Fidelity*** expresses the intensity with which a species is bonded to a biocoenosis type. This concept, essential in phytosociology, enables definition of the characteristic species of a plant association.

Richness

- ***Richness*** represents the number of species (***specific richness***), of synusia types (***synusial richness***) or of phytocoenosis types (***phytocoenotic richness***) counted in a survey.

Diversity

- ***Diversity*** is applied basically to the species of a biocoenosis (***specific diversity***). It is expressed as a relation between the number of individuals (abundance) and number of species (specific richness). But it also pertains to other organizational levels of ecosystems: genetic, phytocoenotic, tesela, etc. diversity. The ***Shannon-Wiener index***, one of the most commonly used, gives a measure of it. Based on information theory, it is independent of a hypothesis of distribution of the data, which makes it very flexible in application. Applied to species it is given by

$$H' = -\sum_{i=1}^{S} \frac{n_i}{N} \cdot \log_2 \frac{n_i}{N}$$

where H' is the Shannon-Wiener specific diversity index; S the total number of species; n_i the number of individuals of the species i; N the total number of individuals.

Thus calculated, the diversity indices range from 0 (no diversity, one single species) to 4.5 bits (very great diversity). This latter value is only exceptionally exceeded: in nature, a still greater diversity actually results in loss of stability of the ecosystem (Frontier and Pichod-Viale, 1991).

Regularity

- ***Regularity*** (or ***equitability***) of a sample represents the ratio of observed specific diversity and the maximum theoretical diversity that could be obtained with the same number of species:

$$R = \frac{H'}{H'_{max}} = \frac{H'}{\log_2 S}$$

where R is the regularity or equitability index; H' the observed diversity (= Shannon-Wiener index); H'_{max} the maximum theoretical diversity; S the total number of species.

Regularity measures the distribution of individuals by species. A biocoenosis in which just a few species provide most of the individuals has low regularity; on the contrary, regularity is high if all the species have nearly the same number of individuals. It is highest (=1) if all the species show the same number of individuals.

7.5.3 From the square metre to hectare, the same process

In the Jura, the pair 'spruce forest with blechnum — podzolized soil' occupies very different areas (Fig. 7.11):

From the square metre ...

- In Raimeux, podzolized soils, in fact true PODZOSOLS, often occur only directly below spruce trees taken individually, surrounded under their crown by a belt of whortleberry and other acid-tolerant plants. Between trees is developed a ***megaphorbia*** with marsh marigold *Caltha palustris*, lady's smoke *Cardamine nemorosa*, chervil *Chaerophyllum cicutaria*, yellow pimpernel *Lysimachia nemorum*, sanicle *Sanicula europaea*, etc., which colonize RÉDOXISOLS. According to the grain of the mosaic, spruce forest and megaphorbia are rated here in the same phytocoenosis.

Megaphorbia: plant formation of tall forbs often broad-leaved, developed on moist soils rich in bioelements, particularly nitrate (Delpech et al., 1985, modified).

... to the hectare ...

- At La Vattay, on the other hand, many hectares of a very homogeneous spruce forest extend between veritable

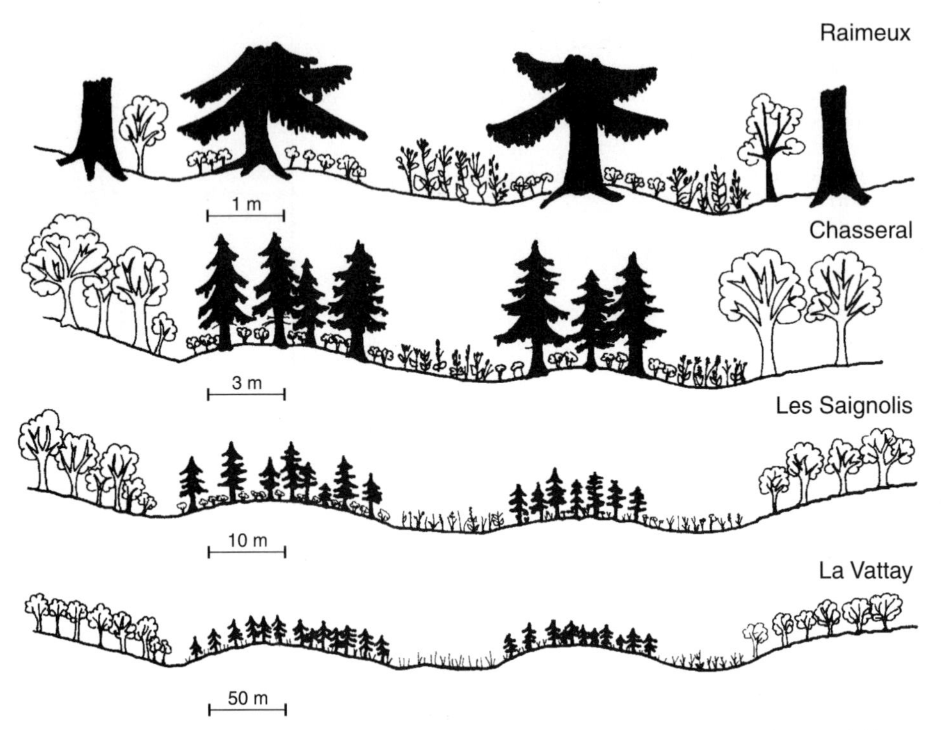

Fig. 7.11 *Podzolized soils and independence of spatial scale: four locations of the spruce forest with blechnum in the Swiss Jura.*

'rivers' of megaphorbia. Here PODZOSOLS are the best developed because of very high rainfall and the crystalline parent material formed of quartz sand and flint pebbles. Calcite appears only at 2-m depth! A mosaic of two extensive, often well-delineated phytocoenoses—spruce forest and megaphorbia—is seen combined in one ecocomplex.

... passing through the are.

- Chasseral and Les Saignolis illustrate intermediate cases.

The spruce forest with blechnum of the Jura leads to the following ideas on relationships between soil and vegetation:

- of the entire phytocoenosis, just three species suffice as bioindicators of podzolized soils compared to HISTOSOLS;
- the very close correspondence between vegetation and soil and the rarity of the situation is extremely informative, because they are highly improbable;
- organization of the vegetation in a mosaic of two very different but linked environments (spruce forest with

blechnum on podzolized soils and megaphorbia on RÉDOXISOLS) is expressed at many spatial scales.

7.6 POPULATION AND THE EDAPHIC FACTOR: WET GRASSLANDS OF LAKE NEUCHÂTEL

Vegetation units of the banks of lakes, subject to intense hydromorphy by subterranean infiltration or inundation, are often dominated by just one or two species.

While the phytocoenoses of wooded pastures, subalpine lawns and spruce forests are rich in species, others have far fewer species and even tend, in some cases, to become monospecific. Thus their soil often presents a very strong ecological constraint that drastically culls species.

7.6.1 A catena governed by groundwater

Example of the south bank of Lake Neuchâtel in Switzerland (Buttler et al., 1985; Buttler and Gobat, 1991; Gobat, 1991).

To illustrate the relationship with their soils of these units with low diversity and regularity, let us consider the four species-poor phytocoenoses constituting the catena:

- offshore, a girdle of common club-rush *Schoenoplectus lacustris* is established in sediments of gyttja type, under more than a metre of water;
- closer to the bank, the reed-bed with *Phragmites australis* grows in zones subject to permanent but shallower inundation;
- where flooding is temporary, grassland with tall sedge (*Caricetum elatae*) colonizes HISTOSOLS or RÉDUCTISOLS; this sedge dominates, often mixed with reed (Fig. 7.12).
- still higher, the flora diversifies in the fen with black bog-rush (*Orchio-Schoenetum*).

In the marsh of Lake Neuchâtel, pedologic differentiation of the phytocoenoses is done at the level of the humus form.

Pedologic differentiation among the phytocoenoses is established at the level of the humus form, very sensitive to the hydric regime. It is particularly noted that three humus forms separate the variants of the grassland with tall sedge (Chap. 6, § 6.4.2).

7.6.2 ...but the soil does not explain everything!

Furthermore, in all the zones not excessively flooded, *Cladium mariscus* forms discontinuous monospecific colonies from which all competition is excluded. This is not a phytocoenosis *per se* but a particular facies. Even though hydrodynamic properties of the soil primarily

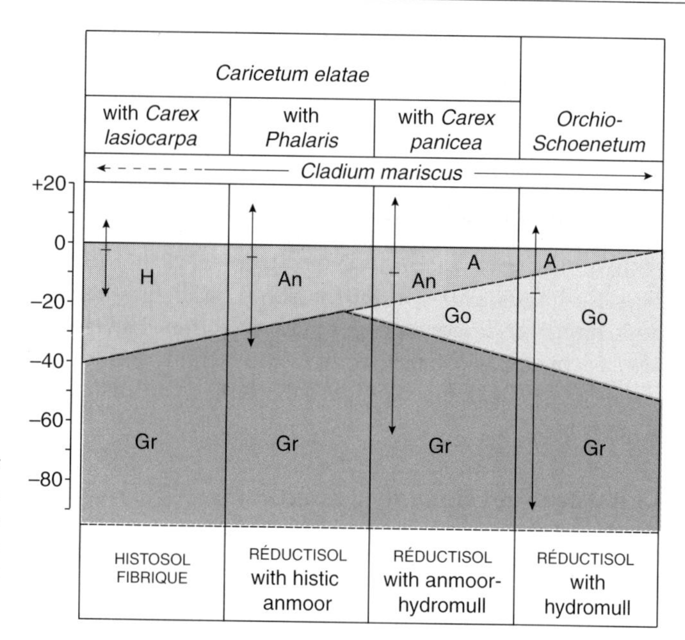

Fig. 7.12 *Partial catena of vegetation units and soils on the south bank of Lake Neuchâtel. Vertical arrows indicate fluctuations of the water table.*

determine the distribution of vegetation (Buttler and Gallandat, 1989), they do not explain why *C. mariscus* very powerfully overruns certain fen areas. Moreover, this species is found further upslope on less wet soils of RÉDOXISOL and Redoxic RENDISOL types.

Dominance of *C. mariscus* combines external edaphic and internal physiological causes.

Two causes integrating edaphic and physiological aspects explain the greatly intrusive behaviour of this species (Buttler, 1987):

- On very wet soil, *C. mariscus* is inhibited in absorption of phosphorus, the limiting bioelement, because of poor adaptation to excess of water; but it is very competitive on drier soil. Interspecific competition explains the resistance of tall sedge to *C. mariscus* in less wet zones, close to reed-beds.

- Outside the wettest zones, a genetic-physiological predisposition enables *C. mariscus* to dominate the tall sedge. Capable of photosynthesis and of growing year round, which the latter cannot, it accumulates certain bioelements during the winter rest. In spring, when the water table drops again and mineral nitrogen is again made available to plants by nitrifying bacteria of soil, *C. mariscus* is ready to gain from it before the other species. Also, it is capable of very rapid translocation of bioelements.

This particularly instructive example underscores the need for suitably adjusting the levels of observation between the two partners according to the aim of the comparison:

Let us adjust the levels!

- Do we want to quickly characterize the pedological conditions of marshes compared to forests? The level of the Référence is adequate for overall understanding of the ecological situation of the catena.
- Do we desire greater precision regarding the functioning of the soils of each phytocoenosis or synusia? The humus form provides this.
- Lastly, does the behaviour of a particular species interest us? The answer is given by a specific ecological factor combined with a specific physiological disposition, depending on autoecology.

7.7 CONCLUSION: RELATIONSHIPS BETWEEN SOIL AND VEGETATION THAT VARY ACCORDING TO CIRCUMSTANCES

7.7.1 Need for organization of knowledge: better understanding of the relationships between soil and vegetation

The few examples reviewed above—hundreds can be found in pedological or ecological literature!—suggest that understanding of the relationships between vegetation and soil is not an easy problem in edaphology. Drouineau (in Callot et al., 1982) emphasized the great complexity of relationships between plant and soil in much simpler systems, those of cultivated soils. Passing from the plant to the plant community, the processes can only be still more complex...

A bit of ***epistemology***...

Epistemology: that aspect of philosophy, which studies the history, methods and principles of the sciences; theory of scientific understanding (*Petit Larousse*).

7.7.2 Some milestones for a future paradigm

The examples in §§ 7.2 to 7.6, supplemented by instances taken from the literature or cited elsewhere in this book, lead to discussion of some conceptual aspects of the relationships between soil and vegetation (see also Van Breemen,1998; Zanella et al., 2001).

'One of the major obstacles that must be surmounted (in ecology) is related to the fact that problems to be resolved result from multiple processes operating and to be understood at different scales of time, space and analysis (...). To resolve the problems of change of scale—biological, temporal and spatial—is certainly one of the priorities in ecological research.' (Barbault, in Auger et al., 1992).

Spatial levels of organization, hierarchy and scale transfers

Relationships between soil and vegetation are established at all levels of organization of ecological systems, from great biomes to the rhizosphere. Correspondences link integrating concepts (e.g. the Reference in soil science or the phytocoenosis in phytosociology) or elementary (e.g. a soil bioelement or a plant genetic character) (Fig. 7.13).

The relationship between soil and vegetation, and the processes linked to them, are nested one within the other. Some exist only at specific levels of organization while others reflect ecological processes independent of spatial scale, such as podzolization under spruce forest with blechnum. These concepts, invariant with scale, are fundamental to development of ***fractal geometry*** in mathematics (Mandelbrot, 1975), in which the same law

Towards fractal geometry

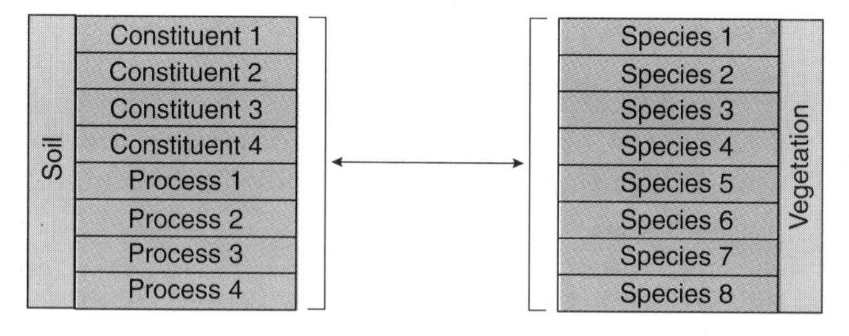

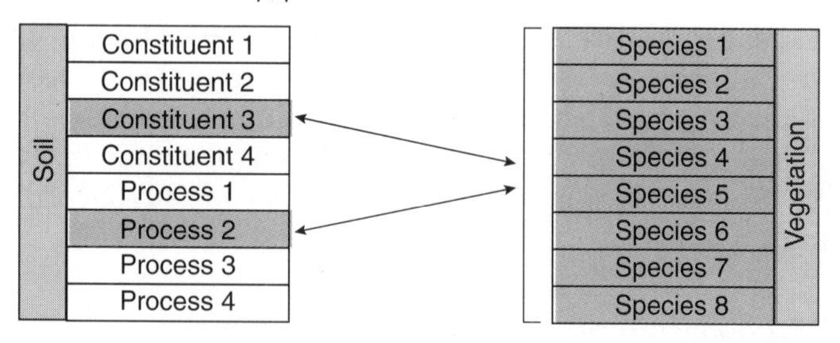

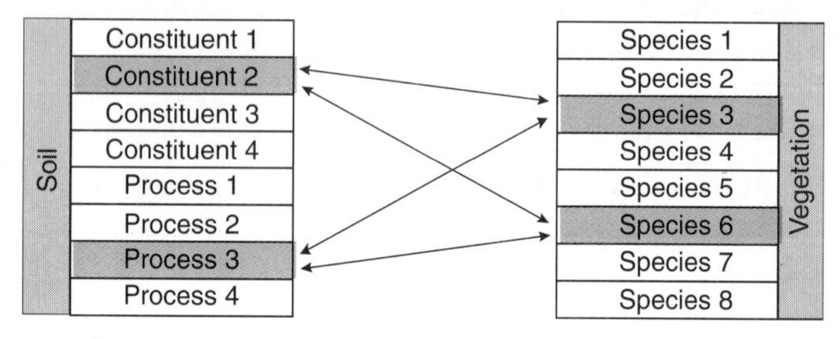

Fig. 7.13 *Possible correspondences between soil and vegetation. The link is not always made between objects at the same organizational level in pedological and phytosociological domains: A. An integrating phytosociological concept may correspond to an integrating pedological concept (e.g. solum and phytocoenosis of the Creux du Van); B. An integrating concept may correspond to an elementary concept (e.g. range of fluctuation of the water table and the phytocoenosis in marshes); C. An elementary concept may correspond to another elementary concept (e.g. available phosphorus and growth of* Cladium mariscus*).*

of transformation determines changes of form at successive different levels of organization. Applications are appearing little by little in ecology (de Foucault, 1986; Milne, 1988; Sugihara and May, 1990; Auger et al., 1992; Schneider, 1994) and in soil science (Rieu and Sposito, 1991; Zanella et al., 2001).

'One of the important problems we have in studying soil systems is to reach the level of resolution at which the processes are occurring.' (Coleman et al., in Edwards et al., 1988).

Amount and accuracy of information

In practical terms, the methods based on calculation of information can be used to quantify the value of the observed bioindication, for example in the case where multivariate analysis is not efficient. In the soil, relationships with the vegetation it supports can be highlighted by directly acting ecological factors or by simple ecological descriptors that do not act on organisms but often indicate many active factors. For example, latitude *per se* does not control distribution of biomes; it simply reflects the zonation of an assemblage of truly influential macroclimatic factors. Determination of the amount of information may, for example, highlight specific microregions within the biomes.

According to information theory, rare or hardly probable relationships are more informative than those corresponding to the general environment.

Relative importance of edaphic factors in soil-vegetation relationship

The action of edaphic factors on vegetation may be:

Edaphic factors do not always *per se* determine the distribution of vegetation, even if this seems to be the case.

- truly determinant (presence of spruce forest with blechnum on podzolized soils, growth of acid-tolerant lawn with spikenard on loess);
- located at the same level as other factors (dominance of *Cladium mariscus* due to its intrinsic physiology as well as soil phosphorus content);
- masked by the action of another ecological factor (presence of different synusiae in the wooded pasture, due chiefly to cattle action); the latter case shows that the fauna in particular may be a very important factor modifying soil-climate-vegetation equilibria.

Within these limits, the soil-vegetation relationship is nevertheless subject to climate. In this context, the correspondence observed between vegetation and soil may be direct causal relationships:

Climate generally retains the upper hand.

- Unidirectional in the soil→vegetation direction. Aluminium released by weathering of silicates during podzolization selects species according to their tolerance of its toxicity (Chap. 4, § 4.3.5).

- Unidirectional in the vegetation→soil direction. In spring, evaporation of water by plants, even though plants are not the prime mover for this (Chap. 3, § 3.4.3), regulates the level of groundwater in the riverain forests of the lake at Neuchâtel (Cornali, 1992).

- Bi-directional, with feedback. Establishment, then development of sphagnum in a minerotrophic peat increases the acidity of the substrate and makes it still more favourable for sphagnum (Chap. 9, § 9.2.2). By continuing to grow, sphagnum transforms and acidifies the substrate more and more (positive feedback), until a negative feedback takes place. At this moment, physicochemical and general climatic conditions intervene and enable attainment, and then preservation of the climactic equilibrium.

But these relationships may also be just simple correspondences or coincidences, such as the almost perfect superposition of the distribution of soils and plant formations in Africa, actually due to their common subjection to the macroclimate and to their different reaction inertias.

'Time' factor and soil-vegetation relationships

The relationship between soil and vegetation is dependent on the time factor, which is not 'seen' the same way by the two protagonists.

Vegetation reacts roughly ten times faster to change than soil. For example, many pastures of the Alps rich in grasses and Leguminosae grow on very well developed PODZOSOLS (Gobat, in Vittoz et al., 1995). But this soil type cannot form under this vegetation since addition of chelating phenolic compounds originating from an acidifying litter is essential for the process. It is also because of these inertial differences that the 'already' climactic beech forest of the alluvial zone of the Sarine 'still' grows on a FLUVIOSOL TYPIQUE, while the general climax of the biogeocoenosis would want a CALCISOL or a BRUNISOL.

Vegetation reacts chiefly to the functioning of the soil, and much less to its morphology.

These two cases illustrate the fact that the vegetation primarily reacts to the functioning of the soil, often to a single limiting edaphic factor, and not to its general morphology indicated by the horizons. Horizons require an adaptation time for inscribing in themselves visible marks, pedofeatures typical of the new conditions. In the same way, certain species adapt themselves very quickly; others more slowly. Good knowledge of their strategies (Chap. 12, § 12.6.2) is thus useful for understanding the dynamics of soil-vegetation relationships.

A non-compliance with the rule... which confuses the issues and the minds!

Because they have forgotten the laws governing the differential speeds of reaction of the compartments of the ecosystem, many authors have thought that they could deny any bioindicative role whatsoever to vegetation units described by phytosociology. Surely actual situations seem to contradict the ideas, preconceived or verified, that we may occasionally have on the correspondences between soil and vegetation. Yet, in most cases, no serious discussion has been undertaken on the spatiotemporal organizational levels actually participating in the soil-vegetation relationship. Faced with paradoxical situations, investigators will have to become more interested in exceptional or non-standard cases.

'The dependence on scale has long been neglected, even totally ignored; this has been the source of many sterile debates on the 'causes' of particular phenomena and their modelling (Allen and Starr, 1982).

The reaction speeds of soil and vegetation are differentiated still more subtly at the lower organizational level, synusia for vegetation and horizon for soil. Figure 7.14 shows that the surface of the terrain, in particular the litter, which is concomitantly vegetation and soil, is a kind of mirror plane between the plant strata and the pedological layers. Thus it is interesting to study the soil not only at the level of the solum in its entirety, but also as a function of the horizons themselves considered as independent functional units (Chap. 5, § 5.6.1).

The 'comparative reaction speeds approach' illustrates all the interest there is in considering the phytocoenosis not only as a whole, but also as a system formed of relatively autonomous synusiae.

Lastly, according to the time-scale chosen, certain changes that act on soil-vegetation relationships reveal themselves to be reversible (an abandoned pasture with spikenard on PODZOSOL returns without difficulty to primitive heath as long as the macroclimate does not change) or irreversible (anthropically induced acceleration

Reversible or irreversible changes! The concept of reversibility is essential in all considerations concerning protection of ecosystems and soils.

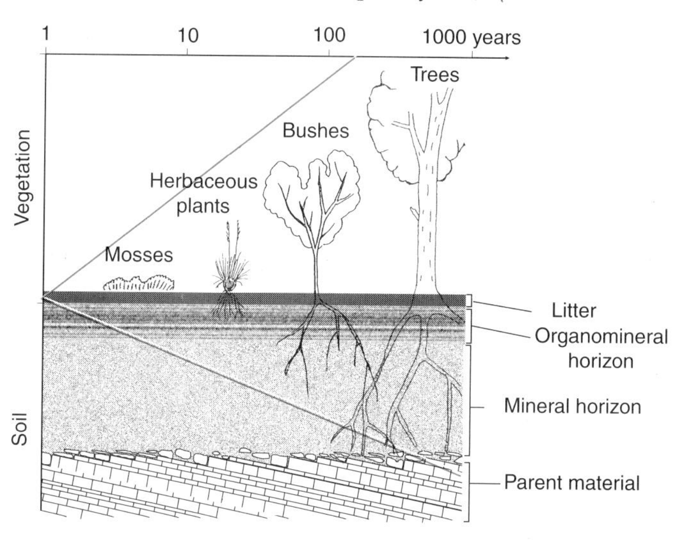

Fig. 7.14 Comparative reaction speeds (blue lines) of plant strata and pedological horizons, on either side of the mirror that the litter represents.

of podzolization observed in the Vosges by Guillet, in Duchaufour, 2001) made recolonization of the sites by the old beech-fir forest impossible).

7.7.3 Some key words for a future general explanatory theory

Some bases for a future paradigm.

Based on the ideas developed above, establishment of a future paradigm of relationships between soil and plant should be articulated around the following themes:

- dependence or independence of a phenomenon with regard to the organizational level;
- effects and changes of scale (Perrier, 1990; Auger et al., 1992; Schneider, 1994);
- comparative speeds of reaction to changes in vegetation and soils;
- adaptive r–K or C–R–S strategies of the species, determining strategies occasionally different at the synusial level (Grime, 1979; Grime et al., 1988);
- exceptional situations;
- spatial or temporal transition zones: ecotones (Holland et al., 1991; Lachavanne and Juge, 1997), sudden disturbance, etc.;
- spatial relations (Collet, 1992; Wilding et al., in Bryant and Arnold, 1994; Legros, 1996);
- reversibility and irreversibility of phenomena;
- space-time relations (Schneider, 1994);
- the relation between morphology and functioning in all compartments of the ecosystem;
- role of biodiversity in stabilization of the soil-plant system.

'In agronomy, the soil-plant relationships domain is one of the most difficult and for researchers, the least perfect one. (...) But teachers and practical engineers require syntheses that present the whole problem of soil-plant relationships' (Drouineau, in Callot et al., 1982).

The ambition to establish an explanatory theory of relationships between soil and vegetation is indeed a great one. However, by a meticulous study of typical situations, it should be possible to define the fundamental laws governing these relations and to compare them with the basic laws of functioning of ecosystems (Odum, 1971, 1996; Frontier and Pichod-Viale, 1991; Dajoz, 1996; Beeby and Brennan, 1997). By this way, we should even tend towards a paradigm concerning not only the soil and vegetation, but even the fauna and microorganisms...

CHAPTER 8

DEAD WOOD, EXCREMENTS, CARCASSES AND STONES: SOIL ANNEXES

This chapter discusses an aspect often neglected in books on soil science, that of such structures as dead trees or dry stone walls that variegate the soil surface yet retain a very marked individuality. While they have a definite relationship on the biological plane, ***soil annexes*** contain specialized animal and fungal communities. The invertebrates that live in them, however, may be considered mostly forming part of the pedofauna.

Soil annexes, underestimated in soil science...

Soil annexes (dependances of the soil, according to Delamare-Deboutteville, 1951): simple or complex structure that variegates the surface.

Evolution of organic matter constitutes the red thread of the second part of this book. In this perspective, the organic annexes (excrements, dead animal or plant tissues) represent an essential transition step between living organisms and the humiferous episolum. Mineral annexes contribute to the maintenance of an abundant and varied epigeal fauna by providing them temporary shelter and favourable habitats.

8.1 MINERAL AND ORGANIC ANNEXES OF SOIL

8.1.1 Types of annexes

Soil annexes can be distinguished according to their nature (mineral or organic) or their location:

A stony layer is a PEYROSOL and litter is a pedological horizon (OL). These are not soil annexes.

- direct annexes of mineral nature (more or less permanent): isolated stones, accumulations of pebbles, dry stone walls; they shelter a cryptozoic fauna (§ 8.2);
- rapidly evolving direct organic annexes (disappear in less than one year): carcasses of small animals, faeces, droppings, dung (§ 8.3); certain fruits fallen on the soil senescent carpophores of certain fungi are sometimes considered as very rapidly evolving indirect annexes (§ 8.8.2);
- slowly evolving organic annexes (disappearance in more than one year): dead wood, trunks (decaying trees), stumps, garden composts in contact with soil (§§ 8.4 to 8.7; Chap. 10, § 10.8);
- indirect organic annexes: epiphytic or epilithic soils, tree holes (§ 8.8).

8.1.2 The degradative successions

The degradative succession: squads of microorganisms and detritivorous animals that follow each other.

Organic annexe: soil structure serving as semi-permanent or temporary habitat as well as nutrient reserve, sometimes only one, sometimes the other. This is a milieu composed of dead, well-differentiated, impermanent organic matter that shelters a characteristic rapidly **evolving biocoenosis** different from the soil biocoenosis.

All biocoenoses are transformed more or less rapidly in response to changes in their biotopes: these are *ecological successions* (Dajoz, 1996; Odum, 1971). This notion does not give a full account of the status of ***organic annexes*** of soil. These rapidly evolving (weeks, months, years) media are inhabited by microorganisms and invertebrates that occupy them in successive waves of different compositions. They together form ***degradative successions*** (Begon et al., 1986), also called 'destructive successions' (Dajoz, 1996), 'faunal successions' (Smith, 1986) or 'rapidly evolving biocoenoses' (by us). They are in fact decomposition food chains in which the several actors do not act in concert but succeed each other in groups, one preparing for the arrival of the other. The process culminates, as in every food chain of decomposition, in mineralization of the carcass, dung or dead wood and in their return to the cycle of bioelements.

When the waves of decomposers succeed each other on a carcass or on dead wood, the activity of the first wave sets in motion physicochemical modifications of the medium such that it becomes more and more hostile for the first wave, thus favouring a second; this itself will be replaced by a third and so on until the annexe disappears, in a way 'digested by the soil'.

Organic annexes are not ecosystems!

Organic annexes are simultaneously more or less temporary nutrient sources and habitats. But it would be wrong to see small ecosystems in them, as is sometimes done (Chap. 4, § 4.4.1). Actually, when they evolve more or less rapidly towards an equilibrium state, the climax, these annexes are destined to disappear. They constitute the starting point of often specialized decomposition food chains (Chap. 13, § 13.6) and actually represent only steps, stopping points in the cycle of bioelements.

8.2 DIRECT ANNEXES OF MINERAL NATURE

Piles of pebbles and dry stone walls constitute temporary or permanent habitats and refuges much used by the epigeal pedofauna. Their internal microclimate is buffered compared to that of the surface of cultivated areas and pastures with short grasses (Tab. 8.1).

Annexes of great importance to the epigeal fauna!

Table 8.1 Microclimate of **mineral annexes**: pasture walls

Ecological factor	Stone surface	Interior of wall
Temperature	Wide variations, rapid changes	Attenuated abrupt variations; thermal inertia
Relative humidity	Very wide variations, rapid changes	Permanently high, very stable
Wind	Marked on side exposed to wind currents	Very low
Light	Strong on side exposed to sun	Low to none

Figure 8.1 highlights the microclimatic characteristics of these mineral annexes. In high mountains, where climatic conditions are severe, the microclimate of these habitats stay relatively constant even if the soil is bare. In winter, they are insulated by snow and the temperature in them does not fall below 0°C.

Invertebrates living under stones or in the milieu under fallen trunks or below bark are ***cryptozoans*** (Coleman and Crossley, 1996). Their habitats, termed ***cryptozoic***, acquire considerable importance in high mountains. Because of the microclimates offered by the mineral medium, survival of invertebrates, most often cold ***stenotherms***, becomes possible under the generally severe condition. These communities react to confinement and to macroclimatic (abundance of UV radiation, gusting winds) and microclimatic (high humidity) factors by

Stenothermal: having a narrow tolerance range with respect to temperature. There are stenothermal species dependent on cold regions ***(oligothermal)*** or hot regions ***(polythermal)***. The opposite of stenothermal is ***eurythermal*** (with wide thermal tolerance range).

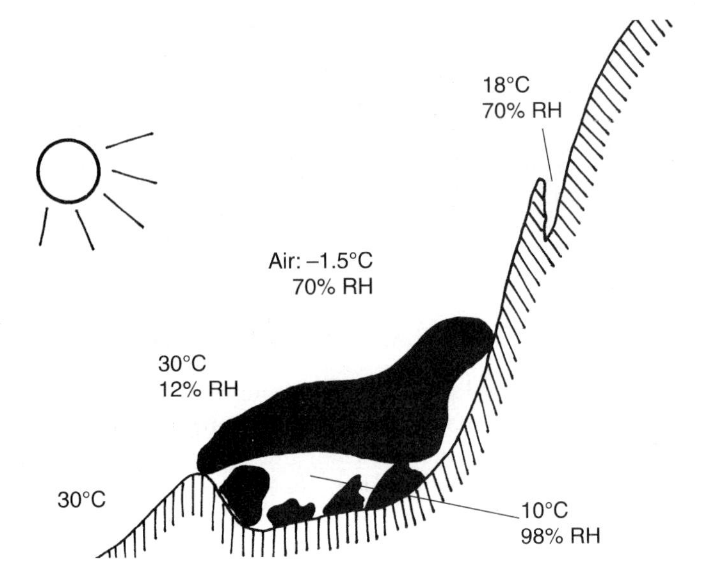

Fig 8.1 *Temperature and humidity in a cryptozoic habitat at high altitude. The mineral substrate is strongly heated on the surface by the sun. The heat wave slowly diffuses through the pebbles and is maintained even after the sun has set and the temperature occasionally falls below 0°C on the surface. The relative humidity (RH) stays at a constantly high level because the small interstitial habitats are hardly accessible to the wind (after Mani, 1962).*

Melanism: development of black pigmentation in animals.

reduction in size or of wings, as well as by ***melanism*** or other physiological adaptations.

In the alpine zone, the cryptozoic communities comprise Acarina, Collembola, Dermaptera (= earwigs), Coleoptera and Diptera. They are assemblages of species with no particular links, which feed mostly outside the cryptozoic habitat and reassemble to spend the night and unsuitable periods in bearable microclimatic conditions.

The importance of mineral annexes is no less at lower altitude. Thus in the mountain stage, dry stone walls and piles of pebbles are wintering sites, refuges during dry periods and shelters for many vertebrates and invertebrates.

8.3 RAPIDLY EVOLVING DIRECT ORGANIC ANNEXES

We shall describe here two major types of rapidly evolving direct organic annexes: carcasses and dung and other excrements.

8.3.1 Decomposition of carcasses of vertebrates

From the carcass to the soil, a 'grey' litter

Despite the abundance in nature of birds and rodents, even of larger mammals, it is rather rare that their carcasses are found. The carcasses quickly enter the decomposition chains and disappear under the combined action of the trio of bacteria, fungi and invertebrates. Their existence is more or less ephemeral according to their size, nature of the soil that supports them, the climate and the composition of the evolving biocoenosis that develops there. On the other hand, the species (of the dead animal) has little importance in taxonomic structure and succession of waves.

Carcasses answer to the definition of direct organic annexes of soil. The products of their decomposition form part of the 'grey' litter reaching the soil (Chap. 2, § 2.2.1).

How can a carcass be recycled? It is 'worked upon' by groups of arthropods that colonize it in a specific order, the ***waves*** or ***squads*** (Smith, 1986). A wave is always formed of the same species under a given climate. The invertebrates that compose it are attracted by the odours emitted during decomposition of the carrion. The number of waves, their species composition and their efficiency vary with region, season, meteorological conditions, size of the carcass and its location on the land.

'If the environmental conditions were always perfectly regular, such that the succession of putrefactions is perfectly regular, the law of succession would be, so to say, mathematical for carcasses comparable between themselves as much by their own characteristics as by the environments in which they are found' (Leclercq, 1978).

Stages of decomposition

In Australia, the decomposition of carcasses of guinea-pigs consists of five putrefaction stages corresponding to an equal number of waves of necrophages:

- initial decomposition of the carcass: 0 to 2 days;
- stage of internal putrefaction: 2 to 12 days;
- stage of black putrefaction: 12 to 20 days;
- stage of butyric fermentation: 20 to 40 days;
- stage of dry decomposition (mummification): 40 to 50 days.

In Central Europe, eight waves can be seen succeeding each other (theoretically) on the carcass (Leclercq, 1978):

In Central Europe, eight waves take the carcass to skeletonization.

- The first wave consists only of Diptera (Calliphoridae and Muscidae, to which the housefly belongs). They appear immediately after death and deposit a large number of eggs on the skin and in orifices of the carcass. During good weather, growth is very rapid, of the order of a couple of weeks. Along with bacteria, the larvae that feed on dead tissue modify the physicochemical characteristics of the carcass and hence of the effluvia it emits.

- As soon as the odour of death is given out, other species of Diptera forming the second wave appear and lay eggs (Sarcophagidae and Calliphoridae). Their larvae replace those of the first group that have meanwhile hatched

Necrophagous Diptera, perfect examples of r strategists (Chap. 12, § 12.6.2).

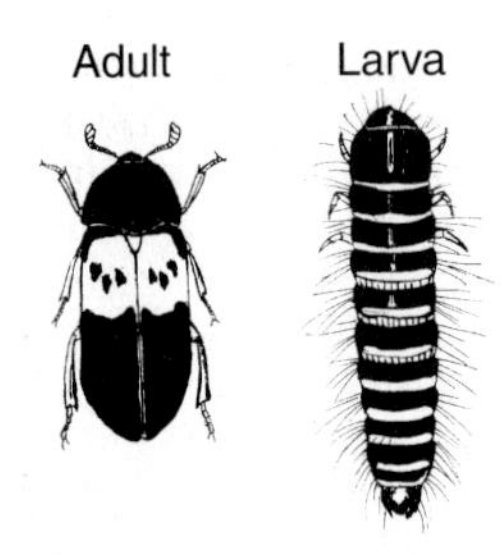

Fig. 8-A. *Bacon* or *larder beetle* (Dermestes lardarius). *Length: adult 7-9 mm, larva 10-12 mm.*

and whose adults have departed in search of other substrates for laying eggs. The necrophagous Diptera, r strategists, are good flyers, equipped with very sensitive sensory organs enabling them to detect carcasses quickly.

- The third wave appears with butyric fermentation causing rancidity in fats, that is three to six months after death in the case of a large carcass (pig, humans). It comprises Dermestidae (Coleoptera) (Fig. 8-A) and Pyralidae (Lepidoptera).

- The fourth wave comes at the end of butyric fermentation. Diptera such as skipper flies (Piophilidae) arrive on carrion, while the putrid liquids that exude from it are colonized by the larvae of other Diptera such as Drosophilidae and Syrphidae (*Eristalis*).

- Flies and beetles of the fifth wave respond to the fermentation of proteinic substances accompanied by production of ammonia (ammonification, Chap. 4, Fig. 4.21).

'A pair of *Calliphora* eats the carcass of a donkey as fast as a lion does' (Linnaeus, after an Arabic proverb).

- From here on, after about a year, the corpse is desiccated. The sixth wave, formed of populations of Acarina (Astigmata) absorb the last fluids and thus contribute to mummification.

Different parts of a large carcass do not evolve at the same speed; hence many groups can coexist.

- The last two waves comprise beetles (skin beetles family, e.g., *Attagenus* spp., *Anthrenus* spp.) and moths (Tineidae such as clothes moths). They feed on skin, hairs and dried flesh. After their passing, in other words after about three years, a pig carcass is skeletonized. It can still attract carnivores and rodents.

In hot desert climate, desiccation of carcasses is very rapid. In this case, the beetles of the third wave, then of the seventh and eighth, are the principal agents of skeletonization (Smith, 1986).

Numerous representatives of the last two waves have found alternative environments in human habitations, for example in stores and warehouses. They attack curtains, carpets, leather and woollen garments and even collections in museums of natural history... They also live in bird nests (§ 8.8).

The above scenario may be interrupted at any time by the intervention of carrion-eating vertebrates.

The guild of carrion feeders (e.g., jackals, vultures and crows) also plays an important role in the disappearance of carcasses. For example, in the Russian steppe two thirds of ground-squirrel carcasses are eaten before their decomposition begins (Elton, 1966). The remains of small animals can also be buried by necrophores *(Necrophorus* spp.*)*. These burying beetles are mostly attracted by the fermentation of proteins. By burying the dead flesh, they create a food reserve for their larvae.

Organization and diversity of the necrophagous community

The necrophagous community comprises numerous predators of carrion-eating arthropods: Coleoptera (Carabidae, Staphylinoidea, Histeridae), Diptera (Empididae, Calliphoridae), Acari (Gamasida). Most of them feed on the maggots.

The necrophagous community is organized in functional compartments like those of the decomposition food chain (Chap. 13, § 13.7).

Also, opportunistic species such as wasps, scarabaeids, ants, carrion beetles, slugs and even shrews regularly exploit this particular environment and its inhabitants.

In Europe, 56 species of necrophages have been associated with fox carcasses and 36–38 with those of rabbits. In USA, more than 300 species have been verified on pig carcasses. On the contrary, on the carcasses of African elephants only an astonishingly small number of species, for example, only two of Calliphoridae, have been verified, against half a dozen on a fox carcass in Europe. On the other hand, they form very large populations: more than 26,000 larvae on a single carcass (Smith, 1986).

Use of necrophages in criminology

Fungi, along with bacteria, are the principal agents for putrefaction of carcasses of animals... or of human beings. They also appear according to a succession, which in other respects only imperfectly matches what has been described above. In 1930, Bianchini, a pioneer in criminology, distinguished three waves or squads of insects corresponding to gaseous putrefaction, transformation of fats and skeletonization, respectively (Leclercq, 1978). Knowledge of the progressive interaction and coordination of animals, bacteria and fungi in the decomposition of carcasses is used in medicolegal investigations, when it is a question of the date of death of an individual once the rigor and lividity of the corpse have disappeared (Leclercq, 1978; Smith, 1986).

Crimes and necrophages...

8.3.2 Cowdung and other excrements, densely populated media

Disappearance of dung, an ecological... and economic necessity

Among the excrements of large mammals, those of herbivores attract a particularly rich fauna. Dung and other excrements are colonized by successive waves of ***coprophilous*** invertebrates, chiefly insects. They feed and

"Le village ne connaît guère ces chalets à odeur d'ammoniaque où dans les villes vont se soulager nos misères. Un petit mur pas plus haut que ça, une haie, un buisson, c'est tout ce que le paysan demande au moment où il désire être seul. C'est assez dire à quelles rencontres pareille sans-façon vous expose... Revenez le lendemain. La chose a disparu, la place est nette: les bousiers ont passé par là." (Fabre, 1925).

often reproduce there, leading by their activity to modification of the physicochemical characteristics of this clearly demarcated environment. Coprophilous animals occupy a series of specialized but ephemeral ecological niches that belong to the decomposition webs. The faecal matter transformed by coprophages and by activity of the microflora is later integrated in the soil (Chap. 5, § 5.2.4). Disappearance of the dung leads to disappearance of the specialized community living in it.

Coprophilous: said of all living things, coprophagous or not, attracted to excrements or living in them. Muscidae, which lay eggs in dung, are coprophilous.

Let us take as an example cowpats that have all the characteristics of direct organic annexes. According to season and altitude, they attract coprophilous insects during a period lasting a few weeks to a few months; on drying they gradually become inhospitable to these insects. They are then colonized by a cryptozoic fauna similar to that which colonizes the underside of fallen wood, then by the true pedofauna, especially earthworms, which reintegrate them in the soil.

'Long dried' cowpats: a shelter similar to fallen dead wood.

The work of decomposition done by the necrophagous fauna is of prime importance in the ecology of pastures. Actually one adult cow drops a dozen cowpats every day. With an average diameter of 30 cm, they altogether cover one square metre of soil with a 5-cm-thick layer, this deposit obviously reducing the area available for grazing. Also the soiled grass around the points where they have fallen is avoided by cattle, leaving a protective border 10–30 cm wide, the ***refuse***, in which inedible plants establish. Theoretically, a herd of 100 cattle thus covers one are of pasture daily, to which must be added 4 to 12 ares for the refuses (Gittings et al., 1994).

Integration of cowpats in the soil is concomitantly a great economic necessity for agriculture and an essential process in the functioning of terrestrial ecosystems.

Evolution of the coprophilous community of cowdung

In the evolution of dung, the fauna provides the essence of mechanical action, but clay-leaching, cryoturbation, evaporation and gas circulation also contribute. Bacteria and fungi are most efficient in biochemical transformation of faecal matter (Chap. 4, § 4.4.1) but their action is still only poorly understood. The coprophilous animal community consists mainly of Coleoptera and Diptera, along with nematodes, enchytraeids, earthworms, mites and springtails. Eight phases have been recognized in the evolution of dung (Fig. 8.2a–h).

A necessary study: that of the parallel physicochemical and biochemical evolution of the 'dung' environment.

- Phase 1 (Fig. 8.2a). Hydrophilidae of the subfamily Sphaeridiinae are the first arrivals; they penetrate the dung and lead a semi-aquatic life there. Diptera (families of Calliphoridae, Muscidae, Psychodidae, Sarcophagidae,

Fig. 8.2a Phase 1. Evolution of a cowpat. When it exits the alimentary canal, the dung represents an anoxic medium rich in water (on average 88% in the case of bovines) and plant debris. It forms the site of intense bacterial activity accompanied by considerable release of methane. Diptera: left, adult Psychodidae; right, adult Sepsidae.

Scatophagidae and Sepsidae) follow them immediately, laying a large number of eggs below the surface of the dung. Staphylinid beetles, predators of eggs and larvae of dipterans, also arrive soon. They will be present right through the evolution of the dung.

• Phase 2 (Fig. 8.2b). The egg-laying of dipterans is stopped by formation of a crust their ovipository organ cannot penetrate. Larvae of Diptera will not survive in this still almost anoxic medium without the work of Hydrophilidae, who promote oxygenation of the medium by their movements.

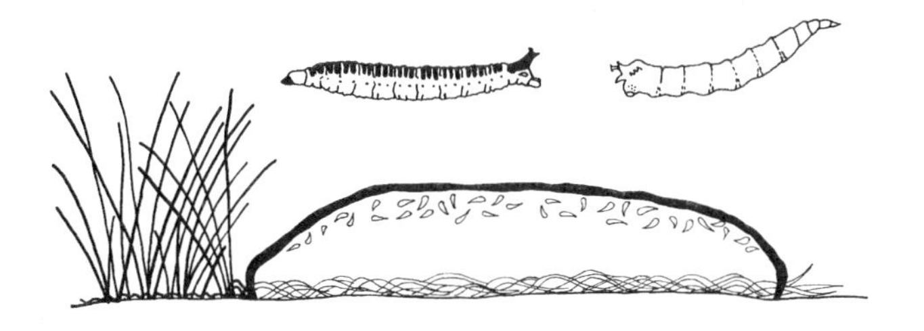

Fig. 8.2b Phase 2. In the hours following deposition of the dung, a surface crust is formed by drying. Methane forms pockets in the mass. Diptera larvae: left, Psychodidae; right, Sepsidae.

• Phase 3 (Fig. 8.2c). Hydrophilidae are still abundant; drilling round holes in the crust. Larvae of dipterans of the first wave grow and Histeridae, predators of these maggots, fly in as do necrophagous and mycetophagous Ptilidae. Hydrophilidae and Histeridae improve aeration of the faecal mass by their movements in it and accelerate its drying. New waves of dipterans come to lay eggs in the tunnels of these coleopterans. The abundance of Ptilidae could be the consequence of a proliferation of fungi.

• Phase 4 (Fig. 8.2d). Arrival of Scarabaeidae. They mix the faecal matter, feed on it, bury it under the dung or, in southern regions, roll it in the form of pellets into their burrows (e.g. the ball-roller scarabs). The larvae of Diptera

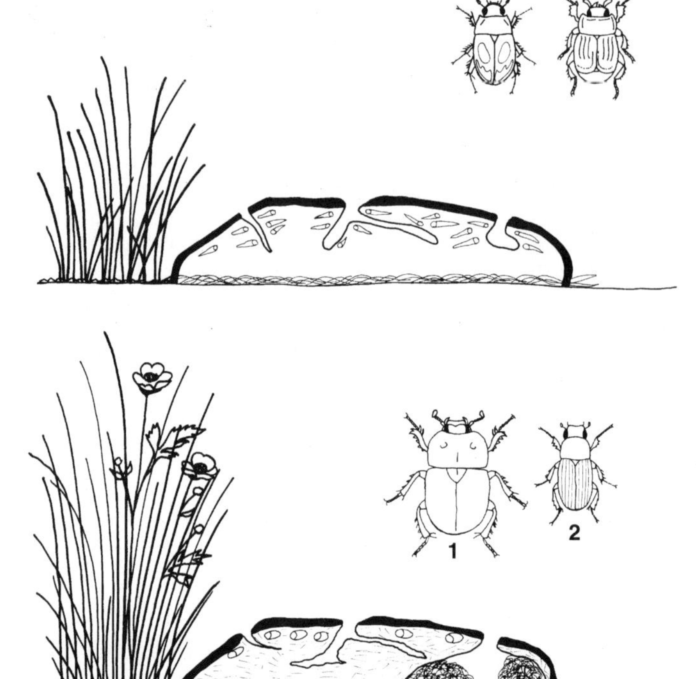

Fig. 8.2c *Phase 3. The crust hardens and grows thicker. The interior of the mass is condensed but remains soft. Coleoptera: left, adult Hydrophilidae; right, adult Histeridae.*

Fig. 8.2d *Phase 4. The cowpat dries out more and more. Coleoptera: (1) larvae, adult and burrow of* Geotrupes stercorarius; *(2) adult and burrow of* Aphodius.

are partially destroyed by their activity. The food regime of certain maggots that issued from the first eggs has changed: from coprophagous it has become carnivorous. Epigeal earthworms appear at the soil-cowpat interface, occupied by crushed plants.

- Phase 5. (Fig. 8.2e). Numerous Microcoleoptera and edaphic larvae of Diptera colonize the excrement. The population of epigeal worms increases and anecic worms feed on the pat from below. The refuse formed is constituted of nonconsummate plants.

- Phase 6 (Fig. 8.2f). The cowpat is fragmented by birds, often Corvidae in search of larvae and earthworms (carrion crows at low altitude or yellow-billed choughs in alpine meadows).

Fig. 8.2e *Phase 5. The cowpat is now dry. The crust is further thickened and cracked. The porous interior is a felting of fibrous nature (plant remains that have resisted digestion until now). Earthworms.*

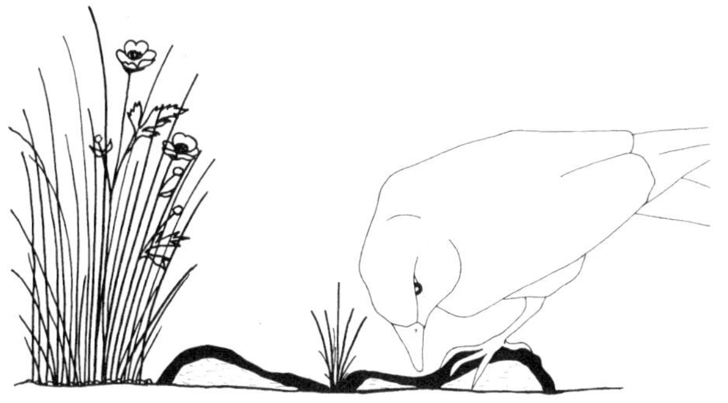

Fig. 8.2f *Phase 6. The cowpat cracks and breaks up.*

- Phase 7 (Fig. 8.2g). Earthworms achieve its burial.

- Phase 8 (Fig. 8.2h). The site of the cowpat is colonized by meadow vegetation during the following year. The refuses stay visible for a long time.

The time taken for decomposition of cowpats varies with season, location and the autochthonous entomological fauna; periods of 60 to 240 days have been noted in the Swiss Jura, 100 to 150 days in Great Britain, 300 to 450 days in Japan, 520 days in New Zealand, 360 to 1000 days in California and several years in British Columbia.

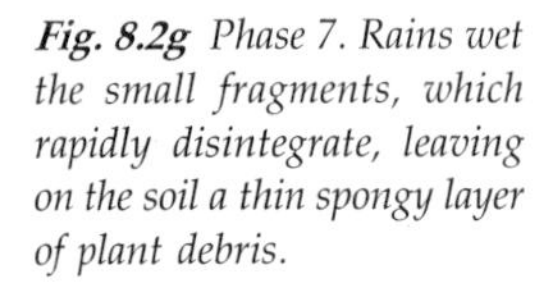

Fig. 8.2g *Phase 7. Rains wet the small fragments, which rapidly disintegrate, leaving on the soil a thin spongy layer of plant debris.*

Fig. 8.2h *Phase 8. The cowpat has disappeared; only traces remain on the soil.*

Flies and scarabaeids, secret scavengers

Two insect orders are particularly well represented in cowpats: Diptera and Coleoptera.

Coprophagous Diptera.

The public at large readily associates Diptera with excrements... and it is true that these insects are often coprophilous. For example, 161 species of flies have been counted in human faeces and 172 in cowpats (Stubbs and Chandler, 1978). The larvae of Diptera play an important role in the ecology of cowpats, the adults being just 'inseminators' and 'disseminators'. The number of eggs laid is very large, as shown by the 2469 dipterous larvae belonging to 16 families extracted from a cowpat drawn from a pasture of the Jura mountains and placed in a Berlese-Tullgren extractor (Chap. 12, § 12.2.2) for a month (W. Matthey, unpublished). Identification of these larvae is delicate. In spite of the availability of some manuals (Hennig, 1948–52; Smith, 1986; Ferrar, 1987), the most certain method is to rear the larvae and identify the adult that emerges. Many species of coprophilous Diptera can proliferate in cowpats and become a nuisance (e.g. the biting horsefly *Stomoxys calcitrans*, or the face fly *Musca autumnalis*).

The face fly is quite similar to the housefly; it torments cattle and occasionally invades houses.

Coprophagous Coleoptera.

Four families of coleopterans are particularly important in this kind of soil annexe. In order of their arrival on the dung these are Hydrophilidae, Staphylinidae, Histeridae and Scarabaeidae. But others can be abundant too, Ptilidae for example. The Scarabaeidae are the best known and also the most efficient in the evolution of cowpats.

Fabre's Geotrupes

'I propose this time to evaluate what a common dor beetle is capable of hiding in one session. Towards sunset I serve my twelve captives with an entire heap left just then by a mule at my doorstep. There is enough in it for one basketful. The next morning, the heap has disappeared underground. Nothing or very little of it is seen above. A very close evaluation is possible, and I find that each of my beetles... has stored very nearly one cubic decimetre of material!' (Fabre, 1925).

The coprophagous scarabaeids represent an ensemble of greatly diversified families and their subfamilies in the tropics. The European species belong to two of them: Geotrupidae (Fig. 8-B, some ten species) and Scarabaeidae (110 species in Central Europe, 140 in France). These insects are good flyers and detect odorous faeces up to 10 km windward. Once they have reached the site, they reveal themselves to be remarkable diggers, because of their enlarged crenellated legs, their flattened head that serves as a shovel and their surprising strength.

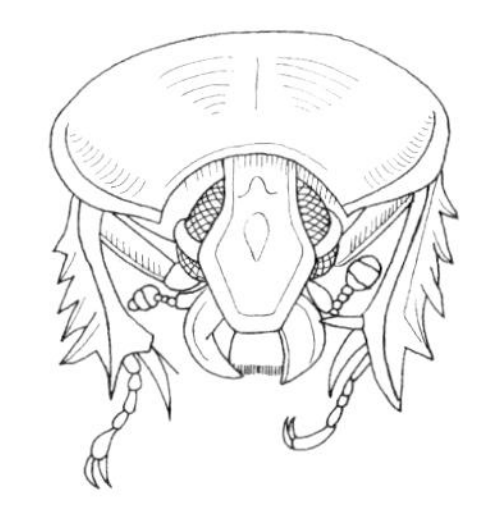

Fig. 8-B *Common dor beetle* (Geotrupes stercorarius), *front view.*

The Australian scarabaeids: an episodic story

- Bovines were introduced into Australia at the end of the eighteenth century. Today they number nearly 25 million and eject almost 300 million cowpats annually in the continent.
- Indigenous coprophagous scarabaeids had been accustomed to marsupial droppings, dry and of small size, such as for example those of kangaroos. Consequently they were not ready to utilize the large liquid cowpats.
- The pats not integrated in the soil by the autochthonous scarabaeids, which were quite incapable of doing so, remained two to five years in pastures, reducing the grazing area by about 10,000 km^2 every year. Furthermore, a fly (*Musca vetustissima*), similar to the housefly, and the biting horn fly (*Haematobia irritans*) became veritable scourges in certain regions of the continent, because they grew several generations annually in the pats that remained in place.
- In comparison, southern Africa, where large herds of wild herbivores live, does not experience these problems because of the activity of more than 2000 species of coprophagous scarabaeids adapted to the faeces of herbivores.

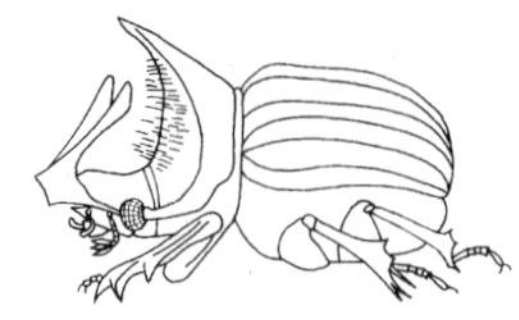

***Fig. 8-C** The giant scarab (*Heliocopris gigas*), ultimate weapon in the war against cowpats in Australia. Length up to 55 mm.*

The problem related to medication of cattle is found not only in Australia; it exists world wide in all regions of intensive cattle rearing (Lumaret, 1993; Lumaret et al., 1993).

- Australian entomologists decided in 1967 to introduce African scarabaeids in the continent to resolve the problem of cowpats and flies. Success was immediate, chiefly due to two species *Onthophagus gazella* and *Heliocopris gigas* (Fig. 8-C). The latter, as big as a golf ball, showed itself capable, working as a pair, of burying up to one kg of cowdung in one night!

- To intensify the war against the undesirable flies, Histeridae and Gamasida were also introduced. These two groups of arthropods indeed feed on the larvae of Diptera.

- Equilibrium is seen today. But entomologists are confronted with a new situation. Chemical remedies have been used in the fight against cattle parasites. Most of these medicines are eliminated through the faeces and are found in the dung. Coprophages in it are affected directly or by reduction in their fecundity. The most recent of these products, the ivermectines, are used as vermifuges. They persist in the dung for two weeks to several months, causing lowering of fecundity and leading to massive death of the newly hatched Coleoptera. Also the dung of treated animals survives much longer on the soil! To see the splendid results obtained by soil biologists thus vitiated by rashly used chemical treatments must lead us to reflect on problems in coordination!

Scarab beetles exploit faecal matter in many ways. Some species establish themselves in it; others hide it in tunnels. Many species can breed in it, each one occupying a niche of its own. Thus, in an African elephant's droppings weighing seven kilograms, more than 7000 scarab beetles have been counted, all simultaneously exploiting the medium in many ways and at different times of the day. In Central and Northern Europe, Scarabaeidae comprise essentially the genera *Geotrupes, Aphodius* and *Onthophagus.* These Coleoptera fill their tunnels, occasionally deeper than twenty centimetres, with pellets of the dung or droppings, which constitute reserves for their larvae.

8.3.3 Coprophilous fungi

The scientific names of coprophilous fungi often leave no doubt regarding their gastronomic preferences: *Ascobolus stercorarius, Sordaria fimicola, Coprinus sterquilinus, Hypocopra merdaria...* Tell me who you are and I shall tell you what you eat...

The 'manna' represented by faecal matter is useful not only for insects. Fungi, often specialists in this rather special dish, follow at different stages of evolution of the substrate. The high nitrogen content is particularly favourable for them. The succession of fungi nearly follows the ascending taxonomic order: first of all Zygomycetes, rapidly-growing primitive species with non-septate mycelium, for example *Pilobolus.* Then come Ascomycetes such as *Ascobolus stercorarius* (Fig. 8.3), while coprophilous Basidiomycetes (particularly *Coprinus*) accompany the last phases in the evolution of dung.

The spores of many coprophilous fungi, in particular the early specialist species, are already present in the dung when it is deposited by herbivores. The animals have ingested them while grazing and the spores have passed through their alimentary canal. This passage has a double effect: besides the expected inoculation of the dung, it is indispensable for ***activation*** of the ***dormant*** spores of these fungi. As long as these spores adhere to vegetation they do not have to germinate, even if the moisture and/or temperature conditions are favourable for them. If they do germinate, the fungus will die, lacking the ability to pursue its growth. The dormancy is then removed only during passage through the digestive tract, due to rise in temperature and/or chemical or biochemical factors. The spores then germinate and the fungus begins its growth once aerobic conditions return, that is, when the dung is dropped.

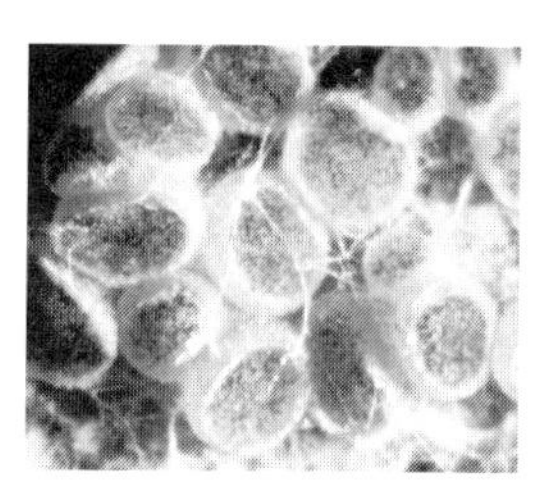

Fig. 8.3 Ascobolus, *a coprophilous fungus with relatively slow growth (photo J.-P. Hertzeisen).*

Activation: Removal of the dormancy of a spore or an organ by a physical (e.g. thermal shock) or chemical signal. The activated spores can germinate immediately when placed in a medium favourable for growth of the fungus.

Dormant: said of a cell or organ that, even when placed in conditions favourable for its vegetative development, remains inactive. Its metabolism is then kept at a very low level to ensure its survival.

But herbivores do not eat their excrement from which the fruit bodies of fungi are formed. It is then necessary that the spores be projected to a good distance and adhere to plants beyond the refuse. The method found by many coprophilous fungi is to expel their spores in a mass. Such masses have a lower surface area to volume ratio than the spores taken individually; also the friction of the air is smaller; for the same initial impetus, the mass travels a greater distance.

In the case of *Ascobolus stercorarius*, the eight ascospores in each ascus are surrounded and bound to each other by a sticky mucilage, and are projected to a good distance in a single mass. The mucilage also enables it to stick to plants. In the case of *Pilobolus*, the entire sporangium, containing tens of thousands of spores, is ejected to a horizontal distance of nearly two metres (Fig. 8.4)! In both cases, the catapulting organs (asci or sporangiophores) exhibit positive ***phototropism***, thus ensuring ejection of the spore mass into the open.

Phototropism: reaction of displacement or orientation of a living organism relative to light. Positive phototropism (turning towards the light source) is distinguished from negative phototropism (turning away).

8.4 DECOMPOSITION OF WOOD: GENERAL PRINCIPLES

A natural forest in which the forester does not intervene comprises a large number of withered or dead trees, upright or fallen. For example, in the vast coniferous

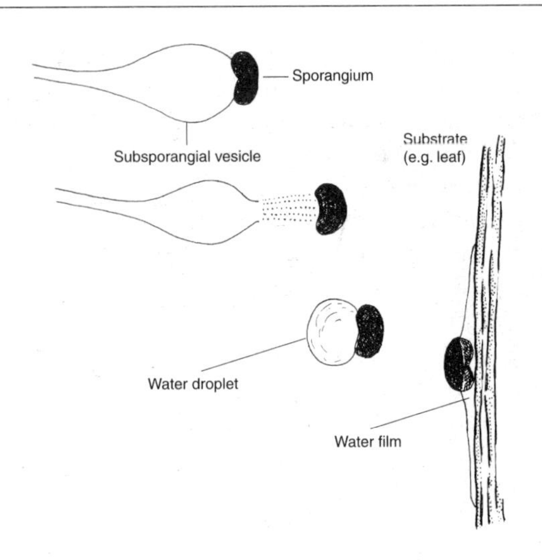

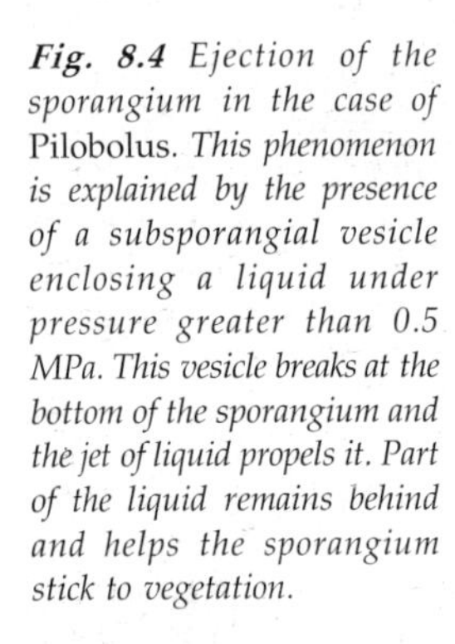
Fig. 8.4 *Ejection of the sporangium in the case of* Pilobolus. *This phenomenon is explained by the presence of a subsporangial vesicle enclosing a liquid under pressure greater than 0.5 MPa. This vesicle breaks at the bottom of the sporangium and the jet of liquid propels it. Part of the liquid remains behind and helps the sporangium stick to vegetation.*

In the undergrowth of a chestnut grove (Switzerland, Tessin), more than 1600 kg ha^{-1} of large dead wood was measured.

forests of the Swiss National Park, comprising some 20 million trees, it is estimated that one-fourth of them are dying or dead (Elton, 1966). Even in maintained forests a large quantity of dead wood falls on the soil: more than 1000 kg ha^{-1} y^{-1} in an oak forest with hornbeam in Belgium. Decomposition releases in the soil the bioelements contained in the wood. This is a part of the biological cycle of substances. From this viewpoint, the concept of 'clean forest', an expression signifying that the undergrowth is rid of its dead wood and its withered trees, is an ecological absurdity.

The 'clean forest', an ecological absurdity.

Dead wood, soil annexe or constituent of litter?

A trunk and a stump undergoing decomposition can certainly be considered soil annexes, a twig as a constituent of litter. While the latter is humified at nearly the same rate as the leaves, floral debris or fruits, more than a century is often necessary for the trunk of a dead tree or the stump to be completely integrated in the humiferous episolum! Even buried, a branch a few centimetres in diameter persists many years in the soil with a distinct shape, 'dictating its law' to the organisms that live in it and decompose it. In this sense, it is a transitory annexe, destined to 'become soil' in a period of time that depends on species, environmental conditions, location of the dead wood (in contact with the soil or not), characteristics of the soil (wet or dry, shallow or deep) and on the structure of the forest.

Wood, that tyrant!

Complete destruction of the tree takes a period almost as long as its life.

Not all parts of the tree evolve at the same speed: different stages of decomposition are simultaneously seen. Most studies having been done on the above-ground portion of trees, the decomposition of their root system is

I-1. *Campodea* sp. (Diplura) (→ 5 mm). These primitive insects, close to the Collembola, live in litter and rotten wood and under stones (Chap. 2, § 2.6.3; photo J. Zettel).

I-2. Springtail of the euedaphon, *Onychiurus* sp. (1 to 2 mm) (Chap. 12, §§ 12.4.9, 12.5.3; photo J. Zettel).

I-3. Springtail of the epiedaphon, *Orchesella villosa* (→ 5 mm) (Chap. 12, §§ 12.4.9, 11.5.3; photo J. Zettel).

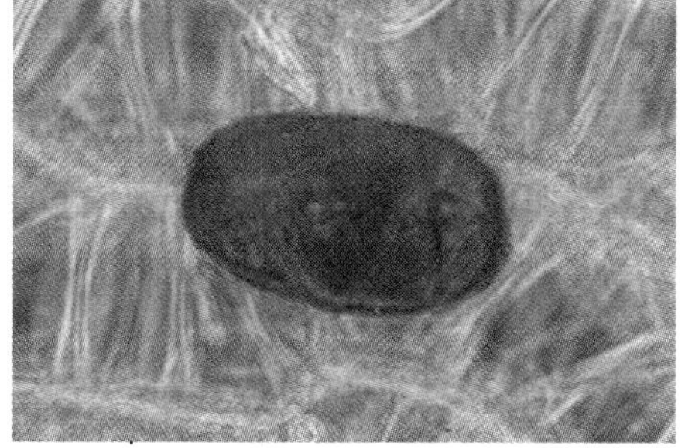

I-4. A typical testate amoeba of raised bogs, the 'Keg of peat bogs' *Amphitrema flavum* (→ 75 μm) on *Sphagnum papillosum* (Chap. 9, § 9.1.3; Chap. 12, § 12.4.1; photo E. Mitchell).

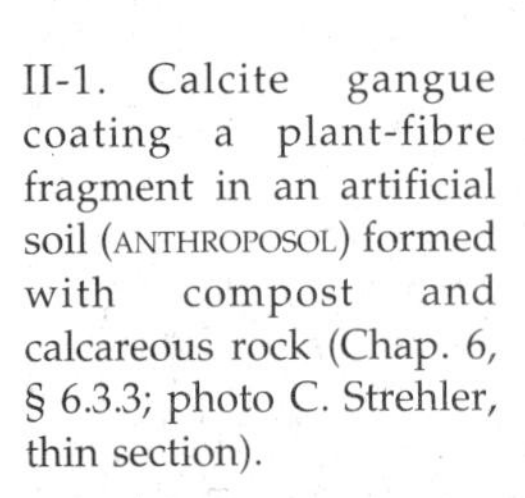

II-1. Calcite gangue coating a plant-fibre fragment in an artificial soil (ANTHROPOSOL) formed with compost and calcareous rock (Chap. 6, § 6.3.3; photo C. Strehler, thin section).

II-2. Argillans deposited by clay-leaching in a small crack in a massive silty soil (Chap. 3, § 3.2.4; Chap. 5, § 5.3.2; photo J.-M. Gobat).

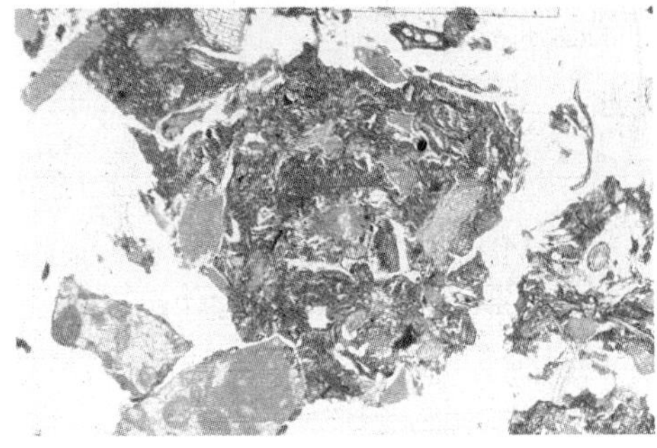

II-3. Cross-section of an earthworm cast showing the mineral particles agglomerated by humified organic matter, during passage through the gut of the animal (Chap. 4, § 4.5.1; photo C. Strehler, thin section).

II-4. Humified organic matter deposited by lateral transport in a sandy horizon downstream of a PODZOSOL. Harz, Germany (Chap. 3, § 3.2.4; Chap. 5, § 5.3.2; photo J.-D. Gallandat).

II-5. Coating of ferric oxide (Fe_2O_3) deposited on the walls of a channel formed at the site of an old root. The blue-grey colour of the surrounding sandy matrix is due to ferrous carbonate ($FeCO_3$). Bottom of a PODZOSOL, La Vattay, French Jura (Chap. 3, § 3.2.4; Chap. 5, § 5.3.2; photo J.-M. Gobat).

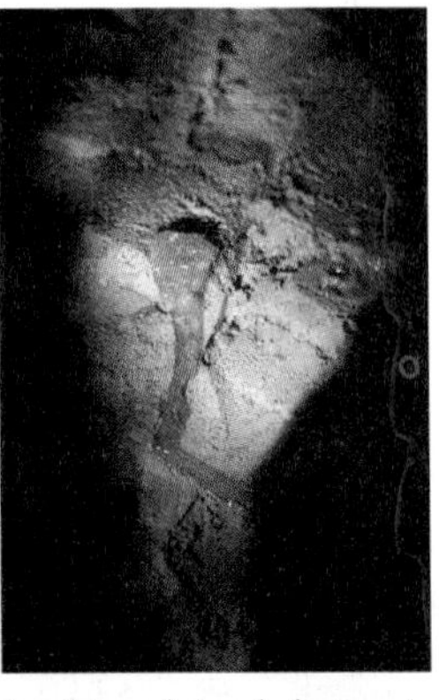

II-6. Liquid 'pedofeature': at the bottom of a PODZOSOL, drainage path of soil solution very rich in ferric chelates. La Vattay, French Jura (Chap. 5, § 5.3.2; photo J.-M. Gobat).

III-1. Solifluction terraces in a CALCOSOL-CALCISOL of the alpine stage, at the base of a dolomite cliff. Moving soil layers slowly cover the scree but are themselves covered by new scree detached from the cliff. The system appears to be at equilibrium. Val de Moiry, Valais, Switzerland, altitude 2700 m (Chap. 5, § 5.5.4; photo J.-M. Gobat, July 1997).

III-2. Section through a solifluction terrace. During summer, solifluction returns the upstream soil over the downstream greensward, resulting in the two herbaceous strata facing each other. The end-product is thus a soil composed of three superposed sola, clearly visible on the right in the photograph: at the top, the surface solum of the solifluction terrace, not yet overturned; in the middle, the old surface solum of the solifluction terrace, overturned; just under the old solum in place, at present buried. Réchy Valley, Valais, Switzerland, altitude 2600 m (Chap. 5, § 5.5.4; photo J.-M. Gobat, August 1989).

III-3. Garden compost in temperate climate, after about two years' addition of household and garden wastes. Enclosed in a cylindrical wire-cage (0.8 m in diameter and 1 m tall), it has been regularly aerated by turning over starting from the surface. The three layers of waste with different degrees of transformation roughly correspond to the natural humiferous OL, OF and OH horizons (Chap. 10, § 10.8.1; photo J.-M. Gobat).

III-4. Process of formation of sphagnum peat. Four layers are seen: green, living sphagnum (here *Sphagnum fallax*); brown, dead sphagnum; loose brown peat with chemical composition very close to that of sphagnum; dense black peat, rich in humified organic matter (Chap. 9, § 9.2.2; photo A. Buttler).

IV-1. Colluvial RENDOSOL with carbonate eumull. Beechgrove with tooth-wort. Le Pâquier, Neuchâtel, Switzerland (photo J.-M. Gobat).

IV-2. CALCOSOL with mesomull. Sedges-beech forest. Le Landeron, Neuchâtel, Switzerland (photo J.-M. Gobat).

IV-3. NÉOLUVISOL with dysmull. Wood rushes-beech forest. Neuchâtel, Switzerland (photo J.-M. Gobat).

IV-4. PODZOSOL HUMIQUE with eumoder. Fescue-lawn. Vallemaggia, Tessin, Switzerland (photo J.-M. Gobat).

IV-5. PODZOSOL MEUBLE with mor. Rhododendron heathland. Saint-Luc, Valais, Switzerland (photo J.-M. Gobat).

IV-6. Truncated FERSIALSOL ÉLUVIQUE with hemimoder. Scrub with arbutus and heather. Villeneuvette, Hérault, France (photo J.-M. Gobat).

IV-7. Polyphasic FLUVIOSOL TYPIQUE with dysmull. Alpine alluvial fen with small sedges. Binntal, Valais, Switzerland (photo J.-M. Gobat).

IV-8. RÉDUCTISOL TYPIQUE with anmoor. Black-alder forest. Worben, Berne, Switzerland (photo J.-M. Gobat).

IV-9. HISTOSOL FIBRIQUE with mor. Pine forest with sphagnum and ericaceous plants. Les Pontins, Berne, Switzerland (photo J.-M. Gobat).

V-1. Carbonate mull. Dentaria-beech forest. Le Pâquier, Neuchâtel, Switzerland (Chap. 6, § 6.3.3; photo J.-M. Gobat).

V-2. Mesosaturated mull. Beech-fir forest. Le Pâquier, Neuchâtel, Switzerland (Chap. 6, § 6.3.4; photo J.-M. Gobat).

V-3. Unsaturated mor. The unsaturated mor actually forms the solum, here an ORGANOSOL INSATURÉ. Spleenwort-spruce forest. Creux du Van, Neuchâtel, Switzerland (Chap. 6, § 6. 6.3.5; photo J.-M. Gobat).

V-4. Unsaturated mor; horizons taken from the solum (from top to bottom): OL, OF, OH, A. Spleenwort-spruce forest. Creux du Van, Neuchâtel, Switzerland (Chap. 6, § 6.3.5; photo J.-M. Gobat).

5 10 15 20 25 30

80 70 60 50 40 35

VI-1. Decomposition index using sodium pyrophosphate. Low indices indicate scarcely decomposed peats, high indices well-humified peats (Chap. 9, § 9.1.4; (photo J.-M. Gobat).

VI-2. A sphagnum species typical of peatlands under regeneration, on HISTOSOL FLOTTANT, *Sphagnum teres* (Chap. 9, § 9.2.2; photo F. Matthey).

VI-3. A sphagnum species of hollow rims and lawns, *Sphagnum recurvum* s.l., with sporophytes (Chap. 9, § 9.2.2; photo J.-M. Gobat).

VI-4. A sphagnum species of hummocks, in open area, *Sphagnum magellanicum* (Chap. 9, § 9.2.2; photo F. Matthey).

VI-5. A sphagnum species of peaty spruce forests, *Sphagnum girgensohnii* (Chap. 9, § 9.2.2; photo J.-M. Gobat).

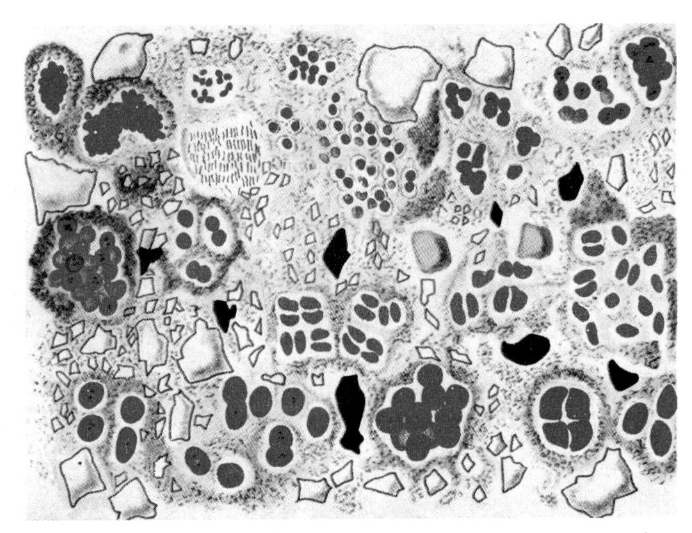

VII-1. The microflora *in situ* in a cultivated field on fallow land in Brie-Comte-Robert, France; after Winogradsky (1949), who described it as follows: 'The Brie soil is relatively rich in colloids, both colourless and brownish-yellow, covering the entire background with their flocs. Moderate quantity of black particles; some bright yellow particles.' With permission of Éditions Masson et Cie, Paris (Chap. 2, § 2.5.2).

VII-2. Extension of an ectomycorrhizal fungus (*Suillus bovinus*) related to the root system of the plant (*Pinus sylvestris*) (Chap. 16, § 16.1.1; photo D.J. Read, in Smith and Read, 1997; with permission of the author).

VII-3. Nodule with definite growth on the legume *Vigna luteola*. The pink colour is due to leghaemoglobin (Chap. 16, § 16.2.1; photo W. Broughton).

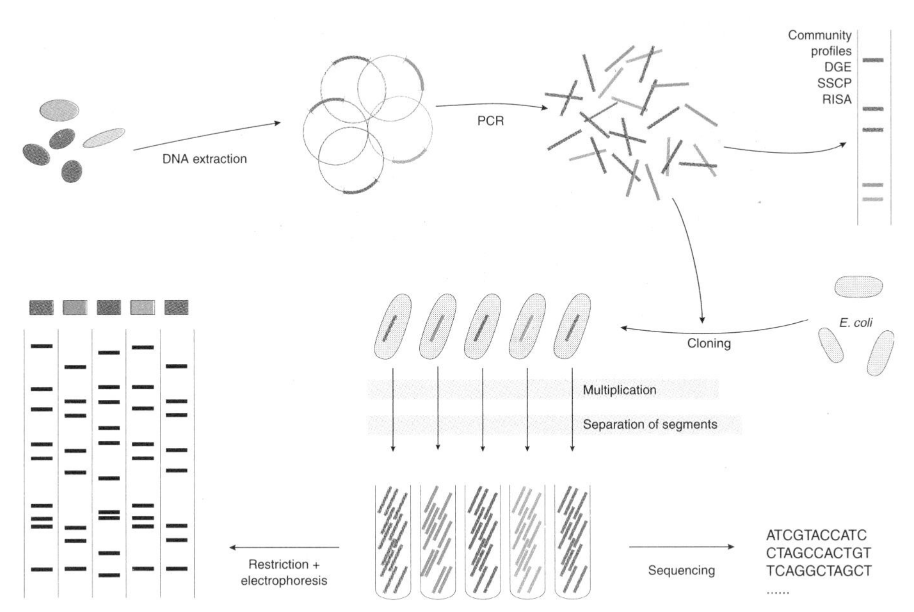

VIII-1. Principle of molecular characterization of bacterial communities. Explanations in the text, Chapter 15, § 15.5.5.

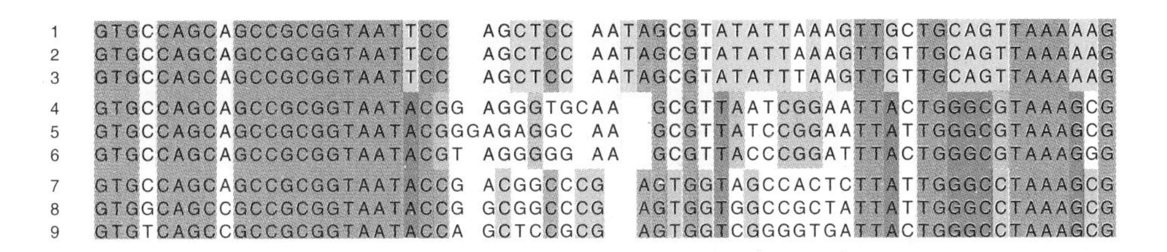

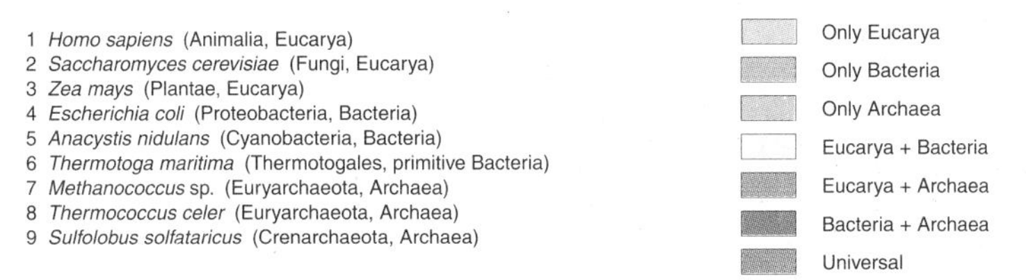

VIII-2. Comparison of a homologous segment of the gene coding for 16s RNA (18s for eukaryotes) in three eukaryotes (*Eucarya*), three bacteria (*Bacteria*) and three *Archaea*. Explanations in the text, Chapter 15, § 15.5.5 (boxed text, *Review of molecular biology*).

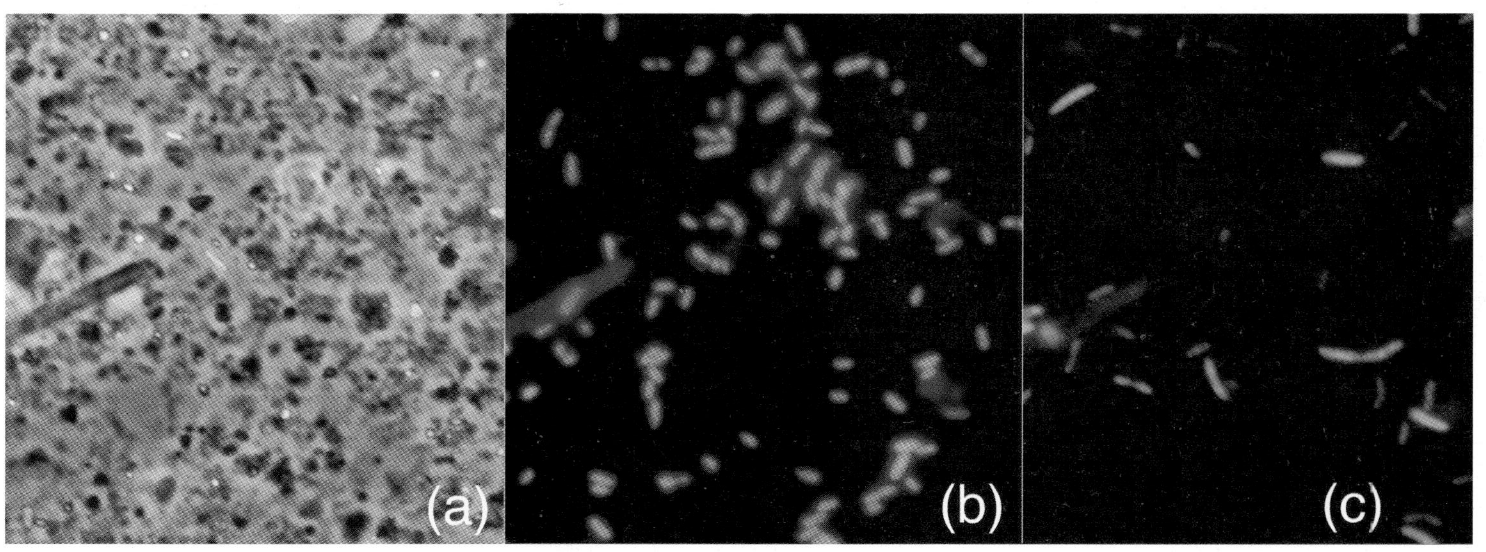

VIII-3. Highlighting the taxonomic groups in a bacterial community by means of molecular-fluorescence probes. The three images represent the same microscopic field in a soil suspension. (a) Phase-contrast photograph showing all the components (cells and organic and mineral particles); (b) and (c) Epifluorescence photographs with filters to highlight: (b) the γ phylum of Proteobacteria (yellow fluorescence) and (c) the β phylum of Proteobacteria (green fluorescence). Magnification 1000 × (photos D. Hahn) (Chap. 15, § 15.5.5; boxed text, *Fluorescent-molecular probes: direct observation of biodiversity*).

poorly understood (Chap. 2, § 2.2.1; Chap. 4, § 4.1.7). Thus there are no precise data available on the duration of the phenomenon because, to follow it from beginning to end, several generations of investigators are required! Climatic conditions also influence it greatly.

During decomposition, wood is profoundly modified in its chemical composition and its physical and mechanical properties. It is attacked by a succession of insects and by fungi, which together constitute a degradative succession.

The process of decomposition of wood follows the same pattern as that of other soil annexes, but with different players.

Decomposition of wood can be 'viewed' at two levels: that of mechanical actions at the scale of the millimetre or centimetre, which is that of invertebrates; and that of biochemical reactions at the scale of extracellular enzymes and fungal hyphae, therefore of the lignolytic fungi. The two following sections will successively cover the degradation of wood at the scale of invertebrates (§ 8.5) and of fungi (§ 8.6). Lastly, the profound synergism between these two levels of attack on wood will be shown through a few examples (§ 8.7).

8.5 DEGRADATION OF WOOD AT THE SCALE OF INVERTEBRATES

8.5.1 From dead wood to saproxylic complex: example of oak

In the first stage, the tree is colonized by a fauna of xylophagous insects composed 90 per cent of coleopterans, and by fungi. Together they transform it into a mass of rotten wood, the ***saproxylic complex***. The second stage sees the latter progressively invaded by the pedofauna, its introduction in the decomposition chains of the soil and its disappearance as an annexe. These two periods are subdivided into four better defined phases.

In the first phase, the withering tree (attacked by fungi, undergoing desiccation, victim of pollution) is colonized by coleopterans of the first wave.

• Phase 1. Coleopterans invade the cortical zone, cutting their galleries in the phloem and interrupting sap circulation. This first wave chiefly comprises Buprestidae and Cerambycidae, mostly billeted under the bark. A few Scolytidae are also found. At this stage, only the larvae of *Rhagium* (Cerambycidae) penetrate deep into the wood. The latter retains the same consistence as in the living tree but takes on a reddish tint. The xylophages of the first phase are not part of the pedofauna but are linked to it in the dynamics of the system, since they prepare the wood in some way for the soil organisms. On conifers, the Scolytidae (e.g. *Ips typographus*) are very important in this first phase. When they proliferate, they kill the withering trees and can even attack healthy ones. The first phase lasts about one to three years (Dajoz, 1996).

In the second phase, mould develops under the separated parts of bark and mycetophages proliferate.

- Phase 2. In the next three to four years, Cerambycidae become more abundant in species and in individuals; Anobiidae (e.g. death-watch beetles) establish themselves right to the heart of the trunk. Predators are more numerous. The bark loses its adherence. The wood is always sufficiently hard and resistant, but its colour becomes darker than in Phase 1. The holes made in bark and wood by woodpeckers in search of larvae are also points of attack by fungi.

The saproxylic complex: an original structure midway between dead wood and the soil

There is only indirect faunistic affinity between the fauna of the saproxylic complex and the first xylophages. They end their development and some mycetophages and carnivores continue there for a while (Histeridae, Colydiidae). The saproxylic communities are thus no more characteristic of different species of trees but are uniform, for example those of beech and oak.

The saproxylic complex offers microclimatic conditions very similar to those of the O and A horizons.

The fauna of the humiferous episolum has no difficulty in colonizing this soil annexe. Mites, springtails and earthworms are essentially edaphic species, while isopods and gastropods are littericoles. According to Dajoz (1996), of 25 coleopteran species captured from litter, 11, essentially carnivorous, are found in the saproxylic complex. Also, numerous larvae of mycetophagous and carnivorous dipterans live in the medium if its water content is high (e.g. Cecidomyiidae, Mycetophilidae, Sciaridae).

In number of individuals the groups dominating the saproxylic complex are mites (67%), springtails (17%) and coleopterans (8%). Other insects (silverfish, diplurans, ants, larvae of dipterans) represent 3.5% and other invertebrates (worms, snails and slugs, woodlice, centipedes, spiders) 4.5%. Snails and slugs (34%) and earthworms (25%) are the most important in the zoomass.

In the third phase, the wood is slowly invaded by rot and becomes friable.

- Phase 3. From the fifth year up to ten years, the bark sloughs and falls in slabs, often revealing a mycelial pad formed underneath. The Cerambycidae gradually diminish. Specialized saproxylophagous coleopterans can then become abundant. Lucanidae mostly exploit the wood already degraded by other insects and by fungi. Elateridae, simultaneously saproxylophages and predators, play an increasingly important role.

In the fourth phase, the structure of the dead wood is profoundly modified in various ways according to the fungi that have invaded it.

- Phase 4. The last waves then transform the wood to a more or less friable mass, the saproxylic complex. This is progressively invaded by elements of the soil fauna, who contribute to completing its transformation to humus and ensure its integration with the soil.

8.5.2 Three scenarios of wood decomposition in nature

The tree that dies standing

Cavity-dwelling birds...

...fungi and insects.

In a tree weakened or withered erect, decomposition initiated by insects and fungi is activated by cavity-dwelling birds who scoop out their nests in the trunks. Starting from cavities and cracks in large branches, fungi colonize the wood, advancing towards the centre, which they gradually rot, and are followed by new insect species.

In the Fontainebleau beech forest (Lemée, in Lamotte and Bourlière, 1978), 14 to 22 upright dead trees were counted per hectare. Decomposition is realized in several phases. According to Kelner-Pillault (1967), there are four phases in the formation of cavities in a tree such as the chestnut (Fig. 8.5).

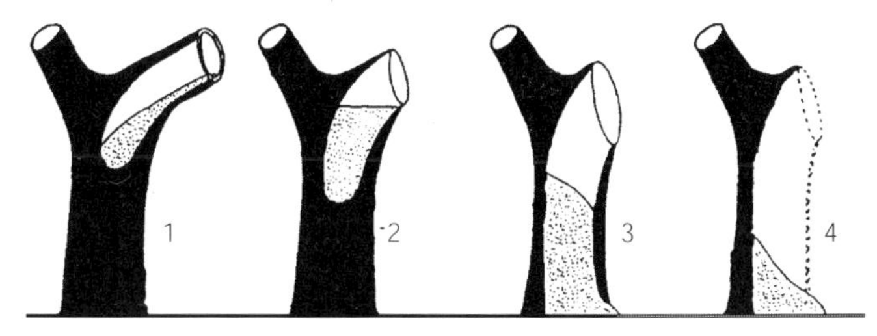

Fig. 8.5. *Phases in the rotting of an erect chestnut tree (after Kelner-Pillault, 1967). Explanations in the text.*

- Phase 1. The attacks start with the upper branches whose centre is decomposed by fungi and insects; initially, these cavities are confined to decayed wood inhabited mostly by larvae of coleopterans, whose faeces gradually make a mycelium-rich vegetable mould;

- Phase 2. The cavities extend from the large branches to the trunk; the hollowed branches collapse, leaving a large opening two or three metres from the ground; the quantity of vegetable mould is augmented;

- Phase 3. The cavities, connected together, are extended gradually to ground level and a second opening is hollowed out at the base of the trunk; a portion of the mould falls out onto the soil;

- Phase 4. A portion of the trunk collapses and the cavity is largely open to the outside; thus weakened, the tree's survival is limited.

The cavities in the chestnut-tree are excellent examples of indirect soil annexes, colonized by an astonishingly

The cavities in the chestnut-tree: an abundant specialized fauna mixed with edaphic animals

abundant saproxylic fauna mixed with elements of the pedofauna. Senescence of this tree and decomposition of its wood are very slow and greatly exceed a hundred years. The characteristic biocoenosis of the vegetable mould may thus subsist in the same tree over several human generations. A study done in Tessin, Switzerland (W. Matthey, unpublished; Tab. 8.5) shows that the perched cavities contain numerous microarthropods (Acari, springtails), proving thereby their communication with the soil. But they are also veritable 'nurseries' that shelter the young larvae of millipedes, dipterans and coleopterans. Small arthropods, more rarely found in the soil (Protura, Diplura, Pauropoda) also proliferate in it (§ 8.8.3).

The fallen tree

'Classic' waves of xylophages whose activity leads to loss of the bark, weakening and then breaking of the branches and contact of the trunk with soil.

The trunk, limbs and branches of conifers fallen on the soil are attacked by different species of Scolytidae, whose size is appropriate to the part affected. Formation of the saproxylic complex is accelerated on contact with moist soil and the lower part of the trunk and large branches are invaded by the pedofauna. The following sequences illustrate the process (Fig. 8.6a–e, after Delamare-Deboutteville, 1951).

- Phase 1 (Fig. 8.6a). The dominant groups are Scolytidae, Cerambycidae, Buprestidae and Curculionidae.

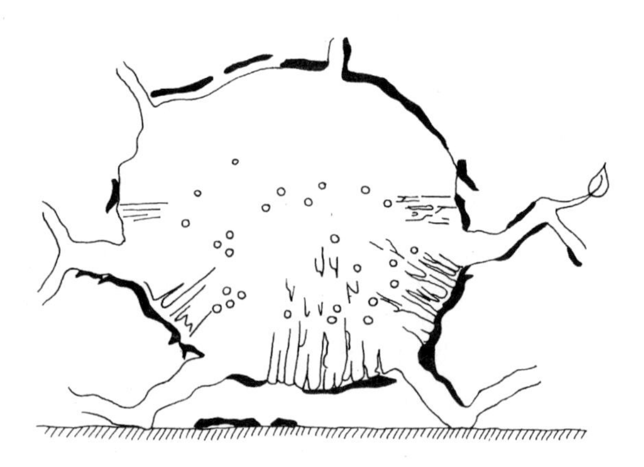

Fig. 8.6a *Phase 1. Start of the attack. Duration 1 to 4 years.*

- Phase 2 (Fig. 8.6b). The branches give way, thereby bringing the trunk in contact with the soil. Dominant groups: Cerambycidae, Anobiidae (death-watch beetles).

- Phase 3. (Fig. 8.6c). The soil fauna starts penetrating the base of the trunk: Collembola, Acari, Diplura,

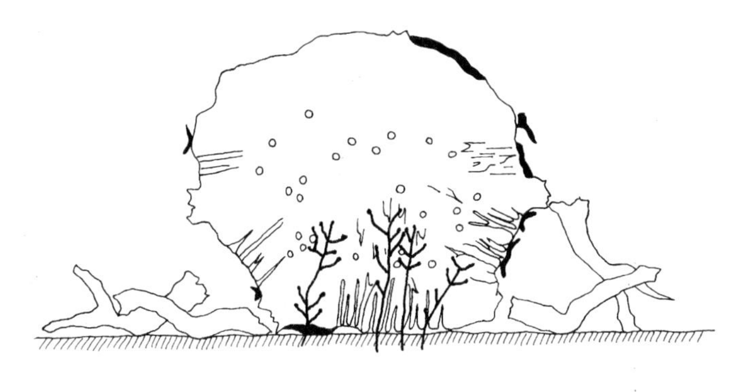

Fig. 8.6b Phase 2. Moist decomposition. Duration 3 to 7 years.

Fig. 8.6c Phase 3. The bark has sloughed. The saproxylic complex forms on the underside of the trunk. Duration 6 to 10 years.

Myriapoda, Isopoda, Lumbricidae. Furthermore, this assemblage of fallen bark, large branches and the trunk itself acts as shelter for numerous epigeal macroinvertebrates (slugs, ground beetles, centipedes, millipedes), which belong to the cryptozoic fauna.

- Phase 4 (Fig. 8.6d). The underside of the trunk is 'absorbed' by the soil. The remainder is occupied by the saproxylic complex protected by an outer layer of more resistant wood. A diversified soil fauna invades it: Acari, (mites, mesostigmatic mites), Collembola, Isopoda, Myriapoda, Lumbricidae, ground beetles and larvae of coleopterans and dipterans. The decayed trunk is a refuge for the fauna of the litter during the dry period since it stays moist longer.

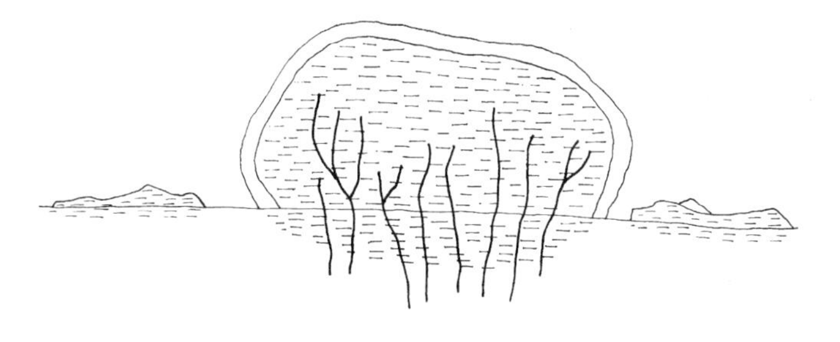

Fig. 8.6d Phase 4. Phase of collapse. Duration greater than 10 years.

• Phase 5 (Fig. 8.6e). The fauna is the same as in the preceding phase, but is less abundant. The debris of the trunk forms shelters for the cryptozoic fauna.

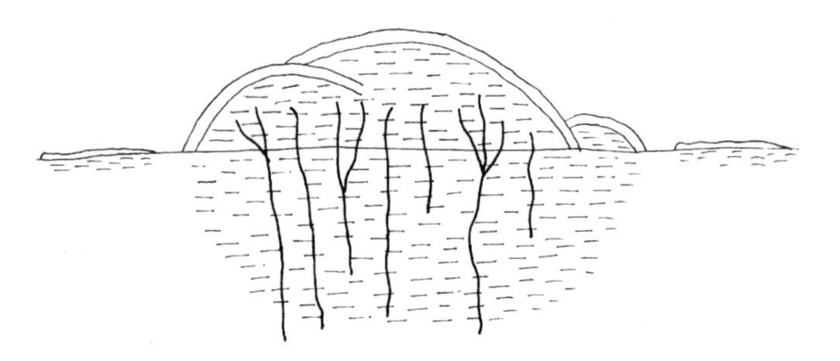

Fig. 8.6 e *Phase 5. Annihilation phase. The fragmented trunk is for the most part integrated in the soil.*

Termites and wood

The destructive power of termites is feared in wooden buildings and with respect to all that contains cellulose, paper for instance (Chap. 12, § 12.4.9). In nature, termites are primarily recyclers of dead wood and litter, extremely active and more efficient in the tropical zone than other xylophagous insects and even fungi. Trunks 20 m long and 1 m in diameter can be completely hollowed out, the wood being replaced by clay brought from depth (Chap. 4, § 4.5.1).

The stump

Stumps: remnants at times hard to digest! Their water content determines three decomposition paths: dry, moist and wet (Wallwork, 1976).

Stumps are the remnants of trees cut or broken off a little above ground level, still held by roots. The composition of the fauna accompanying decomposition of the stump passes from that of a community of saproxylophages to that of an ensemble of saproxylophagous, mycophagous and mycetophagous invertebrates. The composition of the waves is not unchanging. It varies from one species of tree to another, according to the sites and even from one part of the stump to another (microclimate).

In a very sunny location and on a soil with little moisture: dry decomposition.

In dry decomposition, the wood is colonized mainly by xylophagous insects that do not belong to the pedofauna. These are specifically coleopterans, whose larvae drill galleries (families of Anobiidae, Buprestidae, Cerambycidae and Scolytidae). Among the Hymenoptera, Siricidae seek dry wood for laying their eggs (§ 8.7), while solitary bees, for example xylocopids, dig their nests in it.

Ants (e.g. *Camponotus herculeanus*) frequently establish themselves in well-dried stumps, more rarely in fallen or erect trunks. They lay out a nest in it by digging in spring wood, which is softer. Thus only the autumn wood, which forms partitions separating chambers and galleries, survives. Formic acid impregnates the wood and makes it almost impervious to decay. In southern Europe, it is the termite *Calotermes flavicollis* that builds nests in pine stumps.

In moist decomposition—the most favourable for the various decomposers (Wallwork, 1976)—the stump is invaded in three ways:

Moist decomposition: between dry decomposition and saturation of the wood with water.

• Through the periphery: the partially detached bark and cracks are colonized by Elateridae, Lucanidae and Scolytidae. Elements of the soil fauna insinuate themselves under the half-detached bark: xylophagous mites, earthworms and slugs as well as seasonal arthropods that colonize these refuges in autumn (ground beetles, silphids, woodlice, centipedes).

An attack through the periphery...

• Through the heart of the trunk, the attack starting from the cut. Algae, carpophores of various fungi, lichens and mosses grow here, modifying the microclimate and creating new ecological niches for the fauna. The sawdust resulting from the work of xylophages is abundantly colonized by nematodes, enchytraeids and larvae of dipterans; it also favours establishment of bryophytes. Decomposition of the central portion of the stump, which retains moisture better, is often more rapid than that of the periphery. Under the influence of xylophages (mites, springtails, larvae of Lucanidae and Elateridae) and fungi, the centre is hollowed out into a cup filled with woody debris rich in mycelia.

...one through the heart of the trunk...

• Through the roots, under the action of invertebrates and fungi of the soil.

...and one through the roots.

The final result is reduction of the stump to small, partially decomposed particles. They are ingested by earthworms and gastropods and intimately mixed by them in the underlying soil. Occasionally, only the outer wall, very dry, survives. It is slowly colonized by plants and often provides support for a hill nest of *Formica* spp.

8.6 DECOMPOSITION OF WOOD AT THE SCALE OF FUNGI

8.6.1 Fungi, essential players in decomposition of wood

Fungi play an essential role in the decomposition of wood. But wood is one of the most selective of biological media: let not those who wish venture into it!

Because of their filamentous structure and ability in translocation, fungi are vigorous colonizers of the porous structures of wood and soil (Chap. 2, § 2.5.3). But wood knows how to defend itself! Oak stakes have been found buried in the mud of lake shores nearly 3000 years ago by our ancestors of the Bronze Age. Also, almost intact beams remain in thousand-year buildings, the chief reason for their rareness being rather the frequency of fires or unskillful renovation than their transformation into humus!

Soft, white or brown rots sooner or later (rather later), reduce the lofty trunk to the state of humus! The annexe has become soil.

Except under extreme conditions of water saturation or dryness, fungi play various tricks and strategies to attack the wood, feeding and fructifying in it; from entering the trunk, from the top or from the bottom, to invasion of the fallen limb by fungi already existing in the soil; from entry through natural openings to invasion through wounds or galleries cut by larvae of coleopterans; from selective destruction of lignin to that of polysaccharides. Numerous synergistic and antagonistic relationships exist between the action of fungi and that of the fauna. We shall discuss them in § 8.7.

Study of the decomposition of wood by fungi is an important domain, as much by its implication in general ecology as in understanding degradation of timber and methods to prevent it. We shall only touch upon the subject in this book; the reader wishing to learn more can usefully consult the books by Rayner and Boddy (1988) and Dix and Webster (1995).

Greek roots for animals and Latin roots for fungi. Lignivorous and lignolytic are equivalent to xylophagous, used for animals.

Lign- and the rest... a complicated terminology!

In the area of wood fungi and their action, terminology is not always very clear. We shall use the following terms in this book:

- ***lignicole:*** living in wood, without necessarily degrading it significantly;
- ***lignolytic*** (or ***lignivorous):*** which significantly degrades wood by modifying its properties, but not necessarily attacking lignin; lignolytic fungi are divided into ***cellulolytic*** (cellulose-degrading) and ligninolytic;
- ***ligninolytic:*** which degrades or solubilizes lignin significantly.

8.6.2 Wood, a selective medium

Wood, a very selective environment, accommodates a rich community of fungal species, often highly specialized.

Even though the anatomical and chemical complexity of wood gives rise to a wide variety of microbial ecological niches, wood remains very selective. This is related to anatomical, biochemical (lignin, tannins, soluble inhibitory compounds), physicochemical (hydric stress, CO_2 and oxygen contents) and nutritional (particularly available nitrogen) factors.

Attack on lignin

'Prefabricated humin' (Chap. 2, § 2.2.3; Chap. 14, § 14.3.4), lignin is a macromolecular structure simultaneously very complex and disordered, with composition varying according to plant categories. Apart from its intrinsic biochemical stability, it embeds the polysaccharides of cell walls and combines with them chemically into lignocellulose, preventing saccharolytic enzymes from reaching their substrates. Because of this, attack on or modification of lignin is often the *sine qua non* for decomposition of wood. Only fungi, chiefly Basidiomycetes, succeed in this significantly (Hammel, in Cadisch and Giller, 1997).

Lignin is a tough molecule, attacked with difficulty by enzymes!

Presence of aromatic inhibitory compounds

Certain constituents of wood exercise an inhibitory effect on most microorganisms. This holds true for tannins and other soluble polyphenolic compounds, the 'extractables', among which phenolic acids (e.g. vanillic, gallic, caffeic, ferulic acids) and flavonoids (e.g. catechin, quercetin, pyrogallol, catechol). Only organisms resistant to these compounds can grow on undegraded wood. Certain fungi, especially Basidiomycetes and soft-rot fungi, are capable of degrading these substances by means of polyphenol-oxidases (Chap. 14, § 14.3.3; Lyr, 1962), allowing other more sensitive organisms to be established in turn. Lixiviation of wood also has the effect of diminishing the concentration of extractables and therefore the selective effect.

Specific inhibitors of microbial life that often work together synergistically.

Heart of the wood (or ***heartwood):*** in a trunk, the central portion of the wood through which water is not transported, which does not have living tissues, but serves as support for the plant. Heartwood is impregnated with tannins and other anti-decay substances, which protect it against degradation by microorganisms and give it a dark colour (after Lüttge, et al., 1996). However, in spite of this protection, it is often attacked before the sapwood because of its lower water content.

Importance of moisture content

Water is a major factor in orienting the decomposition of wood. If there is too much of it and the pores of the wood are filled, diffusion of oxygen is greatly reduced, limiting the growth of fungi. Under these conditions, bacteria are often at an advantage and their activity can give rise to complete anoxia. This is what happens in certain living trees in which the ***heartwood*** is saturated with water (wet wood). A deeply anaerobic bacterial community is established, the methanogenic syntrophy (Chap. 4, § 4.4.2; Zeikus and Ward, 1974; Schink et al., 1981). Biogas is then produced in the trunk!

Trunks that produce biogas!

It is thought that the often preferential attack on the heartwood is related to the fact that the ***sapwood,*** especially in its portions active in conduction of the sap, exhibits an oxygen deficit unfavourable to fungi (Rayner, in Ayres and Boddy, 1986).

Sapwood: peripheral zone in a wood trunk comprising active functional tissues, the xylem vessels in particular, transporters of water and mineral salts from the roots to the leaves. The tannin-free sapwood is light coloured.

Provided its moisture content does not exceed 20%, sound timber is completely protected against mycelial attack.

On the contrary, lack of water often prevents development of lignolytic fungi. The limit of hydric potential permitting their growth is between –5 and –7 MPa (Chap. 3, § 3.4.3). In undegraded wood the hydric potential is –5 to –6 MPa for a water content of 30% and –7 to –8 MPa for content of 25% (Dix, 1985).

Under natural conditions, dead wood (standing or on the ground) undergoes large variations in moisture content. On drying, the latter can fall well below the critical point permitting growth of fungi. In order to maintain themselves under such conditions, many lignicole species develop a strategy of survival at very low hydric potentials, without showing active growth. While it requires a potential higher than –6 MPa for growth, *Schizophyllum commune* (Basidiomycetes, Aphyllophorales) survives more than 27 weeks in wood at –20 MPa (Fig. 8.7; Dix and Webster, 1995).

Fig. 8.7. Schizophyllum commune, *champion of drought resistance (photo J. Keller).*

But watch out! For the same moisture content, wood degradation raises the hydric potential considerably (Tab. 8.2). Even if the fungus starts its attack under conditions of extreme hydric stress, it improves them later. This results in a positive feedback, which permits acceleration of fungal development and allows ***carpophore*** formation.

Carpophore (literally 'fruit-bearer'): morphologically differentiated macroscopic structure formed by certain fungi of which it constitutes the fruit-body. This is the 'mushroom' in its popular sense. The carpophore is formed from condensed mycelia and carries the organs (asci or basidia), which are the site of sexual phenomena and formation of 'sexual' spores, the latter ensuring multiplication, dispersion and often conservation of the species.

Table 8.2 Variation in hydric potential in undegraded and degraded oak wood as a function of moisture content (after Dix, 1985) 276

Moisture content (%)	Hydric potential (MPa)	
	Undegraded wood	*Degraded wood*
22	–9.5	–3.8
11	–18.2	–6.1

In the tissues of decomposing wood, the carbon dioxide content is often high, reaching 15 to 20%. The growth of lignicole fungi is optimum at 10% CO_2 and it may still reach half its maximum under 50% CO_2. As a rule, non-lignicole fungi are strongly inhibited under more than 15% CO_2.

The fungus is not only a fruit-body... and all fungi do not wear a hat!

Macromycetes (e.g. morels, polyporus, boletus) exhibit large fruit-bodies or ***carpophores***. ***Micromycetes*** (e.g. plant-parasitic rusts and mildews, moulds) have minute carpophores or do not form them, in which case their sexual organs appear directly on the vegetative mycelium. In numerous micromycetes the sexual phase is unknown and, in many cases, is probably non-existent. These are grouped under the name 'imperfect' fungi or ***Deuteromycetes***.

Lignicole fungi do not let themselves be asphyxiated by carbon dioxide!

Frequent nitrogen deficiency in dead wood

Sound wood contains only 0.03 to 0.3% combined nitrogen (average of 0.09%), which corresponds to very high C/N ratios, between 300 and 1000. This low N content is a limiting factor of fungal growth. Furthermore, most of this nitrogen is in an insoluble form or combined in aromatic complexes and is therefore not directly available to microorganisms. Lignicole fungi often exhibit a very high affinity for nitrogenous compounds and are thus active at very high C/N ratios. In addition, they have developed a very efficient strategy for conservation and recycling of nitrogen with intense turnover of proteins and translocation of nitrogen from senescent hyphae to actively growing regions of the mycelium (Levi et al., 1968).

Nitrogen: the (nearly) great absentee from wood.

Certain fungi *(Hoehenbuehelia, Pleurotus)* are concomitantly lignolytic and nematophagous (Chap. 4, § 4.4.7; Thorn and Barron, 1984, 1986). Therefore they participate in a predatory chain, bacteria-eating nematodes in turn becoming the prey of fungi. *Pleurotus ostreatus* is even capable of directly lysing bacterial cells and feeding on the cell contents (Barron and Thorn, 1987). Bacteria produce ureases (Chap. 14, § 14.3.2) that transform urea excreted by certain animals to ammonia and CO_2. Association of fungi with nitrogen-fixing bacteria has been postulated but still not proved.

The 'nitrogen economy' of wood fungi is also related to biotic factors.

8.6.3 Attack and colonization of wood

Standing trees

The wood of a healthy living tree is not a sterile medium. It harbours a flora often rich in ***endophytic*** fungi, which are ***saprophytic*** and appear to be neutral with respect to the tree (Petrini, in Andrews and Hirano, 1992). Commensals or mutualists, their precise role is not clear. Their spread is very limited, often a few mycelial filaments confined to a single cell. Certain endophytes isolated from healthy wood are also known as decomposers of wood as, for example, Xylariaceae (Ascomycetes). It is therefore a matter of latent infections, comparable to the presence of potentially pathogenic bacteria on the skin *(Staphylococcus aureus)* or in the lungs *(Pneumococcus)*. Some of these endophytes exert a protective effect on the wood, for example by producing metabolites toxic to xylophagous insects (Carrol, in Barbosa et al., 1991; Dix and Webster, 1995).

Trees die standing up!

Endophytic: said of an organism, in particular a microscopic fungus, living within the tissues of a plant without visible effects on the latter. In the living wood of certain trees, the presence has been noted of endophytes that attack, as decomposers, only other tree species.

Saprophytic: said of a microorganism that feeds on dead organic material.

From the top and from the bottom...

The attack on a living tree is the work of parasites or 'aggressive saprophytes'. The ***rots*** that infect the tree from the top (top rots) and developing from the top downwards are often distinguished from those that penetrate from the foot (butt rots) and invade the trunk by moving up. The former are essentially caused by spores that germinate and penetrate the wood through wounds (pruned or broken branches), through natural orifices (lenticels, foliar scars) or holes drilled by xylophagous insects, or through tissues weakened by desiccation or by microbial attacks. The latter have different origins: transmission of the mycelium from the roots of a diseased tree to those of a healthy one, directly in the case of the dangerous parasite *Heterobasidion annosum*, or indirectly through rhizomorphs. Certain fungi establish first in the root sapwood before invading the heartwood (e.g. *Armillaria mellea*).

Rot: this term is used here in a specific sense. It refers to visible symptoms (colour, modification in texture and resistance) associated with the development of one or several saprophytic or parasitic fungi on wood.

Certain species enter in the form of spores or mycelial fragments through minor wounds, from where they are transported in a diffuse manner by the circulating sap. Damage can then be manifested in every physiologically weakened organ with no apparent way of entry.

To and fro!

Once the fungus has entered, axial colonization of the trunk can be rapid, the hyphae growing in the vessels without being stopped by cellular barriers. Radial colonization is slower as it involves prior rupture of the septa of ***pits***, colonization of walls of the non-lignified parenchyma and penetration by ***medullary rays***. The mode of colonization thus depends closely on the anatomy of the wood.

Pit: communication path between cells, forming where the cell wall is thin. It allows radial intercellular transport of cytoplasm elements by virtue of micropores through the cell wall and membrane.

Medullary rays: chain of parenchymal cells that radially link the pith, located at the centre of the stem, and the peripheral bark.

Basidiomycetes, in particular *Polyporus* (Tab. 8.3), cause the greatest damage, resulting in brown or white rot of heartwood of the trunk or major branches. Other destroyers are opportunistic invaders of the sapwood. Functioning sapwood is normally saturated with water, therefore too poorly aerated to permit mycelial growth (Rayner, in Ayres and Boddy, 1986). Following a major injury, the water column in the active vessels is ruptured; thus the tissues dry up and become aerated, thereby greatly facilitating colonization.

Even if they have attacked the living tree as parasites, fungi generally continue to grow as saprophytes after the host dies.

Non-specific fungi colonize fallen wood and stumps.

Colonization of fallen wood and stumps

Stumps of felled trees, large branches and trunks broken off by wind are rapidly colonized by

Table 8.3 Examples of Basidiomycetes common on trees (after Dix and Webster, 1995)

Species	Type of rot
Fungi growing on standing trees	
Heterobasidion annosum	White rot of heartwood of conifers
Phaeolus schweinitzii	Brown rot of heartwood of conifers
Armillaria mellea	White rot of heartwood and sapwood of hardwood trees
Armillaria ostoyae	White rot of heartwood and sapwood of conifers
Oudemansiella mucida	White rot of sapwood of beech
Fistulina hepatica	Brown rot of heartwood of oak
Fomes fomentarius	White rot of heartwood and sapwood of beech and birch
Ganoderma applanatum	White rot of heartwood of beech
Pleurotus ostreatus	White rot of heartwood of hardwood trees, rare in conifers
Polyporus squamosus	White rot of heartwood of hardwood trees, chiefly maple and elm
Piptoporus betulinus	Brown rot of sapwood of birch
Laetiporus sulphureus	Brown rot of hardwood trees, chiefly oak and chestnut
Fungi growing on stumps and fallen wood	
Bjerkandera adusta	White rot of hardwood trees
Coriolus versicolor	White rot of hardwood trees
Stereum hirsutum	White rot of hardwood trees, chiefly oak
Stereum sanguinolentum	White rot of conifers
Chondrostereum purpureum	White rot of hardwood trees, chiefly beech and birch
Daedalea quercina	Brown rot of hardwood trees, chiefly oak
Pseudotrametes gibbosa	White rot of beech
Coniophora puteana	Brown rot of all types of wood, common in timber

Basidiomycetes with relatively low specificity (Tab. 8.4), from spores or rhizomorphs originating from mycelial networks already present in the forest soil. The strands are fed by pre-existing nutrients accumulated in the mycelium. This facilitates invasion of new substrates. Examined superficially, the fruit-bodies of the latter fungi may seem to be formed in the soil, but the nourishing wood is always found by carefully following the strands.

Other colonizers, linked to subsequent decomposition stages, appear later. These are mostly Agaricales. Also found on moist, highly decomposed wood are representatives of Dacrymycetales (Basidiomycetes) and Discomycetes (Ascomycetes); the latter form masses of small ***apothecia*** (Fig. 8.8).

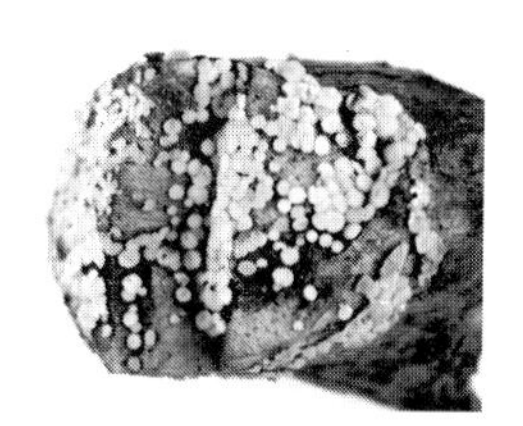

Fig. 8.8 *Humariaceae (Ascomycetes) on moist, decomposing wood (photo J. Keller).*

Apothecium: Fruit-body characteristic of Discomycetes, often cup-shaped.

Besides the major leaders in decomposition of wood, successions of Micromycetes are also seen at all the stages of decay.

The first decomposing Micromycetes are r-strategists (Chap. 4, Table 4.3) incapable of attacking lignin and the complexes it forms with polysaccharides. They do not cause significant structural changes. Other more aggressive Micromycetes responsible for soft rots and lignolytic Basidiomycetes arrive later. Finally, late cellulolytic Micromycetes develop, benefiting from the

Table 8.4 Some fungi colonizing stumps and fallen wood

Macromycetes		
Infection from spores	Infection from mycelial strands	Late colonizers (advanced stages of decomposition of wood)
Chondrostereum purpureum *Coriolus versicolor* *Corticium* spp. *Pseudotrametes gibbosa* *Phlebia radiata* *Stereum hirsutum* *Ascocoryne sarcoides* (Ascomycetes)	*Armillaria* spp. *Hypholoma fasciculare* *Tricholomopsis (Collybia) platyphylla* *Phallus impudicus*	*Mycena galericulata* *Pluteus cervinus* *Psathyrella hydrophila* (only deciduous) *Paxillus atrotomentosus* (conifers) *Tricholomopsis rutilans* (conifers) *Lycoperdon pyriforme*
Micromycetes		
Early saprophytes *Cylindrocarpon* *Fusarium* *Penicillium* *Cladosporium* *Trichoderma* *Aureobasidium* *Mucor* *Mortierella*	Soft-rot fungi *Phialophora* *Chaetomium* *Humicola* *Doratomyces*	Late cellulolytic fungi Various moulds

access provided to cellulose of the walls following destruction or solubilization of lignin by Basidiomycetes.

Figure 8.9 shows an example of a succession of microorganisms on buried beech-wood.

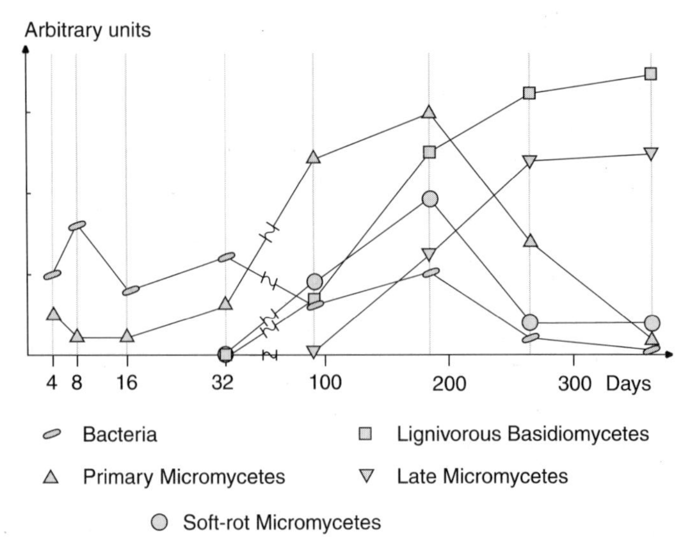

Fig. 8.9 *Frequency of colonization (arbitrary units) of buried beech-wood (after Clubbe, 1980). The first important colonizers are bacteria, followed by primary Micromycetes. Without themselves causing great damage, these organisms facilitate subsequent invasion by mycelial filaments of Basidiomycetes, by rupturing the septa of pits for example. The 'big guns' of destruction of wood, the soft rots (transiently) and mostly Basidiomycetes, make their appearance only much later. Late or secondary Micromycetes (generally cellulolytic moulds) follow last. These last two groups, K-strategists (Chap. 4, Table 4.3) persist a long time whereas the importance of the others diminishes considerably after several months.*

8.6.4 Types of rot

Soft rot, brown rot or white rot?

Biochemical characteristics of the enzymes released to ensure decomposition of wood are indicated by expression of very typical symptoms, the rots. In general, three types of rot are distinguished:

- ***Soft rot*** is achieved by Micromycetes (Ascomycetes and Deuteromycetes) that often have transitory growth. Several hundred species cause soft rot (Seeham et al., 1975): the most represented genera are *Acremonium, Cephalosporium, Chaetomium, Doratomyces, Phialophora* and *Sporocybe*. These organisms decompose cellulose and hemicelluloses. Some of them decompose lignin superficially (degradation of aliphatic sidechains and demethylation). Soft rots are more active on wood of hardwood trees and, unlike Basidiomycetes, tolerate a relatively high moisture content. The mechanical strength of wood is diminished by their action and the wood becomes soft and spongy.

- ***Brown rot*** (or ***cubic rot***) is provoked by about 10% of the lignolytic Basidiomycetes, essentially Aphyllophorales (Tab. 8.3, Fig. 8.10). Hemicelluloses are attacked first, followed by cellulose; lignin is only affected superficially through demethylation. The extracellular enzymes are diffusible. Selective degradation of polysaccharides leaves a lignin skeleton, coloured brown, which exhibits fractures perpendicular to the direction of fibres.

Fig. 8.10 *Brown rot. The general appearance of wood attacked by brown rot recalls to mind that of carbonized wood. This should not be surprising: when wood is burnt, the polysaccharides are degraded faster than lignin (photo D. Job).*

- ***White rot*** is characterized by bleaching of the wood, which acquires a fibrous or spongy consistence (Fig. 8.11). The fungi attack polysaccharides as well as lignin; the latter could be considered a cosubstrate. It is never eliminated and/or dissolved alone: cellulose or hemicelluloses are always degraded concomitantly with lignin. One type of white rot is represented by *Coriolus versicolor*, which accomplishes simultaneous oxidation of lignin and cellulose. Another, like *Ganoderma*, oxidizes lignin along with hemicelluloses. Cellulose is decomposed only very slowly (Otjen and Blanchette, 1985).

Fig. 8.11 *White rot. This is the work of most lignolytic Macromycetes (mostly Basidiomycetes) as well as of certain Ascomycetes (Xylariaceae). Extracellular enzymes of the white-rot fungi are essentially localized on mycelium walls or impregnate the mucilaginous sheaths. Thus they do not diffuse far and the attack takes place at the point of contact of the mycelium (photo D. Job).*

Actually, the main goal of lignin degradation is not the production of molecules assimilable by the fungus. Rather, it removes the protection that impeded polysaccharide degradation. The chemical radicals and quinones resulting from oxidative decomposition of lignin mainly enter processes of synthesis of insolubilization humin H2 (Chap. 5, § 5.2.3; Chap. 14, § 14.4.3). White-rot fungi are thus much more ligninolytic than 'ligninophagous'!

Trees and fungi: indissolubly bonded, for better or for worse!

The tree has need of mycorrhizal fungi for its growth and nutrition (Chap. 16, § 16.1). Often it dies from attack of other fungi and continues as an organic annexe of soil, to shelter and regulate a succession of very characteristic fungal communities, accompanied by a procession of bacteria, Myxomycetes and invertebrates. It is a very different life from that of the soil that, in a period of time comparable to what was needed for constituting the tree, returns it to the soil. And in that time a seed fallen from a nearby tree may germinate there, its rootlets may encounter a mycorrhizal fungal filament, and...

8.7 COMBINATION OF FUNGI AND INSECTS IN DECOMPOSITION OF WOOD

Interactions with invertebrates: the combination gives strength!

Colonization of wood by fungi is often modulated by invertebrates, in particular insect larvae that cut galleries in it and feed there. Just as in a litter, there is a mutually beneficial association between the fauna and mycoflora. These relations are very complex, going from predation to symbiosis, passing through transport and dissemination (Anderson et al., 1984).

- Insects, described fungivorous, mycophagous or mycetophagous, are predators of fungi (e.g. the dipterous Mycetophilidae whose larvae are mycetophagous and mycophagous, and the Cisidae, coleopterans that live in *Polyporus*.

- Fungi are half-parasitic, half-predatory on insects that they kill more or less slowly (e.g. *Beauveria* spp., used in biological control, grow at the expense of larvae of the common cockchafer).

- Many fungi are intestinal symbionts of xylophagous invertebrates. Mites and springtails, by moving in the galleries cut by coleopterous larvae, carry spores and mycelial fragments into the wood in their pellets and on their bodies, favouring their dissemination (Swift and Boddy, in Anderson et al., 1984).

A concrete example: the Siricidae, astonishing wood wasps

The females of these Hymenoptera deposit their eggs deep within the wood (one or two centimetres) by means of a sting that is a drill as well as an ovipositor. This perfected apparatus carries at its base two small pouches filled with spores and mycelial filaments bathed in a glandular secretion. Each egg is introduced into the wood wrapped in a droplet of this liquid. The mycelium quickly invades the channel where the eggs are laid and maintains an environment favourable for egg survival during incubation, in particular by lowering the moisture content of the wood. It then serves as sole nourishment for young larvae that during their first two stages are incapable of gnawing wood.

Older larvae cut ascending or descending galleries, feed on wood and mycelium and finally fill up their galleries with compacted sawdust in which fungi proliferate. Old larvae already have a pouch with fungi in which mycelial fragments float. These fragments survive the metamorphoses of the insects, passing from larvae to nymphs, then to adults. This association of xylophagous insect and fungi (Basidiomycetes, family Thelephoraceae) is essential for survival of the larvae.

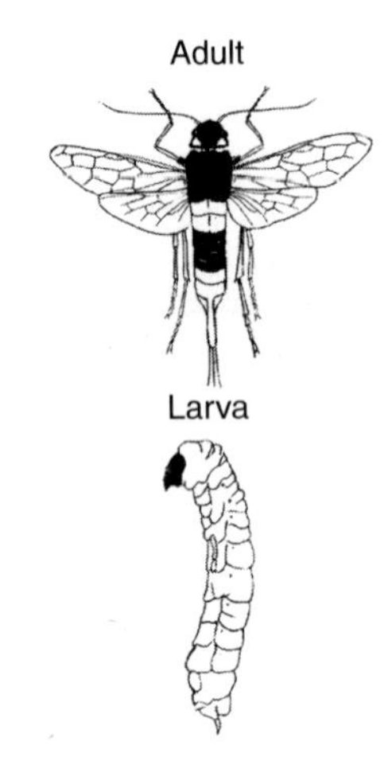

Fig. 8-D *Giant wood-wasp* (Uroceras gigas). *The female is recognized by its ovipositor. Length: adult up to 44 mm, larvae up to 35 mm.*

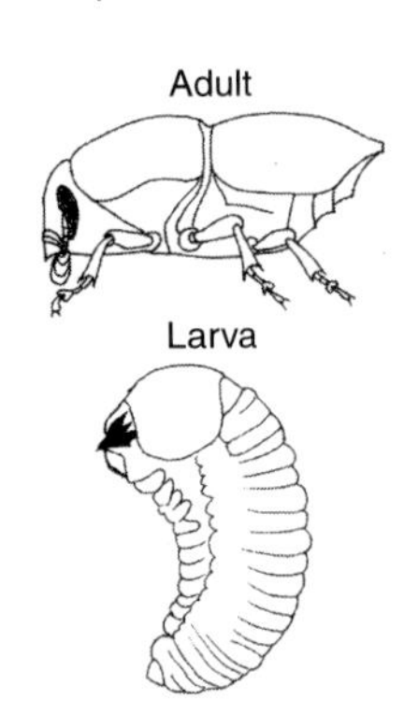

Fig. 8-E *Large elm-bark beetle* (Scolytus scolytus). *Length: larvae and adults 3–6 mm.*

- Specialized fungi form a mutually beneficial association with higher termites and leaf-cutter ants. The fungi develop in 'fungus gardens' composed of wood or leaves collected and prepared by insects. They have a predigestive action necessary to enable termites and ants utilize this food; they themselves enter the alimentary regime of their host (e.g. *Termitomyces* and Macrotermitinae, *Attamyces* and *Atta*).

- Xylophagous fungi modify the structure of wood transforming it into a food more suitable for invertebrates. Their enzymes oxidize toxic phenols and weaken cell walls; the C/N ratio diminishes. The coleopteran *Xestobium rufovillosum*, which lives in dead wood of oak and willow, has been well studied. Raised in fresh willow-wood it attains a size of 3 mm in 9.5 years; in wood decomposed by fungi it completes its cycle in 10 to 17 months (Swift and Boddy, in Anderson et al., 1984).

- In turn, xylophagous insects (e.g. Siricidae, Fig. 8-D) facilitate colonization of trunks by fungi. They augment the specific surface area of the wood by pulverizing it and thereby promote attack by extracellular enzymes. Locally, production of faecal matter provides a supplementary source of nitrogen. Fungi thus find a favourable environment in the droppings that fill the galleries, more so because the microclimate of this milieu is particularly favourable for them.

- Fungi are transported by insects, possible vectors of cryptogamic diseases (pathogenic association), e.g. large elm-bark beetle *Scolytus scolytus* (Fig. 8-E), which transports spores of *Ceratocystis ulmi*, causal agent of Dutch Elm disease.

8.8 INDIRECT ORGANIC ANNEXES

Indirect annexe: this is the term used for a habitat that, without being in contact with soil, shelters a diversified microfauna and mesofauna exhibiting considerable similarity with those of the soil.

Indirect annexes of soil are grouped under three categories (Fig. 8.12, which also presents other structures giving variety to the wood surface):

- suspended soils of epiphytes and epiliths;
- carpophores of fungi;
- cavities in trunks.

Cavicole: which lives in cavities.

With the exception of those of ***cavicole*** species, bird nests constructed on trees do not appear to be true soil annexes. Actually, the pedofauna hardly colonize them,

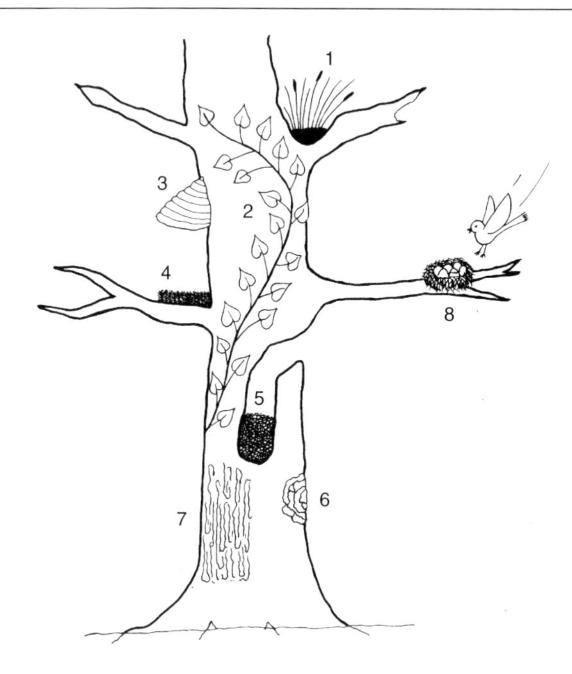

Fig. 8.12 *The majority of indirect annexes are located on trees (here, a hardwood tree in the temperate zone).*
1. *Soil of epiphytes*
2. *Creeper (ivy)*
3. *Carpophore of polypore*
4. *Mosses*
5. *Cavity in trunk*
6. *Lichens*
7. *Wrinkles in bark*
8. *Bird nest (see text)*

even though certain elements of the last two waves in carcass decomposition are seen (§ 8.3.1) feeding on remains of eggs, feathers or even on dead fledglings.

8.8.1 Suspended soils of epiphytes and epiliths

Plant debris that accumulates and decomposes on large horizontal branches and in forks forms organic 'microsols' termed *suspended soils* (Paulian, 1988). In temperate climate, mosses that develop there as thick pads or mantles function as dust traps and accumulate mineral particles. In tropical forests, suspended soils have still greater importance. ***Epiphytic*** plants (Bromeliaceae, Orchidaceae, ferns) germinate there and grow to a height of forty metres. They form what Delamare-Deboutteville (1951) terms soils of epiphytes.

Epiphyte: plant, lichen or fungal carpophore growing on a living woody plant.

The colonizing pedofauna of suspended soils and cavities located at height climbs up the trunks, the way being facilitated by the presence of creepers, for example ivy.

Ivy and other creepers (wild vine, clematis) open up the upward path for the soil fauna.

Epilithic mosses and lichens are the initiators of microsols on mineral substrates, natural or not. In a pad of the moss *Grimmia pulvinata*, the mineral fraction represents 42% on dry weight basis (Matthey et al., 1984). This capacity of mosses to filter fine particles can prime pedogenesis on unweathered substrates such as alluvial

Where soils begin in moss!

Epilith: plant or lichen growing on a crag, cliff, wall, roof, etc.

sandbanks (Arnold and Gobat, 1998). If the slope permits accumulation of organic debris, a slow evolution takes place; its stages are marked by a succession of epiliths: lichens-mosses-ferns or flowering plants.

Moss pads are well-structured environments, exhibiting several microhorizons (Fig. 8.13):

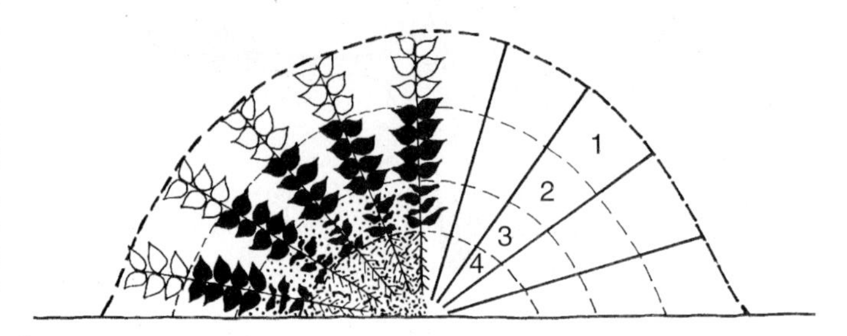

Fig. 8.13 *Structure of a moss pad. In a 5g (dry weight) pad of* Syntrichia ruralis, *located at 3m height on a southwest-facing limestone wall under oceanic temperate climate, 303 living Acari (299 mites, 2 mesostigmatic mites, 2 prostigmatic mites), 102 dead mites, 2 snail shells (*Pupilla muscorum*) as well as the remains of Coleoptera and ants were counted (Matthey et al., 1984).*

- a living outer zone, aerated, with green leaves (1);
- dead but still whole leaves (2);
- fragmented and decomposed leaves mixed with mineral debris (3);
- a central part, fixation zone, rich in rhizoids (Fig. 8.14), with low porosity; this is also the richest in mineral debris from the substrate or aerial dusts retained by the moss (4).

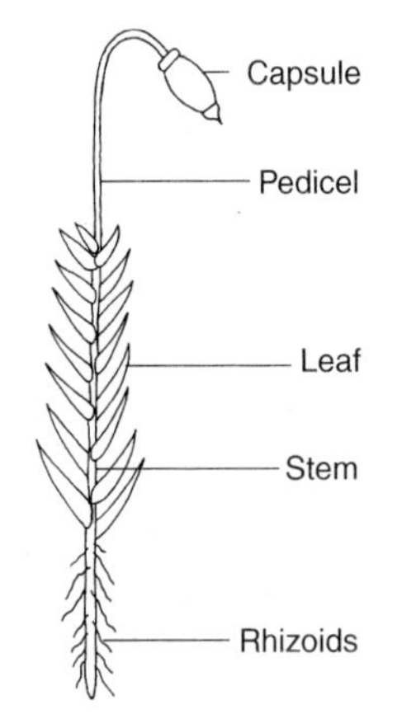

Fig. 8.14 *A moss shoot, with rhizoids.*

8.8.2 Fungal carpophores

Carpophores of fungi are also indirect annexes of soil, but not all of them!

The carpophores of certain polypores, epiphytic fungi that occasionally grow in large numbers on tree trunks (e.g. *Piptoporus betulinus* on birch) may also be considered indirect annexes of soil. They are colonized by a mycetophagous fauna composed mostly of coleopterans. Specialized coleopteran species complete their entire growth cycle there (e.g. Cisidae). An old boletus that remains accessible to the fauna for several days is also another example of annexe, unlike the Shaggy cap whose sudden and rapid liquefaction is the work of an autolysis that does not require soil organisms.

8.8.3 Tree holes

Trunks and large branches have cavities of varying size with walls gradually decomposing under attack by fungi and xylophagous invertebrates. According to the process illustrated by Kelner-Pillaut (1967) and Arpin et al. (2000), the cavity is extended little by little into the trunk which finally collapses (Fig. 8.5; § 8.5.2). Some birds (nuthatches, tits) and rodents (edible dormice, garden dormice and possibly squirrels) set up favourable holes by carpeting them with grass, hair and feathers; they leave behind droppings and occasionally even dead bodies of their young. These materials attract new species of invertebrates that enrich the cavicole community.

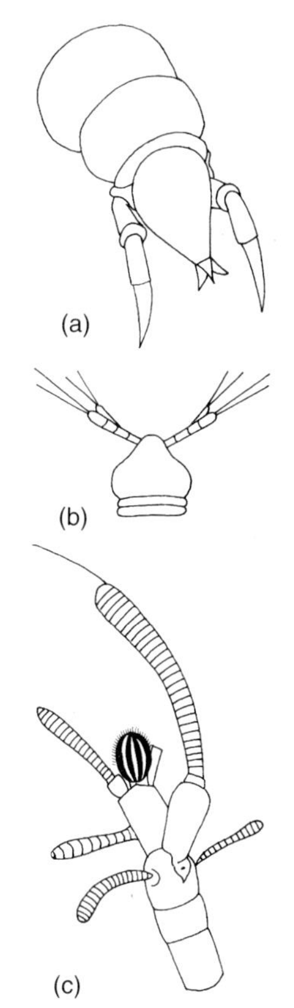

Fig. 8.15 *Protura and Pauropoda: (a) Head and first two thoracic segments of a Protura; the first pair of legs replaces the antennae. (b) Head of a Pauropoda. (c) Complexity of a Pauropoda antenna.*

An amazing small world swarms in the vegetable mould of a chestnut tree

One kilogram (moist weight) of vegetable mould was taken from a stage-2 cavity (Fig. 8.5) in the trunk of an old chestnut tree (Tessin, southern Switzerland) and placed in a Berlese-Tullgren extractor (Tab. 8.5). More than 12,300 arthropods belonging to 39 taxa emerged from it! Microarthropods were the most abundant, chiefly mites (88.2%). Larvae of dipterans, most often mycophagous, were less numerous but very varied (ten families). Abundance of taxa rather rare in soil extracts, such as the Protura (Insecta, 0.5–2 mm) and Pauropoda (Myriapoda, 2–5 mm) (Fig. 8.15), was also striking. This closed environment was also favourable for young stages of Diplopoda (220 young Iulidae) and small species of other Myriapoda: 130 Pauropoda and 300 Symphyla were counted in it.

Table 8.5 Comparative abundance of arthropods living in cavities of old chestnut trees in Switzerland (W. Matthey, unpublished) and in USA (Park et al., 1950).

Taxonomic groups	Switzerland (Tessin)	USA
	in % of total number of individuals	
Oribatid mites, Acari	88.2	61
Mesostigmatic mites, Acari	1.8	(all Acari)
Collembola	1.9	27.2
Coleoptera (larvae and imagos)	0.1	3.2
Diptera (larvae)	0.3	3.2
Other insects (e.g. Protura, Coccidae)	1.9	3.2
Other arthropods (e.g. Arachnida, Myriapoda)	5.8	2.0
Total number of individuals (per kg vegetable mould)	12,306	?

8.9 CONCLUSION

In soil science, the importance of soil annexes is often underestimated. However, they diversify the environment, providing the pedofauna and mycoflora with numerous specialized habitats and shelter the epigeal edaphic fauna.

Soil annexes: essential diversifying elements of ecosystems...

...that illustrate all the great themes of soil biology.

The biologist and the soil scientist see in them the operation of decomposition chains essential to recycling of dead organic materials and to the functioning of biogeochemical cycles. They find in them prey-predator relationships, demography and competition and, above all, ecological niche, biological diversity and bioindication. By studying the disappearance of the dead body of a mole, a cowpat or a chestnut-tree trunk, we touch upon the functioning even of the ecosystem, in particular its soil subsystem.

CHAPTER 9

JAMMED DECOMPOSITION: FROM SPHAGNUM TO PEAT

Peat soils are kept a little apart from the great variety of the Earth's soil mantle. Sometimes neglected in books on soil science, they owe their relative isolation to the actual characters of the material constituting them—peat. Neither truly soil nor truly litter but a little bit of both, this baffling material cannot be easily approached by usual methods of soil science or plant ecology.

From choice, this chapter is centred on temperate boreal bioclimatic conditions, more particularly on those of acid peat soils, typical of raised bogs with ***sphagnum***; only some data are given for other situations. Furthermore, it does not present raised mires as such, but focuses on peat as a material, its formation and evolution as well as on the soils corresponding to it, the HISTOSOLS. However, to better understand the statements on formation and evolution of peat and HISTOSOLS, a quick survey of the bog ecosystem is necessary; this will be found again in its totality at the end of the chapter, after its progressive reconstruction over the pages.

This chapter within the second part of the book illustrates a case in which decomposition of organic matter is greatly slowed by environmental constraints, contrary to dead wood with moderate evolution studied in Chapter 8 or compost, whose accelerated evolution forms the topic of Chapter 10. It lays out the picture of peat through the organizational levels of the soil system (Chap. 1, § 1.2.2):

- composition of peat (§ 9.1),

Sphagnum (or peat moss): generic name of a type of bryophyte characterized by a leafy stem surmounted by a more or less spherical head, the ***capitulum*** (Fig. 9.1). Sphagnum mosses are the chief agents responsible for formation of peat in an oligotrophic environment (§ 9.2.2). When conditions are suitable for it, an individual sphagnum grows indefinitely, its upper part continuing to push on while the lower part decomposes and is imperceptibly transformed into peat.

Fig. 9.1 *Anatomy of sphagnum* (Sphagnum palustre).

'Never has anyone, priest or poet, described Hell as well as peat-bogs do!' (Linnaeus, quoted by Terasmae, in Brawner and Radforth, 1977).

- formation (§ 9.2) and evolution (§ 9.3) of peat,
- histic horizons (§ 9.4),
- HISTOSOLS (§ 9.5) and their functioning (§ 9.6),
- utilization and conservation of peats (§ 9.7).

What is a mire, a raised bog, a fen?

A ***mire*** or ***peatland*** is a marshy biogeocoenosis in which the soil is composed of peat, comprising specialized animal and plant communities. All mires are bogs or fens, but the converse is not true: some fens formed on hydromorphic non-peaty soils (e.g. floodplain fens on FLUVIOSOLS) do not answer to the definition of mires (Manneville et al., 1999).

Also called raised mire, the ***raised bog*** bulges out progressively by elevation of its surface following accumulation of organic matter in the form of peat produced essentially by sphagnums. Its vegetation belongs to the phytosociological classes of *Oxycocco-Sphagnetea* and *Vaccinio-Piceetea* (Chap. 7, § 7.1.4), and the soil is a HISTOSOL. Figure 9.2 summarizes the principal characteristics of a bog.

The reverse of raised bog, ***fen*** may be acid (e.g. fen with common sedge, *Caricion fuscae* alliance) or neutral to alkaline (e.g. fen with Davall sedge, *Caricion davallianae* alliance), peaty or non-peaty. ***Limnogenic, topogenic*** and ***soligenic*** fens are distinguished (Bridgham et al., 1996). Fens are common at the edges of raised bogs.

Certain mires, intermediate between bogs and fens, are termed ***transitional mires*** (Ge. *Zwischenmoor, Übergangsmoor;* Fr. *marais de transition*); they shelter vegetation hardly sensitive to pH (*Caricion lasiocarpae* and *Rhynchosporion* alliances).

Raised bog (or raised mire; *Ge. Hochmoor;* Fr. *haut-marais, marais bombé*): mire with hydric and mineral nutrition originating essentially from precipitation, without terrestrial additions. See Manneville et al. (1999) for further details about mires and their specialized terminology, particularly correspondences between French, German and English terms.

Fen (Ge. *Niedermoor;* Fr. *bas-marais*): marshy biogeocoenosis fed with water both from precipitation and by terrestrial additions. Some less or more peaty fens are called swamps.

Limnogenic: qualifies a marsh formed on the banks of a lake or water-course.

Topogenic: qualifies a marsh situated in a basin and fed by lateral seepage or by phreatic groundwater.

Soligenic: qualifies a marsh situated on a slope and fed by lateral seepage.

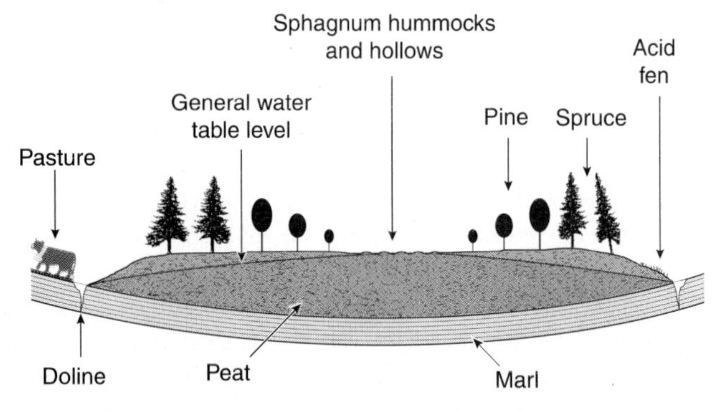

Fig. 9.2 *Some characteristics of a raised bog, in particular of its phreatic groundwater. It should be noted that 'the' phreatic groundwater is actually triple: a water table suspended in the acrotelm (see § 9.6.2), a water table of capillary imbibition in the catotelm and the true groundwater located at the base of the peat mass and in the first mineral layers (see § 9.6.2, where another possibility developed by Bridgham et al., 1996, is presented). In fens the three water tables are generally intermixed.*

9.1 PEAT, AN ALMOST TOTALLY ORGANIC MATERIAL

9.1.1 Peat under the hand lens or microscope

Observation of peat of bogs under the hand lens or microscope reveals varied plant debris in different states of preservation: nearly intact woody twigs and pine needles, more or less unravelled ocreae of Cyperaceae and whole or fragmented stems and leaves of sphagnum (Fig. 9.3).

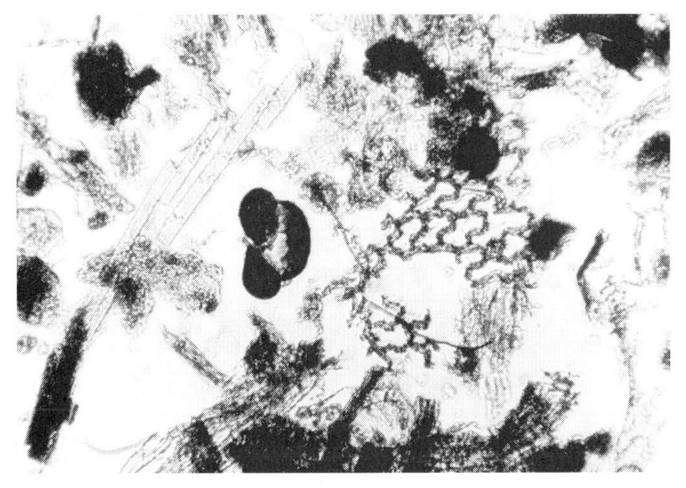

Fig. 9.3 *Raised-bog peat viewed under the microscope (50–200 μm particle-size fraction). According to peat type, such a study can reveal highly fragmented plant tissues with open cells, pollen grains, carcasses of small invertebrates, testate amoebae for example, fungal mycelia, diatoms or masses of humified organic matter (photo J.-M. Gobat).*

Many constituents of peat cannot be brought out by such study. Water, and also gases such as CO_2 and methane, organic molecules and mineral salts dissolved in the soil solution are not visible. Others do not occur in this type of peat, such as molluscs whose shells are typical of fen peats.

Peat: material formed by the accumulation under anoxic hydromorphic conditions of more or less decomposed organic matter. This definition allows qualifying as peat a material that is no longer presently under anoxic hydromorphic conditions, for example following drainage that would have dried it out. The conditions prevailing at its formation are determinant, even if they still exist only in the form of morphological memory in the soil (Chap. 7, § 7.7.2).

9.1.2 Definition of peat

Visual observation and knowledge of the hydric regime of peat soils (§ 9.6) permit a definition of ***peat*** and its differentiation from aerated hologanic horizons of certain soils in which the organic matter accumulates under oxic conditions (O horizons, Chap. 5, § 5.4.3). Starting from the peat material, many criteria permit definition of histic horizons (§ 9.4).

The water content of a histic horizon may attain 95% of the fresh weight.

The solid bulk of peat is formed principally of plant debris in varying degrees of decomposition.

Fig. 9-A *Ancient trunks and stumps found buried in peat (photo W. Matthey).*

9.1.3 Constituents of peat

In the strict sense, peat is a solid material, almost totally organic and of plant origin. But it is actually preserved with its attributes only when it is very highly saturated with water that it strongly retains.

Plant constituents

The thoroughly vegetal character of peat is due to the very great predominance of primary production compared to secondary production. Mason and Standen (in Gore, 1983) reported that the total animal production in an English raised bog with *Calluna* and *Eriophorum* is 100 kJ m^{-2} y^{-1} as against 13,336 kJ m^{-2} y^{-1} for plant production, that is 133 times less! It may be recalled that production can be expressed equally in mass or energy per unit area in unit time (Chap. 2, § 2.2.1). According to pedoclimatic formation conditions, three principal types of plant debris constitute peat or are preserved in it (Tab. 9.1). The largest are ancient trunks and stumps of trees (Fig. 9-A).

Table 9.1 Principal types of plant debris constituting peat (after Grosse-Brauckmann, in Göttlich, 1990)

Type of residue	Plant types, families	Genera and species
Woody (parts of leaves, needles, twigs, stalks, roots, stumps, trunks)	• Coniferous trees (Pinaceae) • Hardwood trees (Betulaceae) • Dwarfshrubs (Ericaceae)	• *Pinus* spp., *Picea abies* • *Alnus glutinosa*, *Betula* spp. • *Calluna vulgaris*, *Erica* spp., *Vaccinium* spp.
Herbaceous (ocreae, leaves, roots)	• Trees and dwarfshrubs • Cyperaceae (*Carex*) • Cyperaceae (other genera) • Poaceae • Various families	• As above • *Carex elata, C. gracilis, C. lasiocarpa, C. limosa, C. rostrata* • *Cladium mariscus, Eriophorum angustifolium, E. vaginatum, Schoenus nigricans, Trichophorum caespitosum* • *Molinia coerulea, Phragmites australis* • *Menyanthes trifoliata, Scheuchzeria palustris, Equisetum fluviatile*, etc.
Herbaceous (stalks, leaves)	• Bryophyta (*Sphagnum*) • Bryophyta (other genera)	• *Sphagnum cuspidatum, S. rubellum, S. angustifolium, S. magellanicum*, etc. • *Aulacomnium palustre, Hypnum* spp., *Polytrichum* spp., *Scorpidium scorpioides*

Before their residues are incorporated into the HISTOSOL, higher plants influence its functioning by their activities of mineral nutrition, evapotranspiration and root secretion (Chap. 4, § 4.1.5). The last is particularly high in acid peats compared to neutroalkaline peats, as proved with ^{14}C study of *Eriophorum angustifolium* (Saarinen et al., in Laiho et al., 1996). This could contribute to sustaining microbial activity in the upper layers of HISTOSOLS.

Animals

Various groups of animals live in peatlands; according to their affinity for this environment, they are grouped into ***tyrphobionts***, ***tyrphophiles***, ***tyrphotolerants*** and ***tyrphoxenes***.

Macroinvertebrates of decomposition are rare in raised bogs because of absence of calcium salts for millipedes, woodlice and snails, and soil anoxia and low nutritive value of the litter for earthworms. As a result, saprophagous nematodes, enchytraeids and certain dipterous larvae (Tipulidae in particular) become the principal actors in detritus chains in peat (Coulson and Whittaker, in Heal and Perkins, 1978).

Springtails, oribatid mites and other dipterous larvae, algophagous and mycophagous, are greatly abundant in the moss stratum (Moore and Bellamy, 1974), as also are predators (gamasid mites, spiders, ground beetles, staphylinid beetles, larvae of dipterans, ants) (Burmeister, in Göttlich, 1990). Sphagnum on lawns accommodates, apart from rich populations of algae and fungi, a surprisingly diverse microfauna: protozoans, rotifers, nematodes, bear-animalcules, crustaceans, acari, springtails, larvae of dipterans (non-biting midges) are assembled here into a complex zoocoenosis (Hingley, 1993).

Although the taxonomic groups are often few, the animal populations are large in accordance with the rule that regularity of a biocoenosis is low in extreme environments (Chap. 7, § 7.5.2; Matthey and Borcard, 1996). In ombrogenic bogs of the Swiss Jura for example, the oribatid mites constitute 71 to 85% of the animal individuals obtained by coring of peat and extraction in the Berlese-Tullgren extractor, that is, 11 to 19 times more than springtails (Borcard, 1988, 1991). This represents between 140,000 and 245,000 individuals m^{-2} for oribatid mites, 13,300 and 32,400 individuals m^{-2} for other acari and 12,300 and 18,300 individuals m^{-2} for springtails.

Tyrphobiont: animal that reproduces exclusively in mires.

Tyrphophile: animal that sharply prefers mires to other environments.

Tyrphotolerant: animal that can reproduce in mires but whose optimum is located elsewhere.

Tyrphoxene: animal that is only transitory in mires.

Acid peat: an extreme environment for animals!

Despite their abundance, the part of animals in the biomass is negligible compared to that of plants.

According to environment, the animal biomass comprises between 0.95 and 2.20 g m^{-2} (Borcard, 1988). This confirms the data of Mason and Standen (in Gore, 1983), who compared them to 1564 g m^{-2} of the aboveground phytomass of a heath with *Eriophorum* and *Calluna*!

In bogs, soil animals are the movers of a slightly peculiar bioturbation when they transport far and rather deep humified substances and other compounds held on their surface. Their cuticle, rich in proteins and chitin, actually represents one of the few possible sites for precipitation of certain organic acids in peat environments. In turn, the products thus adsorbed play a part in controlling the cuticular respiration of animals (Mathur and Farnham, in Aiken et al., 1985).

Palaeoecological: pertaining to study of past ecological conditions. Palaeoecology seeks to reconstruct ancient ecosystems and their functioning from testimonies from various witnesses that have subsisted (pollen grains, testate amoebae, grains, growth rings in fossil wood, etc.).

Frequent in bogs, thecamoebians (Chap. 12, § 12.4.1) are very efficient ***palaeoecological*** indicators, their siliceous shells being quite resistant to acids. Each layer of accumulated peat hides a certain panel of species (Fig. 9.4; Plate I-4), reflecting the conditions prevailing during its formation. The information given expresses long-term climatic modifications (Blackford, 2000), and also recent changes due, for example, to exploitation (§ 9.7; Warner, 1990; Buttler et al., 1996; Mitchell et al., 2000a, b).

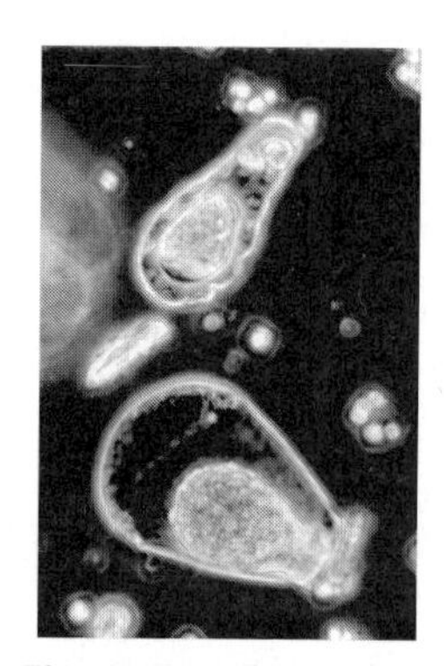

Fig. 9.4 *Two thecamoebians of acid mires: below,* Hyalosphenia papilio; *above,* Hyalosphenia elegans *(photo E. Mitchell).*

Microorganisms

Even without attaining the density that they have in other soils, microorganisms are present in large quantities in peats, to a depth of many metres in thick HISTOSOLS. The dominant group is fungi (Favre, 1948), especially in acid peats, with numerous species of Ascomycetes (*Hymenoscyphus, Oïdiodendron;* Smith and Read, 1997) forming mycorrhizae with the roots of ericaceous plants (*Calluna, Empetrum, Vaccinium*) (Chap. 16, § 16.1.3; Burgeff, 1961). Others are the principal agents of organic matter decomposition in these environments: *Collybia* spp., *Lyophyllum palustre, Galerina paludosa* (Lütt, 1992). The length of mycelial filaments is 15 to 180 m cm^{-3} in peats against 160 to 580 m cm^{-3} in a meadow on limestone (Moore and Bellamy, 1974).

Because of the weak general biological activity of peat, the biomass of microorganisms has long been underestimated (Küster, in Göttlich, 1990). However, nearly two-thirds of the CO_2 released by peat originates from their respiration.

Bacteria are most abundant in minerotrophic and eutrophic peats with neutroalkaline pH. In acid peat, a number of still poorly understood—because of culturing difficulties (Chap. 15, § 15.5.4)—genera and species assert their presence: *Arthrobacter* spp., *Clostridium* spp., *Cytophaga* spp., *Micrococcus* spp., etc. Despite this

'Pyromaniac' fungi?

By supposition, certain fires in dried-out bogs have been attributed to auto-ignition of the peat, which would have been taken to 70°C by the combination of hyperactive thermophilic fungi (maximum possible temperature = 61.5°C) and consecutive exothermic chemical reactions! This succession of biological processes followed by chemical reactions is analogous to that which creates will-o'-the-wisps (Chap. 15, § 15.4).

'In general, however, there is still relatively little known about the activities of microorganisms and particularly about microbial interactions in peat' (Clymo, in Gore, 1983)... and this has scarcely changed since!

taxonomic limitation, the functions of bacteria are varied because of numerous aerated and anoxic microsites that characterize peat (William and Crawford, 1983). Anaerobic bacteria, still present in the surface layer, increase in number with depth.

Actinomycetes, Cyanobacteria and algae are generally low in quantity and are scarcely active (Chap. 2, § 2.5), with the exception of diatoms, which may be abundant in submerged living sphagnum (Wüthrich and Matthey, 1978).

Organic compounds

Overall, the organic acids of peats are similar to those of mineral soils (Mathur and Farnham, in Aiken et al., 1985). Mainly their proportions differ, reflecting at the same time the botanical composition of the debris and the degree of transformation of the material. In peats still containing living tissues, this renders tricky the separation for analytical purposes of products actually originating from humification and the usual constituents of the cytoplasmic liquid. Although this difficulty exists in all soils, it is greater in organic materials in which scarcity of the mineral fraction limits stabilization of humic compounds in the clay-humus complex and therefore their separation by physical means (Chap. 2, § 2.1.3).

Besides compounds inherited from plant material and constituents of cell walls—cellulose, lignin as well as polymers of uronic acids which themselves constitute 30% of the dry mass of the walls of certain sphagnum mosses (Clymo, in Parkyn et al., 1997)—peat contains the bulk of organic constituents of soil.

Peat contains numerous phenolic compounds from degradation of lignin, which modify the growth of higher plants. In experiments *in vitro* their effect on plant growth was determined to be positive or negative according to environmental conditions and the species (Flaig and Söchtig, 1972). But their toxicity *in situ* with respect to microorganisms tends to curtail their beneficial effect on higher plants as the general functioning of the rhizosphere gets disturbed. Sphagnums produce other compounds toxic to bacteria such as sphagnic acid (Fig. 9.5), which may attain a concentration of 300 mg kg^{-1} of dry weight of sphagnum (Naucke, in Göttlich, 1990).

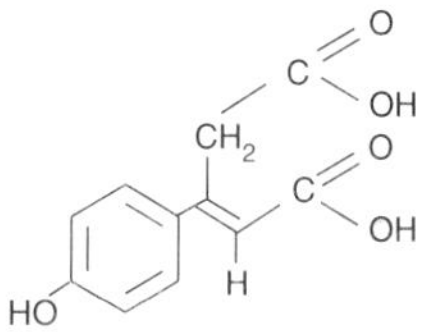

Fig. 9.5 *Molecular structure of sphagnic acid (after Naucke, in Göttlich, 1990).*

And the rhizosphere of bog plants?

Opinion is divided on the existence of an actual rhizosphere around the roots of certain raised-bog plants. Besides ericaceous plants that are strongly mycorrhized or associated with decomposing fungi (Wallén, 1986), studies often show no fungus associated with the roots of Cyperaceae. One possible cause is the very high rate of renewal of rootlets (Chap. 4, § 4.1.6), which does not leave bacteria and fungi sufficient time to colonize them in an environment with strong ecological stress (Dickinson, in Gore, 1983).

Von Post humification scale: scale from 1 to 10 to indicate the degree of decomposition of peats. It is established by pressing in the hand a sample of peat and noting the colour of the solution thus extracted and the appearance of the material left behind on the palm. Low values indicate a poorly decomposed peat.

A large quantity of products resulting from humification colours the water extract of peat dark brown or black, whereas the extract is colourless or yellowish with less humified matter. Jointly with fibre content this property is used in describing peats and qualifying histic horizons, in the form of the ***von Post humification scale***.

Sphagnum mosses and the pharmacopoeia

The presence of antibiotic products in sphagnum mosses has long been known since these mosses were used as aseptic dressing, particularly during the war of 1914–1918 to compensate for the shortage of medical supplies.

Mineral constituents

The mineral content of peats is low because, according to definition, this material must not contain more than 50% minerals on dry weight basis (AFES, 1995, 1998).

Fen peats are richer in minerals (generally 5 to 20% of dry weight) than acid peats (1 to 10%). Silica SiO_2 predominates (10 to 45% of the total minerals) followed by oxides of calcium (2 to 45%), magnesium (1 to 20%), aluminium (1 to 11%) and iron (1.5 to 5.5%). The low values of these estimates are those of raised bogs (Naucke, in Göttlich, 1990). The very low quantities of calcium in acid peats explain the absence of Diplopoda—their cuticle is impregnated—at the centre of raised bogs, and also that of shell molluscs.

The cation exchange capacity (CEC) of peat is high (between 150 and 250 cmol (+) kg^{-1}, greater than that of clay minerals) (Chap. 3, § 3.8.1). It depends on the botanical origin—sphagnums have a much higher CEC than *Eriophorum vaginatum* for example, because of their high content of uronic acids—and on the degree of humification, young peats having very low values. Basic cation saturation percentage is very low in peat of raised bogs, lower than 25%, whereas it may reach 100% in calcic peats.

Botanical composition (of peat): relative proportion of different taxonomic categories found in the plant debris constituting a peat. The categories may be simple (e.g. mosses, herbaceous debris, woody debris) or more detailed, going to a species of sphagnum, sedge or ericaceous plant.

9.1.4 Types of peat

Different criteria have been proposed for classification of peats—botanical, physical, biochemical, trophic, chemical (acidity) and hydrological (Succow, 1988; AFES, 1995, 1998; Bridgham et al., 1996; Grosvernier et al., 1999):

• As the first criterion, Lévesque et al. (1980), Clymo (in Gore, 1983) and Grosse-Brauckmann (in Göttlich, 1990) suggest, not without logic, consideration of ***botanical composition***. As peat is a material formed almost totally from plant residues that confer its major features on it, it is relevant to retain this aspect as primary. Peats are then differentiated according to the residues they contain: sphagnum peats, sedge peats, reed peats, woody peats, etc. (Tab. 9.1). Proportions are determined by mass or by number counted on a grating using a binocular eyepiece.

• Physical nature. This criterion groups peats into three categories according to their ***fibre content***: ***fibric peats*** (> 40% of ***rubbed fibres*** on dry weight basis), ***mesic peats*** (10 to 40%) and ***sapric peats*** (< 10%). It forms the basis for separating histic horizons (§ 9.4).

• Quantity of total organic matter. Evaluated by the ***loss on ignition***, it separates peats from anmoors (Chap. 6, § 6.2.1), richer in minerals, and divides them into three categories: *Halbtorfe, Volltorfe* and *Reintorfe* (Succow, 1988).

• Biochemical nature of the organic matter. This is evaluated by a semi-quantitative estimation of the humified organic matter extracted by an alkaline reagent (Chap. 2, § 2.2.4). The ***pyrophosphate index*** thus obtained enables grouping of peats along a gradient of humification (Plate VI-1; Kaila, 1956; Lévesque and Dinel, 1982; Clymo, in Gore, 1983).

• Trophic level of the bioelements controlling plant growth (NO_3^-, NH_4^+, PO_4^{3-}, K^+). It separates ***eutrophic*** peats rich in them from ***oligotrophic*** peats.

• Concentration of mineral elements other than N, P and K, in particular calcium. This separates ***ombrotrophic*** peats, acid with low basic cation saturation percentage, from ***minerotrophic*** peats, neutroalkaline with high percentage.

• Hydrological regime. This criterion specifies the origin of the water responsible for peat formation. It is mostly applied to the mire in its entirety (see introduction to this chapter).

Fibre content: proportion in per cent of dry weight of fibres relative to the entire sample. ***Fibres:*** in a peat or similar material (litter, compost, etc.), ensemble of debris, generally from plants, retained on a 200-µm sieve (Bascomb et al., 1997). Sieving is done under a current of water on the fresh sample (Chap. 2, § 2.1.3).

Rubbed fibres: fibres that have, prior to sieving, undergone a mechanical treatment designed to break up their masses (agitation by turning over or light crumbling between the fingers) (Lévesque and Dinel, 1997).

Loss on ignition: 100 – ash content (%). ***Ash content:*** ratio in % of the weight of material that remains after calcination at 650°C and that of the dry material before calcination. In peaty materials, loss on ignition is a handy estimate of total organic matter.

Pyrophosphate index (or '***pyro index***'): absorbance at 550 nm, multiplied by 100, of a solution of organic matter extracted by a 0.025 *M* solution of sodium pyrophosphate $Na_4P_2O_7$. The pyro index does not reflect the whole of the humified matter because it is significantly correlated only with fulvic acids and not with humic (Schnitzer, 1967). It is therefore always necessary to interpret it with other descriptors.

Ombrotrophic: qualifies a peat essentially fed by precipitation.

There is trophy... and then trophy!

Minerotrophic: qualifies a peat nourished by precipitation *and* by lateral additions (surface or deep seepage, capillary rise).

Nitrogen nutrition... and the others...

As Bridgham et al. (1996) and Grosvernier (1996) remarked, in mires it is erroneous to assimilate the trophic levels in nitrogen, phosphorus and potassium to those of other bioelements such as calcium or magnesium. They are too often confused in the literature, probably for historical reasons linked to conditions in Scandinavia where the first descriptions were made and where these two types of trophy are correlated. In Central Europe for example, there are fens with *Carex davalliana*, which colonize peats concomitantly minerotrophic, rich in calcium, and oligotrophic, hence poor in nitrogen. This linguistic pitfall had already been discussed and avoided ever since pedological horizons were defined in the *Référentiel pédologique* (Chap. 5, § 5.4.3; AFES, 1995, 1998).

These criteria, characterizing present-day peat, are often used for evaluating its past evolution (Bascomb et al., 1977; Lévesque et al., 1980; Gobat and Portal, 1985). Such a course is admissible to the extent that comparison limits are clear, in particular with respect to homogeneity of botanical composition, the essential parameter.

9.2 FORMATION OF PEAT

9.2.1 Two conditions and three stages

Excess of water and excess of carbon

Two conditions are necessary and sufficient for peat to accumulate and, consequently, for a HISTOSOL to form (Fig. 9.6; Mitsch and Gosselink, 1993):

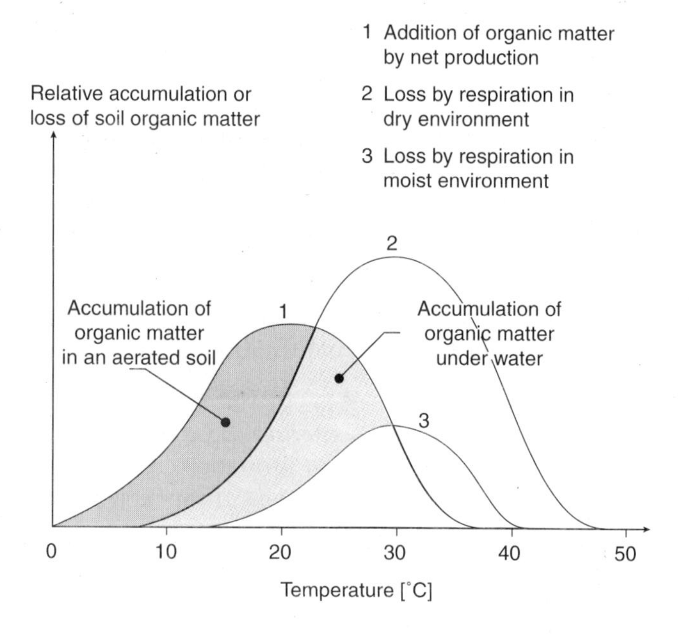

Fig. 9.6 *Budget of accumulation and loss of soil organic matter as a function of mean temperature and hydromorphy (after Mohr and van Baren, 1959).*

Peats, mires and the global carbon cycle: some data

Organic carbon stored in mires represents one-third of the total organic carbon content of soils of the world, estimated at $1395 \cdot 10^9$ t (Post et al., 1982); it is also three-fourths of the atmospheric carbon, evaluated at $600 \cdot 10^9$ t (Clymo, in Laiho et al., 1996). With a global area of $346 \cdot 10^6$ ha or 2.3% of the land area, boreal and Arctic mires, most widespread on the Earth, lock up nearly $455 \cdot 10^9$ t carbon (Gorham, 1991). 98.5% is contained in peat against just 1.5% in living vegetation of mires; about one-half of the total is in sphagnums or their residues (Clymo, in Parkyn et al., 1997). This enormous quantity of C fixed in peats makes them an essential regulator of the global cycle; it is even thought that their evolution plays a role in the initiation of glacial periods (Franzén et al., 1996; Klinger et al., 1996).

Each year, about $0.096 \cdot 10^9$ t carbon is trapped in peat, or less than 10% of the net primary production of peatlands. The remainder is immediately redistributed, in the short or medium term, to the atmosphere as CO_2 or CH_4. The latter, a more effective greenhouse gas than the former, is produced at the rate of $19.5 \cdot 10^6$ t y^{-1} by northern mires (Crill et al., in Vasander and Starr, 1992); this represents nearly 4% of the total methane released into the atmosphere, estimated at $535 \cdot 10^6$ t y^{-1} (Smith and Bogner, 1997; Tuittila, 2000; Reynolds and Fenner, 2001).

The carbon of peat can be rapidly transferred to the atmosphere by drainage of the peat or due to any other cause that speeds up biological activity, such as rise in temperature. A significant correlation has been found between temperature of peat soils and release of methane (Crill et al., in Vasander and Starr, 1992). But, in the opposite direction, warming of climate also augments production of ***turfigenic*** plants that fix more carbon in their tissues and thus abstract it from the atmosphere.

So, in a warming situation, the effect of HISTOSOLS on the global carbon cycle depends on the balance between increased mineralization and an even stronger carbon fixation. Today it is impossible to tell to which side the equilibrium will be shifted.

Francez (2000) gives a general review of this set of problems.

The carbon of mires: more than three times the carbon of tropical forests (Malthy, in Parkyn et al., 1997)! Because of antagonism between production and decomposition, peat environments play an important role in regulating the carbon cycle at the global level (Chap. 5, § 5.2).

Compared to that of other soils, the carbon of peat, chiefly that of the surface layer, is rather labile, quickly released by oxidation or reduction.

Turfigenic: prone to form peat.

'Given the diversity of possible responses by boreal and subarctic peatlands to climatic warming, it is impossible at present to predict their future contributions to the global carbon cycle' (Gorham, 1991). Confirming this, Lal et al. (1998) and Arp et al. (in Cadisch and Giller, 1997) point out the present-day paucity of research on carbon and nitrogen in HISTOSOLS.

'Thus, instead of viewing peat as an immediate result of a specific fermentation, we must rather consider it formed because of an obstacle to this fermentation and this essential obstacle is presence of water' (Lesquereux, 1844).

- A positive water balance in which the sum of precipitation and terrestrial additions is greater than that of evapotranspiration and export losses (lateral seepage, drainage, etc.). Seasonal distribution of precipitation is important, as is its type: snow, rain, dew, etc.
- A net primary production greater than organic matter decomposition. In other words, the carbon fixed by photosynthesis in the biomass must be greater than that released by decomposition of organic matter as CO_2 in oxic conditions or $CO_2 + CH_4$ in an anoxic environment.

In raised bogs, three stages succeed one another (Fig. 9.7):

Three stages, from production of sphagnums to humified peat.

- from photosynthesis to production of sphagnums (§ 9.2.2);

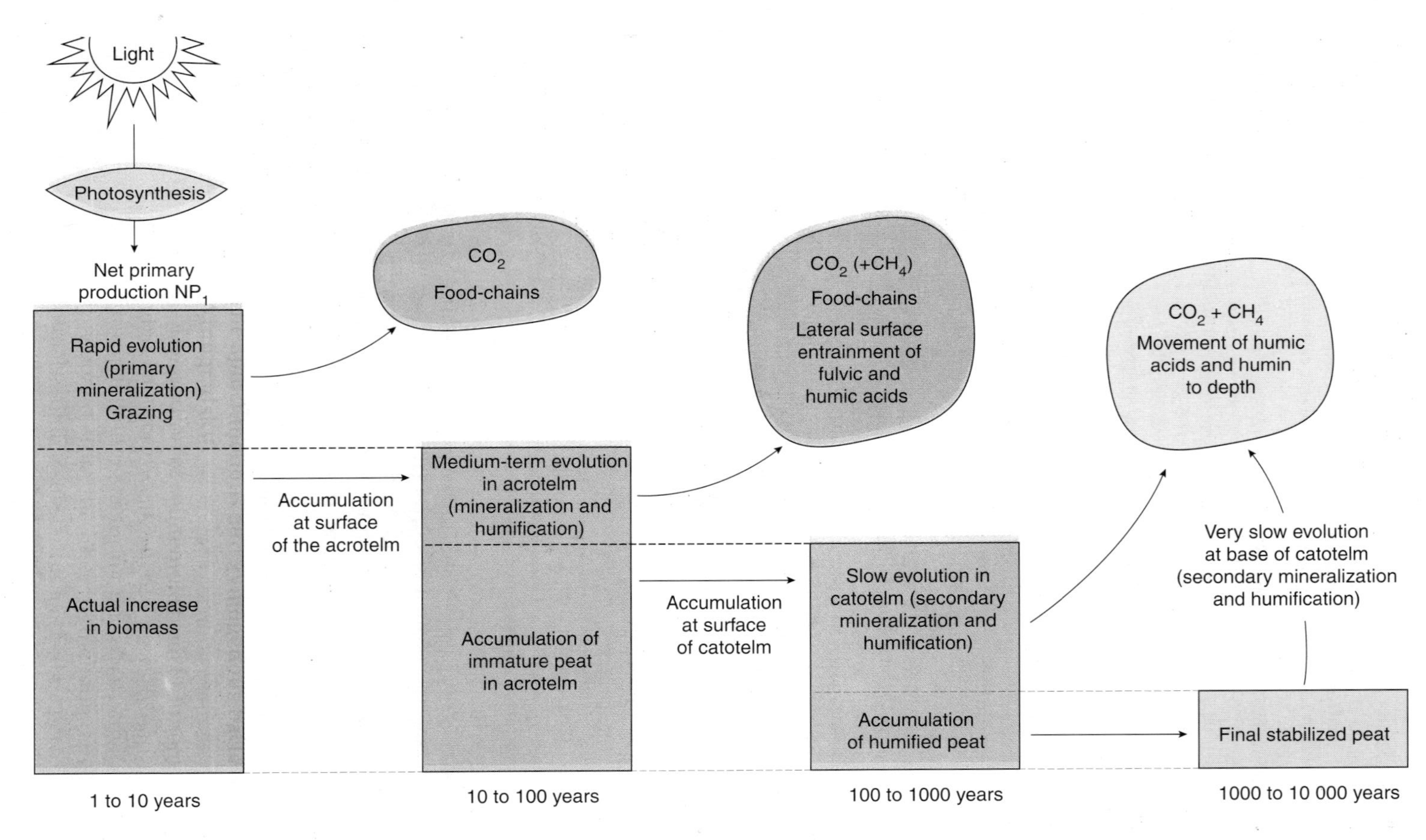

Fig. 9.7 *Stages in accumulation of peat.*

- from sphagnum production to accumulation of fresh organic matter (§ 9.2.3);
- from fresh organic matter to humified peat (§ 9.2.4).

9.2.2 Net production of sphagnum

Sphagnum, motor and fuel of acid peats

It was thought for a very long time that peat was a purely mineral substance, 'a soil mixed with resin, oil, sulphur and other substances that render it combustible'. Many authors of the eighteenth century suggested that peat was a particular kind of underground vegetation, a web of roots that continued to grow, to raise themselves without surface plants exerting any influence on this growth. Varied causes explained its origin: survival of an organic world drowned in the Flood and returned to the surface; fertility at the bottom of the seas which during great floods would have been thrown out on their shores and even into lands with plants whose decomposition would have provided the fuel. By studying the peatlands of the Swiss Jura, Lesquereux (1844) adduced the first totally convincing proof of the plant origin of peat. He distinguished peats with sedges and reeds, formed in ***infra-aquatic*** environment from those constituted in ***supra-aquatic*** environment by sphagnums, ericaceous plants and cotton grass.

'If the vertical section of a peat layer after exploitation is studied from top to bottom, it is seen that living plants that still preserve their shapes gradually lose them by imperceptible degrees and finally reach the peat stage' (Lesquereux, 1844).

Infra-aquatic: situated or formed below the mean water level of a pond or lake shore, for example.

Supra-aquatic: located or formed above the mean water level.

In boreal and temperate regions that accommodate acid peats taken as example here, accumulation of peat is ensured chiefly by growth of specialized mosses, the sphagnums (Plates III-4 and VI-2 to VI-5).

Net production and growth of different sphagnum species were studied in detail by Clymo (1970), Clymo and Reddaway (1971), Hayward and Clymo (1983), Francez (1992) and, more recently Grosvernier (1996) and Grosvernier et al. (1997). In a greenhouse experiment in which other potential factors such as light, precipitation and trophic level were kept constant, the last-mentioned authors showed that:

All other conditions being the same, three priority factors determine the net production of sphagnums: species, groundwater level and physicochemical nature of the peat.

- intensity of growth, measured by increase in length or in biomass, depends primarily on the species, which explains 49% of the ***variance***; globally, *Sphagnum fallax* is the most productive, followed by *S. magellanicum* and *S. fuscum*;
- groundwater level is the second important factor, with 11% of the variance; a high water table promotes

Variance: measure of dispersion of a variable around its mean (Legendre and Legendre, 1984).

longitudinal growth and increase in mass in *S. fallax*, like *S. magellanicum*, whereas specific differences are clearer with low water table;

- physicochemical nature of the peat support intervenes only in the third place, explaining 6% of the variance; its effect is more prominent with a low water table at 40 cm than with a surface water table at 1-cm depth.

Compared to that of many ecosystems the net production of ombrogenic peatlands is small. On the other hand, that of soligenic mires places them among the most productive ecosystems of temperate regions (Moore and Bellamy, 1974).

Lütt (1992) and Mitsch and Gosselink (1993) provided many quantitative data on net production of sphagnums and ombrogenic peatlands. The orders of magnitude are very variable chiefly between species, but also because of ecological conditions and non-uniform methods of determination! The values are thus distributed on average between 100 and 300 g m^{-2} y^{-1} with minimum of 10 g m^{-2} y^{-1} and maximum reaching 1000 g m^{-2} y^{-1}. Production of soligenic mires is much greater than that of ombrogenic mires, reaching 4600 g m^{-2} y^{-1} in marshes dominated by *Scirpus lacustris*.

Sphagnums not only are very efficient manufacturers of peat, but also play an important role in biogeochemical cycles, in particular that of nitrogen.

Paradoxically, sphagnums love nitrate!
Contrary to all expectations, sphagnums are the principal consumers of nitrogen brought by rain. They react to rise in nitrate concentration by a rapid induction of their nitrate-reductase, the enzyme responsible for assimilative reduction of nitrogen (Chap. 4, § 4.4.4). This enzymatic action is the strongest in the capitulum, where nitrate 'lands'; it decreases progressively along the stem (Woodin and Lee, 1987). The nitrogen consumed is not always indicated by an increase in biomass or in length. It sometimes accumulates in the tissues (luxury uptake) or is eliminated in the environment as amino acids, in what is probably a detoxifying action. Excess nitrogen may also be leached down before being immobilized by microorganisms (assimilation). With time, excess nitrogen in the acrotelm disrupts the equilibrium of the plant communities, favouring vascular plants to the detriment of bryophytes (see also Mitchell et al., 2002).

'Sphagnum is not only the most successful of the bryophytes, it is a notably successful plant by any standards.' (Clymo, in Parkyn et al., 1997).

Specific strategy of sphagnums

Sphagnums may be considered simultaneously C- and S-type strategists, competitive in environments with great stress (Grime et al., 1988). By their increase in biomass they gradually replace the pioneer R species, such as *Eriophorum vaginatum* and *Polytrichum strictum*, which initiate recolonization of a new peat (Buttler et al., in Wheeler et al., 1995; Matthey, 1996).

Living sphagnums: the 'rich' compartment of the HISTOSOL, with higher concentration of bioelements than found in other layers (Damman, 1978; Gobat, 1984).

Sphagnums create their own chemical environment by virtue of their ability to concentrate basic cations from very dilute soil solution or rain-water (high cation exchange capacity) and to abstract them from other plants. Sphagnums accumulate basic cations in their mineralomass by continually releasing protons (Chap. 4, § 4.2.1); they thus contribute to acidification of the proximate soil they modify for their own benefit. Klinger (1996) proved their role in podzolization of soils in Alaska. Very low pH values, going to 3, have been measured on hummocks after a long rain-free period (Clymo, in Parkyn et al., 1997).

But sphagnums also create their own physical environment by retaining water strongly in their specialized cells and in micropores resulting from progressive compaction of the peat during its formation. This leads to formation of a perched water table as well as to very strong water retention; this water becomes unavailable to other species (Chap. 3, § 3.4.4).

9.2.3 From sphagnum production to accumulation of fresh organic matter

Sphagnum production should not be confused with accumulation of fresh organic matter, still less with that of peat (Fig. 9.7). In a few years considerable quantities of matter may be degraded by microorganisms and by grazing (Chap. 13, § 13.3.1), the latter however being less active in ombrogenic peatlands. The matter thus attacked, mineralized or transmitted through the food-chains, ranges between 50 and 200 g m^{-2} y^{-1}, but limited to a maximum of one-half the net production (Lütt, 1992).

The balance of accumulation of sphagnum depends on microtopography, itself determined by, among other phenomena, feedback of production of plants and kind of species.

Following the differentiated growth of sphagnum populations, three topographic microstructures are formed in an ombrogenic bog (Fig. 9.8):

- ***hummock*** (Ge. *Bult;* Fr. *butte*), on which grow plants that can better endure the physiological drought in peat (see *The desert of peat bogs* in Chap. 3, § 3.4), such as *Andromeda polifolia, Calluna vulgaris, Vaccinium* spp.;

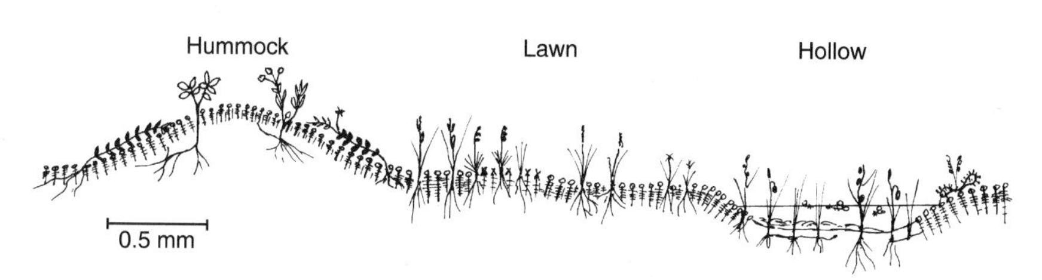

Fig. 9.8 *Complex of hummocks, lawns and hollows in an ombrogenic bog (after Gobat et al., 1986).*

• ***lawn*** (Ge. *Rase;* Fr. *replat*), often dominated by Cyperaceae such as *Eriophorum angustifolium, E. vaginatum* and *Trichophorum caespitosum;*

• ***hollow*** (Ge. *Schlenke;* Fr. *gouille*), almost permanently flooded, in which *Carex limosa, C. rostrata, Rynchospora* spp., *Scheuchzeria palustris* and *Sphagnum cuspidatum* grow.

Linked to each other, hollows constitute the major route for elimination of bioelements evacuated by superficial seepage.

Hollows sometimes form a veritable hydrographic network on the surface, collecting the seepage water of the entire acrotelm (§ 9.6.2). With slope, however gentle, losses of bioelements can go up to 90% of the potassium and sodium, and 70% of the calcium and magnesium added by precipitation, the only 'wet-nurse' of raised bogs (Streefkerk and Casparie, 1989).

9.2.4 From accumulation of fresh organic matter to humified peat

The annual or multi-annual accumulation of fresh plant material on the surface of HISTOSOLS still does not represent the long-term accumulation of peat.

'The most important components in modelling peat accumulations are the decay parameters for the component fractions and not the productive parameters' (Wallén, in Vasander and Starr, 1992).

Between its immediate deposition and its stabilization as peat, organic matter undergoes physical changes, mineralization and humification under the influence of climate, microorganisms and, to a lesser degree, of the pedofauna (Fig. 9.7). The balance of these modifications is estimated by the rate of net accumulation of peat; this depends on sphagnum production but more so on the speed of decomposition. Two general methods are followed for estimating the accumulation (Clymo, in Laiho, et al., 1996):

• measurement in controlled cultures or *in situ* of the gas flows between peat and the atmosphere (chiefly CO_2 and CH_4); these fluxes reveal the losses of organic matter;

• modelling the processes of accumulation on the basis of historical observations and the nature of peats of different ages, for example by utilizing ^{14}C dating.

Accumulation of peat depends on the functioning of two hydric subsystems of the peat-bog.

To evaluate the actual accumulation of peat it is also necessary to take into account the functioning of two hydric subsystems of HISTOSOLS of the raised bog (§ 9.6.2). Although peat is elaborated in the top layer, the acrotelm, it is stabilized below, in the catotelm. And it is at the interface of the acrotelm with the catotelm that the actual net accumulation occurs. Transformation continues in the catotelm, but here it is about 100 to 1000 times slower (Clymo, in Parkyn et al., 1997). Clymo (in Gore, 1983) estimated that 16 to 28 years are required for the organic matter to reach the catotelm (or, more rightly to be joined up § 9.6.2) and to evolve in a more constant and slower manner.

'The acrotelm... is not itself a peat accumulator but acts as a selective preprocessor of the plant material before passing it on as peat to the catotelm: the true site of peat accumulation' (Clymo, in Laiho et al., 1996).

Peat and O horizons

A rough functional equivalence may be drawn with aerated holorganic horizons: the OL horizon (litter) corresponds to living and recently dead sphagnums, the OF horizon (fragmentation layer) to the acrotelm/catotelm interface and, lastly, the OH horizon of humification and accumulation of stabilized organic matter to the catotelm.

Final accumulation of peat is low compared to the original production. Wallén (in Vasander and Starr, 1992) quoted values of 30 g m^{-2} y^{-1} finally transferred to the catotelm from 800 g m^{-2} y^{-1} produced. This represents only 4% of the net production being transformed into peat. Heal et al. (in Rosswell and Heal, 1975) gave precise figures: of 635 g m^{-2} y^{-1} total net production of a bog in great Britain, 540 were decomposed by the microflora in different stages, 6 were grazed by herbivores and 64, that is, 10% accumulated as peat in the catotelm. The fate of the remainder was not known; it might have been evacuated in soluble form through the outlet of the bog. It is interesting to note here that 1% of the energy circulates in grazing-predation chains against at least 85% in decomposition chains (Chap. 13, § 13.5.1), most of it in microorganisms, and this in spite of the strong restrictions the peat imposes on their activity.

Accumulation of peat represents just 4 to 10% on average of the net primary production.

Accumulation of peat is not infinite. The average thickness of peat for all boreal and Arctic peatlands has been evaluated at 2.3 metres (Gorham, 1991). This is the result of an average net accumulation of 0.53 mm y^{-1} (Gorham, 1991), the values ranging from 0.2 to 1.6 mm y^{-1} according to conditions (AFES, 1995, 1998). The small impermeable subcircular depressions in valleys of the Haut-Jura generate peats thicker than those of large, gently undulating shelves (blanket bogs) formed in Ireland, for example.

Peat accumulation is limited by climatic factors and by the substratum underneath the HISTOSOL, as a function of its permeability and its horizontal extent.

In the bogs of temperate regions, the average final augmentation of peat in the catotelm is estimated at between 0.5 and 1 mm y^{-1} for an organic-matter accumulation of 1 cm y^{-1} in the acrotelm. In specific situations, such as ancient mined pits, one to two metres of peat may occasionally form in a few decades (Matthey, 1996). This is then a very poorly decomposed fibrous peat with loose structure that may still be considered a recent accumulation of biomass. This observation is compatible with the average speed that takes into account decomposition and compaction.

Organic matter: a sequestration of energy in chemical form (Chap. 2, § 2.2.1; § 9.7).

In all ecosystems the organic matter is a stage in the energy flow, during which the latter is sequestered for a longer or shorter period. Peat is no exception, it is even a perfect example of this... for ages understood by those who use peat as fuel and, by doing so, simply release the trapped chemical energy and transform it into heat energy (Chap. 5, § 5.1.1).

9.3 EVOLUTION OF PEAT: PROCESSES, FACTORS, SPEED

9.3.1 Highlights

When conditions of the water and carbon budgets permitting formation of peat are fulfilled (§ 9.2.1), the particulate plant litter accumulates and is slowly transformed in such a way that peat is formed.

The pathways of transformation of peat are similar to those of other organic materials: primary and secondary mineralization, humification (Chap. 5, § 5.2). During maturation, the products such as cellulose and hemicelluloses inherited from the plant material evolve in the reverse direction to humic acids, bitumens and lignin (Fig. 9.9). These transformations are under the principal control of water content and aeration of soil, as well as temperature, and their speed obeys the same laws linking soil development to time (Chap. 5, § 5.5.5; Lütt, 1992).

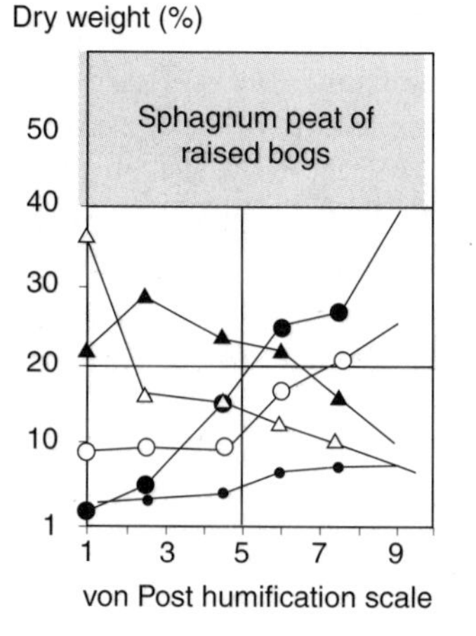

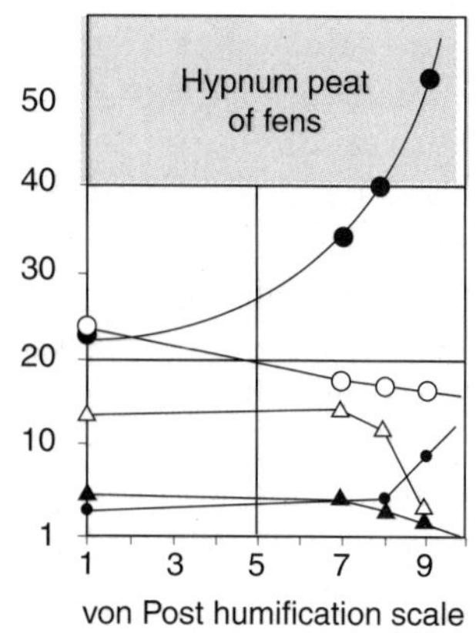

• Bitumens
● Humic acids
△ Hemicelluloses
▲ Cellulose
○ Lignins with other residues

Fig. 9.9 Evolution of constituents of peat during its humification (after Naucke, in Göttlich, 1990).

Production of humic acids is only one of the causes of acidity in ombrotrophic peats (Mitsch and Gosselink, 1993). Four other explanations exist: release of protons by sphagnums (principal cause, § 9.2.2), cation nutrition of plants (Chap. 4, § 4.2.1), oxidation of sulphides to sulphates (Chap. 4, § 4.4.4) and acid rain.

Many biochemical indicators, from the roughest to the most sensitive, enable evaluation of the degree of humification:

- organic carbon content, which distinguishes peats from non-peaty materials;
- the extractability of organic matter which provides a general estimate of the intensity of humification (Chap. 6, § 6.4.2);
- the pyrophosphate index, which reflects the concentration of certain humified compounds;
- the fulvic acid/humic acid ratio (***FA/HA ratio***), which reveals the first stages of humification by polycondensation H2 (Chap. 5, § 5.2.3);
- the H/C and O/C atomic ratios, which indicate the mean 'packing' of molecules; they diminish with polycondensation of humins and increase with mineralization, whether anaerobic to CH_4 or aerobic to CO_2 (Fig. 9.10);
- certain ratios between functional chemical groups of humic acids brought out by infrared spectrometry; they show, for example, the progressive replacement during humification of alcoholic (–OH) groups by carboxylic (–COOH) groups, diminution of aliphatic structures and even the progress of keto (>C=O) groups.

By way of example, determinations done on five ombrotrophic peats of the Swiss Jura, each representing a stage in dynamics of the vegetation, are synthesized in Table 9.2. The ensemble of criteria, measured on the whole peat, on particle-size fractions or just on the humic acids reflects the evolution of peaty material from the young peat of *Sphagno-Caricetum rostratae* association to denuded peat of heath with *Calluna vulgaris*.

9.3.2 Importance of botanical composition and other factors

'The diversity of botanical origin causes these properties (humification, degradation, etc.) to vary widely' (Mathur and Farnham, in Aiken et al., 1985).

The high correlation observed between plant species and peat type confirms the dominant role of the original botanical composition in formation of characteristics of

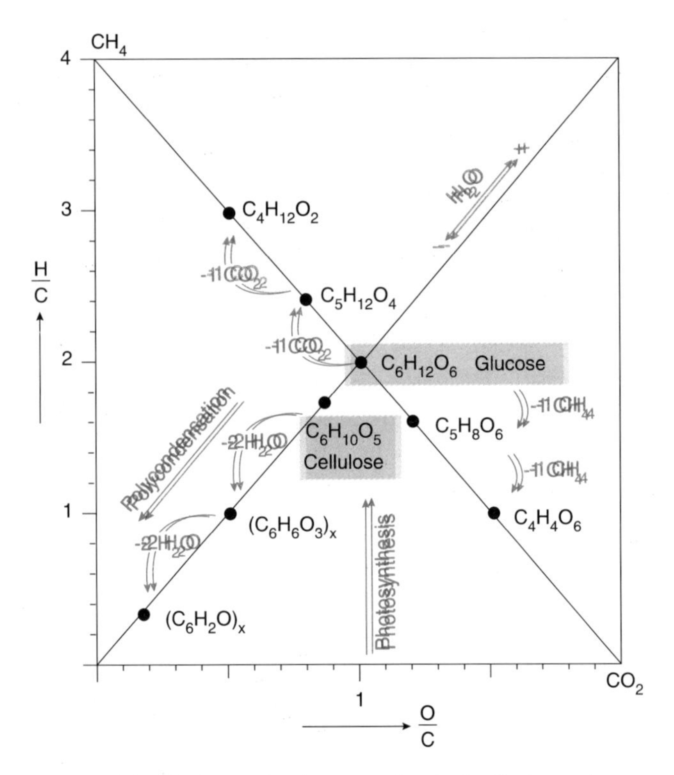

Fig. 9.10 *Evolution of humification according to the H/C and O/C ratios of different compounds in organic matter. According to the conditions, losses of CO_2, CH_4 and H_2O modify these ratios, in number of atoms of carbon, hydrogen and oxygen constituting organic molecules (after Naucke, in Göttlich, 1990).*

peats and in their evolution, particularly their resistance to degradation. Clymo (1965) and Lütt (1992) observed differences in speed of degradation between sphagnum species, some being more resistant than others to mineralization and humification, for morphological and biochemical reasons; thus, *Sphagnum magellanicum, S. cuspidatum* and *S. rubellum* are more resistant than *Sphagnum riparium, S. fimbriatum* and *S. papillosum*. Curiously, nitrogen content is scarcely differential, while this is of major importance in aerated litter (Chap. 2, § 2.2.1).

The importance of the qualities linked to species in degradation processes has generally been shown in sphagnum cultures. Conditions prevailing *in situ* have scarcely been studied and perhaps they hold surprises (Chap. 17, § 17.1.3). Lütt (1992) pointed out by example that a high summer temperature provokes a total drying of the head of sphagnum, making the simple sugars very susceptible to leaching by late summer rains. This thus benefits microbial activity immediately.

For the same botanical composition, the speed of evolution of peat is controlled by other ecological factors:

Beware of results from experimental cultures... without verification in the field!

'Once formed, a bog is remarkably resistant to conditions that alter the water balance and accumulation of peat. The perched water table, the water-holding capacity of the peat (...) create a microclimate that is stable under fairly wide environmental fluctuation.' (Mitsch and Gosselink, 1993).

Table 9.2 Physical and biochemical characteristics of five peats representative of changes in vegetation in the Swiss Jura (from Gobat and Portal, 1985)

Plant formation	**Transitional bog with sedge and sphagnum**	**Unwooded raised bog**	**Pine-land of raised bog**	**Spruce forest of raised bog**	**Bare peat of a heath with ericaceous plants**
Vegetation unit	***Sphagno-C. rostratae***	***Sphagnetum magellanici***	***Pino mugo-Sphagnetum***	***Sphagno-Piceetum***	**Heath with *Calluna vulgaris***
Soil (HISTOSOL)	*HISTOSOL FLOTTANT*	*HISTOSOL FIBRIQUE*	*HISTOSOL FIBRIQUE*	*HISTOSOL FIBRIQUE*	*HISTOSOL FIBRIQUE*
General analysis of the surface Hf horizon (0–5 cm)					
Ash content (%)	1.8	3.1	1.4	6.0	3.3
pH_{H2O}	4.0	4.1	3.8	4.1	3.6
Rubbed fibre content (%)	99.0	99.0	83.2	68.1	41.5
Pyrophosphate index	3.7	4.6	4.4	14.7	20.7
Monophenols (mg g^{-1})	2.5	3.3	2.6	10.3	9.8
Detailed analysis of the 200–2000 μm size fraction					
Organic C (%)	43.4	n.d	50.1	n.d.	48.6
Extractability (%)	4.5	n.d	7.0	n.d.	12.2
FA/HA ratio	1.8	n.d	1.1	n.d.	0.3
C/N ratio of humic acids	17.9	n.d	19.9	n.d.	22.1
H/C atomic ratio of humic acids	1.39	n.d	1.35	n.d.	1.11
Slope of basic line of IR analysis (%)	3.1	n.d	17.2	n.d.	32.8
COOH/OH ratio of humic acids (by IR)	0.35	n.d	0.42	n.d.	1.01

pH, chemical nature of the water, aeration, temperature, seasonal fluctuations in climate, topography, etc.

Once formed, peat contributes to its own stability by various homoeostatic processes (Chap. 7, § 7.1.4): blocking of water by increase in matric potential, retention of bioelements by increase in cation exchange capacity, diminution of bacterial activity by toxic compounds, creation of a well-buffered microclimate, etc. Schmeidl (1978) gave a good proof of the exceptional thermal buffering of peat, studied during a very hot day. With a surface temperature of 71.1°C, a bare peat, black, attained only 31°C at 5-cm depth; its buffering power was thus 40°C in 5-cm thickness!

'At first an acid is formed—which will be named humic or ulmic by Sprengel in 1821—which hinders rapid decomposition of plants' (Einhof, 1804, quoted by Lesquereux, 1844). The release of toxic products that restrict decomposing activity was very soon accepted as explanation for accumulation of peat.

9.3.3 Biological activity and evolution of peat

The slow evolution of peat does not signify absence of biological activity!

Biological activity in a peat bog is mostly concentrated in the surface layer of the HISTOSOL, in the upper part of the acrotelm where conditions are most variable (§ 9.6.2) and oxygenation is favourable for the fauna. In raised bogs of the Jura, Borcard (1988) counted about 75% of the biomass of oribatid mites in the first 35 mm of a core to 130 mm.

The idea of 'hot spots of activity', defined for other soils (Chap. 6, § 6.2.1; Chap. 13, § 13.5.2; Coleman and Crossley, 1996) could be applied to HISTOSOLS as well.

Acid peat contains no earthworms, very sensitive to pH, but it shelters enchytraeid worms such as *Cognettia sphagnetorum*, which attacks the residues of *Calluna vulgaris* (Heal et al., in Heal and Perkins, 1978). Enchytraeids and their associated mesofauna prefer nitrogen-rich microhabitats; such microhabitats are provided in the acrotelm by roots of higher plants (Coulson and Butterfield, 1978).

Greatly reduced bacterial activity.

In acid bogs, bacterial activity is reduced by bacteriostatic compounds released by sphagnum but also by sequestration by sphagnum of metallic cations necessary for bacteria (Clymo, in Parkyn et al., 1997). Slow decomposition of the peat is also due to inactivity or absence of nitrifying and nitrogen-fixing bacteria. This results in a very poor supply of combined nitrogen to plants; in some cases, this is compensated for by a partially heterotrophic N nutrition (carnivorous plants, Chap. 4, § 4.3.1). Furthermore, a large part of the nitrogen is removed from the biogeochemical cycle by fixation in the biomass and necromass.

The greatest role in decomposition then belongs to fungi, well adapted to acid environment and successful in penetrating the cell walls to digest the cytoplasm (Lütt, 1992). They are abundant mostly in the first 20 centimetres of soil, like the mycophagous fauna often bound to it.

Autolytic: said of the destruction of a cell or organism by enzymes it itself produced. It is often a controlled process.

Some decomposition activities are ***autolytic***, due to the action of enzymes released by long-dead cells and preserved 'in running condition' by association with humified substances or metals (Chap. 14, § 14.2.4); copper is particularly complexed (Mathur and Farnham, in Aiken et al., 1985; Logan et al., 1997). Long-term maintenance of enzyme activity, frequent in other soils, thus happens in HISTOSOLS too. These autolytic reactions may even predominate in peats in which biological activity is greatly reduced by acidity, anoxia or deficiency in bioelements.

'A better understanding of the role of enzymes in the humification of organic soils is prevented by the lack of proper criteria and methods for distinguishing between functional enzymes that are ephemeral (transient) in the soils and those which are accumulated in the stabilized (immobilized) form' (Mathur and Farnham, in Aiken et al., 1985; Chap. 14, § 14.2.4).

UFOM: Unidentified Flying Organic Matter!

When mineral soils undergo erosion by water or wind, their material remains on the Earth although laterally displaced over very great distances. Erosion of peat is different because it is due to mineralization. Organic matter transformed into water, carbon dioxide and methane 'takes to the skies' and disappears in the atmosphere. In the physical sense of the term, we may almost speak of a gigantic ***sublimation***!

Following the lysis of plant tissues and their mineralization, loss in height of one to two metres in less than a century is caused in peats whose evolution has been greatly accelerated following drainage. Eggelsmann (in Göttlich, 1990) quoted the record of Kehdinger Moor in Germany, which lost 2.80 metres thickness in 100 years. Similar cases are known on the Swiss plateau.

Sublimation: physical process by which a body in the solid state passes directly to the gaseous state without going through a liquid phase.

9.3.4 Time-scales

Plant production is basic to formation of peat. This implies that conditions determining its speed may change more rapidly than for another kind of soil, at least in the acrotelm; actually, in view of the changes, vegetation and its organic matter exhibit a weaker inertia than the mineral constituents (Chap. 7, § 7.7.2).

In a rather thick HISTOSOL about 10,000 years old, as found on perialpine massifs, some peat layers were formed rapidly while at other depths (other epochs), the process was very much slower. Even the opposite phenomenon is found, in which ancient peats were temporarily subjected to conditions permitting their degradation before conditions again became favourable for accumulation.

But peat, by its ***tanning*** power, very well preserves the residues incorporated in it. The most resistant, such as pollen grains, spores or dead testate amoebae, serve as chronological and ecological markers for reconstructing the history of the peat. The ^{210}Pb and ^{137}Cs deposited by nuclear tests since the 1960s (Clymo, in Gore, 1983) have also been used as markers of time. The concentrations of Pb and Sc have even enabled observation of the pre-industrial periods of the Roman Empire in a peat bog of the Swiss Jura (Shotyk et al., 1998). Also, considering the layers of ash deposited by great volcanic eruptions, peat-bogs are irreplaceable archives for understanding climatic changes (Blackford, in Parkyn et al., 1997).

The distinction between age of a soil and its degree of development is still trickier to apply in a HISTOSOL than in another kind of soil (Chap. 5, § 5.5.5)!

Tanning (power): said of the capacity of certain materials containing tannins to be able to block the evolution of substances with aromatic radicals and thereby their decomposition. The tanning power of peats has enabled preservation of plant and animal residues almost intact despite their great age. More than 700 very well preserved mummies dating from the Iron Age have been discovered in the peat bogs of northern Europe. The Tollund Man of Denmark (200 BC) is the best known case.

9.4 HISTIC HORIZONS

Although some soil scientists continue to use the term 'peat' to designate the material and also a pedological horizon, a soil type and a humus form (!), a consensus is gradually emerging to clearly separate the different cases:

- peat is a biohydrogeological material just like any other substratum;

Several types of H horizons are distinguished on the basis of their fibre content, pyrophosphate index and/or ecological changes undergone (Aandahl et al., 1974; USDA, 1999).

Histic horizon (H horizon): holorganic horizon formed in an environment saturated with water for long periods (more than six months in the year) and formed chiefly from debris of hygrophilic and subaquatic plants (AFES, 1995, 1998).

• the pedological horizon corresponding to certain conditions in the presence of peat is termed the ***histic horizon*** (from Gk *histos* = tissue, referring to the plant debris constituting peat);
• a solum formed of histic horizons is a HISTOSOL;
• because of the organic character of HISTOSOL, the latter term may also be applied to the corresponding humus form; occasionally however, for example in case of drying, another humus form may develop in the surface layer of the HISTOSOL, such as mor or moder (Chap. 6, § 6.3.2).

The *Référentiel pédologique* (AFES, 1995, 1998) defines five types of histic H horizons (Tab. 9.3).

Table 9.3 Characteristics of histic horizons (after AFES, 1995, 1998)

Descriptor	Hf horizon	Hm horizon	Hs horizon	Ha horizon	LH horizon
Qualifier	Fibric	Mesic	Sapric	Muck	Ploughed muck
Rubbed fibre content	>40%	10–40%	<10%	Not detectable	Not detectable
Pyrophosphate index	About 0–20	About 20–40	>about 40	>50	>50
Von Post humification scale	0–3	4–7	8–10	Gen. 8–10	Gen. 8–10
Hydraulic conductivity	High	Moderate	Low	Low	Low
Bulk density (g cm^{-3})	<0.10	0.07–0.18	>0.18	Gen. >0.18	Gen. >0.18
Humified amorphous organic matter	Absent	Moderate to high proportion	Very high proportion	Very high proportion	Very high proportion
Ecological conditions (at similar botanical composition)	Permanently very wet, increasing peat	Permanently wet, stabilized peat	Possible temporary drying, decomposing peat	Lowered water table, uncultivated	Lowered water table, cultivated after ploughing

9.5 HISTOSOLS

9.5.1 Definition

The taxon HISTOSOLS is recognized by almost all soil classifications, even though the nomenclature may occasionally differ (Chap. 5, § 5.6.1).

Composed of organic materials and water, a HISTOSOL is mostly made up of histic H horizons, at least to a certain depth. Below this is generally found a permanently water-saturated, anoxic, reductive Gr or Cg horizon. The grouping of the solum as a HISTOSOL or RÉDUCTISOL depends on the horizons predominating in it.

9.5.2 Categories of HISTOSOLS (Références)

Seven Références, from HISTOSOL FIBRIQUE to HISTOSOL FLOTTANT.

Just as for the material peat or histic horizon, it is always the criterion of physical and botanical nature of the material, veritable red thread through the organizational levels, that enables differentiation of HISTOSOLS. Here it is the predominance in the first 120 centimetres of such or such histic horizon that 'cross-checks' this criterion through the characteristics of each type of H horizon. Seven Références are identified (AFES, 1995, 1998):

- HISTOSOL FIBRIQUE: Hf predominant between 40- and 120-cm depth, more than 60 cm thick; corresponds to *Fibrist* of *Soil Taxonomy* (USDA, 1999);
- HISTOSOL MÉSIQUE: Hm predominant, more than 40-cm thick *(Hemist)*;
- HISTOSOL SAPRIQUE: Hs predominant, more than 40 cm thick *(Saprist)*;
- HISTOSOL COMPOSITE: No specific type of H horizon predominant between 20 and 120 cm depth;
- HISTOSOL LEPTIQUE: Solum less than 60 cm thick if Hf predominates or less than 40 cm thick if Hm or Hs predominates; mineral substratum appearing at these depths;
- HISTOSOL RECOUVERT: a terric material (e.g. added topsoil) or mineral horizons (e.g. alluvial deposition) 10 to 40 cm thick covering the H horizons;
- HISTOSOL FLOTTANT: it floats on the free water surface that appears at 40 to 160 cm depth; typical of transitional mires, it is often formed of sphagnums caught up in pioneer plants with long floating rhizomes, such as *Carex chordorrhiza, Menyanthes trifoliata* or *Potentilla palustris* (Fig. 9-B).

Fig. 9-B *Above,* Menyanthes trifoliata; *below,* Potentilla palustris *(after Hess et al., 1967ff). With permission of Éditions Birkhäuser, Basle.*

9.5.3 HISTOSOL: also a humus form

One more peculiarity of HISTOSOLS...

Because of their special character, the holorganic layers of a HISTOSOL may also be considered a humus form (Chap. 6, § 6.1.2).

Furthermore, in a 'normal' soil, with perhaps the exception of ORGANOSOLS INSATURÉS that resemble thin HISTOSOLS (Chap. 5, § 5.5.1), the humiferous episolum only sits on the remainder of the solum, generally mineral. In this situation, it is perfectly legitimate, even necessary, to distinguish two separate typologies, one for the solum in its entirety and the other just for the humus form (Chap. 6, § 6.1.1). Obviously this is hardly valid in peat soils in which the 'humiferous episolum' *is* the solum! It is necessary to accept this small exception in the basic principle of pedological nomenclature, which looks forward to distinctly separating the typology of humus forms and that of sola (Green et al., 1993).

9.6 HYDRIC REGIME OF HISTOSOLS

9.6.1 Hydraulic conductivity and nature of peat

The hydric regime of HISTOSOLS depends largely on nature of the organic matter, itself a consequence of the botanical composition and degree of decomposition.

The still scarcely degraded peats are relatively porous and exhibit a higher hydraulic conductivity than other more decomposed peats (Fig. 9.11). The hydric flux is also influenced by topography but only feebly. The very great water-holding power of peat means that the

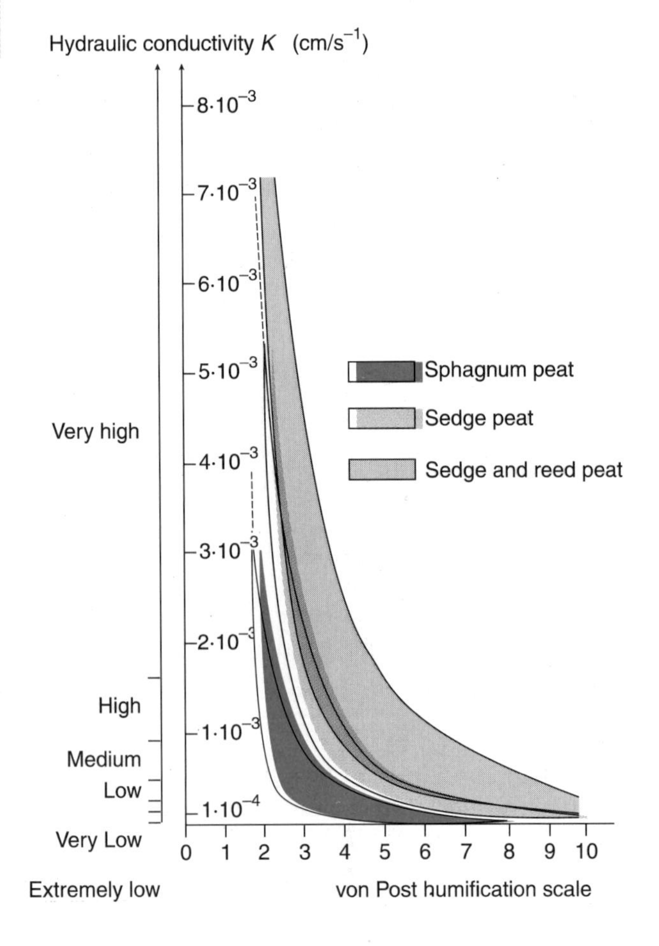

Fig. 9.11 *Hydraulic conductivity of three types of peat as a function of the von Post humification scale (after Schweikle, in Göttlich, 1990). Verry and Boelter (in Mitsch and Gosselink, 1993) measured hydraulic conductivity* K *greater than* $1.50 \cdot 10^{-3}$ *cm* s^{-1} *in a fibric peat and lower than* $0.012 \cdot 10^{-3}$ *cm* s^{-1} *in a sapric peat. In comparison, the conductivity of a clay is about* $0.0005 \cdot 10^{-3}$ *cm* s^{-1} *and that of a sand* $50 \cdot 10^{-3}$ *cm* s^{-1}.

gravitational potential is often negligible compared to the matric potential (Chap. 3, § 3.4.5).

In turn, the speed of hydric fluxes acts on vegetation and biological activity. Where water circulates a little faster, the oxygen concentration increases, favouring the soil fauna, fungi and aerobic bacteria. On the other hand, the deep compacted zones, where the water is almost stationary, are colonized only by anaerobic bacteria. Drainage with attendant drying out of peats also favours the dynamics of certain tyrphotolerant animal populations with r strategy to the detriment of K strategists typical of intact peats (tyrphobiont animals).

9.6.2 Acrotelm and catotelm, the dual hydric regime of ombrogenic peatlands

Definition and comparison of acrotelm and catotelm

If the bottom water table in contact with the mineral substratum (Fig. 9.2) is neglected, a dual hydric regime appears in HISTOSOLS of raised bogs. Two zones are distinguished according to the very abrupt drop in porosity and hydraulic conductivity from the first decimetres to the lower layers, on both sides of an interface just a few centimetres thick (Figs. 9.12, 9.13).

HISTOSOLS of bogs, soils with two sup-erposed water tables..., which almost ignore each other!

The layers close to the surface are inhabited by active aerobic organisms and plants preferentially develop their feeding roots there. In the ccontext of raised bogs they undergo the greatest changes and alternations in ecological conditions. Their totality constitutes the ***acrotelm.***

Acrotelm (*acro* = raised; *telm* = peat): all the layers of a HISTOSOL close to the surface, with numerous exchanges between water and the atmosphere as well as large fluctuations in the water table, directly related to precipitation. The acr-otelm and catotelm were defined by Ingram (1982) and Ingram in Gore (1983) based on the work of Ivanov (1953).

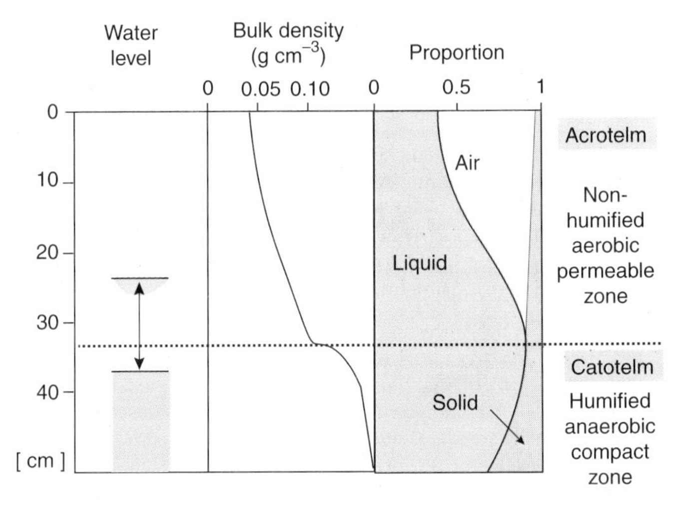

Fig. 9.12 *Some properties of the acrotelm and catotelm in a raised bog (various sources).*

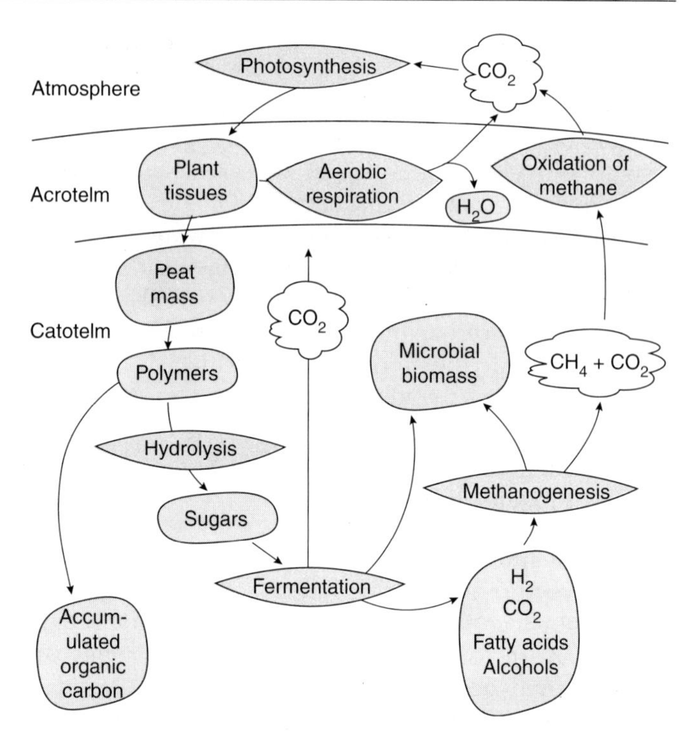

Fig. 9.13 *Evolution of organic matter in acrotelm and catotelm (various sources).*

Catotelm (*cato* = below, *telm* = peat): all the deep, permanently water-saturated, poorly permeable layers of a HISTOSOL. This is the zone in which the peat is progressively humified and mineral elements leached down from the base of the acrotelm accumulate (Damman et al., in Vasander and Starr, 1992).

Below this, between 20 and 60 cm, permeability decreases by humification and the pressure exerted by the upper layers (Streefkerk and Casparie, 1989). Plant debris is gradually decomposed and humified, which reduces its attractiveness to decomposers. The fauna is non-existent here and the highly reductive water is almost immobile. This is the ***catotelm***.

The distinction between acrotelm and catotelm is non-existent in fen-peats, in which the deep groundwater and that nourished by precipitation are mixed up into a single one.

What should be thought of humification in the catotelm?

A still poorly understood process: humification in the catotelm

Which humification is active in the catotelm? In aerated soils, humification is due to inheritance, polycondensation or bacterial neosynthesis (Chap. 5, § 5.2.3; Chap. 14, § 14.4). None of these pathways seems applicable singly to the slow process that takes place in the anoxic catotelm:

- 'Classic' inheritance and the weak transformations pertaining to it, as observed in a mor (Chap. 6, § 6.3.5), are aerobic processes; thus we should envisage fermentations here (Chap. 4, § 4.4.2).
- Humification by polycondensation is possible because of activity of the enzymes (oxidases and peroxidases) that require oxygen for their formation. But whence comes this oxygen? Salts of Mn(III)

could be formed in the oxic acrotelm by manganese-peroxidases (Chap. 14, § 14.3.3) and leached down into the catotelm.

- Lastly, humification by bacterial neosynthesis, with formation of polysaccharides by anaerobic bacteria, is theoretically possible; but physicochemical conditions in the catotelm, notably the pH, prevent the bacteria from acting, even surviving!

Furthermore—but this is perhaps related—similar unknowns subsist regarding methane formation in the catotelm, because the bacteria that are responsible, the methanogenic bacteria, are eliminated at $pH < 6.5$! But new species are perhaps yet to be discovered...

Many mysteries remain in the microbiology of bogs...

Essential role of the interface between acrotelm and catotelm

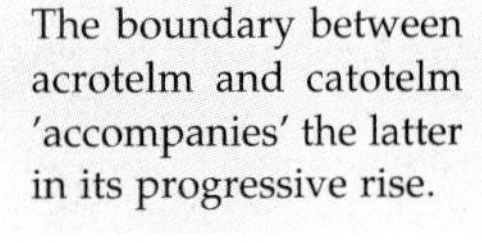
The boundary between acrotelm and catotelm 'accompanies' the latter in its progressive rise.

Evolution of a HISTOSOL is scarcely marked by a change in the vertical sequence of horizons as in another soil (Chap. 5, § 5.3), but rather by a progressive 'rising' of its surface and the acrotelm/catotelm boundary. Because of the two-layer hydric regime and accumulation of peat at the acrotelm/catotelm interface, the boundary between these two parts of the HISTOSOL stays at a relatively constant distance from the topographic surface; it thus preserves the thickness of the acrotelm. This is of course not valid for cultivated HISTOSOLS in which the surface layer has been ploughed or scraped off, destroying the interface. Similarly, raised bogs of small diameter exhibit a water table with very strong natural seepage because of the slope of their edges; the acrotelm may be non-existent at the periphery and the catotelm may be dried out at the sides (Fig. 9.14).

Water as old as the peat that contains it?

The hydric relationships between the acrotelm and catotelm are not intense, the water of the former mixing only slightly with that of the latter. We may occasionally read that the latter is as old as the peat, that is, thousands of years old. Although isotope tracing has put these figures into perspective, this water is much older than that of the acrotelm (Warner, pers. comm.).

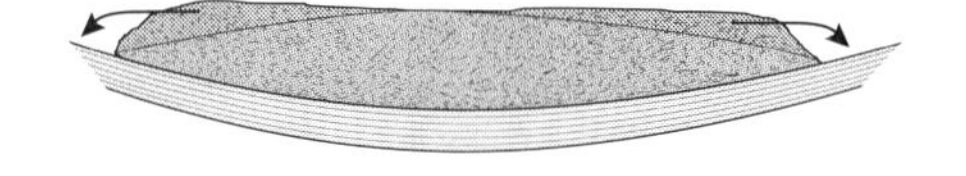

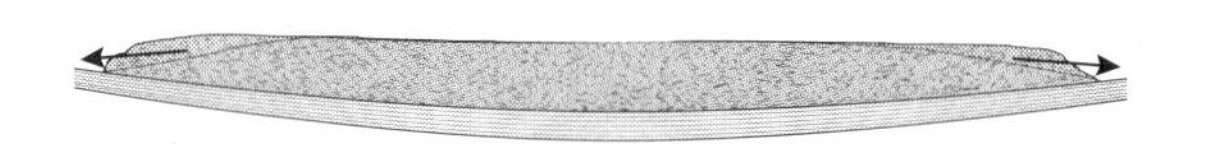

Fig. 9.14 Relationship between shape and extent of peat bogs and water table (vertical scale greatly exaggerated!).

Because of its low porosity, the catotelm constitutes a veritable floor impermeable to water of the acrotelm, which therefore tends to flow laterally. A gentle slope suffices for this lateral runoff if the catotelm is scarcely porous (Schneebeli, 1989). The surface water may thus leave the raised bog through the network of hollows (§ 9.2.3) and feed the surrounding acid fens; if the natural zonation of vegetation and soils is preserved, it may even rejoin the alkaline fens farther downstream (Gallandat, 1982; Gobat, 1984).

Another hydric-regime hypothesis

Some workers doubt the very sharp separation between the waters of the acrotelm and catotelm, which could explain some of the mysterious microbial processes of which there is some talk...

Contrary to general opinion—but this may be only a question of scale of observation (Chap. 7, § 7.7.2)!—Chason and Siegel (1986) and Siegel and Glaser (1987) measured higher hydraulic conductivity than expected in certain zones of the catotelm. Preferential paths will exist through which the water of the upper layers will move vertically downwards 'pushing' the water of deeper layers to the bottom by hydrostatic pressure; this water will come out on the sides of the bog through the bottom (Bridgham et al., 1996). Similarly, methane pockets freed of their gas under the pressure of water could also modify the hydraulic conductivity of the catotelm (Buttler et al., 1991), allowing, for instance, evacuation of the water-soluble organic matter through the base of the HISTOSOL. But this is yet very poorly described and still less quantified (Clymo, in Laiho et al., 1996).

'We call *découverte* or *bourin* the upper layer of a peat with very loose texture, too little compacted to be usefully exploited. This layer is ordinarily a foot to a foot-and-a-half (30 to 45 cm) thick. It is below this layer that the peat starts to become useful. If it is admitted that growth is two feet per century, it will require seventy-five years to produce this upper layer and consequently to lead the part it covers to an adequate stage of decomposition. After an equal length of time, what was formerly the surface will be found sunken by a foot-and-a-half or more, to the peat state.' (Lesquereux, 1844).

Lesquereux, the Agassiz of peatlands!

More than a century ahead of his time, the naturalist and watchmaker of Neuchâtel, Switzerland, Léo Lesquereux (1806–1889) highlighted the subdivision of ombrogenic peat bogs into two functional layers. By observing the work of peat-cutters, he was able to foresee the existence of the acrotelm and catotelm (Lesquereux, 1844). He also understood the transfer of fresh organic matter from the acrotelm to the catotelm and attempted to estimate its speed. Furthermore, it is interesting to note that even today peat-cutters use a vocabulary that well reflects the chief functional layers highlighted by scientists: the *découverte* (or blond peat) corresponds to the acrotelm, 'brown peat' to the upper part of the catotelm where the peat is not yet highly humified and 'black peat' to its deeper layers, humified and very old.

After his researches on the mires of the Jura, Lesquereux emigrated to the USA, where the great scientist Louis Agassiz appealed to him, for he was also a recognized palaeontologist. He also worked in the search for coal seams and petroleum deposits and was one of the precursors of the concept of zonality of soils (Boulaine, 1989).

9.7 UTILIZATION AND PROTECTION OF PEATS AND PEATLANDS

Peatlands merit absolute, unconditional protection, looking to their value for biodiversity and their fundamental role in regulating the global carbon cycle. But the nature of the material constituting them, namely peat, makes them very readily vulnerable to human actions. Curiously, it is perhaps man who will restore their fullness by his well-chosen 'boosts'...

Peatlands, whether ombrogenic or soligenic, constitute an irreplaceable pedological and biological patrimony at the global scale.

Numerous uses...

By virtue of its composition, peat is a material of choice for man, particularly in regions where HISTOSOLS abound, such as Canada, Scandinavia, Scotland, Ireland, Russia, Poland and the Alpine countries. On global average, about 22% of the HISTOSOLS have economic use (Mathur and Farnham, in Aiken et al., 1985). This proportion is much higher in places, reaching 75% in Germany and Poland, and nearly 85% in Switzerland (Grünig et al., 1986). Numerous direct or indirect uses have then been verified (Göttlich, 1990):

Since the end of the eighteenth century, nearly 30,000 megatons of CO_2 have been released to the atmosphere following drainage and cultivation of peatlands (Malthy, in Parkyn et al., 1997).

- extraction of peat for heating houses or for distillation, as practised in Ireland for whiskey;
- roasting of malt by smoking in peat fires in the Scottish islands;
- extraction for horticultural purposes; these first three types of utilization cover $4.5 \cdot 10^6$ ha in the world;
- drainage and cultivation (10^6 ha);
- silvicultural use ($0.5 \cdot 10^6$ ha);
- medicinal use in peat-heated baths as certain hydropathic establishments (Fig. 9-C) suggest;
- manufacture of electricity in thermal power plants.

Fig. 9-C *Without comment... (advertisement for a hydropathic establishment, Salzbourg, Austria; photo J.-M. Gobat).*

'Development' of peat soils by and for agricultural use in particular, directly or consecutive to another use, has caused disappearance of vast areas of peatlands all over the globe. It is a matter of a net loss of stored organic matter because it is generally destroyed by drainage required for agricultural utilization (§ 9.3.3). Well, whatever the utilization of peat, it always results in disappearance of a valuable ecosystem non-renewable at the time-scale of the destructions!

'The destruction of the world's peatlands must stop now...' (Bellamy, in Parkyn et al., 1997). Over the entire globe, mining of peat exceeds its accumulation; the balance is thus negative (Moore and Bellamy, 1974).

...that necessitate increased protection

Utilization of peat poses very acute problems of protection of nature, at least in certain regions where peatlands have almost totally disappeared. In Switzerland, for example, it is estimated that not more than 15% of the original peatlands remain; still more serious, of this 15% only 10%, or 1.5% of the total, are still primary, undisturbed, the remainder generally having been dried out or deforested (Grünig et al., 1986).

Restarting growth of sphagnums by voluntary techniques (Akkermann, 1982; Poschlod, 1990).

But the situation is improving and many peatland areas exploited long ago are returning to the wild state... occasionally with a small human priming necessary for triggering plant production. In Switzerland, for example, the canton of Neuchâtel has launched a big research programme on revitalization of areas of peat denuded for horticultural purposes. Attempts at restarting growth of sphagnum in the shade of planted or seeded pioneer plants (in particular *Eriophorum vaginatum*) have been put in operation, in a manner to again prime vegetation dynamics towards the hummock (§ 9.2.3; Buttler et al., in Wheeler et al., 1995; Grosvernier et al., 1999).

Thus Man alone occasionally has the power to re-establish a favourable balance in peat accumulation, thereby safeguarding not only this very special material, but also peat soils and peatlands in their entirety, rich in an extremely original biodiversity.

CHAPTER 10

COMPOSTING: A VALUE ADDITION TO OUR WASTES

A speeded-up imitation of the natural aerobic process of waste management, as takes place in a litter (Chap. 5, § 5.2), ***composting*** allows treatment of the fraction of biological origin of human wastes. The resultant ***compost*** has dual nature: ***amendment*** because it comprises organic compounds that are precursors to humus, and ***fertilizer*** by virtue of its content of bioelements. Compost thus enables compensating for the humic deficit of overexploited soils and ameliorating their long-term fertility.

This chapter presents an artificially accelerated path of evolution in the logic of the decomposition-formation chain of soil organic matter; it is the reverse of what takes place in peat, which results in an extreme slowing-down of the dynamics of organic matter (Chap. 9). Compost also concerns the *living soil* as a source of amendments and fertilizers, enabling improvement of fertility of cultivated soils and ensuring renewal of their humic reserve. This chapter is just an introduction to the general problems of the composting process and compost utilization. The reader who wishes to know more may refer to the works of Hoitink and Keener (1993) and de Bertoldi et al. (1996), who discuss compost in greater detail.

Composting, accelerated humification!

Composting: method of intensive treatment of organic wastes, which carries out biological aerobic processes of degradation and stabilization of complex organic substances while optimizing them.

Compost: prehumified material resulting from the composting process, exhibiting amending as well as fertilizing character.

Amendment: humigenic organic materials added to a soil for reconstituting humus and improving soil structure. ***Humigenic:*** said of substances resistant to primary mineralization M1 and liable to be transformed by enzymatic and chemical processes to insolubilization humin H2 or to generate residual humin H1.

10.1 IMITATING NATURE?

Fertilizer: bioelement added to the soil in the form of mineral salts or organic compounds containing the element; it is designed for plant nutrition.

Natural disposal of wastes by ecosystems, except where the latter are in decline, does not lead to increase in atmospheric CO_2 content. Actually, fixation of this gas by plants in the net primary production is equal to (in ecosystems at equilibrium) or higher than (in growing ecosystems) its release by consumers.

In nature, plant residues (leaves, branches, flowers, fruits, rain-leachates, litter, root secretions) and animal residues (faecal matter, moults, carcasses) are transformed progressively and finally oxidized to CO_2 (Fig. 10.1).

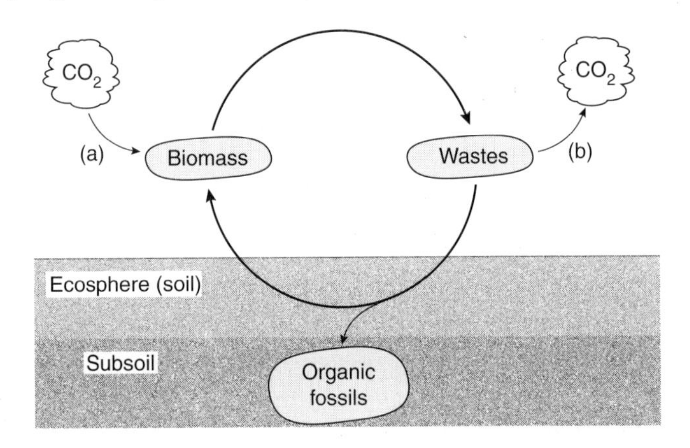

Fig. 10.1 *Natural waste-management cycle. In an ecosystem at equilibrium, the quantity of CO_2 fixed in the biomass by net primary production PN_1 (a) is equal to the quantity produced by decomposition of biomass wastes (b) but for a small fraction possibly fossilized.*

Biogas: mixture of methane (50–65% by volume) and carbon dioxide (35–50% by volume) produced by biological methanization.

Biological methanization (or ***biomethanization, anaerobic digestion***): method of intensive treatment of organic wastes that carries out, while optimizing them, anaerobic biological processes of degradation and stabilization of complex organic materials. Biological methanization generates biogas; it leaves a residue rich in bioelements, containing unaltered organic materials, part of which may later be degraded under aerobic conditions, for example in a composting process.

However, part of the organic matter in the process of transformation and/or neosynthesis—humus—exhibits great resistance to degradation (Chap. 5, § 5.2.2). This confers on it a residence time that can be measured in years, centuries or millennia (Chap. 5, § 5.5.5; Balesdent, 1982; Paul, in Grubb and Whittaker, 1989; Paul and Clark, 1996).

If the organic wastes reach an anoxic environment such as a sediment or a marsh soil (RÉDUCTISOL, HISTOSOL), their evolution is different. A large portion, comprising lignified materials, is not degraded under these conditions or is humified very slowly (Chap. 9, § 9.2.4): depending on the case, it could later undergo fossilization. This, for example, occurred in the Carboniferous when coal seams were formed. This accumulation is linked to that of oxygen in the atmosphere (Chap. 4, § 4.4.4). The fraction degradable under anaerobiosis gives a mixture of methane and carbon dioxide—marsh gas or ***biogas***.

In a way, composting reproduces the process of aerobic degradation of plant litter while accelerating it. Likewise, ***biological methanization*** (Glauser et al., in Wise, 1987;

Wheatley, 1990; Hobson and Wheatley, 1993) is similar to the natural anaerobic process that takes place in sediments and marshes (Chap. 4, § 4.4.2). Both processes may be combined in an approach aimed at the cyclic management of wastes of biological origin (Aragno, 1985). The key idea of such management is to replace, whenever possible, the strategy of nuisance minimization by that of maximization of ecological benefit, in this case renewal of humus.

10.2 HUMAN WASTES

Production of wastes by living organisms is considerable.

It is estimated that in Europe forests produce ten times more waste than the human population. The problem of human wastes is thus not so much a matter of their quantity as of their quality and the way they are managed (Fig. 10.2).

Technological improvements are aimed primarily at minimizing the short-term and long-term consequences of return of wastes to the environment.

These wastes are produced partly from substances drawn from primary fossil materials and modified by industry. Industry itself produces wastes, often very toxic, resulting from manufacturing processes. Although treatment of human wastes increasingly consists of concentrating them and confining them at great expense, a large portion is nevertheless dispersed into the air, soils and waters. This management is essentially acyclic and prejudicial to the environment.

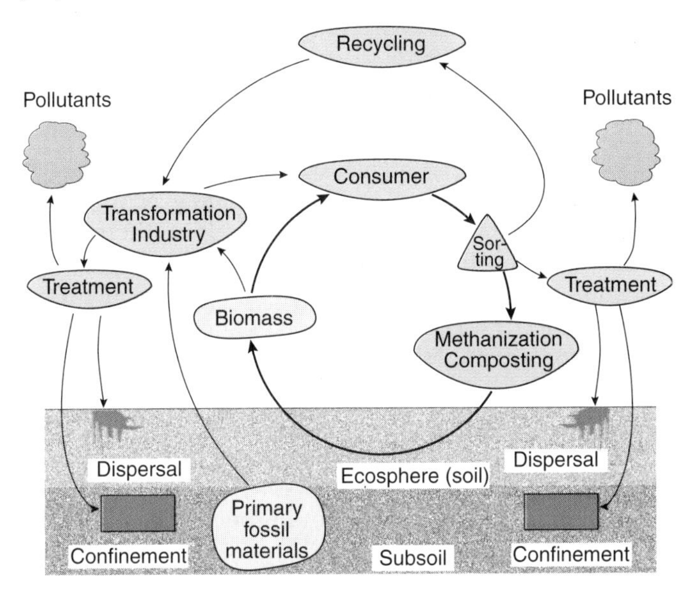

Fig. 10.2 *Management of human wastes. To the natural waste cycle, man has added a non-cyclic process, which draws from non-renewable primary fossil materials and generates polluting wastes and/or those that are not recycled by the ecosphere. The ecological consequences are dispersal of these materials in the environment (air, water, soil), the costly solution being to concentrate and confine them.*

Recycling of paper...

As in the case of paper, recycling and reuse of primary materials should be encouraged. They increase useful life of these materials and thereby reduce their turnover by man. However, it should be noted that these materials are not indefinitely recyclable. Reused paper, for example, is as much contaminated as new paper and cannot be recycled once again: during treatment the cellulose molecules are shortened and the mechanical strength of the paper reduced.

About two-thirds of solid domestic wastes are of biological origin (Fig. 10.3). However, part of these, mostly paper and cardboard, are strongly contaminated at the source. Their return to the environment through the biological pathway is accompanied by dispersal of the pollutants they contain—heavy metals and ***xenobiotic*** organic compounds (see Chap. 11). On the other hand, some types of paper are often used to manufacture recycled paper.

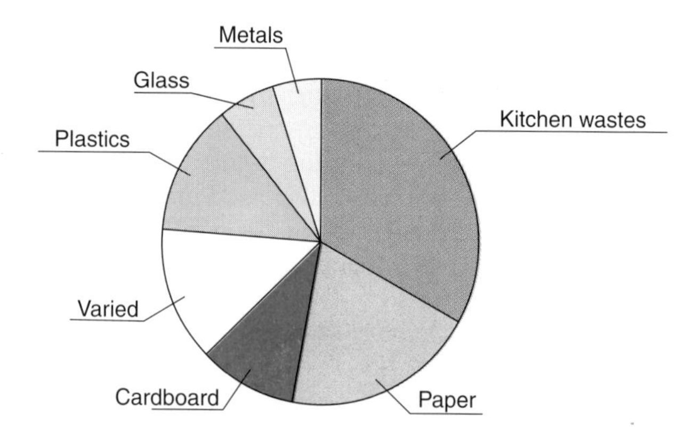

Fig. 10.3 *Composition of domestic wastes in Switzerland, from an investigation of the Federal Office of Environmental Protection, Berne* (Cahier de l'Environnement no. **27**, 1984).

...and other wastes of biological origin.

Other wastes of biological origin should be managed in a cyclic and beneficial manner, imitating Nature (Fig. 10.4). For this, it is necessary to avoid their prior contamination with toxic substances such as heavy metals, solvents, etc. This is achieved by sorting them at the source. Indeed, separate collection of kitchen and garden wastes is increasingly organized.

The different kinds of wastes do not lend themselves similarly to treatment through composting or methanization.

Rich in woody and structured materials (forest wastes, fallen leaves, tree cuttings, part of garden wastes), certain wastes are predetermined for composting: after crushing or shredding they form a porous, easily aerated substrate, while their high content of lignin-impregnated substances makes them poor methanization substrates. Contrarily, water-rich substrates, scarcely lignified and hardly

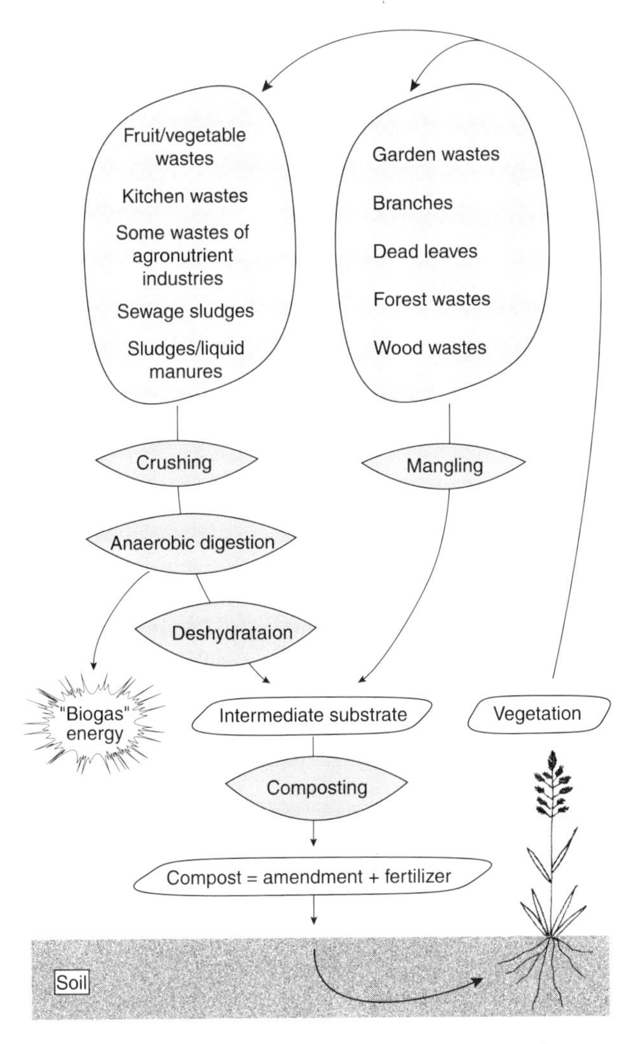

Fig. 10.4 *Principle of cyclic management of organic wastes. Explanations in the text.*

structured, such as fruits and vegetables and sewage sludges, give after crushing a compact mass that is difficult to aerate. On the other hand, they are richer in substances that can be transformed under anaerobic conditions (pure cellulose, hemicelluloses, pectins, starch, proteins, lipids, free sugars) and are therefore particularly suitable for methane digestion.

For an integrated waste management

Composting and methanization are complementary rather than competing processes. Also, biogas generated by methanization may be used as an energy source for production of electricity, motive power and hot water. The ensemble of processes of waste management is thus rendered energy-wise autonomous, either partly

A good imitation of the natural cycle!

or entirely. Such would be the situation in a plant that combines water purification and treatment of organic wastes. Furthermore, after partial dehydration the residues from methanization can be combined with woody residues for composting. Rich in bioelements, they enable optimization of the biological process, while adding a fertilizer character to the final compost. Thus integrated management of treatment of 'green' wastes is achieved (Fig. 10.4; Aragno, 1994).

Such a system of integrated waste management could be combined with greenhouse production of vegetables or fruits. Composting would provide heat for moderating greenhouses, carbon dioxide for enriching the atmosphere and stimulating plant growth and, of course, the compost itself as substrate for growing plants.

10.3 COMPOSTING PROCESSES

10.3.1 The aerobic process and production of heat

When the volume of the substrate to be composted is sufficient, a large rise in temperature occurs (Fig. 10.5) because aerobic microbial activity is an ***exothermic*** process. The heat produced is proportional to the mass and hence to the volume of the compost, whereas the heat is dissipated into the air through its surface area. Rise in temperature is thus greater the higher the volume/area ratio of the compost heap.

Exothermic: qualifies a physical or chemical process that results in release of energy as heat. The opposite is ***endothermic***.

Microorganisms: very economical under anaerobic conditions.

Under anaerobic conditions the amount of energy a microorganism derives from degradation of organic matter is very low. It therefore has to be thrifty with the little energy available to it. The heat released by its anaerobic metabolic activities is thus very modest compared to that generated by aerobic processes (Chap. 4, § 4.4.2). In the anoxic core of a landfill, despite the extremely low dissipation of energy, the temperature rarely exceeds 30–40°C.

A substrate subject to composting processes manifests a thermal cycle: the temperature first rises rapidly under the effect of the heat produced (thermogenic phase). In this stage, two factors limit the production of heat—insufficient aeration and the maximum temperature permitting activity of the microorganisms present:

- Limitation by aeration. With increasing temperature the solubility of oxygen diminishes while its aerobic consumption increases. If the compost is inadequately ventilated, anoxic zones appear at the levels of both microstructure and general structure (Chap. 3, § 3.2), in which much less thermogenic anaerobic activity is manifested.

• Limitation by the maximum temperature. Every microorganism, once it has reached its temperature of maximum activity, is inhibited in respiration and growth; it no longer produces heat. Other more ***thermotolerant*** or ***thermophilic*** organisms take its place. But compost microorganisms must be present in relative abundance in the substrate right from the start. The speed of heating is thus limited by the potential of the microflora of the original substrate to produce a succession of increasingly thermophilic organisms. The maximum temperature attained is that supported by the most thermophilic organisms present (Beffa et al., 1996a). The system then functions as a 'biothermostat'!

Thermotolerant: said of a microorganism that generally lives or grows best at medium temperatures (up to 35–40°C), but tolerates a maximum temperature that may exceed 50°C.

Thermophilic (literally 'loving high temperatures'): every microorganism has a minimum temperature below which it does not grow, an optimum temperature at which its growth is the fastest, and a maximum temperature above which it does not grow. A strictly thermophilic bacterium has in principle an optimum temperature equal to or higher than 60°C. Certain bacteria isolated from compost have a maximum temperature a little higher than 80°C. The record is held by bacteria from submarine geothermal sources, which grow at temperatures up to 123°C! The most thermophilic fungi do not function above 62°C.

This 'cruising' temperature is maintained as long as easily metabolizable substrates are available. But these are gradually consumed and the remaining substances are more and more difficultly and therefore slowly degradable. Because of this, production of heat drops, resulting in a progressive lowering of temperature. When the latter is brought close to the ambient temperature, the most important phase of the maturation process is terminated.

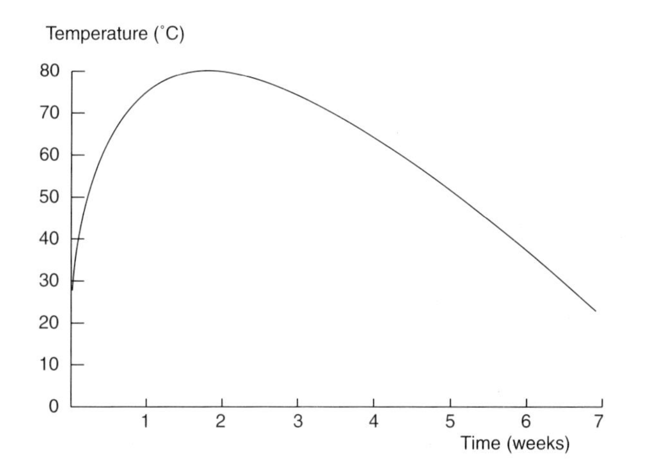

Fig. 10.5 *Change in temperature at the centre of a compost heap.*

10.3.2 Structure of the substrate and aeration

A porous, permeable structure of the substrate (Fig. 10.6) is essential to ensure good progress in composting. Circulation of air in the compost depends on it. Circulation intervenes passively by a 'thermosiphon' phenomenon (Fig. 10.7) caused by heating up of the mass.

Compost behaves somewhat like a biological furnace, the 'chimney' of which ensures a draft.

A compact substrate, highly hydrated and poorly structured, has low permeability to air and is therefore

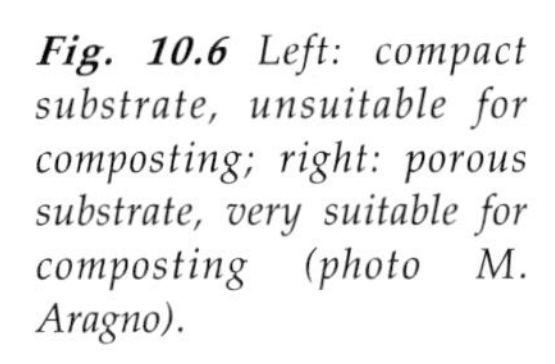

Fig. 10.6 *Left: compact substrate, unsuitable for composting; right: porous substrate, very suitable for composting (photo M. Aragno).*

Aeration can be increased by blowing air or by providing passive aerators in the mass, for example perforated tubes pushed in.

difficult to compost. Also, to be able to compost fruit and plant wastes or sewage sludges, it is necessary to mix them with large quantities of woody materials.

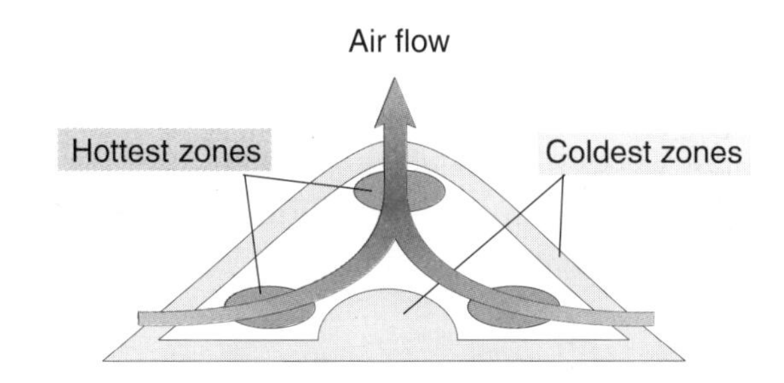

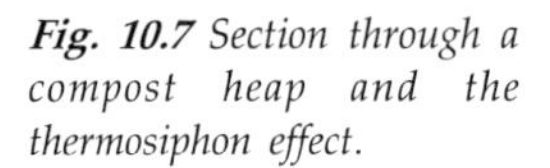

Fig. 10.7 *Section through a compost heap and the thermosiphon effect.*

10.3.3 Biological process

Microorganisms eat their white bread first.

Faced with the variety of foods offered by the materials to be composted, microorganisms first assimilate the small soluble molecules (simple sugars, amino acids, alcohols) that they can absorb without prior modification in the external environment. Then they attack, by means of their extracellular enzymes, more or less easily degradable biopolymers (proteins, nucleic acids, starch, hemicelluloses, pectins, cellulose) (Lynch, in Hoitink and Keener, 1993). The constituent elements of these substances, chiefly carbon, nitrogen, sulphur, phosphorus and metals, are then released as CO_2 and mineral salts. The process is similar to primary mineralization M1 of natural soils (Chap. 5, § 5.2.2).

Lignin, scarcely affected by composting.

Transformation of less degradable materials takes place in composts only during the phase of cooling and stabilization, leading to prehumification. However, certain very resistant polymers such as lignin are scarcely degraded or not at all during composting. In natural litter, ligninolysis is achieved essentially by certain fungi that do not grow at high temperatures (Chap. 14, § 14.3.4) and thus does not take place during the thermogenic phase of the process. This leads to preservation of lignin, very favourable for compost quality: it should not be forgotten that it is an essential precursor of inherited humins H1 and insolubilization humins H2 (see Chap. 14, § 14.4).

During the thermogenic phase, thermophilic sulphur-oxidizing bacteria have been observed, some of which are capable of growing at temperatures up to 80°C (Beffa et al., 1996b). It was thought until now that such bacteria live exclusively in hot springs of volcanic regions!

A volcano in our compost!

Nitrogen is mineralized to the ammoniacal form and sulphur to hydrogen sulphide. Ammonia is not often directly utilizable by plants while hydrogen sulphide is a respiratory poison. In soils these compounds are oxidized respectively to nitrate and sulphate by chemolitho-autotrophic nitrifying and sulphur-oxidizing bacteria (Chap. 4, § 4.4.2). These phenomena have been very little studied in composts.

The nitrogen cycle: very poorly understood in composts.

Nitrification does not occur at high temperatures as confirmed by the relatively high concentration of ammonia in the air of composts during the thermogenic phase (between 60 and 80°C). During the cooling phase, ammonia is oxidized by nitrifying bacteria when the temperature drops below 45°C. If nitrate reaches temporarily anoxic zones of compost, it may be reduced to elemental nitrogen by denitrifying bacteria (Chap. 4, § 4.4.4). Loss of combined nitrogen from the compost follows. If the anoxic zones become large and permanent, later reduction of sulphate to H_2S may ensue, or even formation of methane (Chap. 4, § 4.4.2).

Nitrification benefits from lowering of temperature.

10.4 HYGIENE PROBLEMS AND SOLUTIONS

We have emphasized the importance of composting in a global approach to management of organic wastes. Under certain conditions however, composting presents a health hazard that must be understood in order to control it. First of all, the wastes to be composted are liable to be contaminated by pathogenic organisms parasitic on man, plants or animals. The composting process, if conducted at sufficiently high temperature, will eliminate these microbes.

It is wrong to think that all that comes from Nature is good and not dangerous to man.

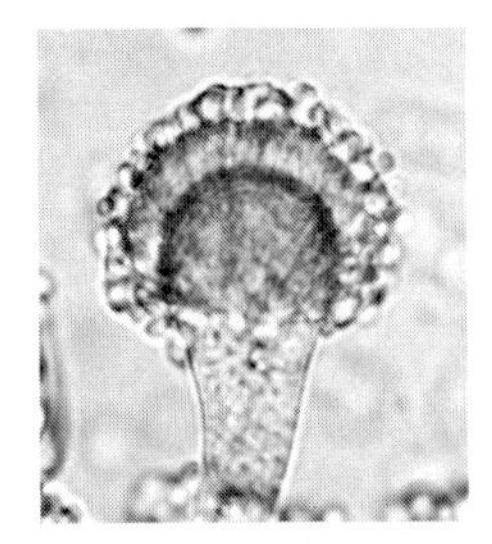

Fig. 10.8 Aspergillus fumigatus. *Capable of decomposing cellulose, this microscopic fungus is a usual inhabitant of compost (photo J. Lott Fischer).*

Another hazard is represented by organisms indigenous to the composting process. This is particularly true of the mould *Aspergillus fumigatus* (Fig. 10.8), which tolerates relatively high temperatures and develops particularly well between 35 and 50°C. It has few competitors in this range and greatly dominates over other moulds. Unfortunately, it is also a potent agent of allergic diseases, similar to those caused by forking of hay ('farmers lung disease'). Also, it causes serious internal mycoses in some immune-deficient persons: it is thus an opportunistic parasite as well. However, it is eliminated at temperatures higher than 60°C.

It is mostly found in conventional composting in open-air heaps, in which a temperature gradient is established between the hot centre and the surface (Figs. 10.7, 10.9).

Fig. 10.9 *Mycelial layer due mainly to* Aspergillus fumigatus *in a compost heap (photo T. Beffa).*

The temperatures most favourable for *Aspergillus fumigatus* are attained in a layer close to the surface, which appears white due to the presence of mycelia. There can be up to 10 million spores per gram of dry compost in this layer! Deeper down, at temperatures higher than 60°C, this number is greatly reduced. During maturation also the number is reduced and rarely exceeds a few thousands per gram of mature compost.

The normal content of *Aspergillus fumigatus* spores in the atmosphere is less than a few tens per cubic metre.

The pathological hazard linked to composting (Lott Fischer et al., 1995) is mainly related to dispersal of spores in the air when the heaps are turned (Fig. 10.10). Compost workers are thus the most exposed.

To avert the hazard in composting operations, certain precautions must be taken:

- compost workers: wear a protective mask during the turning operations, ensure medical follow-up; allergic or immune-deficient persons should not be hired;
- the neighbourhood: do not install composting sites very close to hospitals or convalescent homes.

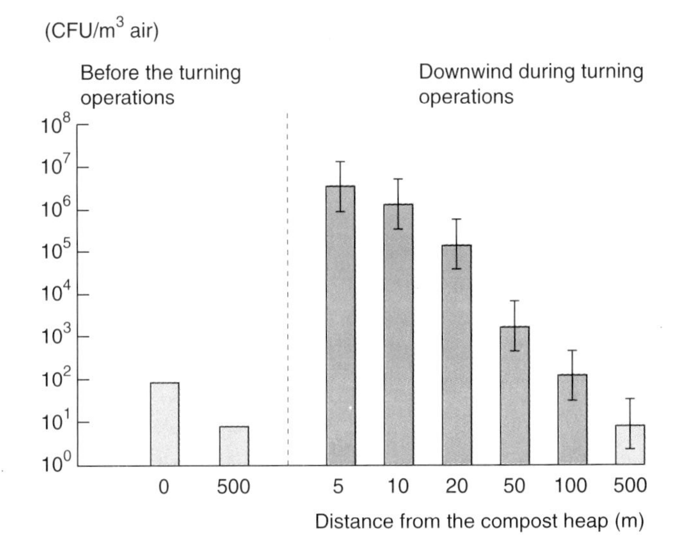

Fig. 10.10 *Dispersal of* Aspergillus fumigatus *spores in the air from a compost heap. During turning of the heap, up to 10 million spores per cubic metre of air have been counted 5 metres downwind! Yet, before the turning, the air at the site contained only about a hundred. At more than 500 metres downwind, the content in the air does not exceed the background level.*

But more can be done. Growth of allergenic moulds, inhibited or killed at temperatures higher than 60°C, is strongly repressed if composting techniques are modified for this purpose (Lott Fischer et al., in de Bertoldi et al., 1996).

10.5 COMPOSTING TECHNIQUES

Composting can be accomplished at different scales (from 0.1 to 100,000 t y^{-1}) and following different techniques (Stentford, in Hoitink and Keener, 1993), among which are individual composting, composting in heaps in the open, trench composting and bioreactor composting:

Individual composting in the garden: close to natural decom-position of litter.

- ***Individual composting*** of domestic and garden wastes is close to natural decomposition of litter (§ 10.8). Taking place just as in soil, it allows the edaphic fauna to establish itself and participate in compost maturation (§ 10.8.1; Plate III–3). Except when large quantities of material are deposited at one time on the heap, rise in temperature is small. The process is slow: one to two years are required to obtain an equilibrated compost, with one or two annual turnings. Individual composting should be encouraged as long as the compost is to be used at site; it is therefore tied to owning a garden. It poses no risk to persons in normal health. Patients under immunosuppressive treatment and also those suffering from immuno-depressing diseases (e.g. AIDS, certain cancers) should, however, avoid turning garden compost.

Composting in heaps: a relatively simple and economical procedure.

• ***Composting in heaps*** in the open consists of establishing heaps in a suitable area—elongated piles with triangular cross-section, about one to two metres high (Fig. 10.11). These heaps are regularly turned by means of a digger or more specialized machines (Fig. 10.12). This type of composting is applied at different scales, from a few tons per year in the case of a small community to tens of thousands of tons per year for a region. In Europe, most composting installations of a certain size use this system.

An investigation conducted in the Microbiology laboratory at the University of Neuchâtel (Lott Fischer et al., in de Bertoldi et al., 1996) showed an interesting relation between frequency of turning of compost heaps and growth of *Aspergillus*. With intensive management (turning every two days during the first 2 to 3 weeks,

Fig. 10.11 *Composting in heaps (photo T. Beffa).*

Fig. 10.12 *Turning of a compost in heaps (photo T. Beffa).*

then about twice a week), growth of the fungus was a hundred to a thousand times less than with extensive management (fortnightly turning). Intensive management enables higher temperatures to be attained, while each turning displaces the mould colonies to the hot part of the heap where they are killed by the very high temperature.

- In ***trench composting***, the compost laid out in concrete trenches is mixed by machines moving along these trenches (Fig. 10.13). Air is also blown through it. Such systems can be established in open or covered sheds with entirely automated turning. Thus workers do not come into contact with spores and dust, and the air in the composting sheds could be directed through a filter.

Trench composting: a more sophisticated variant than composting in heaps.

Fig. 10.13 *Trench composting (photo T. Beffa)*

- Although different systems of ***bioreactor composting*** have been commercially set up, few are presently functioning (Fig. 10.14). They occupy a small area and their environmental impact (e.g. emission of odours) is minimal. In principle, the temperature attained in the bioreactor should be high and uniform: thermal hygienization will then take place in the entire mass. However, depending on the system, aeration is difficult to control: occasionally preferential air-circulation channels are formed, other zones then becoming anoxic. Mixing or turning is often impossible.

Bioreactor composting: the most intensive of all; but it involves high investment.

Fig. 10.14 *Bioreactor composting (photo M. Aragno).*

10.6 CHARACTERISTICS OF MATURE COMPOSTS

A great variety of composts.

Composts are very varied because they are the product of a combination of several factors: nature of the original material, degree of optimization of the process, degree of maturity, texture, addition of minerals or other additives.

Certain elements are concentrated, others reduced.

During composting (Chen and Inbar, in Hoitink and Keener, 1993) the substrate essentially loses carbon (in the form of CO_2) and hydrogen and oxygen (as water). To a lesser extent it also loses nitrogen in the form of ammonia and, by denitrification, as elemental nitrogen. Compared to the original substrate, mature compost exhibits higher concentrations of other bioelements—phosphorus, sulphur, potassium, calcium, magnesium and trace elements. Considering the rather minor losses, the relative content of nitrogen also rises (C/N ratio is lowered).

The content of inorganic salts and humigenic materials in compost depends on the composition of the original substrate.

The characteristics of a mature compost of green wastes are given in Table 10.1. Compounds such as sewage sludges also contain many nutrient elements but little humigenic material. Contrarily, wood and bark wastes are poor in bioelements but rich in polyphenolic compounds (lignin, suberin), precursors of humus.

Table 10.1 Average characteristics of a mature compost of domestic wastes

Dry matter (as % of fresh material)	55–70
Bulk density (g L^{-1} fresh material)	500–800
Water capacity (% by volume)	45–65
Organic matter (% of dry matter)	20–40
C/N ratio	10–20
pH	7–8
Total nitrogen (N, % of dry matter)	0.5–1.8
Phosphorus (P_2O_5, % of dry matter)	0.4–1.0
Potassium (K_2O, % of dry matter)	0.6–1.8
Magnesium (MgO, % of dry matter)	0.7–3.0
Calcium (CaO, % of dry matter)	3–12

Two extreme types of compost... and intermediates.

A first type of compost, rich in inorganic nutrient elements, essentially functions as fertilizer. A second, primarily designed for improving soil structure, acts as amendment. Composts obtained from kitchen and garden wastes generally have intermediate properties and serve as fertilizer as well as amendment.

According to requirement, composts are sieved to fine (up to 10 mm), medium (up to 20 mm) and coarse (up to

40 mm) texture. For specific applications or for correcting the nutrient-element content of composts, 'vegetable mould' or certain minerals are added.

Compost and heavy metals
The level of contamination of the starting materials by toxic substances directly influences the quality of the end compost and its possible applications. Requirements are higher for composts designed for cultivated soils (Tab. 10.2) than for those used for greening in road building or for urban management, for example.

Table 10.2 Heavy metal contents of composts of green wastes (mg kg^{-1} dry matter)

Metal	Composts of green wastes (average contents)	Permissible limit
Cadmium	0.1–1	1.5
Chromium	25–60	100
Copper	30–50	100
Mercury	0.1–0.5	1
Nickel	10–30	50
Lead	50–100	150
Zinc	150–350	400

10.7 USE OF COMPOST

Composting has two objectives:

- treatment of organic wastes that represent an environmental load because of their putrescibility and the nuisances they cause—smells, hygiene problems, attracting certain animals (rats, gulls, crows, flies, etc.);
- production of amendments and fertilizer for long-term conservation of cultivated soils, for intensive garden crop cultivation and horticulture, and for home gardening.

Each partner in composting benefits from it.

Why some reservation about use of compost?
The aims of composting involve two partners: the person in charge of disposal of wastes and the farmer; in other words, the producer of compost and the user. Very often, unfortunately, it is only the producer that assumes promotion and development, the consumer requiring pressure. However, there would be no reason for sorting and composting organic wastes if the product were not profitably usable for improvement of soils and have finally to be eliminated by incineration or dumping. However, bad experiences with poorly prepared composts that have undergone insufficient maturation or are loaded with undesirable materials (glass or plastic debris, heavy metals, organic pollutants) have led the farmer,

The fertile structure of soil is weakened; compost helps to restore it.

agricultural adviser or even the agricultural scientist to distrust compost. Bulk composting of poorly sorted or unsorted urban wastes, as sometimes done in the past, has obviously reinforced this distrust.

Following compaction by very heavy agricultural machinery and over-fertilization, the organomineral complex is broken up (Chap. 3, § 3.6.2), rendering the humin more vulnerable to secondary mineralization and erosion. Most cultivated soils are threatened with a humic deficit caused, among others, by intensive cultivation methods. It is thus necessary to promote recycling of humic materials instead of incinerating or dumping them.

10.7.1 Effects of compost application on soil properties

Effect on physical properties

What benefits do the farmer, market gardener and horticulturist derive from using compost?

Application of compost to a soil greatly modifies its physical, chemical and biological properties in both the short and long term (Dick and McCoy, in Hoitink and Keener, 1993). For example, it reduces bulk density and increases porosity of the soil, promoting aeration (Fig. 10.15; Chap. 2, § 2.4.2; Chap. 3, § 3.3.3).

Water retention capacity and availability of water to plants are augmented (Fig. 10.16; Chap. 3, § 3.4.3), as is structural stability of the soil.

A strict restriction on use of peat.

Improvement of physical properties of soil by compost is slower but longer lasting than that provided by addition of peat. Generally poor in humigenic polyphenolic substances, peat has mainly an immediate mechanical

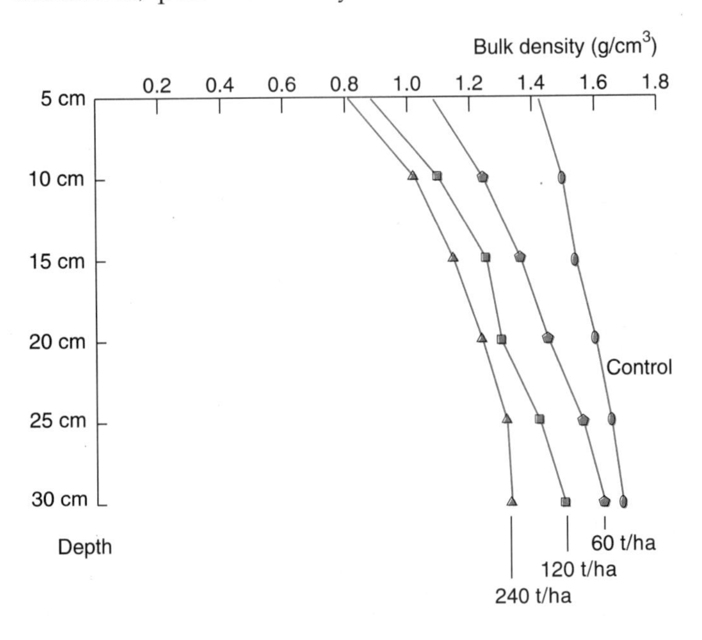

Fig. 10.15 *Reduction of bulk density of soil after application of different quantities of compost or inorganic fertilizer (after Tester, 1990). In the control soil without compost application, addition of bioelements was achieved through application of inorganic fertilizer. It is seen that the effect of compost extended to more than 30 cm below the amended layer five years after a surface application, because of the fauna that buried the organic matter. Soil compaction decreased considerably, promoting penetration of roots... and implements.*

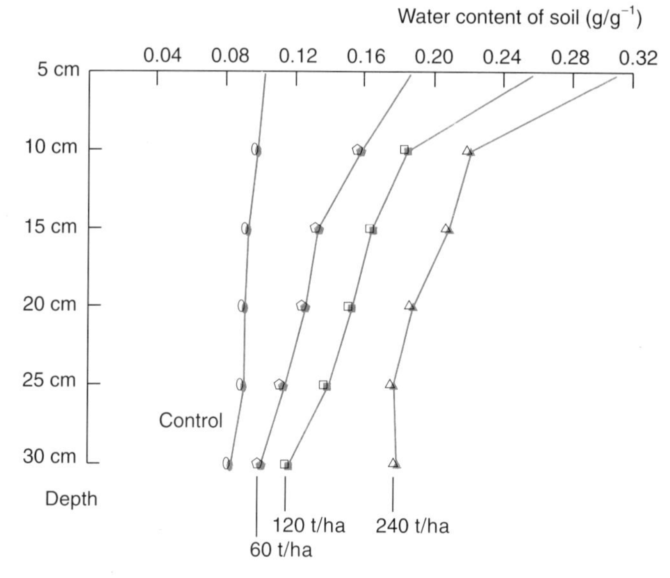

Fig. 10.16 *Water retention in a soil with and without compost application (after Tester, 1990). Same conditions as in Fig. 10.15. Addition of compost improved water retention particularly in the topsoil layers.*

effect on soil. On the other hand, under normal conditions of use, it degrades rapidly. Its use as amendment should be severely restricted because of the ecological importance of the ecosystems (peat bogs) from which it comes (Chap. 9, § 9.7).

Effect on chemical properties

Cation exchange capacity is augmented by addition of humigenic substances in compost: the soil retains more inorganic salts, which promotes root nutrition and prevents leaching of ions. This effect attains its maximum only several years after application has begun, when the humification process has integrated the materials added through the compost. The humic substances bind certain xenobiotic aromatic compounds (Chap. 14, § 14.4.4; Bollag, 1991; Bollag and Dec, in Grimwell and de Leer, 1995) and chelate heavy metals. This allows use of composting and its products in bioremediation of wastes contaminated with pesticides and other xenobiotic compounds (Chap. 11, § 11.3.2).

Compost application raises the concentration in soil of organic matter and nutrient elements.

A detoxifying effect on polluted soils.

The inorganic-salt content of composts confers on them high buffering capacity: they stabilize pH and neutralize very acid soils. Unlike inorganic fertilizers they add, apart from macroelements, a complement of trace elements very useful to plants (Chap. 4, § 4.3.4).

Composts are neutral to slightly alkaline.

Effect on biological properties

Except if they have undergone thermal post-treatment (pasteurization), mature composts contain a large and

Enzymatic and nutritional effects

varied community of mesophilic microorganisms. Their use leads to a significant increase in enzymatic activity in soil (Chap. 14, § 14.2), which depends, however, on the type of compost used.

We still have little information on the impact of addition of compost microorganisms on life in the soil and their ability to maintain themselves in the long term. Furthermore, similar to litter, compost also represents a food source for indigenous soil organisms and can modify food webs. Compost releases CO_2 to the extent of its mineralization by soil microorganisms. Concentration of the gas increases not only in the soil air, but also in the layer of air immediately above, to the advantage of photosynthesis in lower plant strata.

10.7.2 Effects of compost application on plant growth

Modifications brought about in soil by application of compost are beneficial to plant growth:

- better retention and availability of water;
- better aeration;
- easy root penetration in the soil;
- stabilization of soil structure;
- increase in ionic retention and exchange;
- addition of nutrient trace elements and macroelements;
- maintenance of a favourable pH.

The effect of compost? Even in your plate!

Use of compost often significantly improves the quality of food crops (Vogtmann et al., in Hoitink and Keener, 1993). Tomato, for example, tastes better. Vegetables keep better; their nitrate content diminishes, that of vitamin C increases.

Patience! The effect of added compost on plants is manifested only several years after it is first applied.

Long-term studies conducted in Japan (Suzuki et al., 1990) showed that for the same addition of nitrogen, accumulation by plants was initially higher with inorganic fertilizers. After several years, however, compost gave markedly better results. In the best case, the beneficial effect of compost was seen after two to three years. Addition of certain elements by compost, nitrogen in particular, may not be sufficient in certain periods to ensure maximum growth. It is then useful to supplement the effect of compost by inorganic additions.

Application of compost often enables reduction of the impact of parasitic fungi and bacteria that attack plant roots. In some cases, however, the opposite phenomenon has been seen.

10.7.3 Addition or suppression of phytoparasites

By itself, compost can be a source of pathogens if ***propagules*** (spores, ***sclerotia***) originating from the plant material used for composting have not been killed during the composting process. Optimization of the thermogenic phase enables reduction, even elimination, of this hazard. Modifications caused in the soil or rhizosphere environment by application of certain types of compost sometimes favour existing pathogens or their action. For example, salt-rich composts stimulate development of some *Phytophthora* (Hoitink et al., in Hoitink and Keener, 1993). Parasites (*Phytophthora, Fusarium, Erwinia*) are favoured by very low C/N ratios, such as those of composts obtained exclusively from sewage sludges.

However, composts generally contribute to ***suppression*** of soil parasites by promoting the development or activity of antagonistic organisms. They may also play a more direct role by inducing at the root level resistance to penetration of the pathogen. The effects of microbes antagonistic to phytoparasites are of three orders:

- competition for bioelements, especially organic carbon—and therefore energy—and iron (see text box, Chap. 15, § 15.3.3);
- effect of antibiotic substances secreted by the antagonist;
- ***hyperparasitism***.

Competition results in widespread suppression, barely specific or not at all. In the other two cases, the phenomenon is often more targeted due to a specific antagonist and a specific parasite; this is specific suppression (Baker and Cook, 1974).

Compost can add suppressor organisms to the soil. Occasionally, suppressors develop spontaneously when the compost is recolonized by mesophiles after the thermogenic phase. The possibility of adding specific suppressor organisms to compost has also been suggested. The 'biological vacuum' created at the end of the thermogenic phase, before spontaneous recolonization of compost by mesophiles could be particularly favourable for establishing the inoculated organisms in the long term (Hoitink et al., in Hoitink and Keener, 1993).

Also, indigenous suppressing agents in the soil are occasionally favoured by the addition of compost. This occurs with general suppression, wherein the large biomass added with compost results in severe competition among the organisms. This competition often acts against pathogens such as *Pythium* and *Phytophthora* (Fuchs, 1995).

Propagule: in the case of a fungus, any structure that enables dispersal and multiplication of the organism or preservation of the fungus under unfavourable conditions (drought, cold, nutrient deficiency).

Sclerotium: in fungi, mass of poorly differentiated cells, generally hard and pigmented, dormant, through which the organism can face adverse conditions. The sclerotium germinates by occasionally giving fruit bodies directly. Ergot of rye is a sclerotium.

Suppression (of a parasite): intervention with a view to obstructing attack by a parasite. A suppressive organism may be an antagonist of the parasite in the soil; it may also induce in the plant a mechanism of resistance to the parasite.

Hyperparasitism: type of parasitism in which the antagonist is a parasite on the phytoparasite. Such are certain *Trichoderma*, fungi that invade and kill the sclerotia of *Sclerotium rolfsii* and *Rhizoctonia solani* (also see Chap. 13, § 13.3.2).

Biostatic: said of a chemical compound that inhibits growth of an organism at low concentrations without killing it. A prefix can specify the group to which the targeted organism belongs (e.g. bacteriostatic). The suffix *-cide* signifies that the compound kills the targeted organism (e.g. fungicide: compound that kills fungi).

Another effect of induction of this microflora is to stimulate production of pathogen-inhibiting ***biostatic*** compounds.

Generally speaking, better knowledge of the complex interactions among the host plant, pathogens, soil microorganisms and amendments added is necessary (Hoitink et al., in Hoitink and Keener, 1993).

10.7.4 Conditions of compost application

Few general rules...

The variety of composts makes it difficult to establish general rules for their use. First of all, their chemical composition (total organic carbon, nitrogen, phosphorus and potassium contents) must be examined, then their degree of maturity and the characteristics of the soil to be amended.

The time of application of compost is critical. Application of insufficiently matured compost should be avoided at sowing: there is risk of introducing toxic volatile organic acids in the soil and causing intense microbial growth that monopolizes the mineral salts. An autumn application results in loss of part of the nutrient elements following the nitrification (Chap. 4, § 4.4.4) and also in leaching of fine particles during the crop-free season.

...and fine, precise directives.

It is difficult to give directives regarding the quantities of compost to be used. According to the nature of the latter, application of 20 to 100 t ha^{-1} y^{-1} is recommended. Excessively large quantities result in very high soil-moisture content, reduction in oxygen concentration and lack of soil cohesion.

The mode of application is important.

Many specialists recommend broadcasting compost on the surface rather than mixing it into the soil, especially in the case of a one-time application of a large quantity (e.g. 200 to 300 t ha^{-1}). This avoids competition for nitrogen between roots and compost microorganisms. Furthermore, a superficial layer of compost counters the effects of rain such as leaching of mineral salts, soil erosion (Chap. 17, § 17.2.1) and crust formation. On the other hand, incorporation in the soil is preferable for repeated application of small doses.

Simultaneous application of compost and inorganic fertilizers: a synergistic effect on soil fertility.

Application of 100 kg ha^{-1} nitrogen as a mixture of compost and chemical fertilizer gives better results than application of the same quantity of nitrogen through compost or fertilizer alone (Dick and McCoy, in Hoitink and Keener, 1993).

10.7.5 Other uses

From fertilization of cultivated soils to creation of artificial soils.

As noted earlier application of compost contributes to amelioration of degraded soils and detoxification of contaminated soils. Compost may be used in place of peat as additive for establishment of reconstituted ANTHROPOSOLS. Sometimes composts mixed with crushed rock are used; for example, the materials resulting from tunnel cutting or debris from cement factories can be used to create artificial soils that can be greened. Such soils are used for highway or urban construction sites (Strehler, 1997) or for covering rubbish dumps (Galli and Parenti, 1990).

10.8 GARDEN COMPOST: A RESERVOIR OF ANIMAL BIODIVERSITY

For the zoologist, garden composts formed from kitchen wastes, cut grass and dead leaves are soil annexes (Chap. 8, § 8.1). Only detritus chains , quite appropriately named in the circumstances, are found in it. Two major conditions render garden composts very attractive environments for invertebrates:

- the variety and abundance of continually renewed food resources;
- the moist and moderately warm microclimate, which promotes growth of many species.

10.8.1 Some faunistic data

The first year's compost is a stratified environment in which 3 layers can be distinguished (Fig. 10.17; Plate III-3).

Top layer

A top layer densely colonized by insects with r strategy, efficient at locating and colonizing new environments (Chap. 6, § 6.1.3; Chap. 12, § 12.6.2).

Organic materials of diverse nature, kitchen leavings and garden wastes accumulate here to a thickness of about 30 cm. The fauna that feeds on this highly mixed litter is quite heterogeneous. It comprises phytosaprophages (larvae of Sphaeroceridae), microphages (nematodes, springtails, larvae of dipterans), coprophages (Hydrophilidae, Scarabaeidae) necrophages (larvae of Calliphoridae) and carnivores (gamasid mites, larvae of Empididae). Bacterial and fungal activity is particularly intense in this layer; microphages constitute abundant populations. Their predators, staphylinid beetles and

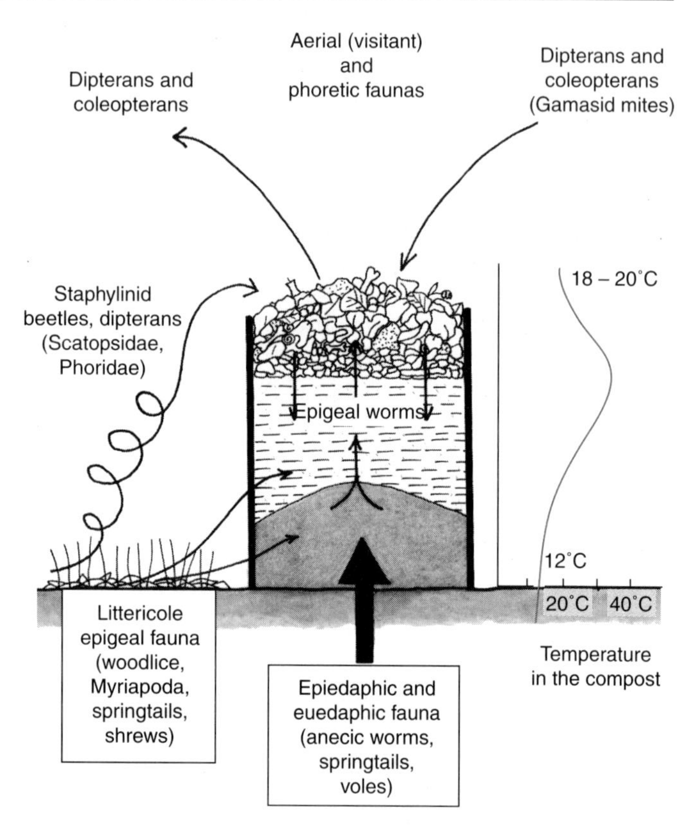

Fig. 10.17 *Cross-section of a garden compost. Arrows indicate origin of the fauna, and the curve on the side the general distribution of temperature.*

phoretic mites (Gamasida and Uropoda), are also abundant.

In one year, 12,400 dipterans were captured on a single compost by means of four yellow water traps (W. Matthey, unpublished).

In one year, more than a hundred species of adult dipterans were enumerated on compost, of which at least 25 to 30% developed there. In the same period of time, 1800 coleopterans of 14 families were captured, mostly staphylinid beetles that hunt in the upper few centimetres of the top layer, and also Hydrophilidae and Scarabaeidae of dung. The dipterans belonged to 45 families, 11 of which could be considered characteristic of these environments: Anthomyiidae, Chironomidae, Drosophilidae, Empididae, Muscidae, Phoridae, Scatopsidae, Sciaridae, Sepsidae, Sphaeroceridae and Trichoceridae. Outside composts, their larvae live in litter, on rotted fruits, in dung or on carcasses (Chap. 8, § 8.3).

Middle layer

The animal population of the middle layer is less typical.

Woodlice, pseudoscorpions, springtails and various larvae dominate this layer. This littericole fauna is composed mostly of species that come from the top (springtails, larvae, compost worms) and bottom (myriapods, diplurians) layers.

Bottom layer

The pedofauna, chiefly anecic worms, but also myriapods and woodlice, do considerable mixing of this layer. The most spectacular animals are vertebrates: shrews, voles and slow-worms often take up their abode here, mostly in winter.

The bottom layer is colonized by the soil fauna.

Species of the middle and bottom layers are found in the more uniform composts of the second year. Those of the upper layer will have disappeared or become rare.

10.8.2 Importance of garden compost in animal biodiversity

Apart from the species that live in them, garden composts form a continually renewed food source for many visitant insects: bees in spring, various wasps in summer and autumn, blue- and greenbottle flies throughout the year. Even birds (blackbirds, magpies, sparrows, tits) have learnt to sustain themselves on them.

The presence of an abundant decomposing fauna in urban and suburban environments often depends on garden composts.

Composts help to maintain or to extend the species that have become rare by disappearance of their natural habitats. An example is given by the rose chafer. The larvae of this superb beetle, which usually live in vegetable mould and rotten wood of old broad-leaved tree trunks, find an alternative habitat in garden composts.

What is known of the zoology of composts?

The zoology of composts and garbage has aroused the interest of investigators. The high faunal diversity of these environments, particularly of dipterans and microarthropods, the question of detritus food chains and physiology of organisms that inhabit them in spite of high temperatures and semi-anoxic conditions have stimulated numerous more or less site-specific studies. Among those devoted to microarthropods can be mentioned Gisin (1952), Streit and Roser-Hosch (1982) and Roser-Hosch and Streit (1982). Werner (1996) studied the dipterans of these environments. Lastly, Topp (1971) authored a paper still without equal on the ecology or composts. A general study on the zoology and ecology of different types of composts is lacking till today.

In the framework of this book, we only shall mention the broad lines of observations, yet unpublished, on a garden compost at an altitude of 750 m in the Jura mountains, in a semi-urban environment.

10.9 CONCLUSION

In a reactionary view, economics are often posed against environmental protection and ecology. Composting is a good example of the opposite.

Globality and long term: two key words for harmonious integration of economics and ecology, environment and sustainable development (Chap. 17, § 17.4).

Starting from a waste of negative value (it costs the community to dispose of it), a product of positive value useful for soil conservation is manufactured by composting: thus there is value-addition for the manager of wastes as well as the user. However, for inclusion of composting in improvement of cultivated soils, agricultural economics must necessarily take the long-term view: the compost applied today will give the full measure of its effect only some years later. It is also necessary that the community and the private sector view the waste problem in its totality, from source to treatment to application: more general recourse to the practice of composting involves measures right from the beginning of the process (conception and choice of primary materials, sorting of wastes) to the process itself (optimization, sanitation) and its end (utilization).

CHAPTER 11

BIOREMEDIATION OF CONTAMINATED SOILS

11.1 INTRODUCTION

Management of biosphere wastes has taken place almost cyclically and without problems in the history of life (Chap. 10, § 10.1, Fig. 10.1). Only a very small fraction was fossilized in each cycle, but its accumulation resulted in the formation of deposits of fossil fuels (coal, asphalt, oil, natural gas) as well as of diffuse compounds in sedimentary rocks.

When Man stopped following the rules of Nature.

Man has modified this cycle considerably through his activities and industry (Chap. 10, § 10.2, Fig. 10.2). Industrial wastes and consumer's wastes have led to dispersal throughout the environment of transformed or unchanged compounds that originated chiefly from organic fossilized materials (oil, coal) or inorganic compounds (ores). Obviously, the best means to counter this pollution consists of avoiding it! But, realization and implementation of this simple truth is recent. Furthermore, it is not always applied, far from it! Mankind must now manage millions of hectares of gravely contaminated soils.

Diffuse contamination: a problem that can be solved only at the source.

Soil contamination may be diffuse and widespread, or intense and localized. Diffuse contamination is perhaps the most problematic, because it is difficult to manage: pollutant concentrations, although significant, are relatively low; they can often be concentrated along food chains and create serious problems at the end of the food chain (Chap. 4, § 4.6.4; Chap. 13, § 13.3). On the other hand, the soil volumes affected are much too large, the diversity of contaminants too high and their concentrations too low to permit designing an appropriate treatment.

Intense, localized contamination often results from industrial activities and from uncontrolled dumping of hazardous industrial wastes. It may also result from an accident or from an act of war. Contaminant concentration is high, but only a limited number of chemical species occur, so that specific processes for treatment can be designed.

Landfilling is *not* a decontamination.

Polluted soils may be confined in appropriately designed landfills; this is not actually a decontamination process but rather a temporary evasion of the problem! The cost of landfilling is high, hence long-term monitoring and assessment of the environmental hazard are required.

Chemistry and physics are more expensive than biology!

Bioremediation: treatment of a milieu by a biological process for eliminating contaminating and toxic substances from it (see O'Connell et al., 1996, for example).

Soil remediation is possible by physical, chemical and biological means. Incineration and chemical extraction are very expensive and often lead to the destruction of soil structure. On the contrary, ***bioremediation*** generally allows the treated soils to recover, at least for certain applications. In principle, costs are lower overall.

Soil contaminants may be considered to be of two types: heavy metals, and organic compounds, which entail completely different biological processes for their elimination or inactivation.

11.2 BIOREMEDIATION OF SOILS CONTAMINATED BY HEAVY METALS: PHYTOREMEDIATION

11.2.1 Plants with high affinity for heavy metals

Unlike organic pollutants, heavy metals [see box on next page] are chemical elements, therefore, indestructible compounds. Radioactive isotopes may also be considered the same way. All treatments must therefore aim at extracting these substances from the milieu and to concentrate them for their final confinement. Although chemical extraction processes have been suggested, they are very costly and it is understandable that one turns to biologists to find other solutions.

Heavy metals

The ***heavy metals*** are chemical elements with relatively high atomic mass. Some of them such as iron (Fe), manganese (Mn), copper (Cu), nickel (Ni), cobalt (Co), zinc (Zn), molybdenum (Mo), tungsten (W), vanadium (V), are necessary in small amounts for living things and thus form part of the trace elements (Chap. 4, Tab. 4.2). They often function as inorganic *cofactors* of enzymes such as of nitrogenase (Fe, Mo, V), or even as constituents of organic cofactors, such as vitamin B_{12} (Co) or haemes (Fe). We may also mention the utilization of Mn as the oxidation-reduction intermediate of manganese peroxidase (Chap. 14, § 14.3.3) and of iron as electron donor in iron-oxidizing bacteria and as electron acceptor in iron-reducing bacteria (Chap. 4, § 4.4.2).

Other heavy metals, such as lead (Pb), cadmium (Cd), mercury (Hg) and silver (Ag) do not participate in structures or functions in living things. They are however present in small amounts in natural environments.

At higher concentrations, however, all of them exhibit a more or less elevated toxicity (Chap. 4, § 4.3.4) depending on ambient conditions. In fact, these elements manifest their toxicity in solution form, whereas their solubility depends on numerous environmental factors such as pH, redox potential, and presence of anions and of chelating molecules.... At the same total concentration, a heavy metal may show higher or lower toxicity according to the nature of the soil.

Relatively high amounts of a heavy metal in a soil do not necessarily result from anthropogenic pollution. It can originate from high contents in the parent material.

Establishment of norms, even if more or less arbitrary, is however necessary in order to define legal limits for toxicity in a soil (Tab. 11.1) as in composts (Chap. 10, Tab. 10.2). Ideally, each case should be considered individually rather than applying these legal norms blindly and uniformly. But this remains a politico-scientific challenge at present still insurmountable.

Heavy metals: a little bit is good...

...but too much is definitely not!

Table 11.1 Principal heavy metals contaminating soils and their indicative limiting concentrations, according to Swiss laws

Heavy metal	Level (mg kg^{-1} of dry matter)	
	Total	**Soluble**
Chromium (Cr)	50	—
Nickel (Ni)	50	0.2
Copper (Cu)	40	0.7
Zinc (Zn)	150	0.5
Molybdenum (Mo)	5	—
Cadmium (Cd)	0.8	0.02
Mercury (Hg)	0.5	—
Lead (Pb)	50	—

Indeed, many living organisms are able to solubilize and accumulate heavy metals and to concentrate them.

The most efficient in this are bacteria, followed by fungi and plants but they are difficult to extract from soil. Fungi can be more favourable *a priori*. It is known that fruit-bodies of mushrooms grown in contaminated soils are loaded with heavy metals. But it is difficult to control growth of fruit-bodies in inappropriate milieus except in rare cases. Plants, even though their accumulation and concentration capabilities are lower, remain our most useful helpers for extraction of heavy metals. Soil decontamination by plants is called ***phytoremediation***.

Plants to 'pump out' heavy metals...

It must be recalled, however, that plants are not alone in soils. They are associated in the rhizosphere (Chap. 15) with bacteria and mycorrhizal fungi (Chap. 16, § 16.1). Certain bacteria (PGPR, Chap. 15, § 15.3.3) promote plant growth by solubilizing inorganic nutrients, including heavy metals, from minerals (Chap. 4, § 4.4.5). Mycorrhizal fungi are also efficient mineral solubilizers and extractors. Moreover, they are able to take up and mineralize metal-organic compounds and transferring these metals to the plant as inorganic ions. By their extension in the soil (see Plate VII-2), the large surface area of their extraradicular mycelium and their great concentrating power they greatly increase the ability for heavy-metal uptake in the plant accommodating them. Indeed, arbuscular mycorrhizal fungi are often seen associated with plants growing on contaminated sites: they certainly play a major role in the capacity of these plants to accumulate heavy metals.

... with the help of microorganisms!

Plants that may be used for phytoremediation should exhibit a combination of several characteristics:

Superplants!

- they must be resistant to the heavy metals (and also, if need be, to other toxic compounds) present in the site to be treated;
- they must have high capacity to concentrate and accumulate the heavy metals to be extracted (***hyperaccumulating plants***);
- they must have rapid growth and high biomass production.

Fig. 11-A Thlaspi caerulescens, *the Alpine pennycress, an accumulator of Cd and Zn.*

Investigations in this field comprise selection of naturally hyperaccumulating plants, identification of the genes responsible for hyperaccumulation and lastly, modification by means of such genes of plants till now less efficient in accumulation but having rapid growth. However, the widespread use of genetically modified plants for phytoextraction is at present severely restricted by law in many countries. Research on optimization of phytoremediation should therefore give more emphasis to the benefits to be drawn from the use of mycorrhizal fungi and rhizosphere bacteria. Indeed, plants behave quite differently in a soil with all its microflora than in a sterile, hydroponics solution (Chap. 4, § 4.2.4)!

Two variants of phytoremediation will be considered here: phytoextraction and rhizofiltration.

11.2.2 Phytoextraction

Briefly, ***phytoextraction*** consists of cultivating, on contaminated soils, plants selected for their capacity to concentrate one or several heavy metals. The plants will then be cut, incinerated and the ash confined or the heavy metals extracted.

Among the heavy metals that can be accumulated by plants are lead, uranium, caesium, strontium, chromium, zinc, selenium, manganese....

A difficult choice.

It is not possible to apply a single given plant species to all situations. Choice of plant(s) (Fig. 11.A and 11.B) will depend in particular on:

- the heavy metal(s) to be extracted, the plant's capacity to accumulate it and its tolerance limit;
- the characteristics of the soil to be decontaminated;
- the depth which the soil has been contaminated;
- the climate.

If the soil has to be treated *in situ*, the zone of accumulation of heavy metals should not extend below the rooting zone of plants or the zone of extension of mycorrhizae (mycorrhizosphere, Chap. 4, § 4.1.3). Contamination from the surface is particularly suitable for this kind of treatment because most heavy metals are not easily leached down. It has also been shown that addition of certain amendments such as compost improves the concentration of heavy metals by plants. Small-scale trials should be attempted before making a choice of the best combination of accumulator plants and soil-preparation techniques, the whole process perhaps extending over several consecutive years.

Fig. 11-B Alyssum saxatile, *the Basket of Gold, a nickel accumulator.*

11.2.3 Rhizofiltration

Artificial marshes for natural elimination.

Rhizofiltration consists of creating 'artificial marshes', beds formed of flooded soils planted with marsh species selected for their capacity to accumulate heavy metals (Fig. 11.1). Contaminated water passes through these rhizofilters, the plants taking up the heavy metals and accumulating them. The plants are regularly cut,

Fig. 11.1 Artificial marshes for rhizofiltration. Umwelt-schutz Nord installations, near Bremen, Germany (photo M. Aragno).

incinerated and their ash containing the heavy metals in high concentration is confined. Such a system can also be applied to the treatment of sewage from a small community.

11.2.4 Case studies

Lead in the leaf!

A process of lead removal by phytoremediation was recently developed by Phytotec Inc., a spin-off company of Rutgers University (USA): plants accumulated up to 2% (dry matter basis) of lead, which is enormous compared to the usual content in plants. Such a treatment would cost just 15-25% of a conventional chemical treatment. A trial conducted in a residential area at Trenton (NJ) enabled the concentration of lead to be lowered by 75% (for example, from an original concentration of 1500 ppm to less than 400 ppm) after three plant generations.

Plants for Tchernobyl... and for nuclear power!

Phytoextraction has also been successfully tested for the removal of radioactive isotopes. In particular, tests were conducted in the Tchernobyl region, Ukraine, for extraction of ^{90}Sr and ^{137}Cs from heavily contaminated soils. Similar success has been achieved in extracting uranium from contaminated soils in regions where nuclear weapons testing was done. Rhizofiltration has also been successfully used to extract uranium from underground water in mining regions and in the vicinity of nuclear plants. Varieties of sunflower are particularly efficient in concentrating uranium: ratios of 30,000:1 between the concentrations in root biomass and water have been attained. The residual concentration in the water was brought to very low values, below the norm of 20 $\mu g\ L^{-1}$.

11.3 BIOREMEDIATION OF SOILS CONTAMINATED BY ORGANIC COMPOUNDS

In this case, very different from that of heavy metals, the aim is not to extract and concentrate the pollutants to confine them elsewhere, but rather to promote microbial activities in order to destroy and mineralize them. The players will then essentially be heterotrophic microorganisms, fungi and, most of all, bacteria. The soil

Organic pollutants

Different categories of organic compounds may be considered soil pollutants:

- Compounds occurring as such in nature. These are as a rule completely biodegradable under certain conditions. They are to be considered polluting if:
  — their concentration is abnormally high and induces great stress in the soil biocoenosis, which can lead to inactivation of potential biodegradational mechanisms and to loss of the homoeostatic properties of soil;
  — their chemical nature prevents them from being degraded under the existing soil conditions, for example decomposition of many hydrocarbons is impossible under anoxic conditions.
- ***Xenobiotics***: artificial compounds invented by man and differing greatly in their chemical structure from compounds synthesized by living things. We may distinguish here:
  — inert, non-degradable, non-toxic compounds that do not interact with living organisms and behave as fillers. Most plastics belong to this group;
  — toxic, more or less degradable compounds that pass the cell barrier when plants draw their nutrients (Chap. 4, § 4.2.1; Table 11.2). These substances may get concentrated along food chains, leading to physiological harm, to severe poisoning or to death. Many examples are known in carnivorous animals, and this even if the concentration of toxic compounds in the soil is not itself damaging. For example, the very harmful effects of certain PCBs (polychlorobiphenyls) on super-predator birds such as peregrine falcons, whose eggs develop thin shells that crack during brooding (Chap. 4, § 4.6.4).

'Natural' pollutants comprise petroleum derivatives, which often leak from industrial establishments. Most are aliphatic or aromatic compounds whose degradation requires molecular oxygen, through the mediation of oxygenases (Chap. 4, § 4.4.4, oxygen cycle).

Microbial biodegradation mechanisms have adapted over geological time to the *invention* of new, natural organic structures: this evolution occurred in 'small steps', so that the probability of appearance of mutations allowing an enzyme to degrade a 'new' substrate was high. This rule was broken since the invention by organic chemists of structures greatly differing from those of biotic compounds—the *xenobiotic* compounds (literally, 'foreign to life'). The probability of mutations able to produce enzymes capable of degrading them may be extremely low, particularly in a short time scale.

Nature produces pollutants...

...but man has invented worse than that!

to be decontaminated is considered a culture medium, in which microbial activities have to be optimized towards biotransformation and/or biodegradation of the compounds to be eliminated.

Table 11.2 Examples of xenobiotic compounds and their aerobic and anaerobic biodegradability

Group	Example	Possibility of degradation	
		Aerobic	Anaerobic
Aromatic monochlorine compounds	Chlorobenzene	+	–
Benzene, toluene, xylene		+	(+)
Non-halogenated phenolic compounds and cresols	2-methyl phenol	+	(+)
Aromatic polycyclic hydrocarbons	Creosote	+	–
Alkanes and alkenes	Petroleum	+	–
PCBs (polychlorobiphenyls)	Trichlorobiphenyl	+	+
Chlorophenols	Pentachlorophenol	+	+
Heterocyclic nitrogen compounds	Pyridine	+	+
Chlorinated solvents:			
Alkanes	Chloroform	+	+
Alkenes	Trichloroethylene	(+)	+

11.3.1 Optimizing biodegradation by promoting microbial activities related to it

To let be or to act, that is the question!

In certain cases, decontamination is spontaneous; just a follow-up is required by analysis of the concentration of contaminants in the soil *in situ*. Most often, however, it will be necessary to promote biodegradation by more or less intensive practices. Here too, as in phytoremediation, there is no general 'recipe'. The most suitable techniques must be tested and selected in each case. Among the different points to be considered are:

Supermicrobes for sale!

Bacterial strains: the production of strains specialized in decontaminating soils is an interesting development..., sometimes economically fruitful. In some cases, adaptation to xenobiotic substrates has been obtained by accelerated 'artificial evolution' at the laboratory scale in specially designed bioreactors. But such organisms are not necessarily capable of competing with native organisms for being acclimatized in the complex community of soil. In most cases, the organisms most capable of decontamination are already present in the contaminated soil, which represented for them a milieu of spontaneous enrichment. Most often, the decontamination process consists of optimizing the conditions for growth and activity of these 'spontaneous' bacteria rather than addition of preselected strains. We have to distrust sellers of 'magic potions'!

A welcome oxygenation!

Processes of aerobiosis/anaerobiosis: degradation of polluting compounds often requires the presence of oxygen (Tab. 11.2). In fact, their degradation could involve oxygenases, enzymes that catalyse a direct attack on the organic molecule by molecular oxygen (Chap. 4, § 4.4.4, oxygen cycle).

Improvement of oxygenation of a soil may be achieved by turning, by addition of structure-forming organic amendments that have the effect of lightening the soil, or by insufflation of air.

However, in some cases, in organochlorine compounds, anaerobic reduction of the C—Cl bonds is more rapid and efficient:

$$-\overset{|}{\underset{|}{C}}-Cl + 2[H] \rightarrow -\overset{|}{\underset{|}{C}}-H + HCl$$

Such a reaction occurs in an anaerobic respiratory process (Chap. 4, § 4.4.2, Fig. 4.19), in which the chlorinated compound acts as electron acceptor. This phenomenon has been particularly well studied in the anaerobic reduction of tetrachloroethylene (see box and Fig. 11.2).

Degradation of tetrachloroethylene

Tetrachloroethylene is a xenobiotic compound with low toxicity and low volatility. It is, therefore, widely used as a solvent, for example in laundries. It is frequently found as a soil contaminant. Under anaerobic conditions it undergoes reductive dechlorination leading to the successive formation of trichlorethylene, the three isomers of dichloroethylene, and finally of the monochlorinated compound, vinyl chloride.

The more the dechlorination the greater the volatility and toxicity, vinyl chloride being the most toxic and volatile of all. If the conditions remain anoxic, this compound accumulates and problems can then arise. Further degradation can, however, take place under oxic conditions; aerobic bacteria accomplish total mineralization of vinyl chloride, and thus its detoxification. Complete elimination of tetrachloroethylene therefore requires a succession of anoxic and oxic conditions.

Give them this day their daily nutrients!

Nutritional aspects: The soil to be decontaminated must supply to the bacteria responsible for the biotransformations the essential bioelements at a non-limiting concentration. The mineral salts in short supply must be added to soils with deficiencies.

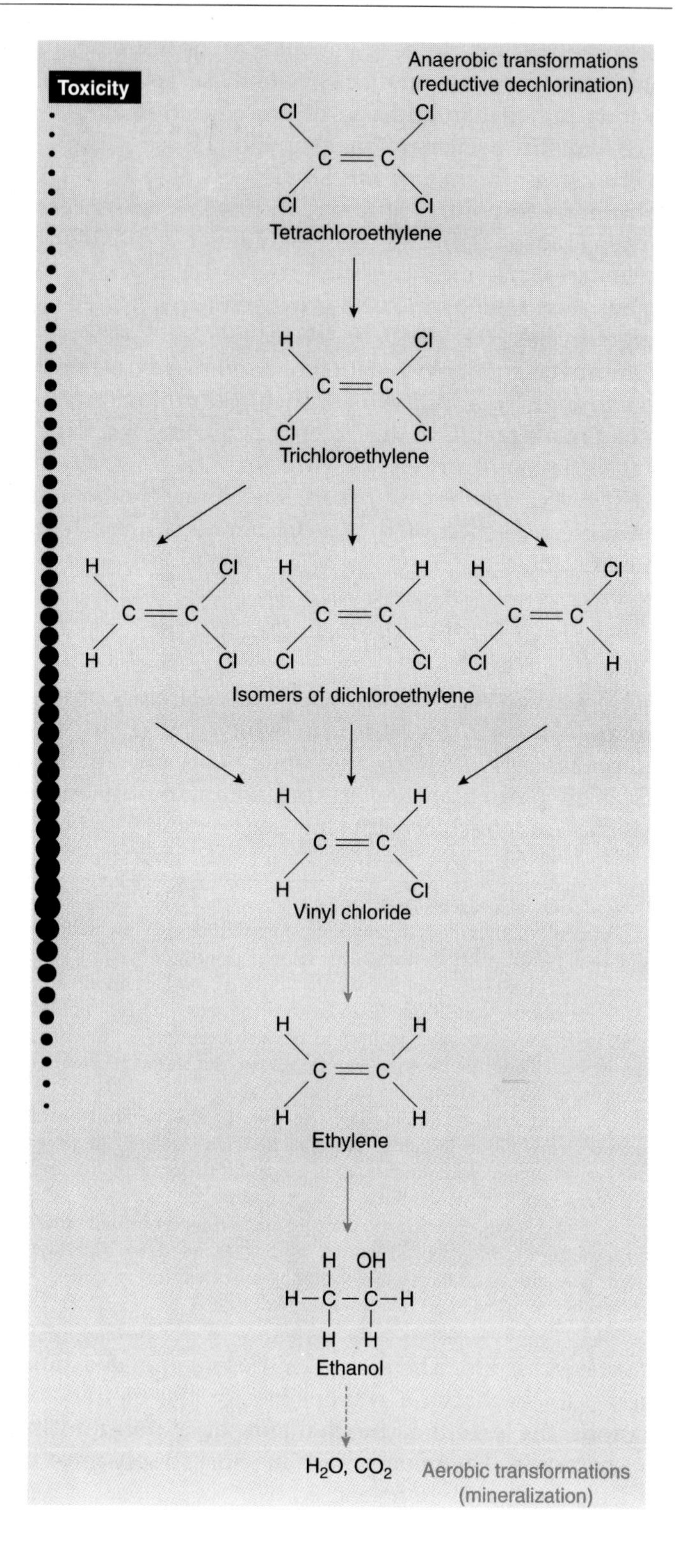

Fig. 11.2 Biodegradation of tetrachloroethylene.

Generally, the concentration and/or nature of organic contaminants does not permit fast and vigorous growth of the organisms responsible for their elimination. Addition of non-toxic organic substrates may promote bacterial growth and increase bacterial biomass; this practice, however, may lead to a few problems:

- it may stimulate growth of organisms different from (and concurrent with) those that are sought to be stimulated;
- it may induce ***diauxic*** behaviour in the organisms, that is, repression of the utilization of the target substrates when the more easily degradable added substrates are utilized.

On the other hand, addition of complementary substrates (***co-substrates***) is necessary when the compound to be eliminated is not used as food by the bacterium, but may be degraded by broad-spectrum enzymes induced (Chap. 14, § 14.2.5) during the degradation of another compound (***cometabolism,*** Fig. 11.3).

Like a sugarcoated pill!

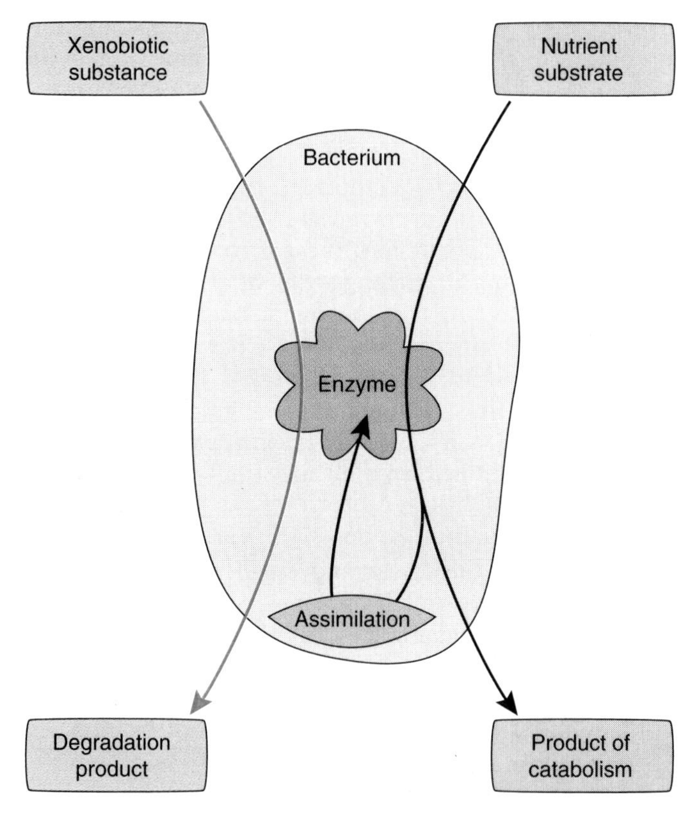

Fig. 11.3 *Principle of cometabolism.*

Broad-spectrum oxygenases
Oxygenases catalyse the direct oxidation by molecular oxygen of aliphatic or aromatic hydrocarbons. They often have a broad spectrum of activity, which permits them to attack a wide variety of compounds, in particular xenobiotic compounds that are then degraded by ***cometabolism*** (Fig. 11.3). Tests of injection of gaseous hydrocarbons such as methane, propane and butane into contaminated soils have been successful; the treatment induced synthesis of the related oxygenases, which were also able to oxidize xenobiotic contaminants as co-substrates.

Where plants provide the logistics!

The rhizosphere (Chap. 15) may promote cometabolism of xenobiotics by supplying suitable substrates or broad-spectrum enzymes through root secretions. This would then be a sort of indirect phytoremediation, in which the plant does not function as an accumulator of the toxic compounds but as supplier of 'tools of degradation'.

11.3.2 Immobilization of xenobiotics in humic matter

Put the xenobiotics in jail!

Another way of detoxification of some xenobiotics consists of immobilizing them very firmly in the humified organic matrix of soil. This prevents their leaching into ground waters, ultimately impeding their entry into living cells and subsequent concentration in food chains. A number of aromatic derivatives of phenols (parathion, oxadiazon, 2,4-D, 2,4,5-T) and anilines (phenylureas, phenylcarbamates, acylanilines, dinitroanilines, 2,6-dichloro-4-aniline) are currently used as pesticides. Like natural phenols and anilines, these xenobiotic compounds, or the products of their partial degradation by soil organisms, may undergo enzymatic oxidation by broad-spectrum polyphenol-oxidases (Chap. 14, § 14.3.3), leading to the formation of reactive radicals and quinones. These are then integrated in humin molecules by oxidative coupling (Chap. 4, § 4.4.3) , in the same way as the natural compounds. In this form they are completely inactivated (Bollag and Loll, 1983).

Fixation of xenobiotics in humic matter: a time bomb?

But what about the long-term behaviour of these compounds? In the cases studied, release of xenobiotics during degradation of humin was very slow, and the xenobiotics were mineralized to the same extent (Bollag, 1991). These results are encouraging, but more studies should be attempted with a wide spectrum of potential contaminants susceptible to such a cycle of fixation and release (Chap. 17, § 17.2.2; Rivière, 1998).

11.3.3 Decontamination procedures

No panacea for decontamination!

As stated earlier, in bioremediation there is no recipe universally applicable to any type of contamination and of soil. Each case is unique, and companies dealing with soil decontamination must have available an analytical laboratory and pilot-trial systems in order to select the optimum procedure before acting on a large scale. At the same efficiency, it is obvious that the cheapest and the least 'heavy' for the environment is the best. In order of increasing 'interventionism' may be mentioned:

Ambulatory treatment of the soil...

- *Soil treatment in place.* Treatments *in situ* are the most desirable, but they are mainly applicable to surface contaminations.
  - — *Letting Nature do it* by spontaneous decontamination: monitoring the changes in concentration and distribution of the contaminants by regular analyses.
  - — *Compensating for nutrient deficiencies* by addition of inorganic salts (fertilizers).
  - — *Improvement of soil structure* to promote aeration in it, by turning and/or application of amendments (e.g. compost, see Chap. 10, § 10.7.1).
  - — *Soil active aeration* to boost *in situ* oxidation of organic contaminants. Depending on their volatility, this would also lead to extraction by volatilization.
  - — *Saturating the soil with water* to make it anoxic, at least temporarily, so that anaerobic degradation processes are promoted.
  - — *Establishing a system for percolation through the soil* (Fig. 11.4) recycling the percolate through a **bioreactor**. Here, the main role of water consists

Bioreactor: vessel in which a biological reaction (e.g. culturing a micro-organism) or biochemical reaction (e.g. enzymatic) is conducted. If need be, it is provided with systems for stirring the medium and controlling the physico-chemical environment.

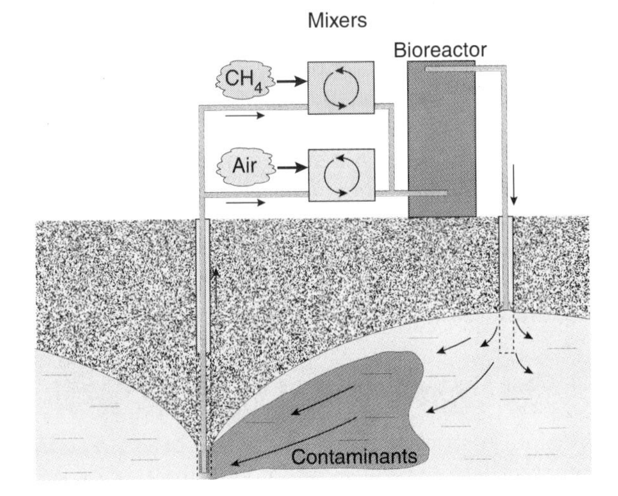

Fig. 11.4 *Percolation system for extraction and treatment of organic contaminants of soil with a bioreactor on the surface.*

of extracting the compounds, which will then be eliminated in the bioreactor installed at the surface. The system works continuously, and the decrease of concentration in the aqueous phase of the compounds to be eliminated is monitored.

... or hospitalization?

• *Treatment on an area.* In this case, the soil is transported from its original location to an intensive-treatment area.

— *Treatment in beds or heaps* (Fig. 11.5): the soil is spread over the treatment area in beds of varying dimensions. The area is often covered, and caution is taken to avoid dispersal of volatile toxic or foul-smelling compounds to the environment. Water, mineral salts or organic amendments may be added to increase biological activity, and the beds are regularly turned. A percolate-collection system may also be installed.

Fig. 11.5 *Treatment on an area of a contaminated soil (Umweltschutz Nord, Germany; photo M. Aragno).*

— *Composting:* treatment similar to the preceding, but the soil is mixed with a fresh compost substrate. The process is then thermogenic and thermophilic, as in a normal compost (Chap 10, § 10.3.1). High temperatures may accelerate the degradation and cometabolic processes, but the variety of organisms active at high temperature may be small.

• *Treatment in bioreactor.* This is the costliest mode of treatment but often also the most efficient. A treatment in the solid phase may be though of, with a bioreactor with biofiltration or percolation, or treatment in the stirred and aerated liquid phase in the form of a suspension (mud).

11.3.4 Case studies

Two examples, among so many others!

We shall restrict ourselves here to a brief presentation of two studies dealing with bioelimination of organic contaminants. Bioremediation systems are coming more and more into use, as much for their efficiency as for their lower cost compared to a physical or chemical treatment. The interested reader is exhorted to consult specialized literature or to make a search on the Internet, which is full of information in this domain.

1. Degradation of polycyclic aromatic hydrocarbons by a bacterial consortium

Polycyclic aromatic hydrocarbons (PAH) are frequent contaminants of soils and underground water. They come from accidental leaks of petroleum distillates or industrial solvents. Accumulation of PAH in the environment can cause great damage to ecosystems and pose serious health risk to humans, animals and plants. Mixed bacterial cultures (*Achromobacter* spp. and *Mycobacterium* spp.) have been found to be particularly efficient for rapid biodegradation of PAH. Indeed, the first organism breaks down the large PAH molecule into simpler organic molecules, which are completely mineralized by the second.

PAH: aromatic core, but hard to swallow!

2. Treatment of a soil heavily contaminated with crude oil in a treatment cell in the vicinity of the polluted site

A leak of crude oil had occurred in a petroleum storage facility in southeastern Texas. More than 1000 m^3 of clayey soil thus had to be treated after having been scraped off (according to D. Vance on the Internet).

Oil runs, but bacteria are going to recapture it!

In order to test feasibility and to optimize bioremediation of this site, a study was first of all taken up to find out if the soil to be treated contained autochthonous bacterial populations capable of degrading the components of oil. The soil properties that could be affected by the treatment were determined, as also the nutrient supplements to be applied. Studies in microcosms at laboratory scale enabled evaluation of the effects of

The microcosm, a technique for getting closer to natural conditions

Studies in soil biology are more and more taking recourse to the experimental technique of the microcosm, an apparatus that permits us reconstruct, more or less faithfully, the real soil existing in the field. A microcosm is generally formed from a plastic cylinder fitted with a grill at its base and various appendages depending on the experiments planned. Sometimes microcosm is distinguished from mesocosm according to the size of the cylinder. Figure 11.6 shows an installation designed for the study of the first stages of structuration of alluvial soils by earthworms (Vadi et al., 2002).

The microcosm, or the soil at home.

To learn more about microcosms, their possibilities and their limits, see for example Kampichler et al. (2001), Chen et al. (2001) and Martinez and Medel (2002).

aeration and addition of nutrients. In less than six months, with aeration and application of ammonium and orthophosphate, degradation was optimum.

Fig. 11.6 Microcosms containing reconstituted alluvial soils (photo F. Hainard and G. Vadi).

The selected system of treatment consisted of a treatment area of 35 m × 35 m, covered with a thick polyethylene sheet and surrounded by a 1-m enclosure. This area was given a gentle slope to permit collection of the leachate. Aeration was provided by slotted PVC pipes laid along the bottom of the bed and connected to an aspiration system. The soil to be treated was deposited in a layer 90 to 120 cm in thickness. The leachate was collected and recycled on the layer after addition if needed of nitrogen and phosphorus salts.

Follow-up was ensured by analyses done at regular intervals on soil moisture, nutrient concentrations and concentration of hydrocarbons (Fig. 11.7). Bacterial counts

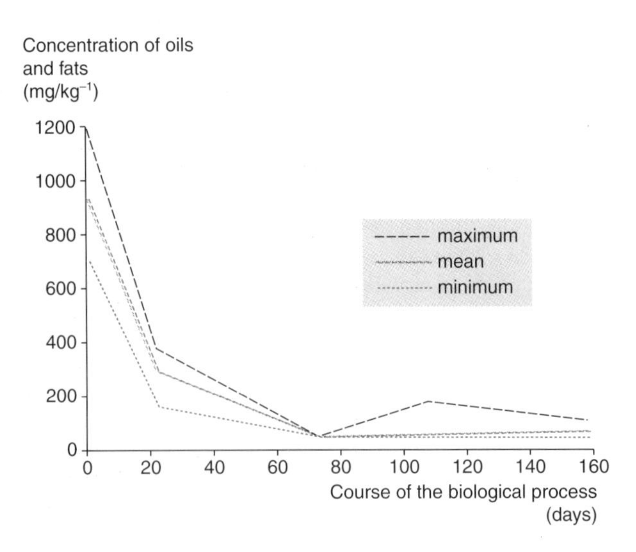

Fig. 11.7 Decrease in hydrocarbon concentration in a contaminated soil treated close to the site.

(ordinary saprophytic flora and hydrocarbon-degrading flora) were also taken. The treatment was stopped after six months, when the mean concentration of hydrocarbons had dropped from about 1000 mg kg^{-1} to <50 mg kg^{-1}. The soil could then be put back in place.

11.4 CONCLUSION

Bioremediators: firemen of the soil!

The bioremediation of contaminated soils is an expanding domain of applied research. But it is not a panacea. Often many years will be needed for decontaminating a soil which required but a few minutes to be polluted; also, biological techniques are not applicable to diffuse pollution. Bioremediation is a partial remedy for the irresponsibility of polluters; it should not be considered a 'license to pollute'. Bioremediation companies are the 'firemen of soils'. However, just as for fires, the main thing is the action has to be done in advance, by preventive measures regarding pollution.

CHAPTER 12

ANIMALS AND ECOLOGICAL NICHES

Many animal species have colonized the soil. They live together in it in hundreds, forming a sort of 'skeleton' of chains of decomposers (Chap. 13, § 13.6). But the efficiency of invertebrates would appear derisory if they were not associated with bacteria and fungi.

An inseparable trio, veritable engine for the evolution of soil organic matter.

Evolution of organic substances forms the red thread of the second part of this book. From this perspective, this chapter presents the animals responsible for transformation of these substances and their functional role in the ecosystem. General information, in particular methodological, forms the subject of §§ 12.1 to 12.3. The soil fauna is then presented according to two complementary approaches:

- first, systematics (description of the organisms) with illustrations of major groups of the pedofauna (§ 12.4);
- second, based on functional specificity of the animals, it follows the idea of ecological niche, which enables us understand why and how so many species coexist in the same soil (§§ 12.5, 12.6).

12.1 AT WHAT STAGE IS SOIL ZOOLOGY?

Animals that live temporarily or permanently in the soil, at its surface or in its annexes (Chap. 8) constitute the ***pedofauna***. Their study is not so recent as is often said. For example, as early as 1837, Darwin studied the

Darwin, one of the fathers of soil zoology.

role of worms in the formation of 'vegetable mould'—which he elsewhere preferred to call 'animal mould' (Boulaine, 1997)—at the same time as Ehrenberg drew attention to the possible role in the soil of protozoans and other microorganisms (Kevan, 1962).

Thanks to improvement and miniaturization of scientific equipment, we have in one-and-a-half centuries progressed from mere description of soil organisms to sensory physiology and synecology, through fundamental as well as applied research. But the so-called conventional approaches, such as systematics and 'natural history' of animals, remain as indispensable as in the past and support all the others.

Three attitudes towards soil zoology:

The missing link...

...a sometimes exaggerated perfectionism...

...uneven knowledge but quite genuine.

Three attitudes

Three attitudes towards soil zoology exist:

- That which consists of saying that it is the link which, if not missing, is at least the weakest in our understanding of the soil system. This is the view of those who, working most often in other domains of soil biology, are not in the know of the very many new studies on the pedofauna and its role.
- That of highly specialized systematists who consider that so long as some species are still under study in a taxon, nothing is known about the totality of that particular group.
- Lastly, that of informed generalists which is ours as well as. They aver that knowledge is not non-existent but uneven, some groups being much better understood than others at the level of systematics. Earthworms, isopods, mites and termites are examples. Contrarily, other important taxa are definitely less explored. Such is the case of dipterans with edaphic larvae, whose importance is great in the soil. Knowledge of them requires increased efforts.

'The status of soil can be assessed by using soil fauna (especially ciliates, nematodes and microarthropods) as bioindicators, and it is important to present the techniques available to estimate soil faunal biodiversity' (Benckiser, 1997).

The ensemble of these approaches results in the degree of ignorance in soil zoology not being higher than in other biological disciplines. Indeed, as in the entire scientific domain, there is much to discover; but it is not correct to say that everything remains to be done. In our opinion, it is necessary to:

- conduct a broad synthesis of the numerous publications dispersed in the literature;
- integrate this knowledge with those of soil science, mycology, bacteriology, etc., an effort that demands an interdisciplinary approach such as the one attempted by the authors of this book.

12.2 TOOLS OF THE ZOOLOGIST

Methods used for enumerating and studying the soil fauna should be adapted for different groups of soil animals: protozoans, nematodes, enchytraeids, earthworms and arthropods. This methodology would itself fill one large chapter and so we redirect the reader to some reference works: Lamotte and Bourlière, 1969; Vannier, 1970; Phillipson, 1971; Bachelier, 1978; Southwood, 1978; Dunger and Fiedler, 1997. Three methods frequently used to study arthropods illustrate the general procedure:

Study of the soil fauna consists in studying the contents of a black box!

- the Barber trap,
- the Berlese-Tullgren extractor,
- the emergence trap.

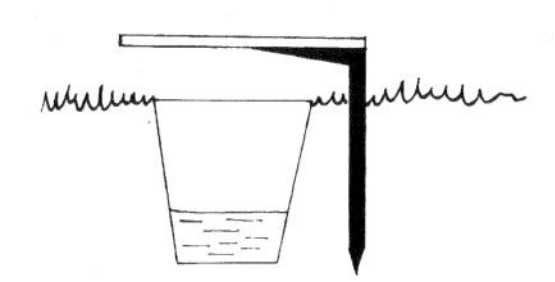

Fig. 12.1 The Barber trap. Each trap consists of a plastic cup buried in the soil with its rim a little below the soil surface. The receiver is one-third filled with a non-volatile liquid preservative (ethylene glycol or picric acid) Arthropods fall into the trap neither attractive nor repellent to them, by their random movements; they are killed and fixed by the liquid preservative.

12.2.1 How to capture epigeal arthropods?

Epigeal macroarthropods (spiders, insects, myriapods and woodlice) are captured with the help of ***Barber traps*** or ***pitfall traps*** (Fig. 12.1), named for their 'inventor'. In principle, the content of the traps should be taken every week. Their number and distribution vary depending on the purpose but in practice ten traps per site suffice to capture the very great majority of mobile species in the area. The Barber trap is an activity trap. It does not allow estimation of the exact number of ground beetles or staphylinid beetles living permanently in a given area but indicates the quantity of these insects active during the trapping period. On the other hand, by combining the Barber trap with marking techniques, the population strengths can be determined.

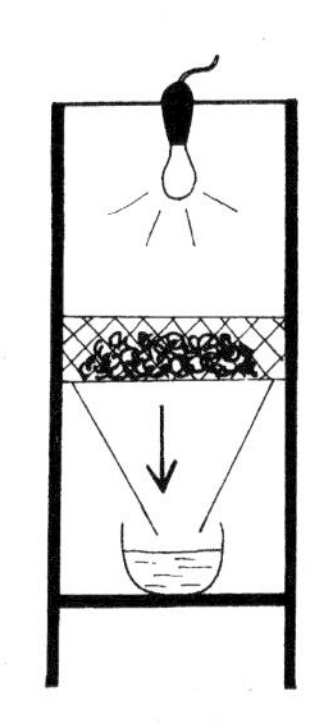

Fig. 12.2 The Berlese-Tullgren funnel. Driven out by dryness, the fauna leaves its home and falls in the receiver. High precision can be attained if the sample size and number of samples are correlated with the size of the organisms considered, macroarthropods or microarthropods.

12.2.2 How to capture endogeal arthropods?

The ***Berlese-Tullgren funnel*** (Fig. 12.2) is particularly convenient for studying endogeal arthropods. Its principle consists of placing in it a known volume of soil for slow drying from top to bottom. To estimate the populations of springtails, ticks and mites, several small samples are drawn by means of a probe (5-cm dia., 2.5-cm high), Macroarthropods are extracted from a smaller numbers of larger samples (e.g. 20 cm × 20 cm × 5 cm) (Fig. 12.3).

The ***emergence trap*** (Fig. 12.4) enables capturing aerial insects with almost immobile edaphic larvae. This is the

Fig. 12.3 Result of an extraction using the Berlese-Tullgren apparatus. The Swiss 5-centime coin (17-mm dia) provides the scale (photo Y. Borcard).

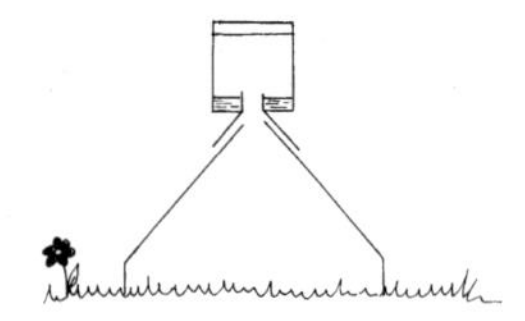

Fig. 12.4 The emergence trap isolates a 0.1 to 0.25 m^2 area of the soil.

case with numerous dipterans and coleopterans. Designed to collect insects when they hatch, it enables sufficiently precise estimation of abundance of the organisms considered.

12.3 AFTER CAPTURE, IDENTIFICATION

The history of biology shows that it is necessary to identify with greatest exactness the animals, plants, fungi and bacteria that are studied. This is a fundamental rule and that is why systematics, the earliest discipline in natural sciences, retains its relevance.

'Aristotle actually contributed by his methodical approach to development of taxonomy (systematics)... While describing a group he gave priority to morphological-functional properties he himself had observed' (de Wit, 1997).

Since antiquity, with Aristotle, naturalists have been sorting and comparing living things according to different criteria. They built classification systems that were perfected as new techniques, based on molecular biology and genetics, provided supplementary comparative criteria (Chap. 15, § 15.5.5). It is thus logical that classification of living things evolves continuously. But it is also impossible to present a system that could bring about unanimity. Contact between phylogenetic nomenclature (Ridley, 1997; Lecointre and Le Guyader, 2001) and the 'utilitarian' nomenclature frequently used today by practitioners of biology seems difficult for the time being. In this book we shall hold on to the latter and the systematic levels used remain ***phylum, class, order, family, genus*** and ***species***. Intermediate categories such as superfamily, subfamily, subgenus, etc., complete this hierarchy (Table 12.1).

'Insects are numerous and belong to many species... Some are winged such as bees, others winged as well as wingless such as ants; some lack wings and legs' (Pliny, in *Naturalis Historia*, Book XI, Chaps. 1 and 2).

Table 12.1 Examples of classification of soil animals

Systematic category	Mole	Field cricket	One of the earthworms
Phylum	Vertebrata	Arthropoda	Annelida
Class	Mammalia	Insecta	Oligochaeta
Order	Insectivora	Orthoptera	Haplotaxida
Family	Talpidae	Gryllidae	Lumbricidae
Genus	*Talpa*	*Gryllus*	*Lumbricus*
Species	*europaea*	*campestris*	*terrestris*

Taxonomy: a tool more than an intangible biological reality

The idea of taxonomic level remains a mental view, a tool enabling the taxonomist to 'fix' a classification and hierarchize it. Biological reality exhibits an evolutionary continuum in which the taxonomist establishes, often arbitrarily, hierarchical levels capable of expressing in discrete units the information provided by comparison of living things. Homonymous hierarchical levels thus have very different meanings according to the group of organisms considered, or even according to the investigator who studies it...

In biology as in soil science (Chap. 5, § 5.6.1) or in phytosociology (Chap. 7, § 7.1.4), the art of classification is difficult.

12.4 TOWARDS A LITTLE MORE KNOWLEDGE OF SOIL ANIMALS

This section presents in a condensed manner the principal groups of the pedofauna under the following headings (with occasional exceptions): taxonomic position, morphological characteristics, size, number of species, geographical distribution, habitat, location in soil, life mode, nutrition, abundance, significance in the soil, reference books. Drawings in the margins illustrate some examples of organisms. The nomenclature used is widely scattered in books on soil zoology. Although it does not always correspond to the present-day fully evolved taxonomy, it retains its value from a functional viewpoint (see also Lavelle and Spain, 2001).

This section initiates discussion of the ecological niche, which will be developed in §§ 12.5 and 12.6.

For an introductory overview of forms of soil arthropods, see for example Coineau et al. (1997) and Zettel (1999).

12.4.1 Protozoa, midgets of the pedofauna

All the Protozoa are grouped together here for convenience in the single kingdom of Protista. But recent studies based on molecular criteria have shown that there are in fact more than ten distinct kingdoms (Patterson, in Hausmann and Hulsmann, 1994). They also differ from the animal and plant kingdoms and fungi. The terms Protozoa and Protista have even disappeared from recent

Protozoa: very diverse organisms probably belonging to several kingdoms.

Zooflagellates: 1-2 flagellae, 3-10 μm

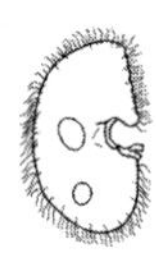

Fig. 12-A *Soil Protozoa.*

books on phylogenetic classification (e.g. Lecointre and Le Guyader, 2001).

Taxonomic position. Kingdom of Protista. Subkingdom Protozoa. 11 classes, of which three are important in the soil: the Zoomastigophora or Zooflagellata, Rhizopoda (naked amoebae and thecamoebians or testaceans; Chap. 9, Fig. 9.4; Plate I-4) and Ciliata (Fig. 12-A).

Morphological characteristics. Protozoa are unicellular eukaryotes. Their reproduction by ***fission*** gives them great ***biotic potential***.

Size of species. Protozoa form part of the microfauna, from 3 to 250 μm, occasionally up to 3 mm.
Number of species. Nearly 50,000 known.
Geographic distribution. Worldwide.
Habitat. Interstitial soil water and film water on surface of aggregates (Chap. 3, § 3.4.2).
Vertical distribution in soil. OL, OF and A horizons.
Life mode. Free-living, but parasitic and symbiotic soil species exist. ***Cysts*** are resistant dispersal forms.
Nutrition. Bacteriophagous, saprophagous, predators on other protozoans.

Fission: asexual reproduction of the organism by division. Each part again forms a complete animal.

Biotic potential: maximum proliferating capacity of a species.

Abundance. Except for bacteria, these are the most abundant organisms in soil. Estimates, most often extrapolated from number of individuals per gram dry soil, should be prudently considered: they range for all the Protozoa taken together between 10^7 and 10^9 ind. m^{-2} (Lavelle and Spain, 2001). Up to 823,000 Zooflagellata, 823,000 naked Amoebae, 73,000 Thecamoebae and 133,000 Ciliata have been counted per gram dry soil (Dunger and Fiedler, 1997; Bamforth, in Sumner, 2000).

Reference works. Bamforth, in Benckiser, 1997 (modern introduction to soil ecology and methods of study); Lousier and Bamforth, in Dindal, 1990 (presentation of the group and identification key up to genus; cites American and Canadian work); de Puytorac et al., 1987 (general introduction to Protozoa).

Significance in soil. Very great in the biological equilibria at the level of microorganisms. Protozoa are the principal predators of bacteria. Among them, naked amoebae are the most active bacterivores because, with their pseudopodia, they reach the finest pores (<6 μm) from where they expel their prey. Protozoa also have an important symbiotic role, degrading cellulosic tissues or stimulating bacterial activity in the gut of their host (e.g. ciliates in earthworms and flagellates in lower termites; § 12.4.9).

A soil nematode, *Coenorhabditis elegans*, has been selected as model animal in many current studies in fundamental biology!

12.4.2 Nematodes, small worms in great abundance

Taxonomic position. The phylum Nematoda comprises 2 classes, 20 orders and nearly 200 families.

Morphological characteristics. These non-segmented, fusiform or vermiform worms show great structural uniformity (Fig. 12-B). The mouthparts and position of sensory organs on the body are used to identify species. The sexes are separate. Growth takes place through moults of the cuticle.

Size of species in temperate regions. Soil nematodes usually range in length from 0.5 to 3 mm, with diameter varying between 1/20 and 1/50 of length.
Number of species. 20,000 known globally; there may perhaps be 100,000.
Geographic distribution. Worldwide.

Habitat. Nematodes are aquatic animals living in film and interstitial water, rhizosphere and decomposing material (they are very plentiful in garden composts; Chap. 10, § 10.8). They form cysts capable of ***reviviscence*** after 10 to 30 years' inactivity in soil.

Vertical distribution in soil. Most (85%) individuals live in the top three centimetres of the rooting zone. In unvegetated soils, their maximum abundance is located deeper.

Life mode. (a) Free forms (e.g. *Rhabditis* spp.); (b) free forms parasitic on plants (rhizoparasites) at one stage of their growth (e.g. *Heterodera carotae*); (c) strict parasites in vertebrates (filaria, ascarids) and invertebrates (e.g. Mermithidae).
Nutrition. Microphagous, (micro)phytophagous, carnivorous.

Abundance. Nematodes are the most abundant animals in soil after the protozoans. Density: 100 to 1,000 individuals per gram soil, or 1 to 30 million ind. m^{-2} (= biomass of 1-30 g m^{-2}). More than 10,000 individuals belonging to about a hundred species can be counted in one cm^3 of soil. Some data: soil of beech forest, 12 million ind. m^{-2}; meadow soils, 2–20 million ind. m^{-2}; cultivated soil, 1 million ind. m^{-2}. In an apple rotting on the soil, 90,000 nematodes of several species were counted.

Significance in soil. Very great. Nematodes are the functional intermediates between bacteria, fungi, protists, and the mesofauna.

Small phyla for minuscule invertebrates

Other than protozoans and nematodes, the microfauna includes other less abundant invertebrates such as Tardigrada and Rotifera (Fig. 12-C). By means of ***anhydrobiosis*** these animals are adapted to moist environments susceptible to desiccation, conditions often found in soil.

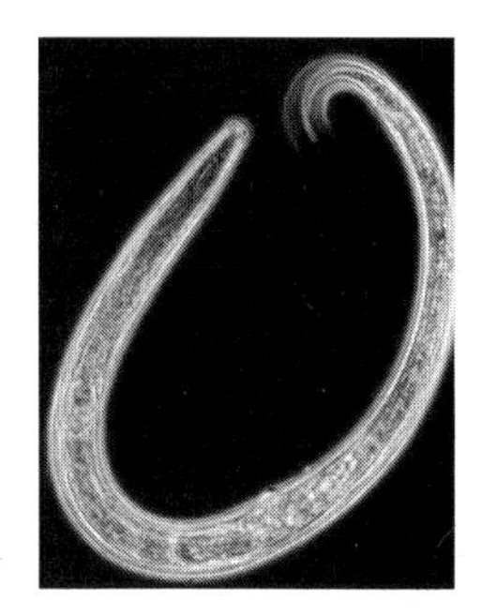

Fig. 12-B *Soil nematode (photo E. Mitchell).*

Reviviscence: resumption of activity after a period of encystment.

Nematodes have their predators: gamasid mites for example which seize their prey with chelicerae that function somewhat like spaghetti tongs. Fungi predatory on nematodes also exist (Chap. 4, § 4.4.7).

Reference works. Grassé, 1965 ('compendium' on nematodes); Dunger, 1983 (generalities); Freckmann and Baldwin, in Dindal, 1990 (group presentation and identification key up to genus; refers to American and Canadian work; Zunke and Perry, in Benckiser, 1997 (modern introduction to nematodes, exhaustive bibliography); Lavelle and Spain, 2001.

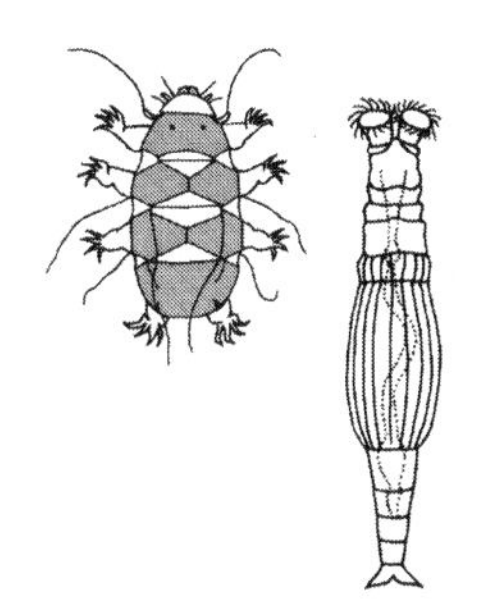

Fig. 12-C *A tardigrade (left) and a rotifer (right).*

Anhydrobiosis: resistant state that permits animals to survive loss of more than 95% of the water from their bodies for long periods without forming cysts. Anhydrobiosis is a state of latent life enabling animals to bear normally lethal temperatures, close to absolute zero and to that of boiling water. Animals in anhydrobiosis rapidly regain their activity when rehydrated (reviviscence).

Smaller than 1 mm, ***Tardigrada*** exhibit a mixture of the features of arthropods and annelids. Nearly 550 species are known in the world, characterized by four pairs of non-articulated legs carrying large claws. The mouth is furnished with styles enabling the animal to puncture plant cells or animal prey at their scale. Tardigrada live mainly in mosses and film water of soil. When dried mosspads kept in a herbarium for more than 60 years were rewetted, Tardigrada were simultaneously revived. Density in soil: 10,000 to 200,000 ind. m^{-2}. Example: *Macrobiotus* spp.

Rotifera or Rotatoria (wheel animalcules) measure 0.04 mm to 2 mm in length. About 1800 species are known, mostly planktonic. An order, the Bdelloida, groups Rotifera of the pedofauna. Aquatic like other microfaunal organisms, they are active in interstitial and film water. A crown of cilia surrounds their mouth; their movements, reminiscent of those of a wheel (Rotatoria, from Latin *rota*, wheel), create an eddy that leads food particles to the mouth. The food thus filtered comprises algae, fungi, bacteria, protozoans and fine organic debris. Rotifera are abundant in soil interstices, litter, mosses and lichens. Density: 33,000 to 450,000 ind. m^{-2} depending on the soil. Example: *Rotatoria* spp.

'Earthworms are the intestines of soil' (Aristotle).

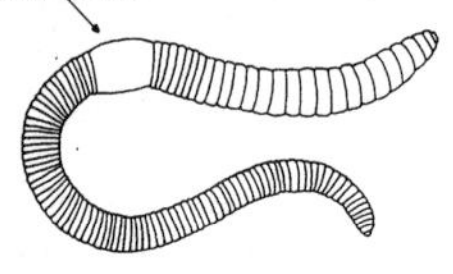

Fig. 12-D *Earthworm.*

Clitellum: Pad 4–10 segments long formed in the skin of sexually mature earthworms. It is located in the anterior part of the body and secretes cocoons in which eggs are laid.

An Australian earthworm, *Megascolides australis*, attains 3 m length for a diameter of 4 cm!

Epigeal and endogeal, used here as substantives, are also qualifiers applied to other soil animals or to the environment itself.

12.4.3 Lumbricidae or earthworms, kings of the soil

Taxonomic position (after Edwards and Bohlen, 1996). Phylum Annelida, class Oligochaeta, order Haplotaxida, suborder Lumbricina, comprising 20 families including the Lumbricidae.

Morphological characteristics. Worm divided into many segments or ***metameres*** separated by dividing membranes. ***Clitellum*** present (Fig. 12-D↓). Reproduction is hermaphroditic with cross-fertilisation and the development is direct, without larval stages.

Species size. Central Europe: up to 35 cm. Southern Europe: up to 75 cm.

Number of species. 3700 worldwide (up to 6000 in Lavelle and Spain, 2001) of which 19 are common in Europe.

Geographic distribution. Earthworms are ubiquitous in all the continents, absent only in hot or cold deserts.

Habitat, vertical distribution in soil, life mode and nutrition. According to Bouché (1972), three ecological categories of earthworms are distinguished (Fig. 12.5):

- ***epigeals*** (or ***epigeics***) live on the surface; they are linked to litter, manure, composts, dead wood; their anatomy predisposes them to rapid movement (e.g. *Eisenia* spp.).
- ***anecic species*** are the large, 'vertical' tunnelling worms (Chap. 4, § 4.5.1), which build a network of galleries throughout the solum, even if they feed mainly on the

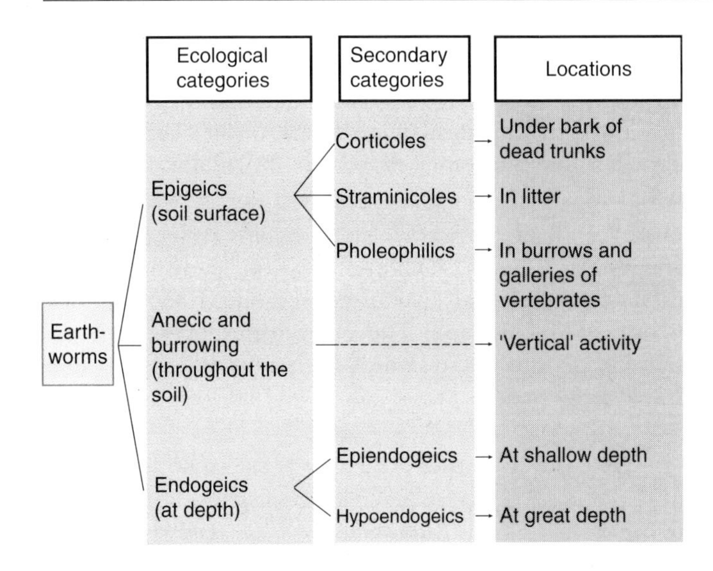

Fig. 12.5 *Ecological categories of earthworms.*

surface; they are the species having maximum impact on the soil (e.g. *Lumbricus terrestris*).

- ***endogeals*** (or ***endogeics***) live at depth, 'horizontally'; they are geophagous and rhizophagous (e.g. *Allolobophora icterica*).

Epigeal or epigeic; endogeal or endogeic!

Abundance. Live biomass commonly ranges from 30 to 100 g m^{-2} (Lavelle and Spain, 2001). Some determinations of density and/or biomass m^{-2} have been done in different environments of western Switzerland (Ducommun, 1989; Matthey et al., 1990; Cuendet et al., 1997):

Earthworms form the largest animal biomass on the continents.

- permanent grassland of the Swiss Plateau: 130–515 g m^{-2};
- directly seeded crop: 200 ind. m^{-2} weighing 152 g m^{-2};
- maize field with bare soil: 100 g m^{-2};
- deep ploughing: 4–18 ind. m^{-2} weighing 6–21 g m^{-2}.

Significance in the soil (see also Chap. 4, § 4.5.1). The work of earthworms has important effects on soil:

- porosity is increased by 20 to 30%, improving natural drainage and soil aeration;
- their mixing action brings up mineralized earth and buries organic matter;
- their galleries constitute preferential routes for roots (Chap. 4, Fig. 4.37) and can partly counteract compaction by agricultural machinery;
- earthworms constitute an important food source for numerous carnivores and insectivores, for example moles, little owls, seagulls and foxes;
- they accumulate soil micropollutants (bioaccumulation, Chap. 4, § 4.6.4).

Reference works: Bouché, 1972 (reference book on ecology and systematics); Bouché, 1984 (good popular paper); Edwards and Bohlen, 1996 (modern book giving current status); Dindal, 1990 (Chaps 11–13. Earthworms from North America); Lavelle and Spain, 2001 (functional roles).

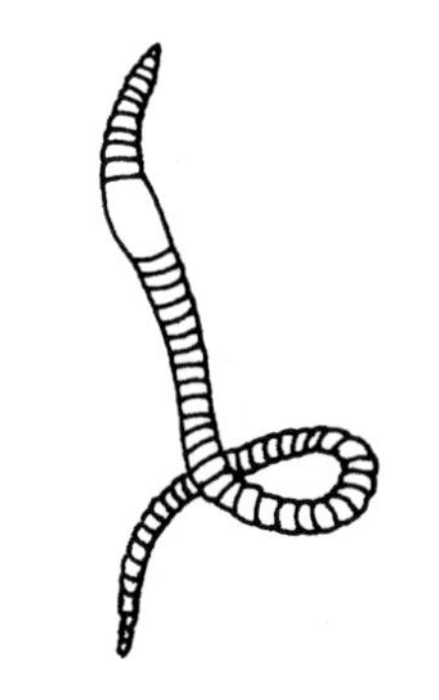

Fig. 12-E *Enchytraeid* (Fredericia spp.)

12.4.4 Enchytraeidae, small cousins of earthworms

Taxonomic position (after Edwards and Bohlen, 1996). Phylum Annelida, class Oligochaeta, order Haplotaxida, suborder Enchytraeina, of which the major family is Enchytraeidae.

Morphological characteristics. Enchytraeids resemble small earthworms (Fig. 12-E). Identification is difficult and based on anatomical structures magnified 400 × under an optical microscope. The chief differences between earthworms and enchytraeids are summarized in Table 12.2.

Table 12.2 Comparison of earthworms and enchytraeids

Descriptor	Earthworms	Enchytraeids
Diameter	≥ 5 mm	≤ 2 mm
Length	60–350 mm (Central Europe)	10–20 mm (up to 45 mm)
Number of segments	100–180	15–70
Colour	Reddish, brownish, bluish, greenish	Whitish
Reproductory organs	2 pairs of testicles	1 pair of testicles

Size of species in temperate regions. 2–35 mm (occasionally up to 45 mm).

Number of species. About 600 worldwide, of which more than 120 occur in Europe.

The very plentiful and stable casts of enchytraeids occasionally form actual layers in soil and are incorporated in aggregates, contributing to structural stability (Chap. 3, § 3.2.1).

Geographic distribution. Abundant in temperate, Arctic and Antarctic regions.

Habitat. In forest litter, mostly of conifers, and in moist soils including peat bogs (Chap. 9, § 9.3.3). Most species are terrestrial but some live in aquatic environment. Enchytraeids abound in sludge from used-water purification plants.

Vertical distribution in soil. Enchytraeids live mostly in the top ten centimetres of the soil, migrating according to moisture content and temperature.

Nutrition. Microphagous and phytosaprophagous. Enchytraeids feed on faeces of springtails and other phytosaprophages.

Abundance. From 10,000 to 290,000 ind. m^{-2}, representing a fresh biomass of 3–53 g m^{-2}. They are distributed in patches.

Reference works. Dunger, 1983 (generalities); Dash, in Dindal, 1990 (presentation of the group and identification key up to genus; cites American and Canadian work); Didden et al., in Benckiser, 1997 (modern introduction to ecology and investigative methods).

Significance in soil. Enchytraeids are efficient phytosaprophages and can consume 2–31% of the organic matter. They exert a stimulatory effect on bacterial and fungal populations by their predation. They also increase porosity in the surface layer by their tunnelling activity (Chap. 4, § 4.5.1).

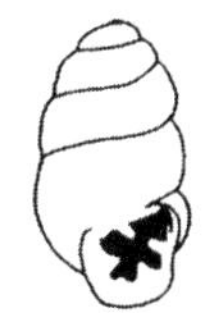

Very small land-shell (*Vertigo pygmaea*). Height 2 mm

Garden snail (*Cepaea hortensis*). Diameter 16-22 mm

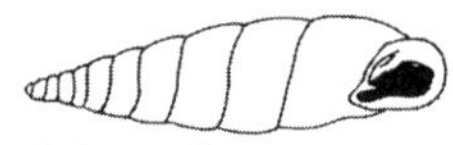

Spire snail (*Macrogastra ventricosa*). Height 17-19 mm

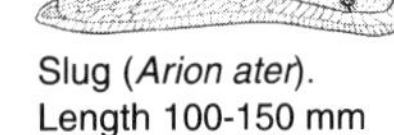

Slug (*Arion ater*). Length 100-150 mm

Fig. 12-F *Some land molluscs.*

12.4.5 Gastropoda, soil animals often forgotten

Taxonomic position. Phylum Mollusca, class Gastropoda. Land molluscs belong to subclass Pulmonata (air-breathing), order Stylommatophora. They are snails (with external shell) and **slugs** (with internal or no shell) (Fig. 12-F).

Morphological characteristics. Land molluscs are recognizable through their 'foot'. Discharge of mucus, which marks their path, is required for creeping. The head is armed with two pairs of retractile tentacles, the upper with eye-spots at the tip.

Size of species in temperate regions. Snails: shell 1–50 mm diameter and 2–50 mm height. Slugs: 20–200 mm length for the largest, *Limax maximus* and *L. cinereoniger*.

Number of species. Worldwide, more than 100,000 Mollusca. In Central and Northern Europe, nearly 400 species of Gastropoda.

Geographic distribution. These animals occur throughout the continents except in hot and cold deserts, from the sea coast to the subalpine zone.

Habitat. Snails and slugs, mostly dependent on high moisture levels, are most plentiful in moist soils and litter. Snails prefer calcareous environments, calcite being a fundamental constituent of their shell; they derive this calcium from food.

Vertical distribution in soil. They are active in the top 10 cm of the soil, migrating according to moisture content and temperature. They overwinter by burying themselves more or less deeply, from 10 cm to more than 2 m depending on species.

Life mode. In winter, these animals enter ***diapause***. Snails close the entrance to their shell with an operculum of hard mucus, while slugs surround themselves with a mucous layer forming a sort of cocoon. They also undergo periods of summer inactivity during which they retreat to their shelters in order to reduce evaporation (soil cracks or under stones: cryptozoic habitat, Chap. 8, § 8.2).

Diapause: Winter sleep or deep torpor of invertebrates during which their metabolism is greatly slowed.

Radula: tongue of gastropods, furnished with small sharp teeth that rasp the food.

Cecilioides acicula is the only truly edaphic snail of Central Europe. Its shell measures 1.5 mm in diameter and 5 mm in height. White and blind, it leads a totally subterranean life, mostly in the first 40 cm of highly porous calcareous soils, feeding on mycelia. It winters at up to 90-cm depth.

Reference works. Lutman, in Heal and Perkins, 1978 (oft-cited ecological study); Godan, 1983 (very good introduction to ecology of Gastropoda, applied aspects); Bogon, 1990 (well-illustrated book useful for identification); Hommay, 1995 (introduction to knowledge of slugs, good illustrations).

Nutrition. Gastropods feed by means of their ***radula***. Two major nutrition modes are seen:

- non-specialist phytophages and phytosaprophages (for example, slugs and snails are abundant in garden composts; Chap. 10, § 10.8);
- some species are exclusively carnivorous. Most slugs are necrophagous as well and eat fresh carrion of other gastropods.

Abundance. Some densities recorded in the temperate zone:
- slugs: in beech forest, 14 ind. m^{-2}; in open grassland, 1 to 30; in cultivated fields, 110 to 150;
- snails: in birch forest, 1–490 ind. m^{-2}.

Gastropods often form masses containing more than 1000 ind. m^{-2}.

Significance in soil. Land gastropods are often associated with reworked or immature soils. They produce large quantities of excrement composed of finely fragmented green or dead leaves, impregnated with enzymes (molluscs secrete a quite complete spectrum of these biological catalysts, including cellulases, Chap. 14, § 14.3.1). Furthermore, the mucus they produce contributes to formation of microaggregates, like that of earthworms (Chap. 3, § 3.2.1).

12.4.6 Terrestrial Isopoda or woodlice; discreet but efficient littericoles

Taxonomic position. Phylum Arthropoda, class **Crustacea**, order Isopoda, suborder Oniscoidea.

Morphological characteristics. Oval shape, dorsoventrally flattened and provided with 7 pairs of legs. Colour often mottled grey, occasionally white. Woodlice have 2 pairs of antennae, 4 pairs of mouth parts and jointed abdominal appendages (Fig. 12-G).

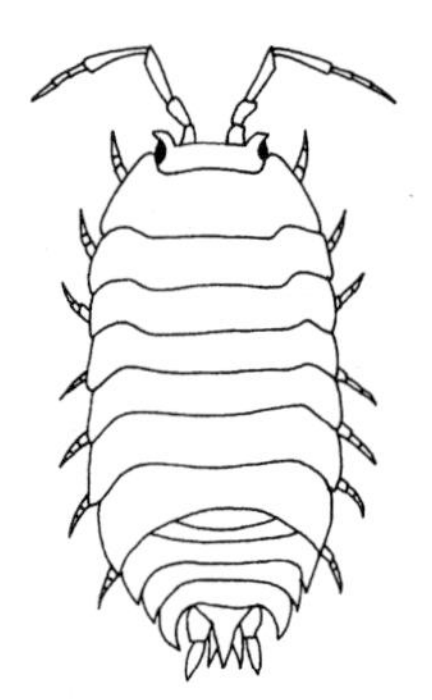

Fig. 12-G *Woodlouse* (Oniscus asellus). *Length up to 18 mm.*

Size of species of temperate regions. From 5 to 20 mm. Woodlice belong to the macrofauna.

Number of species. Worldwide nearly 2000. In France, 166. In Switzerland, 51.

Geographic distribution. From tropical forest to the tundra and from the sea coast (halophilic species) to the snowline.

Habitat. Litter, soil annexes (rotted wood, composts, cryptozoic habitats, fissures). Some species are commensals of ants.

Vertical distribution in soil. Woodlice are permanent surface geophiles, belonging to the epi- and hemiedaphon.

Life mode. Isopoda form a very varied group. Numerous species are marine or freshwater-living. Of nine suborders, just one, Oniscoidea, brings together the terrestrial species, most of which are closely associated with high moisture content. Isopoda are active during the summer months and undergo a winter diapause.

Nutrition. These animals are phytosaprophagous (leaves and dead wood), mycophagous, rhizophagous and phytophagous (young shoots). Their gut accommodates an abundant and varied microflora that degrades cellulose.

Abundance. Temperate zone, 50 to 200 ind. m^{-2} in forest and 500 to 7900 in grassland. Another figure: 4000 ind. m^{-2} in an Australian forest.

Woodlice can proliferate and become harmful in mushroom beds and greenhouses.

Significance in soil. Woodlice are fragmenters belonging to the first compartment in the detritus food chain (Chap. 13, § 13.7.2). When earthworms are absent they can, together with diplopods, cut up 30 to 50% of the annual litter. Certain semi-desert species of south-eastern Russia (*Hemilepistus* spp.) dig galleries that reach 90 cm deep, mixing up to 5 t ha^{-1} y^{-1} of soil and burying 1.5 t ha^{-1} y^{-1} of excrements (data corresponding to a density of 600,000 ind. ha^{-1})

Reference works. Vandel, 1960, 1972 (reference work for Francophone isopodologists). Muchmore, W.B. Terre-strial Isopoda. In Dindal, 1990 (key to genus of North America).

12.4.7 Myriapoda: millipedes, centipedes and others

Taxonomic position. Phylum Arthropoda. Superclass Myriapoda, to give a very expressive name, comprises four clearly differentiated classes: Pauropoda, Symphyla, Diplopoda (millipedes) and Chilopoda (centipedes).

Millipedes or millipeds; Centipedes or centipeds!

Morphological characteristics. The head bears one pair of antennae and one pair of mandibles (Chap. 13, Fig. 13.3). The body comprises 12 to 180 segments, each bearing one pair of legs (two in the case of Diplopoda).

Reference works. Wallwork, 1970 (well-presented general coverage); Bachelier, 1978 (ditto); Demange, 1981 (presentation of the superclass with identification keys); Eisenbeis and Wichard, 1985 (elements of ecology and scanning-electron micrographs); Dindal, 1990 (broad coverage of the group with identification keys; refers to the fauna of North America).

Diplopoda, two pairs of legs per segment

Morphological characteristics. These animals have a large number of segments (11 to 100), each bearing two pairs of ventrally articulated legs (Fig. 12-H). The body is cylindrical or semi-cylindrical, occasionally with lateral growths. The cuticle is impregnated with mineral salts. Four principal types have been described: juliform millipede, pill millipede, polydesmid millipede and polyxenid millipede.

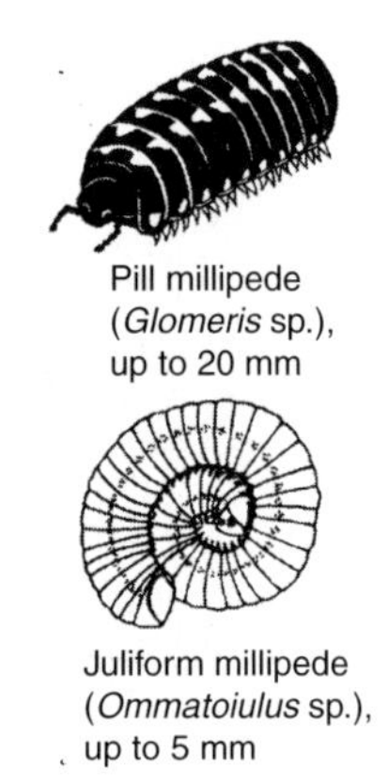

Fig. 12-H *Some examples of millipedes.*

Species size. Worldwide: 2 to 300 mm. Most European species are from 5 to 50 mm long.
Number of species. Worldwide about 10,000.
Geographic distribution. All continents except in frozen deserts. From the sea coast to the Alpine zone.

Habitat. Litter, dead wood and other soil annexes.
Vertical distribution in soil. OL to A horizons. Diplopods belong to the epi- and hemiedaphon.

Life mode. Diplopods exhibit morphological (§ 12.5.3) and functional adaptations: flattening of littericoles (polydesmid millipedes), volution (pill millipeds, juliform millipedes), cylindrical shape of tunnellers (juliform millipedes). They are permanent geophiles, active during summer and undergoing diapause in winter. They can be long-lived, *Glomeris marginata* having a life span of eleven years when reared.

The term myriapod evokes the abundance of appendages of locomotion, always in excess of 9 pairs, and even as high as 375 pairs in a tropical species *Illacma plenipes*.

Nutrition. These animals eat litter (they are cutters) and dead wood and are occasionally phytophagous and algae-grazing (polyxenid millipedes). They show preference for certain foods depending on their nitrogen, sugar and water contents and degree of infestation with microflora. They are also coprophagous, reworking several times their bacteria-rich faeces. They harbour a cellulose-degrading intestinal microflora.

Abundance. Several hundred diplopods may be found per square metre where earthworms are few. They occasionally multiply and then move *en masse*.

Reference work. Hoffman, in Dindal, 1990 (introduction to the taxon and identification key); Pedroli-Christen, 1993 (general and Swiss faunistics).

Significance in soil. Diplopods are important in the first compartment of the decomposition food chain (Chap. 13, § 13.7.2). They consume up to 25% of the annual litter and proceed to the first cutting of litter and moist dead wood, which promotes microfloral activity. Occasionally they ravage crops.

Chilopoda: one pair of legs per segment

Chilopoda, efficient predators.

Morphological characteristics. Each segment bears one pair of laterally articulated legs. Shape dorsoventrally flattened. The cuticle is sclerotinized, softer than in diplopods. The first pair of appendages is modified into poison claws (forcipules). Four types of chilopods exist: Geophilomorpha (25–177 pairs of legs), Scolopendromorpha (21–23 pairs), Lithobiomorpha (15 pairs), and Scutigeromorpha (15 pairs of long legs, the tarsi of which comprise 500 to 600 segments!) (Figs. 12-Ia, 12-Ib).

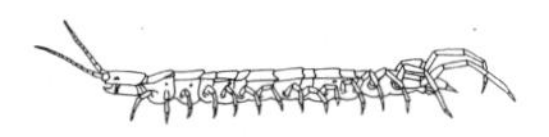

Fig. 12-Ia *Lithobiomorph centipede* (Lithobius forficatus), *18-32 mm.*

Size. In temperate zone, from 5 to 100 mm. Tropical species attain 30 cm length and 1 cm width.
Number of species. Worldwide more than 3500. In France, more than 120.

Geographic distribution. All continents, from the sea coast to high mountains, but rarer at altitude. Lithobiomorphs are restricted to temperate regions.

Habitat. Chilopods are associated with rather moist environments (litter, composts, cryptozoic habitats).

Vertical distribution in soil. These animals belong to the epi- and hemiedaphon. The short chilopods (lithobiomorpha, scolopendromorpha) live in litter and cryptozoic habitats. The geophilomorpha, long, flexible and blind, go down to the A horizon.
Life mode. Chilopods are active during summer and undergo a winter diapause. Life span: 3 to 4 years.

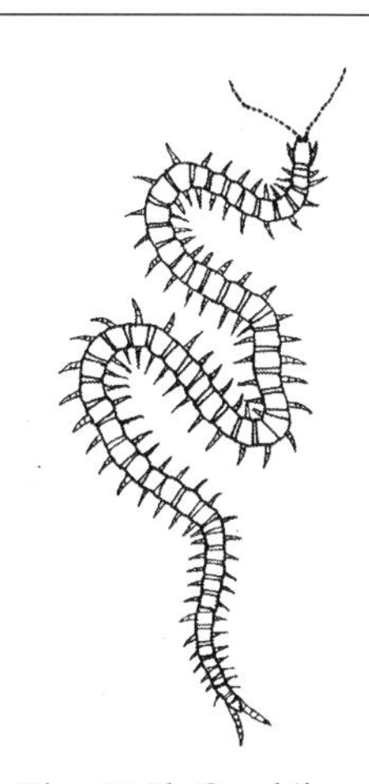

Fig. 12-Ib *Geophilomorph centipede* (Necrophloeophagus longicornis). *Length up to 44 mm, common in grassland soils (from Broleman, 1930).*

Nutrition. Chilopods belong to the first compartment of the detritus food chain (Chap. 13, § 13.7.2). Primarily carnivorous, they feed on other invertebrates that they kill by injecting venom through their poison claws. Lithobiomorphs mostly lie in wait for their prey while Geophilomorpha, depending on size, actively search out earthworms, larvae of dipterans or enchytraeids. Chilopods also consume varying quantities of litter.

Abundance. Density: from 40 to more than 400 ind. m^{-2} in thick forest litter. Their biomass varies according to the environment: 180 mg m^{-2} for 265 individuals in forest soil, 60 mg m^{-2} for 140 individuals in a meadow and 100 mg m^{-2} in an agroecosystem have been reported in the temperate zone.

Significance in soil. Like all predators, they contribute to the demographic balance of prey populations.

Reference works. Broleman, 1930 (still *the* reference work for Western Europe despite its great age); Eason, 1964 (excellent introduction to the group). Mundal, in Dindal, 1990 (broad coverage of the class, with identification key to orders and families of the world).

12.4.8 Arachnida, from scorpions to mites

Taxonomic position. Phylum Arthropoda, subphylum Chelicerata. Class Arachnida comprises 11 subclasses or orders that all belong, totally or partly, to the soil fauna: Scorpiones, Pseudoscorpiones, Araneae, Opiliones, Acari, Solifugae, Palpigradi, Uropygi, Schizomida, Amblypygi and Ricinulei. The first five are present in temperate regions.

Morphological characteristics. Body composed of two parts, ***cephalothorax,*** bearing four pairs of legs, and the

Depending on the author, Acari constitute an order containing 8 suborders or a subclass containing 8 orders. Krantz (1978) estimated that more than 30,000 species are known, but the number could be as high as half-a-million.

Phoresy: behaviour consisting of being transported from one habitat to another by an animal of another species. Bumblebees and scarabaeids often transport mesostigmatic mites. Thus these scarcely mobile mites have found the means of travelling far by insect hitch-hiking!

abdomen. One pair of ***chelicerae,*** pincer-shaped prehensile appendages, (Chap. 13, Fig. 13.3); stylet shaped or ending in articulated poison claws (spiders) present.

Only Acari, the most important arachnids in the functioning of the soil system, are covered here.

Acari, a vast group still insufficiently understood

The Acari have very variable morphology and habits; among them are found phytosaprophages, predators, sapsuckers and haematophages, often vectors of human and animal diseases, and also animal parasites. Three orders are plentifully represented in the second compartment of detritus food chains (Chap. 13, § 13.7.3): Gamasida, often ***phoretic,*** **Actinedida** and Oribatida. Characteristics of the first two are summarized in Table 12.3. Oribatida, important decomposers, are presented later in greater detail.

Oribatida

Morphological characteristics. Variously shaped bodies, light brown to black. Mouth parts are crushers, with pincer-shaped chelicerae. Two pairs of ventral shields conceal the genital and anal openings. The front part of the body bears two long sensory setae (pseudostigmatic organs) (Figs. 12-Ja, 12-Jb). Identification of oribatids, like that of other Acari, is a matter for specialists.

Table 12.3 Characteristics of Gamasida and Actinedida

Order	Size (mm)	Nutrition	Habitat
Gamasida (suborders Parasitiformes and Uropoda)	0.5 to 2	Carnivorous (springtails, larvae of dipterans, other Acari, nematodes, enchytraeids); haematophagous tendency; fungivorous	O horizon, S layer (mosses, according to Green et al., 1993); galleries, burrows; dung, droppings; many phoretic species.
Actinedida (=Prostigmata or Trombidiformes)	0.2 to 3	Carnivorous (springtails, other Acari); sap-sucking; ectoparasitic	O horizon, S layer (mosses); one aquatic family

Reference works. Krantz, 1978 (very comprehensive coverage of all the Acari, centred on systematics); Travé et al., 1996 (excellent introduction to the Oribata, but the book does not contain identification keys); Larink, in Benckiser, 1997 (microarthropods in agroecosystems, large bibliography).

Size. Oribatid mites are microarthropods 0.1 to 2 mm in length (average 0.3 to 0.6 mm) and form part of the mesofauna.

Number of species. More than 6000 have been described till now but the total number is estimated to be between 30,000 and 50,000.

Geographic distribution. Over the entire land mass above sea level, from the sea coast to the snowline.

Habitat. Soil (euedaphic species) and soil annexes. Oribatid mites are found in extreme habitats such as lichens growing on bare rocks or the mosses of Antarctica.

Fig. 12-Jb *Two species of oribatid mites seen through the scanning electron microscope. Left:* Nothridae *sp., size about 1 mm; right:* Galumnidae *sp., size about 0.5 mm (photos D. Borcard).*

Vertical distribution in soil. In grassland, 50% of the strength live in the first two centimetres of soil; under forest, one-third of the oribatids are located in the OF horizon. Species size decreases with soil depth.

Nutrition. Oribatids are generally phytosaprophagous and microphagous but may be coprophagous or pollen-eating. Their digestive tube contains a varied microflora that degrades cellulose.

Abundance. Up to 425,000 ind. m^{-2} in soils of coniferous forest in the temperate zone. They may represent up to 2% of the zoomass in soil.

Significance in soil. Important players of the second compartment of detritus food chains, oribatids contribute to microfragmentation and mixing of organic materials and to dispersal and regulation of the microflora. Because of their microphagous nutrition they intervene in the dynamics of fungal populations. They are efficient at cutting needles of conifers. They have an essential role in the polar and circumpolar zones devoid of macroarthropods and earthworms.

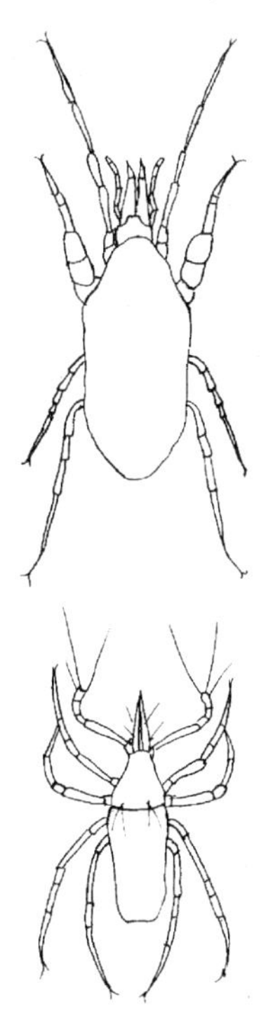

Fig. 12-Ja *Above: gamasid mite (*Pergamasus *sp.). Length 1.25 mm Below: prostigmatic mite (Bdellidae). Length 1 mm.*

Insects constitute the most varied zoological group on the continents; it is not possible to estimate the number of their species, which is between 3 and 30 million! Novotny et al., 2002)

12.4.9 Insects, more than 90% of animal species!

Insects are Arthropoda provided with one pair of antennae, one pair of mandibles and three pairs of legs. They occupy all imaginable ecological niches, including soils. Of 31 to 34 orders, 5 participate more specifically in the functioning of soil: Collembola, Isoptera (termites), Coleoptera (e.g. ground beetles, cockchafers, staphylinid beetles), Diptera (flies, horseflies, crane flies) and Hymenoptera (ants, wasps, bees).

Reference works. (introductory or reference). Grassé, 1949, 1951; Grandi, 1984 (2000 pages of well-illustrated general entomology); Chinery, 1986 (profusely illustrated guide); Gullan and Cranston, 1994 (good book on general entomology); Resh and Cordé, 2003 (a remarkable modern reference book on general entomology)

Collembola, omnipresent in soils

Taxonomic position. These are primitive insects belonging to subclass Apterygota (primitively wingless in the adult stage).

Morphological characteristics. Abdomen in 6 segments, bearing three specific organs (transformed appendages): ***ventral tube***, which enables absorption of water and dissolved inorganic salts, ***retinaculum*** and ***furcula*** (furca), which together form the organ for jumping (Figs. 12-Ka, 12-Kb). The euedaphic species are white, others of varying hues. Springtails exhibit a morphological adaptation for living at depth (§ 12.5.3; Plates I-2, I-3).

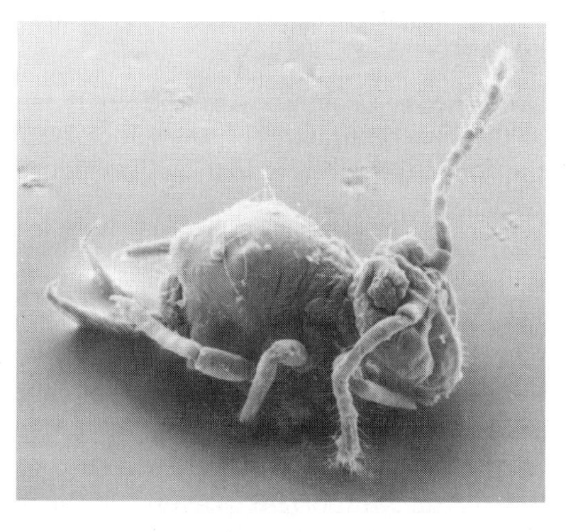

Fig. 12-Ka *Springtail (Sminthuridae) viewed under the scanning electron microscope; about 2 mm; photo D. Borcard.*

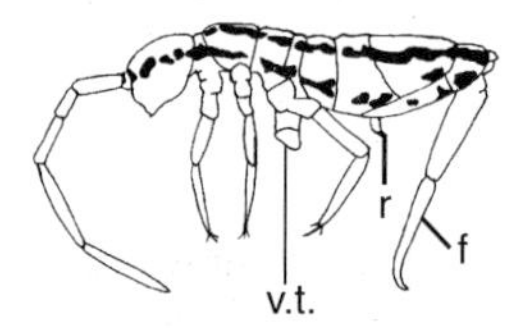

Fig. 12-Kb *Springtail (Entomobryidae).*
v.t. ventral tube
r retinaculum
f furcula (furca)

Size. Length 0.25 to 5 mm, rarely 10 mm. Springtails belong to the mesofauna.

Number of species. To date about 3000 species have been described throughout the world, 820 of them from Europe (Gisin, 1960).

Geographic distribution. Springtails are present in all kinds of soils between the equator and polar regions, from the sea coast to the snow zone.

Habitat. Eu-, hemi- and epiedaphon, soil annexes.

Vertical distribution in soil. Greatest abundance in the first three centimetres, then to 10 cm. Species size diminishes with increasing depth.

Nutrition. Collembola are mainly fungivorous, but phytosaprophagous, coprophagous, pollen-eating, carnivorous (nematodes) and phytophagous (living roots and germinating plants) species also exist.

Abundance. From 2000 to 200,000 (occasionally 500,000) ind. m^{-2}. In the grassland soils of western Switzerland, from 95,000 to 180,000 ind. m^{-2}. Springtails are r-strategists unlike oribatids, which are K-strategists.

Significance in soil. Their role is similar to that of oribatids. Like the latter, Collembola are important actors in the second compartment of the detritus chain. They contribute to microfragmentation and mixing of organic matter, as well as to dispersal and regulation of the microflora. Because of their microphagous habit they influence fungal population dynamics considerably, as their nibbling stimulates growth of fungi. They can become harmful when they multiply; they have caused great havoc in sugar-beet cultivation.

Reference works. Gisin, 1960 (classic book on identific-ation); Massoud, in Pesson, 1971 (introduction to the ecology of Collembola); Christiansen, in Dindal, 1990 (general introduction and key to general levels); Larin, in Benckiser, 1997 (micro-arthropods in agroeco-systems, large biblio-graphy).

Termites, innumerable residents of tropical soils

Taxonomic position. Order Isoptera. The family Termitidae groups the so-called higher termites, the most evolved (nearly 60% of known species). Five other families contain the lower termites.

Morphological characteristics. Termites live in communities comprising different types of sexed (male and female) and sexless (workers and soldiers) individuals forming castes. Young sexed isopterans (*iso* = equal) have four nearly identical membranous wings that lie flat on their backs and are detached after swarming. Termites have chewing mouth parts; mandibles of soldiers can be hypertrophied. Growth is ***hemimetabolous.***

Hemimetabolous: qualifies the growth of an insect whereby it gradually develops, by moulting without metamorphoses, from the larval stage to the imago. Cockroaches, crickets, and bugs are, like termites, hemimetabolous insects.

Size. From 2 to 20 mm without wings. Elderly queens of certain Termitidae grow to 14 cm in length and 3.5 cm in abdominal diameter.

Queens of the Termitidae are obesity champions. During their life, which could reach a century, their weight increases 300-fold. They lay 25 eggs per minute non-stop, or 11 million eggs per year!

Number of species. About 2300 worldwide, 630 in Africa, 60 in North America and 3 in Europe: *Calotermes flavicollis, Reticulitermes lucifugus* and *R. flavipes*, the last introduced from North America.

Geographic distribution. Termites are present on all continents in forest and semi-grassland (savanna, steppe) environments, and even in desert regions. They are primarily tropical insects but a limited number of species live in temperate regions. Termites are found from sea level to 2000-m altitude (exceptionally 3000 m in the Himalayas).

Habitat: Termites are ***lucifugous*** insects (Fig. 12-L) and build nests of many types:

- haphazard assemblages of galleries and chambers dug in the soil, attaining a total length of several hundred metres;
- built nests in dead wood;

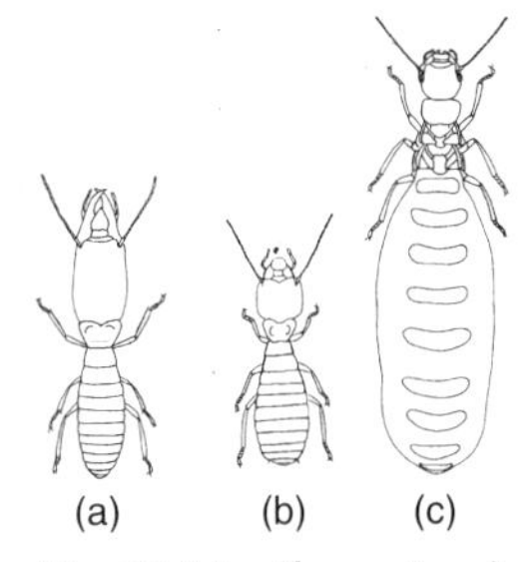

Fig. 12-L *Lucifugous termite* (Reticulitermes lucifugus); *(a) Soldier, 5-5.5 mm; (b) worker, 5-6 mm; (c) queen, 10-15 mm.*

Lucifugous: avoiding light.

• imposing edifices shaped as mounds or 'cathedrals that can go up to 6 m in height and 30 m in diameter at the base, undoubtedly forming the most complex structures of the animal world (except for man);
• boxed tree nests built by forest species.

Caste: in social insects, category of individuals differing in shape and function, although belonging to the same species.

Pheromone: external chemical compound secreted by glands. Its release constitutes a kind of message that sets up a physiological or behavioural reaction in individuals of the same species. There are different kinds of pheromones: aggregating, alarming, marking, sexual and social.

Vertical distribution in soil. Colonies are located a few decimetres to several metres below the surface. Some galleries go down 55 m towards the groundwater.

Life modes. Termites are exclusively social insects. Colony size varies with species: 1000 to 2000 individuals at the most in the European species *Calotermes flavicollis*; up to 10 million in the case of the African *Bellicositermes*, which build large mounds. Colonies of social insects comprise several ***castes***, numbering three to seven in termite societies. Social ***pheromones*** determine the demographic equilibrium of the castes, between workers and soldiers, for example.

Nutrition. Cellulose forms the basic food of termites, which devour all materials containing it, from timber and other wood to paper. But they are incapable of digesting it themselves (Chap. 13, § 13.6.4). In the case of lower termites this is made possible with the help of zooflagellata and symbiotic bacteria living in the posterior intestine (Chap. 13, Fig. 13.8). In fact termites utilize the metabolic wastes of their symbionts. In higher termites, termed fungus farmers, plant material is accumulated in fungus combs in the termitarium and predigested by the fungi growing there. Termites eat this wood and fungi, completing digestion by means of cellulases secreted by their gut and by their symbiotic bacteria. Contrarily, lignin is scarcely attacked or not at all and is found in large part unaltered in the excrements (Swift et al., 1979; Breznak, in Anderson et al., 1984).

'The loss of 75% of printed paper on termites will scarcely diminish our knowledge of these insects' (Grassé, 1982).

Abundance. In Australia, but chiefly in Africa, colonies contain several hundred thousand or several million individuals forming one single family, descended as they are from the same royal couple. In Africa, the number of termites ranges from 1000 to 10,000 ind. m^{-2}. In one hectare of savanna in Côte d'Ivoire, Josens (1974) in Dajoz (1996) counted:

Foraging: collecting dry grass for bringing to the termitarium.

Humivorous: eating litter and humified organic matter.

• 1.6 million **foraging** termites weighing 7.9 kg (fresh weight) and consuming 30 to 40 kg grass per year;
• 4.5 million **humivorous** termites, weighing 10 kg and consuming 30 kg cellulose from various sources per year;
• 5 million fungus-farming termites, weighing 6.3 kg and bringing 1.4 t dry litter per year into their termitaria.

Significance in soil. Termites are, with ants, the most important elements of the tropical pedofauna. They play an essential role in chemical and physical changes in soils (Chap. 4, §§ 4.5.1, 4.5.2), in recycling dead wood and plants and also of dry dung. In the southern United States, they annually eliminate from the soil surface nearly 80 kg wood and 450 kg litter per hectare.

Reference works. Grassé, 1982 (all researches on termites in the nineteen-eighties); Netting, in Dindal, 1990 (modern synthesis); Lavelle and Spain, 2001 (recent general review).

Termites and man

Although termites have a favourable effect on tropical soils, they often are a considerable menace for human activity. Their nutrition based on cellulose means that they direct their attack against timber, goods depots and stored paper (which endangers preservation of archives, for example). They are thus economically very important destroyers difficult to battle against. Their havoc is not restricted to hot regions of the world because the climate of large cities promotes their spread in the temperate zone.

In Europe, *Reticulitermes flavipes* established itself in many large ports (Hamburg, Bordeaux), then extended its coverage to the urban environment up to Paris and Lyons, where it has since been very damaging.

Coleoptera, elytra of all sizes

Morphological characteristics. These insects, often with thick hard cuticle, are provided with a pair of tough wings, the ***elytra***, which protect the membranous wings used for flying (Fig. 12-M). The mouthparts are of biting or chewing type, with well-developed mandibles (Chap. 13, Fig. 13.3).

Size of species in temperate regions. Adults: from 0.5 mm (Ptilidae) to 75 mm (Lucanidae); fully grown larvae: length from <1 mm (Ptilidae) to 100 mm (Lucanidae).

Number of species. Coleoptera probably comprise more than one million species worldwide and more than 6000 in Central Europe. The principal edaphic families of temperate regions are given in Table 12.4. However this list is far from complete: it lacks among others a dozen families of endogeal Coleoptera whose biology is poorly understood (Coiffait, 1960).

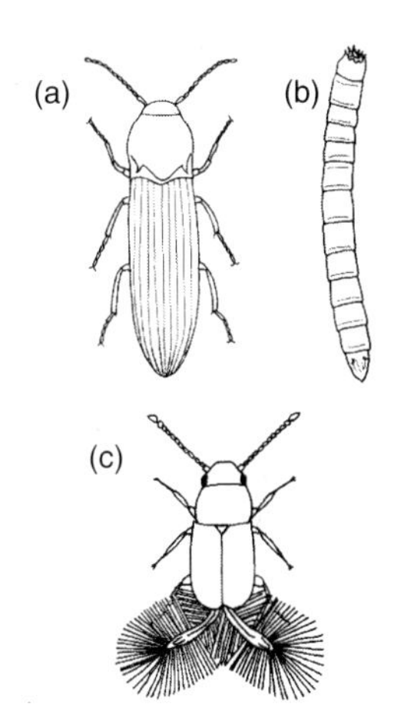

Fig. 12-M *Above: click-beetle* Agriotes *(Elateridae). (a) adult 6-10 mm; (b) larva 15-18 mm.*
Below (c): an endogeal coleopteran (Ptilidae), 1.2-2 mm.

Geographic distribution. Coleoptera are found in all terrestrial ecosystems.

Vertical distribution in soil. Surface and OL to OH horizons, soil annexes; to 1-m depth for hibernation (cockchafer larvae).

Life mode in soil. Temporarily inactive and active geophiles, periodics, geobionts (§ 12.5.5). Many families (e.g. Carabidae, Catopidae, Ptilidae, Staphylinidae) have endogeal species exhibiting adaptations to edaphic living, such as small size, thin or flattened body, shortening of legs, apterism, reduction of eyes, depigmentation.

Endogeal: said of an organism (or species) that spends its entire life within the soil. Caution: endogeal has a specific meaning in the case of earthworms (§ 12.4.3)!

Table 12.4 Location and principal nutrition regimes of edaphic Coleoptera

Family	Location	Nutrition regime in soil (L = larvae, A = adults)					
		1	2	3	4	5	6
Cantharidae	I-II-III	L					
Carabidae	I-II-III	L + A					
Cerambycidae	III						L
Curculionidae	II					L	L
Elateridae	II-III	L				L	L
Scarabaeidae	generally			L + A	A	L	L
Cockchafers	II					L	
Earth-boring dung beetles	II-III			L + A	A		
Silphidae	I-III	L + A	L + A				
Staphylinidae	I-II-III	L + A			A		

Location
I OL to H horizons, S layer (mosses, see Green et al., 1993)
II A horizon, rooting zone
III Soil annexes (more or less rotten dead wood, faeces, carcasses, etc.)

Nutrition regime
1. Predation 2. Necrophagy 3. Coprophagy
4. Mycetophagy 5. Rhizophagy 6. (Sapro)xylophagy

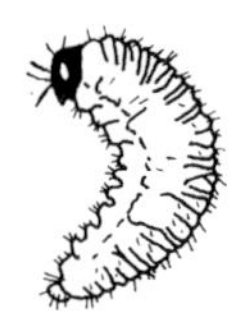

Fig. 12-N *Rhizophagous larva of weevil, 6 mm.*

Reference works. du Chatenet, 1985 (illustrated guide, from Carabidae to Scarabaeidae); Klausnitzer, 1978 (book on identification of coleopteran larvae); Crowsen, 1981, and Paulian, 1988 (excellent general books on Coleoptera, full of facts).

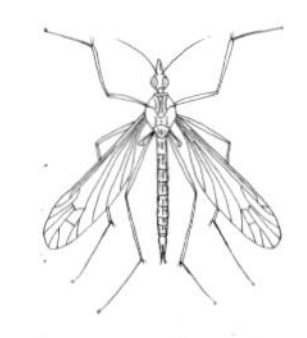

Fig. 12-O. *A fully extended dipteran. Adult female crane fly* (Tipula maxima). *Length up to 40 mm.*

Nutrition. Rhizophagous (Fig. 12-N), saprophagous and coprophagous larvae and adults contain a rich intestinal microflora, occasionally located in 'bacterial chambers' (Scarabaeidae). Endogeal coleopterans are 80% predatory but may be mycophagous and commensal with ants.

Abundance. A few tens to some hundreds per square metre.

Significance in soil. Coleoptera play an important role as decomposers. Numerous predatory and mycophagous species contribute to the biological equilibrium in soils. The rhizophages may become important ravagers of crops, such as larvae of cockchafers, click-beetles ('wire worms') and weevils.

Diptera, larvae and still more larvae ...

Morphological characteristics. Adults have only a single pair of wings, the second pair having been transformed to halteres. Some species have wingless adults that live in the soil (Sciaridae, Limoniidae). Mouthparts are sucking, occasionally stinging. The crane fly, mosquito and house fly are very representative examples of the entire order so far as adults are concerned (Fig. 12-O). But it is essentially the larvae that form part of the active

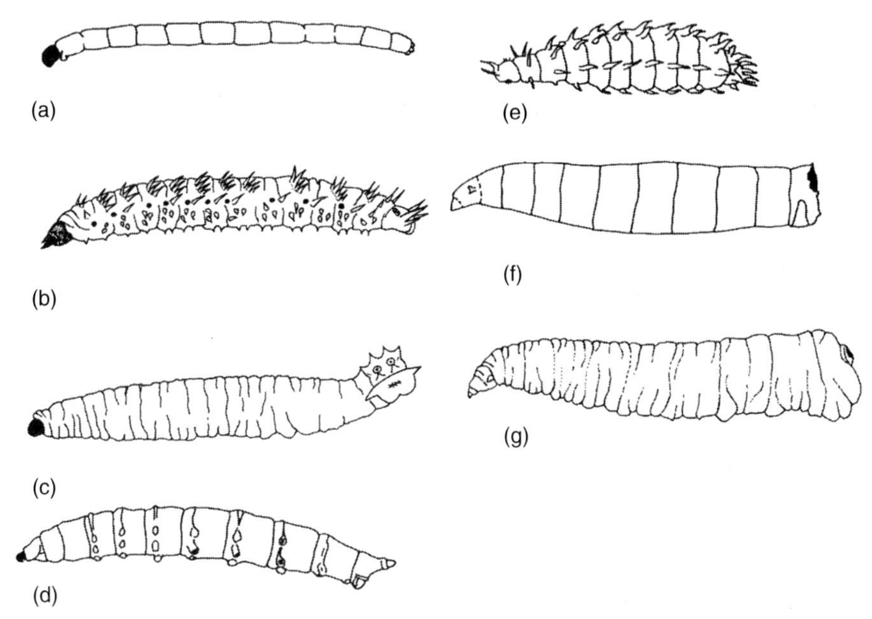

Fig. 12.6 *Larvae of dipterans: (a) Chironomidae, 4 mm; (b) Bibionidae, 20 mm; (c) Tipulidae, 30–40 mm; (d) Tabanidae, 15 mm; (e) Fanniidae, 10 mm, saprophagous; (f) Scatophagidae, 14 mm, (g) Calliphoridae, 6–14 mm (other nutrition regimes, see Tab. 12.5). Head of larvae to the left.*

pedofauna. Beneath an apparent uniformity, they exhibit great morphological, ecological and behavioural diversity (Fig. 12.6; Chap. 13, Fig. 13.3). They are still very poorly understood (Chap. 17, § 17.1.1).

Size of species in temperate regions. Adults and fully developed larvae, 2 to 40 mm long.

Number of species. About 125,000 species worldwide (certainly many more); in Switzerland, more than 6000. According to Brauns (1954), 44 families of a total of 113 have edaphic larvae.

Geographic distribution. All terrestrial ecosystems.

Habitat. Chiefly moist soils, in litter. Each type of soil annexe (carcasses, dead wood, compost, burrows, etc.) harbours a different larval population.

Vertical distribution in soil. OL to OH horizons, S layer of bryophytes (Green et al., 1993; Chap. 6, § 6.2.2), soil annexes.

Life mode. Temporarily active geophiles. Larval growth lasts a few weeks to several years.

Nutrition. Larvae of dipterans can utilize the different resources offered by soil and its annexes. Their gut

Table 12.5 Location and nutrition regimes of major families of Diptera with edaphic larvae

Family	Location	Nutrition regime of larvae in soil							
		1	2	3	4	5	6	7	8
Bibionidae	I-II-III			×		×	×	×	
Calliphoridae	I-(II)-III	×	×	×		×			
Cecidomyiidae	I-II-III	×			×				(×)
Chironomidae	I-(II)-III					×	×		
Dolichopodidae	I-II-III	×							
Drosophilidae	I-II-III	×	×	×	×	×	×	×	
Empididae	I-II-III	×							
Limoniidae	I-II-III	×				×	×		
Muscidae	I-(II)-III	×	×	×		×			
Mycetophilidae	I-(II)-III	×			×				
Phoridae	I-II-III	×	×	×	×				
Scatopsidae	I-II-III			×	×	×	×		×
Sciaridae	I-II-III			×	×	×	×		
Sphaeroceridae	I-(II)-III		×	×		×			
Stratiomyiidae	I-III	×		×		×		×	
Tabanidae	(I)-II-(III)	×							
Tipulidae	I-II-III	×				×		×	×

Location
I OL to OH horizons, S layer (mosses; see Green et al., 1993)
II A horizon, rooting zone
III Soil annexes (more or less rotten dead wood, faeces, carcasses, etc.)

Nutrition regime
1. Predation
2. Necrophagy
3. Coprophagy
4. Mycophagy + mycetophagy
5. Phytosaprophagy + microphagy
6. Microphagy
7. Phytophagy (roots + living leaves)
8. Saproxylophagy (+ microphagy)

contains a very active microflora. The principal nutrition regimes are given in Table 12.5.

Abundance. From a few tens to some thousands of larvae per square metre, often distributed in patches. Larvae of Sciaridae may constitute more than 75% of the strength; thus, in a European forest, up to 20,000 ind. m^{-2} were counted in a mull and 12,000 in a moder.

Significance in soil. Dipteran larvae fragment litter efficiently: of them alone, those of the Bibionidae cut up to 15% of the leaves fallen in autumn. Their activity becomes all the more important in destruction of animal carcasses (Chap. 8, § 8.3.1) and in recycling of vertebrate excrements (Chap. 8, § 8.3.2). Many species preying on macroarthropods and others, mycophagous, contribute to the biological equilibrium in soils.

Reference works. The literature of dipterology is vast but more than 90% of the publications are concerned with adults: McAlpine et al., 1981-1989 (all the data on North American Diptera); Smith, 1989 (presents the status of knowledge of larvae of Great Britain); Matile, 1993, 1995 (very good, concise introduction to Diptera).

Ants, principal enemies of termites

Taxonomic position. Order Hymenoptera, suborder Aculeata. The Hymenoptera comprise three families of social insects: Formicidae (**ants**), Vespidae (wasps) and Apidae (bees).

Morphological characteristics. Four membranous wings, the second pair smaller and coupled to the first by a series of hooks (characteristic of Hymenoptera). In ants, only male and female sexed individuals bear wings, which are used only for the nuptial flight. Wings of females drop before establishment of the nest (males die soon after coupling). The thorax is separated from the abdomen by a constriction, the ***petiole***; the antennae are segmented (Fig. 12-P). Mouthparts are of the chewing-licking type. Females and workers have a sting connected to poison glands which, however, have disappeared in the case of Formicinae (red, yellow and black ants). In this subfamily, only the poison glands survive: they secrete formic acid that is sprayed on prey and enemies. Development is ***holometabolous***.

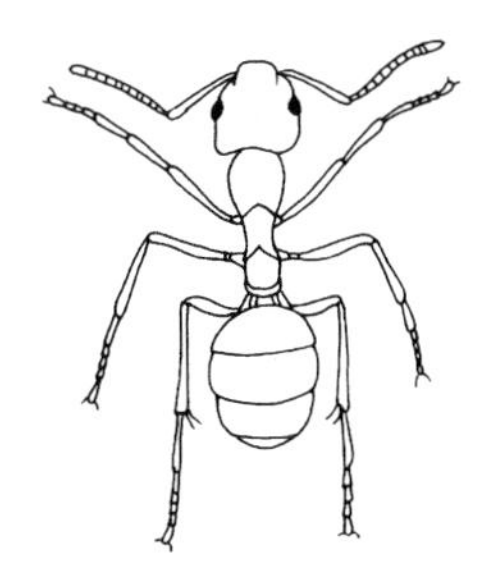

Fig. 12-P *Black ant* (Lasius niger) *worker. Length 2-5 mm (from Kutter, 1977).*

Holometabolous: said of development interrupted by metamorphoses during which the insect changes shape totally. The larva is transformed to nymph, pupa or chrysalis, then to ***imago*** or full-grown insect. The classic example is butterflies: egg–caterpillar –chrysalis–adult.

Size of species in Western Europe. In the case of workers, between <2 mm (*Leptothorax* spp., *Diplorhoptrum fugax*) and 14 mm (*Camponotus herculeanus*); sizes vary considerably in the same colony. Queens are much larger: that of *Camponotus* attains 18 mm.

Number of species. About 9500 worldwide, 192 in Western Europe, 180 in France, 110 in Switzerland.

Geographic distribution. Throughout the continents except in the coldest regions. Three-fourths of the species live in the tropical zone. Ants are present up to 3000 m in the Alps.

Habitat. Life mode very varied, leading to wide variety in construction of ant nests. Most of these are established underground, occasionally raised into a mound (*Formica* spp., *Lasius* spp.). Soil annexes also furnish many favourable sites, dead wood in particular. Certain ants establish tiny ant nests in chestnuts and acorns on the soil. Others dig enormous nests: the *Atta* (Chap. 4, § 4.5.1) or the granivorous *Messor*, whose nests contain up to 200 subterranean granaries dug to two metres depth (Bernard, 1968). In tropical forest, numerous tree-living ants build boxed nests from wood or leaves. The formidable driver ants (*Dorylus, Anomma*) of Africa and the army ants (*Eciton*) of tropical America are nomadic. Their colonies with many million individuals establish rudimentary nests or encampments from where raids are mounted for collecting food. The colonies change location periodically in accordance with the reproductive cycle of the queen and development of the larvae.

Wilson, in Hölldobler and Wilson (1996) observed 43 ant species living at the same time on a single tree in the Amazon forest.

Reference works. Forel, 1920 (work of a pioneer in myrmecology; still usefully read with interest;) Bernard, 1968 (book containing identification keys and information on the biology of species of Western Europe); Cherix, 1986 (excellent short book, very lively, presenting the biology of the *Formica* or red ants); Hölldobler and Wilson, 1996 (exciting presentation of ants of the world); Lavelle and Spain, 2001 (global picture).

Vertical distribution in soil. From the surface to several metres deep, depending on species and nest size.

Life mode. Ant societies comprise three castes: females or queens, males and workers. The last, sterile, are occasionally divided into subcastes according to size (minor, media and major).

Nutrition. Phytophagous, granivorous and often omnivorous. Many species take advantage of honeydew. *Lasius flavus*, one of the most edaphic ants of temperate regions, rears plant-lice on plant roots and scarcely leaves its anthill (Chap. 4, § 4.5.1). Many species have very specialized nutrition regimes, for example capturing only springtails.

Abundance. Ants form one of the most important groups of the pedofauna. In the Amazon forest for example, they are eight times more numerous than termites and their biomass represents four times that of all the vertebrates living there (Hölldobler and Wilson, 1996). Ant societies at times have impressive strengths:

- those of driver ants *(Anomma)* attain 20 million individuals (a single queen gives rise to 300 million descendants);
- Cherix (1986) described a supercolony of *Formica lugubris* in the Swiss Jura mountains: 1200 nests distributed over 70 ha were interlinked by 100 km of tracks and sheltered a population of 150 million ants (density 2×10^6 ind. ha^{-1});
- in Japan, one supercolony of *Formica yessensis* covered 270 ha and numbered 300 million individuals in 45,000 nests;
- in the African savanna, 7000 nests were counted per hectare, accommodating 20 million individuals.

The literature devoted to ants is very plentiful, probably more than 50,000 publications!

Significance in soil. The abundance of ants allows us to imagine how great the impact of these predators is on the remainder of the soil fauna: it is accepted for example that one colony of driver ants destroys 1.6 million insects per ha in ten days and that small ants are the chief consumers of microarthropods. Furthermore, their bioturbation activity is considerable (Chap. 4, § 4.5.1; Chap. 5, § 5.3.3).

"Les fourmis,
Chacune d'elles ressemble au chiffre 3
Et il y en a! il y en a
Il y en a 333333333333...
jusqu'à l'infini."
(Jules Renard, *Histoires naturelles*).

12.4.10 Vertebrates, or the soil megafauna

Are there birds, fishes and amphibians in soil?

On all continents, vertebrates (classes Pisces, **Amphibia**, **Reptilia**, **Aves** and **Mammalia**) count as edaphic species forming part of the megafauna (Chap. 2, § 2.6.3) Birds with underground nests can be grouped in the category of temporarily active geophiles: the kingfisher and sand martin are good examples. Reptiles also have edaphic representatives; in Central Europe, the slowworm (20 cm) is of this category. There even are edaphic fishes (lungfish) and amphibians (toads, frogs and newts) that survive difficult climatic periods by burying themselves in the soil; they are temporarily inactive geophilides (§ 12.5.5).

In Europe, only a small number of mammals significantly influence functioning of soils. Voles (Rodentia) and moles (Insectivora) can have considerable effect through bioturbation, the former in particular when breeding. Burrowing animals such as rabbits (Lagomorpha) and badgers (Mustelidae) locally move large quantities of earth.

Other mammals not belonging to the pedofauna may modify the surface layer of soil through ripping (e.g. wild boar) or trampling (e.g. large herbivores).

Reference works. Kevan, 1962; Wallwork, 1970 (two general books with brief coverage of vertebrates); Paton et al., 1995 (particularly on the role of birds).

12.5 THE FAUNA IN SOIL, ECOLOGICAL NICHE

12.5.1 Introduction to the concept of ecological niche

A considerable number of animal species coexist in the soil without eliminating one another because of the fact that they occupy different ***ecological niches***. To define a niche one uses several criteria relating to the morphology of the species considered, its behaviour, physiology, diet, abiotic environment, etc. The descriptors presented later (§§ 12.5.2–12.5.8) are relatively easy to define. Others, pertaining to behaviour or physiology, are less obvious, for example the reactions of animals to abiotic factors or their demographic strategies (Chap. 4, § 4.4.6; § 12.6.2). But all of them contribute to delineation of the niche. In a single biocoenosis, coexistence of two or more species is possible only if their niches differ in at least one important parameter (diet, period of activity, site of reproduction, etc.). A partial overlap of two niches is thus possible, but not complete coverage, following the principle of ***competitive exclusion***. This principle is well illustrated among other cases by the marsupials of Australia that have been ousted by exotic mammals introduced into the continent. But unoccupied niches may also exist in an ecosystem, allowing establishment of new species in a biocoenosis without causing disappearance of species already existing.

To each species its niche... and to each niche its species!

Ecological niche: place, function and specialization of a species within a community (from Ramade, 1993).

Competitive exclusion: principle postulating that durable coexistence of two species having exactly the same ecological niche is not possible in a single ecosystem and that, in this case, the better performing species eliminates the other (Dajoz, 1996).

The term ***guild*** is used to group species close to each other from the taxonomic viewpoint and occupying neighbouring ecological niches, in other words exploiting a single category of resources (Chap. 16, § 16.1.6) Separation of niches is then done on very precise criteria (periodicity, size of food, etc.). Some examples of guilds:

- woodpeckers that exploit bark of old tree trunks in a single forest;
- sphagnum mosses on hummocks in raised bogs (Chap. 9, § 9.2.3);
- stonecrops colonizing an old wall;
- decomposing coprophagous coleopterans in a cowpat (Chap. 8, § 8.3.2).

12.5.2 Classification of the pedofauna according to size

Size: the simple first approach, very informative nonetheless.

The approach to ecological niche by size of organisms is very informative. Many ecological characteristics depend on the species size: choice of food, life mode, aquatic or terrestrial, colonization of soil at depth as a function of soil porosity, mode of locomotion and abundance. The size of certain euedaphic invertebrates diminishes with increasing depth: this is seen in the mesofauna (springtails and oribatid mites) and even in the microfauna (thecamoebians). Anecic earthworms are an exception because the largest species (20 to 30 cm) occasionally dig their galleries two or three metres deep.

Three... or four size categories?

English-language authors divide soil animals into three size categories: <0.2 mm, microfauna; from 0.2 to 10 mm, mesofauna, and >10 mm, macrofauna. We prefer a division into four categories that appear to take into account ecology and the impact of animals on soil (see Chap. 2, § 2.6.3 for the relevant taxa and Chap. 13, §§ 13.6 and 13.7 for their roles in the food chain):

Micro-, meso-, macro- and megafauna.

- ***microfauna***

Size: length 4–200 µm, diameter <100 µm (0.1 mm).
Ecological characteristics: aquatic life, formation of cysts and forms resistant to desiccation.
Nutrition: microphagy predominant.
Abundance: expressed in 10^6 ind. m^{-2}.
Third compartment of the decomposition food chain.

- ***mesofauna***

Size: length 0.2 to (3)4 mm, diameter 0.1 to 2 mm.
Ecological characteristics: terrestrial life.

Nutrition: microphagy, saprophagy, zoophagy.
Abundance: expressed in 10^4–10^5 ind. m^{-2}.
Second compartment of the decomposition food chain.

- ***macrofauna***

Size: length (2) 4 to 80 mm, diameter 2 to 20 mm.
Ecological characteristics: terrestrial life.
Nutrition: zoophagy, saprophagy; microphagy less widespread.
Abundance: expressed in 10^2–10^3 ind. m^{-2}.
First compartment of the decomposition food chain.

- ***megafauna***, whose size is still greater, comprises vertebrates and the large tropical earthworms.

Entirely relative limits between categories.

In Chapter 2, § 2.6.3, fig. 2.12 highlights the relationships of these classes strictly according to size. For example, a 5-cm shrew actually belongs to the megafauna, even though it measures just one-fourth the length of *Lumbricus terrestris*, which itself belongs to the macrofauna! In tropical regions, macroarthropods attain very large size: carnivores (Scolopendromorph and geophilomorph centipedes, scorpions, spiders) and saprophages (diplopods) greatly exceed 10 cm in length. Lastly, numerous insect larvae during growth change from mesofauna to macrofauna.

12.5.3 Classification based on morphological adaptations

This second approach to the ecological niche postulates the existence of a relationship between the shape of organisms and their life mode. This provides some very spectacular examples of adaptations to life in soil. Five of them are examined here.

Are morphological adaptations really useful?
Animals apparently poorly adapted to life in the soil coexist there with species that seem to be greatly specialized (e.g. the field-cricket and the mole-cricket). But wait, we should not commit the 'refrigerator mistake' in considering the subject of morphological adaptations! We so term such an error in understanding which leads to thinking that when two-thirds of humanity have no refrigerator but manage well without it, this domestic appliance is of no use!

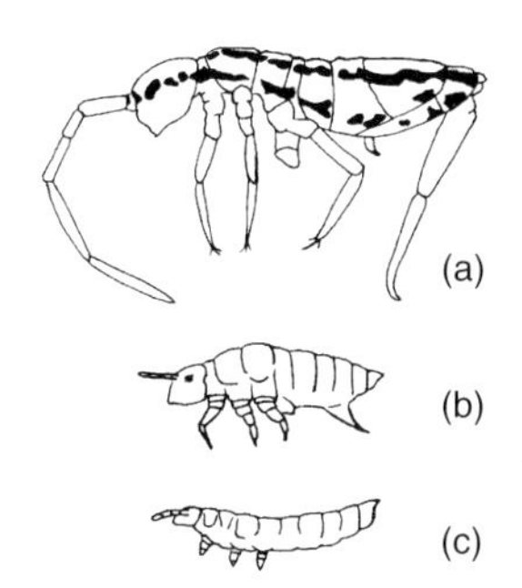

Fig. 12.7 *Types of springtails.*

Shortening of appendages and pigmentation
Edaphic springtails show that small size is often accompanied by reduction of legs, antennae and the

furcula. Hairiness disappears. Darkness causes diminution in pigmentation; forms living at depth and in hollows are generally white (Fig. 12.7; Plates I-2 and I-3):

- springtails of the epiedaphon (a): developed eyes, pigmented cuticle, long antennae and furcula, large size (up to 6, rarely 8 mm);
- springtails of the hemiedaphon (b): developed eyes, pigmented cuticle, moderately long antennae and furcula, medium size (2 to 4 mm);
- springtails of the euedaphon (c): smaller eyes and antennae, white cuticle, furcula small or absent, small size (1 to 2 mm),

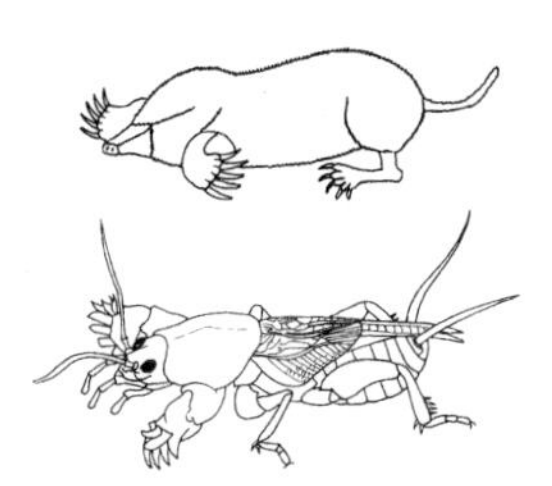

Fig. 12.8 *Convergence of shapes in the mole and mole-cricket, brought to same size.*

'Pedodynamic' body

Different species of moles of the five continents have the same characteristics: fusiform body, powerful forelimbs enabling the animal to dig, regression of the eyes, long sensory setae. The same convergence of shapes is found in the mole-cricket, with similar ecology (Fig. 12.8).

Hydrostatic skeleton

Haemolymph: liquid filling the interstitial spaces in the body of invertebrates such as the earthworm. The same term is used for the blood of arthropods.

Earthworms and dipteran larvae exhibit a remarkable adaptation for movement in the soil. Their soft, elastic and watertight cuticle is maintained under tension by a constant volume of ***haemolymph***. Beneath the cuticle, two layers of opposing muscles, circular and longitudinal, take support from the body fluid. When the circular muscles contract, the body lengthens; the body is again shortened under the effect of the longitudinal muscles (Fig. 12.9).

Flattening of littericole organisms

Littericole: living in litter.

This shape allows larvae of dipterans (e.g. *Fannia*) or certain ***littericole*** diplopods (polydesmid millipedes) and chilopods (lithobiomorphs) to insinuate themselves between layers of dead leaves more or less stuck together. Borcard (1981) noted that among the Carabidae, those of medium size and flattened shape (*Pterostichus melanarius, Abax* spp.) greatly dominate in acidophilic beech forests with thick litter.

Volution: behaviour of certain animals of rolling themselves into a ball (Fig. 12.10). This behaviour is seen in mites (Phthiracaridae), isopods (Armadillidiidae), myriapods (Glomeris) as well as in mammals (hedgehogs, armadillos, scaly anteaters).

Volution

Certain animals protect themselves against predators through ***volution***. In the case of pill millipedes, the defence is reinforced by secretion of repellent substances. Volution is also a protection against desiccation: a voluted pill millipede *(Glomeris)* evaporates up to 12 times less than an unrolled active animal, because volution masks

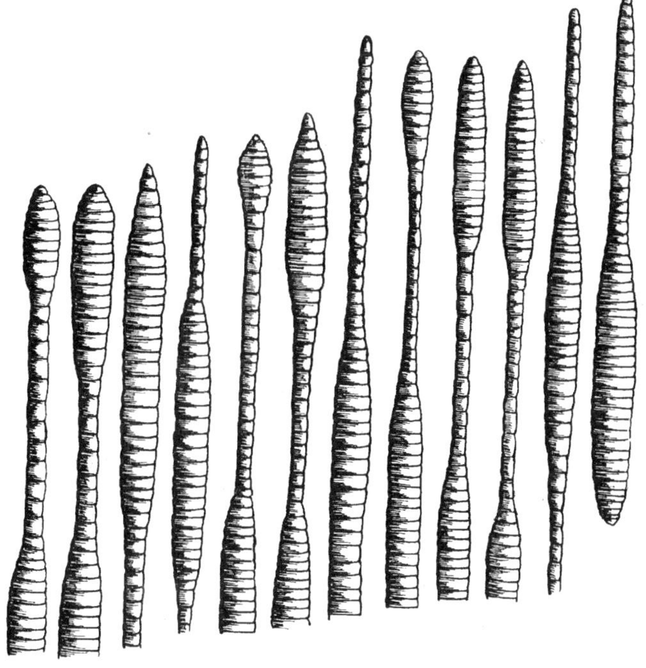

Fig. 12.9 *Movement of earthworms. Coordinated by nerves, a wave of contraction runs through the body and enables the earthworm to move in the soil.*

its respiratory openings located on the ventral surface, through which most of the water loss takes place (Eisenbeis and Wichard, 1985).

12.5.4 Classification according to mode of movement

Microfauna, aquatic, swim in film and interstitial water (Chap. 3, § 3.4.2; Chap. 13, § 13.7.4). The small organisms of the mesofauna move by using the available space, without much digging. On the contrary, the larger animals (macro- and megafauna) must open out a path in the soil. These movements contribute greatly to bioturbation of soil (Chap. 5, § 5.3.3):

- ***Burrowers*** dig chiefly with their limbs, whether modified or not (e.g. mole, marmot, vole, mole-cricket, cricket, scarabaeid).
- ***Miners*** dig their underground home with their mandibles or teeth. They carry out the earth (e.g. ants, termites) or push it behind them with their limbs (e.g. mole-rats, cockchafer larvae).
- ***Tunnellers*** drill galleries in two ways. They either force their way by enlarging the interstices (e.g. juliform

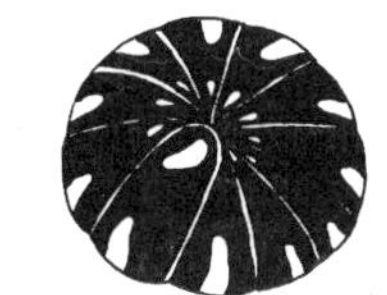

Fig. 12.10 *Volution in* Glomeris.

There is an obvious relationship among body shape, the 'tools' available to animals and their modĕ of locomotion. For example, by virtue of its hydrostatic skeleton an earthworm 20 to 30 cm long, such as *Lumbricus terrestris*, can exert a pressure of 1 kg cm^{-2} with its anterior end.

diplopods, earthworms) or ingest the soil and evacuate it in the form of excreta they line their galleries with or jettison (earthworm casts).

12.5.5 Classification based on more or less continued presence in soil

According to the ancient Jacot classification (in Wallwork, 1970), still valid, soil animals are divided into four categories:

- ***Temporarily inactive geophiles*** spend time in the soil to overwinter or for metamorphosis. They exert no mechanical force on it. Thus caterpillars of many butterfly families (Noctuidae, Geometridae, Sphingidae) bury themselves in the ground at the time of ***nymphosis***. Their chrysalides constitute, at certain times of the year, an important food source for edaphic carnivores, parasitoids, necrophages and decomposers.

Nymphosis: period of transformation of the larva to imago. During nymphosis, insects in the nymph form are mostly immobile. They pass this more or less extended period in a small chamber hollowed out in the soil by the larva or in locations protected by vegetation. The nymph (termed chrysalis in the case of butterflies and pupa in dipterans) is characteristic of holometabolous (§ 12.4.9) insects.

- ***Temporarily active geophiles*** accomplish their development from eggs laid in the soil. Aerial adults generally have a short life outside the soil, during which they reproduce and disperse their eggs. It is larvae and nymphs that actually belong to the pedofauna. Many dipterans with edaphic larvae (e.g. Tipulidae, Bibionidae, Sciaridae) belong to this group.

- ***Periodic geophiles*** exhibit a still higher degree of bonding with soil because their entire life is spent in it. But these organisms, winged or not, can change location when their habitat becomes unfavourable or rise in population density results in very tough intraspecific competition (e.g. scarabaeids, mole-crickets, certain diplopods).

- ***Geobionts*** are permanently present in the soil. Their dispersal ability being low, they expand their area of distribution by 'contact', that is, by moving short distances in the soil or on its surface (e.g. earthworms, mites, springtails).

What is the speed of propagation of anecic earthworms?

Allolobophora caliginosa, a European earthworm, was introduced in New Zealand to improve the output from grasslands. The capacity for expansion from the point of release of 25 earthworms was measured. The result of their impact on the soil was easily seen because the grass became definitely taller and greener where the worms prospered than elsewhere. After four years, a patch several metres in diameter was visible around each inoculation. And these patches, signifying the presence of *A. caliginosa*, attained a diameter of 200 metres after eight years of exponential growth (Edwards and Lofty, 1977)!

12.5.6 Classification of animal communities according to their stratification

Animal communities are differentiated according to their vertical stratification:

- the ***hyperedaphon*** is the community of low plants, enriched with several elements of the pedofauna;
- the ***epiedaphon*** occupies the OL, OF and OH horizons and S (layer of bryophytes, in the sense of Green et al., 1993; Chap. 6, § 6.2.2);
- the ***hemiedaphon*** corresponds to the A horizon;
- the ***euedaphon***, at depth, corresponds to the S (as defined by AFES, 1995, 1998), B and sometimes the C and D horizons.

Hyper- and epiedaphon are diversified by the presence of soil annexes such as dead wood, stumps, isolated or heaped stones, moss pads and mats, cowpats, dung and carcasses (Chap. 8).

12.5.7 Classification according to preferred habitats

Table 12.6 gives a classification of different edaphic environments used by the soil fauna and is equally useful for identification of the ecological niche.

12.5.8 Classification according to nutrition regime

Based on understanding of trophic functioning of the ecosystem, with its cycles of matter and energy fluxes,

Table 12.6 Habitats of the soil fauna

General category	Subcategory	Location
Epigeal edaphic milieu	Film-dwelling milieu	Lichens, mosses on bare rock
	Littericole milieu	In litter
	Stone-dwelling milieu	Beneath rocks
	Moss-dwelling milieu	Mosses (S layer—Green et al., 1993) and plants in pads
	Saproxylic milieu	In dead wood
Hypogeal edaphic milieu	Humicole milieu	In the top part of soil (A horizon)
	Endogeal milieu	In the deeper part of soil (S horizon—AFES, 1995, 1998; B, C and D)
	Terrestrial phreatic milieu	Cracks in rock occasionally filled with water, which cannot be penetrated by man
	Cavernicole milieu	Caverns, large fissures

this classification requires detailed presentation and forms the subject of Chapter 13.

By combining the seven groups of criteria that have been discussed, we get a picture, an outline of the ecological niche of a species, still rough and incomplete, but rich in facts.

12.6 SUMMARY OF THE POSITION AND ROLE OF SOIL ANIMALS

12.6.1 Five examples of ecological niche

Five organisms illustrate this attempt at defining 'synthetic ecological niche': mole-cricket, dusky cockroach, an oribatid, snow scorpion fly and earthworm:

- Mole-cricket (*Gryllotalpa gryllotalpa*, Orthoptera, Insecta; Fig. 12.8): macrofauna; periodic active geophile; burrowing insect; humicole environment; carnivorous, possibly rhizophagous; belongs to the hemiedaphon; activity during summer.

- Dusky cockroach [*Ectobius lapponicus*, Dictyoptera, Insecta; Fig. 12-Q(a)]: macrofauna; temporary inactive geophile; littericole epigeal environment; omnivorous; belongs to the epiedaphon; activity early spring to autumn.

- An oribatid of family Phthiracaridae [Acari, Arachnida; Fig. 12-Q(b)]: mesofauna; geobiont; littericole epigeal and saproxylic environments; saprophagous (occasionally microphagous); belongs to the epi- and hemiedaphon; activity during summer.

- Snow scorpionfly [*Boreus hiemalis*; Mecoptera, Insecta; Fig. 12-Q(c)]: macrofauna; periodic geophile; epigeal moss-dwelling and littericole environment; muscivorous; belongs to the epiedaphon; active year-round (adult found on snow in peak winter).

- Earthworm (*Lumbricus terrestris*, Oligochaeta, Annelida, Fig. 12-D): macro- and megafauna; geobiont; endogeal environment; phytosaprophagous and geophagous; belongs to the euedaphon and hemiedaphon (anecic worm); tunneller active chiefly in spring and autumn.

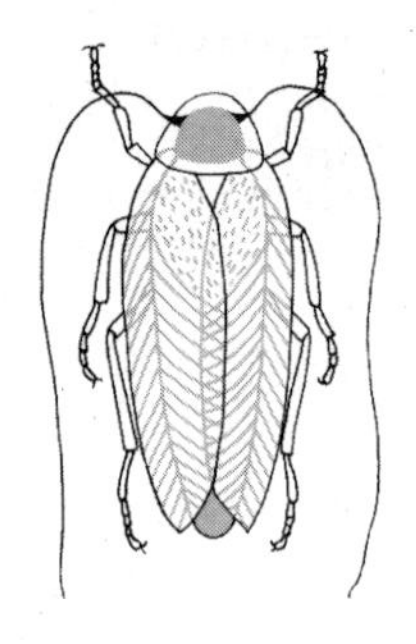

(a) Dusky cockroach, 8-11 mm

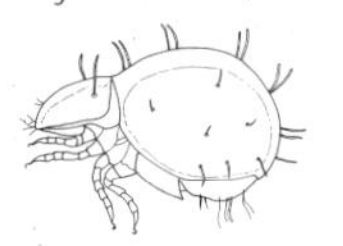

(b) Phthiracaridae oribatid, 0.5-0.8 mm

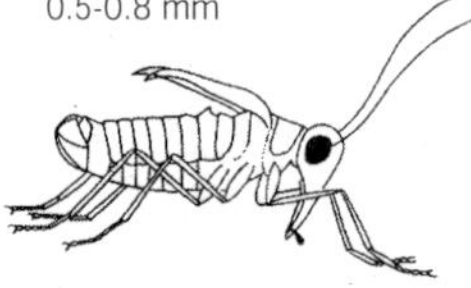

(c) Snow scorpionfly, 2-3 mm

Fig. 12-Q *Three animals with different ecological niches.*

'Sixty years of controversy for not arriving at the correct definition, such is the more or less puzzling balance of the concept of ecological niche. Although if it has such a tough life, it should indicate a profound theoretical necessity' (Lachaise, 1979).

12.6.2 Future of the concept of ecological niches

The concept of ecological niche, one of the oldest in ecology, defined as early as 1927 (Elton, 1927)—based on early papers of Grinnel dating from 1917—then refined in 1957 (Hutchinson, 1957) has not finished being discussed, improved, questioned again or completed.

Although argued through numerous criteria, this first synthesis of ecological niches of some animals in the soil is yet incomplete. Current discussions on the ecological niche are actually integrating little by little other extremely fruitful criteria:

- recent or ancient history of the ecosystem in which the species lives; the niche forms part of a dynamic system and its present boundaries depend on past conditions also of the ecosystem (Chap. 5, § 5.1.1; Chap. 6, § 6.1.3);
- role of this system and its emerging properties in delimitation and homoeostasy of the niches (Chap. 1, § 1.2.2, Chap. 7, § 7.1.4);
- probability of realization and occupation of a particular ecological niche;
- role of interspecific competition (Chap. 4, § 4.5.3) in separation of niches;
- comparison between potential and real ecological niches (Chap. 4, § 4.2.4);
- 'evolutionary strategies' (life-history strategies, Blondel, 1986, 1995);
- population dynamics, with r–K or C–R–S demographic strategies (Chap. 4, § 4.4.6; Chap. 6, § 6.1.3).

Adaptive demographic strategies

Species are morphologically and physiologically adapted to life in certain environments. They are also adapted at the level of their demography, a concept integrating among others the birth and death rates. Two demographic 'behaviours' of populations can be distinguished: the r and K strategies or selections.

The r ***strategists*** exhibit strong aptitude to colonize new unstable and ephemeral environments because of a high rate of reproduction, rapid growth and a high capacity to disperse. The selection that operates in this case favours taxa with high fertility (e.g. certain springtails, corpse flies, plant colonizers of wastelands, fungi of the genus *Mucor*).

The **K** ***strategists*** have more specialized ecological niches that they retain because of their competitive abilities. Survival of these species rests on their specialization. Their fertility is limited by the capacity of the environment. K-selection favours efficiency in exploitation of the environment (e.g. numerous earthworms, trees such as beech and oak in temperate regions, lignivorous fungi such as *Trametes* spp.).

When these ideas are extended to all the biocoenoses, we have ***coenotic strategies***, with the letters r and K replaced by **i** and **s**.

Proposed by MacArthur and Wilson (1967), the theory of r–K strategies is an essential concept in ecology. It explains the biodemographic adaptation of species to the environments they exploit. But its essentially bipolar character does not always allow accounting for the complex dynamics of the population-environment

By convention, in population dynamics, **r** designates the natural growth rate of populations and **K** the maximum capacity of the environment containing them (Chap. 4, § 4.4.6).

R.H. MacArthur (1930–1972) is one of the great names in modern ecology. He is known for his studies on insular populations (theory of islands) in which are included those with r and K strategies. E.O. Wilson is a renowned myrmecologist who would later be recognized as the founder of animal sociobiology and defender of biodiversity (Wilson, 1992).

Other authors have attempted to improve the descriptive power of the theory of r–K strategies.

system (Barbault, 1990). For this reason, Grime (1979) and Grime et al. (1988) (also see Vanden Berghen, 1982; Begon et al., 1986) propose another classification based on the concept of perturbation (= causes of destruction of the biomass) and on constraint (= factors limiting production). Grime's theory, restricted to plants, is subdivided into three categories. Two correspond roughly to those of MacArthur and Wilson; these are **C *strategists***, as in Competitive, analogous to K strategists (e.g. numerous forest species, *Rumex* spp.) and **R *strategists***, as in Ruderal, analogous to r strategists (e.g. white goosefoot). One category is new, that of **S *strategists***, as in Stress-tolerant, which are eliminated by competition from environments with moderate conditions but manage to survive in difficult conditions (e.g. mountain pine, bog rosemary, saxifrages in rock fissures).

In Chapter 4, § 4.4.6, Table 4.3 illustrates the similarities between different approaches with respect to similar concepts developed in microbiology (zymogenous and autochthonous species).

CHAPTER 13

FOOD CHAINS AND WEBS IN SOIL

Every transformation of organic matter takes place through soil organisms. The action of plants is more specifically covered in Chapter 4, that of microorganisms in Chapters 4, 14, 15 and 16; that of the fauna is detailed in Chapter 4 and the present one.

Chapter 13, half-way along the logic of the second part of this book, presents the 'food paths' in soil. It covers the foundations of the trophic-dynamic principle of the ecosystem (§ 13.1) with an outline of the methodology corresponding to it (§ 13.2). Sections 13.3 and 13.4 are devoted to two levels of organization of trophic relations, food chain and food web respectively, then § 13.5 puts the soil at the centre of these relations as the 'recycling compartment' of the ecosystem. The dominant food chain in the soil, that of decomposers, forms the subject of a detailed picture in § 13.6 (general aspects) and § 13.7 (functional compartments).

This chapter is closely related to Chapter 12, which covered the players in the food chains from the viewpoint of systematics (taxonomic groups) and functioning (ecological niches).

13.1 TROPHIC-DYNAMIC PRINCIPLE OF THE ECOSYSTEM

How to detect a structure in the profusion of plants, animals, fungi and bacteria that constitute the biocoenosis (Chap. 4, § 4.2.4)? There is one method: it is based on the concept of food chain, that is, on the study of ***trophic*** relationships among species. For the biologist, this idea is an indispensable tool for understanding not only the biocoenosis, but also the ecosystem in its entirety.

A biocoenosis of the temperate zone is formed from hundreds, even thousands of species; it is very difficult to discern *a priori* a structure in this vast assemblage.

Trophic: from Gk. *trophé*; pertaining to nutrition.

13.1.1 Everything starts in the tundra... and outside the soil

Naturalists of Oxford University have studied life in the tundra in Spitzbergen.

In 1920, young Charles Elton took part in an expedition to Bear Island, near Spitzbergen. As zoologist, Elton observed Arctic foxes, clearly visible in the treeless environment, and specifically their food habits. In summer their diet consists chiefly of tundra birds, their young and their eggs. Of this prey, the ptarmigan feeds on berries and green leaves, and the snow bunting chiefly on phytophagous insects. Thus the foxes and these birds, like all other animals of the tundra, are bound among themselves by trophic relationships that Elton termed ***food chains***. A chain comprises several consumption levels: vegetation, herbivores, carnivores of the first level, carnivores of the second level; for example:

- berries → ptarmigan → fox,
- leaves → insects → snow bunting → fox.

Elton and Lindeman, two fathers of functional ecology

Charles Elton (died 1991) was a pioneer in ecology. He clearly defined for the first time the very productive principles of food web and ecological niche, which ecologists have still not fully explored. These ideas were published in 1927 in a surprisingly modern book, *Animal Ecology*, reprinted many times until 1996.

Raymond L. Lindeman (1915–1942) was an American limnologist known mainly for his studies on Cedar Bog Lake, a small peat bog lake in America. Despite the briefness of his career, Lindeman laid the foundation of quantitative functional ecology by using energy (the calorie) as the common denominator for all levels of the trophic chain. The Lindemanian approach enables us conceive of the ecosystem in its entirety.

Three of Elton's principles: food web, ecological niche and pyramid of numbers.

'Analysis of the cycles of trophic relationships shows that a biotic community cannot be clearly separated from its abiotic environment. The ecosystem must therefore be considered the most fundamental ecological unit' (Lindeman, 1942).

Although originally centred solely on the tundra, these observations have gained general value, proved since by many investigators working in all parts of the world. Also, Elton deduced from them a series of principles that enable understanding the trophic framework of living communities. We mention just three:

- the ***food web,*** which shows the arrangement of the chains connected to one another in the ecosystem;
- the ***ecological niche,*** which defines the role of each species in the community based on its food regime and its relationships with the other organisms of the biocoenosis;

• the ***pyramid of numbers***; Elton showed that organisms are as a rule more numerous in the first levels of the chain than at the top; he thus introduced the quantitative aspect of food chains, which would later be developed by Lindeman (establishment of the ***trophic-dynamic principle*** of functioning of the ecosystem).

13.1.2 Nature has said: eat each other... and no leftovers!

In the ecosystem, energy and nutrients circulate through the biocoenosis, following the food pathways. In this complex whole, trophic or nutritional relationships distribute the living things, from bacteria to mammals, not in one but in a multitude of chains, generally quite short and interconnected. Thus they form a vast web, the food web, which defines the ***trophic niche*** of each species in the biocoenosis (Fig. 13.1).

An intense and complex circulation.

The trophic ladder rests totally on autotrophic production (net primary production, Chap. 2, § 2.2.1) that is consumed by herbivores who themselves serve as food for carnivores. The leftovers, carcasses and excrements feed detritus chains that recycle the dead organic matter, making it again assimilable by plant roots. This is the food cycle (Fig. 13.2).

13.1.3 A specific food regime

Feeding is an essential activity of heterotrophs, animals in particular, which fuel their energy motor in this manner. Each organism has a nutrition strategy that enables it to find, overcome, absorb and digest certain foods but not others. This strategy is a function of size, shape of mouthparts, enzymes in the alimentary canal, behaviour of the hunt and activity rhythm. These characteristics differ from one species to another, thereby permitting each one to occupy its own food niche in the biocoenosis.

The energy contained in the food is often the principal limiting factor for animal populations, unlike for plants for which the sun provides unlimited energy. From the largest animal to the smallest, all must eat to live.

In the animal world, a great variety of regimes is seen, sometimes very specialized, sometimes very broad; this variety permits exploitation of all the food resources present on the surface of the globe (Tab. 13.1). Just like size, the mode of reproduction or temperature requirements and food regimes are ecological characteristics of the species.

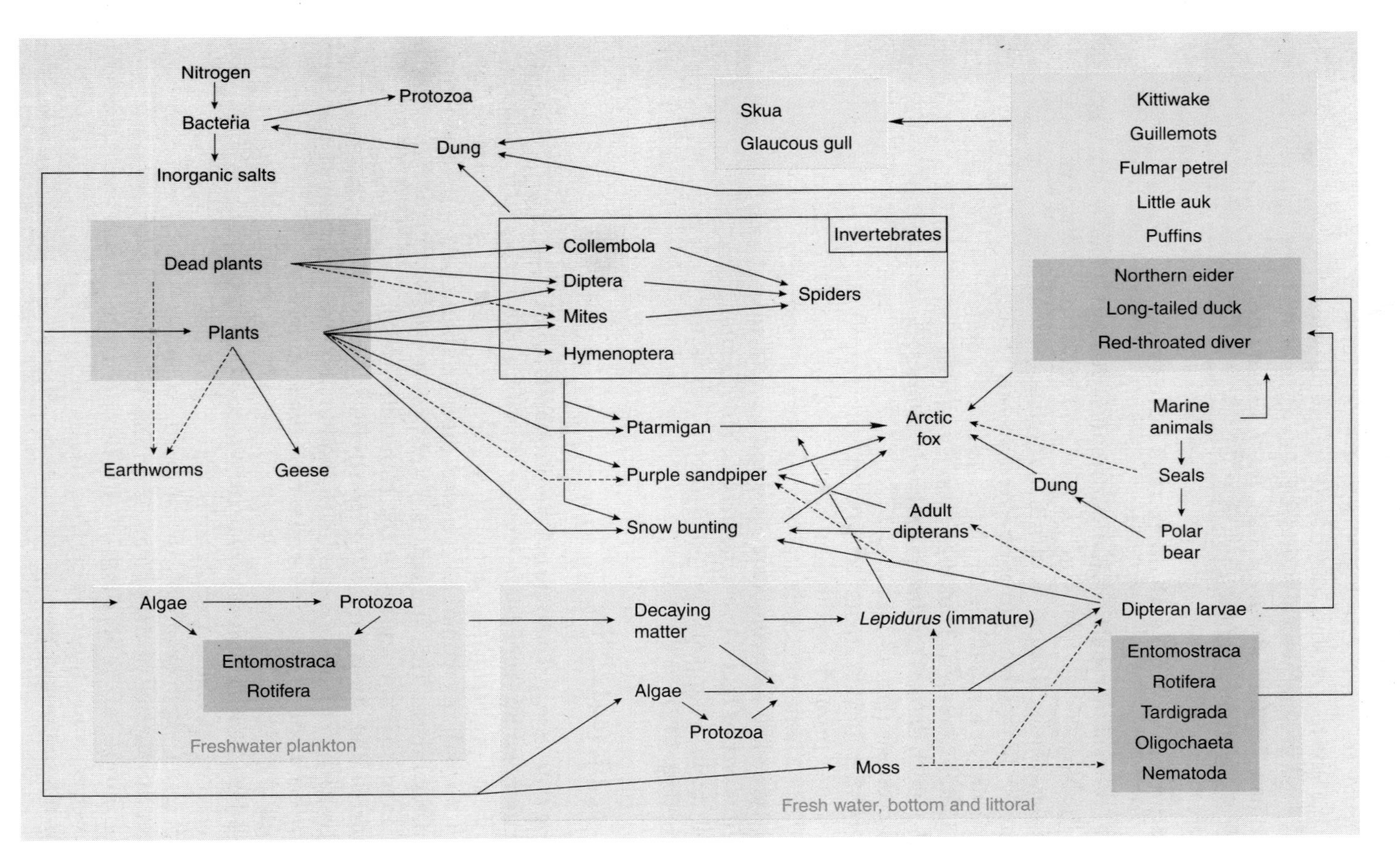

Fig. 13.1 *Food web in the tundra of Bear Island (from Summerhayes and Elton, 1923). The historical importance of the scheme, summarized still more, justifies its being chosen here. The totality of the food web rests on the primary production of autotrophic bacteria, lichens and plants with chlorophyll, the latter being capable of capturing solar energy for conversion into biochemical energy.*

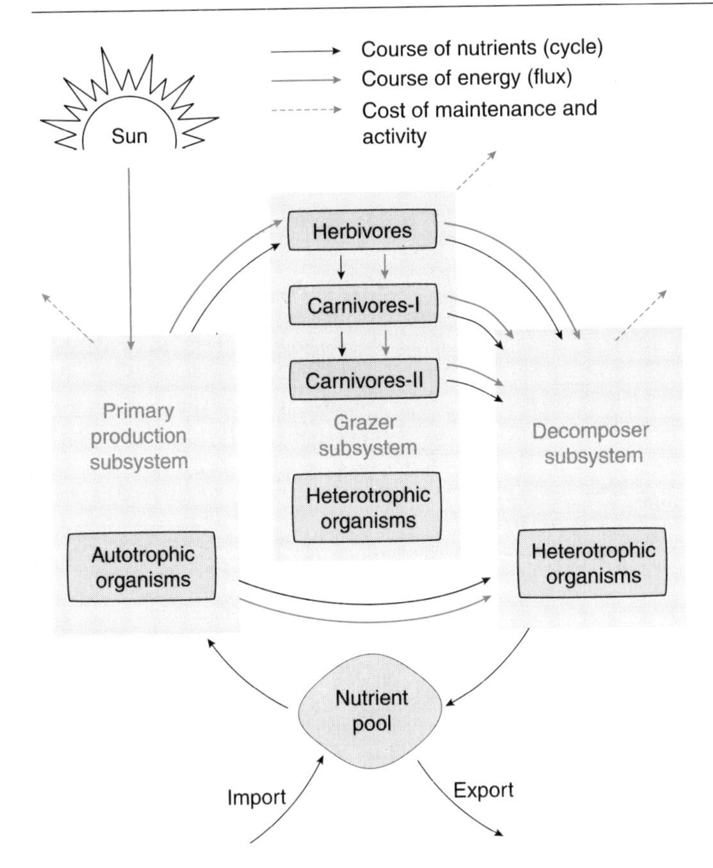

Fig. 13.2 Grazing food chain and food cycle.

Table 13.1 Diet of the pedofauna in the temperate region: some definitions

Category	Food	Examples of organisms
A In the grazing food chains		
A1 Living food of plant origin		
Algophagous	algae	springtails, larvae, nematodes
Fungivorous = mycetophagous	fungi (generally including carpophores)	larvae of dipterans: Mycetophilidae
Frugivorous	fallen fruits	ants, larvae of elaterid beetles
Granivorous	grains	ground beetles, weevils
Herbivorous	mosses, higher plants, lichens	caterpillars
Lichenivorous	lichens	oribatids, silverfish
Muscivorous	mosses	*Boreus* (Mecoptera), tardigrades
Mycophagous	moulds (micromycetes)	Liodidae (Coleoptera), springtails, oribatids
Opophagous	sap (sucking)	root aphids
Pollinovorous	pollen	coleopterans, flower flies (Syrphidae)
Rhizophagous	living roots	larvae of weevils, cockchafers

(Contd.)

Table 13.1 (*Contd.*)

Category	Food	Examples of organisms
Xylophagous = lignivorous	living wood	coleopterans, larvae and adults
A2 Living food of animal origin		
Carnivorous (vertebrates)	herbivores (vertebrates)	weasel, fox
Drilophagous	earthworms	mole
Entomophagous	insects	mole-cricket, hister beetles
Haematophagous	blood (sucking)	gamasid mites, ticks, fleas, female horseflies
Insectivorous (vertebrates)	insects	shrews, woodpeckers
Myrmecophagous	ants	woodpeckers, foxes
Termitophagous	termites	ants, various vertebrates
A3 Living food of microbial origin		
Bacterivorous	bacteria	protozoans, nematodes
Microphagous	protozoa, bacteria, algae, fungi	larvae of dipterans, oribatids, springtails, nematodes
B In the detritus food chains		
Carrion-eating = diversivorous (vertebrates)	carcasses of vertebrates	various birds and mammals
Coprophagous	dung	scarabaeids, fly larvae
Foraging	litter	termites, ants
Geophagous	earth	earthworms
Humivorous	humus	termites, earthworms
Necrophagous	carcasses	larvae of dipterans, burying beetles
Saprophagous	unspecified dead organic matter	numerous invertebrate groups
Saprophytophagous = phytosaprophagous	dead leaves	woodlice, diplopods
Saprorhizophagous	dead roots	larvae of coleopterans, earthworms
Saproxylophagous	dead wood	diplopods, larvae of dipterans and coleopterans
C Specificity in diet		
Monophagy	very specialized food	
Oligophagy	diet restricted to a single family of plants or to a very definite range of prey	
Omnivory = euryphagy	very varied plant, animal and microbial material	
Polyphagy	different kinds of food, insects, plants, etc.	

Sometimes nutrition becomes cannibalism

Nutrition is perhaps the most widespread relationship among species in the animal world. It is even an internal factor of demographic regulation when it becomes ***cannibalism***, the process in which individuals eat other individuals of their own species, most often in the form of eggs, larvae or the young, but occasionally adults as well. Cannibalism is widespread among most metazoans, from earthworms to primates, and may be manifested at all stages of development (Elgar and Crespi, 1992).

Cannibalism is seen in many groups of edaphic insects, mostly among the larvae (coleopterans: *Aphodius*, ground beetles, staphylinid beetles; dipterans: horseflies) as well as in vertebrates soil-associated (shrews, rodents).

13.1.4 Digression on functional morphology

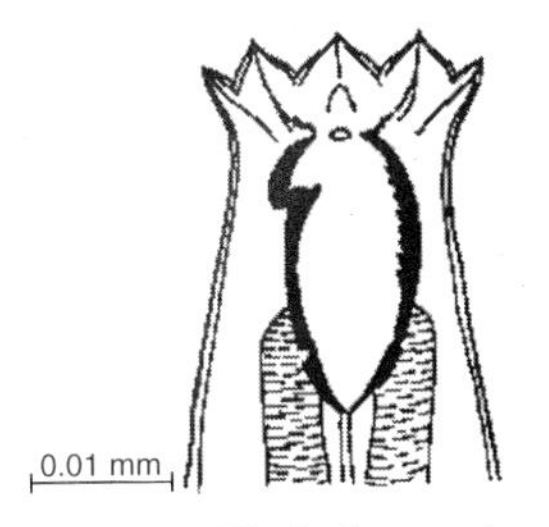

Fig. 13-A *Head of a carnivorous nematode* (Prionculus muscorum). *After Brauns (1968).*

Animals, according to their size, capture a certain range of prey that can be neither too large nor too small. The structure of the mouthparts indicates the mode of feeding of the animals (Fig. 13.3).

The violet ground beetle is a carnivore, its powerful and pointed mandibles are provided with sharp teeth that enable it to kill and cut up earthworms, insect larvae and molluscs. These mouthparts are categorically the

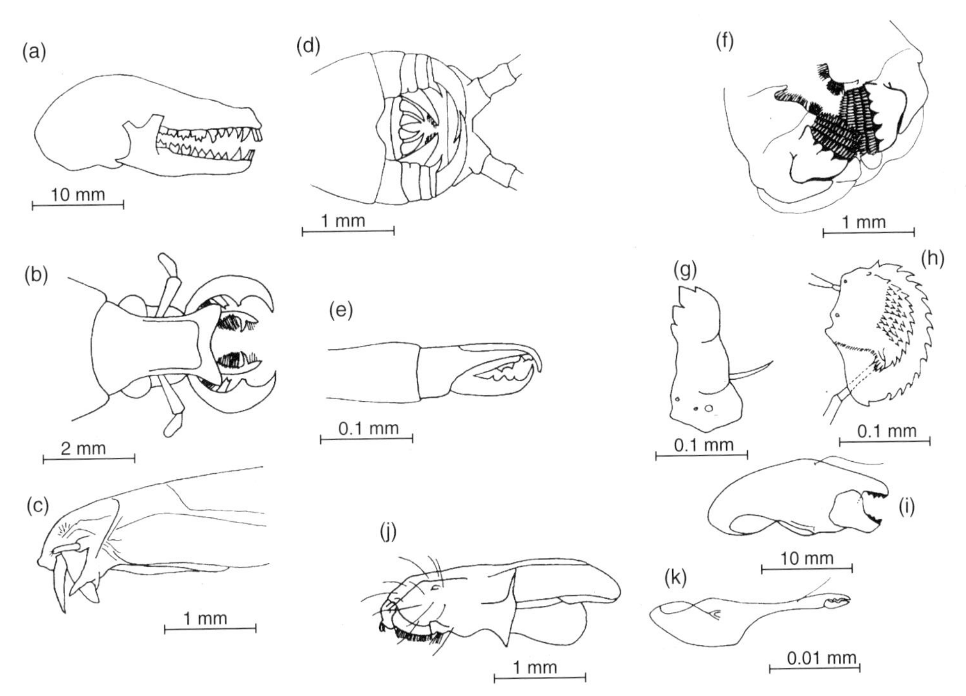

Fig. 13.3 *Comparative morphology of mouthparts of soil animals. Carnivores: (a) Skull of mole (Insectivora). (b) Head of ground beetle (Coleoptera) viewed from above. (c) Head of larva of horsefly (Diptera), side view; (d) Head of lithobiomorph centipede (Chilopoda), viewed from below. (e) Chelicera of gamasid mite (Acari). Phytosaprophages: (f) Mandibles of juliform millipede (Diplopoda). (g) Mandible of St Mark's fly (Diptera). (h) Mandible of midge (Diptera). (i) Chelicera of an oribatid mite (Acari). Microphages: (j) Head of soldier fly (Diptera), side view. (k) Chelicera of an oribatid mite (Acari).*

Predator: organism that kills and eats a heterotrophic organism.

same as those of the mole, centipedes, gamasid mites and nematodes (Fig. 13-A). True, these animals do not belong to the same taxonomic group and differ greatly in size; nonetheless they perform a similar function, scaled to each specifically, in the biocoenosis: namely, that of ***predator***.

Emulating Sherlock Holmes, the zoologist deduces from the morphology of animals several indices that enable prediction of their trophic niche.

Body shape and mode of locomotion show an adaptation to certain environments. The ground beetle, with long legs, hunts on the soil surface (Chap. 12, § 12.4.9); the long, thin, flexible geophilomorph (one European species reaches 10 cm), provided with many pairs of legs, is comfortable in soil interstices from where it moves with ease (Chap. 12, § 12.4.7). The same is true for the much smaller (1 to 2 mm) gamasid mites, which hunt for prey at their scale in the same environment (Chap. 12, § 12.4.8).

13.2 HOW TO STUDY THE FOOD REGIMES?

The study of food regimes of animals is not easy.

Naturalists have long been interested in this aspect of animal ecology, clearly less easy to approach than it seems here, in spite of many methods available. None of the methods described later suits all species, each one must be adapted for different taxa.

13.2.1 Three apparently simple methods...

Three 'simple' methods.

Three methods, requiring only optical equipment, will be discussed first:

- direct observation,
- analysis of faeces,
- examination of stomach contents.

The last two are not applied to species with liquid nutrition, which aspirate the contents of their prey (e.g. spiders, ground beetles, predatory larvae of dipterans).

Direct observation of animals

Direct observation is hard to apply systematically because, by definition, the endogeal fauna lives hidden from sight while the epigeal, visible on the surface, is often nocturnal. Occasional observations, for example that of plant lice adhering to plant roots may, however, set the investigator on the right track. Then research can be done on reared animals. For example, putting a portion of the soil in a ***rhizotron*** enables us define the nutritional

Rhizotron: vertical glass-panelled box in which a section of the rooting zone is placed.

behaviour of rhizophagous larvae of cockchafers and weevils, or that of earthworms.

Experiments on food selection, familiarly termed ***cafeteria-tests***, often provide interesting information. They are applicable to different taxa of the pedofauna (Acari, diplopods, isopods, earthworms) and consist of offering to oribatids of forest litter, for example, a choice of dead leaves of different species (Tab. 13.2). The relative disappearance of the fragments is measured, which indicates the preferences in choice of food. But this method is not always very reliable because the subjects sometimes make different choices in the laboratory (confined conditions) than in nature (Reutimann, 1987).

Meet me at the cafeteria!

Table 13.2 Cafeteria-test with three species of phytosaprophagous oribatid mites (Berthet, 1964)

Species	Hornbeam	Hazel	Oak	Beech
Nothrus palustris	1	n.c.	n.c.	n.c.
Steganacarus magnus	1	2	4	3
Phthiracarus borealis	1	4	2	3

Note: The numerals correspond to the order of preference; n.c. = not consumed. Hornbeam leaves are eaten first by all three species, which then show different tastes.

Study of faeces

The droppings of different groups of edaphic arthropods, often with characteristic shape and size, have been the object of precise description. For example, the excrements of isopods are rectangular, those of diplopods more or less cylindrical; their size is in keeping with that of the animals dropping them.

The following terms are considered synonymous: faeces, droppings, faecal pellets.

Example of a very characteristic faeces

The faeces of *Glomeris hexasticha* has the shape of a truncated cone 2 mm in height and 1.2 to 1.4 mm in diameter (Fig. 13-B). Its dry weight comprises between 0.40 and 0.45 mg. It contains fragments of dead leaves (maximum area 0.17 mm^2), pollen grains, pieces of mycelium, algal cells, unrecognizable very fine organic matter and mineral particles (maximum diameter 0.07 mm) (Tajovsky et al., 1992).

Fig. 13-B *Faeces of* Glomeris spp.

The method valid for herbivorous invertebrates, phytosaprophages, eaters of algae and spores, consists of:

- establishing, from rearing chambers, an atlas enabling recognition by comparison of the faeces found in soil, chiefly in the litter (Zachariae, 1979);

• making a microscopic analysis of these faeces to identify the plant residues they contain.

Biological activity of the humiferous episolum, essential classification criterion for humus forms, is revealed by faeces in particular (Chap. 6, § 6.2).

Analysis of stomach contents

Analysis of stomach contents is conventional in the study of vertebrates (moles, voles, woodmice). But it has also been successfully applied to phytosaprophagous macroinvertebrates (e.g. earthworms, isopods, diplopods) and even to oribatid mites. In the case of woodmouse, it shows that the regime changes during the year. In summer, this rodent becomes an insectivore very closely bound to the soil from which it extracts chrysalides for eating.

Squashing: pressing of material between a glass slide and cover slip for observation under an optical microscope.

Applied to oribatid mites, this analysis has proved on the basis of four identifiable materials (plant residues, fungal hyphae, fungal spores and structureless material) that the food regime varies considerably from one species to another. To examine the intestinal content of animals as well as oribatids, we proceed by ***squashing***.

13.2.2 ...and some others to the rescue

Three more sophisticated methods.

More sophisticated laboratory methods complement the above:

- highlighting digestive enzymes;
- the immunological approach;
- tracing the food chains with radioactive and stable isotopes.

Highlighting digestive enzymes

When enzymes direct the biologist.

The panoply of enzymes encountered in the alimentary canal of an animal point to its current food regime. These enzymes are generally inducible (Chap. 14, §§ 14.1, 14.2.5); in principle, there is no gratuitous production of enzymes. Traces of enzymes released into the soil by the pedofauna (in excreta and regurgitations) should enable tracing or quantifying the activity of different functional groups in the decomposition chains.

Enzymes secreted into the alimentary canal enable digestion of ingested materials. Thus amylases make hydrolysis of starch possible and chitinase that of chitin. There is a fairly clear-cut relationship between the food regime and the enzymatic equipment of the alimentary canal. For example, in the case of ground beetles predatory on arthropods, chitinases and proteases are secreted (Jasper-Versali and Jeuniaux, 1987). On the other hand, in the case of larvae of the necrophagous blowfly, production of tryptase, collagenase, peptidase and lipase enables digestion of dead tissues of vertebrates.

This analysis requires rather large investment of time and equipment and to date has produced almost no important results. For example, Ghilarov and Semenova (1978) attempted to establish a correlation between the anatomy of the alimentary canal in edaphic larvae, the enzymes produced therein and their food regime. But their results fell short of those obtained by more conventional methods.

The study of digestive enzymes is laborious and often deceptive; the difficulties are of the same order as those met with in soil enzymology (Chap. 14, § 14.2).

This laborious method includes an assumption. But it has already given interesting results.

Immunological approach or serological method

The ***serological method*** is used for invertebrates and enables us to ascertain whether prey A forms part of the food regime of predator B. The following are done:

- antigenic fraction is prepared from prey A;
- this ***antigen*** is repeatedly injected intramuscularly into a rabbit; this antigen generates antibodies (highest sensitivity level reached after about three months);
- a predator B is captured from nature, killed and its alimentary canal taken out;
- the serum from the rabbit ***immunized*** against prey A is added to the contents of this alimentary canal;
- if ***agglutination*** occurs, it is indicated that B has ingested A.

Antigen: molecular structure foreign to the organism, determining antibody production.

Antibody: protein (immunoglobulin) produced in response to the presence of an antigen, to which it bonds, causing agglutination of the antigen.

Immunize: to provoke formation of antibodies by injection of an antigen (e.g. vaccination).

Used in the study of larvae of *Hybomitra bimaculata* (Diptera: Tabanidae), the serological method has helped show that the food regime of these carnivores comprises rhizophagous larvae of *Donacia* spp. (Coleoptera: Chrysomelidae); detritivorous larvae of *Prionocera turcica* (Diptera: Tipulidae); carnivorous larvae of *Phylidorea* spp. and *Tricyphona* spp. and detritivorous larvae of *Erioptera* spp. (Diptera: Limonidae). A fragment of the food web in a moist peaty soil could thus be reconstructed (Affolter et al., 1981).

Tracing the food chains with radioactive substances

After ***labelling*** the organisms, the progress of radioactivity through the biocoenosis traces the path of the food chain:

The method consists of labelling plants or prey with radioactive isotopes. The herbivores and predators that feed on them become radioactive in turn.

- isotopes emitting β-rays are very suitable for study of predator-prey relationships on the surface; according to duration of the experiment ^{32}P (half-life 12 days), ^{35}S (half-life 87 days) or ^{14}C (half-life 5730 y) is used;
- isotopes that emit the more penetrating γ-rays are used in soil or in wood. The chains are then labelled with ^{137}Cs (half-life 30 y).

Chernobyl, a gigantic tracing 'experiment'

An experiment with radioactive labelling was 'conducted' at a continental scale at the time of the Chernobyl accident on 26 April 1986. Many food chains were thus labelled in the Russian and Scandinavian tundra, in Scotland and even in the lakes of the southern Alps, leading at times to banning of consumption of foods such as fish and mushrooms.

Several tens of field experiments have been done, especially in Great Britain and the United States.

In use, it has been noted that the analytical determination of elements, unequal partitioning of radioactivity among organisms and diffusion in the environment render quantitative interpretation of the results quite tricky. But when well conducted, the method gives interesting information, as shown by the work of Frank (1967) on predation of chrysalides of the winter moth (Lepidoptera: Geometridae) in forest soils. This author highlighted a complex of predators comprising shrews, fieldmice, ground beetles, staphylinid beetles and elaterid larvae. Other studies have been carried out on ants, termites, oribatids and earthworms. A good review of the method has been written by Southwood (1978). Odum (1971) and Ramade (1995) have approached radioactive pollution of the environment in a more general manner.

13.3 FOOD CHAINS

Five categories of food chains.

According to the food source that forms their starting point, five categories of food chains can be defined:

- grazing or biting food chains;
- parasitoid food chains;
- ectoparasite (haematophage) food chains;
- endoparasite food chains;
- detritus food chains.

13.3.1 Grazing or biting food chains

A great loss of energy.

Grazing food chains mostly comprise three or four trophic levels, although chains with five or six levels are thermodynamically possible (Tab. 13.3). According to Lindeman (1942), each trophic level converts only about 10% of the energy originating from the preceding level and kept available for the following one. Thus, after five levels, including that of primary production, only 1/10,000 of the energy started with remains available for possible utilization. Understandably a fifth level, though possible for superpredators, is rare.

Table 13.3 Number of trophic levels in 101 food chains studied in different ecosystems (from Pimm, 1982).

Number of trophic levels	Number of cases	Expressed in %
2	2 (23)	3.5 (51)
3	23 (19)	41 (42)
4	24 (1)	43 (2)
5	5 (2)	9 (5)
6	2 (0)	3.5 (0)
Total	56 (45)	100 (100)

Note: Numbers in parentheses pertain to cases wherein the predators at the top of the chain are small invertebrates, liable to be only a link in a longer, but unknown chain.

The first level is that of photosynthetic plants, which convert light energy to biochemical energy and the latter to plant organic matter (leaves, wood, sap). This net production (Chap. 2, § 2.2.1) is exploited by heterotrophic organisms, the herbivores, who themselves support one or more levels of carnivores. The size of consumers increases from one level to the next, but their number diminishes [e.g. nettle leaves → snails (1–2 cm) → ground beetles (2 cm) → shrews (6 cm) → barn owls (34 cm)].

At the first level, the net primary production sustains the trophic ladder.

Importance of predators in the detritus food chains

The diversity of predators is wide in each compartment of the detritus food chains. Though they do not participate directly in decomposition processes, they are nevertheless quite important because they are the chief regulators of growth of populations of decomposers and microphages. By predation they maintain the strengths of these two groups compatible with the capacity of the environment (Chap. 4, § 4.5.3; Chap. 15, § 15.3.2). They are the guardians of the subtle equilibria among species, enabling proper functioning of the biting and detritus food chains.

13.3.2 Parasitoid food chains

Parasitoids are specialized insects belonging mainly to the Hymenoptera (e.g. Ichneumonidae, Braconidae,

Parasitoid: insect that lays its eggs upon or in the body of other invertebrates (e.g. caterpillars, chrysalides, earthworms). Its larvae develop by slowly devouring their host, but the latter does not die until the parasitoid has attained maturity. The parasitoid may in turn become victim of a ***hyperparasitoid*** whose larvae develop at the cost of those of the primary parasitoid.

Chalcidoidea) and Diptera (e.g. Tachinidae). They most often attack specific species of caterpillars and chrysalides or larvae of coleopterans above and in the soil. The ***parasitoid food chains***, particular instances of grazing chains, comprise just one or two levels, the size of organisms diminishing from one to the other [e.g. larvae of cockchafers parasitized by larvae of *Dexia* sp. (Tachinidae)]; chrysalides of the winter moth (Lepidoptera: Geometridae) parasitized by larvae of *Cratichneumon culex* (Hymenoptera: Ichneumonidae).

13.3.3 Ectoparasite food chains

In the case of fleas, adults are haematophagous and larvae saprophagous.

'Big fleas have little fleas
Upon their backs to bite 'em
And little fleas have lesser fleas
And so, ad infinitum.'

(J. Swift cited by Odum, 1971)

The poem in the margin by J. Swift, author of *Gulliver's Travels*, picturesquely illustrates the ectoparasite food chain to which fleas belong.

Ectoparasite food chains are also a specific case of biting food chains but here the organisms consume only blood or lymph from their prey, some will say host. Soil vertebrates (e.g. fieldmice, voles, moles) are carriers of fleas, lice, etc. Part of the development of parasites takes place in the soil, such as that of flea larvae in the burrows of their hosts. The nests of cavicole birds, indirect appendages of soil (Chap. 8, § 8.8.3) also shelter these ectoparasites to which should be added gamasid mites, predatory Acari that show all intermediate stages between predation and ectoparasitism.

13.3.4 Endoparasite food chains

Endoparasites have 'the best possible coexistence' relationship with their host, a vital requirement because the death of the host leads to that of the parasite.

Endoparasite food chains pertain to organisms that use the biting food chains for closing their evolutionary cycle. The typical case is that of helminths (e.g. solitary worms, liver flukes, parasitic nematodes). They accomplish the various stages of development in intermediate hosts, passing, before predation, from one level to another for reaching the adult stage in a final host. Soil animals are often involved as intermediate or final hosts (e.g. insect larvae, fieldmice, moles).

Are endoparasite food chains truly trophic chains?

Although often considered at par with other trophic chains, endoparasite chains appear more to us as evolutionary cycles of organisms, linked to biting or detritus food chains.

13.3.5 Detritus food chains

The organisms constituting the ***detritus food chains*** (or ***decomposition food chains***) are agents of mineralization and humification of organic matter (Chap. 5, § 5.2). Closely tied to the soil, these food chains are complex, formed of functional modules related to animals successively belonging to the macrofauna (e.g. woodlouse, 1 cm), mesofauna (e.g. mites, 1 mm) and microfauna (e.g. amoeba, 10 to 100 µm). Fungi and bacteria (1 to 10 µm diameter) are found upstream from the chain, then through the modules and finally downstream (Fig. 13.9). Such a functioning is very different from that of the grazing or biting food chains.

Along with biting food chains, detritus food chains are the most numerous and also most fundamentally necessary in the functioning of the ecosystem.

Detritus food chains show three general characters, which will be covered in §§ 13.6 and 13.7:

Detritus food chains recycle dead organic matter and render the mineral constituents available to plants.

- three groups of living things are involved in decomposition: bacteria, fungi and animals;

- at different levels of the chains, detritivores eat and reingest the same leaf, each time a little more fragmented and chemically modified by its passage through different digestive tracts;

- the size of the animals diminishes and their number rises from level to level, contrary to what is seen in grazing food chains.

13.4 FOOD WEBS

The five types of food chains are represented in and on the soil. They are not simple, linear or isolated. For example, the detritus food chains bristle with predation, parasitoid, endoparasite and ectoparasite sidechains. The trophic structure of the edaphic community thus takes the shape of a very dense ***food web*** of interconnected food chains.

It is appropriate to speak of food web or trophic web rather than of chains to illustrate the circulation of energy and bioelements in the soil.

In this apparent jumble, soil biologists distinguish three functional subsystems that structure the biocoenosis and are contained in the biotope (abiotic environment); (Swift et al., 1979; Coleman and Crossley, 1996) (Fig. 13.4).

One such structure is clearly seen in a pasture grassland of the Swiss Plateau (Fig. 13.5).

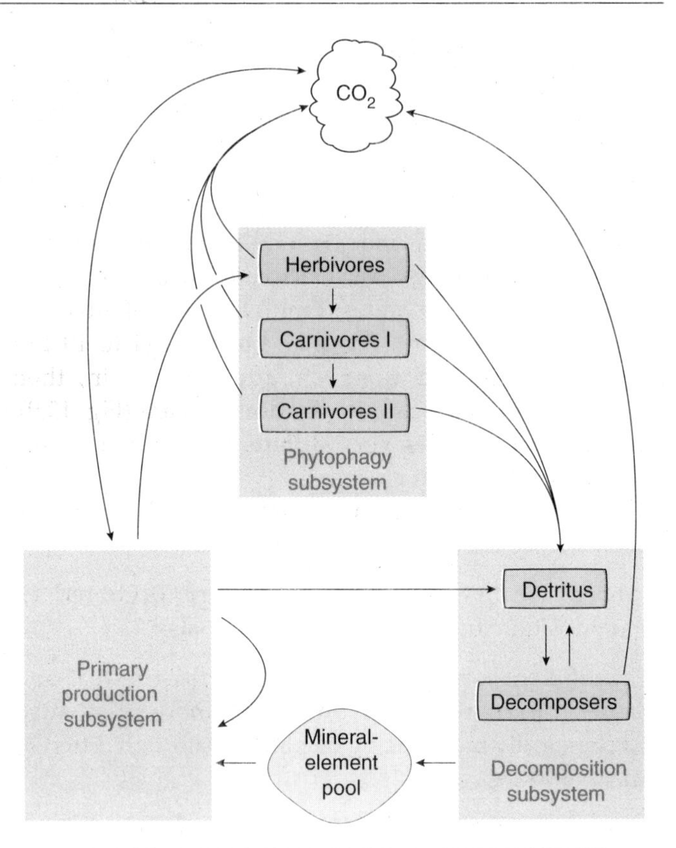

Fig. 13.4 *Organization of the ecosystem and subsystems. The primary production system feeds the grazing food chains (phytophagy subsystem) and the detritus food chains (decomposition subsystem).*

13.5 SOIL, RECYCLING COMPARTMENT OF THE ECOSYSTEM

13.5.1 Large quantities of organic wastes feed the detritus food chains

In grassland environment, one part of the net primary production NP_1 (Chap 2, § 2.2.1) is consumed by large herbivores (livestock or wild vertebrates) and another part by invertebrates (insects in particular). The uneaten plant parts, left in place, constitute the necromass and part of the litter. It enters the detritus food chains as herbivore faeces and dead bodies of animals.

Very variable utilization of the net primary production.

In agricultural systems, estimates of the utilization of NP_1 by the fauna and quantification of matter returned to the soil vary according to the type of crop (biological, conventional), the species of herbivore and grazing pressure. In a semi-arid grassland put to extensive grazing, only 5% of the NP_1 is consumed by cattle and 75% forms litter. At the other extreme, 90% of the NP_1 may be taken up by intensive grazing (Curry, 1994).

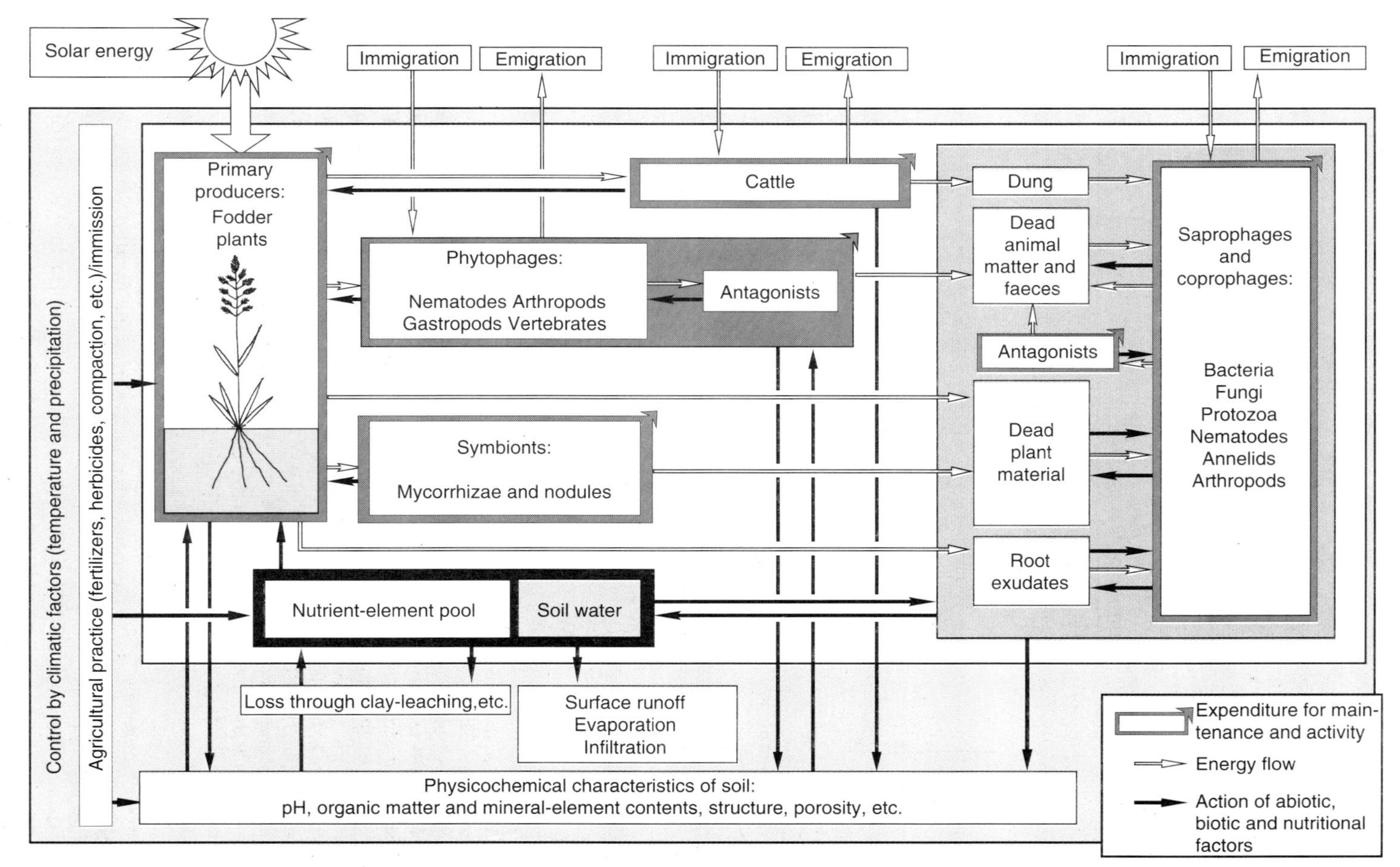

Fig. 13.5 *Example of the overall functioning of an ecosystem: organization of a pasture grassland of the Swiss Plateau (from Bieri, Matthey and Zettel, unpublished).*

A large part of browsed grass is returned to the soil as dung (Tab. 13.4).

Table 13.4 Proportion of the net epigeal primary production entering detritus food chains (in %)

Type of detritus	a. Intensive grazing	b. Extensive grazing
Litter	7	37
Cattle dung	23	18
Invertebrate droppings	5	21

a. Grazed grassland, Normandy (from Ricou, in Lamotte and Bourlière, 1978)
b. Semi-natural grassland, England, with free-grazing beef cattle; grazing intensity comparable to that of game (from Duvigneaud, 1984)

In meadows for hay, successive cuttings export 85 to 90% out of the ecosystem; the only return of dung to the soil is assured by invertebrates, unless livestock is put to graze in autumn or the cultivator spreads manure.

Wood, a temporary energy reserve.

In forest ecosystems, herbivores consume only 10% of the NP_1 under normal conditions. Another part (20 to 60%) is removed from the detritus food chains by storage in the form of wood. In unexploited forests, this is a temporary reserve because dead trees are decomposed in place. Contrarily, export of trunks out of the system by woodcutters, systematic elimination of old trees and often burning of woody debris lead to a large food deficit for wood-decomposition chains (Chap. 8, §§ 8.4 to 8.7). A visible consequence is the reduction of density of many saproxylophagous species (Speight, 1989).

IBP: International Biological Programme. Research sponsored by UNESCO which from 1963 to about 1980, mobilized ecologists from 57 countries to study production of the ecosystems at the global scale. Extremely fruitful, the IBP exerted a profound influence on ecology at the close of the twentieth century.

To sum up, whatever the mode of exploitation might be, the unused fraction of the above-ground vegetation—future litter—is considerable: the results of ***IBP*** enabled estimation of the total quantity of litter on the land mass at 111×10^9 t. This mass, according to Whittaker (1975), represents only part of what may be called the 'necrosphere'. Carrion and faeces constitute two other additions to the decomposition food web but are difficult to quantify. To this above-ground litter is added the root necromass, often still larger (Chap. 2, § 2.2.1; Chap. 4, § 4.1.7).

In soil, the humiferous episolum concentrates most of the detritus food chains (Chap. 6, § 6.1.3).

Detritus chains are mostly located in the soil. This environment, which may be termed 'decomposition subsystem' (Fig. 13.4), is very important in the functioning of ecosystems. It is here that an essential part of the biogeochemical cycles operates and most of the energy circulates (Odum, 1971, 1996).

13.5.2 How can we structure this complex subsystem?

Seven approaches are possible:

- Determining fragments of the web 'nourished' from different additions: litter, dead wood, excrements or carcasses.

- A more pedological approach consists of determining the quantity of dead organic matter and biological activity in the various horizons. It can be ascertained when both diminish with increasing depth. To the extent that they are sunk in the soil, invertebrates of the macro- and mesofauna, which act powerfully in fragmentation of surface material, give place to a microfauna that feeds on fine debris washed down from the surface. However, burrows of moles, voles and earthworms (Chap 4, § 4.5.1; Chap. 12, § 12.4.3) and roots of living and dead trees provide access for the surface fauna to horizons below. For example, 120 arthropod species have been enumerated in mole 'nests'.

- Beare et al. (1995) approached the organization of the decomposition system very differently. They sought sites of intense biological activity in it, which they termed ***hot spots of activity*** (Chap. 6, § 6.2.1; Chap. 17, § 17.1.3). They listed
  - — the ***porosphere***, comprising all soil voids (Chap. 3, § 3.3);
  - — the ***drilosphere***, the portion of soil influenced by secretions of earthworms (see details in Lavelle and Spain, 2001);
  - — the ***aggregatosphere***, aggregates and their immediate neighbourhood (Chap. 3, § 3.2; Chap. 4, § 4.5.1);
  - — the ***detritusphere***, masses of organic matter;
  - — the ***rhizosphere***, immediate vicinity of rootlets (Chap. 4, § 4.1.3; Chap. 15).

These seats of intense biological activity will occupy only 10% of the soil volume but will concentrate more than 90% of the biological activity in most soils of the world.

The approach through 'hot spots of activity', a little out of the ordinary for the vocabulary it generates, gives a very interesting picture of the compartment of decomposition.

- Lavelle and Spain (2001) described four 'biological systems of regulation' (BSR) involved in the decomposition processes, not equally important in all ecosystems:

- The litter system, litter as food, lateral surface roots, epigeal invertebrates and microflora dominated by fungi.
- The rhizosphere, living subterranean roots, soil and microflora that they influence.
- The drilosphere, part of the soil influenced by earthworm activities.
- The **termitosphere**, volume of soil and organic resources influenced by termites.

A fifth BSR could be defined in the future, the *myrmecosphere*, concerning ants and their sphere of influence.

• Taxonomic criteria may be used (populations of mites and ticks, larvae of dipterans, etc.).

• Communities individualized in the soil at different depths may be investigated: epiedaphon, hemiedaphon, euedaphon, phreatic land fauna (Chap. 12, § 12.5.6).

An approach with three functional compartments.

• Lastly, it is possible to combine modular and faunistic approaches by relating the entire decomposition subsystem to a detritus food chain model (with three modules or functional compartments) and by selecting for reference scale the size of dominant invertebrates in each module (macro-, meso- and microfauna; Chap. 2, § 2.6.3; Chap. 12, § 12.5.2). We shall adopt this viewpoint in § 13.7.

13.6 HOW DO DETRITUS FOOD CHAINS FUNCTION?

13.6.1 Principle

In the grazing food chain (§ 13.3.1), each trophic level is nourished at the expense of the preceding level or levels. The detritus food chain functions differently.

In the detritus food chain, dead organic matter and the microflora form a sort of red thread along which the waves of phytosaprophagous invertebrates succeed one another (Chap. 8, §§ 8.3, 8.5). These invertebrates eat, therefore fragment; they digest, hence chemically modify. They discharge faeces, a product different from the original material. The nutrient elements and energy the original material contains is far from exhausted by the first level of the chain because macroarthropods utilize at the most 40% of the nourishment available. The waves that follow benefit from it and are nourished by excrements and the still substantial debris, in turn modifying them physically and chemically (Tab. 13.5).

Table 13.5 Characteristics of the animals and protists composing the three functional compartments of the detritus food chain

Descriptor	Compartment 1	Compartment 2	Compartment 3
Length	4 to 80 mm	0.2 to 4 mm	< 0.2 mm
Diameter	2 to >20 mm	0.1 to 2 mm	< 0.1 mm
Individuals m^{-2}	10^2	10^4 to 10^5	10^6 to 10^7
Utilization coefficient	20–40%	50%	95% (bacteria, 100%)

The organization is roughly the same for various detritus food chains but there are subtle differences depending on the type of waste and environment. These chains show peculiarities according to the nature of the litter or dead wood that feeds them and the site where they are located (surface of soil or at depth, altitude, climate).

Detritus food chains and degradative succession.

Decomposition of vertebrate carcasses (Chap. 8, § 8.3.1) follows the same scheme. Several waves of necrophagous insects succeed one another on dead flesh, in a constant sequence, modifying the flesh chemically and physically until it disappears. The succession of coprophages, particularly spectacular in dung (Chap. 8, § 8.3.2) exhibits the same organization. Waves of coleopterans, larvae of dipterans, worms and fungi succeed one another until the faecal matter is dried and buried.

The type of material to be decomposed decides the composition and behaviour of the chains.

Plant and animal materials and excrements are also distinguished. Among the first, rapid detritus food chains decompose in a few years the annual litter of dead leaves and grasses, while slow chains degrade certain resistant plant residues (Chap. 2, § 2.2.1) and dead wood (Chap. 8, §§ 8.5, 8.6); the time-scale here is the decade or century. Figure 13.6 presents a classification of these chains.

13.6.2 A partnership for decomposition

Is the fauna really necessary in the decomposition processes?

The importance of fragmentation? Divide a cube...

The nature of plants, mineralogical composition of the soil and climatic conditions play an important part in the rapidity of functioning of the decomposition food chain. And the degree of fragmentation of wastes as well. The reader will be convinced of the importance of

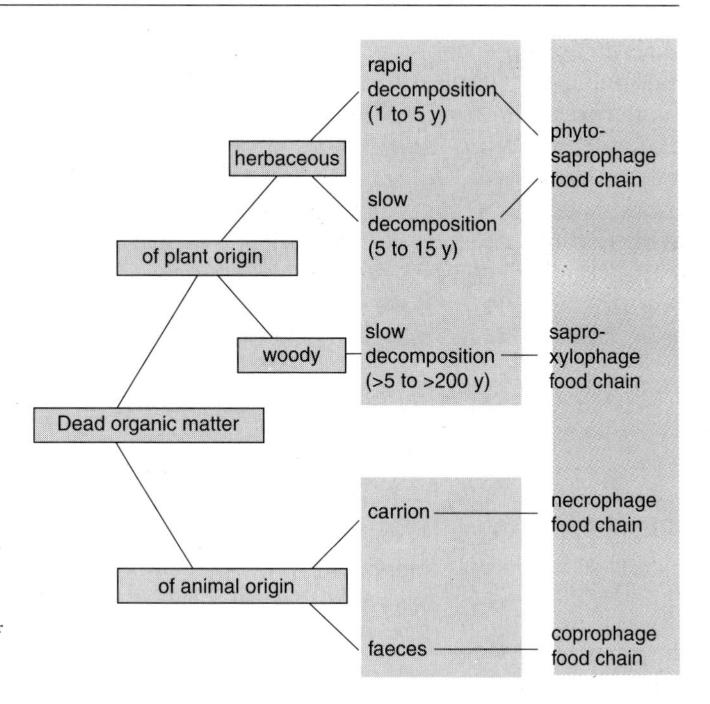

Fig. 13.6 *Classification of detritus food chains.*

fragmentation by successively dividing a cube of 1 cm^3 into cubes of 1 mm, 1/10 mm and 1/100 mm sides and calculating the increase in total surface area obtained (10, 100 and 1000 times).

It is known that soil invertebrates play an essential role in the mechanical attack on litter, hence in its fragmentation (Chap. 4, § 4.5.1; Chap. 5, § 5.2.4). Almost intact fragments of leaves, apparently unattacked during passage through the gut, can be recognized under the microscope in the faeces of phytosaprophagous macroinvertebrates (§ 13.2.1). Nevertheless, crushing of the cuticle has enabled access to the interior of the leaf and to cell contents. This cutting up significantly multiplies the surface area of the substrate that serves as 'culture medium' for the microflora, itself largely responsible for biochemical attack.

Eliminate the fauna to understand its action!

To demonstrate the action of the fauna in the natural environment, attempts have been made to eliminate this action from experimental areas by spreading naphthalene, an insecticide with repellent properties that suppresses up to 90% of the mesofauna. Various authors (cited by Bachelier, 1978) have concluded from these rather rough attempts that in the presence of the pedofauna, decomposition processes are accelerated overall 2.5 to 8 times depending on the nature of the litter, or even 50 times (Van der Drift, 1965). These figures should be considered with caution, however, given the difficulties in interpretation of such experiments.

Van der Drift and Witkamp (1960) showed by measuring release of CO_2 that the intensity of microbial attack is about seven times higher in the faeces of a land-living trichopteran (*Enoicyla pusilla*) than in whole leaves that serve as its food. A similar increase is seen in the case of leaves ground in a blender, which proves the importance of fragmentation for this process.

...and is the microflora necessary for the fauna?

Plants represent overall a food of rather mediocre quality for animals because of the low content of available mineral elements (Swift et al., 1979) and because most animal species are incapable of digesting the most abundant constituents of plant tissue—cellulose, hemicelluloses and lignin. Microorganisms are the most efficient in utilizing these compounds. They transform them into microbial biomass, a food of better quality for animals, rich in nitrogen and phosphorus, digestible and obtained without great expenditure of energy. The association of scarabaeid larvae with bacteria or of higher termites with fungi (Chap. 8, § 8.7; Chap. 12, § 12.4.9) may be mentioned to illustrate this remark.

Importance of the associations to the mutual benefit of bacteria, fungi and rhizophagous and saprophagous soil invertebrates (Chap. 16, § 16.1.6).

13.6.3 Detritus food chains in action

Outline of methodology

Detritus food chains may be reduced schematically to three functional compartments (§ 13.7.1). The role of each in degradation of litter is highlighted by the ***bag method*** with calibrated meshes, a laboratory or field experiment that enables determination, by elimination, of the respective roles of the macro-, meso- and microfauna.

Litter in bags!

Litter bag or soil bag method, how to proceed with it?

Bolting-cloth bags of size (often 10 cm × 10 cm) and mesh sizes (5 mm, 1 to 1.2 mm and 45 to 80 µm) suitable for the aims of the experiment, are filled with previously sterilized litter, then sealed. Placed at various sites in the soil (below litter, in the rooting zone, etc.), they are withdrawn after three, six, nine or twelve months. The effect of detritivores and decomposers on the litter (fragmentation, chemical modification) is measured after each of these periods. A somewhat different method, termed the nested-bag method, has also been used with success (Fig. 13.7).

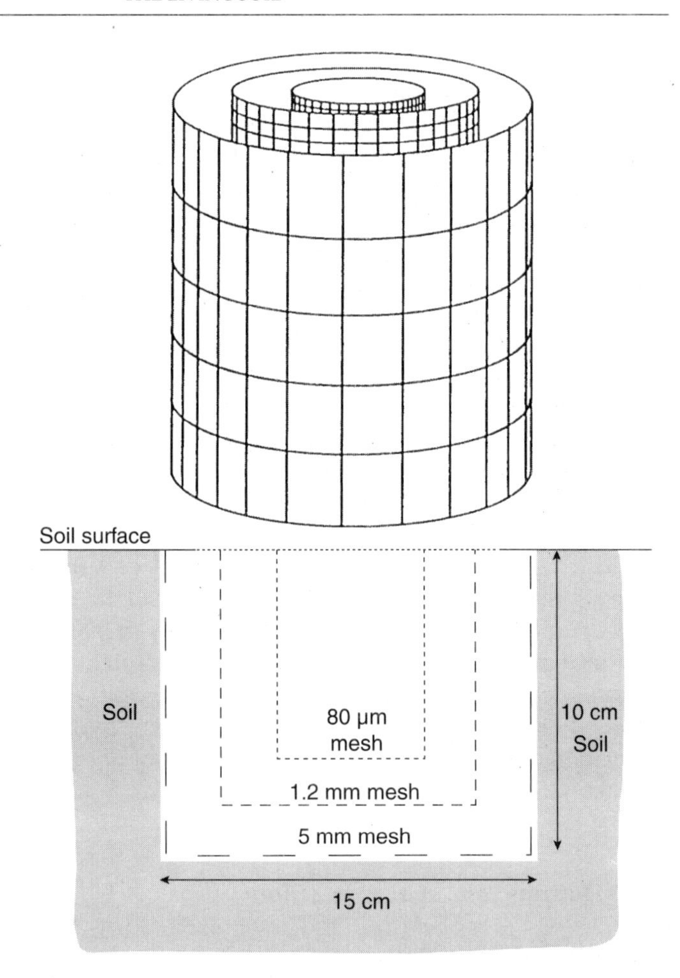

Fig. 13.7 *Nested-bag method (from Maire, Borcard and Matthey, unpublished).*

Data in ecophysiology of nutrition of saprophagous invertebrates are deficient.

From the soil... to the soil!

Efficiency of the pedofauna in detritus chains

Results pertaining to overall efficiency of detritus food chains, often variously expressed in publications, give little information about the actual role of the pedofauna in decomposition processes. Only some groups of saprophagous invertebrates, easier to rear than others, have been taken for studying the ecophysiology of nutrition: earthworms, isopods, diplopods, springtails. However, some data are available. In forest, soil animals generally consume 20 to 100% of the litter in one year. Depending on the faunistic composition and climatic conditions, 60 to 90% of the material gulped down returns to the soil as faeces.

From autumn, the litter of grey-alder forests is cut up and buried in six weeks by Bibionidae larvae, often abundant in these forest environments. In their absence, diplopods and isopods do the same work in close to a year. In oak, beech and chestnut forests, the litter is more leathery and lasts longer. Three conditions must be fulfilled for litter to become attractive to soil animals:

- it should have absorbed water;
- it should have lost, by leaching, any repellent chemical substances it might contain (ketones, tannins, phenolic acids);
- the microflora should have modified the substances and microtopography of the epidermis (Chap. 17, § 17.1.3), thus favouring the first attack on the leaf.

Microarthropods and small dipteran larvae (Cecidomyiidae, Chironomidae) then pierce the protective tissue and allow fungi and bacteria to invade the interior of the leaf (Chap. 5, Fig. 5.5). Resistant leaves undergo these changes during winter which follows their arrival in the soil. It is thus clear why the littericole pedofauna is particularly active in spring and beginning of summer (Dunger, 1958; Garay et al., 1986).

In grassland in Sweden, according to Persson and Lohm (1977), phytosaprophagous arthropods and earthworms consume only 2 kg dry matter m^{-2} y^{-1}, while the above-ground and below-ground additions constitute 41.5 kg m^{-2} y^{-1} in the first 40 centimetres of the soil.

The efficiency in grassland environment seems lower.

When constantly unfavourable conditions of oxygenation, pH and temperature put a stop to functioning of the food chains, plant matter accumulates as peat (Chap. 9, § 9.2.4).

The seasons of detritus food chains

The community of detritivores is structured in time and space.

The community of detritivores follows a seasonal rhythm particularly visible at the level of macroinvertebrates. In winter, the composition of detritivorous populations is dominated by larvae of dipterans as also larvae and imagos of coleopterans, for the most part in diapause. These species revive their activity in spring. During summer, isopods, diplopods and spiders are the most abundant and most efficient. Ants are omnipresent (inactive in winter) as are earthworms (generally inactive in summer).

Seasonal climatic factors influence quality of the food as well. For example, in a cafeteria-test, isopods ingested equal quantities of fresh autumn litter and spring litter, soaked and leached. On the contrary, diplopods prefer the latter and consume four to eight times more of it.

13.6.4 'Big boys' of the soil fauna and detritus food chain

'Big guns'. 'teeth' and 'alimentary canals' on welfare!

Among the very numerous phytosaprophagous organisms of the detritus chains, three groups of invertebrates have a very significant impact on soils: earthworms, ants and termites, the last present chiefly in tropical regions (Tab. 13.6).

Table 13.6 Comparative importance of the most significant invertebrates in soils

Invertebrate group	Density (ind. m^{-2})	Biomass (g m^{-2})	Principal nutrition regime
A. Lamto savanna, Côte d'Ivoire (from Lamotte, in Lamotte and Bourlière, 1978)			
Earthworms	230	30	Saprophagous, geophagous
Ants	500	2	Predatory, omnivorous, granivorous
Termites	880	1.4	Phytosaprophagous, geophagous, xylophagous, herbivorous
Other macroinvertebrates	7 to 21	0.2 to 1.5	Rhizophagous
B. Temperate soils (from Bachelier, 1978)			
Earthworms	50 to 400	20 to 250	Phytosaprophagous, geophagous
Ants	No comparative data		
Termites	Absent		
Other macroinvertebrates	>900	15	Phytosaprophagous, rhizophagous, xylophagous, predatory

Big guns of temperate soils

Earthworms: a digest of the detritus food chain!

Earthworms are by themselves a digest of the detritus food chain because they fragment leaves, ingest them and pulverize them in their crop. Their alimentary canal contains an abundant microfauna and microflora that work on the intestinal contents. Their casts are composed of a mixture of finely fragmented litter and soil (Plate II-3).

Module 3 of the decomposition food chain is already partially present in the gut of earthworms.

Earthworms represent the largest edaphic zoomass (Chap. 2, Tab. 2.11; Chap. 12, § 12.4.3). Anecic worms in particular have great mechanical impact on the soil by their mixing action. They short-circuit the second compartment of the chain by burying still-intact whole dead leaves in their tunnels, consuming them and mixing them with earth in their alimentary canal. The alimentary canal, a veritable bioreactor (Chap. 14, § 14.4), contains very active protozoans and bacteria.

The role of ants in the detritus chain is essential in regulating decomposer communities.

The teeth of the earth

Ants exhibit a wide variety of habits (Chap. 12, § 12.4.9). Like termites, some are fungus-farmers (*Atta*);

others rear aphids (*Lasius*) or are granivorous (*Messor*). But most are predators and exert great pressure on the soil fauna, capturing in temperate regions about 30% of the dipterans with edaphic larvae when they hatch and 40% of young spiders. Among the ants, small endogeal species are found, for example *Epitritus argiolus* (2 mm long), which feed on springtails.

'In the greater part of the continents, the vegetation cycle and life of soils would be profoundly changed if ants were absent, above all in hot regions and the cultivated and forested portions of temperate countries.' (Bernard, 1968).

The work of Pavan (1976) on wood ants, *Formica* spp., showed their impact on the invertebrate fauna in coniferous forests of the Italian Alps, after reintroduction of colonies into zones from where they had disappeared. Pavan estimated after a few decades up to one million *Formica* nests spread over 579,000 ha, or about 300 thousand million individuals. Each weighing 8 mg on average, the total weight of *Formica* ants reaches 2400 tons. If an individual ingests 1/20 of its weight in food each day, 120 tons are consumed daily or, in a season of 200 days of activity, 24,000 tons of food. Sixty per cent of this, or 14,400 tons, comprise invertebrates captured from plants and the soil surface. Even if these values pertain to only 20 to 30% of the pedofauna, the impact of *Formica* at this level is considerable.

2400 tons of ants eat 24,000 tons of food in 200 days of activity!

In tropical regions, where up to 25,000 ants and 10,000 termites can be counted per 100 m^2, the equilibrium between the two groups controls forest-soil life, termites recycling dead wood and litter, while certain ants feed on termites.

Ants of the Lamto savanna, Ivory Coast

The variety and density of ant populations are often considerable. Thus the preforest savanna of Lamto in Ivory Coast, accommodates more than 150 species (Lamotte et al., in Lamotte and Bourlière, 1978). The majority are soil-dwelling and build from 3000 to 7000 nests per hectare. Predation and utilization of animal wastes predominate in the nutrition regime, and maximum exploitation of the environment is made possible by the trophic specialization of the species. Thus, among others, *Aenictus* ants feed essentially on... ant larvae, *Amblyopone* ants on chilopods, *Hypoponera caeca* on microarthropods, *Leptogenys conradti* on woodlice, *Megaponera foetens* on termites and *Pheidole* spp. on seeds and dead arthropods.

Termites, or how to get food digested by others

Termites play an essential role in hot regions (Chap. 12, § 12.4.9). They exploit dead plant matter, chiefly wood. In the forests of the Congo, they consume from 6 to 7 t ha^{-1} y^{-1} of organic matter fallen on the soil. But they attack living plants too, becoming major ravagers of citrus and mango trees in some regions of Africa. Lastly,

About disappearance of litter (and of some table legs!) in tropical countries.

The termite-zooflagellate-bacteria relationship: a perfect example of the principle of symbiosis (Chap. 16).

cannibalism is practised at the expense of dead or injured individuals of the colony.

All the species have the same food: cellulose. Paradoxically, termites, but for a few exceptions, are themselves incapable of digesting this polysaccharide and harbour in their alimentary canal symbiotic microorganisms that take care of it: bacteria and zooflagellates (Fig. 13.8) digest cellulose and produce waste products that can be utilized by their host. These flagellates themselves contain symbiotic bacteria that attack cellulose or contribute to completing the decomposition path (methanogenic bacteria, Chap. 16). If the alimentary canal of the termite is sterilized, it becomes incapable of digesting the wood it ingests and dies of starvation in two or three weeks.

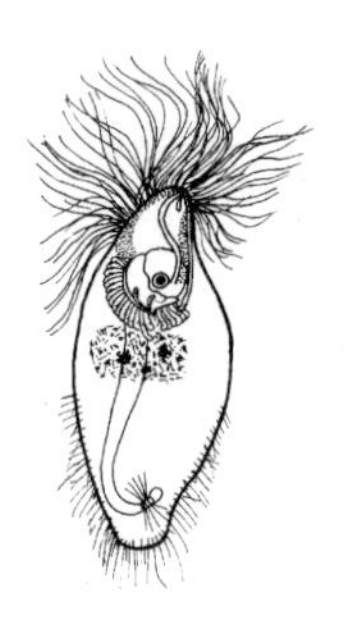

Fig. 13.8 *Flagellate* Joenia *sp. from the rectal pouch of a yellow-neck termite* Calotermes flavicollis *(size 0.07 mm).*

Three types of nutrition regimes function by means of bacteria-protozoa-termites symbiosis:

- xylophagous termites dig colonies in dead branches and trunks, which serve both as shelter and as food source;
- humivores ingest soil and litter already partially decomposed by the microflora. Their galleries are many in the upper horizons;
- foragers feed on litter, herbaceous plants (mostly graminaceous) and seeds.

Three nutrition regimes for termites with protozoans.

Fungus-farming termites (Macrotermitinae) bring sawdust and litter into the termitarium and make them into more or less spherical balls. Specialized mushrooms—thirty species of the genus *Termitomyces*—develop there and wrap these fungus combs with mycelial velveting and small fruit bodies. The latter, ***sporodochia***, constitute a source of vitamins and cellulolytic enzymes for the termites. The termites consume the material of the old termite combs chemically modified by the mycelium; most of their food thus consists of wood and litter predigested by the mushrooms. Digestion is achieved by cellulolytic bacteria and enzymes present in the alimentary canal of the insects.

The *Termitomyces* are much valued as edible mushrooms in the tropical regions. Attempts are being made to cultivate them industrially.

Sporodochium (or ***hyphal tip***): spherical structure composed of a bundle of conidiophores giving rise to conidia.

13.7 MODULAR EXPRESSION OF THE DETRITUS FOOD CHAIN

13.7.1 Three functional compartments

Three functional compartments, three categories of edaphic fauna.

Viewed in a theoretical and simplified manner, the detritus food chain is formed of three functional

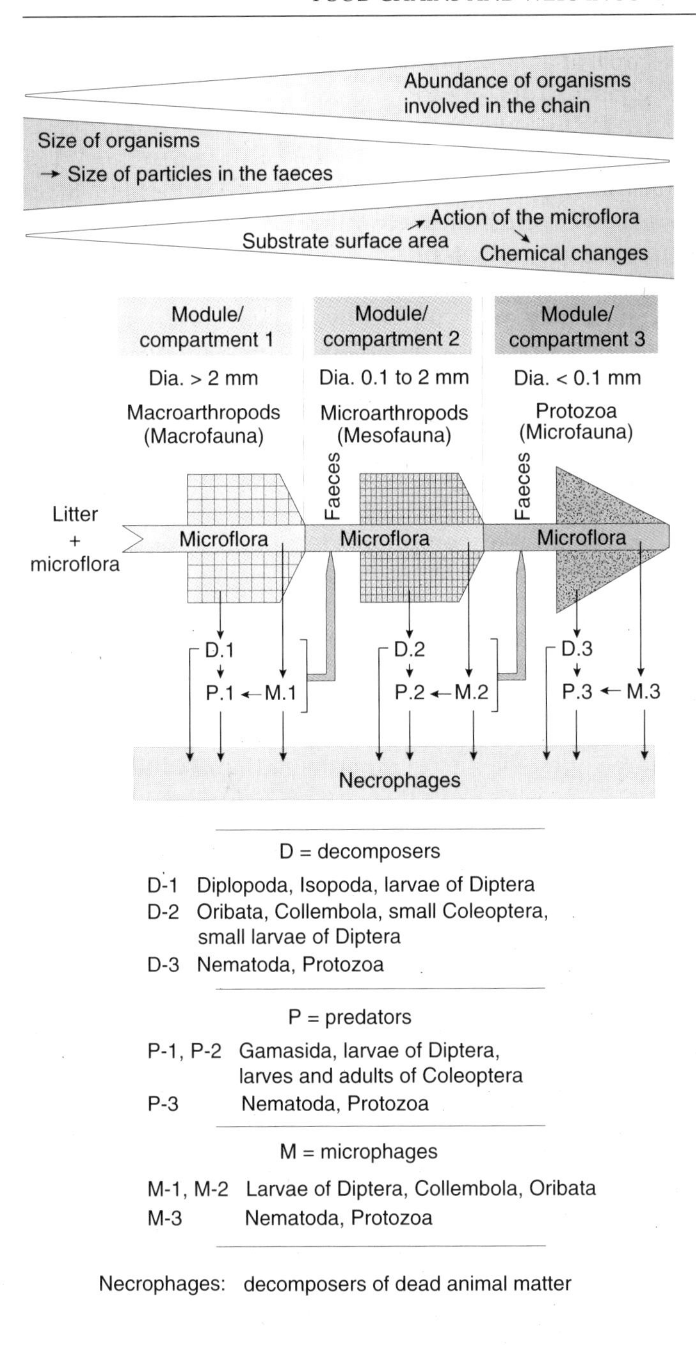

Fig. 13.9 *Modular organization of the detritus food chain. Explanations in the text.*

compartments or modules, corresponding respectively to the macrofauna, mesofauna and microfauna (Fig. 13.9).

Physical fractionation and selection in the first two compartments. Biochemical processes in the third compartment.

In the first two, the essential activities are cutting, burial and also dispersal and selection of the microflora. The microflora present on the food is selected in the alimentary canal of invertebrates, all of them remaining very active; aerobic and anaerobic microfloras intermingle in the intestinal environment. Lastly, in the faeces, where the original material is physically and chemically modified, bacterial and fungal populations explode. In the third compartment, the biochemical processes largely overtake physical processes because of the microflora. The microfauna, chiefly composed of bacterivorous and fungivorous species, contribute above all to maintaining biological equilibrium in soil.

Each module comprises detritivores consumed by predators, both liable to become victims of parasitoids and parasites. Microphages come to enrich the web; dead bodies are consumed by necrophages. The faeces are often reingested several times by the detritivores that eject them. Then they, as well as the uneaten fragments (remains or leftovers of the 'meals' of macroarthropods), are taken charge of by the next compartment (Fig. 13.10).

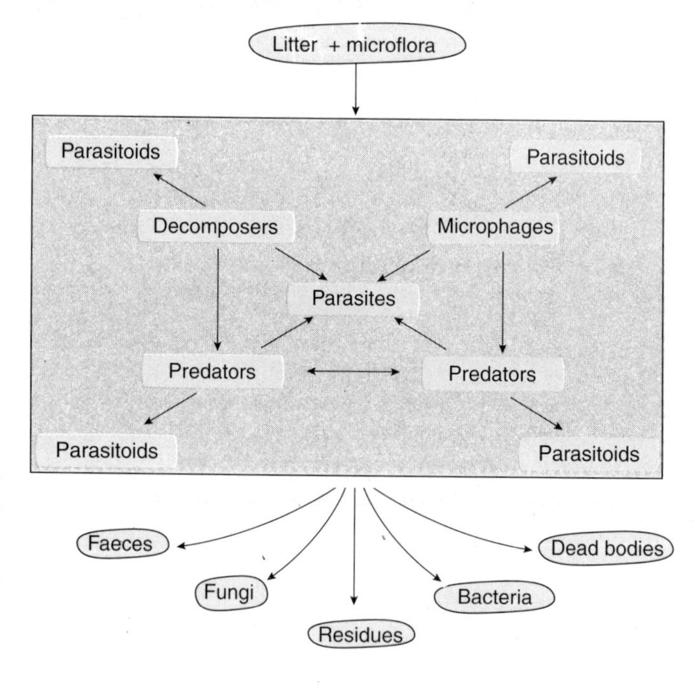

Fig. 13.10 *Break-up of the functioning of the first two compartments of the decomposition chain. Litter passes through a veritable biological 'blender' that pulverizes and chemically modifies it.*

Figure 13.9 allows us to conclude that from the first compartment to the third:

- the microflora is always present,
- size of the invertebrates diminishes,
- that of the plant particles 'treated' by the fauna also diminishes,
- abundance of organisms taking part increases considerably,
- the proportion of microphages increases from module to module,
- the ***utilization coefficient*** increases (Tab. 13.5).

Utilization coefficient (or ***assimilation efficiency***): depending on author, this refers to the ratio between gross production and ingested energy, or between net production and ingested energy.

13.7.2 First compartment: a community at the centimetre scale

Many papers concern the enumeration, phenology and systematics of macroarthropods typical of this compartment but very few relate to its functional organization (Dunger, 1958; Garay et al., 1986).

Typical of this compartment, macroarthropods (Chap. 2, § 2.6.3; Chap. 12, § 12.5.2) colonize chiefly the upper soil horizons. The OL layer, well aerated, suits mobile species (e.g. woodlice, diplopods, spiders), while dipteran larvae, more static, exist in large numbers in the OF layer, often wet and compact.

Macroinvertebrates harbour numerous parasites: worms, protozoans (***Apicomplexa***) and fungi. These organisms can be very widespread among the macrofauna and mesofauna (10 to 80% of the individuals examined); they can kill their host in soils subject to immissions of SO_2 (Purrini, 1982). Classified according to their nutrition, the principal groups of the first compartment are:

Apicomplexa (formerly termed Sporozoa): phylum of the kingdom of Protista (Chap. 12, § 12.4.1). All Apicomplexa are parasitic on animals. Some of them cause grave diseases in man, for example the genus *Plasmodium*, cause of malaria. Gregarines develop in the case of arthropods.

- detritivores: termites, larvae of Diptera (Bibionidae, Cecidomyiidae, Chironomidae, Scatopsidae, Sciaridae, Tipulidae and Trichoceridae), Isopoda, Diplopoda and molluscs;
- predators: larvae and adults of Coleoptera [chiefly Carabidae (Fig. 13-C) and Staphylinidae], ants, spiders (e.g. Dysderidae, Lycosidae, Linyphiidae);
- microphages: scarcely diversified in this compartment; chiefly larvae of Diptera (Stratiomyidae, Fanniidae);
- parasitoids: Hymenoptera (e.g. Chalcidoidea, Braconidae, Ichneumonidae) that attack eggs, larvae and nymphs of other arthropods; Diptera (e.g. Tachinidae, Calliphoridae), some species of which develop in the bodies of earthworms, often leading to increased mortality.

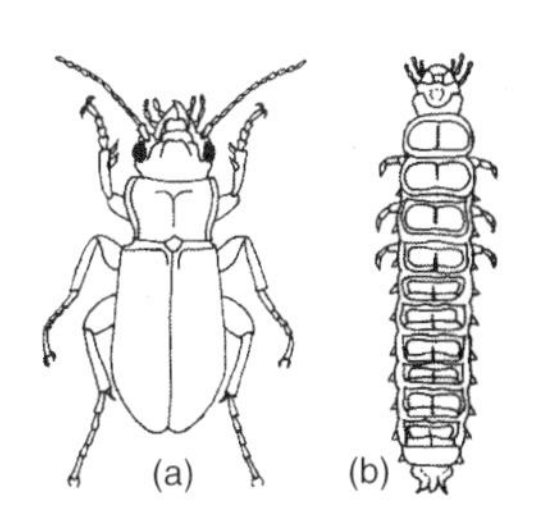

Fig. 13-C *Predatory Carabidae:*
(a) Adult Molops *(9–15 mm)*
(b) Larva of Carabus *(20 mm)*

These faunistic and functional groups are entered in the schematic diagram in Fig. 13.10.

Because of constant interactions between microarthropods and macroarthropods, quantification of the impact of a single compartment on litter is difficult.

Data on the overall efficiency of the detritivorous community of the first compartment are rather rare. It is generally agreed that 40% of the litter is consumed by the macrofauna. According to Dunger (1958), of 150 g m^{-2} of litter in an alluvial forest near Leipzig, 50 g were transformed into droppings by isopods and diplopods, although these taxa represent only 10% of the biomass of saprophages. Diplopods are twice as efficient here as isopods.

13.7.3 Second compartment of the detritus food chain: passing to the millimetre scale...or the world of microarthropods

A good example of pulverization of material: the Phthiracaridae.

This module functions according to the same principles as the first, but attack on the whole leaf consists more of perforation than of fragmentation. Thus conditioned by mites and springtails, litter decomposes 2.5 times faster. Springtails are also capable of skeletonizing leaves leaving behind only the veins. Phytosaprophagous microarthropods, by consuming the faeces of macroarthropods, subject the vegetal particles they contain to a supplementary fragmentation. For example, the Phthiracaridae (Oribata) eject faeces about 2000 times smaller than those they feed on (faeces of *Glomeris*, § 13.2.1). Microphages are very important here because many springtails and mites are specialized mycophages that influence the evolution of fungal populations. Thus a moderate browsing optimizes growth of fungi (Wardle, 1995) but 'overgrazing' caused by a high density of microarthropods may lead to diminution of the mycelial biomass.

Caution! *Micro*arthropods belong to the *meso*fauna!

Microarthropods (mites and springtails above all) are very plentiful, with densities of 16,000 to 386,000 ind. m^{-2} depending on the soil (André et al., 1994). Their biomass is almost equal to that of macroarthropods, as is their overall metabolic activity measured by their respiration. They are so varied that they are found in practically all types of ecological niches, except in that of parasitoids. The following are found in the second compartment:

- detritivores: numerous Acari (Oribata, Uropoda), springtails;
- predators: Acari (Gamasida, Actinedida);
- microphages: Acari (Oribata, Acaridida), springtails.

Beare et al. (1992) applied the ***method of exclusion*** in cultivated fields to study the respective impact of the different taxa in the litter. It was found that the action of saprophytic fungi associated with microarthropods was predominant on the surface of the soil, whereas in litter buried by the plough, bacteria associated with nematodes were the most efficient.

Method of exclusion: method of eliminating certain categories of species from a community in order to evaluate the efficiency of the surviving species. In soil biology, different treatments are carried out with bactericides, fungicides, nematicides, acaricides or insecticides (§ 13.6.2).

Many interactions between microarthropods and the microflora

In the second compartment, according to Lussenhop (1992), two types of interaction are seen between microarthropods and the microflora, which affect the distribution and abundance of fungi:

- control of their distribution by selective browsing often has the effect of accelerated decomposition;
- dispersal of fungal propagules by microarthropods seems particularly important in the colonization of substrates and successions of fungi.

Three other categories of interaction influence the metabolism of microorganisms:

- return of bioelements through faeces might be an important stimulant of the microflora;
- growth of bacteria is temporarily favoured by browsing of fungi (diminution of competition);
- in the absence of microarthropods, the speed of decomposition is inversely proportional to the number of fungal species; by their browsing, mites and springtails reduce competition between species of fungi, thus speeding up decomposition of the substrate.

13.7.4 Third compartment: an aquatic microcosm in the terrestrial milieu

Even if this compartment is defined at the scale of the microfauna, it plays only an indirect role in it because crushing of plant material has practically ceased. The microflora is now in the forefront, by controlling biochemical processes such as enzymatic breakdown of long polysaccharide molecules (Chap. 14, § 14.3.1). It thus ends the recycling of litter started in compartments 1 and 2 (Chap. 4, § 4.4.1) with a utilization coefficient 2 to 5 times higher than that of invertebrates (Tab. 13.5). Protozoa and nematodes intervene by predation; they regulate and stimulate fungal and bacterial populations, thus indirectly influencing the mode and rapidity of decomposition processes (Chap. 14, § 14.3.2).

The last compartment in the chain is that of the microfauna... and microflora.

Minuscule size but high biomass.

Compared to the preceding compartments, there are great differences in ecology of the invertebrates. Their smaller size (0.01 to 0.5 mm), morphology and physiology bind them to an aquatic life. In consequence they are

Nematodes: particularly important because of their predatory activity (10–15% of the edaphic animal respiration) and their influence on soil stability (abundant deposition of excrements in the finest cavities).

active only in film and interstitial water, in soil solution and in organic liquids of decomposing materials. Their capacity for active movement is limited; also, while remaining in place when the soil dries, protozoans, tardigrades, rotifers and nematodes survive only because of their very efficient physiological mechanisms, anhydrobiosis or encystment (Chap. 12, § 12.4.2). Thus they withstand the strongest desiccation and the most extreme temperatures for long periods. Dehydrated or encysted, they are dispersed by runoff water, wind, animals and man.

Under the control of abiotic factors, the predation relationships and interspecific competition among organisms of the third compartment play an essential part in the dynamics of their populations. Even at the scale of microorganisms, there can be choice of a certain food, hence a specialized nutrition régime. Thus exudates from bacteria attract certain amoebae and repel others.

Nematodes and protozoans on the one side and fungi and bacteria on the other maintain relationships of the predator-prey type. At this scale, animals, fungi and bacteria are found in 'direct trophic relationship'.

At the scale of the microfauna, excluding parasitoids, the same nutrition regimes are found as in the other compartments:

- detritivores: nematodes, protozoans, especially amoebae and thecamoebians;
- predators: nematodes, tardigrades, protozoans, nematophagous fungi;
- microphages: nematodes, rotifers, tardigrades, protozoans, **cytophagous** bacteria (myxobacteria).

13.8 CONCLUSION

And thus nature leaves no residue, in accordance with the principle stated at the beginning of this chapter.

Thus ends the 'journey' of litter through the detritus food chain. The fauna and microflora have worked together in mineralizing it and reintroducing it in the nutrient cycles. The principle of recycling of animal matter (decomposition of dead bodies) and that of faeces are similar (Chap. 8, § 8.3).

Decomposition food chains are active in every ecosystem. Everywhere litter, dead wood, dead bodies and faeces are worked upon by armies of decomposers along paths similar to those explained in this chapter, with changes only in their modalities. In all soils (and also in aquatic ecosystems), the 'decomposition part' of biogeochemical cycles function because of these innumerable scavengers. Functioning of the biosphere rests on them... also we should better ensure that their work is not disturbed, for example with noxious substances spread in excess and by introducing non-degradable or very slowly degradable organic molecules in soils.

CHAPTER 14

SOIL ENZYMES

Soil enzymes: a conceptual and methodological headache.

Enzymes play an essential role in the functioning of soil. Yet they are too rarely studied in depth. Few textbooks and reviews are specifically devoted to this theme (Burns, 1978; Schinner and Sonnleitner, 1996; Gianfreda and Bollag, 1996; Bollag, in Stotzky, 1997). This is surely because of the difficulties involved in their study.

Understanding the importance of soil enzymes, without looking for completeness in their enumeration.

In this chapter, we present soil enzymes in general, while the biochemistry of humification, perhaps the most pedological of soil phenomena involving enzymes, is treated in a more detailed manner. We shall also define certain concepts of enzymology, advising the reader who wishes more in-depth knowledge to consult specialized books in this area.

Enzymes are located at an important pivot in the evolution of organic matter because their functions lead them as much to degradation of organic matter (covered in Chaps 8 to 13) as to its synthesis and transformation (principal topics in Chaps 15 and 16).

14.1 WHAT IS AN ENZYME?

An ***enzyme*** is a protein endowed with the properties of a ***catalyst***. Proteins can be irreversibly modified (denatured) by certain treatments, in particular by high temperature (e.g. egg albumin) or by changes in their physicochemical environment (e.g. pH, surface tension). They are destroyed by proteases, enzymes that hydrolyse peptide bonds.

Catalyst: substance that speeds up a chemical or biochemical reaction and is found unchanged at the end of the reaction.

All proteins are not necessarily endowed with catalytic properties...

Certain proteins have a structural function such as those that enter in the composition of ribosomes or those that surround the cell envelope of certain bacteria. Others are, for example, transporters of ions (symporter, Chap 4, § 4.2.1), energy or small molecules; toxins that form part of the weaponry of certain parasites; contractile proteins with mechanical function, such as ***actin*** and ***myosin*** of muscular fibres; immunoglobulins involved in immune-protection.

...and certain biocatalysts are not proteins!

Ribozymes are ribonucleic acids (RNA) having catalytic properties similar to those of enzymes. An interesting hypothesis was recently put forth according to which ribozymes would have been the first biological catalysts, appearing much before enzymes, in the evolution of life on Earth. To date, nothing is known about the presence of ribozymes in soils.

It is difficult sometimes to distinguish between enzymatic and non-enzymatic reactions.

Conversely, certain chemical transformations that occur in soils are not catalysed by enzymes. They take place under the control of non-biological catalysts or are spontaneous, as for example the polymerization of insolubilization humin H2 (Chap. 5, § 5.2.3; § 14.4.3).

Of the suffix *ase*!

Names of enzymes almost always end in the suffix *-ase*. Their popular name often qualifies their substrate, whereas the more 'official' name refers to the type of reaction or the chemical bond they form, modify or rupture. Thus a cellulase (enzyme that hydrolyses cellulose) is formally a β-1,4 glucosidase.

An enzyme has one or several ***catalytic sites*** that constitute a 'molecular environment' favouring the catalysed reaction. Occasionally, an enzyme cofactor, a protein or a non-protein molecule, participates directly in the reaction. Many vitamins, such as those of the B group, are precursors of organic non-protein cofactors.

Haeme: prosthetic group consisting of a tetrapyrrole nucleus associated with an iron atom (e.g. haemoglobin, cytochromes, catalase, peroxidases).

Some cofactors are permanently integrated with the concerned enzyme. They may be metal ions: they are then called ***metalloenzymes***. Although indispensable to enzymatic activity, these metals are required in very low concentrations: they are trace elements such as iron, copper or molybdenum (Chap. 4, § 4.3.4). They may also be ***prosthetic groups***; the ***haeme*** group (Fig. 14.1) of haemoproteins is a good example.

Other cofactors are mobile, associated only temporarily with the enzyme. ***Coenzymes*** are mobile non-protein cofactors. Examples are nicotinamide-adenine dinucleotide (NAD) that takes part in oxidation-reduction

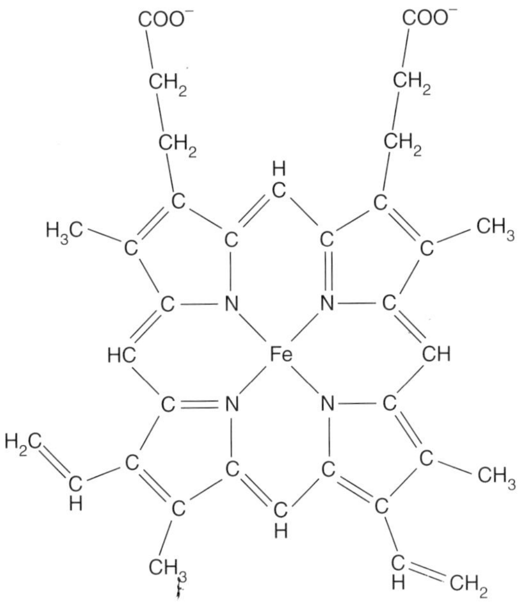

Fig. 14.1 Chemical structure of a haeme group.

reactions by associating itself with dehydrogenases, and ***ATP***, the 'energy currency' of the cell. Unlike prosthetic groups, mobile coenzymes can pass from one enzyme to another. In the cell, they transfer electrons, protons, energy or chemical groups from one reaction site to another. They may also be considered ***cosubstrates*** of the enzymatic reaction. Coenzymes belong to the intracellular environment; the function of their related enzymes is thus strictly bound to the cell. On the other hand, because of their permanent bonding with the cofactor, enzymes with a prosthetic group may be found as extracellular enzymes in soils, as for example peroxidases (§ 14.3.3).

The association of an enzyme with a ***ligand*** may alter its stability and its catalytic properties, by reducing or augmenting its activity, for example.

An organism does not synthesize all the enzymes coded for in its genes but, in principle, only those it needs at the time. For example, a pectinolytic bacterium will synthesize pectinase only if its environment contains pectin and it has no substrate available that is easier to utilize. ***Induction,*** the activation of synthesis of a specific enzyme (which is then termed ***induced enzyme***), responds to a molecular signal linked to the presence of the substrate. This signal is received at the level of DNA, more precisely that of a ***promoter***. On the other hand, enzymes needed permanently by the cell—those that control protein

ATP (adenosine triphosphate): nucleotide formed from an organic base (adenine), a sugar (ribose) and a chain of three phosphoric acid units. Hydrolysis of the bonds between the phosphoric acid molecules releases a large amount of energy that may be recuperated by an endergonic reaction coupled to it.

Ligand: chemical structure that attaches to an enzyme at a site different from the catalytic site. The ligand can actually be a regulator (for cellular enzymes) or stabilizer, particularly in extracellular soil enzymes (§ 14.2.4).

Promoter: regulatory site on DNA located near the region coding for the enzyme or enzyme group thus induced.

Constitutive (enzyme): enzyme permanently present under all conditions.

synthesis for example—are generally ***constitutive enzymes***.

Repression of an enzyme is inhibition of its synthesis by a molecular signal indicating to the cell that it does not require it or that the quantity present is sufficient. A good example is the repression of the nitrogenase complex by ammonium (Chap. 16, § 16.2.1).

14.2 THE HEADACHE OF SOIL ENZYMES

14.2.1 Of what are we speaking?

Cellular enzymes

Synthesis of all enzymes takes place in a living cell.

Synthesis of an enzyme is coded for by one or several genes of a living cell. 'Free' soil enzymes do not escape from this rule though the ambiguous term 'abiotic enzymes' is often used when referring to them!

When the rate of a biochemical reaction is measured in a soil sample, it may have been catalysed by an 'extracellular' enzyme (cell-free enzyme) or by a cellular enzyme bound to a cell or an organism. In theory, the distinction appears simple and the two categories seem well defined. But this conflicts with a major methodological difficulty: how to practically distinguish these activities? By considering these categories more carefully, it can be stated that each conceals a much more complex reality.

Endoenzyme: enzyme located strictly in the cell environment (cytoplasm, organelles, cellular and internal membranes).

Strictly speaking, ***cellular enzymes (endoenzymes)*** are enzymes whose synthesis and activity are controlled by the cell or the organism, and which are actually situated within the cell. In the case of bacteria, certain proteins are located in the ***periplasmic space***, where they play a part in the transport of nutrients or as ultrasensitive sensors informing the cell of the concentration of certain compounds in its environment.

Periplasmic space: space between the cell membrane and the bacterial wall.

Ectoenzyme: enzyme bound to the cell wall but catalysing a reaction outside it. It is at the limit of the definition of cellular enzyme because it is associated with the cell and not located inside it, and because the reaction catalysed takes place outside the cell.

Other cellular enzymes, in the broad sense this time, ***ectoenzymes,*** are located on the outer surface of the bacterial wall where they participate in the hydrolysis of macromolecules by direct contact between the cell and the enzyme substrate (e.g. cellulolysis in the case of *Cytophaga* (Fig. 14.2; Chap. 4, § 4.4.1)).

We shall not consider here the cellular enzymes as soil enzymes since their activity is integrated with that of the organisms containing them.

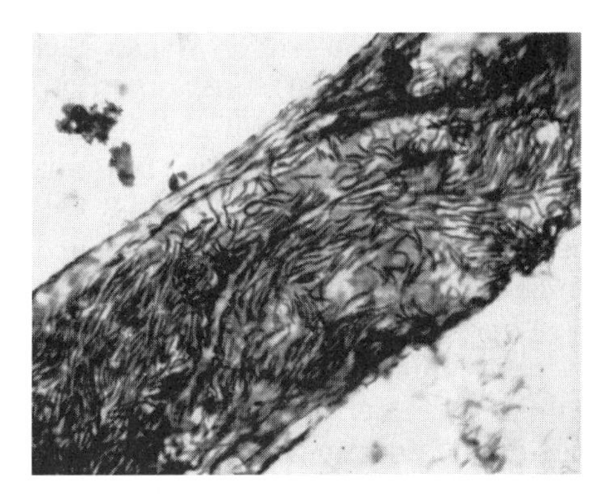

Fig. 14.2 Cells of Cytophaga hutchinsoni *bound to a cellulose fibre that they are digesting (from Winogradsky, 1949). With permission of Éditions Masson et Cie, Paris.*

Extracellular enzymes

Extracellular enzymes or ***cell-free enzymes*** or ***exoenzymes*** are described as enzymes that have left the cell or organism that produced them. Two mechanisms enable their release:

- the enzyme is actively secreted into the environment by the living cell that controls its synthesis;
- it is released by lysis of the cell, a phenomenon not controlled by the organism producing the enzyme.

Extracellular enzymes are separated in space and time from the organisms that have produced them (Schinner and Sonnleitner, 1996). They are found in the free state in soil or are bound to an organic, inorganic or mixed ligand (e.g. excrements of enchytraeids, Chap. 4, § 4.5.2).

The terms endoenzyme and exoenzyme can result in confusion because they also qualify depolymerases that attack a macromolecule, the first at any place in the chain, the second through one of its ends (Fig. 14.6). For example, an α-amylase (§ 14.3.1) is concomitantly an exoenzyme (in the sense of extracellular enzyme) and an endoenzyme (because it attacks the starch molecule in the interior of the chain).

14.2.2 Which then are the 'true' soil enzymes?

Certain edaphic enzymes are functionally linked to the extracellular soil environment. For example, depolymerases of bacteria or fungi must work in the extracellular environment because these organisms are incapable of ingesting particles or macromolecules into their cells. Similarly, phenol-oxidases directly attack lignin and its degradation products, prior to non-enzymatic polymerization of insolubilization humin.

Even if it excludes cellular enzymes, the term soil enzyme is, if not inexact, at least unsatisfactory.

Enzymes released by lysis of cells are cellular in origin. Some cannot function in the soil, either because they require a coenzyme or an activator whose presence is linked to the cellular environment, or because they do not find in the soil their specific substrates or the

After all is said and done, one could consider as 'true' soil enzymes those that are quite stable there and whose function has significance in the soil extracellular environment.

physicochemical conditions necessary for their activity. Others fulfil no significant function or are rapidly degraded or denatured in the soil.

But it is simplistic to exclude from 'true' soil enzymes those that are released by lysis of cells. Adsorbed on ligands of the soil, some proteins can acquire new properties significant in the soil's functioning. Activity of constitutive ***hydrogenase*** with kinetic characteristics (affinity, activity threshold) completely different from those of cellular hydrogenases described for bacteria was detected in soils (Schuler and Conrad, 1990). This activity could have been due to enzymes not necessarily having the same function in the cellular environment.

Hydrogenase: enzyme that transfers electrons from molecular hydrogen to a substrate or conversely from a substrate to protons with formation of molecular hydrogen.

Certain cellular activities are sometimes erroneously considered activities of soil enzymes.

Despite its name, dehydrogenase is not one enzyme!

As is usually heard in soil biology, ***dehydrogenase*** is not one enzyme but the reaction budget of a membrane complex of enzymes that transfers respiratory electrons to artificial acceptors such as triphenyl-tetrazolium chloride (TTC). This compound is irreversibly reduced to a coloured derivative, formazan, easily determined. As this activity involves the participation of coenzymes (NADH, quinones), it takes place only in the cellular environment, which alone can supply them. It very likely has no significance in the extracellular milieu... but it is easy to determine! It actually reflects the presence of a cellular metabolism and should rather be considered an indirect means of evaluating the biomass (Chap. 15, § 15.5.3).

14.2.3 Source of soil enzymes

How to know from which organism soil enzymes originate?

To know from where soil enzymes originate, ideally one would establish their amino-acid sequence and compare it with that of the corresponding genes in the organisms presumed to be the source. But such an approach is faced with currently insuperable methodological difficulties.

High concentrations of free enzymes are found in the faeces of soil animals and in the rhizosphere; thus it is difficult to identify the original source.

Microorganisms—fungi and bacteria—are the major source of soil enzymes. It is possible to support this hypothesis in the case of extracellular enzymes peculiar to certain groups of microorganisms and secreted in pure culture (Fig. 14.3). Plants and animals are also a source of soil enzymes. It is necessary to distinguish here direct and indirect production of enzymes. The latter is the work of microorganisms inhabiting the alimentary canal of animals (Chap. 4, § 4.5.2) or those feeding on the rhizodeposition (Chap. 4, § 4.1.5; Chap. 15, § 15.2.3).

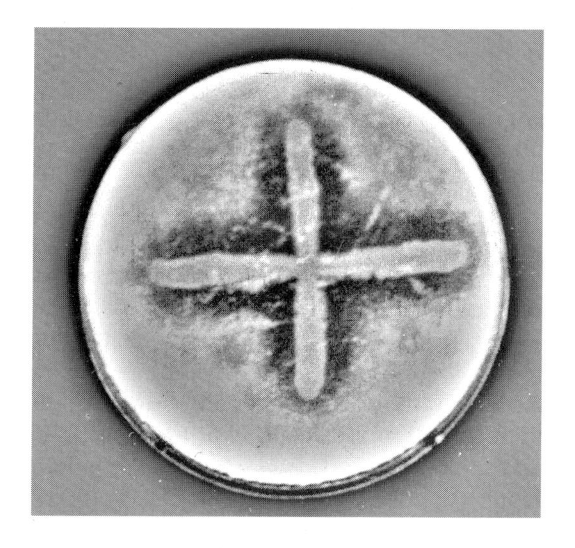

Fig. 14.3 Diffusion of the extracellular enzyme xylanase. The bacterium (a thermophilic Bacillus*) is cultured on agar medium in a Petri dish containing a suspension of xylan, partly insoluble. The free zone where the black background is visible around the bacterial colony indicates that the insoluble xylan has been hydrolysed to a distance by an extracellular enzyme secreted by the bacterium (photo P.-F. Lyon).*

14.2.4 State of soil enzymes

As a general rule, constitutive soil enzymes are adsorbed on colloidal ligands, clay minerals or humic compounds, which act as protectors and stabilizers. Enzymes in this form are resistant to:

- changes in the physicochemical environment that would denature free enzymes; the life of adsorbed enzymes is considerably longer (Lähdesmäki and Pilspanen, 1992);
- degradation by proteases because access is rendered difficult or the ligand inhibits the protease.

Where excess is of no use.

A given soil can adsorb, and thus stabilize, only a certain quantity of enzymes: the excess is degraded rapidly (Nannipieri et al., 1983). It is no use adding free enzymes to a soil if it lacks structures able to adsorb and stabilize them.

Organic amendments promote soil enzymes.

One means of augmenting in the long term the enzyme content of a soil is to apply organic amendments, for example compost (Chap. 10, § 10.7.1). These amendments are slowly transformed to stable humic compounds, which in turn can 'welcome' an addition of enzymes. Specific pre-immobilized enzymes may also be added to soil for the purpose of bioremediation.

Supplying a substrate (glucose or grass) to a soil provokes microbial proliferation, which in turn leads to an abundant synthesis of enzymes. After some time, however, the microbial populations and enzyme activity return to their initial values (Nannipieri et al., 1983). The

Isoelectric point (pI): in a protein, certain amino acids have a basic terminal group liable to accept a proton (e.g. $R{-}NH_2 + H^+ \rightarrow R{-}NH_3^+$) and others have acid terminal groups capable of releasing a proton and thus acquiring a negative charge (e.g. $R{-}COOH \rightarrow R{-}COO^- + H^+$). At low pH positive charges dominate, at high pH negative charges. The isoelectric point is the pH at which the number of positive charges on the enzyme equals that of the negative charges and the enzyme is electrically neutral. Most proteins have pI values between 4.5 and 8.5. At a given pH therefore certain enzymes are positively charged and others negatively. Direct electrostatic bonding to a negative colloid is possible only below its pI. However, electrostatic bonding between two negatively charged molecules may occur through bridging divalent ions (e.g. Ca^{2+}).

same phenomenon is observed when purified enzymes such as urease are added to a soil (Zantua and Bremner, 1977). Similar to a living organism, the soil exhibits homoeostatic properties, which limit interest in additions of bacterial cultures or enzymes, short-term measures designed to augment fertility.

Adsorption of an enzyme on a soil ligand depends on three groups of factors:

- characteristics of the ligand: surface charge, cation exchange capacity, nature of exchangeable ions, hydration;
- characteristics of the enzyme: molecular mass, ***isoelectric point***, number of binding sites, solubility, concentration;
- characteristics of the environment: pH, redox potential, properties of the soil solution, composition and concentration of ions.

Binding of an enzyme to a ligand induces changes in its molecular conformation; these changes alter its activity and kinetic properties. Most often the bonding of an enzyme to a soil colloid lowers its specific activity and its affinity for the substrate. It can also raise its optimum pH (McLaren, in Burns, 1978) and modify its response to change in temperature.

14.2.5 Role of constitutive soil enzymes

What controls soil enzymes? As in the cell (§ 14.1), constitutive and inducible enzymes are distinguished:

Constitutive soil enzymes: ready to act!

- ***Constitutive enzymes*** are present in the soil even when the substrate is absent *at the time* and are ready to start working when the substrate is added. Their concentration does not increase following the addition.

Inducible enzymes: answer only when called!

- ***Inducible*** (or ***induced) enzymes*** appear after the corresponding substrate is introduced. They are synthesized by the producing organisms following their induction by a signal consequent upon the presence of the substrate, then secreted. The activity then increases over time, often after a latent period, necessary for the cell to complete the cycle of induction, synthesis and secretion of the enzyme.

Feedback processes are very important in biological systems (Chap. 7, § 7.7.2)... but less so for constitutive soil enzymes.

There are limited possibilities for regulation of constitutive enzymes. Their activity depends of course on concentration of available substrates and physico-chemical factors of the environment, but one cannot speak of a true regulation. This involves processes of intelligent feedback, positive or negative, a cybernetic behaviour characteristic of the cell machinery. Contrarily, synthesis of inducible enzymes is regulated by the secreting organisms. Although their activity proceeds outside the cell, their concentration is under the control of the living organism. Within the limits of available adsorption sites in soil, induced enzymes can be retained on them and become constitutive.

If microorganisms can secrete high concentrations of inducible enzymes into the soil as a function of current needs, what then is the role of the same enzymes when they are fixed and become constitutive in the soil?

Depolymerases are often inducible.

Depolymerases are actively secreted by living microorganisms that can be nourished only by monomers or dimers resulting from their activity. Their induction requires a molecular signal reaching the interior of cell. For example, the ***xylose*** molecule is the molecular signal that will induce synthesis of xylanase in the cell. How can a macromolecule send such a signal when it cannot pass the barrier of the cell envelope? Constitutive soil enzymes act (Fig. 14.4), triggering an attack on the macromolecule and detaching a few soluble monomers and dimers from it. These are then taken up by the cells and act as a signal inducing synthesis of a considerably larger quantity of the enzyme (Burns, 1982).

Xylose: sugar (pentose) produced by hydrolysis of xylan by the enzyme xylanase.

Constitutive soil enzymes: 'Always prepared!

Fig. 14.4 *Constitutive soil enzymes, the 'scouts' of enzymatic hydrolysis!*

The various mechanisms of degradation of a macromolecule have been well summarized by an experiment by Durand (in Davet, 1996) on degradation of uric acid in soils (Fig. 14.5):

- in a sample sterilized by heat (organisms killed and enzymes denatured), a low but constant non-enzymatic degradation activity (chemical catalysis) is seen;
- in a sample treated with toluene, the living cells are killed but the enzymes (uricases) are not denatured; although clearly faster, degradation of the substrate takes place at constant speed;
- lastly, in an untreated sample, the activity is similar during the first five hours to that in the toluene-treated sample; it then speeds up progressively under the influence of freshly secreted uricases.

When a substrate is introduced, constitutive enzymes start degrading it. The products of their activity induce

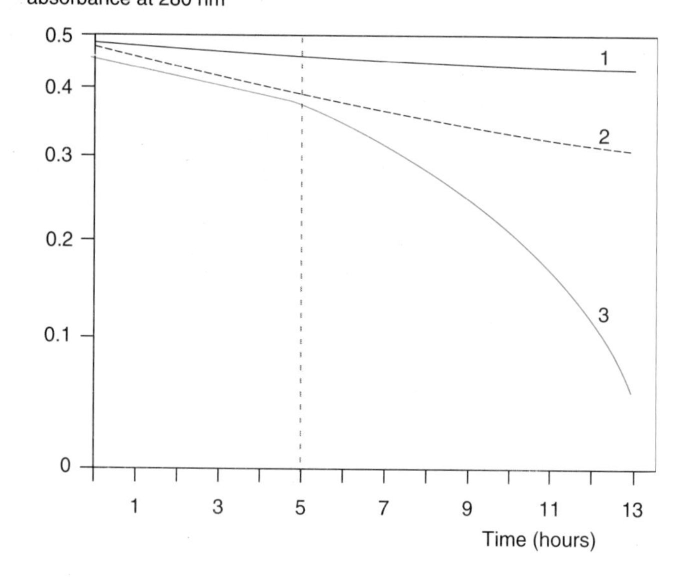

Fig. 14.5 Activity of uricase. Addition of toluene (2), which kills the cells while leaving the enzymes active, keeps the activity constant throughout the period of study; untreated soil (3) shows an immediate similar activity during the first five hours and then an increase; sterilized soil (1) shows only a considerably reduced activity (from Durand, in Davet, 1996). The difference between a constitutive activity and an induced activity is clearly seen here. The former is manifested when the substrate is introduced and stays constant. The latter appears only after a latent period (here 5 hours) and then increases because of continued synthesis of the enzyme and microbial growth induced by uptake of the substrate.

synthesis and secretion in the microorganisms present of a considerably larger quantity of enzyme, active as long as the substrate is not exhausted. Following this, homoeostatic mechanisms of soil intervene, reestablishing the initial enzyme concentrations and microbial populations.

'Soil enzymology is enzymology in the worst possible conditions!'

The difficulties in study of soil, a very heterogeneous medium, compared to those in the case of homogeneous media are evoked in Chap. 15, § 15.5 also.

While it is rather easy to measure overall activity in a soil sample (Chap. 15, § 15.5.3), it is clearly more difficult to distinguish activity of extracellular enzymes from that of living cells or organisms.

14.2.6 Methodological difficulties

Enzymologists are habituated to working with purified enzymes and in homogeneous buffer solutions that provide the enzyme with optimum surroundings. Under these conditions, many enzymatic activities can be monitored on line by means of a spectrophotometer. It is also possible to prepare large quantities of highly purified enzymes that can often be crystallized in such a way as to study their three-dimensional structure and to make it possible to obtain their amino-acid sequence.

Soil enzymologists study a heterogeneous medium in which biochemical reactions take place at the interface of the solid, liquid and gas phases. Conditions are very remote from optimum, subject to gradients, transitions and limitations. Some activities pertain to living cells, others to extracellular enzymes, most often bound to organic or mineral particles.

The methodological difficulties of soil enzymology are many and pertain to determination of activity and fractionation, extraction and purification (also see Schinner et al., 1996).

Determination of activity

A few treatments designed for sterilizing soil (that is, to kill all the living cells) without denaturing the enzymes have been proposed. ***Autoclaving*** destroys the enzymes as well as living cells. Certain treatments have the effect of sterilizing the entire soil more or less sparing the enzymes. This is what happens with high-energy radiation (e.g. γ-rays) or with solvents. Toluene is surely the least harmful of these agents and also the most often used. However, while it kills the cells, it does not eliminate the activity of some of their cellular enzymes. On the contrary, these enzymes are brought into contact with the external environment by the effect of the solvent, which makes the membranes permeable!

Autoclaving: sterilization in steam under pressure at 120°C.

Treatments with radiation or solvents can disturb the activity of some enzymes.

A totally different approach consists of measuring, on untreated soils, the *immediate* activity of transformation of the introduced substrate during a very brief incubation before the induced secretion of new enzymes resumes. The results of determination of uricase presented in Fig. 14.5 illustrate the two approaches.

Fractionation, extraction and purification

Generally, fractionation and extraction treatments (Chap. 2, § 2.2.2) result in organic or organomineral fractions associated with enzymes, often with several enzymes simultaneously. The applied treatments occasionally result in extraction of cellular enzymes. If solubilization and purification of soil enzymes are achieved, they are found in a quite different state and their properties likewise altered.

Fractionation of soils has often been attempted for locating enzymatic activities and for purifying and characterizing the enzymes in them.

Conversely, the behaviour of enzymes isolated from cultures of soil organisms and placed in contact with various types of colloids (e.g. humic compounds, tannins, melanin, lignin, synthetic polymers) has been experimentally studied (Schinner and Sonnleitner, 1996). This approach is perhaps more promising than analytical study for understanding the relationships of enzymes with soil particles.

14.3 PRINCIPAL TYPES OF SOIL ENZYMES

Most constitutive soil enzymes can be grouped into three functional categories (Tab. 14.1):

Three broad categories of soil enzymes.

Table 14.1 Principal types of soil enzymes and their functions (not exhaustive!)

Depolymerases	*Splitting of polymers into monomeric or dimeric units*	
Enzymes catalysing hydrolysis of glycoside bonds	Cellulases Amylases Pectinases Xylanases Laminarinases Chitinases Invertases	Hydrolysis of cellulose (cellulolysis) H. of starch (amylolysis) H. of pectin (pectinolysis) H. of xylans H. of laminarin and lichenin H. of chitin H. of sucrose
Esterases (h. of ester bonds)	Lipases	H. of lipids
Proteases (h. of peptide bonds)		H. of proteins
Nucleases		H. of nucleic acids, DNA and RNA splitting
Mineralization enzymes	*Mineralization of organic monomers or oligomers*	
Mineralization of amino acids	Asparaginases Glutaminases	Asparagine → aspartic acid and ammonia Glutamine → glutamic acid and ammonia
Mineralization of urea and uric acid	Ureases Uricases	Urea → $NH_3 + CO_2$ Uric acid → allantoin
Mineralization of phosphorus compounds	Phosphatases Polyphosphatases, pyrophosphatases	Organic phosphorus → orthophosphate Polyphosphates → orthophosphate
M. of sulphur compounds	Sulphatases	Organic sulphates → inorganic sulphates
Enzymes related to humification	*Degradation of phenolic polymers and synthesis of humin*	
Phenol-oxidases (oxidation of phenolic compounds)	Peroxidases • e.g. manganese-peroxidase	Catalysis of oxidation-reduction reactions between peroxides and phenols, aromatic amines, metals, etc. Catalysis of oxidation of Mn(II) to Mn(III)
	Polyphenol-oxidases • Tyrosinases • Laccases	Catalysis of oxidation of phenolic compounds by means of oxygen Orthohydroxylation of monophenols and oxidation of orthodiphenols to orthoquinones Oxidation of mono-, di- and polyphenols and aromatic amines (anilines).

Permease: transporter protein located in the cell membrane, controlling passage of specific substances across the latter (from Widmer and Beffa, 2000).

- depolymerases,
- enzymes associated with mineralization of certain organic compounds,
- enzymes linked to humification.

We shall not detail these categories here; they form the subject of an exhaustive review in a recent publication (Schinner and Sonnleitner, 1996).

14.3.1 Depolymerases

Depolymerases are enzymes responsible for breakdown of biopolymers into monomeric or oligomeric

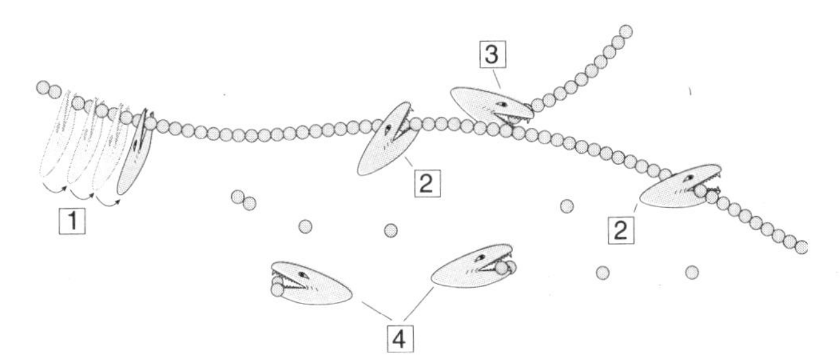

Fig. 14.6 Mode of action of depolymerases. Exodepolymerases (1) attack the molecule from one end. Endodepolymerases (2) attack within the chain. Some (3) cut the side-chains. 'Oligomerases' (4) split oligomers.

units (Fig. 14.6). Soil microorganisms are then able to transport these to their cytoplasm by means of specific ***permeases*** (Chap. 4, § 4.4.1). Typically, depolymerases are inducible enzymes and are actively secreted by living cells of microorganisms. Three types are distinguished:

- ***exodepolymerases,***
- ***endodepolymerases,***
- ***'oligomerases'.***

Hydrolysis of polysaccharides

Cellulose is a polymer of β-glucose. The most abundant organic molecule on Earth, it forms by itself about one-third of the plant litter (Chap. 2, Fig. 2.6). ***Cellulases*** catalyse ***cellulolysis***. They are produced by specialized bacterial species (e.g. *Cytophaga, Sporocytophaga, Cellvibrio*), by many fungi and also by some protozoans (amoebae, testate amoebae...) and molluscs. The ability to hydrolyse cellulose depends on its form, with microcrystalline cellulose the most resistant (Saddler, 1986).

Starch, compound of ***amylose*** and ***amylopectin,*** is the principal carbon reserve of plants. ***Amylases*** are responsible for hydrolysing it ***(amylolysis)***. They are produced by a wide variety of microorganisms, by plants and by certain animals. Four types of amylases are recognized: α-amylase (α-1,4 endoglucanase), β-amylase (α-1,4 exoglucanase), maltase (α-1,4 glucosidase) as well as an α-1,6 glucosidase responsible for hydrolysis of the side-chains of amylopectin.

Caution, possible confusions

α-1,4 *endo*glucanase is an *exo*enzyme (in the sense of extracellular enzyme), while β-amylase is an α-1,4 *exo*glucanase. The authors disown all responsibility in these regrettable terminological conflicts!

Through ***pectinolysis, pectinases*** degrade ***pectin,*** which constitutes the intercellular cement that binds plant cells together. Pectinases include poly-D-galacturonase

Exodepolymerase: depolymerase that attacks the polymer through one of its ends and successively detaches mono- or dimeric units (e.g. β-amylase).

Endodepolymerase: depolymerase that attacks the polymer at any point in the chain, which it cuts into shorter and shorter oligomeric units (e.g. α-amylase). Its action is often faster than, and also complementary to, that of exodepolymerases.

Oligomerase: depolymerase that splits the di-, tri- or tetramers into monomeric units (e.g. maltase).

Cellulolysis: hydrolysis of cellulose. Cellulolysis involves the three categories of depolymerases: cellulase C_x, which is a β-1,4 endoglucanase, cellulase C_1, a β-1,4 exoglucanase and cellobiase (β-glucosidase), an oligomerase that breaks down the dimer cellobiose or the trimer cellotriose into glucose molecules.

Amylose: one of the components of starch, formed of linear chains of α-glucose.

Amylopectin: the second component of starch, formed of branched chains.

Pectinolysis: hydrolysis of pectin.

Pectin: polymer essentially composed of D-galacturonic acid units whose carbonyl group is more or less esterified by methanol.

Xylan: polymer formed essentially from D-xylose; the most abundant form of hemicelluloses.

that breaks the glycoside bond between two galacturonic acid units, and also pectin-esterase, which demethylates the molecule with release of methanol.

Some enzymes hydrolyse other polysaccharides:

- ***xylanases*** hydrolyse ***xylans***;
- ***laminarinases*** (β-1,3 glucanases) hydrolyse polymers of the laminarin and lichenin type; these substances are plentiful in fungi and some brown algae (e.g. *Laminaria, Fucus*);
- ***chitinases***, specific for chitin;
- ***invertase*** hydrolyses sucrose into D-glucose and D-fructose.

Hydrolysis of lipids: lipases

Lipases are esterases that break down lipids (tri- and diglycerides) by hydrolysing the ester bonds between glycerol and fatty acids. Phospholipid degradation requires a glycerophosphatase also.

Numerous microorganisms and animals secrete proteases.

Hydrolysis of proteins: proteases

Extracellular ***proteases (peptidases)*** are responsible for 'digestion' of proteins ***(proteolysis)*** to oligopeptides and amino acids. They break the ***peptide bond*** between two amino acids (Fig. 14.7). Endopeptidases and exopeptidases are distinguished according to mode of action. Exopeptidases are of two kinds that attack the peptide at its carboxyl end or its amine end.

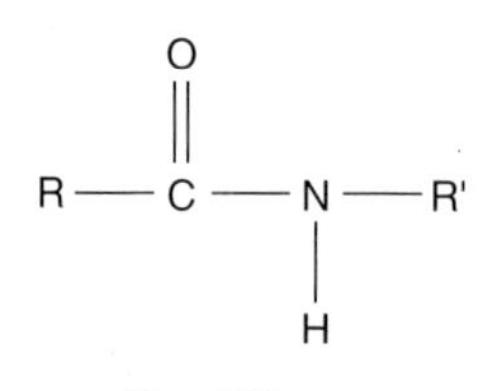

Fig. 14.7 *Peptide bond between two amino acids.*

Hydrolysis of nucleic acids: extracellular nucleases

The nucleic acids DNA and RNA represent a substantial portion of the living biomass. Bacteria contain 3–4% DNA and up to 20% RNA (dry weight basis). Some microorganisms secrete extracellular ***nucleases*** that attack nucleic acids anywhere in the nucleotide chain regardless of the specific base sequence (Chap. 15, § 15.5.5). They thus totally differ from ***restriction endonucleases***. Extracellular nucleases belong to the group of phosphodiesterases. They rupture the two ester bonds between phosphoric acid and the adjacent pentoses.

Restriction endonuclease: strictly intracellular enzyme that recognizes specific base sequences and splits DNA very precisely at these sites. Restriction endonucleases are instruments of 'molecular surgery', one of the basic tools of molecular biology.

14.3.2 Enzymes involved in mineralization phenomena

Some nitrogen, phosphorus and sulphur compounds, often monomers resulting from the activity of depolymerases, undergo further changes before being absorbed and assimilated by microorganisms. Constitutive soil enzymes catalyse these transformations.

Amino acids are deaminated and decarboxylated by soil enzymes. For example, ***asparaginase,*** an amidase, converts asparagine to aspartic acid and ammonia. Similarly, ***glutaminase*** converts glutamine to glutamic acid and ammonia. Some constitutive soil enzymes catalyse decarboxylation of aromatic amino acids and transform tryptophan in particular to ***indoleacetic acid*** (Chalvignac, 1971).

Mineralization of amino acids.

Indoleacetic acid: plant-growth hormone. Its formation in soils can modulate root growth (Chap. 15, § 15.3.6).

Urease, which hydrolyses urea to NH_3 and CO_2, is a common enzyme in plants, animals and microorganisms. In its very stable constitutive form in soils, it has often been used as a test for fertility. Uric acid is transformed to allantoin by ***uricase***. We have earlier defined (§ 14.2.5; Fig. 14.5) the constitutive and inducible activities of this enzyme.

Mineralization of urea and uric acid.

Organic phosphates are not assimilated by vegetation nor by most microorganisms. ***Phosphatases*** are then responsible for enzymatic mobilization of organic phosphorus. Depending on whether the phosphoric acid molecule is esterified with one, two or three organic groups, hydrolysis is achieved by ***phosphomonoesterases, phosphodiesterases*** or ***phosphotriesterases***. The first have been specifically studied because of their importance in soil fertility (Speir and Colwing, 1991).

Mineralization of phosphorus compounds.

Two classes of phosphatases are distinguished according to the optimum pH: acid phosphatases and alkaline phosphatases. The former is often present in root exudates (Chap. 4, § 4.1.5); it is a typical indicator of the rhizosphere environment. The latter is produced only by microorganisms, in particular mycorrhizal fungi (Dighton, 1983). Soil ***pyrophosphatases*** and ***polyphosphatases*** hydrolyse ***polyphosphates*** and release orthophosphate, making the phosphate available to plants.

Two classes of phosphatases

Polyphosphates: linear polymers of phosphoric acid units.

Sulphur is also present in soils in the organic form, which represents 60 to 98% of the total sulphur. Organic sulphates (R–O–SO_3) are found, as well as compounds with direct carbon-sulphur bonds, such as the amino acids cysteine and methionine, which can be abundant in very insoluble proteins. Constitutive ***sulphatases*** have been identified in soils. They are responsible for release from organic sulphates of sulphate ions that can be assimilated by vegetation and microorganisms.

Mineralization of sulphur compounds. The forms of sulphur in the soil are not all known (Chap. 4, § 4.4.4).

14.3.3 Phenol-oxidases

The enzymes that conduct an oxidative transformation of phenolic compounds are grouped under the general term ***phenol-oxidases*** (Filip and Preusse, 1985). Phenol-oxidases are essential for degrading phenolic polymers such as lignin and melanin, as well as for synthesis of insolubilization humin H2 (Chap. 5, § 5.2.3).

Two major categories of phenol-oxidases: peroxidases and polyphenol-oxidases.

Oxidations, reductions, degradations, transformations...

Peroxidases are iron-containing haemoproteins. They are produced chiefly by plants and by microorganisms. They catalyse oxidation-reduction reactions between hydrogen peroxide (H_2O_2) or an organic peroxide on the one hand, and phenols, aromatic amines, metals and other compounds on the other. These reactions often result in very reactive aromatic radicals. Peroxidases act in ligninolysis (ligninases and manganese-peroxidase) and in the oxidative transformation of aromatic monomers, as prelude to their polymerization (§ 14.4). Some peroxidases are cellular enzymes, others are actively secreted. They are abundant constitutive enzymes in soils (Bartha and Bordeleau, 1969).

Some peroxidases, the ***manganese-peroxidases***, have an indirect effect by catalysing the oxidation of Mn(II) to Mn(III). Later the Mn(III) reacts chemically with the phenolic compounds to be oxidized (Chap. 9, § 9.6.2). Such indirect enzymatic oxidation is particularly efficient in microporous substrates. Indeed, manganese is a relatively small ion and can act where big enzyme molecules cannot penetrate.

...biosyntheses, hydroxylations, detoxifications, polymerizations, ...phenol-oxidases are not idle!

Polyphenol-oxidases (monophenol-monooxygenases) are metalloproteins containing copper. They catalyse the oxidation of a broad spectrum of phenolic compounds using molecular oxygen. They include tyrosinases and laccases.

- ***Tyrosinases*** (*ortho*-diphenol-oxidases) are found in animals, plants and microorganisms: they achieve two types of oxidation (Fig. 14.8). *In vivo* they act in many biosyntheses, in particular that of lignin, tannins and melanin. Fungi appear to be the major source of constitutive tyrosinases in soils.

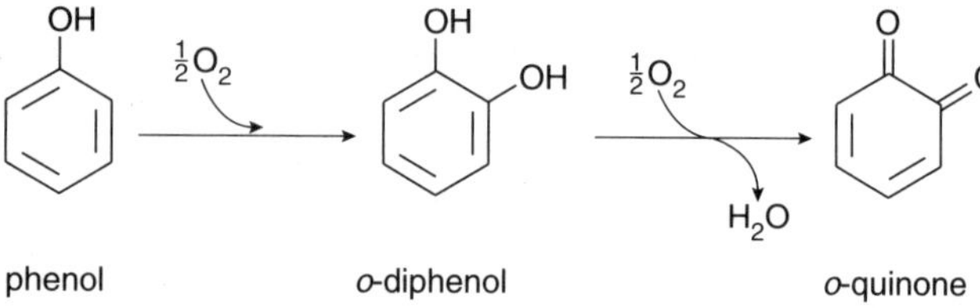

OH
$\frac{1}{2}O_2$
OH
OH
$\frac{1}{2}O_2$
O
O
H_2O
phenol
o-diphenol
o-quinone

Fig. 14.8 *Reactions of tyrosinases. Tyrosinases achieve two types of oxidation: orthohydroxylation of monophenols and oxidation of orthodiphenols to orthoquinones.*

Soil laccases have a detoxifying effect.

- ***Laccases*** (*para*-diphenol-oxidases) are abundant in plants and fungi, the Basidiomycetes in particular.

Oxidation consists of a transfer of electrons from the substrate to oxygen, with formation of two molecules of water (Fig. 14.9). Thus no oxygen atom is introduced in the oxidized molecule. The product is a radical which, because of its instability, enters in non-enzymatic reactions and couples to other phenolic compounds. Laccases can oxidize mono-, di- and polyphenols as well as aromatic amines (anilines). They trigger polymerization of the phenols produced during ligninolysis, whose accumulation would otherwise be toxic to soil organisms.

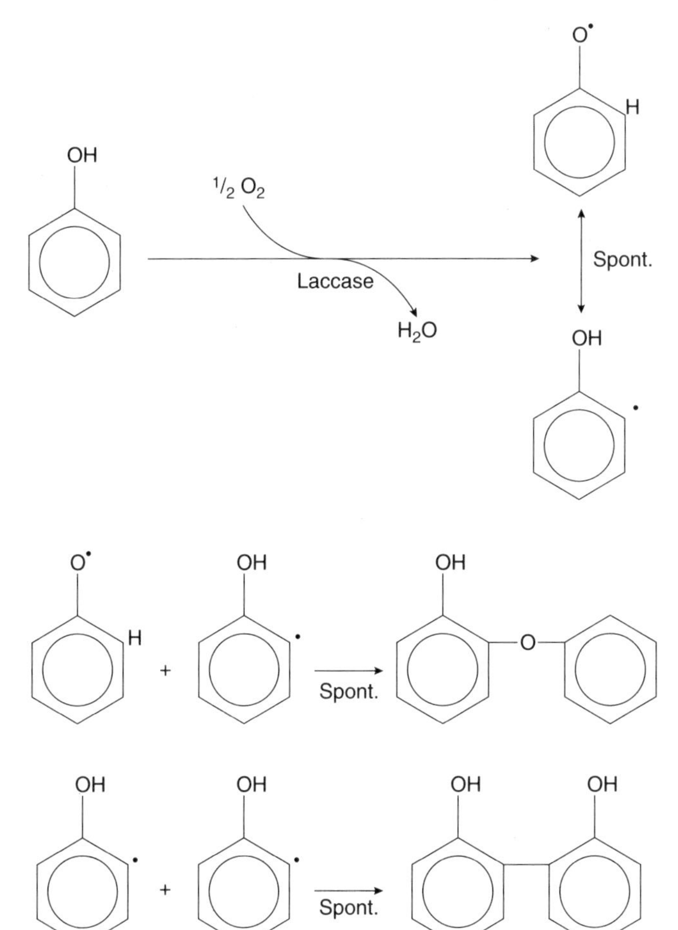

Fig. 14.9 Principle of oxidation reactions catalysed by laccases and resulting spontaneous polymerization reactions between aromatic radicals.

14.3.4 Lignin, its synthesis and degradation

Lignin synthesis takes place from precursors derived from phenylpropane.

Synthesis of lignin is initiated by the enzymatic oxidation of precursors (Fig. 14.10; Chap. 2, Tab. 2.6) to phenoxy radicals. These radicals spontaneously couple in different ways among themselves and with lignin under formation. Enzymes only activate the precursors but do not direct polymerization, which takes place in a random manner (Fig. 14.11).

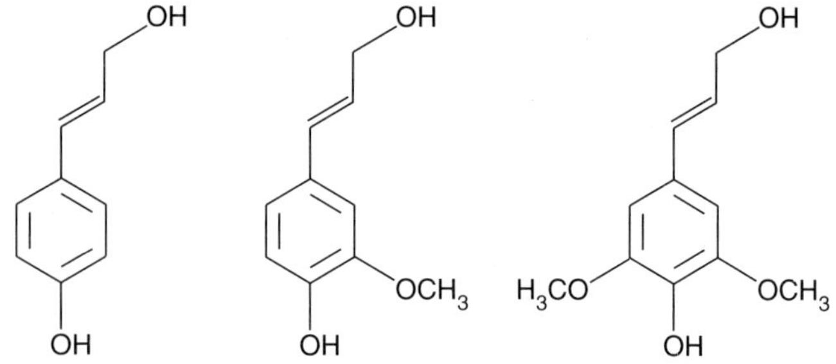

p-Coumaryl alcohol | Coniferyl alcohol | Sinapyl alcohol

Fig. 14.10 *The precursors of lignin.*

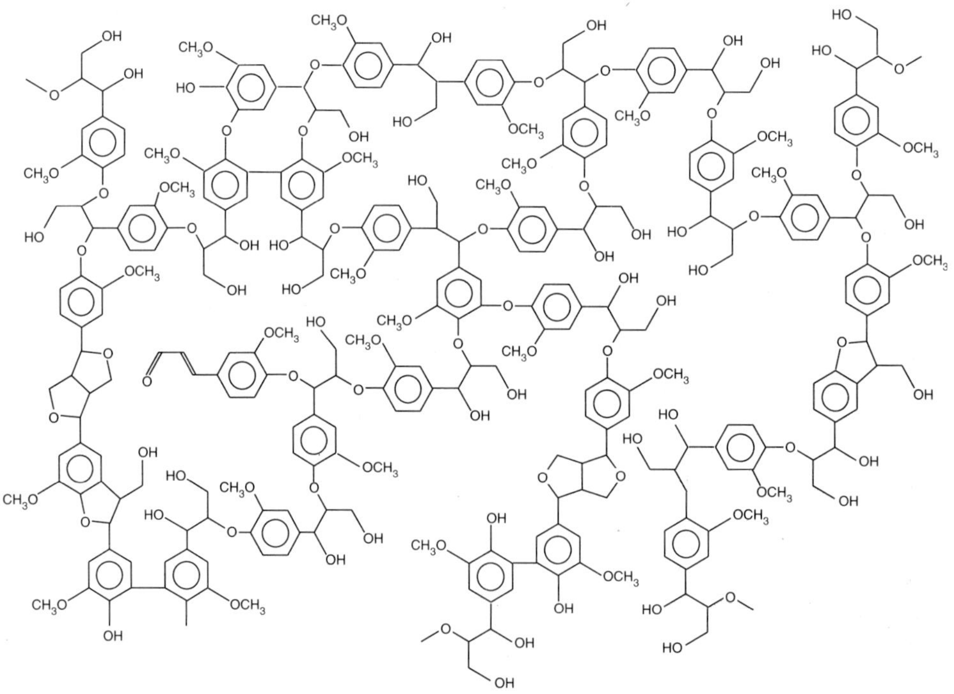

Fig. 14.11 *Schematic diagram of a portion of the lignin molecule (from Reid, 1995).*

Lignin is a very complex phenolic polymer difficult to degrade ***(ligninolysis)***. Only higher fungi, basidiomycetes and ascomycetes can degrade it to any great extent (Chap. 8, § 8.6.1; Hammel, in Cadisch and Giller, 1997).

Lignin: difficult degradation...

Most of the bonds between monomers are of the ether or carbon-carbon type and are not directly hydrolysable. The complexity and irregularity of the molecule makes recognition by enzymes difficult; the latter must therefore have low specificity with respect to different aromatic structures. The enzymes responsible for attack on lignin are chiefly peroxidases (ligninase and manganese-peroxidase) and oxidases (laccase) that initiate the process by activating the lignin. The oxidative fragmentation reaction, exenergonic and therefore thermodynamically favourable (Chap. 5, § 5.1.1), follows spontaneously, without enzymatic control. Such a process is often referred to as ***enzymatic combustion*** (Kirk and Farrell, 1987; Eriksson et al., 1990; Reid, 1995; Tuor et al., 1995).

...essentially of oxidative type.

14.4 BIOCHEMISTRY OF HUMIFICATION

Three fundamental processes act in humus formation; they 'culminate' in the three types of humin defined in Chap. 5, § 5.2.3:

- humification by inheritance H1, which gives residual humin;
- humification by bacterial neosynthesis H3, which results in microbial humin;
- humification by polycondensation H2, which gives insolubilization humin.

Formation of humus integrates degradation and synthesis phenomena (Hayes, in Wilson, 1991).

These processes involve the activity of soil enzymes and microorganisms. But the role of certain animals such as earthworms (Chap. 4, § 4.5; Chap. 13, § 13.6.4) should not be overlooked: an earthworm functions as a veritable ***bioreactor***, which mashes and homogenizes the substrates-enzymes-microorganisms mixture, all by maintaining a physicochemical environment favourable for biological reactions. The worm even secretes a contingent of enzymes into its alimentary canal.

These three processes exhibit in the sequence they are mentioned increasing requirements with respect to environmental conditions. This explains why they participate to various degrees in each form of humus (Chap. 6, § 6.3.2).

14.4.1 Humification by inheritance H1

A 'skeleton' of unaltered or scarcely transformed substances, residual humin.

Humification by inheritance is an essentially degradational, 'negative' process. Inherited materials are those that could not be degraded in the prevailing conditions; they are often phenolic polymers, lignin in particular. The latter is most often degraded by very specialized fungi with precise requirements for their growth. In very extreme conditions (acid pH, anoxia), other depolymerizations are suppressed, as indicated by a considerable accumulation of organic matter including polysaccharides as well; this is what happens in formation of peat (Chap. 9, § 9.2).

14.4.2 Humification by bacterial neosynthesis H3

Polysaccharides very resistant to enzymatic degradation.

In this case, bacteria ensure neosynthesis and secretion of biopolymers, essentially polysaccharides. This synthesis takes place by polymerization of molecules of simple sugars, which requires expenditure of energy. Because of the involvement of coenzymes as energy vectors (e.g. ATP), this synthesis can take place only in living cells. The sugars utilized are derived from monomers absorbed after extracellular hydrolysis of plant polysaccharides or are newly synthesized by the bacterium itself. These very resistant polysaccharides accumulate in soils, notably in the litter neighbourhood and in the rhizosphere (Chap. 15, § 15.3.1).

Example of biosynthesis of a polysaccharide

Addition of one unit of sugar is achieved with consumption of energy through hydrolysis of one molecule of ATP. Transfer of energy is accomplished here through another nucleotide, GTP (guanosine triphosphate).

$$\text{Sugar} + \text{GTP} \rightarrow \text{Sugar–GDP} + \text{orthophosphate}$$
$$(\text{Sugar})_n + \text{sugar–GDP} \rightarrow (\text{Sugar})_{n+1} + \text{GDP}$$
$$\text{GDP} + \text{ATP} \rightarrow \text{GTP} + \text{ADP}$$

$$\text{Sugar} + (\text{Sugar})_n + \text{ATP} \rightarrow (\text{Sugar})_{n+1} + \text{ADP} + \text{orthophosphate}$$

The primary role of polysaccharides is either protection of the bacterial cell (capsule formed of poorly hydrated polymers) or formation of bacterial mats on surfaces or even formation of aggregates of cells and microcolonies (mucilaginous layers formed of highly hydrated polymers; Chap. 3, § 3.2.1).

14.4.3 Humification by polycondensation H2

Humification by polycondensation includes reactions of degradation, transformation of monomers and extracellular polymerization; it puts to work extracellular enzymes and non-enzymatic reactions; it occurs without addition of outside energy to the molecules involved and to their free energy of oxidation. Built at random by the coming together of reactive chemical groups (quinones, radicals) by reactions not controlled by enzymes, insolubilization humin has an indeterminate and very complex structure. This humin and the fulvic and humic acids that lead to it hardly differ in their general organization, but mainly in their degree of polymerization (that is, molecular size), which determines their solubility (Chap. 2, § 2.2.4; Chap. 5, § 5.2.1).

Humification by polycondensation: the most important and most complex humification process.

Chance meeting...

A structure of 'Lego®' bricks built by a madman...

Considering the almost infinite number of combinations among the subunits of insolubilization humin, the fundamental constituents and their bonds form an unlimited number of compounds. What a contrast to the orderly arrangement of sugars in a polysaccharide of microbial humin!

There are probably no two identical molecules of insolubilization humin in the ecosphere!

A model for humic acid (Fig. 14.12), precursor of insolubilization humin, was proposed by Stevenson (1976). It is a small hypothetical molecule showing all the most important constituents and bond types that can be found in humin. Aromatic rings attached by covalent C–C and C–O–C bonds, reactive hydroxyl (–OH), carbonyl (>C=O) and carboxyl (–COOH) groups, quinonoid rings, heterocyclic units as well as peptide chains and linked sugars are seen. Insolubilization humin is also rich in stable free radicals that confer oxidative properties on it and enable attraction of ionized or protonated compounds (Stevenson, 1982).

Insolubilization humin and lignin both result from a non-enzymatic polymerization process in which chance

Insolubilization humin exhibits similarities with lignin in complexity and disordered structure...

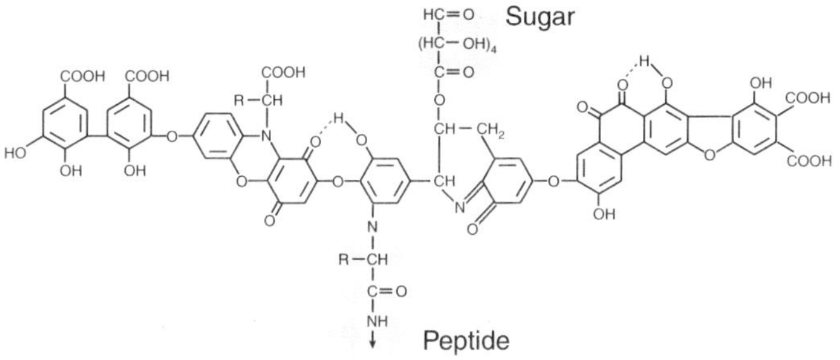

Fig. 14.12 *Model of humic acid, one of the constituents of insolubilization humin (from Stevenson, 1976). Other models have also been proposed (Chap. 2, § 2.2.4).*

plays an essential role. Both are structures particularly resistant to enzymatic degradation. It is surely difficult for an enzyme to find its way around in such a chaos of chemical bonds!

Our understanding of the processes of humin polymerization is still very fragmentary, particularly with respect to fixation of polysaccharide and peptide chains. Polymerization of aromatic molecules is a little better understood. The precursors are phenols and anilines. Most of them originate from degradation of lignin and other phenolic polymers (melanins, tannins) but aromatic compounds (flavones, anthocyanins, catechin) released by living or dead plant tissues should not be omitted.

Ligninolysis yields very reactive oxidized radicals (Fig. 14.13). Phenols and anilines, less reactive, undergo oxidation reactions catalysed by phenol-oxidases (peroxidases, tyrosinases and laccases) or, depending on soil type, abiotic reactions. They are transformed to radicals or very reactive quinones. Thus activated, these compounds polymerize spontaneously. This is the phenomenon of ***oxidative coupling*** (Sjoblad and Bollag, in Paul and Ladd, 1981).

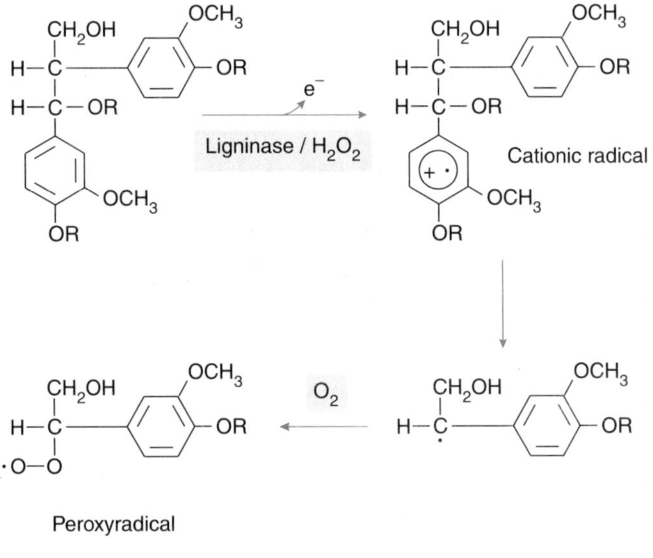

Fig. 14.13 *Action of fungal peroxidases on lignin (from Haider, in Stotzky and Bollag, 1992).*

Enzymes control the formation and activation of precursors, but not polymerization itself.

In the presence of phenol-oxidases, Martin and Haider (1980) obtained synthetic humic compounds experimentally by polymerization of monomers derived from lignin. These compounds had properties very similar to those of natural humic materials. The oxidative-coupling reactions take place essentially at neutral and alkaline pH. This is related to the dominance of residual humin over insolubilization humin in some acid holorganic horizons of OF or OH type found in mors (Chap. 6, § 6.3.5).

14.5 CONCLUSION

Soil enzymes play a major role in degradation of polymers of the litter, in release of mineral salts for the benefit of plants and in synthesis of insolubilization humin. Passively or actively released from the cell generating them, they become stable, autonomous actors in the 'Living Soil'. When constitutive, they work as intermediaries between substances reaching the soil and organisms called upon to destroy them. When induced, they accumulate where their activity is required, ensuring the nutrition of the heterotrophic organisms that have produced them and of their commensals.

The many difficulties in study of enzymes should not conceal the fact that enzymes are essential indicators of fertility of soils.

Chapter 15

THE RHIZOSPHERE: A (MICRO)BIOLOGICALLY ACTIVE INTERFACE BETWEEN PLANT AND SOIL

Chapter 13 described the energy-circulation networks and Chapter 14 the fundamental tools—enzymes in the soil-plant interface. In the sequence of evolution of organic matter, Chapter 15 evokes the functional communication between soil and plant. It precedes the chapter on symbioses, in which the partner organisms integrate so as to form a single organism-like entity.

The rhizosphere, acting as interface between the living plant and mineral matter.

After a brief recapitulation of essential definitions (§ 15.1) this chapter will emphasize the dual functional relationship established in the rhizosphere: effect of the root on its environment (§ 15.2) and response of the microflora to root activity (§ 15.3). Two other aspects are again detailed: the 'inverted' rhizosphere of marsh plants (§ 15.4) and a methodological overview of the most recent techniques in study of the rhizosphere, in particular those of molecular biology (§ 15.5).

15.1 RECAPITULATION OF DEFINITIONS, GENERALITIES

Strictly speaking, the ***rhizosphere*** is the soil zone under the influence of the root (Chap. 4, § 4.1.3). Very poorly

Extent of the rhizosphere.

understood, it must nevertheless be admitted that it represents an essential interface between plant and soil, an interface activated by the presence of micro-organisms—bacteria and fungi—as well as of their predators (Curl and Truelove, 1986; Lynch, in Lynch, 1990; Lavelle and Spain, 2001). Considered the habitat of microorganisms linked to root activities, the rhizosphere *sensu lato* extends to the surface of tissues (rhizoplane) and their interior (endorhizosphere). The zone of soil in the vicinity of the root and influenced by it is then termed the ***exorhizosphere.***

A close dialogue between the plant and the 'workers of the soil', up to symbioses and parasitoses (Chap. 16).

In Chapter 4, §§ 4.1.4 and 4.1.5, we mainly described the physical structure of the root environment and nature of the organic products of the root in its surroundings, the rhizodeposition. The root considerably modifies the physical and chemical characteristics of this environment by these organic products and its metabolic and physiological activity. The microflora—bacteria and fungi—as well as the microfauna accompanying this microflora (§ 15.3.2) respond to these changes. These responses induce other modifications and secretions which, in turn, work on the root through feedback (Bazin et al. and Hedges and Messens, in Lynch, 1990).

Considering the multiple gradients that characterize the rhizosphere, its study is very difficult.

The major obstacle to study of the rhizosphere is related to sampling problems and the always large difference between field conditions and those reproduced in the laboratory, often outrageously simplified (Chapter 17, § 17.1.3). However, recent achievements in very efficient molecular biology techniques enables us throw some light on the so far misunderstood reality of the surroundings of the root (see § 15.5.5).

15.2 EFFECTS OF THE ROOT ON ITS ENVIRONMENT

The root influences in many ways the physical and chemical conditions of the surrounding soil (Drew, in Lynch, 1990).

The effects of the root are manifested in various directions with different intensities depending on the root zone and the characteristics of the surrounding environment. For example, roots of aquatic and marsh plants, immersed as they are in an anoxic sediment, determine a redox gradient opposite to that in the rhizosphere of a terrestrial plant (§ 15.4).

15.2.1 Water absorption, drainage

Water absorption enables compensating for evapotranspiration by aerial parts and nourishes the plant.

This provokes drainage of water from the soil, resulting in an increase in permeability of the latter to air. This drainage is facilitated by the presence on the root surface of a mucilaginous sheath, the mucigel, which maintains a favourable moisture content while eliminating the voids that would otherwise isolate the root absorption surfaces from the soil. Similarly, mycorrhizal fungi participate by translocation in the hydric nutrition of the plant, by drawing soil water sometimes at great distances from the root (Chap. 2, § 2.5.3; Chap. 16, § 16.1.2).

The root constitutes the principal entry point of water into the plant (Chap. 3, § 3.4). The zone of maximum absorption is located at the level of the absorbing root hairs (Chap. 4, § 4.1.4).

15.2.2 Ion absorption

Absorption of ions needed for nutrition of the plant (Ca^{2+}, Mg^{2+}, K^+, Na^+, NH_4^+, NO_3^-, SO_4^{2-}, HPO_4^{2-}, $H_2PO_4^-$, Cl^-, as well as the trace elements Fe, Mn, Zn, Cu, Co, Ni, etc.; see Chap. 4, § 4.3.4) takes place as a rule through the electrochemical potential generated by H^+-ATPase. However, the root alone is not efficient enough in an environment deficient in nutrient elements because of the relatively low affinity of their ion-transport system compared to microbial systems, and the relatively low absorption surface area of the root compared to that of the associated microorganisms (bacteria and mycorrhizae, see Plate VII-2). On the contrary, bacteria and fungi handle this problem by very efficient mechanisms of active transport. These mechanisms do involve an expenditure of energy, but they enable the organisms to concentrate the ions in their cells, often to a considerable extent. A large microbial biomass growing at the periphery of the root can thus corner these ions and form a 'screen' cutting off the plant from its mineral nutrition.

The root absorbs ions in the soil solution along with water.

Two mechanisms can turn the efficiency of ion transport by microorganisms to the benefit of the plant: mycorrhization, which we shall cover in detail in Chap. 16, § 16.1.1, and predation by the microfauna (§ 15.3.2).

Absorption of ions by the root may also have an effect on the pH because of the proton pump (Chap. 4, § 4.2.1). In the case of Dicotyledoneae, the tendency to absorb more cations than anions leads to acidification, whereas in Monocotyledoneae the quantities of anions and cations absorbed are almost equal (Marschner and Römheld, 1983). The dominant form of nitrogen absorbed by the plant has a marked effect on the pH. Absorption of nitrate tends to raise it, that of ammonium to lower it.

The effect of the root on pH varies along the root axis because of different functions (secretion, absorption, lysis) accomplished by different portions of the root (Chap. 4, Fig. 4.3).

15.2.3 The rhizodeposition: carbon and energy fluxes

The plant is a vector between light, the universal form of energy, and soil (Chap. 3, § 3.5.1).

A large portion of the matter photosynthesized by the plant (from 20 to 50%) is secreted into the soil by the living root in the form of secreted polymers, soluble exudates, detached cells and lysates. This does not take into account the source of nutrients represented by dead roots, a veritable underground litter (Chap. 2, § 2.2.1).

Of the elements necessary for constitution of microbial matter (Tab. 15.1) and by abstracting hydrogen and oxygen from it, the rhizodeposition essentially provides organic carbon to the rhizosphere. Although precise data in natural conditions are still lacking with respect to its elemental composition, it is generally estimated that the rhizodeposition contains relatively small quantities of other elements compared to the needs of microorganisms, which then have to draw them from the soil.

Table 15.1 Approximate concentration of major elements in the microbial biomass

Element	Per cent of dry matter	Element	Per cent of dry matter
Carbon	50	Sulphur	1
Oxygen	20	Potassium	1
Nitrogen	14	Calcium	0.5
Hydrogen	8	Magnesium	0.5
Phosphorus	3	Iron	0.2

15.2.4 Signals for the microflora

From attraction to antagonism

Chemotaxis: behaviour of organisms *vis-à-vis* chemical compounds that attract them (positive chemotaxis) or repel them (negative chemotaxis).

Beyond the effects and products of the root, which work non-specifically on the entire rhizospheric biocoenosis, many substances secreted by the plant are important as specific signals for particular organisms. Some species of bacteria are attracted by compounds secreted by a particular group of plants. This behaviour of ***chemotaxis*** explains their specificity for the rhizosphere of these plants. Attractants may have an antagonistic effect (e.g. antibiotic) *vis-à-vis* other microorganisms. Thus some phenolic compounds secreted through wounds caused to roots act as antibacterial defences. *Agrobacterium*, a parasitic bacterium entering through such wounds, is in turn attracted by these compounds. Flavonoids secreted by roots of legumes (Fabales) are simultaneously attractants for *Rhizobium*, inducers of their nodulating genes, and antagonists of other soil bacteria (Chap. 16, § 16.2.1).

14.2.5 Oxygen consumption

The root is a heterotrophic organ (Chap. 4, § 4.1.1) that depends day and night on respiration for obtaining the energy needed for its cellular activities. Although withdrawal of water by absorbing root hairs leads to increased soil aeration, the root consumes part of the oxygen thus diffused by its respiration; this tends to lower the oxygen concentration in the rhizosphere. This sets up at equilibrium a negative oxygen gradient between the bulk soil and the root surface. The rhizosphere may thus accommodate ***microaerophilic*** microorganisms.

Microaerophilic: qualifies a microorganism with aerobic respiratory metabolism, but for which oxygen in high concentrations is toxic. It thus finds its optimum at low oxygen concentrations. Sometimes microaerophilic behaviour is conditioned to expression of a metabolic pathway sensitive to oxygen, for example fixation of molecular nitrogen.

15.3 RESPONSES OF THE MICROFLORA TO ROOT ACTIVITY

15.3.1 Importance of flow of energy and matter from the rhizodeposition

The rhizosphere represents an important part of the 'culture medium' of rhizosphere microorganisms, to which it provides energy and organic carbon (Whipps, in Lynch, 1990). It reproduces, in a way, the experiment of Winogradski (Chap. 4, § 4.4.6), who showed the effects of the addition of organic substrate on soil bacterial communities. Rhizosphere microorganisms as a whole are thus very likely more zymogenous than those of the bulk soil.

One part of the rhizodeposition is transformed to energy, another to biomass.

The mucilages secreted by the root are quite easily hydrolysed by organisms of the rhizosphere microflora feeding on them. In turn, bacteria themselves secrete resistant mucilages into the rhizosphere (microbial humin; Chap. 4, § 4.1.5; Chap. 5, § 5.2.3; Chap. 14, § 14.4.2). The resultant mucigel is the physical support for bacterial growth in the root environment (Fig. 15.1).

Exudates, the soluble fraction of the rhizodeposition, are absorbed without prior action of extracellular hydrolytic enzymes (Chap. 14, § 14.2.1); a wide range of bacteria and fungi can utilize them. The lysates of rhizodermal and cortical tissues, as well as the cells sloughed from the root cap, also form food for saprophytic microorganisms of the rhizosphere. Respiratory activity of aerobic bacteria nourished by the rhizodeposition also

Root exudates: the veritable 'culture broth' of the rhizosphere.

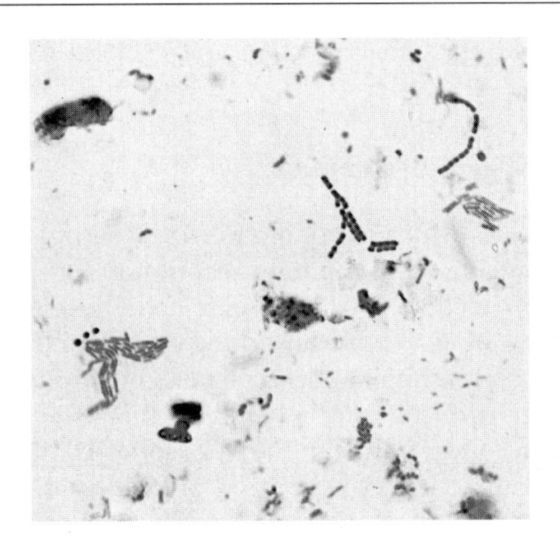

Fig. 15.1 *Bacterial microcolonies in the mucigel (photo M. Aragno).*

results in consumption of oxygen in the root environment. Added to that of the root tissues, this activity intensifies the negative oxic gradient in the rhizosphere.

15.3.2 Bacterial biomass in the rhizosphere: importance of predation

The missing link...

Considering the intensity of the nutrition flux from the rhizodeposition, it may be expected that the biomass of rhizospheric microorganisms formed from very well defined proportions of bioelements (Tab. 15.1) will attain large values. However, it should be noted that an exceedingly large biomass would monopolise much of the elements equally necessary for the plant. This would result in blocking of plant growth… and thus of the root secretions indispensable for the bacteria! Limited to a plant-bacteria interaction, the rhizospheric system will probably be self-blocking. Thus the link for proper functioning is missing.

Clarholm (1985, 1989) identified such a link. Investigating the nitrogen nutrition of wheat, this author cultivated the cereal in an unfertilized soil, sterilized and reinoculated in two ways: with soil bacteria alone and with soil bacteria plus amoebae predatory on the bacteria (Chap. 12, § 12.4.1), The quantity of nitrogen assimilated by the plants was three to four times higher in the presence of amoebae than when they were absent. What had happened?

Where the plant knows its true friends!

The root exudates consumed by rhizosphere bacteria are mostly carbon-containing molecules such as sugars and organic acids (Fig. 15.2). The nitrogen needed for

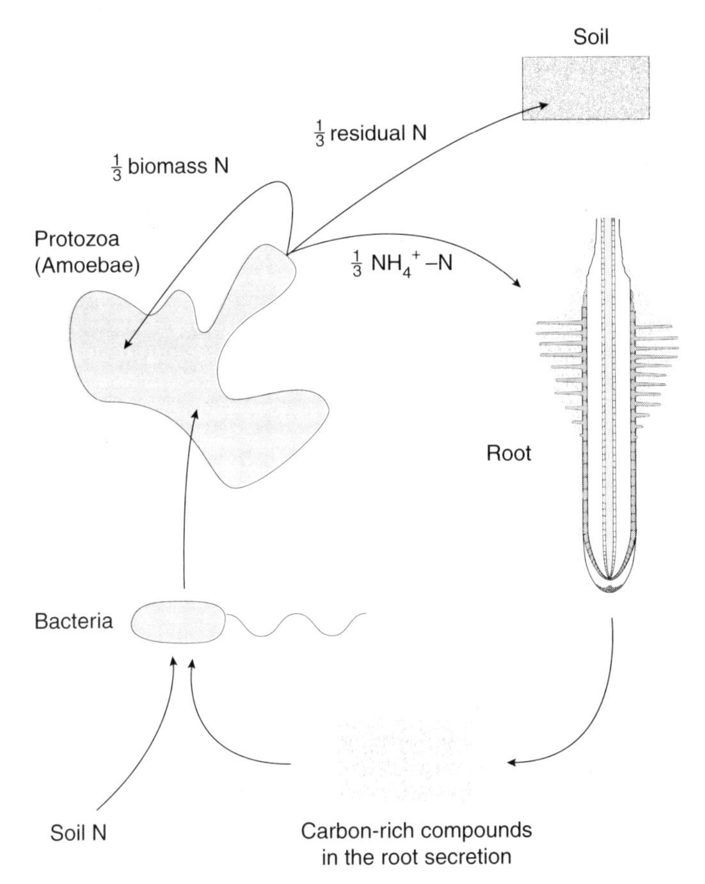

Fig. 15.2 *Effect at the rhizosphere level of the interaction between root, bacteria and protozoa on nitrogen absorption by the root. Explanations in the text.*

bacterial growth being found in the exudates in very small quantities, the major part has to be drawn from the soil. By means of their highly efficient systems of active transport, bacteria are able to considerably concentrate this element (§ 15.2.2). When amoebae feed on bacteria, an estimated two-thirds of the nitrogen contained in this biomass are digested by the amoebae—one-third is assimilated and the other one-third mineralized. The last one-third belongs to the non-digestible compounds rejected as residues and enters the organic-nitrogen pool of the soil.

Higher concentrations of mineral nitrogen.

The nitrogen of the fraction mineralized by amoebae is released as NH_4^+ ions in a concentration sufficient to allow absorption by the root, directly or after nitrification. Thus, through bacteria and protozoa as intermediaries, secretion of carbon-bearing substances by the root makes mineral nitrogen available at high concentration. Similar results were obtained by Ingham et al. (1985) with a system including bacteria, fungi and nematodes.

The same model with three partners can be applied to the other ions assimilated by the bacteriomass conjointly with utilization of the secretions. Predation by microphages (Chap. 13, Tab. 13.1 and § 13.6.2) turns in favour of the plant a process which, without them would have given a competitive advantage to bacteria and fungi: the high concentrating power of their active ion-transport system.

Bacteria preying on bacteria, belonging to the genus *Bdellovibrio* (Jurkevitch, 2000) or to related taxa, were also observed in the rhizosphere of various plants (Jurkevitch et al., 2000). Their significance in bacterial biomass turnover remains to be shown.

The rhizosphere bacteria are found in dynamic equilibrium with cell growth and predation.

Predation thus considerably augments nutrient dynamics (Zwart et al., in Darbyshire, 1994) and turnover of the rhizospheric bacteriomass (Pussard et al., in Darbyshire, 1994). Otherwise, the latter would rapidly attain a threshold imposed by limiting factors in the soil and enter a quasi-stationary state.

In addition to their remarkable ability to concentrate dissolved ions, bacteria have other mechanisms available to augment the availability of certain elements to the benefit of the rhizospheric community: this is particularly true for nitrogen, phosphorus and iron.

15.3.3 Nitrogen

In a soil intrinsically poor in nitrogen, consumption by bacteria for their needs results in a severe shortage of nitrogen in the rhizosphere milieu.

The composition and control of the rhizodeposition are poorly understood, especially under natural soil conditions. If it were particularly rich in organic nitrogen compounds (chiefly amino acids), its nitrogen content could exceed the requirements of the consuming microorganisms and the excess would be released to the medium in the form of ammonium (ammonification) (Fig. 15.3a).

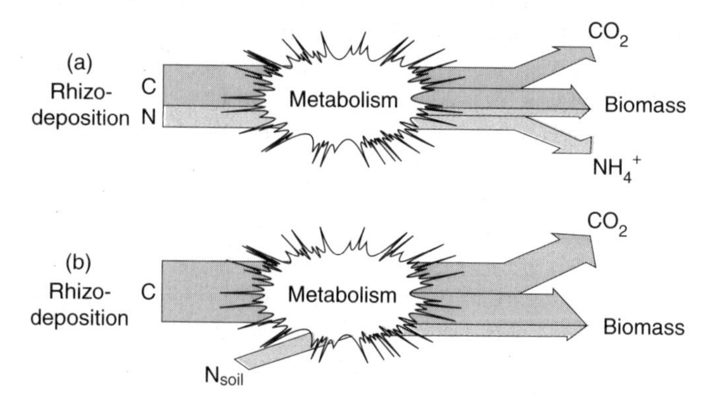

Fig. 15.3 *Effect of nitrogen content of the rhizodeposition on soil nitrogen: (a) high levels; (b) low levels. Explanations in the text.*

Contrarily, if the root production were poor in nitrogen, the bacteria would draw what is lacking for them from the soil (Fig. 15.3b) and thus impoverish the soil.

In the latter case, the specific conditions of the rhizosphere may greatly favour fixation of molecular nitrogen (Chap. 4, § 4.4.3). Actually:

- the fixation reactions, energy-expensive, are promoted by large supplies of organic substrates from the rhizodeposition;
- the concentration of oxygen, a nitrogen-fixation repressor in many bacteria, is lowered by the combined respiratory activities of the root and the primary and secondary consumers of the rhizodeposition;
- deficiency of combined forms of nitrogen (in particular ammonium and nitrate) removes the repression these compounds exert on nitrogenase.

The most spectacular, also the most efficient, example of this fixation is that of root nodules formed in symbiosis with fixing bacteria. But apart from this case pertaining to a limited number of plant groups (Chap. 16, § 16.2), many examples are known of ***associative fixation*** involving exo- or endorhizosphere bacteria. Associative fixation has been intensively studied in tropical plants, particularly forage or fodder grasses, in which it plays a significant economic role.

Associative fixation: nitrogen fixation by bacteria living in association with a root (in the exorhizosphere, on the rhizoplane or in the endorhizosphere) without, however, causing noticeable modification in root morphology as in symbiotic fixation sensu stricto (formation of specialized organs: nodules, rhizothamnia; Chap. 16, §§ 16.2.1 and 16.2.2). Bacteria associated more or less specifically with roots have been isolated: nitrogen-fixing facultatively anaerobic (*Bacillus, Klebsiella, Enterobacter*), microaerophilic (*Azospirillum, Herbaspirillum, Pseudomonas*) and aerobic (*Azotobacter, Beijerinckia, Pseudomonas*) bacteria (Doebereiner and Pedrosa, 1987; Chan et al., 1994). Using molecular methods (§ 15.5.5), Hamelin et al. (2002) detected the presence, in the rhizosphere of *Molinia coerulea*, a grass well suited to nitrogen-poor soils (oligonitrophilic), of a dominant group of nitrogen-fixing bacteria that till then had never been obtained in pure culture.

Stimulating bacteria, but how?

As often stated, numerous bacterial populations in the rhisosphere exert a positive effect on plant growth, health and resistance to parasites. They are often referred to as 'Plant Growth Promoting Rhizobacteria' ***(PGPR)***. They act in different ways and the same population may cumulate different PGP mechanisms. For example:

- Stimulation and control of plant growth by the production of plant hormones § 15.3.6).
- Improvement of plant nutrition by, for instance,
  - concentration of mineral nutriments, due to the high concentration capabilities of the bacteria (§ 15.3.2),
  - solubilization of inorganic phosphate (§ 15.3.4),
  - secretion of plant-compatible siderophores for iron nutrition (§ 15.3.5),
  - associative nitrogen fixation (Chap. 4, § 4.4.3; § 15.3.3),
  - secretion of mucigels that improve root-soil exchange of water and ions (Chap. 14, § 14.4.2; § 15.3.1).
- Protection of the root against parasites by
  - production of compounds repressing the growth of parasites, such as hydrocyanic acid and diacetylphloroglucinol (Fig. 15.4);
  - competition with parasites (e.g. for iron, again through siderophore production, § 15.3.5);
  - stimulation (induction) of plant resistance to parasites.
- Positive interaction with mycorrhiza ('mycorrhiza helper bacteria', MHB, Chap. 16, § 16.1.2).

PGPR: Plant Growth Promoting Rhizobacteria.

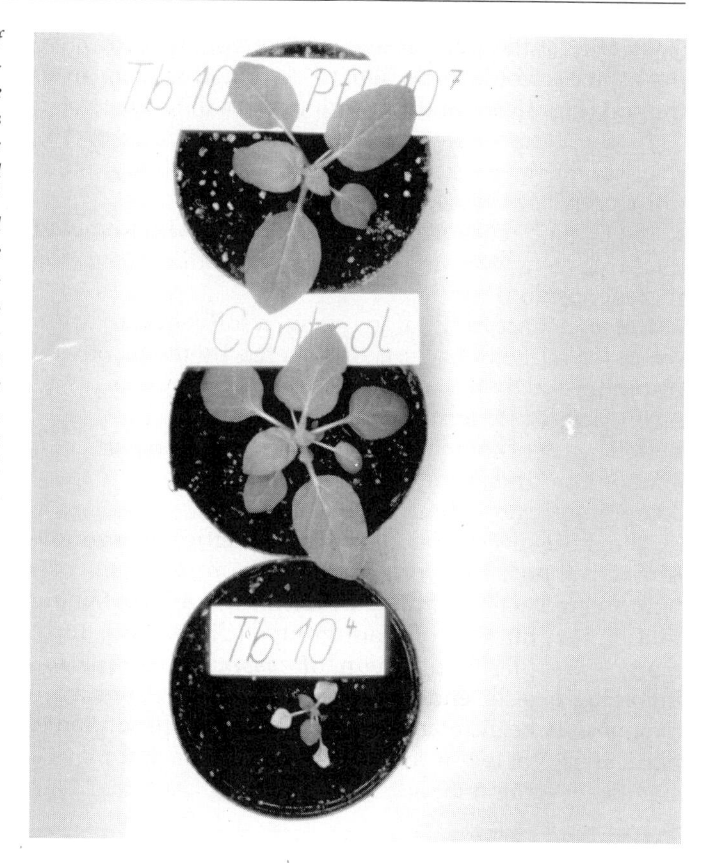

Fig. 15.4 Protection of tobacco (Nicotiana glutinosa) *plants against the root parasite fungus* Thielaviopsis basicola. *Top: tobacco culture inoculated with a total of 10^4 spores g^{-1} soil of the fungal parasite and 10^7 cells g^{-1} soil of the protective bacterium* Pseudomonas fluorescens *CHAO, producing diacetylphloroglucinol. Middle: control culture without the parasite and bacterium. Bottom: culture with the parasite but without protective bacteria (photo G. Défago).*

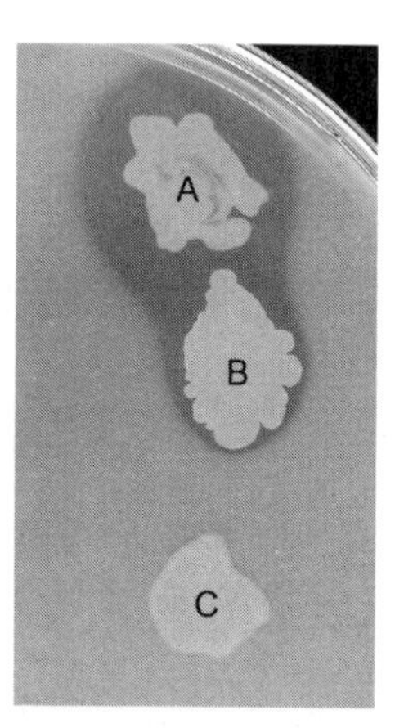

Fig. 15.5 Solubilization of calcium• phosphate around bacterial colonies isolated from the rhizosphere of wheat. A: very active strain; B: scarcely active strain; C: inactive strain (photo D. Roesti and M. Aragno).

The concentration of soluble phosphates in soils is usually very low, the greater part being found in precipitated form.

15.3.4 Phosphorus

Many common bacteria in the rhizosphere solubilize phosphate (Fig. 15.5). This solubilization may occur following overall lowering of rhizosphere pH, itself under the dual control of the root and activity of the heterotrophic microflora of the rhizosphere (§ 15.2.2). However, in most plants, phosphorus nutrition is achieved through mycorrhizae (Chap. 16, § 16.1.2). For example, Piccini and Azcon (1987) showed that only the combination of phosphate-solubilizing bacteria and arbuscular mycorrhizae was able to significantly augment phosphate nutrition of lucerne.

15.3.5 Iron

In an oxic environment, inorganic iron exists in its oxidized form Fe(III), which gives nearly insoluble compounds (e.g. oxides, phosphates). For iron uptake, organisms have to produce small molecules capable of chelating Fe(III) and of being absorbed by the microbial or plant cell through specific transport systems: ***siderophores*** and ***phytosiderophores*** (Figs. 15.6, 15.7)

Siderophore: (literally iron-carrier) molecule capable of complexing iron; it is secreted by an organism to ensure its iron nutrition from nearly insoluble inorganic compounds of trivalent iron.

Phytosiderophore: siderophore produced directly by roots.

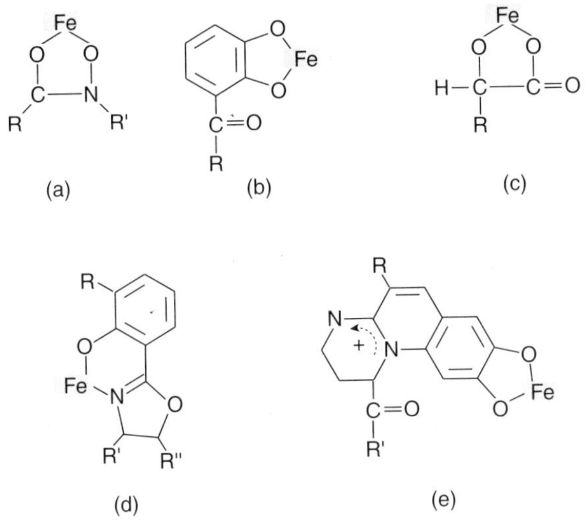

Fig. 15.6 *Usual ligands for Fe(III) in siderophores: (a) hydroxamate (e.g. ferrichromes, rhodotoluric acid); (b) catechol (e.g. enterobactin); (c) hydroxyacids (e.g. citrate, pseudobactin); (d) 2-(2-hydroxyphenyl)-oxazoline (e.g. mycobactins, agrobactin, parabactin, vibriobactin); (e) fluorescent quinolinic chromophore (e.g. pseudobactin).*

Through production of specific siderophores, microorganisms are strong competitors for iron, particularly in the rhizosphere (Schroth and Hancock, 1982). When manifested, such competition could allow, for example, bacteria to block spore germination or attack on roots by pathogenic fungi by depriving the fungi of iron. This competition is analogous to that of sphagnums, which monopolise nitrate in peat bogs before it is made available to other plants (Chap. 9, § 9.2.2).

Iron and siderophores: competition, repression, regulation!

Iron in solution exerts a repressive effect on production of siderophores: these thus intervene only if there is deficiency; here is an important regulatory mechanism, preventing concentrations of iron high enough to be toxic (Chap. 4, § 4.3.5).

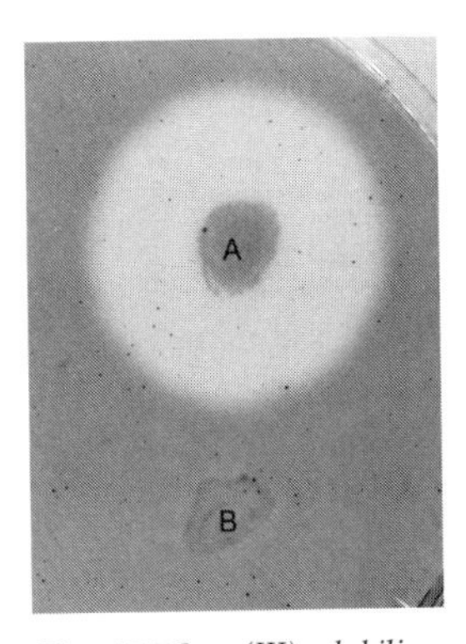

Fig. 15.7 *Iron (III) solubilization following the secretion of siderophores by a bacterium isolated from the rhizosphere of wheat (photo D. Roesti and M. Aragno).*

Does the rhizosphere microflora facilitate iron nutrition of the root by producing, for example, siderophores that can be used by the root, or, on the contrary, does it compete for this element? Both phenomena appear possible (Buyer et al. and Crowley and Gries, in Matthey et al., 1994). On the other hand, the presence of mycorrhizae, thus of their iron-assimilating system, seems to play a major role in the transport of iron towards the root (Chap. 16, § 16.1.1; Cress et al., 1986).

15.3.6 Return signals: effects of the microflora on root growth

Presence of microorganisms in the rhizosphere has an effect on root growth and rhizodeposition.

It is generally accepted that rhizosphere bacteria increase root production, which induces increased secretion to the benefit of the bacteria. Production of ***plant hormones*** by bacteria or fungi is a well-known phenomenon (Lynch, in Vaughan and Malcolm, 1985; Lynch, in Lynch, 1990). These are volatile compounds such as ethylene or non-volatile substances such as gibberellins, auxin (Patten and Glick, 1990) (Fig. 15.8), cytokinins or abscissic acid. The mode of action and the importance of these compounds in the rhizosphere are still poorly understood. Martin and Glätzle, in Klingmüller (1982), demonstrated the positive effect of auxin-producing strains of *Azospirillum* on growth of roots, their extent of ramification and the ratio between biomass of roots and that of aerial parts (Chap. 2, § 2.2.1; Chap. 4, § 4.1.6). These bacteria are often located in the endorhizosphere and it is probable that the most significant interactions take place there (Hedges and Messens, in Lynch, 1990). Signals emitted by microorganisms as a prelude to establishment of symbiosis are the best studied (Chap. 16, § 16.2.1).

***Plant hormones** (or **phytohormones**):* chemical substances of varied nature affecting growth and development of plants at very low concentrations.

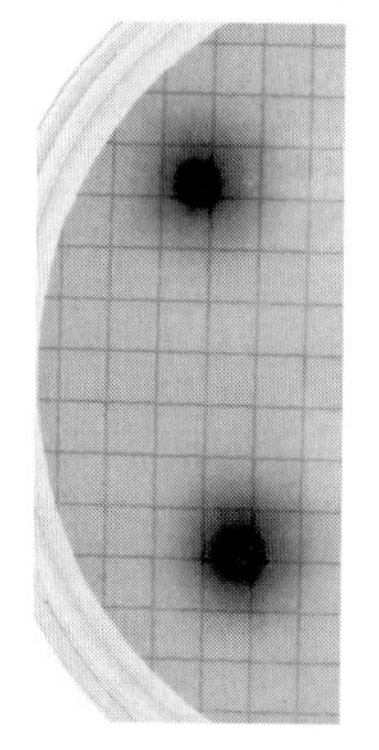

Fig. 15.8 *Production of auxin (indoleacetic acid), a plant hormone, by a bacterium isolated from the rhizosphere of lupin,* Lupinus albus *(photo: L. Weisskopf and M. Aragno).*

15.4 ROOT ENVIRONMENT OF MARSH PLANTS: AN 'INVERTED' RHIZOSPHERE

No diffusion of oxygen to depth.

Rhizosphere conditions of marshes are very different from those of aerated soils. Hydromorphic soils (HISTOSOLS, RÉDUCTISOLS or RÉDOXISOLS) and submerged sediments (sapropel, gyttja, dy; Chap. 6, § 6.2.1) are generally anoxic, the diffusion coefficient of oxygen in the liquid phase being 10,000 times less than that in the gas phase (Chap. 3, § 3.3.3). Furthermore, supply of litter to the surface of the sediments leads to rapid consumption of oxygen, preventing its diffusion to depth.

How to respire in a soil lacking oxygen?

The below-ground organs of marsh plants must respire despite the absence of oxygen in the surrounding soil. For this, in response to anoxic conditions, air-filled ducts are formed in the cortical parenchyma (root aerenchyma,

Fig. 15.9) to lead the oxygen of the air to root tissues (Chap. 4, § 4.2.3). From there, the oxygen diffuses somewhat to the rhizosphere soil, forming an oxic sheath around the root (Armstrong et al., 1994).

This rhizosphere thus exhibits an oxygen gradient as well; this gradient is the reverse of what is established in aerated soils, the highest oxygen concentration being located in contact with the root. The low oxygen levels prevailing here are particularly favourable for microaerophilic organisms, such as many nitrogen-fixing bacteria.

In oxic waters, the dominant inorganic nitrogen form is nitrate, while sulphur is found essentially in the form of sulphate. These ions serve as electron acceptors for anaerobic respirations (Chap. 4, § 4.4).

Nitrate is reduced to elemental nitrogen through denitrification (Fig. 15.10, item 3). This involves facultatively aerobic organisms (Chap. 4, § 4.4.4) and is dominant as a rule in the zone of movement of groundwater (Go horizon), where the conditions are alternately oxic and anoxic. It provokes a loss of combined nitrogen from the medium.

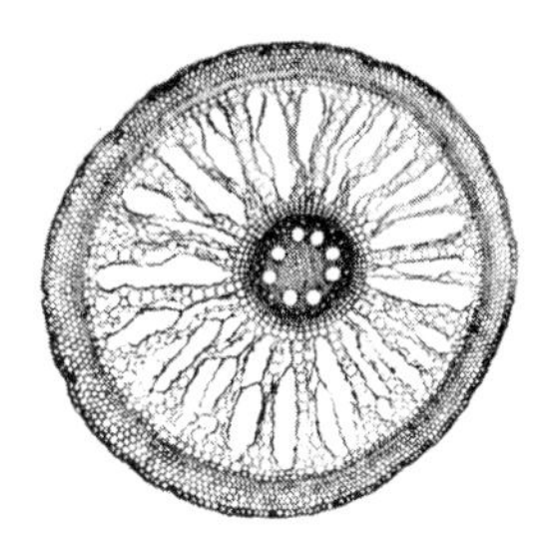

Fig. 15.9 Section through a root of reed Phragmites australis *with aerenchyma (photo P. Küpfer).*

The oxic sheath is a perfect example of pedological feature (Chap. 3, § 3.2.4).

Nitrate and sulphate: ions with relatively high redox potential.

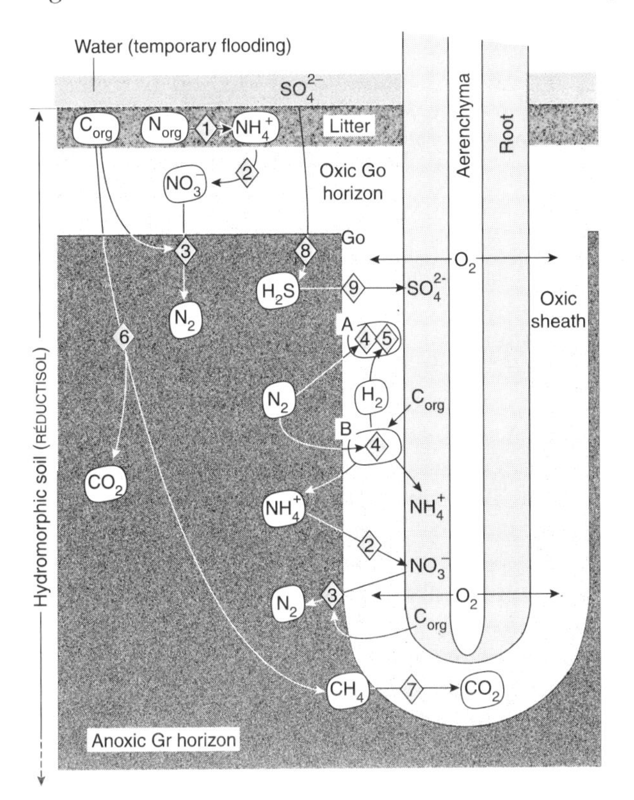

Fig. 15.10 Simplified schematic diagram of the root environment of a marsh plant showing the microbial activities linked to it. 1. mineralization of nitrogen (ammonification); 2. nitrification; 3. denitrification; 4. fixation of molecular nitrogen; 5. hydrogenotrophy; 6. methanogenesis; 7. methanotrophy; 8. sulphate reduction; 9. sulphide oxidation; A: nitrogen-fixing hydrogen bacteria; B: nitrogen-fixing heterotrophic bacteria.

Sulphate is reduced to hydrogen sulphide (Fig. 15.10, item 8), part of which precipitates some metals (iron, manganese); this gives a dark colour, grey or black, to the sediments, as seen very clearly in the TH horizons of THIOSOLS. Anoxic conditions also favour reduction of (hydr)oxides of Fe(III), leading to solubilization of this element in the form of Fe^{2+} ions (Chap. 4, § 4.4.2). In marine sediments, the high sulphate concentration in sea water is generally sufficient for ensuring complete oxidation of organic substances diffusing into the anoxic layers, except in environments such as mangroves with very high plant production. Freshwater milieus are generally poor in sulphate; formation of large quantities of methane, the climactic stage of anaerobic evolution of organic matter (Chap. 4, Fig. 4.20), is often observed in their sediments.

Two kinds of phenomena can occur in the oxic sheath:

fixation of molecular nitrogen...

Nitrogen fixation (Fig. 15.10, item 4) is particularly favoured in this kind of rhizosphere: root secretion provides the necessary energy; the oxygen concentration in it is very low and the denitrification that occurs in the zone of groundwater movement causes a loss of combined nitrogen. These three factors result in a de-repression of microaerophilic fixation of nitrogen (Chap. 4, § 4.4.3). Thus it should be no surprise that environments such as reed-beds, although having a soil solution extremely poor in combined nitrogen—for example those of the south bank of Lake Neuchâtel in Switzerland (Buttler, 1987)—exhibit a very high vegetative production, one of the highest of all plant formations in the world.

... and oxidation of the reduced compounds dominant in anoxic soils (Aragno and Ulehlova, in Lachavanne and Juge, 1997).

Reactions inverse to those taking place during diffusion of oxidized anions in the anoxic zones of the sediment occur in the oxic sheath (Conrad, 1993): these oxidation reactions are the work of chemolithoautotrophic aerobic organisms (Chap. 4, § 4.4.2). Thus, hydrogen sulphide is oxidized to sulphate (Fig. 15.10, item 9) by sulphide-oxidizing bacteria, ammonia to nitrate (Fig. 15.10, item 2) by nitrifying bacteria and methane to carbon dioxide (Fig. 15.10, item 7) by methanotrophs. Molecular hydrogen resulting from the fixation of molecular nitrogen is oxidized by aerobic hydrogen bacteria (Fig. 15.10, item 5). Also, soluble divalent iron may be oxidized to the much less soluble trivalent iron. A large portion of the methane is also oxidized by methanotrophic aerobic bacteria (Fig. 15.10, item 7) before it enters the root aerenchyma, thereby diminishing its transfer to the atmosphere (Conrad, 1993).

Some of these oxidation reactions are particularly important in the root environment: hydrogen sulphide is a powerful respiratory poison, whereas sulphate is the sulphur source needed by plants; nitrate is often better assimilated than ammonium.

Bacteria work as detoxifying agents, at the same time allowing plants to assimilate the elements in the oxidized form.

Marsh gas: greenhouse effect and will-o'-the-wisp

In most cases, the anaerobic transformation of organic matter results in formation of biogas under the effect of an association of bacteria that works in a closely coordinated fashion (Fig. 15.10, item 6; Chap. 4, Fig. 4.20). In marsh sediments, this gas is released in the form of bubbles that may be easily extracted by stirring the sediment with a stick. Part of the dissolved methane diffuses into the roots and also towards the aerenchyma, from where it is easily transported to the atmosphere. Marsh environments, especially paddy fields, are an important source of the methane reaching the atmosphere and thus contribute to the greenhouse effect. We should remember that methane is a gas with considerably greater greenhouse effect than CO_2 (Chap. 5, § 5.2.3; Chap. 9, § 9.2.1).

The will-o'-the-wisp, source of numerous fantasies and legends, has been attributed to self-ignition of marsh gas. But methane does not ignite spontaneously in air. Bacterial production of diphosphane (P_2H_4) has recently been shown to occur by reduction of phosphates. This self-igniting gas could serve as a primer for the ignition of biogas.

Biogas or marsh gas: a mixture of methane and carbon dioxide.

The will-o'-the-wisp? Diphosphane? Unless mischievous spirits...

15.5 METHODS FOR STUDY OF THE RHIZOSPHERE MICROFLORA

The great importance of the rhizosphere microflora in the interaction between root and soil cannot be overstated. There has been considerable development in techniques of microbial ecology in the last few years. These techniques now enable better description of the rhizosphere, not only conceptually but also experimentally. In some cases, a still more exact approach may be achieved in culture systems in the laboratory, but that would be distancing ourselves from the real rhizosphere.

In response to a rather restricted number of publications: developing efficient methods!

Study of the rhizosphere employs a broad spectrum of methods. First those pertaining to sampling and fractionation, which enable spatial definition of the rhizosphere. It is next necessary to measure the biomass and activity of the microorganisms present and to isolate them to determine behaviour and functions. A major obstacle to this approach is the difficulty, even impossibility of culturing in the usual media most of the microorganisms present in a soil. Recent application to

The rhizosphere: where understanding of ecology is totally dependent on the performance of methods!

microbial ecology of methods of molecular biology (so-called 'molecular ecology') enables crossing the limits of culturability in evaluation of biodiversity.

15.5.1 Sampling and fractionation of the rhizosphere

Sampling

One of the major obstacles in study of the rhizosphere? Probably the difficulty of sampling and fractionation!

The minimum sample weight depends on the analysis to be done. Very small samples suffice for some molecular methods, especially those that employ fluorescent probes. But analysis of bacterial-cell compounds may require several tens of grams of material.

Finding a compromise between contradictory requirements.

Samples of adequate volume are required to eliminate the effect of soil heterogeneity. But the gradients controlling microbial life in the soil, particularly in the rhizosphere, are exactly at the scale of this heterogeneity and their description would involve point measurements.

A superposition of gradients in a heterogeneous medium.

The rhizosphere presents a superposition of radial, longitudinal and temporal gradients (Chap. 4, § 4.1.4). These gradients are generally quite pronounced while the soil exhibits, at the scale of the gradient and microorganisms, a very heterogeneous structure. With a large volume, the effect of microheterogeneity of soil is negligible, but the gradients are no more perceptible. In contrast, the reproducibility of a sample of very small volume, even drawn correctly, is poor because of this heterogeneity. The same comments apply to determinations *in situ*, which may be localized or general, and applicable to a small or large soil volume (e.g. determinations of redox potential, Chap. 3, § 3.10.1).

The concept of sample involves a certain quantity of material representative of an ideally homogeneous medium.

In a given volume of soil taken from the root zone, we can try one way or another to measure the root biomass. Thus a relative degree of 'rhizosphericity' is established of the sample which, depending on the case, contains different proportions of roots and hence of rhizosphere soil surrounding them. But in such samples, the fraction of soil subject to high rhizosphere gradients is minor, and this method does not enable highlighting them. Fractionation of samples is therefore recommended.

Fractionation

A rhizosphere fractionated into three: bulk soil, rhizosphere soil, root tissues.

We are often content with dividing the rhizosphere into three fractions. A large sample is drawn from the root zone and the first fraction (bulk soil) is separated from it by moderate agitation. The roots are then left behind with soil particles adhering to them more or less firmly, which are separated by brushing or washing. This second fraction is considered the 'rhizosphere soil'. The washed roots, retaining the microorganisms closely adhering to their surface (rhizoplane) and those that colonize the cortical tissues (endorhizosphere, endophytes), constitute the third fraction. Occasionally a fourth fraction is envisaged, represented by bare soil. This involves artificial elimination of vegetation for a sufficiently long period, which cannot be done without considerably modifying the physical properties of the soil, in particular its structure and porosity.

Sometimes even... a fourth fraction represented by bare soil.

The endorhizosphere fraction in the broad sense, including the rhizoplane, is the best defined. Root tips, the most active zones, are often broken; they are then found in the bulk soil and are not accounted for in the other two fractions, of which they are nevertheless the most representative parts. Moderate agitation is not an easily standardized procedure. The fraction adhering to the root after such a treatment depends on moisture content and texture of the soil and on age of the root (Chap. 4, § 4.1.6). Some roots however, especially of Monocotyledoneae, establish a well-defined sheath of rhizosphere soil (Fig. 15.11).

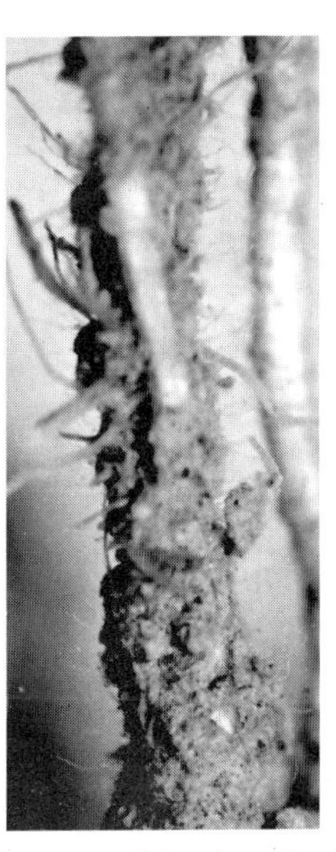

Fig. 15.11 *Rhizosphere sheath of rye-grass* Lolium perenne *(photo L. Marilley).*

The soil samples are generally sieved and after their moisture content is adjusted are incubated for a certain number of days. The aim is to take them to a state that permits comparison, independent of season, weather conditions at the moment of sampling... and root activity. This preconditioning is necessary if we are mainly seeking descriptors drawn from the microflora and enabling comparison of different soils. On the other hand, it results in soils in an artificial state and induces modifications in the microbial communities. The biological parameters measured are then probably only slightly related to the instantaneous reality of the soil.

15.5.2 Determination of the microbial biomass

What does biomass mean?

Biomass is ideally the total mass of living cells in a given area (Chap. 2, § 2.2.1). In a pure culture of bacteria grown in a liquid medium with no chemical precipitate, the cells may be separated by filtration or centrifugation and weighed after drying (dry weight).

Determination of the microbial biomass in the soil is faced with many difficulties.

Compared to a pure culture, direct observation *in situ* is unpredictable: in a soil, which is a mixture of inert particles and living cells, it is sometimes very difficult to distinguish between them, as they might mask each other.

Since culture methods are not applicable, we must develop other methods based on indirect or partial estimation of the biomass.

The method of fumigation by chloroform, which requires large samples, is laborious and gives data difficult to verify.

The method using fumigation by chloroform (Voroney and Paul, 1984; Vance et al., 1987) consists of lysing the biomass by treatment with chloroform. Then the sample is reinoculated with a small quantity of soil containing living bacteria, which will decompose the lysed organisms. After about ten days, the quantity of CO_2 or ammonia produced as a result of this decomposition is measured; this enables estimation of the biomass by using an empirical ratio between biomass and quantity of CO_2 or ammonia.

The most promising methods for estimation of biomass rely on analysis for a cell constituent. The constituent should ideally represent a definite fraction of the total biomass and quickly disappear after the cells die if only living cells are taken into consideration. These conditions are rarely fulfilled. However, the universal energy-rich compound of living cells, adenosine triphosphate (ATP), answers very well to these requirements: its concentration is strictly regulated in living cells and it disappears rapidly after they die (Maire, 1984). On the other hand, it does not enable distinction of the biomass of microbes from that of plant or animal cells.

A universal indicator of life, ATP.

Another indicator: murein, semi-universal compound exclusive to bacterial cell walls.

Murein, the almost universal cell-wall peptidoglycan of bacterial cells, has been used as an index of bacterial biomass. As the cell walls of Gram-positive bacteria contain 10 to 30 times more murein than those of Gram negative bacteria, this descriptor essentially takes the former into account. ***Poly-3-hydroxybutyric acid*** is not so universal and accumulates in cells only under certain conditions. Phospholipids and their fatty acids, on the

Poly-3-hydroxybutyric acid: storage polyester of many bacteria.

Gram-negative and Gram-positive...

Christian Gram, a collaborator of Robert Koch who discovered the causative agent of tuberculosis, gave his name to an empirically found coloration that has retained all its topicality, particularly in medical diagnostics, because it brings out an important difference in the structure of the bacterial wall. When a bacterial smear is successively treated with the dye crystal violet and iodine, a coloured complex is formed in the cell. In ***Gram-negative*** bacteria, washing the smear with alcohol removes this complex and decolorizes the bacteria, which does not happen with ***Gram-positive*** bacteria. This difference in response is due to the thickness of the cell wall, which is much greater in the case of the latter group.

other hand, are good indicators of biomass (Tunlid et al., 1985). Although they require a sample size of several or tens of grams, they represent when taken in their entirety a rather constant fraction of the biomass. Also, these fatty acids are often specific for a given taxonomic group, enabling study of the biodiversity of a sample (Laczkò, 1994; Maire et al., 1994). The disadvantage stems from the cost and the difficulty of this kind of analysis.

15.5.3 Measurement of microbial activity in the field and laboratory

The biomass: dormant or active?

While determination of the biomass defines the amount of living matter in a sample, it does not give information about the specific activity of this biomass or about the functions it fulfils in its environment. Herein lies the aim of activity measurements.

Where the concentration of a metabolic substance is followed.

One measurement of biological activity in a soil generally consists of following the change in concentration of a substrate involved in microbial metabolism. For example, heterotrophic respiratory activity is often estimated by measurement of production of CO_2 by a soil sample. But in this instance, as in many other cases, the descriptor applies to many biotic and abiotic phenomena (Fig. 15.12) and its value is the resultant of

Each determination expresses a budget.

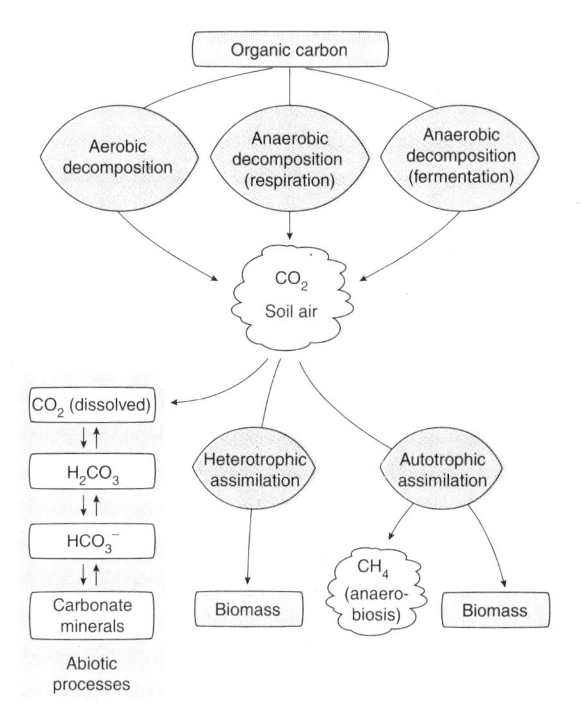

Fig. 15.12 *Carbon dioxide budget of the soil. Explanations in the text.*

these phenomena acting simultaneously. Carbon dioxide is not only a product of the aerobic catabolism of organic matter but also a by-product of numerous fermentation reactions. On the other hand, it is assimilated by autotrophic organisms and to a limited extent by heterotrophs. Release of CO_2 from carbonate minerals can occur by reaction with organic or mineral acids produced by bacterial metabolism or released by plants (Chap. 4, § 4.4.5). Also, it is related to the CO_2 in the soil solution and subject to carbonate equilibria.

Sometimes it is possible to directly and specifically measure an enzymatic reaction indicating an important biogeochemical phenomenon, for example fixation of molecular nitrogen (Chap. 4, § 4.4.3). The enzyme complex responsible, namely, nitrogenase, is able to reduce other compounds with triple bonds, in particular acetylene, which is reduced to ethylene (Fig. 15.13). This reaction is specific to nitrogenase and gas-chromatographic analysis of the ethylene formed is very sensitive.

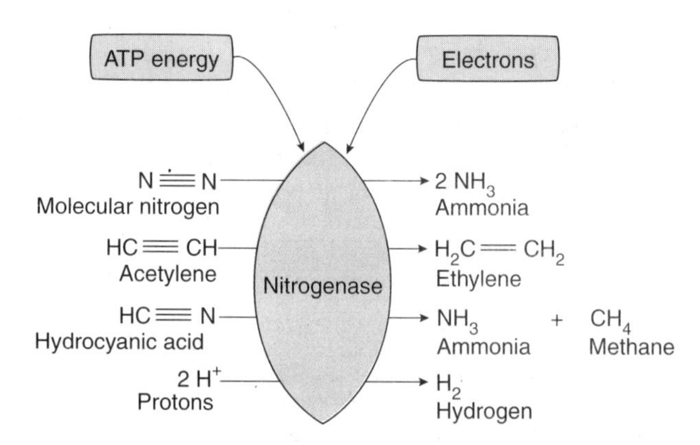

Fig. 15.13 *Reactions catalysed by the nitrogenase complex.*

Study of the change in isotopic ratios to highlight biological activity deserves to be developed.

Isotopes: atoms the nuclei of which contain the same number of protons but differ in number of neutrons.

Some techniques allow an approach a little closer to the reality of microenvironments such as the rhizosphere.

Another promising approach, which however requires considerable infrastructure, is based on ***isotopic selection*** accomplished by living systems: reactions involving compounds of an element occurring in the form of several stable ***isotopes*** (e.g. hydrogen, carbon, nitrogen, oxygen) are faster the lighter the isotope considered.

Unfortunately, most such measurements cannot be done in the field; they require laboratory conditions. It is thus very difficult to know if and to what extent the drawing of samples and their transport to the laboratory modify these activities (Chap. 17, § 17.1.3). Furthermore, measurements of activity often require a rather large sample volume. For certain physical and chemical descriptors, however, microelectrode systems have been developed for almost point measurements in the laboratory and even in the field (Revsbeck, in Stal and Caumette, 1994).

15.5.4 Culture methods

For observing a bacterium, recourse is most often had to experimentation, which involves application of the classic arsenal of culture methods (Pochon and Tardieux, 1962). This includes enrichment cultures, isolation of pure cultures, statistical estimation of the most probable number (MPN) from serial dilutions of the sample serving as inoculum (Fig. 15.14) and taxonomic and functional characterization of the isolated organisms. Molecular methods, while often requiring small samples, may also lead to new perspectives.

While it is quite difficult to study microorganisms in the soil, it is illusory, when observing a bacterium in the soil, to be aware of what it actually does or what it is capable of doing there!

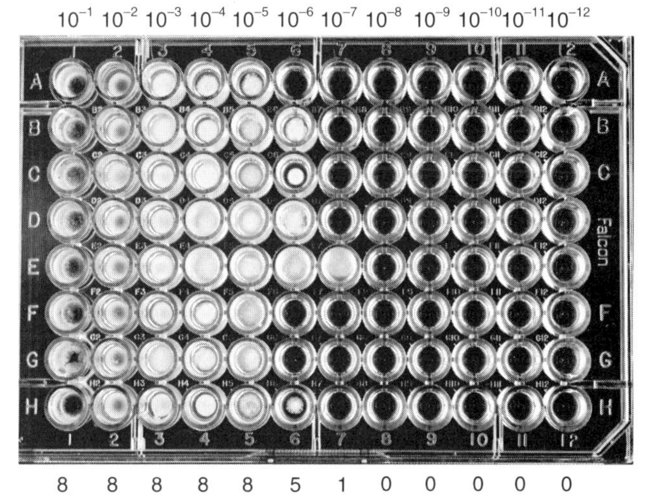

Fig. 15.14 *Determination of most probable number (MPN). Decimal dilutions of a soil sample (10^{-1} to 10^{-12}) served as inoculum (150 μL per cup) for liquid culture media on a 'Nunclon®' plate. For each dilution, 8 replicate cultures were done. The number of positive cultures in each dilution enabled statistical evaluation, using formulae, of the most probable number of microbes in the sample. In this case: 6.8×10^6 microbes per gram (photo N. Jeanneret and M. Aragno).*

The relative density of microorganisms in the rhizosphere compared to that in bulk soil is customarily expressed by the ***rhizosphere coefficient*** or the ***R:S ratio***. Expectedly, if the communities of the rhizosphere are richer than those of bulk soil in zymogenous and culturable organisms (Chap. 4, § 4.4.6), the R:S ratios obtained by culturing methods will inevitably be overestimates. Ratios between 10 and 100 have been reported. However, very rarely has a reliable method for determination of populations and sampling of rhizosphere

Overestimated R:S values?

soil been used; the values have usually been established by counting bacterial colonies on nutrient agar. Maire (1987) obtained much lower (and more realistic!) R:S ratios by comparing ATP contents of rhizosphere and bulk soil.

Elective medium: medium whose composition is adequate to ensure growth of one functional group of bacteria but does not include all the compounds necessary for growth of other groups. For example, a medium without a source of combined nitrogen will permit growth only of nitrogen-fixing bacteria.

Culture methods thus open only a small window on the microbial diversity in soils. However, by means of enrichment cultures on ***elective media***, they enable isolation and counting of specific functional groups of microorganisms (Pochon and Tardieux, 1962; Schlegel and Kröger, 1965). And pure culture is very often an indispensable prelude to experimental study of the behaviour of an organism with respect to its abiotic environment (autoecology, Chap. 4, § 4.2.4).

Culturing does not give an account of nature.

The serious bias in culture methods

Culture methods are vitiated by one serious bias. As seen in Chap. 4, § 4.4.6, only a fraction of soil microorganisms is cultivable on the usual media. Even the most sophisticated broths enable obtaining no more than a few per cent of the total microflora of a soil or a water. An aggravating factor is that organisms that grow best in the laboratory are zymogenic, scant in the soil but responding with 'explosive' growth to sudden nutrient-rich addition (see also Giller et al., 1997). There are several explanations for this phenomenon. Commonly, bacterial cells in soils adhere to form microcolonies (Plate VII-1). Consequently, one 'colony-forming unit' may represent a clump of several associated cells. Counting under the microscope often takes into account intact dead cells as well as living cells. Anyhow, numerous cells in the soil escape culturing by traditional methods (e.g. the classic 'nutrient broth'). This may have several causes:

'It has been estimated that current methods for isolation of bacteria retrieve only 1% of those present in soil (Giller et al., 1997).

- composition of the medium or inappropriate external conditions (e.g. for obligate autotrophs or strict anaerobes).
- inhibitory concentration of nutrients.
- inability to form colonies on agar, or to grow in submerged culture.
- lack of resources, in stationary phase cells, to protect themselves for example against toxic compounds resulting from the partial reduction of oxygen (peroxides, superoxide radical, see Chap. 4, § 4.4.4 oxygen cycle). It has been shown that higher counts are obtained on catalase-containing nutrient media.
- competition by neighbouring colonies or inhibition (e.g. by antibiotics).
- viable but not cultivable cells (VBNC):
  - intrinsically VBNC (e.g. the *Holophaga-Acidobacterium* group) common in soils (molecular clones) and never cultivated.
  - cells of a normally cultivable organism that have entered a dormant state (endospores, VBNC).

'Uncultivable' or 'not yet cultivated'.

Since the nineteen nineties, molecular methods (see below, § 15.5.5) are available that permit identification of organisms in a community avoiding the step of isolation and pure culture. These methods have thrown new light on the 'uncultivable' microbes (Head et al., 1998).

15.5.5 Molecular methods applied in soil ecology

The biology of organisms and their relationships have long been opposed to that related to the study of structures and functions of cells and molecules. Such confrontation is futile and sterile. Biology is one in its diversity, from the molecule to the ecosystem.

All scales combine to elucidate the phenomenon of Life in its entirety.

Development of molecular techniques, in particular those pertaining to characterization of the genome, has given ecologists an instrument that broadens their field of investigation in a manner that would not have been thought possible fifteen years back. This is why we consider it useful to give a broad review of them here. These methods are essentially aimed at identifying the microorganisms, at times even *in situ*, and estimating their genetic and functional diversity. Amann et al. (1995) have critically reviewed them in detail.

'Currently, application of molecular-biology methods is revolutionizing our understanding of the evolutionary relationships between bacteria. (...) molecular-biology methods can be used to analyse diversity in DNA extracted directly from soil and which thus examine diversity across the whole microbial community' (Giller et al., 1997).

Having presented some general principles of molecular biology necessary for understanding these methods (see box below), we shall give some applied examples in soil ecology. Because of the now widely popular method for DNA replication, the polymerase chain reaction (PCR), prior isolation of pure cultures is no longer indispensable in the study of biodiversity.

Recapitulation of molecular biology

The ***genome*** is the heritage of information transmitted by every organism to its descendants. It is composed of chains of a ***nucleic acid***, DNA, the organic-base sequences of which contain many genetic messages **(genes)**, most of them coding the sequence of amino acids of proteins. When synthesis of a protein is required, the related sequence in the DNA is 'photocopied' (transcribed) into a ***messenger ribonucleic acid*** or ***mRNA***. This message is then 'translated' into a specific amino-acid sequence of a protein chain, at the level of a ***ribosome***, making use of the ***genetic code***.

Homologous genes have a common origin and determine similar properties in different organisms. They have identical base sequences in identical organisms. The sequences in two organisms differ more the further apart they are in evolution. The difference between the sequences in two homologous genes may thus be a measure of the evolutionary distance between the organisms carrying them.

In the identification and study of bacterial biodiversity, the genes that code for ***ribosome*** ribonucleic acids ***(rRNA)*** are often made use of, particularly the gene that codes for the small subunit rRNA (16SrRNA). This gene is a segment formed of a sequence of some 1650 bases. But the variability of these sequences changes along the segment. For example, Plate VIII-2 compares a homologous sector of the gene coding for 16S-rRNA (18S for eukaryotes) in three eukaryotes (Eucarya), three bacteria (Bacteria) and three Archaea.

Nucleic acid: macromolecule made up of a double helical chain of molecular units, the ***nucleotides***. The latter are composed of a pentose sugar (ribose in ribonucleic acid ***RNA*** and deoxyribose in deoxyribonucleic acid ***DNA***), phosphoric acid and an organic base bound to the pentose. The organic bases are purines (adenine and guanine) and pyrimidines (cytosine and thymine in DNA or cytosine and uracil in RNA). In double-strand DNA, a purine always faces a pyrimidine, adenine a thymine, and guanine a cytosine; these are complementary bases.

Genetic code: a sequence of three bases ***(codon)*** in the RNA chain determines the insertion of a particular amino acid in the protein chain. For example, the codon CUG (cytosine-uracil-guanine) corresponds to leucine. Some codons (e.g. UAG) do not determine any amino acid but provoke interruption of the protein chain (stop codons).

Ribosomes and ***rRNA:*** ribosomes are cellular organelles at whose level protein chains are manufactured. They are highly precise molecular instruments composed of proteins and ribonucleic acids, rRNA. The latter are copies of genes that do not code for proteins. They themselves have a structural and catalytic function in the ribosome. They can be considered as ribozymes (Chap. 14, § 14.1)

Signature: In a DNA segment, sequence of bases (a few tens usually) exclusive for a given taxonomic group.

DNA-polymerase: enzyme catalysing the formation of a DNA chain from a complementary chain and a primer, an original segment formed of a short sequence of nucleotides.

Primer: small segment of DNA reproducing the complementary sequence of a single-chain DNA chain from which DNA-polymerase starts reconstituting the complementary chain (Fig. 15.15).

Replication: synthesis of a DNA chain starting from a complementary chain serving as template. Separation of the two strands of a double helix

Near both extremities of the gene, there are well-conserved sectors, identical in all the bacteria. Elsewhere in this gene, sectors less well conserved in evolution can be identified, identical only in bacteria of a given species, genus or phylum (Chap. 12, § 12.3). The sequence of bases of one such sector is then a ***signature*** of the group considered. The group may be located at different categorical levels in the taxonomic hierarchy (species, genus, order, etc.)

Extraction and purification of bacterial DNA from soil

To achieve replication of a DNA segment by PCR, it is imperative that this segment not be cut: the DNA should therefore be as intact as possible, a requirement that prohibits application of excessively vigorous extraction treatments. But bacterial DNA is contained in the cytoplasm, itself surrounded by a solid wall that must be ruptured for extraction of the contents. Some bacteria have a particularly resistant wall and their DNA might be scarcely extracted or not at all. This is one of the major obstacles to the molecular approach.

The DNA of 'non-culturable' bacteria as well as of organisms culturable on the usual media can be isolated. Although limited by the extractability of the molecule, the window opened on microbial diversity through soil DNA is therefore very different from and probably wider than that defined by the 'culturability' of organisms. Furthermore, it is not necessary to have large samples, which is a great advantage in the study of the rhizosphere.

The PCR Method and its Applications

Put briefly, the ***PCR method*** consists of multiplying a segment of DNA defined by specific sequences of 15-25 bases situated at each of its extremities. It makes use of ***primers***, oligonucleotides complementary to these sequences, and of an enzyme, ***DNA polymerase***.

The PCR method (Fig. 15.15)

PCR consists in the repetition of a cycle of three stages taking place at different temperatures controlled by an automated system:

- 1 Denaturation, i.e. separation of the two DNA strands at high temperature (e.g. at 94°C).
- 2 Annealing (hybridization) of primers at the 3′ end of both strands of the DNA segment to be amplified. This occurs at relatively low temperature (e.g. 62°C).
- 3 ***Replication***, under the control of a temperature-resistant DNA polymerase. It begins at the level of each primer in the 5′ → 3′ direction at intermediate temperature (e.g. 72°C).

The quantity of the amplified DNA is doubled with each three-stage cycle.

The original segment is replicated up to a million times after twenty or more cycles.

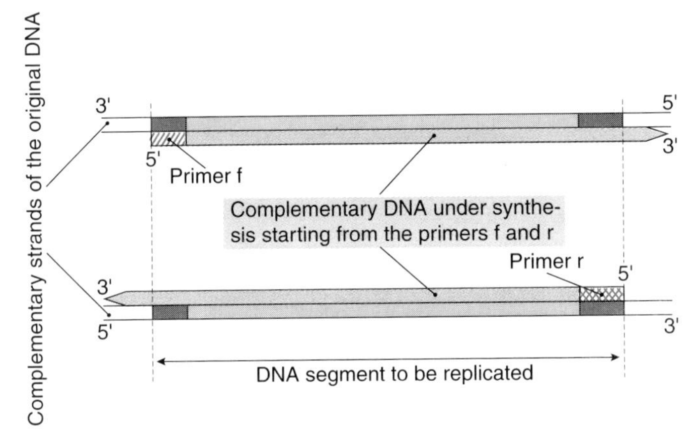

Fig. 15.15 Principle of PCR. Replication of the complementary strands of the original DNA molecule, starting from a pair of primers—f (forward) and r (reverse)—complementary to the specific DNA sequences at the ends of the segment to be replicated.

By choosing for example two primers whose nucleotide sequences correspond to identical regions for all Bacteria and located near each end of the gene coding for the middle ribosomal RNA (Fig. 15.16), a segment representing most part of this gene in all the Bacteria DNA present is replicated. The genes corresponding to the DNÁ of other divisions of organisms (Eucarya and Archaea) on the other hand are not multiplied, because the primers specific to Bacteria do not find their complementary sequences in them.

allows reciprocal replication of each of them, and thus reconstruction of two double helices.

UNI16s-L: 5'-ATTCTAGAGTTTGATCATGGCTCA
UNI16s-R: 5'-ATGGTACCGTGTGACGGGCGGTGTGTA
Bases: A: adenine; T: thymine; G: guanine; C: cytosine

Fig. 15.16 Pair of 'universal' primers used for replication by PCR of most part of the gene coding for ribosomal 16S-RNA in all bacteria.

Other genes can be amplified in a similar way. Ribosomal genes chiefly enable characterization of taxonomic levels above species, while distinction at species and lower levels requires greater variability in the genomic sequences. Such variability is found, for example, in the intergene spacer segment (ITS1) separating the genes 16S and 23S coding for ribosomal RNA, in the ribosomal operon. It comprises genes coding for transfer RNAs, and also non-coding segments. These latter, not being subject to a high selection pressure, have undergone much more evolutionary changes than ribosomal genes. Therefore, ITS1, which differs in length as well as in nucleotide sequences, will be addressed when ribosomal genes show too high a similarity for establishing valid comparisons.

Genes evolve faster when they are not subject to pressures.

Besides, ITS1 can be amplified by means of primers in the well-conserved sequences located in their vicinity, one in the 16S gene and the other in the 23S, so as to get 'universal' amplification of bacterial ITS1 (Fig. 15.17).

Fig. 15.17 Principle of PCR amplification of the intergene between the ribosomal 16S and 23S genes,, by using the conserved primers located close to the neighbouring ends of these ribosomal genes. The different lengths of these intergenes in different bacterial groups enables us to obtain the RISA community profiles (see text box).

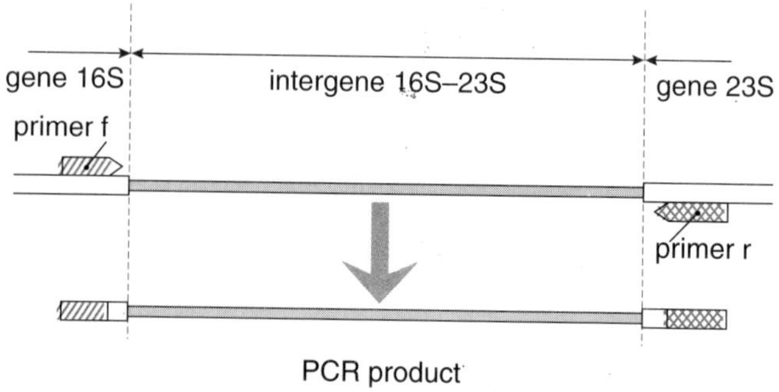

Genes coding for functions can similarly be amplified by PCR, provided they be homologous and have sufficiently similar nucleotide sequences in the different species so much so they are 'recognized' by the common primers. This is the case with nitrogen fixation (Chap. 16, § 16.2.1): the sequence of the *nifH* gene, which codes for a component of the nitrogenase complex, is very well conserved in all nitrogen-fixing bacteria. This allows 'universal' amplification, starting from a sample of DNA extracted from the environment.

In other words, molecular methods enable detection of important populations of nitrogen-fixing bacteria, for example organisms associated with roots, which would have escaped notice in investigations by culture methods.

Molecular profiles of microbial communities

A community of DNA segments representative of the community of bacteria.

After PCR amplification of a given segment from the total DNA extracted from a sample from the environment (e.g. soil), a mixture is obtained of homologous segments representative of all or part of the microbial community in the sample (Plate VIII-1). These segments may then be separated by various techniques of gel ***electrophoresis*** to give a 'profile' characteristic of the community (Fig. 15.18).

Electrophoresis: separation in an electrical field (most often on a gel) of molecules having different electrical charge. If the separated molecules have equivalent charge, they are separated by size, the smallest moving faster than the largest.

Examples of methods of obtaining profiles of microbial communities from DNA PCR amplification products isolated from the environment

- RISA (Ribosome Intergene Spacer Analysis). This is a conventional electrophoresis, based on the differences in length of the ITS1 segments amplified by PCR.
- DGE (Denaturing Gradient Electrophoresis). This approach is based on differences in electrophoretic mobility between double-strand DNA and single-strand ('denatured') DNA. This enables separation of fragments of same length such as the products of

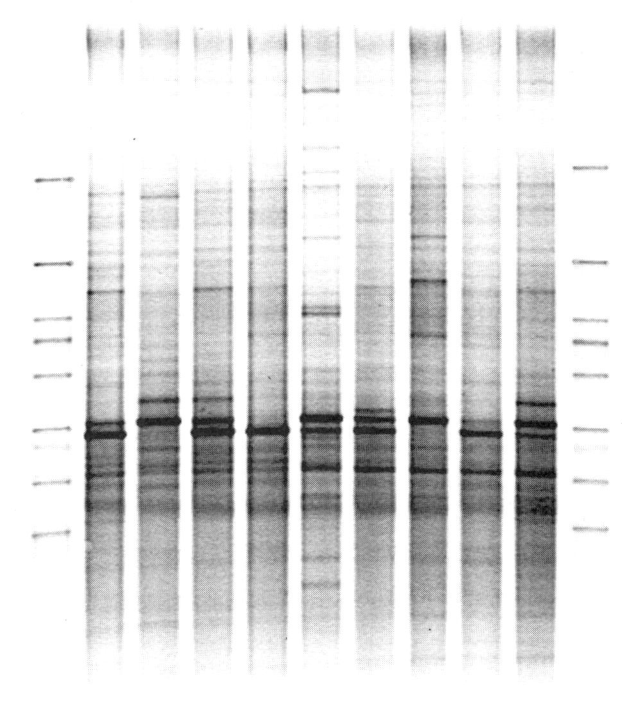

Fig. 15.18 *Profiles of bacterial communities of the rhizosphere of beech* Fagus sylvatica *obtained by DGGE starting from the PCR V3 fragment (a relatively variable fragment) of 16S rDNA. The 'shafts' on the left and the right contain the fragments from a collection of reference bacteria, to give the scale (photo A. Slijepcevic).*

PCR amplification of ribosomal genes, which differ in the contents of the bases guanine and cytosine (% G+C). Three variants exist:

- DGGE, denaturing gradient gel electrophoresis, in which migration takes place along a gradient of chemical denaturating agent;
- TGGE, thermal gradient gel electrophoresis, in which a thermal gradient is applied to the gel;
- TTGE, temporal thermal gradient electrophoresis, in which a slow, linear increase in temperature is applied to the entire gel during the migration.

• SSCP (single strand conformation polymorphism). The PCR fragments, after separation from the DNA strands, are arranged in a three-dimensional (tertiary) structure characteristic of the nucleotide sequences they carry. Different structures show different electrophoretic migration speeds.

• T-RFLP (terminal restriction fragment length polymorphism): End-labelling (e.g. with a fluerescent dye) of PCR product followed by hydrolysis with a restriction endonuclease. The fragments will then be separated according to their length by a conventional electrophoresis, and the labelled end fragments recognised by their fluoresence.

Where 'numerical ecology' catches up with 'molecular ecology'.

Profiles obtained on a gel can be compared by numerical methods so that their degree of similarity can be calculated. Based on the position and intensity of bands, it is also possible to process this information to get indices relating to microbial biodiversity. It must however be kept in mind that an individual band may comprise DNA fragments from different species, with the same electrophoretic characteristics. Besides, subdominant populations (less than 1% of the total community for example) do not give visible bands.

Plasmid: loop or segment of DNA independent of the bacterial nucleus and able to code for some properties of the bacterium (e.g., resistance to antibiotics, ability to utilize certain nutritive substrates). The plasmid can also easily penetrate a bacterial cell or leave it. Segments of foreign DNA can be incorporated in a plasmid.

Clone: bacterial line having a carrier plasmid of a particular DNA segment. Bacterial multiplication leads to that of the plasmid and thus of the incorporated DNA segment.

Certain bands may be cut out and studied in greater depth, for example by cloning and sequencing. This enables us find out if they comprise DNA from one or several populations and to establish their phylogenetic position and their taxonomic affiliations, or in some cases even identify the original organisms.

Cloning and characterization of fragments obtained by PCR

To characterize and attempt to identify individual PCR fragments it is necessary to amplify them individually; for this they have to be cloned, that is, introduced into bacterial cells, generally of *Escherichia coli*, after incorporating the segments in specific ***plasmids*** of this bacterium (Plate VIII-1). Each of the ***clones*** thus obtained reproduces the sequence of the amplified gene of one of the members of the bacterial community of the soil sample. A few of these clones are selected, from which the segment in question is extracted for characterization: This characterization has two major aims:

Two aims in characterizing the individual DNA segments:

- Characteristic profiles can be obtained after separation by electrophoresis of fragments given by restriction enzymes that split the DNA of PCR segments at specific sites. Comparison of these restriction profiles between the different clones and their distribution into ***operational taxonomic units*** (OTUs) provide a picture of the bacterial diversity in the original sample.
- The complete base sequence in the PCR segment under consideration can be established. By comparing such segments with known bases comprising all known homologous sequences, it is possible to place the organism that gave this segment in the global phylogeny of bacteria, and even to identify it if an identical or very similar sequence has been described in a known organism.

Operational taxonomic unit (OTU): ensemble of clones or strains exhibiting a number of identical characteristics without reference to a particular taxonomic level (genus, species...); here ensemble of clones with the same restriction profiles for several different enzymes.

Figure 15.19 shows the frequency distribution of operational taxonomic units (OTUs) obtained after fractionation of the rhizosphere of rye-grass, *Lolium perenne*. Diversity is higher in the bulk soil with one, at the most two, clones for each OTU in about 30 selected at random; some dominant OTUs appear in the rhizosphere soil, a trend still more marked in the endorhizosphere. To date no 'classic' method has provided a comparable picture.

These methods are seen to be particularly promising in the study of biodiversity in the rhizosphere.

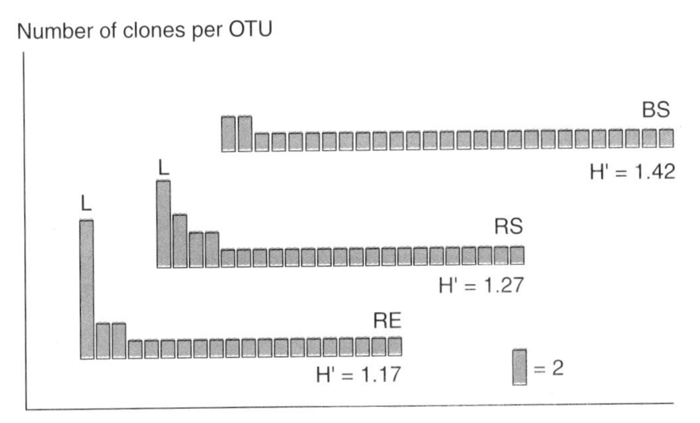

***Fig. 15.19** Bacterial biodiversity in three fractions of the rhizosphere of rye-grass,* Lolium perenne. *Distribution into **operational taxonomic units** (OTUs) of the bacterial clones obtained from each fraction. L, the dominant OTU in the rhizosphere, has been identified by sequencing as Pseudomonas fluorescens. BS: bulk soil; RS: soil of the sheath in rhizosphere soil; RE: rhizoplane and endorhizosphere; H´: Shannon-Wiener diversity index (Chap. 7, § 7.5.2) (from Marilley et al., 1998).*

Fluorescent-molecular probes: direct observation of biodiversity

Direct microscopic observation of a soil or root sample usually gives only very little information on the microflora, which is often confused with inert particles. Certain fluorescent dyes (DAPI, acridine orange) enable distinction of living cells from inert particles. But no method is available for identifying the organisms seen, as cell shape provides only very limited information (Chap. 2, § 2.5.2).

We have seen that the degree of conservation of ribosomal-RNA sequences varies along the molecule and that it is possible to determine regions whose sequences are identical within a given taxonomic group and exclusive to this group (signatures). Complementary DNA segments in the regions considered ***(molecular probes)*** are bound (hybridized) to these regions only in organisms of the group in question.

Ribosomes are very numerous in a bacterial cell. If the selected sequence is located at the periphery of the ribosome and the cell envelope is made permeable, it is possible to hybridize these probes on the ribosome *in situ* in the cells. Moreover, a fluorescent molecule with a given colour may be chemically bound to a given probe.

When such a fluorescent probe is introduced into ***permeabilized*** bacterial cells, those that belong to the taxonomic group considered fix the probe on their ribosomes. These appear coloured when examined under the ***epifluorescence microscope***. By simultaneously using different probes conjugated to different stains, the bacteria belonging to the taxonomic groups considered can be distinguished in the same preparation by their colour (Plate VIII-3). This recently developed technique is promising in the study of bacteria in natural environments, in particular the rhizosphere.

The fully developed technique of fluorescent molecular probes enables identification *in situ* of the taxonomic affiliation of the groups studied.

Permeabilized (cell): cell whose semi-permeable membrane has been destroyed and the wall weakened so that molecular probes are let pass through, without dispersing the cell contents. The cell is of course killed by the treatment.

Epifluorescence microscope: microscope provided with ultraviolet illumination from above the preparation. Fluorescent objects thus emit coloured light on a black background.

Towards a more functional study, the RT-PCR method

To identify a function where it is actually implemented.

Identification of a microorganism does not necessarily provide information on the functions it is able to fulfil (Chap. 17, § 17.1.2). Even if this were true, it will not be known if the function considered is expressed at the instant of sampling. For example, an organism that can fix nitrogen (when conditions permit) does not necessarily fulfil that function where, or when, it is found. The ***RT-PCR method*** opens up a functional window in the application of molecular methods to ecology.

Control of synthesis of an enzyme is done essentially (but not exclusively) during ***transcription***, that is, during the copying of the corresponding gene in the form of a messenger RNA. Thus, the presence of a specific messenger RNA is a strong indication of ***induction*** of the concerned gene and the presence of the corresponding activity. An enzyme isolated from an RNA virus and termed ***reverse transcriptase*** or ***RT*** enables synthesis of a DNA chain ***(cDNA)*** complementary to a given RNA, in this case to a messenger RNA. If primers characteristic of a segment of this complementary DNA are known, this segment can then be specifically replicated by PCR. Obtaining such a replicated product, since it involves the presence of messenger DNA, indicates induction of the corresponding enzyme and hence of the function considered.

Induction (of a gene)***:*** initiation of transcription of a gene or group of genes, usually under the effect of an environmental stimulus.

How to distinguish the workers from the idlers?

A variant of this approach involves amplification of ribosomal RNAs by RT-PCR. Indeed the number of ribosomes in a cell depends greatly on its present activity. Therefore, the comparison of community profiles based on the genes coding for the ribosomal RNA (rDNA, number constant per cell) and on the ribosomal RNAs themselves replicated by RT-PCR (number variable in accordance with activity) gives an indication of the present activity of a population (Fig. 15.20). A small peak in an rDNA profile and a large peak in the corresponding rRNA profile means that the population considered is very active, even if it does not dominate in number. On the other hand, a high peak in the rDNA profile and a low one in the rRNA profile signifies that the population studied, while large in biomass, is only barely active or inactive at the moment of sampling.

We have presented some examples of recent applications of techniques of molecular biology to the study of ecology and microbial diversity. This is a domain in full growth, which also has started to be applied to the *Eucarya* of soil. Out of choice we have not developed this aspect here and refer the reader to Bonhomme et al. in Legay and Barbaut (1995), Redecker (2000) or again to Viaud et al. (2000).

For more in-depth information, the following may be consulted: Campbell and Mathieu (1995); Campbell et al., 1999; Widmer and Beffa (2000).

This molecular approach to soil ecology is not a new science, but rather an instrument that enables opening of

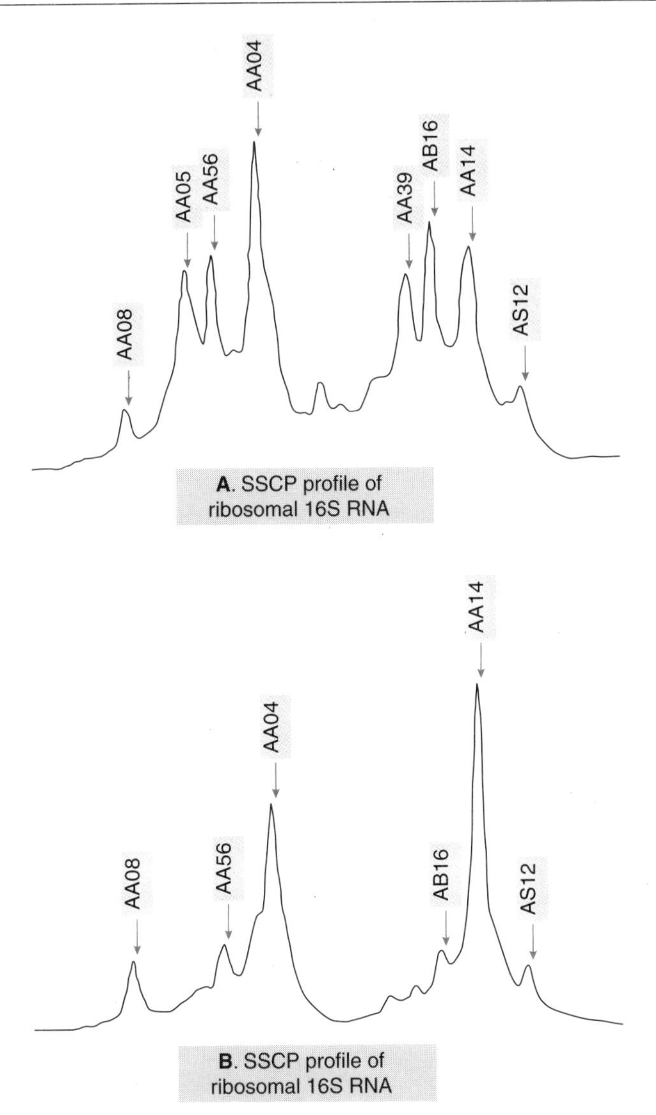

Fig. 15.20 *Comparison of community profiles from a single sample, using SSCP profiles obtained from (above) 16S-DNA PCR products and (below) 16S-RNA RT-PCR products (from Delbès et al., 2002).*

new windows in the understanding of environments difficult to study, such as the rhizosphere. But the 'old' windows should also remain open!

CHAPTER 16

SOIL MUTUALISTIC SYMBIOSES

Here we gain a glimpse into the genetic and molecular mechanisms of establishment and functioning of the most important soil symbioses. Other interactions are evoked in Chapter 13, concerning food chains, and in Chapter 15 on the rhizosphere.

The framework of this book does not allow an exhaustive presentation of all the types of interactions among soil organisms. They are briefly mentioned in the box below, while their effects on the partners involved are summarized in Table 16.1. We are going to concentrate on mutualistic symbioses, certainly the most perfect type of interactions among living organisms. They are also the most intimate among them, along with the true biotrophic parasitoses (parasitic symbioses). This intimacy is revealed more and more with progress in researches at cellular and molecular levels.

Table 16.1 Effect of different kinds of interaction on two partner organisms A and B, A being the more advantaged in this case.

Interaction	Organism A	Organism B
Mutualism (symbiosis, syntrophy)	+	+
Neutralism	0	0
Competition	–	– –
Commensalism	+	0
Parasitism	+	–
Predation	+	– –

This chapter concludes the second part of this book devoted to the evolution of organic matter, by describing the closest links that soil organisms weave among themselves—symbioses. By symbiosis, substances manufactured or transformed in one living organism directly serve another.

Why should a little anthropomorphism not be introduced in the account of interactions among soil organisms, if only to render the discourse less dry?

Mutualism (or ***association for mutual benefit):*** relationship in which each partner derives benefit from the presence and activity of the other partners involved in the interaction. It comprises symbioses and syntrophies.

Commensalism: Relationship favourable to one of the partners who profits from the other without harming it but also without giving it anything in return.

Neutralism, mutual indifference.

Competition, battle for the ecological niche (Chap. 4, § 4.5.3).

Parasitism, unilateral exploitation.

The term ***symbiosis*** is often understood as an altogether physical and trophic mutualistic interaction. Indeed the etymology of 'symbiosis' implies that both partners 'live together', without necessarily a mutual benefit. Biotrophic parasitism, in which the parasite keeps its host alive while diverting the host's function to its unilateral profit, may therefore be considered as a 'parasitic symbiosis'.

Interactions between organisms

Let us compare the soil to a theatre stage on which the characters will be populations of living organisms moving and interacting in a set made up of mineral particles. Between these characters are revealed love, hatred, implacable rivalry, servitude, mendicancy, indifference, theft, crime... as well as trade, management, contract rules. Let us deliberately make this comparison with a wink (but the wrong way round!) like that of Jean de la Fontaine who, to stigmatize human feelings, made use of animals!

Love, friendship, very close relationships, meeting of interests, all these are found in ***mutualism***. Two cases should be distinguished:

- ***Symbiosis***, in which physical union of the partners prevails; this is the absolute bond. The protagonists ***(symbionts)***, in very close contact, exchange nutrients or growth factors exclusively, provided the latter do not appear in the external milieu. They have at their disposal interfaces of physical exchanges [Fig. 16.1(a)].
- ***Syntrophy***, where friendship and neighbourly terms are expressed. Each partner is physically independent, the exchanges being conducted through the milieu as intermediary [Fig. 16.1(b)]. Thus the exchanged substances may also benefit other organisms maintaining a relationship of ***commensalism*** or competition with the members of the syntrophic association. Some associations of this kind are stable, for example the methanogenic syntrophy, typical of anoxic environments (Chap. 4, § 4.4.2).

In ***neutralism***, the organisms involved have no effect on each other, they eat different dishes at different tables, without quarrelling over the available space or getting in each other's way by the substances they secrete or modifications they have brought about in their environment. It is in reality difficult to imagine a perfect neutralism in natural environments subject to manifold nutrient limitations and, very often, to confinement which results in accumulation of metabolites.

Competition is rivalry between two partners. Whatever supporters of boundless ultra-liberalism might say, competition exerts a negative influence on all the shareholders involved. Only when they share the same ecological niche will one eliminate the other... and competition halted. But the winner would have derived no advantage from the momentary presence of the other! If, on the other hand, the two competitors occupy different ecological niches, they must share the cake (nutrition, room) which they are competing for (Chap. 12, § 12.5.1).

Parasitism is the exploitation of one organism, euphemistically termed the ***host***, by another, the ***parasite***, which lives by diverting to its benefit all or part of the activities of the former. In parasitism, the relationship is positive for the parasite and negative for its host. But wait! To be a parasite is an art (Combes, 2001). Just like the woodcutter should not cut off the branch on which he is sitting, the true parasite (qualified as ***biotrophic*** or ***parasitic symbiont***) should not kill the host that provides it with food and tableware. This takes place in some cases when the parasite, then called ***necrotrophic***, continues its growth on the dead host, for example mushrooms parasitic on wood (Chap. 8, § 8.6.3).

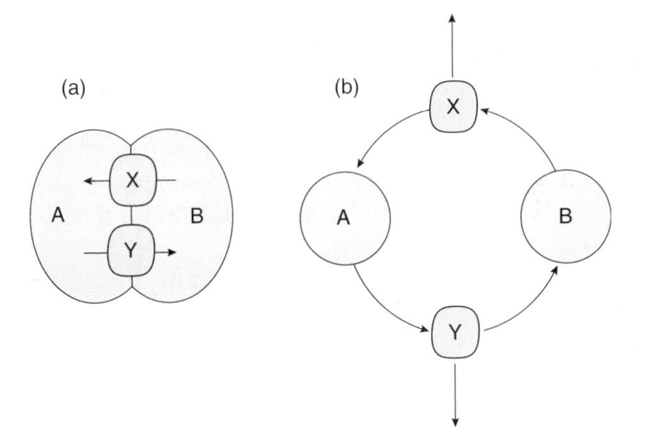

Fig. 16.1 Interactions of mutual benefit between two organisms: (a) symbiosis; (b) simple syntrophy. In these two examples, organism A produces a factor Y indispensable for B, while organism B generates a factor X indispensable for A. The factor may be nutritional or environmental. In the case of symbiosis, X and Y are exchanged exclusively between the two partners; in syntrophy, X and Y appear in the milieu and may influence other commensal organisms.

The killer, the 'assassin', is the ***predator***. But it has the excuse of killing for it to live, to feed itself and then to digest its prey. The assimilation completed, it very often gets back to the hunt.

From this theatrical and anthropomorphic presentation of interactions among organisms, one must not derive a completely black-or-white picture of the personalities involved. It may happen that a normally biotrophic parasite kills its host in specific circumstances. Symbioses are not always the grand passion they seem to be, and a change in external conditions may result in divorce of the partners; sometimes one may even turn against the other! But at bottom, isn't that the way with human relationships?

Predation, 'final' interaction for the one taken (Chap. 4, § 4.5.3).

The great majority of mutualistic symbioses found in soil take place between a photosynthetic organism (bacterium, alga, plant) and a chemoheterotrophic one (Chap. 4, § 4.4.2). The former functions as energy relay between light and the latter, providing it with energy converted to organic carbon compounds, part of which is assimilated. The same link exists with plant pathogens. In mutualistic symbioses, the reciprocal contribution of the heterotrophic organism to the autotrophic organism is more varied and sometimes poorly understood. It is even difficult in some cases to know if we are truly dealing with mutualism. The very same heterotrophic partner (fungus, bacterium) may behave as a mutualist or as a parasite depending on the conditions and degree of compatibility of the two partners.

Diffuse boundary between mutualism and parasitism.

The two principal symbioses of this type found in soil are the mycorrhizal symbiosis between a fungus and a plant (§ 16.1) and the nitrogen-fixing symbioses between a bacterium and a plant (§ 16.2). Through interaction with the partner microorganism, they form specialized organs in which the exchanges characteristic of the

Two broad types of symbiosis: mycorrhizal symbioses and nitrogen-fixing symbioses.

symbiosis take place: mycorrhizae in the former case and nodules in the latter. In both cases the association is almost always established at the level of the roots. In the rhizosphere (including the rhizoplane and the endorhizosphere) are found associations of mutual benefit with nitrogen-fixing bacteria without formation of special organs. This is the case of associative fixation (Chap. 15, § 15.3.3).

A lichen is also a symbiosis!

Other mutualistic symbioses, less closely linked to the soil, also result from the union of one photosynthetic organism with another, chemoheterotrophic organism. We may mention here the well-known union in lichens between a fungus and an alga and in cyanolichens between a fungus and a cyanobacterium.

When the animal benefits from the digestive action of microscopic organisms that live within it.

Another type of symbiosis is found in soil, the digestive symbiosis: often incapable of decomposing certain substances, biopolymers in particular, just by the enzymes it secretes in its digestive tract (Chap. 8, § 8.7), an animal can nourish itself on these substances through the intervention of the microorganisms of its digestive flora (bacteria, protozoans) capable of such decomposition. This type of symbiosis is not a true soil symbiosis; it is also found in ruminants as well as in some protozoans!

The two nested symbioses in termites!

We will here mention only the specific interesting case of xylophagous termites. These insects would be incapable of digesting wood did they not harbour in their digestive tract symbiotic cellulase-producing protozoans (e.g. *Trichonympha*). Very primitive (they contain no mitochondria), these protozoans themselves contain symbiotic methanogenic bacteria. This dual symbiosis achieves the essential phases of methanogenic syntrophy at the scale of the digestive tract of the termite (Chap. 4, Fig. 4.20; Chap. 13, § 13.6.4).

16.1 MYCORRHIZAL SYMBIOSES

16.1.1 Mycorrhizae, preferred partners of plants

Fungi well suited to mycorrhization

Study of mycorrhizae: one of the most promising fields of research in soil biology.

Mycorrhizae constitute essential partners in the plant-soil relationship, particularly in natural environments. The roots of nearly 80 per cent of species of vascular plants harbour or are liable to harbour non-pathogenic fungal partners. Their function is primordial in the entire plant cycle or part of it, chiefly but not exclusively for nutrition. Molecular methods (Chap. 15, § 15.5.5), supplementing conventional approaches, will enable a deeper investigation of these symbioses whose importance has been widely underestimated by the scientific community. A recent book (Smith and Read, 1997) gives an excellent overview of the present status of research and its shortcomings and perspectives.

Four important functional regions: interface with the plant, nutritive zones, rhizomorphs, fruit bodies.

Fungi are particularly well suited to functions linked to mycorrhizal symbiosis. They exhibit four functional 'regions' (Fig. 16.2; also see Plate VII-2):

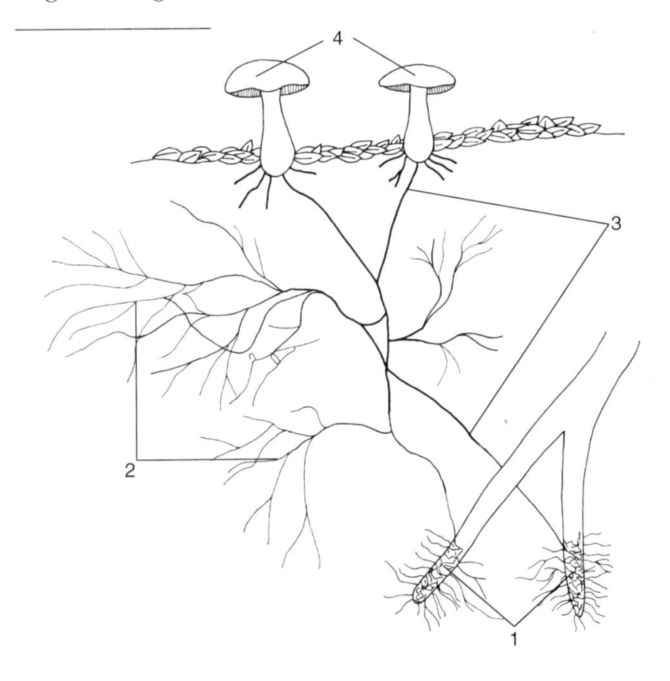

Fig. 16.2 *Functional regions of a mycorrhizal fungus. Explanations in the text.*

- Interface with the plant (1), located around the root, in its tissues or even in the cortical cells, and constituting the actual mycorrhiza.

- Region responsible for nourishing the fungus (2), formed of fine, diffuse hyphae. These hyphae are much less hydrophobic than the rest of the mycelium; their extension into the soil is termed the mycorrhizosphere (Chap. 4, § 4.1.3; Chap. 15, § 15.3.4).

- Region responsible for translocation of water and nutrients (3), which is often distinguished from the preceding by the formation of ***rhizomorphs***, mycelial filaments grouped into bundles. Unlike nutritional hyphae, rhizomorphs are hydrophobic, their impermeability enabling them to function as conduits for water and dissolved substances.

• Fruit bodies (4) or carpophores. In the case of macromycetes, they are large; they are morphologically differentiated and bear the organs in which take place the essential sexual phenomena generating spores (ascospores, basidiospores) responsible for multiplication, dispersal and/or preservation of the fungus under adverse conditions. In the fungi of arbuscular mycorrhizae, vegetative spores are directly formed from the mycelium, without fruit bodies. The fact that fruit bodies of a mushroom appear exclusively close to one particular species or genus of host plant does not necessarily imply that it is mycorrhizal. Often such specificity exists in the relationship of a saprophytic fungus with one type of litter or wood.

Often specific relationships.

Mycorrhizal symbiosis, a living contract

Pass me the sugar and I give you the salt!

Except for the orchid mycorrhizae, the 'symbiosis contract' depends largely on the transfer to the fungus of part of the organic and energy nutrients manufactured by the plant during photosynthesis. In return, the fungus transfers to the plant part of the inorganic salts, above all the phosphorus and nitrogen it has absorbed and translocated. We shall see later that this basic contract may show variations and that trophic exchange, a point common to most mycorrhizae, is not always the most important.

SYMBIOSIS CONTRACT

Between: Mr Fungus from one side, Mrs Plant from the other.
'Mrs Plant is engaged in manufacturing sugars by utilizing atmospheric carbon dioxide and light energy. She provides Mr Fungus with part of the sugars to give him the carbon and energy he needs'.
'Mr Fungus is engaged in return in providing Mrs Plant with water and inorganic salts he has drawn from the soil far from her roots'.
Executed at the office of Mr Mycorrhiza, solicitor, with the consent of the interested parties.

Is the mycorrhizal symbiosis unilateral?

A priori, the advantage of the association is more evident for the fungus than for the plant. By its roots the latter can in principle nourish itself with water and inorganic salts from the soil solution. Thus symbiosis does not appear to be absolutely necessary. However:

• in many natural soils, the available-nutrient reserve is low or scarcely accessible (Chap. 4, §§ 4.2.2, 4.3.1);
• extension of the mycorrhizal fungus into the soil is much greater than that of the root (Plate VII-2); the

absorbing area is thus greatly augmented and so is the soil volume from which inorganic salts are drawn;
- fungi have much more efficient mechanisms for absorption and concentration of mineral nutrients (active transport) than roots (Chap. 4, § 4.2.1);
- fungi, through production of extracellular enzymes, are capable of mineralizing organic compounds containing phosphorus and nitrogen, which are not directly assimilable by the root; they mobilize these compounds and transmit them to the plant in a form it can assimilate.
- through their very thin hyphae, fungi reach regions in the soil not accessible to roots.

The mycorrhizal association assumes very great significance in natural environments and under conditions in which the plant if left to itself might suffer deficiencies. On the other hand, the association does not generally form in over-fertilized soils in which the mycorrhiza is of no use. Attempts at fertilizing forest soils, for example with sludges from water-purification plants, have led to the disappearance of numerous mycorrhizal fungi from treated areas (Horak and Röllin, 1988).

We present here four types of mycorrhizae, particularly important in temperate regions: ectomycorrhizae, ericoid mycorrhizae, orchid mycorrhizae and arbuscular mycorrhizae.

Many types of mycorrhizae are distinguished, formed by various groups of fungi and directed towards various categories of plants or to different ecological conditions.

16.1.2 Ectomycorrhizae

Definition and characteristics

Ectomycorrhizae are seen as a mycelial sheath around a modified rootlet (Fig. 16.3). This sheath or ***coat*** attains 40-μm thickness and constitutes up to 40 per cent of the weight of the rootlet. Filaments from this sheath penetrate between cells of the rhizodermis and the first cortical layers where they form the ***Hartig net***. Development of the root is modified under the effect of the hormones secreted by the fungus: longitudinal growth is less, the roots thicken and ramify strongly (Fig. 16.4; Chap. 4, § 4.1.6). Unlike the ramifications of normal roots, those of mycorrhized roots are very typical, at a right angle.

Ectotrophic mycorrhizae = ectomycorrhizae.

Hartig net: network of intercellular hyphae that forms the two-way nutritional interface between plant and fungus.

The ***extraradicular mycelium*** extends far from the roots, starting from the actual mycorrhizae, most often in

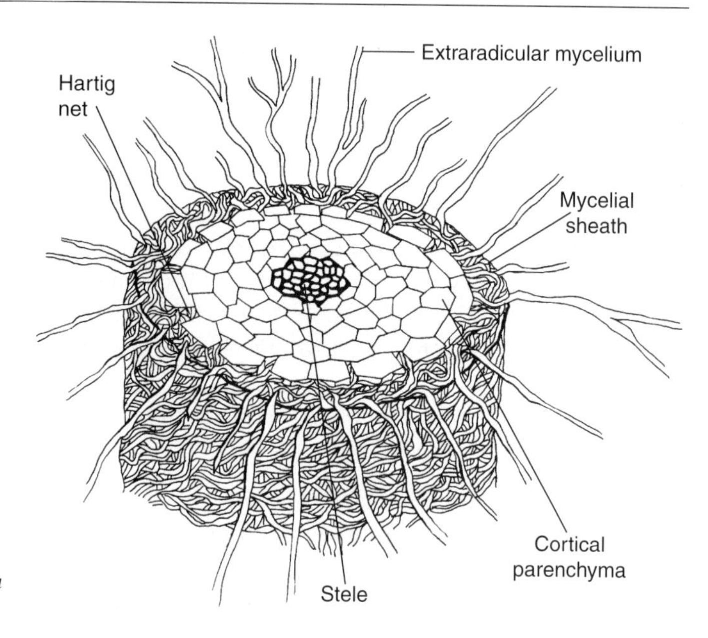

Fig. 16.3 Anatomy of an ectomycorrhiza.

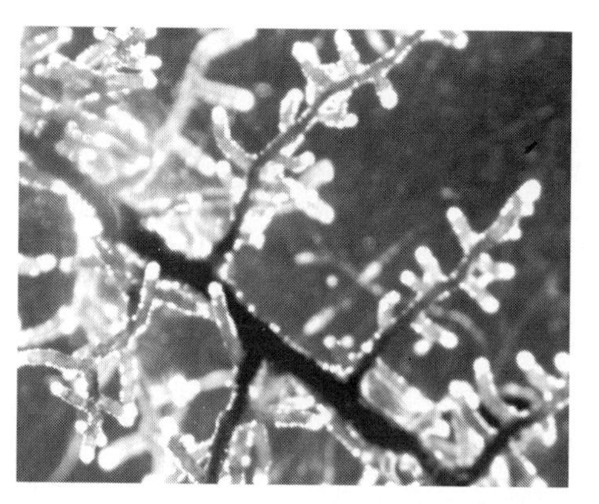

Fig. 16.4 Rootlets transformed by mycorrhization (ectomycorrhizae; photo V. Wiemken).

the form of rhizomorphs through which translocation of water and nutrients is achieved, and in the form of a diffuse mycelium specifically designed to ensure mineral and hydric nutrition (Plate—VII-2). Fruit bodies appear on this mycelium, far from the roots.

Fig. 16.5 Carpophores of Suillus grevillei, *a typical symbiotic fungus of larch (photo J. Keller).*

Botanical families more affected than others

The botanical families most affected by ectomycorrhizae are Pinaceae (Coniferales), Fagaceae, Betulaceae, Rosaceae, Salicaceae and Myrtaceae. Specificity between plants and fungi may be higher or lower. In the case of larch for example, associated only with this genus are fungi such as *Suillus grevillei* (Fig. 16.5), *S. viscidus* or *Boletinus cavipes*. In other cases, a single fungus may form mycorrhizae with several genera of broad-leaved trees or conifers.

Ectomycorrhizae affect woody plants above all. Most often they are formed by higher Basidiomycetes, Agaricales and Boletales and also by some Ascomycetes such as Tuberales (truffles). The fungal partners are macromycetes easy to identify on the basis of the morphology of the carpophores. Culturability of the mycelium facilitates experimentation with this type of organism.

Establishment of an ectotrophic mycorrhiza involves a preliminary biochemical 'dialogue' between the partners-to-be. This starts with secretion by the roots of yet unidentified compounds that more or less specifically attract the mycorrhizal fungi. At the time of establishment of symbiosis, changes are seen in the protein profiles and in the messenger-RNAs formed. Thus, 'mycorrhizines', proteins specific to the association, have been identified. The genetic and molecular aspects of this dialogue still remain largely unexplored.

Presence of fluorescent *Pseudomonas* bacteria in the rhizosphere can stimulate formation of mycorrhizae (mycorrhiza helper bacteria, MHB) (Garbaye, 1994).

Nutritional role favouring the plant

In autumn, when leaves are shed, quantities of inorganic salts reach the soil during a very brief period. In the case of trees with deciduous leaves, rise of sap is stopped at this time. These inorganic salts, not reabsorbed by the plant, cannot remain very long in the episolum and would be rapidly evacuated to depth or to subterranean water. Well, often in the same season the mycelium of mycorrhizal fungi is most abundant. It captures and accumulates a large part of these salts, retains them during winter, mostly in the mycelial coat of the mycorrhizae, and redistributes them to the plant in spring, when the sap rises.

The mycorrhizal fungus, a temporary reserve of inorganic salts in the soil.

The role of ectotrophic mycorrhizae is particularly important in soils of forests of boreal and temperate regions.

Unlike plants, the mycorrhizal fungi are often capable of absorbing organic compounds of nitrogen and phosphorus and mineralizing them before transmitting them to their plant partner. In soils poor in some nutrient elements, they thus enrich the spectrum of substances the plants could benefit from.

European fungi help Australian spruces

Ectotrophic mycorrhizae are very important for development of young trees. In the nineteenth century, it was desired to introduce cultivation of spruce *Picea abies* in Australia. Spruce seeds imported from Europe germinated well but, after some weeks, the seedlings wilted. On the other hand, when such seeds were planted in pots that had contained spruce imported from Europe for Christmas,

Thank you, Christmas trees!

the young plants prospered. Thus, a 'factor' in the soil of Europe not present in the soils of Australia favoured this growth, namely, the specific mycorrhizal fungi of conifers. Later, utilization of mycorrhized spruce plants resolved the problem.

Successions of fungal populations

Precocious (r-strategists) and late (K-strategists; Chap. 12, § 12.6.2) species are distinguished in the successions of populations of mycorrhizal fungi.

Successions of ectomycorrhizal populations have been observed in plantations during the life of a tree. While some fungi appear soon after planting, such as *Hebeloma, Inocybe* and *Laccaria* in the case of birch, others such as *Leccinum, Lactarius* and *Russula* do not develop for several years. These studies are based on the appearance of carpophores, however, and not on identification of the fungi in the mycorrhizae. By comparing the recolonization of pine forests after fires 6, 41, 65 and 121 years back, Visser (1995) showed an increase in diversity over time, then stabilization from 65 years onwards: the pioneer species are conserved and are joined by other, later species.

Direct identification starting from the mycorrhizae.

Use of molecular methods (Chap. 15, § 15.5.5) enables identification of fungi directly in the soil. The first studies published cited a poor correlation between frequency of fungi in roots and that of fruit bodies at the soil surface (Gardes and Bruns, 1996). It is therefore hazardous to make quantitative deductions on the dynamics of fungal populations in the soil just from a study of carpophores on the surface.

16.1.3 Ericoid mycorrhizae

The mycorrhizae of ericaceous plants colonize the rhizodermis of young roots.

Ericaceae (*Calluna, Erica, Vaccinium*), Empetraceae (*Empetrum*) and a few close families accommodate a particular type of root symbiosis: ***ericoid mycorrhizae***. These mycorrhizae are essentially caused by ascomycetes of the genera *Hymenoscyphus* and *Oidiodendron*. The rhizodermal cells are invaded by the mycelium; they do not form root hairs when mycorrhized. Some Ericaceae such as the strawberry tree, *Arbutus unedo*, accommodate ***arbutoid mycorrhizae***, very similar to ectomycorrhizae, instead of ericoid mycorrhizae.

16.1.4 Orchid mycorrhizae

The family Orchidaceae shows mycorrhizae of a very particular kind, formed by fungi with low specificity. Most often the mycorrhizae of orchids are due to fungi belonging to Basidiomycetes (genus *Rhizoctonia*) that form only a sterile mycelium without fruit bodies. In some cases, particularly in plants lacking chlorophyll, they can also be generated by higher Basidiomycetes.

The same *Rhizoctonia* may be a symbiont of orchids, a soil saprophyte and a parasite on other plants.

The very special feature of orchid mycorrhizae is that the fungus does not feed on the carbon provided by the plant. On the contrary, the extraradicular mycelium lives as a saprophyte in the zone surrounding the root and supplies the plant with organic carbon in addition to the inorganic salts it needs. The greater part of the mycelium of the mycorrhiza is seen in the intracellular balls (Fig. 16.6). These balls expand somewhat but are eventually digested by their host cell.

Fig. 16.7 *The heterotrophic orchid* Neottia nidus-avis, *the carbon nutrition of which is ensured in totality by the mycorrhizal fungus (watercolour by Philippe Robert; with permission of the Musée d'histoire naturelle of Neuchâtel).*

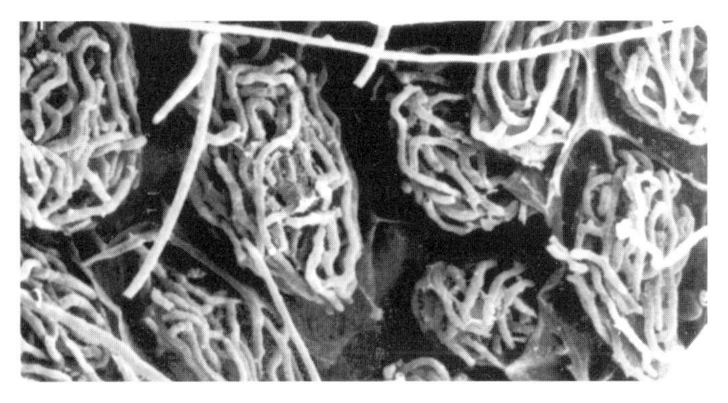

Fig. 16.6 *Mycelial balls* (Rhizoctonia sp.) *in an orchid* (Orchis morio) *mycorrhiza (from Beyrle et al., 1995; with permission from the Canadian Journal of Botany).*

The presence of a mycorrhizal fungus is particularly important during germination of orchid seeds devoid of reserves and active chloroplasts. Although these seeds germinate in the laboratory on rich complex culture media, this is not the case in nature. Presence of the fungal partner is indispensable for the young plant before differentiation of active chloroplasts. The relationship continues in principle throughout the life of the orchid, but is not always obligatory, except the relationship with heterotrophic orchids (e.g. the bird's-nest orchid *Neottia nidus-avis*, Fig. 16.7), in which the fungus ensures the carbon nutrition of the plant throughout its existence.

'Embryos born before term': thus did Noël Bernard, pioneer in their study at the beginning of the twentieth century, describe orchid seeds because of their minute size and absence of storage tissues.

The benefit for the fungal partner of such a symbiosis is not easy to understand. The fungus actively penetrates the tissues of the plant but ends up being digested in the cells it has invaded. It could as well be considered that the plant is parasitic on the fungus!

In orchids the symbiosis contract is very different from that of other mycorrhizae.

16.1.5 Arbuscular mycorrhizae

Where the symbiosis contract is signed again!

It is generally agreed that the symbiosis contract of arbuscular mycorrhizae is the same as that of ectotrophic and ericoid mycorrhizae: the fungus benefits from the sugars supplied by the plant and the plant gains inorganic salts drained by the fungus.

The Glomales are probably of very ancient origin, more than 400 million years old; they have been observed in plant fossils going back to the Devonian.

Arbuscular mycorrhizae are found in most plants that have no other type of mycorrhiza, herbaceous plants in particular. But some plants with other mycorrhizae may also harbour them. The symbionts are lower fungi allied to the Zygomycetes, forming a homogeneous phylogenetic group, the order Glomales (Tab. 16.2). Greatly resistant to desiccation and cold, their extraradicular mycelium forms large multinuclear spores (Fig. 16.8) that can be gathered by sieving the soil.

Table 16.2 Classification and characteristics of the Glomales (from Smith and Read, 1997)

Classification	Characteristics
Order: Glomales	Obligatorily biotrophic symbiotic fungi drawing carbon from the host plant through the intracellular arbuscles.
Suborder: Glominae	Production of arbuscles and vesicles, spores produced in terminal or lateral position on hyphae. No auxiliary cells.
Family: Glomaceae	
Genera: *Glomus* (77 species)	Spores produced singly or in loose aggregates
Genera: *Sclerocystis* (10 species)	Produce sporocarps composed of densely aggregated spores
Family: Acaulosporaceae	Spores formed on or in the constriction of a specialized sporiferous vesicle
Genera: *Acaulospora* (32 species) *Entrophospora* (3 species)	
Suborder: Gigasporineae	Production of arbuscles, never of vesicles. Spores furnished with a bulbous attachment. Auxiliary cells usually formed on the outermost hyphae.
Family: Gigasporaceae	Genera differentiated on the basis of mode of spore germination and ornamentation of auxiliary cells.
Genera: *Gigaspora* (7 species) *Scuttelospora* (23 species)	

Experimental work on these fungi is very difficult because they are not culturable *in vitro* outside their host. A good understanding of their taxonomic position and diversity could be obtained only recently through use of molecular methods (Chap. 15, § 15.5.5; Sanders et al., 1996). In this type of symbiosis, specificity seems to be low. These mycorrhizae comprise an intercellular mycelium that generates two types of very characteristic structures:

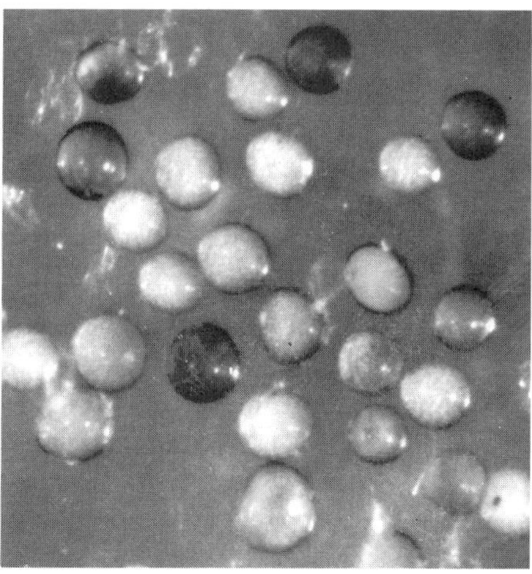

Fig. 16.8 *Spores of* Acaulospora laevis *(Glomales); magnified 3.5 (photo Z. Antoniolli, unpublished).*

- intracellular ***arbuscles*** (Fig. 16.9), highly branched structures that ensure nutrient exchanges between plant cells and fungus, chiefly the carbon nutrition of the latter;
- intra- or intercellular ***vesicles*** whose function is poorly understood yet; they accumulate reserves, lipids in particular; they are formed only in Glominae (Tab. 16.2).

Arbuscular mycorrhizae are the most widespread but also the most difficult to study.

Francis and Read (1995) showed that some plants respond very positively to the presence in their roots of arbuscular mycorrhizae, whereas others (e.g. *Echium vulgare, Reseda luteola*) in which the fungus forms only vesicles and no arbuscles, are diminished in growth and survival capacity. Thus the association is not mutualistic every time.

The presence of a fungus of this type in a root does not necessarily imply that a mutualistic symbiosis operates.

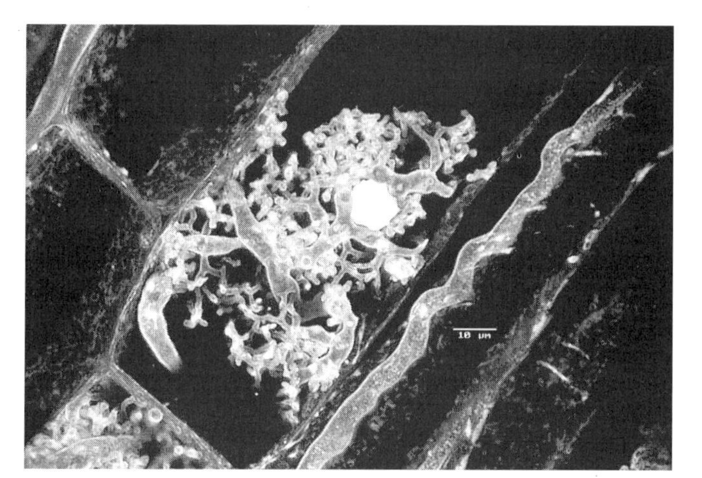

Fig. 16.9 *Arbuscle (photo by confocal microscope by S. Dickson, taken from Smith and Smith, 1997); with permission of* New Phytologist.

Mycorrhizae: major potential factors in relationships of plants among themselves and with their environment.

16.1.6 Multiple ecological role of mycorrhizae

The great age of mycorrhizae and their considerable structural and functional variety should urge us to consider them important factors of survival and competition within plant communities.

'...the detailed investigation directed at understanding mechanisms, which frequently must be done in simplified and ecologically unrealistic experimental systems, must be to an increasing extent focused on answering questions of ecological relevance' (Smith and Read, 1997).

Most studies on mycorrhizae have consisted of experiments done in artificial conditions, in pots for example (an often reductionist approach). However, only a more system-oriented approach to their distribution and above all their functions in natural communities will lead to better understanding of their deeper ecological significance. The 'symbiosis contract' discussed above might be only one aspect of these functions.

Mycorrhizae from the north to the south... or from top to bottom.

Biomes and mycorrhizae: between millimetres and thousands of kilometres

When we visualize a transect through the great biomes of the globe, from the vegetative line in Arctic countries to the equatorial forest (or in the altitudinal dimension, Chap. 7, § 7.2.3), it is seen that the dominant species of mycorrhizae differ depending on the biomes (Read, 1991).

No mycorrhizae in the high latitudes, Arctic or Antarctic, nor in the alpine snow zones!

In extreme zones of semi-permanent snow cover, raw mineral soils (RÉGOSOLS, CRYOSOLS) are scarcely colonized by fungi. The rare plants that live there do not have mycorrhizae while the same species growing in more clement zones are colonized by arbuscular mycorrhizae. For example, hair-grass *Deschampsia caespitosa* lacks mycorrhizae in Antarctic, but has them in the Falkland Islands.

Arbuscular mycorrhizae with *Glomus tenuis* are predominant in the tundra.

The regions below these extremes with a season without snow cover, show a rather continuous vegetative carpet and some accumulation of organic matter on the soil surface (RÉGOSOLS organiques, ORGANOSOLS, RANKOSOLS). The most commonly found mycorrhizae are of arbuscular type with *Glomus tenuis*, a species that, unlike others, forms very fine intercellular hyphae. The kind of relationship these fungi have with plants has not yet been established. Are they commensals, 'gentle' parasites or symbionts?

In moors and peat bogs, place for ericoid mycorrhizae!

Still lower in latitude or altitude, moors and peat bogs of the tundra or taiga are characterized by a vegetation with often-persistent leaves that tolerate the low trophic level of the soils (RANKOSOLS, PODZOSOLS); thus they are determined more by nutritional factors than climatic. The dominant families are Ericaceae, Epacridaceae and

Empetraceae, hosts for ericoid mycorrhizae. Their acidifying litter with high C/N ratio is not easily degradable and accumulates on the soil surface, forming a fibrous matrix in which roots with ericoid mycorrhizae proliferate.

Ectomycorrhizae predominate in the forests of subboreal and temperate zones, in particular on acid soils (BRUNISOLS OLIGOSATURÉS, ALOCRISOLS, LUVISOLS, PODZOSOLS). They essentially form in the superficial holorganic horizons (fragmentation horizon OF) or at the interface with the mineral horizons. Occasionally even an upward growth is seen of lateral roots of trees (Chap. 4, § 4.2.3) allowing them to explore the organic horizons. Plant diversity is relatively low, contrasting with the great diversity of fungi.

Further south (or lower), ectomycorrhizae colonize the subboreal or cool temperate forests.

As the mean temperature and evapotranspiration rise, the basic-cation content and pH of the soils (BRUNISOLS SATURÉS, CALCISOLS, CALCOSOLS) increase, turnover of organic matter speeds up and the C/N ratio falls. Nitrate replaces ammonium and organic nitrogen as the principal nitrogen source and phosphorus gradually takes the place of nitrogen as the major limiting factor among the bioelements of the ecosystem. Under these conditions, plant communities with arbuscular mycorrhizae slowly take the upper hand over those with ectomycorrhizae.

In the warm temperate zones, arbuscular mycorrhizae take the bottom!

Thus the humid tropical forest is largely populated with species with arbuscular mycorrhizae. However, patches formed of guilds of species with ectomycorrhizae live on acid, infertile soils (Ferruginous soils, Ferrallitic soils). These guilds are stabilized by the presence of fungi that favour their host-plants by their specificity. Regeneration of plants of the guilds is thus ensured.

In the tropics, a mosaic of zones with vesicular-arbuscular mycorrhizae and ectomycorrhizae.

Mycorrhizal fungi also attack humic materials

An idea firmly established in the mind of many biologists is that mycorrhizal fungi benefit from the simple carbon-containing molecules supplied by the plant because they would be incapable of producing extracellular enzymes that enable them attack substances of the litter and humus. This simplistic view must be greatly relativized. Many mycorrhizal fungi produce extracellular enzymes of various kinds (peroxidases, polyphenol-oxidases, phospho-monoesterases and phospho-diesterases) that enable them degrade and mineralize complex molecules of soil organic matter (humus peptides, organic phosphates) and, above all, to mobilize them for the benefit of their host-plant.

Mycorrhizal fungi are not disabled!

The assemblages of plant species having mycorrhizal fungi in common are called ***guilds*** (Read, 1991) in a sense slightly different from that given in Chapter 12, § 12.5.1.

A case in point is that of soils of coniferous forests covered with feather-mosses. These mosses form mats that intercept most of the bioelements reaching the soil, just as the sphagnums of peat do (Chap. 9, § 9.2.2). Their senescent parts are often colonized by ectomycorrhizal fungi that can thus redistribute the bioelements from them to their host-trees; Carleton and Read (1991) showed this for example in the moss *Pleurozium schreberi* colonized by *Suillus bovinus*, a symbiont of conifers.

Mycorrhizal fungi are less competitive *vis-à-vis* compounds of the fresh litter than the saprophytic decomposers. But they catch up with them later!

When litter is deposited, the saprophytic decomposers develop first, utilizing the most easily degradable carbon-containing substrates and immobilizing the other elements in their biomass. By doing this they cause a deficiency of easily assimilable carbon. However, well supplied with this element by their plant partner, the mycorrhizal fungi then gain the upper hand, remobilizing the bioelements for the benefit of their host-plant. Such a phenomenon has been observed in pine litter, which shows a phase of nitrogen immobilization up to three years after deposition followed, at the level of the OF horizon, by a phase of remobilization of this element by ectomycorrhizal fungi.

Mycorrhizae and plant diversity

Mycorrhizae, positive or negative 'go-betweens' in relations among plants.

Mycorrhization augments plant biodiversity, particularly on very poor soils. It enables plants occupy different ecological niches, reducing competition among them.

Heaths of the Southern Hemisphere with Epacridaceae show a greater biodiversity than would be expected considering the harsh climatic conditions. Ericaceae *(Calluna, Erica, Vaccinium)* develop roots with ericoid mycorrhizae in the litter on the surface of the soil, while roots of herbaceous plants *(Deschampsia, Molinia, Eriophorum)* go deeper. The two groups thus avoid competition for the same resources. Similarly, the coexistence of Ericaceae, Leguminosae and carnivorous plants, typical of moderately acid heaths, is explained by availability of different forms of nitrogen: that of soil organic matter for the first, via mycorrhizae, elemental nitrogen for the second and partly nitrogen of the animals captured for the last (Chap. 4, § 4.3.1).

Among liverworts and heathers, among arbutus and conifers, they help!

The same fungus can form associations of different types with different plants. This is the case of ericoid mycorrhizae, particularly *Hymenoscyphus ericae,* which form symbioses ***(mycothallus)*** with certain leafy liverworts (families Lepidoziaceae, Calypogeiaceae, Cephaloziaceae and Cephaloziellaceae) living in the same environment (Duckett and Read, 1995). The liverwort mats that sometimes cover moist organic soils are then a source of inoculum for the seeds of Ericaceae that germinate there.

Mycothallus: symbiotic association between a non-vascular plant, thus not having true roots—this is the case of liverworts—and a fungus.

Similarly, some ectomycorrhizal fungi of conifers and Fagaceae form arbutoid mycorrhizae with the genera *Arctostaphylos* (bearberry) and *Arbutus* (strawberry tree). These plants function as reservoirs of these fungi in case of fire or clear-felling, because they are very resistant to these phenomena. Thus regeneration of conifers is much better in the presence of strawberry trees than in their absence.

Mycorrhizae, Internet of the undergrowth!

Because plants of different species sometimes harbour the same mycorrhizal fungi, it can be presumed that the fungi form connections between them. One such communication has been proved between the ectomycorrhizae of the lodgepole pine *Pinus contorta* and the mycorrhizae of the early coral-root orchid *Corallorhiza trifida*, formed by the same mycelium. The nutrition of this heterotrophic species that lacks functional chloroplasts is entirely assured by the fungal partner. Thus the pine nourishes the orchid through the fungus.

A net transfer of carbon has recently been demonstrated between two photosynthetically active plants, namely, a tree in full sunlight and a seedling in the undergrowth (Fig. 16.10; Simard, in Smith and Read, 1997).

Pines, wet-nurses of orchids! Fungi connect plants of different species!

The guilds are not constituted merely of plants showing the same species of symbionts; the plants are physically interconnected by these symbionts.

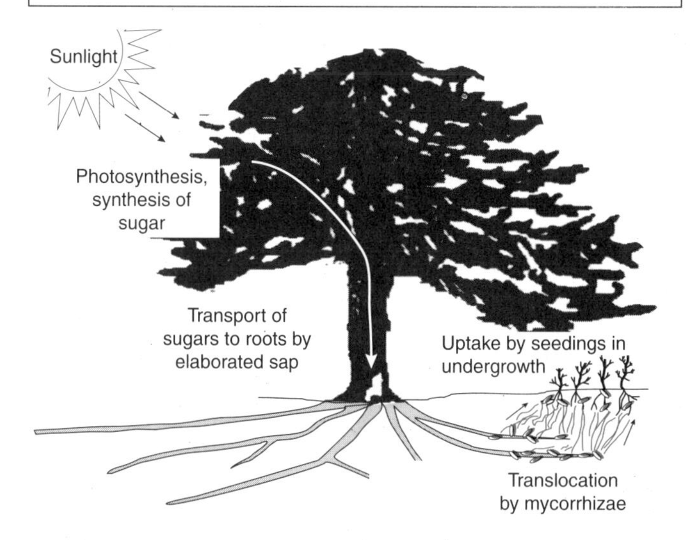

Fig. 16.10 Mycorrhizal symbioses, veritable medicinal drips! Thanks to the mycorrhizae the seedling, although in the shade, gains fully from the sun! The sugars it needs are synthesized by the leaves of the tree 'in sunlight', then transported through the elaborated sap, brought to the rootlets, taken up by the mycorrhizal fungi and finally absorbed by the young plant.

Grime et al. (1987) reconstructed in a growth chamber with and without inoculum of arbuscular mycorrhizal fungi a plant community representing a poor calcareous soil of north-western Europe (Tab. 16.3). This assemblage contained plants forming arbuscular mycorrhizae and others that did not. Three important results were evident:

More generally, mycorrhizae have a deciding effect on competition between plants in the complex communities of natural ecosystems.

- exposing the aerial parts of fescue-grass (mycorrhized grass) to $^{14}CO_2$ enabled demonstration of the transfer of

Table 16.3 Survival rate (%) after 6 months of species of a reconstructed poor calcicole grassland, with and without inoculation by mycorrhizae (from Grime et al., 1987). Explanations in the text.

Plant species	With mycorrhizae	Without mycorrhizae
Species forming arbuscular mycorrhizae		
Centaurium erythraea	64	2
Galium verum	58	11
Hieracium pilosella	49	6
Leontodon hispidus	42	13
Plantago lanceolata	71	10
Sanguisorba minor	53	6
Scabiosa columbaria	84	16
Species not forming arbuscular mycorrhizae		
Arabis hirsuta	8	42
Rumex acetosa	11	60

this carbon only to mycorrhized plants and not to the others;

- in the presence of mycorrhizal fungi, growth of the mycorrhized plants was greater, and that of non-mycorrhized plants less, than when the fungi were absent (Selosse and Le Tacon, 1999);
- survival of plants susceptible to mycorrhization was better when the fungi were present than when they were absent; the reverse was noted for plants not forming mycorrhizae.

Protective mycorrhizae

It was also experimentally proved that the presence of a mycorrhizal partner could reduce the impact of a pathogenic fungus. In the annual grass *Vulpia ciliata*, addition of the pathogenic *Fusarium oxysporum* to the soil resulted in a net diminution of growth, while the simultaneous presence of a *Glomus* restored growth to normal. But addition of *Glomus* alone had no positive effect under the conditions of the experiment (Newsham et al., 1995).

Mycorrhizae, soil fauna, phytosociology...

And the role of the soil fauna in mycorrhization?

The effects of the soil fauna on mycorrhization are not well understood. It is necessary to take into consideration the interactions among many partners, plants, mycorrhizal fungi, pathogenic fungi and fauna, all very complex and difficult to approach. Mycophagous springtails could play a negative role in mycorrhization; certain coleopteran larvae, those of weevils for example, also feed on mycorrhizae.

Conclusion

The study of mycorrhizae, in particular their involvement in the structure and functioning of natural plant communities, has only just begun. Mycorrhizae play a major role and might well explain phenomena such as the existence and floristic composition of certain synusia described by phytosociologists. The future holds better integration of studies in this domain, in which soil science, phytosociology and ecology of mycorrhizae will intervene synergistically.

Mycorrhizae, an explanatory tool for the phytosociologist?

16.2 NITROGEN-FIXING SYMBIOSES

It is estimated that more than two-thirds of the nitrogen fixed in the biosphere—120 million tons per annum—comes from the activity of bacterial symbionts of plants (Paul and Clark, 1996).

120 million tons of nitrogen fixed per year!

Fixation of molecular nitrogen is a reaction very expensive in terms of energy; the fixers able to take in light energy, directly or indirectly, are at a great advantage (Chap. 4, Fig. 4.21). The direct users are certain phototrophic bacteria such as the cyanobacteria (Chap. 4, § 4.4.3). Indirect users are the bacteria living in association with plants, for example the associative fixers (Chap. 15, § 15.3.3) and, above all, symbiotic fixers that form the subject of this section. Symbiosis simultaneously ensures supply of the energy required for the fixation and protection from oxygen.

Fixation of molecular nitrogen, an essential link in the biological nitrogen cycle (Chap. 4, §§ 4.4.3, 4.4.4).

Some nitrogen-fixing associations occur only in aerial parts of plants, for instance that between *Gunnera chilensis*—of tropical Haloragaceae—and cyanobacteria of the genus *Nostoc*, or between some floating ferns of genus *Azolla* and cyanobacteria (*Anabaena azollae*, Chap. 2, § 2.5.2), the second being very important in some rice-paddies.

Many types of fixing associations exist between bacteria and plants, both aerial and below the ground.

In soils, the two most widespread types of symbiosis are associations between leguminous plants (Fabales) and Gram-negative bacteria of family Rhizobiaceae on the one hand, and those between often woody plants belonging to many non-leguminous families and Gram-positive filamentous bacteria of the genus *Frankia* on the other. Other types of fixing root-symbiosis are found in tropical plants, in particular that established between fixing cyanobacteria and roots of *Cycas*, a very primitive genus

In the temperate zone, two broad types of root symbiosis: between leguminous plants and Rhizobia, and between non-leguminous woody plants and *Frankia*. In the tropical zone, an additional type, that of *Cycas* and cyanobacteria.

Fig. 16.11 Coralloid formations in a root of Cycas revoluta. *These formations exist before the invasion by the symbiont but develop considerably and more durably when colonized by bacteria. Nitrogen-fixing symbiosis with cyanobacteria is established at this level.*

of plants. Here symbiosis is established at the level of particular root formations termed coralloid (Fig. 16.11).

The paradoxical situation of symbiotic cyanobacteria.

Nitrogen-fixing cyanobacteria are capable *per se* of capturing and directly transforming light energy. Thus, in epigeal symbioses, the plant partner is not obligatorily the supplier of carbon and energy. In the case of *Cycas*, the paradox is related to the **hypogeal** location (therefore to darkness) of cyanobacteria whose photosynthetic apparatus is nonetheless developed, however. Given the occupancy by a normally phototrophic organism of a habitat without light, is the reason for the symbiotic relationship here primarily non-trophic?

16.2.1 Symbiosis with nodules in leguminous plants (order Fabales)

The nodule, organ differentiated by symbiosis

The ***nodule*** is an organ generated by the interaction between a symbiotic nitrogen-fixing bacterium and a root. Vascularized, this organ harbours bacterial symbionts, provides them with the condition for their nitrogen-fixing activity and favours nutrient exchanges between the partners.

Taxonomy and diversity of symbiotic nodule bacteria

Recent taxonomical studies have dealt with the family Rhizobiaceae (Young et al., 2001). In this book, the ensemble of bacteria belonging to the four genera represented is collectively designated under the term Rhizobium/Rhizobia, without italicization.

Four genera of bacteria, Gram-negative and motile through flagella, form nitrogen-fixing symbioses with leguminous plants:

- *Rhizobium* s. str. (e.g. *R. leguminosarum, R. fredii*);
- *Azorhizobium* (e.g. *A. caulinodans*), the latter forming epigeal nodules on the stalks of *Sesbania grandiflora*, one of the tropical Fabaceae;
- *Bradyrhizobium* (e.g. *B. japonicum*);
- *Mesorhizobium* (e.g. *M. loti, M. ciceri*).

- The genus *Agrobacterium*, which includes the causal agent of crown gall (plant cancer) is now included in *Rhizobium* s. str.

In the order Fabales, the family Caesalpinaceae, very primitive, comprises the smallest proportion of nodulated species. On the other hand, the majority of Mimosaceae and Fabaceae shelter Rhizobia. Herbaceous and arborescent, annual or perennial, arctic or tropical nodulated Fabaceae are found. This almost exclusivity of the Fabales is thus due more to a genetic predisposition common to members of this order than to ecological reasons.

With just one exception (genus *Parasponia*, one of the tropical Ulmaceae), Rhizobia form symbiotic associations with only representatives of the order Fabales.

Leguminosae: a confusion in nomenclature

The terms Leguminosae and Papilionaceae are not correct in current botanical nomenclature because they do not come from names of botanical genera. Here we consider the group of leguminous plants as an order, the Fabales. It is subdivided into three families, one of which, Caesalpinaceae, is primitive and the others, Mimosaceae and Fabaceae are more evolved. Mimosaceae comprise the 'mimosas'—which in fact belong to the genus *Acacia*—and sensitive plants that are the true mimosas (*Mimosa pudica*)! Fabaceae (formerly Papilionaceae) comprises the peas, beans, lucerne, clovers, brooms... and the 'acacias' with white bunches, actually the false-acacia Robinia (*Robinia pseudoacacia*)!

Investigations on the biochemistry and molecular biology of nodulation have multiplied these past years; many recent reviews are devoted to them (e.g. Fisher and Long, 1992; Göttfert, 1993; Fischer, 1994; Mylona et al., 1995; Van Rhijn and Vanderleyden, 1995; Geurts and Franssen, 1996; Long, 2001; Thies et al., 2001; Oldroyd, 2001). This fascination is explained by the usefulness of certain legume-Rhizobium systems as study models, and also by the potential significance of the phenomenon in agronomic applications.

At present, nodular symbiosis is surely best described in terms of establishment and functioning at the molecular scale because of its economic implications.

The formation and functioning of root nodules of leguminous plants exhibit common features and differences, depending on the bacterium-plant system considered. The specificity of the symbiotic relationship is very variable, high or low. Although only a small number of systems has been studied in detail, we shall attempt here to give an overall picture of the totality of the nodulation and fixation process.

Nodulation, a four-stage process: establishment of symbiotic contact, invasion by the symbiont, nodule development and commencement of functioning of the nodule.

Symbiosis establishment: a dialogue in the soil

Rhizobia are soil bacteria. In the absence of legumes they live in the soil in rather modest numbers.

Rhizobia ahoy!

Commencement of a symbiosis involves a 'dialogue' between these bacteria and the future host-plant; this communication starts by exchange of chemical signals enabling the future partners to recognize each other.

Root secretions initiate the dialogue.

The first signals are sent by the root, in the form of more or less specific secretions of sugars, amino acids, carboxylic acids and phenolic compounds (Chap. 4, § 4.1.5) that cause a positive chemotaxis in the Rhizobia. In fact, this phenomenon is not strictly necessary for establishment of symbiosis, but promotes enrichment of Rhizobium populations in the rhizosphere; so it has ecological significance. Among the compounds secreted by the root are also found flavonoids, often in very low concentrations.

nod, nif and fix, nod **genes, Nod proteins and Nod factors: everything in its place!**

Symbols for the genes are written *in italic*, with lower-case initial letter (e.g. the *nodD* gene), while the proteins resulting from their expression are represented by symbols in roman with a capital initial letter (e.g. NodD protein).

Three groups of genes are involved in bacteria in establishment and functioning of nodules:

- ***nod genes*** in the early stages of recognition and nodulation;
- ***nif genes*** in nitrogen fixation; they are homologues of the genes present in the free-living fixers such as *Klebsiella pneumoniae;*
- ***fix genes,*** a heterogeneous group that includes other genes more or less involved in fixation, without homologues in free-living fixers.

Only one of the *nod* genes, the ***nodD gene,*** is transcribed constitutively, that is, permanently (Chap. 14, § 14.6.5). Its product is the **NodD *protein*** capable of binding to regulator sites (*nod boxes*) of the ***operons*** bearing the other *nod* genes. This protein functions in addition as receptor of specific flavonoid compounds secreted by the root. When it is bonded to one of these compounds, NodD activates transcription of the other *nod* genes (Fig. 16.12). The products of these genes are essentially enzymes that participate in the synthesis of a decisive compound in triggering the nodulation process, the ***Nod factor***. This factor is a ***lipo-oligosaccharide*** formed of three to five N-acetyl-glucosamine units with structure similar to that of chitin. The different Nod factors are distinguished by the number of sugar units, nature of fatty acid and presence of other substances in the molecule. These last determine to some extent the specificity and biological activity of the molecule.

Operon: In bacteria, DNA segment comprising a series of adjacent structural genes coding for example for enzymes, and a common regulator site ensuring control (induction, repression) of expression of these structural genes. The latter being transcribed *en bloc* to a single messenger RNA, synthesis of the corresponding enzymes takes place in a coordinated manner (Chap. 14, § 14.2.5).

Lipo-oligosaccharide: macromolecule formed from a short chain of sugar molecules bound to a molecule of long-chain fatty acid.

Sometimes Nod factors induce formation of additional flavonoids in the plant, which in turn act as *nod*-gene inducers: this is a good example of positive feedback!

Some flavonoids are attractant for the bacteria. They are also powerful inducers of *nod* genes, the key determinants of exchange of signals between the two partners because they provoke synthesis of Nod factors. These factors exert a hormonal action on the root at two levels:

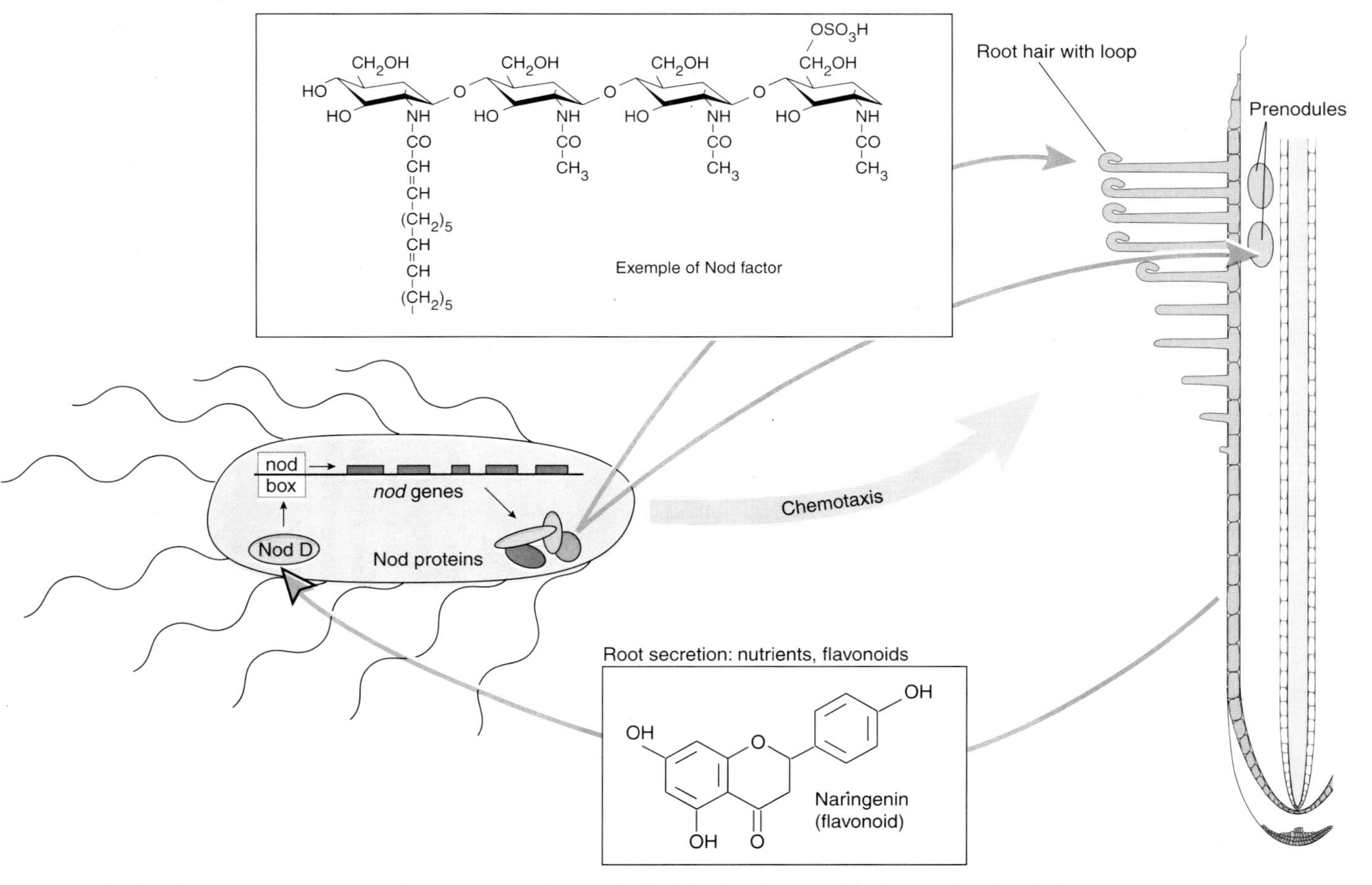

Fig. 16.12 *Molecular dialogue between the root of a leguminous plant and soil Rhizobia prior to nodulation. Explanations in the text.*

Nod factors modify the anatomy of the root.

• on the growing root hairs, they determine deformations, ramifications and curving of the extremity into a characteristic loop or hook (Fig. 16.13);
• in the root cortex (Chap. 4, § 4.1.4), they determine dedifferentiation of some cells which will proliferate and form a pre-nodule ***(primordium)***.

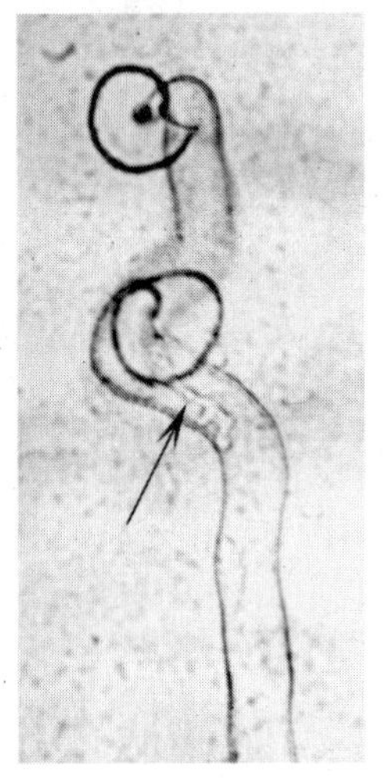

Fig. 16.13 *Deformation of root hairs to a loop or hook prior to nodulation in the legume* Macroptilum atropurpureum. *Formation of an infection thread (arrow) is seen (photo W.J. Broughton).*

These effects may be experimentally obtained in absence of bacteria, just by the action of purified Nod factors at very low concentrations. Deformation of root hairs has been observed starting from 10^{-12} M.

In contrast, plants produce chitinases able to decompose the Nod factors to different degrees depending on the compound they contain. The significance of this phenomenon is still unclear: it may work as a means of defence or regulation of the plant, or of reinforcing in some way the specificity of interaction. Worth watching!

Invasion by the symbiont

Lectin: protein binding specifically to sugars.

Adhesion of the bacteria to the surface of root cells, most often those of root hairs, follows upon this exchange of chemical signals. The adhesive substances are not completely known: specific ***lectins*** formed by the plant, other lectins formed by the bacteria or adhesive molecules such as rhicadhesin, a Ca^{2+}-dependent protein of the bacterial surface have been mentioned.

'Rhizobia caught in a trap! It's in the bag!'

The cells of Rhizobia are found trapped in a pocket formed by the wall—which has been weakened by hydrolysis of its constituents—generally at the bend of the hooks formed at the root-hair ends. The plasmic membrane forms an invagination into which the bacterial cells penetrate. Simultaneously the plant secretes wall material around the bacteria, which proliferate in the form of a linear chain of cells. This chain determines formation of an infection thread directed towards a primordium (Fig. 16.14). The infection threads then release the bacteria into the cytoplasm of cortical cells.

In the case of some leguminous plants such as groundnut, however, the infective Rhizobium does not enter through the root hairs but through rupture of the epidermis provoked by emergence of a lateral root. In this instance it forms no infection thread.

Functioning of the nodule

Cells in a cell!

After the bacterial cells are released from the infection thread into cells of the cortex, they are surrounded by a

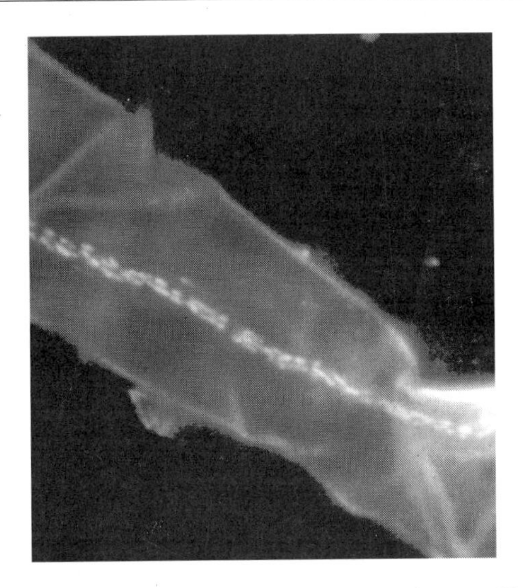

Fig. 16.14 Infection thread in a root hair of clover Trifolium *sp. The bacterial cells, in several rows, were stained by acridine orange and studied under the epifluorescence microscope (photo B.B. Bohlool and W.J. Broughton).*

membrane produced by the plant, ***the peribacteroid membrane***, *PBM*; the ensemble forms the ***symbiosome*** (Fig. 16.15). The PBM contains compounds different from that of the plant cell membrane, with the function of exchange between the bacterium and the cytoplasm of the host-cell. The Rhizobia multiply actively at first in the infected cells. This growth is accompanied by the simultaneous synthesis, by the plant cell, of an enormous area of PBM, about 30 times that of the cell membrane. The Rhizobia cells are then differentiated into their endosymbiotic form, the ***bacteroid***, a larger irregularly shaped cell. This evolution is often irreversible. The

Symbiosome: ensemble formed by one or more bacteroids surrounded by the peribacteroid membrane PBM. The latter provides the bacterium protection against the defence mechanisms the plant could operate to combat an intracellular invasion as well as the mutualistic exchanges between bacterium and plant cell.

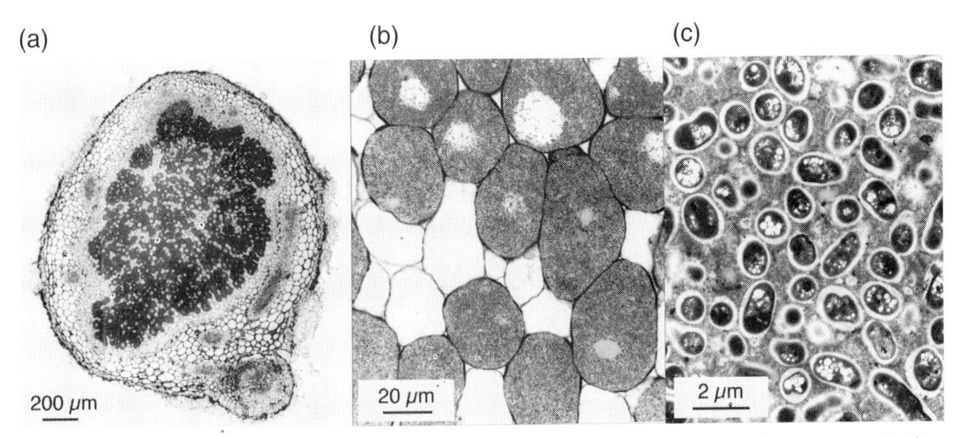

Fig. 16.15 Section, viewed under the transmission electron microscope, of nodules of Vigna unguiculata *infected by* Rhizobium *sp. strain NGR234. (a) The nodule in its entirety; (b) Bacterized (grey) and non-bacterized (white) cells; (c) Bacteroids in a cell, enclosed by the peribacteroid membrane (photos W. Golinowsky, M. Hanin and W.J. Broughton).*

bacteroids can then no longer be dedifferentiated to give back cells capable of autonomous life in the soil.

In nodulation, some strains are only ***infective***, others ***effective***. Infective but non-effective strains form small, light-coloured nodules lacking leghaemoglobin. A common agronomic practice consists of inoculating the seeds of legumes with a selected effective strain in order to optimize the production of active nodules and thus the quantity of nitrogen added to the soil.

Nodule formation does not necessarily lead to the final stage of the fixing nodule.

Infective: qualifies a bacterial strain capable of infecting the root and producing a fixing or non-fixing nodule.

Effective: qualifies a bacterial strain capable of producing an active fixing nodule.

The nodule, guarantor of an environment favourable for nitrogen fixation

In symbiosis, the principal effect of the nodule is to provide a proper environment for ensuring nitrogen fixation by the symbiont and the exchanges between it and the host-plant. The nitrogenase complex is very sensitive to oxygen (Chap. 4, § 4.4.3), while the energy required by nitrogen fixation involves aerobic respiratory activity of the bacteroids. This paradox is resolved by the nodule, which provides a very efficient triple protection:

- by high oxygen demand created by intense oxidative activities that take place in the nodule (respiratory protection);
- by its anatomy (parenchyma scarcely permeable to gases);
- by formation of a molecule peculiar to symbiosis, ***leghaemoglobin.*** This binds strongly to the oxygen, reducing dramatically its concentration in the nodule to as low as 3 to 30 nM O_2, or 10,000 to 100,000 times less than under air saturation.

On the other hand, the peribacteroid membrane regulates transfer of ammonium, the product of nitrogen fixation, towards the plant to be assimilated by it. Ammonium being a repressor of the fixing complex, its active export is necessary to remove this repression.

Lastly, the nodule makes organic carbon available to the bacteroid in an assimilable form. Sucrose is first transported from the leaves to the nodules, where it undergoes partial degradation. Later the organic carbon penetrates the bacteroid essentially in the form of dicarboxylic acids and provides it with the energy needed for fixation.

The foremost ecological role of the nodule concerns oxygen, which should not interfere with the fixation process.

Leghaemoglobin: protein very close in structure and function to the haemoglobin of vertebrates. A transporter of oxygen, it contains a prosthetic group (haeme), synthesized by the bacteroid, and a protein part synthesized by the plant. Leghaemoglobin binds oxygen and conducts it to the terminal oxidase of the bacterial respiratory chain. Thus the oxygen does not interact with the nitrogenase within the cytoplasm. Leghaemoglobin is a product, an emergent property of the symbiotic system (Chap. 1, § 1.2.2).

Nodule development

In most leguminous plants of temperate regions (e.g. peas, lucerne), the primordium is formed from the internal cortex. The nodules are cylindrical (Fig. 16.16) with a continually active meristematic tip at one extremity. Their growth, therefore, is indefinite and they exhibit a development gradient from tip to base. The meristem is followed by the prefixation zone where release of the bacteria into cells takes place, then the zone of fixation where fixing activity is induced and bacteroids differentiated. At the base, in the senescent zone, bacteroids are degraded by the plant.

Two different types of nodules, with indefinite or definite development. But similar anatomy, with peripheral vascular tissues and a central tissue formed of infected and uninfected cells.

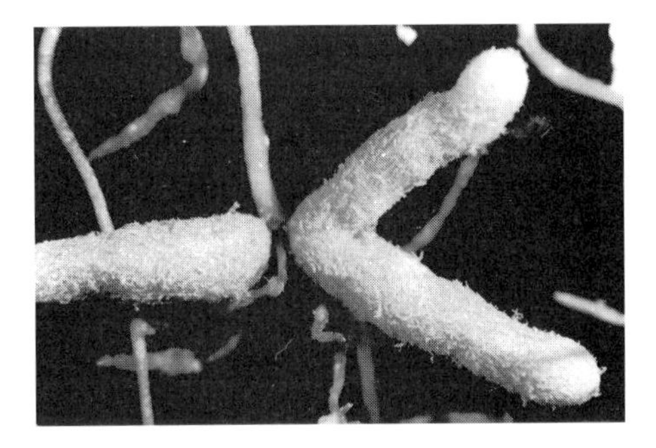

Fig. 16.16 Nodule with indefinite growth in the legume Leucaena leucocephala *(photo W.J. Broughton).*

In many leguminous plants of tropical origin (e.g. soybean, kidney-bean), the primordium appears in the external cortex and proliferation of its cells is limited in time. Bacteria are released from the infection threads during cell multiplication; each daughter-cell receives Rhizobia. After division ceases, the invaded cells simultaneously differentiate to form the central fixing tissue of the nodule. The nodule does not elongate but broadens. In a given nodule, only one stage of differentiation is observed at a given moment. The nodules are at definite growth (Plate VII-3).

Nitrogen-fixing symbiosis involves a gene system of great complexity, complementary and with action synchronized between its partners. The initial phase is characterized by coordinated exchange of chemical signals as prelude to nodulation. The active phase involves coordination of nodule functioning so that optimal conditions for fixation are ensured. These consist essentially of protection against oxygen and its transfer, and also transfer of nitrogen from bacteroid to plant and of energy from plant to bacteroid.

To conclude, nitrogen-fixing symbiosis of legume nodules requires close temporal and spatial coordination between the plant and the bacterium.

16.2.2 Actinorrhizal symbioses

General characteristics

The symbioses that actinomycetes of genus *Frankia* form with many non-leguminous plant genera have only recently been studied intensively, when isolation of symbionts became truly successful (Lalonde, 1978). Many investigations have since been devoted to this topic. The reader wishing to learn more may usefully consult a book entirely devoted to this subject (Schwintzer and Tjepkema, 1990) and also some general review papers (Schwintzer and Tjepkema, in Dilworth and Glenn, 1991; Baker and Mullin, in Stacey et al., 1992), on the physiology of nodules (Tjepkema et al., 1986) or on the application potential of this type of symbiosis (Benoît and Berry, in Schwintzer and Tjepkema, 1990).

Despite their name, actinomycetes are not fungi!

Although not so indicated by their name, ***actinomycetes*** are actually bacteria, not fungi! The confusion originates from their ramified filamentous form (mycelium), often giving rise to ***conidia*** or ***sporangia***. But the resemblance stops there. These are Gram-positive, prokaryotic (Chap. 2, § 2.5.2) organisms abundant in soils and decomposing litter. Numerous actinomycetes are active producers of antibiotics (e.g. *Streptomyces*). Only since 1964 has it been known from electron-microscopic observations that *Frankia* are bacteria. The genus has been so named for a century past, but they had long been considered fungi!

Conidia: external vegetative spores formed singly, in pairs or in chains.

Sporangium: enlarged cell containing endospores (endocellular spores).

Similarities and differences compared to nodule symbioses of leguminous plants.

Actinorrhizal: by analogy with mycorrhizal, qualifies root symbioses with actinomycetes, an exclusive feature of genus *Frankia*.

Stratigraphic palynology: study of evolution of vegetation and climate based on examination of pollen deposited in regularly stratified materials (sediments, peats, etc., Chap. 9, § 9.1.3). The walls of pollen grains, exteriorly constituted of exine, a substance extremely resistant to degradation, have enabled their preservation over thousands of years.

An important difference between ***actinorrhizal*** symbioses and nitrogen-fixing symbioses in legumes is linked to the taxonomic position of the host-plants without specific phylogenetic relationship (Table 16.4). What brings them together is the ecological order: they indeed live in many ecological systems but always on soils with low nitrogen status. The addition of nitrogen by this symbiosis is therefore of great importance, particularly in the Arctic regions where leguminous plants are rare or absent. ***Stratigraphic palynology*** reveals the capital role played by these plants in recolonization of raw mineral soils after the glaciations, for example by the genera *Alnus* and *Hippophae* during the last postglacial period (–12,000 to –8000 y).

Table 16.4 Plant genera forming actinorrhizal nodules (from Schwintzer and Tjepkema, in Dilworth and Glenn, 1991)

Families	Genera
Betulaceae	*Alnus*
Casuarinaceae	*Allocasuarina, Casuarina, Ceuthostoma, Gymnostoma*
Coriariaceae	*Coriaria*
Datiscaceae	*Datisca*
Eleagnaceae	*Eleagnus, Hippophae, Shepherdia*
Myricaceae	*Comptonia, Myrica*
Rhamnaceae	*Ceanothus, Colletia, Discaria, Kentrothamnus, Retanilla, Talguena, Trevoa*
Rosaceae	*Cercocarpus, Chamaebatia, Cowania, Dryas, Purshia, Rubus*

Frankia nodules are found in 25 plant genera belonging to 8 families of 7 different orders.

Frankia, bacteria of actinorrhizal symbiosis

Frankia belongs to a phylum (Gram-positive bacteria) distinct from that of Rhizobia (Proteobacteria).

Subdivision of the genus into species has not yet been established. The mycelium bears ***multilocular sporangia*** that are found at the filament ends or occupy an intercalated position between vegetative cells. In environments with low or no nitrogen, these bacteria also form vesicles 2 to 6 μm in diameter, enclosed in a lipidic capsule, appearing at the filament ends or on short lateral branches. These vesicles are the site of nitrogen fixation and are as well formed in pure culture as in symbiotic nodules.

A unique characteristic of *Frankia* among the actinomycetes: formation of vesicles enclosed in a capsule.

Multilocular sporangium: sporangium subdivided into compartments, each containing several spores.

Unlike Rhizobia, *Frankia* can fix nitrogen in free air (under 20 kPa O_2) in pure culture. The lipids of the vesicle appear to protect the nitrogen-fixing complex from oxygen. Under conditions of nitrogen fixation, 15 monomolecular lipid layers have actually been counted under 4 kPa O_2 and 40 under 40 kPa O_2.

Frankia have an intrinsic means of protection of nitrogenase from oxygen, independent of the structure of the nodule.

Outside the root nodules, *Frankia* are sometimes found, albeit in small number, in soils bearing no plants with which they can associate. On the other hand, they proliferate in the rhizosphere of some non-nodulable plants related to nodulable plants. This is true for birches that belong, like nitrogen-fixing alders, to the family Betulaceae. In this rhizosphere habitat, *Frankia* form vesicles probably capable of fixing nitrogen.

Frankia occupy three ecological niches to varying degrees: bulk soil, rhizosphere and root nodules.

Symbiosis formation

The dialogue prior to infection of the roots is poorly understood. The principal reason is related to the very slow growth of *Frankia* and to the fact that, unlike in Rhizobium, there are no performing genetic tools available for their study.

A dialogue much less studied than that in leguminous plants.

The filaments penetrate through the root hairs or between the epidermal cells and those of the root cortex. The mode of infection chosen depends on the host-plant and not on the bacterial species. The first case recalls what happens in leguminous plants, with branching of the root hairs and formation of a loop. In response to the infection, some cortical cells multiply to form a prenodule later penetrated by the symbiont. The nodule then develops as a modified lateral root, densely branched, sometimes termed ***rhizothamnion*** (Fig. 16.17). The substance or substances—are they related to Nod factors?—responsible for these phenomena have not yet been identified.

Two paths of infection.

At the time of intracellular penetration the plant secretes, as in the case of leguminous plants, a sheath of

Fig. 16.17 Actinorrhizal nodule (rhizothamnion) in Alnus glutinosa. *The Swiss 5-centime coin is provided for scale (17 mm dia.) (photo M. Aragno).*

parietal material surrounding an invagination of the cellular membrane of the plant. However, unlike in Rhizobia, this sheath persists in active nodules.

Anatomy and functioning of the nodule

Nodules very different from those of leguminous plants.

The conductive tissues in *Frankia* nodules are in the middle and infected tissues on the periphery. These nodules have air-filled spaces that therefore do not bring an anatomical protection against oxygen comparable to that of nodules in leguminous plants. However, in addition to the intrinsic protection given by the lipid capsule of vesicles, a haemoglobin has also been observed in actinorrhizal nodules. Its concentration is normally very low, but some nodules without vesicles (for example in *Casuarina* and *Myrica*) contain rather large amounts.

One feature common to *Frankia* in symbiosis and other endosymbiotic fixers is that they do not, under these conditions, form any enzymatic system with high affinity for assimilating ammoniacal nitrogen, like the glutamine synthetase/GOGAT system in free-living nitrogen fixers. Ammonium is thus exported to the plant partner that has such a system available.

As in the Fabales, infective strains capable of causing nodules are here distinguished from others, effective, capable additionally of fixing nitrogen in the nodules.

Strains of *Frankia* may be infective only on certain hosts while they are effective in others. If the number of infective, non-effective strains is higher than that of effective strains, competition may result.

Actinorrhizal symbioses are the object of numerous applications, particularly in improvement and conditioning of marginal soils and deficiencies, or in revegetation (Benoît and Berry, in Schwintzer and Tjepkema, 1990).

16.3 CONCLUSION

The great curtain of the Living Soil Theatre will soon drop. But before beginning the epilogue, we shall have been able to help associations of mutual benefit toward a 'happy ending'. By fusion of their cells and metabolic activities, plants, animals, fungi and bacteria succeed in maintaining themselves and prospering by pooling their resources whereas each group individually cannot.

Equipped with a pickaxe, we went to prepare the photograph for Fig. 16.17, digging out some roots at the foot of hundred-year-old alders. These trees should have been large already at the time of Pasteur and Winogradski, the discoverers of bacteria and their functions in the soil. Plunging our hands into the earth we took out what some would have taken for sticky clods on the roots, but which after superficial washing proved to be nodules, rhizothamnia of the tree's association with *Frankia*.

What better demonstration of the 'Living Soil' than to find in it the expression of such intimacy between the lowliest of the bacteria and the giant among trees, key players in one of the most important biogeochemical cycles of the planet, produced by a wonderful sophisticated regulation at the scale of the cell and the molecule!

CHAPTER 17

IN THE FUTURE... SOIL BIOLOGY!

'We have explored the sun, deserts and fields; we now invite you to a vision of a darker world, that of the soil, of shade. Let us enjoy, all is not lost, all is not destroyed, all is not known. A future takes form in the dark world of the soil, in the liquid world of water, in the shaded world of woods' (Bourguignon, 1996).

At the end of this book but in introduction to a reflection that the reader should pursue, this chapter suggests certain paths along which soil biology could proceed in the next decades. Recent publications (Paton et al., 1995; Wood, 1995; Coleman and Crossley, 1996; Paul and Clark, 1996; Benckiser, 1997; Giller et al., 1997; Smith and Read, 1997), predict its strong development, just as in all of biology.

'Within terrestrial ecosystems, soils may contain some of the last great "unknowns" of many of the biota' (Coleman and Crossley, 1996).

This chapter, detailing some aspects we consider important, discusses four domains in which soil biology has a great future: basic knowledge of soil science and ecology (§ 17.1), applications (§ 17.2), modelling (§ 17.3) and social aspects of soil science (§ 17.4). The last theme, open to human society, gets back to the first pages of the book in which the soil is approached in total relation to mankind (Chap. 1, § 1.1).

'The scenario is speculative, but serves as an example of where we may expect additional breakthroughs occurring in the cryptic and fascinating world of soil ecology' (Coleman and Crossley, 1996).

17.1 SOIL BIOLOGY AND FUNDAMENTAL SOIL SCIENCE KNOWLEDGE

17.1.1 Systematics to be completed and overhauled

A difficult requirement: taxonomic survey of the organisms.

Recent advances in the systematics of living organisms and its techniques, even if very far from being final (Chap. 12, § 12.1; Behan-Pelletier and Bissett, 1993; Coleman and Crossley, 1996; Travé et al., 1996; Lecointre and Le Guyader, 2001), provide a considerable assemblage of information to the soil scientist who knows how to 'read' from bioindicator organisms (Chap. 4, § 4.6). The functions of soil organisms and their effects on pedogenesis are also becoming better understood. When interpreted well, these biological processes are today explanatory of many aspects of soil science and will become still more so in future.

The variation between authors in estimates points to the necessity of continuing research!

Taxonomic identification of organisms is basic to understanding the role of living organisms in the ecosystem, in particular in its subsystem, the soil. Although there is hardly any problem in the case of plants in this respect, at least in temperate regions, we are a long way out in zoology and microbiology. Indeed, the systematics of some zoological groups is well understood (Chap. 12, §§ 12.1, 12.4), but very large deficiencies exist, for example in the oribatid mites. Behan-Pelletier and Bissett (1993) showed that only 30 to 35 per cent of the species of the Northern Hemisphere, though they are the best known, have been described; 100,000 still remain to be discovered. Travé et al. (1996) are more modest in the number of species to be found: between 30,000 and 50,000.

'This difficulty [knowledge of Arthropoda] is compounded by our very poor knowledge of identities of the immature stages of soil fauna, particularly the Acari' (Coleman and Crossley, 1996).

Edaphic dipterans are in a situation similar to that of the acari. While the determination of winged adults is very difficult, what can we say of their larvae, which should have concerned the soil scientist first (Chap. 8, § 8.3.2; Chap. 12, § 12.4.9)? Yet—this perhaps explains it—dipterans are often reduced to a negligible nod, if not totally forgotten, in a large number of publications on soil ecology....

When the concept of species is confusing...

If plants and certain groups of animals are even so well known, it is generally because the concept of species does not pose much problem. It is otherwise for microorganisms. It is known, for example, that arbuscular mycorrhizal fungi (Chap. 16, § 16.1.5) are species capable of merging their mycelia and exchanging their nuclei, resulting in a giant 'networked organism' (Sanders et al.,

1996). In bacteriology, it is necessary in addition to consider both the taxonomic, therefore genetic, identity and the functional identity, which do not always correspond to each other (Chap. 15, § 15.5.4).

Lastly, we should remember that a gram of earth may contain more than a thousand million bacteria belonging to more than 10,000 species (Wood, 1995; Paul and Clark, 1996); and that only 70,000 species of fungi are presently known out of a total presumed to be one-and-a-half million (Coleman and Crossley, 1996)!

At the present rate, 300 to 1500 years to take up an immense challenge...

The ambition, quite justified, of wanting to know all the species of the globe is a utopian challenge practically. For example, the probable number of insect species is estimated at 4 million, of which less than 1 are known. If the present rate of discovery—about 2000 to 10,000 species per years, depending on sources were to continue, at least 300 to 1500 years would be required to describe them all! This estimate should be related with the evolution of the number of systematists, which will be referred to later.

Fortunately, systematics is a science that today has regained a little of its past glory, because it has been recognized as essential to understand environmental problems and, concomitantly, because of the arrival in force of techniques of molecular biology (Chap. 15, § 15.5.5; Giller et al., 1997). The huge research programme DIVERSITAS, initiated by UNESCO and strengthened by the Rio Conference in 1992, has set itself the goal of knowing all species of the globe. Surely one of its great challenges will be the soil fauna!

Though systematics is reborn, the same cannot be said of systematists, particularly in Universities. An American report (Daly, 1995) points out that without strong measures the last University-degreed systematic entomologist of the United States will disappear in 2017, just when the demand for such specialists would be highest to respond to the needs of the DIVERSITAS programme.

17.1.2 From taxonomy to ecological niche

Small biomass, great effect: nitrifying or methanogenic bacteria (Chap. 4, § 4.4.2) are good examples of key-species.

For the ecologist, naming an organism correctly is indeed the first hurdle to cross. But this stage must be followed by research on the role of this organism in the ecosystem: what does it eat, where does it live, how does it reproduce, what is its reaction to physicochemical factors, what is its strategy—in short, which is its

'... 'It thus follows that functions particularly sensitive to disruption will be those performed by a limited number of species' (Giller et al., 1997).

Sometimes the engineers in the soil are also the key-species. Who will replace an earthworm in a grassland soil if not another earthworm of the same species?

ecological niche? Is the studied species dominant, influencing the soil by its abundance or its biomass? Or, on the contrary, despite its low biomass or small population, is it irreplaceable in its exact function, at an essential crossroads of functioning of the soil? Both categories exist, represented by 'ecosystem engineers' (Jones et al., 1994; Chap. 13, § 13.6.4) and 'key-species', respectively, the latter having the formula for a precise operation, even if they are small in numbers.

During the twentieth century, soil science became deeply marked by the concept of zonality of soils and the decisive influence of the macroclimate on evolutionary processes (Chap. 5, § 5.5.1; Boulaine, 1989). Paton et al. (1995) proposed putting into perspective the importance of this paradigm and giving more weight to local physical factors such as cryoturbation, land movements at the scale of catenas, aeolian additions... and bioturbation. When soil evolution is approached in this more holistic manner, the importance of living organisms and their products will undoubtedly be still better perceived!

17.1.3 Interactions among organisms, scale dependence

It is often said: 'Ecology is physiology in the worst possible conditions!' But it can also be said 'Physiology is ecology in the most unrealistic conditions!'

The ecological niche (Chap. 12, § 12.5) is often first studied for a precise taxon, removed from its natural context: the autoecology of the species and its ecophysiology are then studied (Chap. 4, § 4.2.4). It is highlighted in environments where the ecological factors are most controlled: growth chambers, greenhouses, fermenters, etc. Interspecific competition is generally not tolerated.

'Values of these coefficients [of transfer of radioactive isotopes from soil to the plant] obtained under laboratory conditions, however, have little or no field value and progress must be made towards obtaining *in situ* values for these coefficients' (Killham, 1994).

In real conditions, the organisms are indeed permanently interrelated through predation, symbiosis, allelopathy, interspecific competition, rhizodeposition, etc. How then, for example, can we pretend to understand the real functioning of the rhizosphere at its correct organization level and in 'its' soil (Chap. 4, § 4.1.3; Chap. 15), by isolating a root tip, by sterilizing it and by inoculating one species of model bacterium in order to study its interaction with the plant? Indeed root physiologists often cannot do otherwise, unless they do very poor work... consequently greatly displease... field soil scientists!

A wide gap remains between the physiological and ecological approaches to soil organisms (Chap. 4, § 4.2.4; Chap. 15, § 15.5).

The transfer of results and interpretations from one simplified and controlled system to a complex, largely uncontrollable system is only in its infancy... Indeed, in scientific literature, many results are seemingly very convincing; for example, those that translate amounts of nitrate produced in a bioreactor to additions for a cultivated soil. Passing from $\mu g\ mL^{-1}$ to $kg\ ha^{-1}$ is indeed not without danger: it ignores the complexity of the natural system, the phenomena of scale dependence and emergent properties related to it, and also the regulatory mechanisms among organisms, with synergistic and antagonistic effects (Auger et al., 1992; Schneider, 1994).

'Pure cultures (...) are not sufficient and may even lead to erroneous interpretations to the extent that one attempts to attribute to each organism in the natural milieu all the powers that it manifests after isolation' (Pochon et al., in UNESCO, 1969).

Similarly, conclusions drawn from a reciprocal relationship of herbivore-producer type between two organisms often widely differ from those of a triangular relationship, in which the action of the predator or parasite is added (Chap. 4, 4.5.3; Chap. 10, § 10.7.3; Chap. 15, § 15.3.2). But this last is even less understood than the first.... Coleman et al. (in Edwards et al., 1988) proposed to study in particular the soil-plant-microorganisms-fauna interactions in microsites rich in labile carbon, therefore with large energy flux.

Triangular relationship...

The indispensable dialogue between reductionist and holistic approaches

Without ignoring technical problems that are real, the difficulties in comparing results of an observation *in situ* and another *in vitro* are often due to different perceptions of the object studied. The reductionist and holistic approaches are still too often opposed and reciprocally ignored (Chap. 1). Now, a true understanding of ecophysiological phenomena cannot be attained without both of them shared and integrated in a universal but 'plural' perception of scientific research.

Two perceptions to be reconciled.

Of necessity, the reductionist approach simplifies the systems with a view to controlling in them the maximum parameters; it thus will always remain incomplete because of lack of 'inter-object' vision. But the holistic approach, which tackles the problem systemically, will remain just the same because its methods are not applied to explain elementary mechanisms. It is a waste of effort and as much scientific aberration as asking each approach to imitate its counterpart. Leave each one to work with its tasks and responsibilities!

'Ecology must combine holism with reductionism if applications are to benefit society' (Odum, 1977).

But let each one also, in transmitting its results and by its scientific purpose, do what is necessary to enter into contact and have a true dialogue with the other. For that, however, there is still a long way to go...

An interesting approach to reducing this conceptual and methodological 'void' between simplified cultures in controlled media and the complex reality in the field could be employment of certain mapping techniques, or integration of the spatial dimension—we could say microspatial—in the study of plant-microorganism-soil interfaces.

A modern idea: 'minicartography'!

'We need to link specific methods such as noted here with soil thin-section studies. Such means will enable the inclusion of spatial dimensions to soil ecological studies; the addition of temporal ones provides the much-needed aspect of time as well' (Coleman and Crossley, 1996; Chap. 3, § 3.6.1).

The theory of 'hot spots of activity' (Chap. 13, § 13.5.2) is an appealing starting point for integrating microspatial and functional aspects in the soil.

For example, why not invent a sort of mini-GIS (geographic information system, Chap. 1, § 1.2.2; Chap. 7, § 7.1.2) at the scale of microaggregates, of the rhizosphere, of the epidermis of leaves, in other words a mini-GIS that would allow later recourse to the computing power of this methodology? This would be a good dynamization of certain microscopic observations (thin sections of humiferous horizons, point counts of fine pedological structures, scanning-electron microscopy, etc.) that we could realize by superposing the different analytical 'layers', as in a true GIS. Some first attempts have appeared (Binet and Curmi, 1992; Beare et al., 1995).

17.1.4 From edaphic organism to ecocomplex

The 'external' complexity of the soil is still less understood than its 'internal' complexity (Chap. 1, § 1.2.2; Fig. 1.1)!

If despite the difficulties just cited, understanding the relationship between soil and plant draws nearer and nearer, that which exists between soil and vegetation and thus at the level of communities still cruelly lacks fundamental laws and rules (Chap. 7, § 7.7.2). What are the relative speeds of evolution of vegetation and of soil? Is pedogenesis influenced by the vegetation or is it the other way round? Or both reciprocally, in a feedback system? What is the particular role, soil included, of a biogeocoenosis in its drainage basin and in relation to the other ecosystems of the ecocomplex? So many important questions must be better appreciated and established in future, for example with regard to the possible applications of soil science in the domains of environmental protection and agriculture!

Adding a supplementary explanatory and functional dimension to phytosociology through ecological study of mycorrhizae.

With respect to this problem of soil-vegetation relationship, one area appears to us most promising in soil ecology of the future, namely, study of mycorrhizae in relation to the vegetation units described by the phytosociologist (Smith and Read, 1997). The best understanding of fungus-plant relationships at the cell and organism level, the widest knowledge of specific relationships between the partners and the sudden emergence of molecular methods in soil-microbial ecology—these together could give us a glimpse into a field of investigation almost unused but undoubtedly of great potential richness. Mycorrhizae, given their capability for establishing intra- and interpopulation interactions in the soil (transfer of nutrients, promotion of growth, influence on vitality), would be able to add a new explanatory dimension to phytosociological studies.

We should remember that more than 80 per cent of plant species are, or could be, associated with mycorrhizal fungi.

17.2 SOIL BIOLOGY AND APPLIED SOIL SCIENCE

'Fundamental' soil biology often leads to applications; this has long been so in agronomy with the system of crop rotation and the addition of combined nitrogen promoted by leguminous plants (Chap. 16, § 16.2.1). But many other aspects are in full bloom, in particular those in the very vast domain of soil conservation. We shall only briefly mention here the fight against erosion, and detoxification, since the literature on these subjects is abundant (Häberli et al., 1991; Mosimann et al., 1991; Soltner, 1995; Duchaufour, 1997; Gisi et al., 1997) and we shall simply sketch the problems posed in soil science by the use of genetically modified organisms.

Multiple applications of soil biology in agronomy, forestry and environmental sciences.

'Farmers are the group in society most likely to assign a high value to these two functions of soil biodiversity (contribution to the productive capacity and to the resilience to environmental risks) because of their direct effects on production and risk reduction' (Giller et al., 1997).

17.2.1 Fight against erosion

Although the fight against soil erosion depends greatly on agricultural techniques and management of land, a better understanding of many mechanisms linked to soil organisms will enable us to fine-tune and specify the proposed directives (Singer and Ewing in Sumner, 2000):

- the comparative role of fungi and bacteria in aggregation of soil microstructures (Chap. 2, §§ 2.5.2, 2.5.3; Chap. 3, § 3.2.4; Mulder et al., in UNESCO, 1969);
- the positive or negative effects of very strong or very light bioturbation by earthworms or termites (Chap. 4, § 4.5.1; Chap. 5, § 5.3.3; Cuendet and Lorez, 2001);
- the effects of the fauna on undecomposed organic matter left on the surface of cultivated soils (through mulching, fallowing, etc.);
- effects on edaphic fauna and microflora of applications of compost in various conditions (Chap. 10, § 10.7);
- role of the genetic reservoir in anti-erosion landscaping structures (hedges, embankments, trenches, etc.) for edaphic organisms.

'In some regions of the United States and Europe, erosion has crossed two hundred tons per hectare per year and even reached five hundred tons' (Bourguignon, 1996).

17.2.2 Detoxification and bioremediation of soils

Soil pollution by heavy metals, radioisotopes or xenobiotic organic compounds is sometimes considered the principal environmental hazard of the future.

The environmental time-bomb of the twenty-first century?

'To concede that the soil could be polluted went counter to the belief—still very widespread—that soils have a limitless capacity for purification' (Rivière, 1998).

Innumerable compounds are today trapped in the soil more or less finally (Bollag and Loll, 1983; Bollag, 1991; Meyer, 1991; Duchaufour, 1997; Rivière, 1998). A better understanding of bioindication by key-organisms such as bioaccumulators (Chap. 4, § 4.5.3) could help in identifying a danger, very difficult and very expensive to highlight in any other manner (Pillet and Longet, 1989).

Soil organisms are—perhaps—good agents of detoxification. But would not the best solution be to avoid intoxication?

But the living organisms of soil may do more than bioindication! Today experiments are multiplying on the detoxification potential of soils that contain certain organisms, in particular bacteria (Chap. 10, § 10.7.1, Chap 11, § 11.3). The pessimistic opinions of some who think that the microflora will never succeed in destroying xenobiotic compounds retained in soils are opposed to those who see adaptations appear very rapidly in bacterial populations, capable of producing new enzymes to attack new substances. The two views are certainly relevant, but at different time and spatial scales. Another reason to continue research and controlled experiments!

Detoxification and bioremediation of soils, two supplementary domains in which soil organisms excel!

Higher plants are equally sought after, in attempts at phytoremediation for decontamination of the soil of heavy metals (Chap. 11, § 11.2). Certain 'star' species surely start to appear, but we are far yet from having universal extractors available, suitable for the great physicochemical diversity of soils.

17.2.3 Control of genetically modified organisms

Low but not zero risk.

The release, now or later, of genetically modified organisms into ecosystems is an area that has started to concern soil scientists. For example, we wish to multiply the power of degrading xenobiotic substances through transgenic bacteria (Wood, 1995). Unfortunately, this *a priori* positive utilization of the genetic genius still comes up against great ignorance of the ecology of soil organisms; there is danger of a genetically modified bacterium introduced into the soil multiplying uncontrolled or occupying a free or unexpected ecological niche.

'We still do not know what properties confer on a bacterium the ability to compete successfully with other soil organisms, or to colonize the surface of a root. At present, it is impossible (...) to design a successful inoculant for use in the field. This is largely due to the problems in knowing the precise nature of the ecological niche into which the organism is to be introduced' (Wood, 1995).

In addition to the possible modifications of ecological niches, transgenic organisms can also leave in the soil various products, residues of genetic manipulations. Donegan et al. (1997) showed for example that inhibitors of the enzyme protease, synthesized by a modified tobacco plant, remained active in the soil for at least 57 days with marked effects on soil organisms such as protozoans, springtails and nematodes that were exposed to it. Bavage et al. (in Cadisch and Giller, 1997) feel that the litter of manipulated plants could modify the process of evolution of organic matter.

Risk of a catastrophe happening is indeed very small, but the effects of one could be important. It is therefore essential to intensify research on the risks associated with introduction of transgenic organisms into the soil rather than to prohibit their use. Only an adequate combination of traditional methods of soil science with techniques of molecular biology will provide the right answers. Wood (1995) confirmed this view: 'Although there appears to be considerable potential for engineering microorganisms to carry out specific tasks in the soil, the lack of predictive information on the ecological consequences of introducing such organisms may preclude their general use.'

'There is considerable potential for use of genetically modified plants and microbes for agricultural and other purposes. The greater hurdle to overcome for safe and effective use of this technology is developing an understanding of how introduced organisms, genetically modified or otherwise, interact and compete with indigenous soil organisms and how this interaction varies across the full ecological range of soil conditions' (Killham, 1994).

17.3 SOIL BIOLOGY AND SOIL MODELLING

Models that provide scenarios for the future evolution of the planet are multiplying: temperature, concentration of atmospheric carbon dioxide, cycles of water and bioelements, general climatic changes, distribution of agricultural zones, etc., in which the soil appears to be more and more an essential compartment (Bryant and Arnold, 1994). It actually plays an important double role, as regulator of mineral exchanges and as converter and essential accumulator of organic matter, hence of energy (Chap. 5, § 5.1.1; Chap. 6, § 6.1.3). But this role is still not sufficiently understood: what are the proportions of particulate and humified organic matter? What are the respective turnovers of bacterial polysaccharides, humic acids, insolubilization humin, etc.? How do the root secretions, which can represent more than a third of the net production, influence soil organisms and their ability to fix carbon in their biomass?

'Soils are probably the last great frontier in the quest for knowledge about the major sources and sinks of carbon in the biosphere' (Coleman and Crossley, 1996).

All serious modelling of the carbon cycle must take into account the evolution of carbon in soils.

Appearing simple, these questions lead only to complex answers, as much in the field reality as in the establishment of scenarios. This is why recent models tend to improve and raise the number of 'compartments' of organic matter to be considered in modelling (Chap. 5, § 5.2.3; Powlson et al., 1996). But many aspects remain insufficiently studied and cannot still be introduced satisfactorily in the models, often because of lack of field knowledge:

Fortunately we are moving away from the first global models that presented the ensemble of soil as a 'black box' not to be opened under any pretext (!) because of its complexity.

- the precise role of the soil fauna as an often powerful accelerator of organic-matter transformations (Chap. 4, § 4.5.2; Chap. 5, § 5.2.4; de Ruiter et al., in Benckiser, 1997);
- comparison of different soils, in particular of cultivated soils with the virgin soils they are derived from;
- the role of peaty soils that themselves contain one-third of the organic matter blocked in soils (Chap. 9, § 9.2.1).

'More holistic modeling efforts of changes in soil organic matter, particularly ones which include soils, plants, herbivores and detritivores together, are more realistic in their outcomes than those which model the plants, heterotrophs and soil carbon pools separately' (Schimel, in Kareiva et al., 1993).

'In view of the criteria that should be satisfied by C-allocation models, we conclude that today there are no allocation models that satisfy all requirements' (Ågren and Wikström, 1993).

Parallel pursuit of studies *in situ* and improvement of models is indispensable so long as the conclusions of authors are still vague, ambiguous or too limited (Kirschbaum, 2000). Results are particularly contradictory when they relate to the behaviour of the organic-carbon reserve of soil. While the experiments, necessarily simplified, point to an increase in the organic-carbon stock with rise in temperature and/or of atmospheric CO_2 (Van Cleve et al., 1990; Anon, 1992), more global and often also more older models may give the opposite result (Melillo et al., 1982; Gobat et al., in Cebon et al., 1998).

As for field studies on local gradients, they emphasize the impossibility at present of being determined, because of the operation of other ecological factors (for example, the binding capacity of clays for organic matter, Chap. 3, § 3.6.2) and the variety of organic forms in the soil (Johnson, 1993).

17.4 SOIL BIOLOGY AND HUMAN SOCIETY

Gisi et al. (1997) recall in the nick of time the three principal functions of the soil: ecological, socioeconomic and immaterial.

Soils have always carried the mark of human civilizations: eroded red soils of the Mediterranean basin, caused by deforestation and overgrazing; peaty soils destroyed and converted to fuel or horticultural material; alpine soils victim of ski establishments; soils lost as dust in the United States during the Dust Bowl of the 1930s; soils of mining areas overloaded with heavy metals, etc. Besides, in industrialized countries urbanization is the greatest 'eater' of soils (Häberli et al., 1991) as much in quality (the soil is often totally destroyed) as in quantity [hectares disappear one after the other, for example at the rate of one square metre per second in Switzerland (O.F.S., 2001)].

Looking to their future resolution, current environmental problems necessitate a new interpretation of the common past of civilizations and soils. Why, until now, has there been so much insistence in history on the *quantitative* conquest of soil? For centuries, the strength

Which value are you talking of?

It is striking to note that the commercial value of soil is inversely proportional to its biological value: there is no square metre of soil dearer than that—with no biological value because it is totally concreted over—at the commercial centre of large cities and square metre cheaper than an extremely rare peat soil of an ombrogenic bog in a remote area or than a superb PODZOSOL of a forgotten valley in the Alps.

of a lord or a nation was indicated by possession of more hectares or square kilometres of land than the neighbour. Indeed food requirement often has laid down the law, but the simple desire to own vast areas is not behindhand in this cornering of soil. This 'metrical' aspect of the man-soil relationship in the past is rather well known.

But to imagine the renovated links that our civilizations would have to weave with our land, it becomes urgent to better understand the *qualitative* relationship of man with the soil during centuries past. Boulaine (1989) gave some examples of successes or failures of this qualitative relationship, such as the rescue of fertility of cultivated soils through the discovery and moderate application of inorganic fertilizers in the nineteenth century. But we must know more! A good means would be to intensify pedological research in connected disciplines such as archaeology, history, geomorphology and palaeoecology.

In this qualitative-historical rediscovery of soil, biologists are in a position to provide considerable information through the preserved marks of ancient uses: pollens, diatoms, residues of wild or cultivated plants, animal debris, relict organic matter, etc. An immense field is open, which could see historians, soil scientists and systematic biologists converge.... This would be an essential addition to the idea of sustainability, a concept eminently bound to soil conservation and which will succeed in future only if it is (also) inspired by the past.

'We were torn from the bonds to soil—the connections which limited action, making practical virtue possible—when modernization insulated us from plain dirt, from toil, flesh, soil and grave' (Illich et al., 1991).

'As we approach the end of the twentieth century, a century of extraordinary and technological achievements, it is becoming clear that the continued survival of our civilization depends more than ever upon our relationship with the land and soil' (Wood, 1995).

While regional studies or those related to a particular civilization abound, no holistic view of general human history and its biological marks left in soils is yet available.

By way of conclusion ...

As Häberli et al. (1991) put it: 'The problems bound to soil utilization are complex: they show many psychosocial, economic and geographic parameters. As a general rule, analysis shows no simple causes, easy to highlight and correct. Supplementary efforts are therefore necessary to ensure objective information on the life and multiple functions of the soil, and to maintain the public debate on the dimensions of the problem and possible solutions.'

'Therefore, we issue a call for a philosophy of soil: a clear, disciplined analysis of that experience and memory of soil, without which neither virtue nor some new kind of subsistence can be' (Illich et al., 1991).

What better reward for the authors of this book than to see their interdisciplinary will cross the frontiers of soil biology and reach persons and media seemingly not directly concerned with study of the 'earth'! Scientists of other fields, sociologists, politicians, philosophers, practitioners, all of whom would start listening to the soil, especially the *living soil*, perhaps the least known of all...

> 'The aim of sustainable resource management programmes must be to maintain or improve the quality of soil in its broadest sense (...). It is clear that these [environmental] problems will not be solved by scientists alone, but scientists will have an important role to play both as researchers and as teachers. Despite the ever-increasing trend towards specialisation, we shall need to draw on all available knowledge of land and soils. There is a role for everyone to play in the future of our planet' (Wood, 1995).

... or of visual reflection

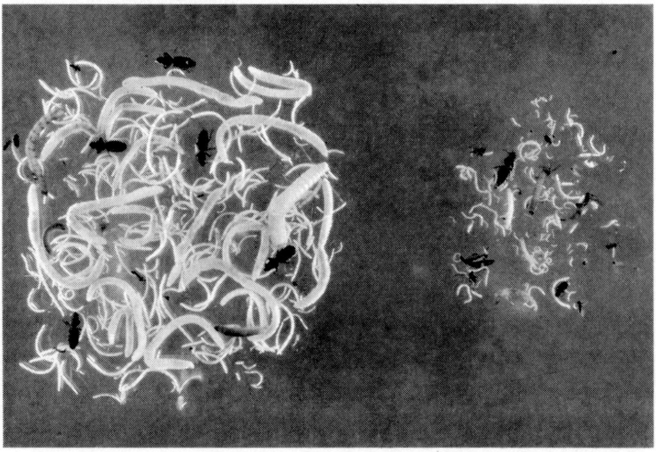

Comparison of two extractions of the soil fauna by the Berlese-Tullgren extractor (0.25 m^2, 0 to 20 cm depth) (photo Y. Borcard): similar soils (CALCISOLS), similar altitudes (500 m), similar climates (oceanic temperate of the Switzerland Plateau), identical crop (wheat). At left, under conditions of biological agriculture; at right, under conditions of conventional agriculture. Life in the soil....

BIBLIOGRAPHY

Aandahl A.R., Buol S.W., Hill D.E. and Bailey H.H. (eds.). 1974. *Histosols. Their Characteristics, Classification, and Use.* SSSA Spl Pub. Ser. 6, Madison WI, U.S.A.

Aber J.D. and Melillo J.M. 1991. *Terrestrial Ecosystems.* Saunders College Publishing, Philadelphia PA, U.S.A.

Acot P. 1988. *Histoire de l'Écologie.* PUF, Paris, France.

Aerts R. and de Caluwe H. 1999. Nitrogen deposition effects on carbon dioxide and methane emissions from temperate peatland soils. *Oikos,* **84**: 44-54.

Aeschimann D. and Burdet H.M. 1994. *Flore de la suisse et des territoires limitrophes.* Éditions du Griffon, Neuchâtel, Switzerland.

AFES. 1995. *Référentiel pédologique.* Coll. Techniques et Pratiques, INRA Éditions, Paris, France.

AFES. 1998. *A Sound Reference Base for Soils. The 'Référentiel pédologique'.* INRA Éditions, Paris.

Affolter F., Auroi C. and Matthey W. 1981. La biocénose des habitąts larvaires de *Hybomitra bimaculata* (Macquart) (Diptères, *Tabanidae). Revue suisse Zool.* **88**(4): 965-975.

Ågren G.I. and Wirkström J.F. 1993. Modelling carbon allocation—a review. *N.Z. Jl Forestry Sci.* **23**: 343-353.

Aguer J.-P. and Richard C. 1996. Reactive species produced on irradiation at 365 nm of aqueous solutions of humic acids. *J. Photochem. Photobiol. A,* **93**(2-3): 193-198.

Aiken G.R., McKnights D.M., Warshaw R.L. and MacCarthy P. (ed.). 1985. *Humic Substances in Soil, Sediment and Water.* Wiley & Sons, New York.

Akkermann R. 1982. Regeneration von Hochmooren. *Inf. Natursch. Landschaftspfl.,* **3**, Wardenburg.

Allen, T.F.H. and Starr T.B. 1982. *Hierarchy: Perspectives for Ecological Complexity.* The University of Chicago Press, Chicago and London.

Amann R.I., Ludwig W., Schleifer, K.-H. 1995. Phylogenetic identification and *in situ* detection of individual microbial cells without cultivation. *Microbiol. Rev.* **59**: 143-169.

Amberger A. 1983. *Pflanzenernährung*. UTB 846, Verlag Eugen Ulmer, Stuttgart, Germany.

Anderson J.M., Rayner A.D.M. and Walton D.W.H. 1984. *Invertebrate-Microbial Interactions*. Cambridge University Press, Cambridge.

Anderson T.A., Guthrie E.A. and Walton B.T. 1993. Bioremediation in the rhizosphere. *Environ. Sci. Technol.* **27**: 2630-2636.

André M.H., Noti M.I. and Lebrun P. 1994. The soil fauna: the other last biotic frontier. *Biodiversity and Conservation*, **3**: 45-56.

Andrews J.H. and Hirano S.S. 1992. *Microbial Ecology of Leaves.* Springer Verlag, New York.

Anon N. 1992. *Soil-warming Experiments in Global Change Research*. Report of a Workshop held in Woods Hole MA, 1991. National Science Foundation, Washington DC.

Aragno M. 1985. Traitement intégré des déchets par méthanisation et compostage. *Bull. ARPEA*, **133**: 13-19.

Aragno M. 1994. Vers une gestion cyclique des déchets d'origine naturelle. In: Analyse et maîtrise des valeurs naturelles. *Actes du Colloque transfrontalier.* Université de Franche-Comté, Besançon, France. pp. 33-36.

Archibold O.W. 1995. *Ecology of World Vegetation*. Chapman & Hall, London.

Armstrong W., Brändle R. and Jackson M.B. 1994. Mechanisms of flood tolerance in plants. *Acta Bot. Neerl.* **43**: 307-358.

Arndt U., Nobel W. and Schweizer B. 1987. *Bioindikatoren: Möglichkeiten, Grenzen und neue Erkenntnisse.* E. Ulmer, Stuttgart.

Arnold C. and Gobat J.-M. 1998. Modifications texturales des soils alluviaux de la Sarine (Suisse) par les Bryophytes. *Écologie*, **29(3)**: 483-492.

Arpin P., Betsch J.-M., Ponge J.-F., Vannier G., Blandin P., Dajoz R. and Luce J.-M. 2000. Les invertébrés dans l'écosystème forestier: expression, fonction, gestion de la diversité. *Les Dossiers forestiers*, **9**. Office national des forêts, Fontainebleau, France.

Aubert G. 1978. *Méthodes d'analyses des sols*. CRDP, Marseille.

Auger P., Baudry J. and Fournier F. 1992. *Hiérarchies et échelles en écologie*. Naturalia Publications, SCOPE France.

Augier J. 1966. *Flore des Bryophytes*. Éditions Paul Lechevalier, Paris.

Ayres P.G. and Boddy L. (eds.). 1986. *Water, Fungi and Plants*. Cambridge University Press, Cambridge.

Bachelier G. 1978. *La faune des sols, son écologie et son action*. ORSTOM, Paris.

Bader K. 1974. Die Bedrohung des humusbildenden Bodenfauna durch Fabrikimmissionen. *Schweiz. Zeitschrift für Forstwesen*, **125**(6): 388-393.

Bailey R.G. 1996. *Ecosystem Geography*. Springer Verlag, New York.

Baize D. 1988. *Guide des analyses courantes en pédologie*. INRA Éditions, Paris.

Baize D. and Jabiol B. 1995. *Guide pour la description des sols*. Coll. Techniques et Pratiques, INRA Éditions, Paris.

Baker A.J.M. 1987. Metal tolerance. *New Phytol.* **106**: 93-111.

Baker K.F. and Cook R.J. 1974. *Biological Control of Plant Pathogens*. Freeman and Co., New York.

Balesdent J. 1982. *Étude de la dynamique de l'humification de sols de prairies d'altitude (Haut-Jura) au moyen des datations ^{14}C des matières organiques*. Doct.-Ing. thesis, University of Nancy, France.

Barbault R. 1990. *Écologie générale*. Masson, Paris.

Barbosa P., Krischik V.A. and Jones C.G. (eds.) 1991. *Microbial Mediation of Plant-Herbivore Interactions*. Wiley and Sons, New York.

Barron G.L. and Thorn R.G. 1987. Destruction of nematodes by species of *Pleurotus*. *Can. J. Bot.* **65**: 774-778.

Bartha K. and Bordeleau L. 1969. Cell-free peroxidases in soil. *Soil. Biol. Biochem.* **1**: 139-143.

Bascomb C.L., Banfield C.F. and Bruton G.O. 1977. Characterization of peaty materials from organic soils (Histosols) in England and Wales. *Geoderma*, **19**: 131-147.

BassiriRad H. 2000. Kinetics of nutrient uptake by roots: responses to global change. *New Phytol.* **147**: 155-169.

Baur B., Joshi J., Schmid B., Hänggi A., Borcard D., Stary J., Pedroli-Christen A., Thommen G.H., Luka H., Rusterholz H.P., Oggier P., Ledergerber S. and Erhardt A. 1996. Variation in species richness of plants and diverse groups of invertebrates in three calcareous grasslands of the Swiss Jura mountains. *Revue suisse de zoologie* **103**: 801-833.

Beare M.H., Coleman D.C., Crossley D.A., Hendrix P.F. and Odum E.P. 1995. A hierarchical approach to evaluating the significance of soil biodiversity to biogeochemical cycling. *Pl. Soil*, **170**: 5-22.

Beare M.H., Parmelee R.W., Hendrix P.F. and Cheng W. 1992. Microbial and faunal interactions and effects on litter nitrogen and decomposition in agroecosystems. *Ecol. Monographs*, **62**(4): 569-591.

Beeby A. and Brennan A.M. 1997. *First Ecology*. Chapman and Hall, London.

Beffa T., Blanc M., Lyon P.-F., Vogt G., Marchiani M., Lott Fischer J. and Aragno M. 1996a. Isolation of *Thermus* strains from hot composts (60-80°C). *Appl. Environ. Microbiol.* **62**: 1723-1727.

Beffa T., Blanc M. and Aragno M. 1996b. Obligately and facultatively autotrophic sulfur and hydrogen-oxidizing thermophilic bacteria isolated from hot composts. *Arch. Microbiol.* **165**: 34-40.

Begon M., Harper J.L. and Townsend C.R. 1986. *Ecology*. Blackwell Scientific Publications, Oxford, U.K.

Behan-Pelletier V.M. and Bissett B. 1993. Biodiversity of nearctic soil arthropods. *Can. Biodiv.* **2**: 5-14.

Benckiser G. (ed.). 1997. *Fauna in Soil Ecosystems*. Marcel Dekker Inc., New York.

Bernard F. 1968. *Les Fourmis* (Hymenoptera Formicidae) *d'Europe occidentale et septentrionale*. Masson, Paris.

Bernier N. 1997. Fonctionnement biologique des humus et dynamique des pessières alpines. Le cas de la forêt de Macot-La-Plagne (Savoie). *Ecologie*, **28**(1): 23-44.

Bernier N. and Ponge J.-F. 1994. Humus form dynamics during the sylvogenetic cycle in a mountain spruce forest. *Soil Biol. Biochem.* **26**(2): 183-220.

Berthet P. 1964. L'activité des Oribates d'une chênaie. *Mémoires de l'Institut royal des Sciences naturelles de Belgique.* Brussels. No. 152.

Berthoud G. 1982. *Contribution à la biologie du hérisson* (Erinaceus europaeus L.) *et applications à sa protection*. Doctoral thesis, University of Neuchâtel, Switzerland.

Beyrle A., Smith S., Peterson R.L. and Franco C.M.M. 1995. Colonisation of *Orchis morio* protocorms by a mycorrhizal fungus: effects of nitrogen nutrition and glyphosphate in modifying the responses. *Can. J. Bot.* **73**: 1128-1140.

Binet F. and Curmi P. 1992. Structural effects of *Lumbricus terrestris* (Oligochaeta: Lumbricidae) on the soil-organic matter system: micromorphological observations and autoradiographs. *Soil Biol. Biochem.* **24**: 1519-1523.

Binet F. and Le Bayon R.C. 1999. Space-time dynamics *in situ* of earthworm casts under temperate cultivated soils. *Soil Biol. Biochem.* **31**: 85-93.

Blackford J. 2000. Paleoclimatic records from peat bogs. *TREE,* **15**: 193-198.

Blandin P. 1986. Bioindicateurs et diagnostic des systèmes écologiques. *Bull. Ecol.* **17**: 215-307.

Bliefert C. and Perraud R. 2001. *Chimie de l'environnement: air, eau, sols, déchets.* De Boek Université, Paris-Brussels.

Blondel J. 1986. *Biogéographie évolutive*. Masson, Paris.

Blondel J. 1995. *Biogéographie. Approche écologique et évolutive*. Masson, Paris.

Blume H.-P., Felix-Henningsen P., Fischer W.R., Frede H.G., Horn R. and Stahr K. (eds.). 1996*ff*. *Handbuch der Bodenkunde*. Ecomed Verlagsgesellschaft, Landsberg/Lech.

Bogon K. 1990. *Landschnecken*. Natur-Verlag, Augsburg.

Bollag J.-M. 1991. Enzymatic binding of pesticide degradation products to soil organic matter and their possible release. In: *Pesticide Transformation Products: Fate and Significance in the Environment*. American Chemical Society Symposium Series Vol. **459**, Washington DC.

Bollag J.-M. and Loll M.J. 1983. Incorporation of xenobiotics into soil humus. *Experientia,* **39**: 1221-1231.

Bonneau M. and Souchier B. (eds.). 1994. *Pédologie 2: Constituants et propriétés du sol*. Masson, Paris.

Borcard D. 1981. Utilisation des pièges Barber dans l'étude des Carabides forestiers sur un transect Grand-Marais - Chasseral. *Bull. Soc. neuchâtel. Sci. nat.* **104**: 107-118.

Borcard D. 1982. Étude des communautés de *Carabidae (Coleoptera)* dans quelques associations forestières de la région neuchâteloise. Aspects statistiques. *Bull. Soc. entomol. suisse,* **55**: 169-179.

Borcard D. 1988. *Les Acariens Oribates des sphaignes de quelques tourbières du Haut-Jura suisse*. Thesis, University of Neuchâtel, Switzerland.

Borcard D. 1991. Les Oribates des tourbières du Jura suisse *(Acari, Oribatei)*: écologie. I. Quelques aspects de la communauté d'Oribates des sphaignes de la tourbière du Cachot. *Rev. suisse Zool.* **98**(2): 303-317.

Bouché M. 1972. *Lombriciens de France. Écologie, systematique*. INRA, Paris.

Bouché M. 1984. Les vers de terre. *La Recherche,* **156**: 796-804.

Bouché M. 1990. *Ecologie opérationelle assistée par ordinateur*. Masson, Paris.

Boulaine J. 1989. *Histoire des pédologues et de la science des sols.* INRA, Paris.

Boulaine J. 1997. Histoire abrégée de la science des sols. *Étude et Gestion des Sols*, **4**(2): 141-151.

Boulet R., Chauvel A., Humbel F.X. and Lucas Y. 1982. Analyse structurale et cartographie en pédologie. *Cahiers ORSTOM, série Pédologie*, **XIX**(4): 309-351.

Bourguignon C. 1996. *Le sol, la terre et les champs*. Sang de la Terre, Paris.

Bouyer Y. 1999. Dynamisme du fer depuis les marais et tourbes de la Vallée des Ponts-de-Martel jusqu'à la résurgence de la Noiraigue. I. Mobilisation et immobilisation du fer dans la pédosphère de la vallée des Ponts. *Bull. Soc neuchâtel. Sci. nat.* **122:** 113-143.

Bouyer Y. and Pochon M. 1980. La migration du fer en milieu marécageux dans le Haut-Jura et dans le Plateau molassique suisse. *Bull. Soc. suisse Pédol.* **4**: 42-48.

Boyd R., Gasper P. and Tront J.D. 1992. *The Philosophy of Science*. MIT Press, Cambridge MA, U.S.A.

Brady N.C. and Weil R.R. 2002. *The Nature and Properties of Soils*. Prentice Hall, Upper Saddle River NJ, U.S.A. 13th ed.

Braun-Blanquet J. 1964. *Pflanzensoziologie. Grundzüge der Vegetationskunde*. Springer Verlag, Vienna-New York.

Braun-Blanquet J., Pallmann H. and Bach R. 1954. Pflanzensoziologische und bodenkundliche Untersuchungen im Schweizerischen Nationalpark und seinen Nachbargebieten. *Ergebnisse der wissenchaftlichen Untersuchungen des Schweizerischen Nationalpark* **4** (Neue Folge).

Brauns A. 1954. *Terricole Dipterenlarven*. 'Musterschmidt' Wissenschaftlicher Verlag, Gottingen.

Brauns A. 1968. *Praktische Bodenbiologie*. Fischer Verlag, Stuttgart, Germany.

Brawner C.O. and Radforth N.W. 1977. *Muskeg Research Conference, 15th, Edmonton*. University of Toronto Press, Toronto, Canada.

Bridgham S.D., Pastor J., Janssens J.A., Chapin C. and Malterer T.J. 1996. Multiple limiting gradients in peatlands: a call for a new paradigm. *Wetlands*, **16**(1): 45-65.

Brolemann H.W. 1930. *Faune des myriapodes de France. Chilopodes*. Imprimerie toulousaine, Paris.

Bruckert S. and Gaiffe M. 1985. *Les sols de Franche Comté*. CUER, University of Franche-Comté, Besançon, France.

Brunner H., Nievergelt J., Peyer K., Weisskopf P. & Zihlmann U., 2002. *Klassifikation der Böden der Schweiz*. FAL Reckenholz, Zürich.

Bryant R.B. and Arnold R.W. (eds.). 1994. *Quantitative Modeling of Soil Forming Processes*. SSSA Special Publication No. 39. Soil Science Society of America, Madison WI, U.S.A.

Bullock P., Fedoroff N., Jongerius A., Stoops G. and Tursina T. 1985. *Handbook for Soil Thin Section Description*. Waine Research Publications, Wolverhampton, U.K.

Bureau F. 1995. *Évolution et fonctionnement des sols en milieu alluvial peu anthropisé*. Thesis No. 1418, EPFL, Lausanne, Switzerland.

Bureau F., Guenat C., Huber K. and Védy J.-C. 1994. Dynamique des sols et de la végétation en milieu alluvial carbonaté. *Écologie*, **25**(4): 217-230.

Burgeff H. 1961. *Mikrobiologie des Hochmoores*. Fischer Verlag, Stuttgart, Germany.

Burns R.G. (ed.). 1978. *Soil Enzymes*. Academic Press, London.

Burns R.G. 1982. Enzyme activity in soils: location and a possible role in microbial ecology. *Soil Biol. Biochem.* **14**: 423-427.

Buttler A. 1987. *Étude écosystémique des marais non boisés de la rive sud du lac de Neuchâtel (Suisse).* Thesis, University of Neuchâtel, Switzerland.

Buttler A. 1990. Quelques aspects climatiques dans les marais non boisés de la rive sud du lac de Neuchâtel (Suisse). *Bull. Soc. neuchâtel. Sci. nat.* **113**: 217-230.

Buttler A. and Gallandat J.-D. 1989. Phytosociologie des prairies humides de la rive sud du lac de Neuchâtel (Suisse) et modèle de succession autogène. *Phytocoenologia,* **18**(1): 129-158.

Buttler A. and Gobat J.-M. 1991. Les sols hydromorphes des prairies humides de la rive sud du lac de Neuchâtel (Suisse). *Bull. Ecol.* **22**(3-4): 405-418.

Buttler A., Bueche M., Cornali P. and Gobat J.-M. 1985. Historischer und ökologischer Ueberblick über das Südostufer des Neuenburger Sees. *Telma* (Hannover) **15**: 31-42.

Buttler A., Dinel H., Lévesque M. and Mathur S.P. 1991. The relation between movement of subsurface water and gaseous methane in a basin bog with a novel instrument. *Can. J. Soil Sci.* **71**: 427-438.

Buttler A., Warner B.G., Grosvernier P. and Y. Matthey. 1996. Vertical patterns of testate amoebae (Protozoa: Rhizopoda) and peat-forming vegetation on cutover bogs in the Jura, Switzerland. *New Phytol.* **134**: 371-382.

Cadisch G. and Giller K.E. (eds.). 1997. *Driven by Nature. Plant Litter Quality and Decomposition*. CAB International, Oxford, U.K.

Calderoni G. and Schnitzer M. 1984. Effects of age on the chemical structure of paleosol humic and fulvic acids. *Geochim. Cosmochim. Acta,* **48**: 2045-2051.

Callot G., Chamayou H., Maertens C. and Salsac L. 1982. *Les interactions sol-racine. Incidence sur la nutrition minérale.* INRA, Paris.

Campbell N.A. and Mathieu R. 1995. *Biologie.* Saint-Laurent, Quebec, Canada.

Campbell N.A., Reece J.B. and Mitchell L.G. 1999. *Biology,* 5th edition. Benjamin Cummings Publishing Co., Inc., Menlo Park CA, U.S.A.

Carles J. 1967. *La nutrition de la plant.* Que-sais-je?, PUF, Paris.

Carleton T.J. and Read D.J. 1991. Ectomycorrhizae and nutrient transfer in conifer-feather moss ecosystems. *Can. J. Bot.* **69**: 778-785.

Carter M.R. (ed.). 1993. *Soil Sampling and Methods of Analysis*. Canadian Society of Soil Science. Lewis Publishers, Boca Raton FL, U.S.A.

Cebon P. Davies H., Dahinden U., Imboden D. and Jäger C. (eds.). 1998. *Views from the Alps: Regional Perspectives on Climate Change.* MIT Press, Cambridge MA, U.S.A.

Chalvignac M.A. 1971. Stabilité et activité d'un système enzymatique dégradant le tryptophane dans divers types de sols. *Soil Biol. Biochem.* **3**: 1-7.

Chamayou H. et Legros J.-P. 1989. *Les bases physiques, chimiques et minéralogiques de la science du sol.* PUF, Paris.

Chan Y.K., Barraquio W.L. and Knowles R. 1994. N_2-fixing pseudomonads and related soil bacteria. *FEMS Microbiology Reviews,* **13**: 95-117.

Chason D.B. and Siegel D.J. 1986. Hydraulic conductivity and related physical properties of peat, Lost River Peatland, northern Minnesota. *Soil Sci.* **142(2)**: 91-99.

Chauvin R. 1967. *Le Monde des Insectes*. Hachette, Paris.

Chen S.K., Edwards C.A. and Subler S. 2001. A microcosm approach for evaluating the effects of the fungicides benomyl and captan on soil ecological processes and plant growth. *Appl. Soil Ecol.* **18:** 69-82.

Cherix D. 1986. *Les fourmis des bois*. Payot, Lausanne, Switzerland.

Chinery M. 1986. *Insectes d'Europe occidentale*. Arthaud, Paris.

Clarholm M. 1985. Interactions of bacteria, protozoa and plants leading to mineralization of soil nitrogen. *Soil Biol. Biochem.* **17**: 181-187.

Clarholm M. 1989. Effect of plant-bacterial-amoebal interaction on plant uptake of nitrogen under field conditions. *Biol. Fertil. Soils*, **8**: 373-378.

Clements F. 1916. *Plant Succession*. Carnegie Institution, Washington DC.

Clubbe C.P. 1980. *The Colonization and Succession of Fungi in Wood*. The International Research Group on Wood Preservation. Doc. IRG/WP/107.

Clymo R.S. 1965. Experiments on breakdown of *Sphagnum* in two bogs. *J. Ecol.* **53**: 747-758.

Clymo R.S. 1970. The growth of *Sphagnum:* methods of measurement. *J. Ecol.* **58**: 13-49.

Clymo R.S. and Reddaway E.J.F. 1971. Productivity of *Sphagnum* (bog moss) and peat accumulation. *Hydrobiologia*, **12**: 181-192.

Coiffait H. 1960. Les Coléopteres du sol. *Actualités scient. et industr.* **1260**. Hermann, Paris.

Coineau J., Cléva R. and du Chatenet G. 1997. *Ces animaux minuscules qui nous entourent*. Delachaux & Niestlé, Lausanne, Switzerland.

Coleman D.C. and Crossley Jr D.A. 1996. *Fundamentals of Soil Ecology*. Academic Press, San Diego CA, U.S.A.

Collet C. *Systèmes d'information géographique en mode image*. Coll. Gérer l'environnement **7**, Presses Polytechniques et Universitaires Romandes, Lausanne, Switzerland.

Collinson A.S. 1988. *Introduction to World Vegetation*. Unwin Hyman, London.

Combes C. 2001. *Les associations du vivant. L'art d'être parasite*. Flammarion Paris.

Conrad R. 1993. Mechanisms controlling methane emissions from wetland rice fields. In: *Biochemistry of Global Change: Radiatively Active Trace Gases*. Chapman and Hall, New York.

Contat F., Claivoz J., Matthey W. and Borcard, D. 1998. Étude de la pédofaune proche de l'usine d'aluminium de Steg. II. Relation entre la bioaccumulation du fluor par quelques groupes d'invertébrés et la teneur en fluor du milieu environnemental. *Bull. Murithienne,* **116**: 33-39.

Cornali P. 1992. *Écologie des pinèdes* (Pinus sylvestris) *de la rive sud du lac de Neuchâtel (Suisse)*. Thesis, University of Neuchâtel, Switzerland.

Coulson J.C. and Butterfield J. 1978. An investigation of the biotic factors determining the rates of plant decomposition on blanket bog. *J. Ecol.* **66**: 631-650.

Coûteaux M.-M., Bottner P. and Berg B. 1995. Litter decomposition, climate and litter quality. *TREE,* **10**(2): 63-66.

CPCS. 1967. *Classification des sols*. Commission de Pédologie et de Cartographie des Sols. Multigraphed.

Crawley M.J. 1997. *Plant Ecology*. Blackwell Science, Oxford, U.K.

Cress W.A., Johnson G.V. and Barton L.L. 1986. The role of endomycorrhizal fungi in iron uptake by *Hilaria jamesii*. *J. Pl. Nutr.* **9**: 547-556.

CRM, CRP, CRC. 1992. *Formulaires et tables. Mathématiques, physique, chimie*. Editions du Tricorne, Geneva, Switzerland.

Crowson R.A. 1981. *The Biology of the Coleoptera*. Academic Press, London.

Cuendet, G. 1985. Répartition des Lombriciens *(Oligochaeta)* dans la Basse Engadine, le Parc national et le Val Müstair (Grisons, Suisse). *Revue suisse de zoologie,* **92**: 145-163.

Cuendet G. and Lorez F. 2001. Effet secondaire néfaste des vers de terre sur les prairies alpines. *Revue suisse Agric.* **33**(6): 253-259.

Cuendet G., Suter E. and Stähli R. 1997. Peuplements lombriciens des prairies permanentes du Plateau suisse. *Cahier de l'Environnement No. 291, Sol*. Office fédéral de l'environnement, des forêts et du paysage, Berne, Switzerland.

Curl E.A. and Truelove B. 1986. *The Rhizosphere*. Springer Verlag, Heidelberg, Germany.

Curry J.P. 1994. *Grassland Invertebrates*. Chapman and Hall, London.

Daget P. and Godron M. 1982. *Analyse de l'écologie des espèces dans les communautés.* Masson, Paris.

Dajoz R. 1996. *Précis d'écologie*. Dunod, Paris.

Daly H.V. 1995. Endangered species: doctoral students in systematic entomology. *Amn Entomologist* (Spring): 55-59.

Damman A.W.H. 1978. Distribution and movement of elements in ombrotrophic peat bogs. *Oikos,* **30**: 480-495.

Darbyshire J.F. (ed.). 1994. *Soil Protozoa*. CAB International, Wallingford, U.K.

Darwin C. 1877. *Les plantes insectivores*. Rheinwald et Cie, Paris.

Darwin C. 1891. *The Formation of Vegetable Mould through the Action of Worms, with Observations of their Habits*. John Murray, London.

Davet P. 1996. *Vie microbienne du sol et production végétale.* INRA, Paris.

Davet P. and Rouxel F. 1997. *Détection et isolement des champignons du sol.* Tec & Doc, INRA, Paris.

Davidson E.A., Trumbore S.E. and Amundson R. 2000. Soil warming and organic carbon content. *Nature, Lond.* **408**: 789-790.

de Bertoldi M., Segui P., Lemmes B. and Papi T. (eds.). 1996. *The Science of Composting*. Blackie Academic and Professional, London.

de Foucault B. 1986. *La phytosociologie sigmatiste: une morphophysique.* Laboratoire Botanique Université de Lille II, Lille, France.

De Lapparent A. 1911. *Abrégé de géologie*. Masson, Paris.

de Puytorac P., Grain J. and Mignot J.-P. 1987. *Précis de Protistologie*. Boubée, Paris.

De Rosnay J. 1975. *Le Macroscope. Vers une vision globale.* Coll. Points, Ed. du Seuil, Paris.

de Wit, H.C.D. 1997. *La vie racontée. Une biographie de la biologie*. PPUR, Lausanne, Switzerland.

Delaloye R. and Reynard E. 2001. Les éboulis gelés du Creux du Van (chaîne du Jura, Suisse). *Environnements périglaciaires,* **8**. Bulletin 26 de l'Association Française du Périglaciaire. Paris-Grenoble.

Delamare-Deboutteville C. 1951. *Microfaune du sol des pays tempérées et tropicaux.* Hermann, Paris.

Delarze R., Gonseth Y. and Galland P. 1998. *Guide des milieux naturels de Suisse.* Delachaux & Niestlé, Lausanne-Paris.

Delbès C., Moletta R. and Godon J.-J. 2000. Monitoring of activity dynamics of an anaerobic digester bacterial community using 16S rRNA PCR-single-Strand Conformation Polymorphism analysis (SSCP). *Environ. Microbiol.* **2:** 506-515.

Delcourt H.R. and Delcourt P.A. 1988. Quaternary landscape ecology: relevant scales in space and time. *Landscape Ecol.* **2**: 23-44.

Deléage J.-P. 1991. *Histoire de l'écologie, une science de l'homme et de la nature.* La Découverte, Paris.

Delhaye L. and Ponge, J.-F. 1993. Étude des peuplements lombriciens et des caractères morphologiques des humus dans la réserve biologique de La Tiliaie. *Bull. Écol.* **24**: 41-51.

Delpech R., Dumé G. and Galmiche P. 1985. *Typologie des stations forestières. Vocabulaire.* IDF, Paris.

Demange J.-M. 1981. *Les Mille-pattes. Myriapodes.* Boubée, Paris.

Dighton J. 1983. Phosphatase production by mycorrhizal fungi. *Pl. Soil,* **71**: 455-462.

Dilworth M.J. and Glenn A.R. (eds.). 1991. *Biology and Biochemistry of Nitrogen Fixation.* Elsevier, New York.

Dindal D.L. (ed.). 1980. *Soil Biology as Related to Land Use Practices.* EPA, Washington DC.

Dindal D.L. 1990. *Soil Biology Guide.* Wiley, New York.

Dix N.J. 1985. *Fungal Ecology.* Chapman and Hall, London.

Dix N.J. and Webster J. 1995. Changes in relationship between water content and water potential after decay and significance for fungal successions. *Trans. Brit. Mycol. Soc.* **85**: 649-653.

Doebereiner J. and Pedrosa F.O. 1987. *Nitrogen-fixing Bacteria in Nonleguminous Crop Plants.* Springer Verlag, Heidelberg, Germany.

Dokouchaev V.V. 1883. *Le chernozem russe.* Thesis, University of St Petersburg.

Donegan K.K., Seidler R.J., Fieland V.J., Schaller D.L., Palm C.J., Ganio L.M., Cardwell D.M. and Steinberger Y. 1997. Decomposition of genetically engineered tobacco under field conditions: persistence of the proteinase inhibitor I product and effects on soil microbial respiration and protozoan, nematode and microarthropod populations. *J. Appl. Ecol.* **34**: 767-777.

Donker M.H., Readeker M.H. and van Straalen N.M. 1996. The role of zinc regulation in the zinc tolerance mechanism of the terrestrial isopod *Porcellio scaber. J. Appl. Ecol.* **33**: 955-964.

Driessen P.M. and Dudal R. (eds.). 1991. *The Major Soils of the World.* Agricultural University, Wageningen and Catholic University of Louvain.

Drouin J.-M. 1994. Histoire et écologie vegetale: les origines du concept de succession. *Ecologie,* **25**(3): 147-155.

du Chatenet G. 1985. *Guide des Coléoptères d'Europe.* Delachaux and Niestlé, Neuchâtel (Switzerland) and Paris.

Duchaufour P. 1976. *Atlas écologique des sols du monde.* Masson, Paris.

Duchaufour P. 1983. *Pédologie 1: Pédogénèse et classification*. Masson, Paris.

Duchaufour P. 2001. *Introduction à la science du sol. Sol, végétation, environnement*. Dunod, Paris.

Duckett J.G. and Read D.J. 1995. Ericoid mycorrhizas and rhizoid ascomycete associations in liverworts share the same mycobiont: isolation of the partners and resynthesis of the association *in vitro*. *New Phytol.* **103**: 457-463.

Ducommun A. 1989. *Influence des boues d'épuration et du fumier sur les macroinvertebrés édaphiques de quelques cultures intensives du Grand-Marais (Plateau suisse)*. Thesis, University of Neuchâtel, Switzerland.

Ducommun A. 1991. Proposition d'une méthode pratique de détermination de la fertilité naturelle globale des sols cultivés au moyen des Macroinvertébrés. *Bull. Soc. entomol. suisse* **64**: 165-172.

Dudal R. 1990. *WRB, World Reference Base*. In: *Gedächtniskolloquium*. E. Schlichting, Hohenheimer Arbeiten. Ulmer. pp. 54-57.

Dunger W. 1958. Ueber die Zersetzung der Laubstreu durch die Boden-Makrofauna in Auenwald. *Zool. Jahrb. (Syst.)* **86**(1-2): 139-180.

Dunger W. 1982. Die Tiere des Bodens als Leitformen für anthropogene Umweltveränderungen. *Decheniana-Beihefte,* **26**: 151-157.

Dunger W. 1983. *Tiere im Boden*. Neue Brehm Bücherei, A. Ziemsen Verlag, Wittemberg-Lutherstadt, Germany.

Dunger W. and Fiedler H.J. 1997. *Methoden der Bodenbiologie*. Fischer Verlag, Stuttgart (Germany) and New York.

Duvigneaud P. 1980. *La synthèse écologique*. Doin, Paris.

Eason T.E. 1964. *Centipedes of the British Isles*. Clowes, London.

Edwards C.A. and Bohlen P.J. 1996. *Biology and Ecology of Earthworms*. Chapman and Hall, London.

Edwards C.A. and Lofty J.R. 1977. *Biology of Earthworms*. Chapman and Hall, London.

Edwards C.A., Stinner B.R., Stinner D. and Rabatin S. 1988. *Biological Interactions in Soil*. Elsevier, Amsterdam.

Eisenbeis G. and Wichard W. 1985. *Atlas zur Biologie der Bodenarthropoden*. Fischer Verlag, Stuttgart, Germany.

Elgar M.A. and Crespi B.J. 1992. *Cannibalism. Ecology and Evolution among Diverse Taxa*. Oxford University Press, Oxford, U.K.

Elhaï H. 1968. *Biogéographie*. Armand Colin, Paris.

Ellenberg H. 1992. Zeigerwerte der Pflanzen in Mitteleuropa. *Scripta Geobot.* **XVIII**.

Ellis S. and Mellor A. 1995. *Soils and Environment*. Routledge, London and New York.

Elton C.S. 1927. *Animal Ecology*. Methuen, London.

Elton C.S. 1966. *The Pattern of Animal Communities*. Methuen, London.

Encyclopedia Universalis. 1985. *Corpus 17*. Encyclopedia Universalis S.A., Paris.

Eriksson K.E.L., Blanchette R.A. and Ander P. 1990. *Microbial and Enzymatic Degradation of Wood and Wood Components*. Springer Verlag, Berlin.

Etherington J.R. 1982. *Environment and Plant Ecology*. Wiley and Sons, Chichester, U.K.

Étienne M. (ed.). 1996. *Western European Silvopastoral Systems*. INRA Editions, Science Update Series, Paris.

Fabre J.-H. 1925. *Souvenirs entomologiques.* Édition définitive illustrée. Second Edition. Delagrave, Paris.

Favre J. 1948. Les associations fongiques des hauts-marais jurassiens et de quelques régions voisines. *Beitr. z. Kryptogamenflora d. Schweiz,* **10.3**. Büchler, Berne, Switzerland.

Ferrar F. 1987. *A Guide to the Breeding Habits and Immature Stages of Diptera Cyclorrhapha.* Brill/Scandinavian Science Press, Leyden (Netherlands) and Copenhagen.

Fierz M., Gobat J.-M. and Guenat C. 1995. Quantification et caractérisation de la matière organique de sols alluviaux au cours de l'évolution de la végétation. *Ann. Sci. For.* **52**: 547-559.

Filip Z. and Preusse T. 1985. Phenoloxidierende Enzyme—Ihre Eigenschaften und Wirkungen im Boden. *Pedobiologia,* **28**: 133-142.

Finck A. 1976. *Pflanzenernährung in Stichworten.* Verlag F. Hirt, Kiel, Germany.

Fischer H.M. 1994. Genetic regulation of nitrogen fixation in rhizobia. *Microbiol. Rev.* **58**: 352-386.

Fisher R.F. and Long S.R. 1992. *Rhizobium*-plant signal exchange. *Nature, Lond.* **357**: 655-660.

Fitter A.H. (ed.). 1985. *Ecological Interactions in Soil. Plants, Microbes and Animals.* Blackwell Scientific Publications, Oxford, U.K.

Fitter A.H. 1987. An architectural approach to the comparative ecology of plant root systems. *New Phytol.* **106**: 61-77.

Flaig W. and Söchtig H. 1972. About the influence of phenolic peat constituents on metabolism of plants. *Proc. 4th int. Peat Cong.,* Otaniemi, Finland, pp. 19-30.

Foissner W. 1987. Soil protozoa: fundamental problems, ecological significance, adaptations in ciliates, and testaceans, bioindicators, and guide to the literature. *Prog. Protistol.* **2**: 69-212.

Forel A. 1920. *Les Fourmis de Suisse.* Second edition. Le Flambeau, La Chaux-de-Fonds, Switzerland.

Forman R.T.T. and Godron M. *Landscape Ecology.* Wiley and Sons, New York.

Foster R.C. and Dormaar J.F. 1991. Bacteria-grazing amoebae *in situ* in the rhizosphere. *Biol. Fertil. Soils,* **11**: 83-87.

Foth H.D. 1990. *Fundamentals of Soil Science.* Wiley and Sons, New York.

Francez A.-J. 1992. Croissance et production primaire des sphaignes dans une tourbière des Monts du Forez (Puy-de-Dôme, France). *Vie et Milieu,* **42**: 21-34.

Francez A.-J. 2000. La dynamique du carbone dans les tourbières à *Sphagnum,* de la sphaigne à l'effet de serre. *L'année biologique,* **39** : 205-270.

Francis R. and Read D.J. 1995. Mutualism and antagonism in their mycorrhizal symbiosis, with special reference to impacts on plant community structure. *Can. J. Bot.* **73**(suppl. 1): 1301-1309.

Frank J.H. 1967. The effect of pupal predators on a population of winter moth, *Operophthera brumata* (L.) (Hydriomenidae). *J. Anim. Ecol.* **36**: 611-621.

Franzén L.G., Chen D. and Klinger L.F. 1996. Principles for a climate regulation mechanism during the late Phanerozoic era, based on carbon fixation in peat-forming wetlands. *Ambio,* **25**(7): 435-442.

Freude H., Harde K.W. and Lohse G.A. 1965*ff*. *Die Käfer Mitteleuropas*. Gustav Fischer Verlag, Jena.

Freuler J., Blandenier G., Meyer H. and Pignon P. 2001. Epigeal fauna in a vegetable agroecosystem. *Bull. Soc. entomol. suisse*, **74**: 17-42.

Frontier S. 1983. *Stratégies d'échantillonage en écologie*. Masson, Paris.

Frontier S. and Pichod-Viale D. 1991. *Écosystèmes: structure, fonctionnement, évolution*. Masson, Paris.

Fuchs J. 1995. Compost: qualité microbiologique et santé des plantes. In: *Valorisation agricole des composts*. Rapport Biophyt S.A.—Compostdiffusion—Sol-Conseil, Nyon and Lausanne, Switzerland.

Gaillard C., Dufaud A., Tommasini R., Kreuz K., Amrein N. and Martinoia E. 1994. A herbicide antidote (safener) induces the activity of both the herbicide detoxifying enzyme and of a vacuolar transporter for the detoxified herbicide. *FEBS Lett.* **352**: 219-221.

Gallandat J.-D. 1982. Prairies marécageuses du Haut-Jura. *Mat. Levé géobot. Suisse*, **58**. H. Huber, Berne.

Gallandat J.-D., Gobat J.-M. and Roulier Ch. 1993. Cartographie des zones alluviales d'importance nationale. *Cahier de l'environnement*, **199**, OFEFP, Berne.

Galli G. and Parenti A. 1990. Land application of industrial residues. *Biocycle*, **31**: 64-65.

Garay I., Mollon A. and Flogaitis E.P. 1986. Étude d'une litière forestière mixte à charme (*Carpinus betulus* L.) et chêne (*Quercus sessiliflora* Smith). II. Succession des macroarthropodes au cours de la décomposition. *Acta Oecol. Oecol. Gener.* **7**(3): 263-288.

Garbaye J. 1994. Helper bacteria: a new dimension to the mycorrhizal symbiosis. *New Phytol.* **128**: 197-210.

Gardes M. and Bruns T.D. 1996. Community structure of ectomycorrhizal fungi in a *Pinus muricata* forest: above- and below-ground views. *Can. J. Bot.* **74**: 1572-1583.

Gaudinski J.B., Trumbore S.E., Davidson E.A., Cook A.C., Markewitz D. and Richter D.D. 2001. The age of fine-root carbon in three forests of the eastern United States measured by radiocarbon. *Oecologia*, online, July 2001.

Géhu J.-M. 1974. Sur l'emploi de la méthode phytosociologique sigmatiste dans l'analyse, la définition et la cartographie des paysages. *C.R. Acad. Sci., Paris*, **279**:1167-1170.

Gerrard J. 1992. *Soil Geomorphology*. Chapman and Hall, London.

Geurts R. and Franssen H. 1996. Signal transduction in *Rhizobium*-induced nodule formation. *Pl. Physiol.* **112**: 447-453.

Ghilarov M.S. and Semenova L.M. 1978. Digestive system of soil insect larvae with different feeding habits. *Rev. Écol. Biol. Sol*, **15**(2): 235-242.

Gianfreda L. and Bollag J.-M. 1996. Influence of natural and anthropogenic factors on enzyme activity in soil. *Soil Biochem.* **9**: 123-193.

Gifford R.M. 1994. The global carbon cycle: a viewpoint on the missing sink. *Aust. J. Pl. Physiol.* **21**: 1-15.

Gigon A. 1971. Vergleich alpiner Rasen auf Silikat- und auf Karbonatboden. *Veröff. Geobot. Inst. ETH, Stift. Rubel*, **48**.

Giller K.E., Beare M.H., Lavelle P., Izac A.M.N. and Swift M.J. 1997. Agricultural intensification, soil biodiversity and agrosystem function. *Appl. Soil Ecol.* **6**: 3-16.

Gillet F. and Gallandat J.-D. 1996. Integrated synusial phytosociology: some notes on a new, multiscalar approach to vegetation analysis. *J. Veg. Sci.* **7**(1): 13-18.

Gillet F., de Foucault B. and Julve P. 1991. La phytosociologie synusiale intégrée; objet et concepts. *Candollea,* **46**: 315-340.

Gillet F., Besson O. and Gobat J.-M. 2002. PATUMOD: a compartment model of vegetation dynamics in wooded pastures. *Ecol. Modelling,* **147**: 267-290.

Girard M.-C. 1989. La cartographie en horizons. *Science du sol,* **21**(1): 41-44.

Gisi U., Schenker R., Schulin R., Stadelmann F.X. and Sticher H. 1997. *Bodenökologie.* Georg Thieme Verlag, Stuttgart, Germany.

Gisin G. 1952. Oekologische Studien über die Collembolen des Blattkomposts. *Revue suisse Zool.* **59:** 543-578.

Gisin H. 1960. *Collembolenfauna Europas.* Muséum d'Histoire Naturelle, Geneva, Switzerland.

Gittings T., Giller P.S. and Stakelum G. 1994. Dung decomposition in contrasting temperate pastures in relation to dung beetle and earthworm activity. *Pedobiologia,* **38**: 455-474.

Gobat J.-M. 1984. *Écologie des contacts entre tourbières acides et marais alcalins dans le Haut-Jura suisse.* Thesis, University of Neuchâtel, Switzerland.

Gobat J.-M. 1991. Recherches scientifiques dans la Grande Cariçaie. La recherche dans les réserves naturelles. *Pub. Acad. suisse Sci. nat.* pp. 71-79.

Gobat J.-M. and Portal J.-M. 1985. Caractérisation de cinq tourbes oligotrophes représentatives d'une dynamique de la végétation dans le Jura suisse. *Science du sol,* **23**(2): 59-74.

Gobat J.-M., Grosvernier P. and Matthey Y. 1986. Les tourbières du Jura suisse: milieux naturels, modifications humaines, caractères des tourbes, potentiel de régénération. *Actes Soc. jurass. Emul.* pp. 213-315.

Gobat J.-M., Duckert O. and Gallandat J.-D. 1989. Relations entre microtopographie, sol et végétation dans les pelouses pseudo-alpines du Jura suisse. *Bull. Soc. neuchâtel. Sci. nat.* **112**: 5-17.

Gobat J.-M., Grosvernier P., Matthey Y. and Buttler A. 1991. Un triangle granulométrique pour les tourbes: analyse semi-automatique et représentation graphique. *Science du sol,* **29**(1): 23-35.

Godan D. 1983. *Pest Slugs and Snails. Biology and Control.* Springer Verlag, Berlin.

Goldsmith E. and Hildyard N. 1990. *Rapport sur la planète Terre.* Stock, Paris.

Gonseth Y. and Mulhauser G. 1996. Bioindication et surfaces de compensation écologique. *Cahier de l'environnement* **261**. Office fédéral de l'environnement, des forêts et du paysage, Berne.

Gore A.J.P. 1983. *Ecosystems of the World, 4A. Mires, Swamp, Bog, Fen and Moor.* Elsevier, Amsterdam.

Gorham E. 1991. Northern peatlands: role in the carbon cycle and probable responses to climatic warming. *Ecol. Appl.* **1**(2): 182-195.

Göttfert M. 1993. Regulation and function of rhizobial nodulation genes. *FEMS Microbiol. Rev.* **104**: 39-64.

Göttlich M. (ed.). 1990. *Moor- und Torfkunde*. E. Schweizebart'sche Verlagsbuchhandlung, Stuttgart, Germany.

Grabherr G. 1997. *Farbatlas. Oekosystem der Erde*. Verlag Eugen Ulmer, Stuttgart, Germany.

Grandi G. 1984. *Introduzione allo Studia Della Entomologia*. Vols. 1 and 2. Edagricole, Bologna, Italy.

Grassé P.P. (ed.). 1949, 1951, 1965. *Traité de Zoologie*. Vols. 4, 9 and 10. Masson, Paris.

Grassé P.P. 1982. *Termitologia*. Vols. I to III. Masson, Paris.

Green R.N., Trowbridge R.L. and Klinka K. 1993. Towards a taxonomic classification of humus forms. *For. Sci. Monograph*, **29**: 1-49.

Greenwood D.J. and Goodman D. 1967. Direct measurement of the distribution of oxygen in soil aggregates and in columns of fine soil crumbs. *J. Soil Sci.* **18**: 182-196.

Grime J.P. 1979. *Plant Strategies and Vegetation Processes*. Wiley and Sons, Chichester, U.K.

Grime J.P., Mackey J.M.L., Hillier S.H. and Read D.J. 1987. Floristic diversity in a model system using experimental microcosms. *Nature, Lond.* **328**: 420-422.

Grime J.P., Hodgson J.G. and Hunt R. 1988. *Comparative Plant Ecology*. Unwin Hyman, London.

Grimwell A. and de Leer E.W.B. (eds.). 1995. *Naturally-produced Organohalogens*. Battelle Press, Columbus OH, U.S.A.

Grodzinski W., Weiner J. and Mayock P.F. 1984. *Forest Ecosystems in Industrial Regions*. Springer Verlag, Berlin.

Grosvernier P. 1996. *Stratégies et génie écologique des sphaignes* (Sphagnum *spp.) dans la restauration spontanée des marais jurassiens suisses. Une approche expérimentale*. Thesis, University of Neuchâtel, Neuchâtel, Switzerland.

Grosvernier P., Matthey Y. and Buttler A. 1997. Growth potential of three *sphagnum* species in relation to water level and peat properties with implications for their restoration in cut-over bogs. *J. Appl. Ecol.* **34**: 471-483.

Grosvernier P., Matthey Y., Buttler A. and Gobat J.-M. 1999. Characterization of peats from Histosols disturbed by different human impacts (drainage, peat extraction, agriculture). *Écologie,* **30**: 23-31.

Grubb P.J. and Whittaker J.B. 1989. *Towards a More Exact Ecology*. Blackwell Scientific Publications, Oxford, U.K.

Grünig A., Vetterli L. and Wildi O. 1986. Les haut-marais et marais de transition de Suisse. *Inst. féd. rech. forest., Birmensdorf, Zurich*, **286**.

Guenat C. 1987. *Les sols forestiers non hydromorphes sur moraines du Jura vaudois. Pédogenèse et relations sol-végétation*. Thesis No. 693, EPFL, Lausanne, Switzerland.

Guillet B., Achoundong G., Youta Happi J., Kamgang Kabeyene Beyala V., Bonvallot J., Riera B., Mariotti A. and Schwartz D. 2001. Agreement between floristic and soil organic carbon isotope ($^{13}C/^{12}C$, ^{14}C) indicators of forest invasion of savannas during the last century in Cameroon. *J. Trop. Ecol.* **17**: 809-832.

Guinochet M. 1973. *Phytosociologie*. Masson, Paris.

Gullan P.J. and Cranston P.S. 1994. *The Insects: an Outline of Entomology*. Chapman and Hall, London.

Häberli R., Lüscher C., Praplan-Chastenay B. and Wyss C. 1991. *L'Affaire SOL. Pour une politique raisonnée de l'utilisation du sol*. Éditions Georg, Geneva, Switzerland.

Hamelin J., Fromin N., Tarnawski S., Teyssier-Cuvelle S. and Aragno M. 2002. nifH gene diversity in the bacterial community associated with the rhizosphere of *Molinia coerulea*, an oligonitrophilic perennial grass. *Environ. Microbiol.* **4(8)**: 477-482.

Hausmann K. and Hulsmann N. (eds.). 1994. *Progress in Protozoology*. Proceedings of the IX International Congress of Protozoology, Berlin 1993. Fischer Verlag, Stuttgart, Germany.

Hausser J. (ed.). 1995. Mammifères de Suisse. *Mém. Acad. suisse Sci. nat.* **103.** Birkhäuser Verlag, Basle.

Havlicek E. 1999. *Les sols des pâturages boisés du Jura suisse. Origine et typologie, relations sol-végétation, pédogenèse des brunisols, évolution des humus*. Thesis, University of Neuchâtel, Neuchâtel, Switzerland.

Havlicek E. and Gobat J.-M. 1996. Les apports éoliens dans les sols du Jura. État des connaissances et nouvelles données en pâturages boisés. *Étude et Gestion des Sols*, **3**(3): 167-178.

Havlicek E. and Gobat J.-M. 1998. Les formes d'humus, révélatrices du fonctionnement de l'écosystème. Un exemple des pâturages boisés du Jura suisse. *Écologie*, **29**(1-2): 367-371.

Havlicek E., Gobat J.-M. and Gillet F. 1998. Réflexions sur les relations sol - végétation: trois exemples du Jura sur matériel allochtone. *Écologie*, **29**(4): 535-546.

Hayward P.M. and Clymo R.S. 1983. The growth of *Sphagnum*: experiments on, and simulation of, some effects of light flux and water-table depth. *J. Ecol.* **71**: 845-863.

Head I.M., Saunders J.R. and Pickup R.W. 1998. Microbial evolution, diversity and ecology: A decade of ribosomal RNA analysis of uncultivated microorganisms. *Microbial Ecol.* **35**: 1-21.

Heal O.W. and Perkins F.D. 1978. *Production Ecology of British Moors and Montane Grasslands*. Springer Verlag, Berlin.

Heller R. 1989. *Physiologie végétale. 1. Nutrition*. Abrégés, Masson, Paris.

Hennig W. 1948-1952. *Die Larvenformen der Dipteren*. Vols. 1 to 3. Akademie Verlag, Berlin.

Hess H.E., Landolt E. and Hirzel R. 1967*ff*. *Flora der Schweiz*. Verlag Birkhaüser, Basle, Switzerland.

Hillel D. 1980. *Fundamentals of Soil Physics*, Academic Press, Orlando FL, U.S.A., 413 pp.

Hingley M. 1993. *Microscopic Life in* Sphagnum. Naturalist's Handbooks 20. Richmond Publishing Co., Slough, U.K.

Hobson P.N. and Wheatley A.D. 1993. *Anaerobic digestion, Modern Theory and Practice*. Elsevier, London-New York.

Hoitink H.A.J. and Keener H.M. (eds.). 1993. *Science and Engineering of Composting: Design, Environmental, Microbiological and Utilization Aspects*. Ohio State University, Wooster OH, U.S.A.

Holland M.M., Risser, P.G. and Naiman R.J. 1991. *Ecotones*. Chapman and Hall, New York.

Hölldobler B. and Wilson E.O. 1996. *Voyages chez les fourmis*. Seuil, Paris.

Hommay G. 1995. Les limaces nuisibles aux cultures. *Revue suisse d'agriculture*, **27**(5): 267-286.

Hoosbeck M.R. and Bryant R.B. 1992. Towards the quantitative modelling of pedogenesis. A review. *Geoderma*, **55**: 183-210.

Horak E. and Röllin O. 1988. Der Einfluss von Klärschlamm auf die Makromycetenflora eines Eichen-Hainbuchenwaldes bei Genf, Schweiz. *Mém. Inst. féd. rech. forest.* **64**: 21-148.

Hutchinson G.E. 1957. Concluding remarks. *Cold Spring Harbor Symp.* **22**: 415-427.

Illich I., Groeneveld S. and Hoinacki L. 1991. Declaration on soil. *ifda dossier* **81**. University of Kassel.

Ingham E.R. and Klein D.A. 1982. Relationship between fluorescein diacetate-stained hyphae and oxygen utilization, glucose utilization, and biomass of submerged fungal batch cultures. *Appl. Environ. Microbiol.* **44**: 363-370.

Ingham E.R., Trofymow J.A., Ingham R.E. and Coleman D.C. 1985. Interactions of bacteria, fungi, and their nematode grazers: effects on nutrient cycling and plant growth. *Ecol. Monogr.* **55**: 119-140.

Ingram H.A.P. 1982. Size and shape in raised mire ecosystems: a geophysical model. *Nature, Lond.* **297**: 300-303.

Isoviita P. Studies on *Sphagnum* L. I. Nomenclatural revision of the European taxa. *Annales Botanici Fennici*, **3**: 199-264.

ISSS. 1998. *World Reference Base for Soil Resources*. FAO, Rome, and ISSS-ISRIC-FAO, Acco, Louvain, Belgium.

Ivanova K.E. 1953. *Gidrologiya bolot*. Gidrometeoizdat, Leningrad.

Jabiol B., Brethes A., Ponge J.-F., Toutain F. and Brun J.-J. 1995. *L'humus sous toutes ses formes*. ENGREF, Nancy, France.

Jamagne M. 1973. *Contribution à l'étude des formations loessiques du Nord de la France*. Doctoral thesis, INRA, Paris.

Jaspar-Versali M.F. and Jeuniaux C. 1987. L'équipement enzymatique digestif des Carabes envisagé sous l'angle de leur niche écologique. *Rev. Écol. Biol. Sol*, **24**(4): 541-547.

Johnson D.W. 1993. Carbon in forest soils—Research needs. *N.Z. Jl. Forestry Sci.* **23**(3): 353-366.

Jolivet C., Guillet B., Karroum M., Andreux F., Bernoux M. and Arrouays D. 2001. Les phénols de la lignine et le ^{13}C, traceurs de l'origine des matières organiques du sol. *C.R. Acad. Sci., Paris, Sci. Terre et planètes*, **333**: 651-657.

Jones C.G., Lawton J.H. and Shachak M. 1994. Organisms as ecosystem engineers. *Oikos*, **69**: 373-386.

Jones R. 1972. Comparative studies of plant growth and distribution in relation to waterlogging. V. The uptake of iron and manganese by dune and slack plants. *J. Ecol.* **60**: 131-140.

Juillard M. 1984. *La chouette chevêche*. Nos Oiseaux, Prangins, Switzerland.

Juillard M., Praz J.-C., Étournaud A. and Baud P. 1978. Données sur la contamination des rapaces de Suisse romande et de leurs oeufs par les biocides organochlorés, les PCB et les métaux lourds. *Nos Oiseaux*, **34**: 189-206.

Jurkevitch E. 2000. The genus *Bdellovibrio*. In: Dworkin M., Flakow S., Rosenberg E., Schleifer K.H. and Stackebrandt (eds.). *The Prokaryotes*. Springer Verlag, New York.

Jurkevitch E., Minz D., Ramati B. and Barel G. 2000. Prey range characterization, ribotyping, and diversity of soil rhizosphere *Bdellovibrio* spp. isolated on phytopathogenic bacteria. *Appl. and Environ. Microbiol.* **66**: 2365-2371.

Kaila A. 1956. Determination of the degree of humification in peat samples. *Maatalous Ackakaoushiva*, **28**: 18-35.

Kampichler C., Bruckner A. and Kandeler E. 2001. Use of enclosed model ecosystems in soil ecology: a bias towards laboratory research. *Soil Biol. Biochem.* **33:** 269-275.

Kareiva P.M., Kingsolver J.G. and Huey R.B. (eds.). 1993. *Biotic Interactions and Global Change*. Sinauer, Sunderland MA, U.S.A. pp. 45-54.

Keller J. 1985. Les cystides cristallifères des Aphyllophorales. *Mycol. Helv.* **I**(5): 277-340.

Kelner-Pillault S. 1974. Étude écologique du peuplement entomologique des terreaux d'arbres creux (châtaigniers et saules). *Bull. Écol.* **5**: 123-156.

Kevan D.C. 1962. *Soil Animals*. Witherby, London.

Killham K. 1994. *Soil Ecology*. Cambridge University Press, Cambridge.

King D., Jamagne M., Chrétien J. and Hardy R. 1994. Soil-space organization model and soil functioning units in Geographical Information Systems. In: *Proc. World Cong. ISSS* (Acapulco), **6a**: 743-757.

Kirk T.K. and Farrell R.L. 1987. Enzymatic 'combustion': the microbial degradation of lignin. *Ann. Rev. Microbiol.* **41**: 465-505.

Kirschbaum M.U.F. 2000. Will changes in soil organic carbon act as a positive or negative feedback on global warming? *Biogeochemistry*, **48**(1): 21-51.

Klausnitzer B. 1978. *Ordnung Coleoptera (Larven)*. Dr W. Junk, The Hague, Netherlands.

Klinger L.F. 1996. Coupling of soils and vegetation in peatland succession. *Arct. Alp. Res.* **28**(3): 380-387.

Klinger L.F., Taylor J.A. and Franzén L.G. 1996. The potential role of peatland dynamics in Ice-Age initiation. *Quaternary Research*, **45**: 89-92.

Klingmüller W. (ed.). 1982. *Azospirillum: Genetics, Physiology, Ecology*. Springer Verlag, Berlin.

Krantz G.W. 1978. *A Manual of Acarology*. Oregon State University Book Store, Corvallis OR, U.S.A.

Krieg N.R. and Holt J.G. 1984ff. *Bergey's Manual of Systematic Bacteriology*. Williams and Wilkins, Baltimore MD, U.S.A.

Krumbein W.E. (ed.). 1983. *Microbial Geochemistry*. Blackwell Scientific Publications, Oxford, U.K. pp. 223-262.

Kuntze H., Roeschmann G. and Schwerdtfeger G. 1988. *Bodenkunde*. UTB 1106, Ulmer Verlag, Stuttgart, Germany.

Kutter H. 1977. *Formicidae. Insecta Helvetica (Fauna)* **6**. Société suisse d'entomologie, Zurich, Switzerland.

Kuzyakov Y. and Domanski G. 2000. Carbon input by plants into the soil. Review. *J. Pl. Nutr. Soil Sci.* **163**: 421-431.

Lachaise D. 1979. Le concept de niche chez les Drosophiles. *La Terre et la Vie,* **33**: 425-456.

Lachavanne J.-B. and Juge J. (eds.). 1997. *Biodiversity in Land/Inland Water Ecotones.* MAB series Vol. 19. Parthenon Publishing, Carnforth, U.K.

Laczkò E. 1994. Neue Ansätze zur Analyse von mikrobiellen Gemeinschaften in Böden: die Phospholipidfettsäuren-Analyse und ihre Anwendung. *Bull. bodenkundl. Ges. Schweiz.* **18**: 23-28.

Lähdesmäki P. and Piispanen R. 1992. Soil enzymology: role of protective colloid systems in the preservation of exoenzyme activities in soil. *Soil Biol. Biochem.* **24**: 1173-1177.

Laiho R., Laine J. and Vasander H. 1996. Northern peatlands in global climatic change. *Publ. Acad. Finland* 1/96. Helsinki, Finland.

Lal R., Kimble J.M., Follett R.F. and Stewart B.A. (eds.). 1998. *Soil Processes and the Carbon Cycle*. CRC Press, Boca Raton FL, U.S.A.

Lal R., Kimble J.M. and Stewart, B.A. 2000. *Global climate change and cold regions ecosystems.* Advances in Soil Science, Lewis Publ., Boca Raton FL, U.S.A.

Lalonde M. 1978. Confirmation of the infectivity of a free-living actinomycete isolated from *Compotonia peregrina* root nodules by immunological and ultrastructural studies. *Can. J. Bot.* **56**: 1621-1635.

Lamotte M. and Bourlière F. 1969. *L'échantillonnage des peuplements animaux des milieux terrestres*. Masson, Paris.

Lamotte M. and Bourlière F. 1978. *Problèmes d'ecologie: écosystèmes terrestres.* Masson, Paris.

Landolt E. 1977. Oekologische Zeigerwerte zur Schweizer Flora. *Veröff. Geobot. Inst. ETH Zürich*, Heft **64**.

Laskowski R. 1991. Are the top carnivores endangered by heavy metal biomagnification? *Oikos,* **60**: 387-390.

Lavelle P. 1987. Interactions, hiérarchies et régulations dans le sol: à la recherche d'une nouvelle approche conceptuelle. *Rev. Écol. Biol. Sol,* **24**(3): 219-229.

Lavelle P. and Spain A.V. 2001. *Soil Ecology*. Kluwer Academic Publishers, Dordrecht, Netherlands.

Lebreton P. 1978. *Éco-logique. Initiation aux disciplines de l'environnement.* InterÉditions, Paris.

Lebrun P. and van Straalen N.M. 1995. Oribatid mites: prospects for their use in ecotoxicology. *Experim. Appl. Acarology,* **19**: 361-379.

Leclerq M. 1978. *Entomologie et médecine légale.* Masson, Paris.

Lecointre G. and Le Guyader H. 2001. *Classification phylogénétique du vivant.* Belin, Paris.

Lefeuvre J.-C. and Barnaud G. 1988. Écologie du paysage: mythe ou réalité? *Bull. Écol.* **19**(4): 493-522.

Legay J.M. and Barbault R. (Eds.). 1995. *La révolution technologique en écologie.* Masson, Paris.

Legendre L. and Legendre P. 1984. *Écologie numérique*. Vols. 1 and 2. Masson, Paris.

Legros J.-P. 1996. *Cartographies des sols*. Coll. Gérer l'environnement No. 10, Presses polytechniques et universitaires romandes, Lausanne, Switzerland.

Lengeler J.W., Drews G. and Schegel H.G. (eds.) 1999. *Biology of the Procaryotes*, Thieme Medical Publishers, Stuttgart, Germany.

Lepart J. and Escarre J. 1983. La succession végétale, mécanismes et modèles: analyse bibliographique. *Bull. Écol.* **14**(3): 133-178.

Leser H. 1991. *Landschaftsökologie*. UTB 521, Eugen Ulmer Verlag, Stuttgart, Germany.

Lesquereux L. 1844. *Quelques recherches sur les marais tourbeux en général*. H. Wolfrath, Neuchâtel, Switzerland.

Levêque C. and Mounolou J.-C. 2001. *Biodiversité. Dynamique biologique et conservation*. Dunod, Paris.

Lévesque M. and Dinel H. 1977. Fiber content, particle-size distribution and some related properties of four peat materials in eastern Canada. *Can. J. Soil Sci.* **57**: 187-195.

Lévesque M. and Dinel H. 1982. Some morphological and chemical aspects of peats applied to the characterisation of Histosols. *Soil Sci.* **133(5)**: 324-332.

Lévesque M., Dinel H. and Marcoux R. 1980. Evaluation des critères de différenciation pour la classification de 92 matériaux du Québec et de l'Ontario. *Can. J. Soil Sci.* **60**: 479-486.

Levi M.P., Merril W. and Cowling E.B. 1968. Role of nitrogen in wood deterioration. IV. Mycelial fractions and model nitrogen compounds as substrates for growth of *Polyporus versicolor* and other wood-destroying and wood inhabiting fungi. *Phytopathology*, **58**: 628-634.

Lindeman R. 1942. The trophic-dynamic aspect of ecology. *Ecology*, **23**(4): 399-418.

Lippmaa T. 1939. The unistratal concept of plant communities (the unions). *Am. Midl. Nat.* **21**: 111-145.

Logan E.M., Pulford I.D., Cook G.T. and Mackenzie A.B. 1997. Complexation of Cu^{2+} and Pb^{2+} by peat and humic acid. *Europ. J. Soil Sci.* **48**: 685-696.

Long S.R. 2001. Gene and signals in the *Rhizobium*-legume symbiosis. *Pl. Physiol.* **125**: 69-72.

Lott Fischer J., Beffa T., Lyon P.-F., Blanc M. and Aragno M. 1995. Development of *Aspergillus fumigatus* during composting of organic wastes. *Recovery, Recycling, Re-integration*, EMPA, **R'95**: 239-244.

Lovelock J. 1992. *GAÏA. Comment soigner une Terre malade?* Robert Laffont, Paris.

Lozet J. and Mathieu C. 1997. *Dictionnaire de science du sol*. Coll. Tec & Doc, Lavoisier, Paris.

Lugon A.M., Weber G., Matthey Y., Gonseth Y. and Wermeille E. 2001. Influence des espèces animales bioindicatrices dans l'élaboration de plans de mesures d'aménagement et d'entretien des milieux naturels. *Bull. Soc. neuchâtel. Sci. nat.* **124**: 198-208.

Lumaret J.-P. 1993. Insectes coprophages et médicaments vétérinaires: une menace à prendre au sérieux. *Insectes*, **91**(4): 2-3.

Lumaret J.-P., Galante E., Lumbreras C., Mena J., Bertrand M., Bernal J.-L., Cooper J.F., Kadiri N. and Crowe D. 1993. Field effects of ivermectin residues on dung beetles. *J. Appl. Ecol.* **30(3)**: 428-436.

Lussenhop J. 1992. Mechanisms of microarthropod-microbial interactions in soil. *Adv. ecol. Res.* **23**: 1-33.

Lütt S. 1992. Produktionsbiologische Untersuchungen zur Sukzession der Torfstichvegetation in Schleswig-Holstein. *Mitt. Arbeitsg. Geobot. Schlesw.-Holst. u. Hamb.* **43**. Kiel.

Lüttge U., Kluge M. and Bauer G. 1996. *Botanique*. Coll. Tec & Doc, Lavoisier, Paris.

Lynch J.M. 1990. *The Rhizosphere*. Wiley and Sons, Chichester, U.K.

Lyr H. 1962. Detoxification of heartwood toxins and chlorophenols by higher fungi. *Nature, Lond.* No. **195**: 289-290.

MacArthur R.H. and Wilson E.O. 1967. *The Theory of Island Biogeography*. Princeton University Press, Princeton NJ, U.S.A.

Madigan M.T., Martinko J.M. and Parker J. 2002. *Brock Biology of Microorganisms*. 10th edition. Prentice-Hall, Upper Saddle River NJ, U.S.A.

Maire N. 1984. Extraction de l'adénosine triphosphate dans les sols: une nouvelle méthode de calcul des pertes en ATP. *Soil Biol. Biochem.* **16**: 361-366.

Maire N. 1987. Evaluation de la vie microbienne dans les sols par un système d'analyses biochimiques standardisé. *Soil Biol. Biochem.* **19**: 491-500.

Maire N., Borcard D., Laczko E. and Matthey W. 1999. Organic matter cycling in grassland soils of the Swiss Jura mountains: biodiversity and strategies of the living communities. *Soil Biol. Biochem.* **31**: 1281-1293.

Mandelbrot B. 1975. *Les objets fractals: forme, hasard et dimension*. Flammarion, Paris.

Mani S. 1962. *Introduction to High Altitude Entomology*. Methuen, London.

Manneville O., Vergne V., Villepoux O. et al. 1999. *Le monde des tourbières et des marais*. Delachaux & Niestlé, Lausanne-Paris.

Manthey J.A., Crowley D.E. and Luster D.G. (eds.). 1994. *Biochemistry of Metal Micronutrients in the Rhizosphere*. CRC Press, Boca Raton FL, U.S.A.

Marilley L., Vogt G., Blanc M. and Aragno M. 1998. Bacterial diversity in the bulk soil and rhizosphere fractions of *Lolium perenne* and *Trifolium repens* as revealed by PCR restriction analysis of 16s rDNA. *Pl. Soil*, **198**: 219-224.

Marschner H. 1995. *Mineral Nutrition of Higher Plants*. Academic Press, London.

Marschner H. and Römheld, V. 1983. In-vivo measurement of root-induced pH changes at the soil-root interface: effect of plant species and nitrogen source. *Z. Pflanzenphysiol.* **111**: 241-251.

Martin J.P. and Haider K. 1980. A comparison of the use of phenolase and peroxidase for the synthesis of model humic acid type polymers. *Soil Sci. Soc. Am. J.* **44**: 983-988.

Martinez G. and Medel R. 2002. Indirect interactions in a microcosm-assembled cladoceran community: implications for apparent competition. *Oikos*, **97:** 111-115.

Martini I.P. and Chesworth W. (eds.). 1992. *Weathering, Soils and Paleosols*. Elsevier, Amsterdam. pp. 349-377.

Martinoia E. Flügge U.I., Kaiser G., Heber U. and Heldt H.W. 1985. Energy-dependent uptake of malate into vacuoles isolated from barley mesophyll protoplasts. *Biochim. Biophys. Acta*, **806**: 311-319.

Mason C.F. 1976. *Decomposition*. Studies in Biology 74, Edward Arnold, London.

Mathys W. 1977. The role of malate, oxalate and mustard-oil glucosides in the evolution of zinc resistance in herbage plants. *Physiol. Plantarum*, **40**: 130-136.

Matile L. 1993, 1995. *Diptères d'Europe occidentale*. Boubée, Paris.

Matthey W. and Borcard D. 1996. La vie animale dans les tourbières jurassiennes. *Bull. Soc. neuchâtel. Sci. nat.* **119**: 3-18.

Matthey W., Dethier M., Galland P., Lienhard C., Rohrer N. and Scheiss T. 1981. Étude écologique et biocénotique d'une pelouse alpine au Parc national suisse. *Bull. Écol.* **12**(4): 339-354.

Matthey W., Zettel J. and Bieri M. 1990. *Invertébrés bioindicateurs de la qualité des sols agricoles/Wirbellose Bodentiere als Bioindikatoren für die Qualität von Landwirtschaftsböden*. Programme national de recherche 22. SOL, Liebefeld-Berne, Switzerland.

Matthey Y. 1996. *Conditions écologiques de la régénération spontanée du* Sphagnion magellanici *dans le Jura suisse*. Thesis, University of Neuchâtel, Neuchâtel, Switzerland.

Mattson W.J. 1977. *The Role of Arthropods in Forest Ecosystems*. Springer Verlag, New York.

Mayeux D. and Savanne D. 1996. *La faune, indicateur de la qualité des sols*. ADEME, Paris.

McAlpine J.P., Peterson B.V., Shewell G.E., Teskey H.J., Vockeroth J.R. and Wood D.M. 1981-1989. *Manual of Nearctic Diptera*. Vols. 1-3. Minister of Supply and Services, Hull, Que., Canada.

Melillo J.M., Aber J. and Muratore J.F. 1982. Nitrogen and lignin control of hardwood leaf litter decomposition dynamics. *Ecology*, **63**: 621-626.

Meyer K. 1991. *La pollution des sols en Suisse*. Programme national de recherche 22. SOL, Liebefeld-Berne, Switzerland.

Meylan A. 1997. Le campagnol terrestre *Arvicola terrestris* (L.): biologie de la forme fouisseuse et méthodes de lutte. *Rev. suisse Agric.* **9(3):** 178-187.

Michalet R. and Bruckert S. 1986. La podzolisation sur calcaire du subalpin du Jura. *Science du sol*, **24**(4): 363-375.

Milne B.T. 1988. Measuring the fractal geometry of landscapes. *Appl. Math. Comput.* **27**: 67-76.

Mitchell E.A.D., Borcard D., Buttler A., Grosvernier P., Gilbert D. and Gobat J.-M. 2000a. Horizontal distribution patterns of testate amoebae (Protozoa) in a *Sphagnum magellanicum* carpet. *Microbial Ecol.* **39**(4): 290-300.

Mitchell E.A.D., Buttler A., Grosvernier P., Rydin H., Albinsson C., Greenup A., Heijmans M., Hoosbeek M.R and Saarinen T. 2000b. Relationships among testate amoebae (Protozoa), vegetation and water chemistry in five *Sphagnum*-dominated peatlands in Europe. *New Phytol.* **145**: 95-106.

Mitchell E.A.D., Buttler A., Grosvernier P., Rydin H., Siegenthaler A. and Gobat, J.-M. 2002. Contrasted effects of increased N and CO_2 supply on two key species in peatland restoration and implications for global change. *J. Ecol.* **90**: 529-533.

Mitsch W.J. and Gosselink J.G. 1993. *Wetlands*. Van Nostrand Reinhold, New York.

Mohr E.C.J. and van Bahren F.A. 1959. *Tropical Soils*. Van Hoeve, The Hague, Netherlands.

Monod J. 1950. La technique de la culture continue. Théorie et applications. *Ann. Inst. Pasteur,* **79**: 390-410.

Mooney H.A., Vitousek P.M. and Matson P.A. 1987. Exchange of materials between terrestrial ecosystems and the atmosphere. *Science, N.Y.* **238**: 926-932.

Moore P.D. and Bellamy D.J. 1974. *Peatlands*. Elek Science, London.

Morard P. 1997. *Les cultures végétales hors sol*. Publications agricoles, Agen, France.

Morel R. 1996. *Les sols cultivés*. Coll. Tec & Doc, Lavoisier, Paris.

Mosimann T., Maillard A., Musy A., Neyroud J.-A., Rüttimann M. and Weisskopf P. 1991. *Lutte contre l'érosion des sols cultivés*. Programme national de recherche 22. SOL, Liebefeld-Berne, Switzerland.

Mückenhausen E. 1985. *Die Bodenkunde*. DLG-Verlag, Frankfurt, Germany.

Müller J. 1994. Diffuse Quellen von PCB in der Schweiz. *Cahier de l'environnement,* **229**. Office fédéral de l'environnement, des forêts et du paysage. Berne.

Müller P.E. 1889. Recherches sur les formes naturelles de l'humus et leur influence sur la végétation et le sol. *Ann. Sci. Agron.* 85-423.

Münch J.C. and Ottow J.C.G. 1977. Modell Untersuchungen zum Mekanismus der bakteriellen Eisenreduktion in hydromorphen Böden. *Z. Pflanzener. Bödenk.* **140**: 549-562.

Musy A. and Soutter M. 1991. *Physique du sol*. Coll. Gérer l'environnement No. 6. Presses polytechniques et universitaires romandes, Lausanne, Switzerland.

Mylona P., Pawloski K. and Bisseling T. 1995. Symbiotic nitrogen fixation. *Pl. Cell,* **7**: 869-885.

Nannipieri P., Muccini L. and Ciardi C. 1983. Microbial biomass and enzyme activities: production and persistence. *Soil Biol. Biochem.* **15**: 679-685.

Nealson K.H. and Saffarini D. 1994. Iron and manganese in anaerobic respiration: environmental significance, physiology and regulation. *Ann. Rev. Microbiol.* **48**: 311-343.

Nef L. 1957. État actuel des connaissances sur le rôle des animaux dans la décomposition des litières des forêts. *Agricultura,* **5**: 245-316.

Newsham K.K., Fitter A.H. and Merryweather J.W. 1995. Multifunctionality and biodiversity in arbuscular mycorrhizas. *TREE,* **10**: 407-411.

Novotny V., Basset Y., Miller S.E., Weiblen G.D., Bremer B., Cizek L., Drozd P. 2002. Low host specificity of herbivorous insects in a tropical forest. *Nature* **416:** 841-844.

O'Connell K.P., Goodman R.M. and Handelsman J. 1996. Engineering the rhizosphere: expressing a bias. *Tibtech,* **14**: 83-88.

Odum E.P. 1971. *Fundamentals of Ecology*. WB Saunders Company, Philadelphia PA, U.S.A.

Odum E.P. 1977. The emergence of ecology as a new integrative discipline. *Science, N.Y.* **195**: 1289-1293.

Odum E.P. 1996. *Ecology. A Bridge between Science and Society*. Sinauer Associates, Sunderland MA, U.S.A.

O.F.S. 2001. *L'utilisation du sol: hier et aujourd'hui. Statistique suisse de la superficie.* Office fédéral de la statistique, Neuchâtel.

Oldroyd G.E.D. 2001. Dissecting symbiosis: Developments in Nod factor signal transduction. *Ann. Bot.,* Lond. 87: 709-718.

Otjen L. and Blanchette R.A. 1985. Selective delignification of aspen wood blocks *in vitro* by three white rot basidiomycetes. *Appl. environ. Microbiol.* **50**: 568-572.

Ozenda P. 1982. *Les végétaux dans la biosphère*. Doin, Paris.

Pancza A. 1988. Un pergélisol actuel dans le Jura neuchâtelois. *Bull. Soc. neuchâtel. Géogr.* **32-33**: 129-140.

Park O., Auerbach S. and Corley G. 1950. The tree hole habitats with emphasis on the Pselaphid beetle fauna. *Bull. Chicago Acad. Sci.* **9**: 19-41.

Parkyn L., Stoneman R.E. and Ingram H.A.P. 1997. *Conserving Peatlands*. CAB International, Oxford (U.K.) and New York.

Paton T.R., Humphreys G.S. and Mitchell P.B. 1995. *Soils. A New Global View.* UCL Press, London.

Patten, C.L. and Glick B.R. 1996. Bacterial biosynthesis of indole-3-acetic acid. *Can. J. Microbiol.* **42**: 207-220.

Paul E.A. and Clark F.E. 1996. *Soil Biology and Biochemistry*. 2nd ed. Academic Press, San Diego CA, U.S.A.

Paul E.A. and Ladd J.N. (eds.). 1981. *Soil Biochemistry*. Vol. 5. Marcel Dekker, New York.

Paulian R. 1988. *Biologie des Coléoptères*. Lechevalier, Paris.

Pavan M. 1976. *Utilisation des fourmis du groupe* Formica rufa *dans la défense des forêts*. Protection de la Nature et des sites du Canton du Vaud, Lausanne, Switzerland.

Pedroli-Christen, A. 1981. Étude des peuplements de diplopodes dans six associations forestières du Jura et du Plateau suisse (région neuchâteloise). *Bull. Soc. neuchâtel. Sci. nat.* **104**: 89-106.

Pedroli-Christen A. 1993. Faunistique des mille-pattes de Suisse (Diplopoda). *Documenta Faunistica Helvetiae*, CSCF, Neuchâtel, Switzerland, **14**.

Perrier E. 1990. Modélisation du fonctionnement hydrique des sols. Passage de l'échelle microscopique à l'échelle macroscopique. In: *Séminfor IV. Quatrième séminaire informatique de l'Orstom. Centre de Brest*, 11 to 13 September 1990, pp. 113-132.

Perruchoud D. and Fischlin A. 1994. The response of the carbon cycle in undisturbed forest ecosystems to climate change: a review of plant-soil models. *J. Biogeogr.* **22**: 2603-2618.

Persson T. and Lohn U. 1977. Energetical significance of the Annelids and Arthropods in a Swedish soil. *Ecol. Bull.* **23**: 1-211.

Pesson P. 1971. *La vie dans les sols*. Gauthier-Villars, Paris.

Phillipson J. (ed.). 1971. *Methods of Study in Quantitative Soil Ecology*. IBP Handbook 18, Blackwell Scientific Publications, Oxford, U.K.

Piccini D. and Azcon R. 1987. Effect of phosphate-solubilizing bacteria and vesicular-arbuscular mycorrhizal fungi on the utilisation of Bayovar rock phosphate by alfalfa plants using a sand-vermiculite medium. *Pl. Soil,* **101**: 45-50.

Pimm S.L. 1982. *Food Webs*. Chapman and Hall, London.

Pochon J. and Tardieux P. 1962. *Techniques d'analyse en microbiologie du sol*. Coll. 'Techniques de base'. Éditions La Tourelle, St-Mandé, France.

Pochon M. 1978. Origine et évolution des sols du Haut-Jura suisse. Phénomènes d'altération des roches calcaires sous climat tempéré humide. *Mém. Soc. helv. Sci. nat.* **XC**.

Polomski J. and Kuhn N. 1998. *Wurzelsysteme in Wald- und Grünlandgemeinschaften*. Paul Haupt Verlag, Berne (Switzerland), Stuttgart (Germany) and Vienna (Austria).

Poschlod P. 1990. Vegetationsentwicklung in abgetorften Hochmooren des bayerischen Alpenvorlandes unter besonderer Berücksichtigung standortskundlicher und populationsbiologischer Faktoren. *Diss. Botan.* **152**. J. Cramer, Stuttgart, Germany.

Post W.M., Emanuel W.R., Zinke P.J. and Stangenberger A.G. 1982. Soil carbon pools and world life zones. *Nature, Lond.* **298**: 156-159.

Powlson D.S., Smith P. and Smith J.U. (eds.). 1996. *Evaluation of Soil Organic Matter Models*. NATO ASI Series. Series I: Global Environment Change, Vol. **38**. Springer Verlag, Berlin.

Purrini K. 1982. *Soil Invertebrates Infected by Microorganisms. New Trends in Soil Biology*. Proceedings of the VIII International Colloquium of Soil Zoology, Louvain-la-Neuve, Belgium. pp. 167-178.

Rabenhorst M.C., Wilding L.P. and Girdner C.L. 1984. Airborne dusts in the Edwards Plateau region of Texas. *Soil Sci. Soc. Am. J.* **48**: 621-627.

Ramade F. 1992. *Écotoxicologie*. Masson, Paris.

Ramade F. 1993. *Dictionnaire encyclopédique de l'écologie et des sciences de l'environnement*. Édiscience, Paris.

Ramade F. 1995. *Éléments d'écologie: écologie appliquée.* Édiscience, Paris.

Rayner A.D.M. and Boddy L. 1988. *Fungal Decomposition of Wood*. Wiley and Sons, Chichester, U.K.

Read D.J. 1990. Ecological integration by mycorrhizal fungi. *Endocytobiology,* **4**: 99-107.

Read D.J. 1991. Mycorrhiza in ecosystems. *Experientia,* **47**: 376-391.

Redecker D. 2000. Specific PCR primers to identify arbuscular mycorrhizal fungi (Glomales) within colonized roots. *Mycorrhiza,* **10**: 73-80.

Reeves H. 1990. *Malicorne. Réflexions d'un observateur de la nature*. Seuil, Paris.

Rehfuss K. 1981. *Waldböden. Entwicklung, Eigenschaften und Nutzung*. Verlag Paul Parey, Hamburg and Berlin.

Reid I.D. 1995. Biodegradation of lignin. *Can. J. Bot.* **73**, suppl. 1: 1011-1018.

Resh V.H. and Cordé R.T. (eds). 2003. *Encyclopedia of insects*. Academic Press and Elsevier Science (USA). San Diego and London.

Reutimann P. 1987. Quantitative aspects of the feeding activity of some oribatid mites (Oribata, Acari) in an alpine meadow ecosystem. *Pedobiologia,* **20**: 425-433.

Reynolds B. and Fenner N. 2001. Export from organic carbon from peat soils. *Nature. Lond.* **412**: 758.

Ribéreau-Gayon P. 1968. *Les composés phénoliques des végétaux*. Dunod, Paris.

Richard J.-L. 1961. Les forêts acidophiles du Jura. *Mat. Levé géobot. Suisse*, **38**. H. Huber, Berne.

Ricou G. 1967. Recherche sur les populations de tipules. Action de certains facteurs écologiques sur *Tipulosa paludosa* Meig. *Ann. Epiphyties*, **18**(4): 451-481.

Ridley M. 1997. *Évolution biologique*. De Boeck Université, Paris-Brussels.

Rieu M. and Sposito G. 1991. Fractal fragmentation, soil porosity and soil water properties, I and II. *Soil Sci. Soc. Am. J.* **55**: 1231-1244.

Ringrose-Voase A.J. and Humphreys G.S. 1994. *Soil Micromorphology: Studies in Management and Genesis*. Elsevier, Amsterdam.

Rivière J.-L. 1998. *Évaluation du risque écologique des sols pollués*. Coll. Tec & Doc, Lavoisier, Paris.

Romell L.G. and Heiberg S.O. 1931. Types of humus layer in the forests of northeastern United States. *Ecology*, **12**: 567-608.

Rorison I.H. (ed.). 1969. *Ecological Aspects of the Mineral Nutrition of Plants*. Blackwell Scientific Publications, Oxford, U.K.

Rorison I.H. 1987. Mineral nutrition in time and space. *New Phytol.* **106**: 79-92.

Roser-Hosch S. and Streit B. 1982. Besiedlungsdichten von Microarthropoden im Verlaufe der Kompostierung von Rindermist. *Schweiz. Landw. Forsch.* **21:** 49-65.

Rosswall T. and Heal O.W. 1975. Structure and function of tundra ecosystems. *Ecol. Bull.* **20**: 295-320.

Ruellan A., Dosso M. and Fritsch E. 1989. L'analyse structurale de la couverture pédologique. *Science du Sol*, **27**(4): 319-334.

Runge E.C.A. 1973. Soil development sequences and energy models. *Soil Sci.* **115**: 183-193.

Saddler N. 1986. Factors limiting the efficiency of cellulase enzymes. *Microbiol. Sci.* **3**: 84-87.

Sala O.E., Jackson R.B., Mooney H.A. and Howarth, R.W. (eds.). 2000. *Methods in Ecosystem Science*. Springer Verlag, New York.

Sanders I.R., Clapp J.P. and Wiemken A. 1996. The genetic diversity of arbuscular mycorrhizal fungi in natural ecosystems—a key to understanding the ecology and functioning of the mycorrhizal symbiosis. *New Phytol.* **133**: 123-134.

Schink B., Ward J.C. and Zeikus J.G. 1981. Microbiology of wetwood: role of anaerobic bacteria populations in living trees. *J. gen. Microbiol.* **123**: 313-323.

Schinner F. and Sonnleitner R. 1996. *Bodenökologie: Mikrobiologie und Bodensystematik. Band I. Grundlagen, Klima, Vegetation und Bodentyp*. Springer Verlag, Berlin.

Schinner F. Oehlinger R., Kandeler E. and Margesin R. 1996. *Methods in Soil Biology*. Springer Verlag, Berlin.

Schlegel H.G. and Kröger E. 1965. *Anreicherungskultur und Mutantenauslese*. Fischer Verlag, Stuttgart, Germany.

Schlichting E. Blume H.P. and Stahr K. 1995. *Bodenkundliches Praktikum*. Pareys Studientexte 81. Blackwell, Berlin and Vienna.

Schmeidl H. 1978. Ein Beitrag zum Mikroklima der Hochmoore. *Telma*, **8**: 83-105.

Schneebeli M. 1989. Zusammenhänge zwischen Moorwachstum und hydraulischer Durchlässigkeit und ihre Anwendung auf den Regenerationsprozess. *Telma*, Beiheft **2**: 257-264.

Schneider D.C. 1994. *Quantitative Ecology. Spatial and Temporal Scaling*. Academic Press, San Diego CA, U.S.A.

Schnitzer M. 1967. Humic-fulvic acid relationships in organic soils and humification of the organic matter in these soils. *Can. J. Soil Sci.* **47**: 245-250.

Schönborn W. 1982. Estimation of annual production of *Testacea* (*Protozoa*) in mull and moder (II). *Pedobiologia*, **23**: 383-393.

Schönborn W. 1986. Population dynamics and production biology of testate amoebae (*Rhizopodea, Testacea*) in raw humus of two coniferous forest soils. *Arch. Protistenkd.* **132**: 325-342.

Schroeder D. 1978. *Bodenkunde in Stichtworten*. Verlag Ferdinand Hirt, Kiel, Germany.

Schroth M.N. and Hancock J.G. 1982. Disease-suppressive soil and root-colonizing bacteria. *Science, N.Y.* **216**: 1376-1381.

Schubert R. 1991. *Bioindikation in terrestrischen Oekosystemen*. G. Fischer Verlag, Jena.

Schuler S. and Conrad R. 1990. Soils contain two different activities for oxidation of hydrogen. *FEMS Microbial. Ecol.* **73**: 77-84.

Schulten H.R. and Schnitzer M. 1997. Chemical model structures for soil organic matter and soils. *Soil Sci.* **162**: 115-130.

Schwartz D. 1988. Some podzols on Bateke Sands and their origins, People's republic of Congo. *Geoderma*, **43**: 229-247.

Schwarz E. 1988. *La révolution des systèmes*. DelVal, Cousset, Fribourg, Switzerland.

Schwarz E. 1997. *Cours de systémique*. University of Neuchâtel, Neuchâtel, Switzerland.

Schwintzer C.R. and Tjepkema J.D. (eds.). 1990. *The Biology of* Frankia *and actinorhizal plants*. Academic Press, New York.

Seeham G., Leise W. and Kess B. 1975. *Lists of Fungi in Soft-rot Tests*. International Research Group, Wood Preservation, Princes Risborough, U.K. Doc. IRG/WP/105.

Selosse M.-A. and Le Tacon F. 1999. Les champignons dopent la forêt. *La Recherche*, **319**: 33-35.

Shotyk W., Weiss D., Appleby P.G., Cheburkin A.K., Frei R., Gloor M., Kramers J.D., Reese S. and Van Der Knaap W.O. 1998. History of atmospheric lead deposition since 12,370 ^{14}C yr BP from a peat bog, Jura Mountains, Switzerland. *Science, NY*, **281**: 1635-1640.

Siegel D.I. and Glaser P.H. 1987. Groundwater flow in a bog-fen complex, Lost River Peatland, Northern Minnesota. *J. Ecol.* **75**: 743-754.

Singer M.J. and Munns D.N. 1996. *Soils: an Introduction*. Prentice-Hall Inc., Upper Saddle River NJ, U.S.A.

Singh V.P. (ed.). 1995. *Biotransformations: Microbial Degradation of Health Risk Compounds*. Elsevier Science Publishers, Amsterdam.

Smit A.L., Bengough A.G., Engels C., van Noordwijk M., Pellerin S. and van de Geijn S.C. (eds.). 2000. *Root Methods. A Handbook*. Springer Verlag, Berlin.

Smith F.A. and Smith S.E. 1997. Tansley Review No. 96: Structural diversity in (vesicular)-arbuscular mycorrhizal symbioses. *New Phytol.* **137**: 373-388.

Smith K.G.V. 1986. *A Manual of Forensic Entomology*. Trustees British Museum (Natural History), London.

Smith K.G.V. 1989. *An Introduction to the Immature Stages of British Flies*. Handbk Ident. Br. Insects, Royal Entomological Society of London. Dorset Press, Dorset, U.K.

Smith K. and Bogner J. 1997. Measurement and modelling of methane fluxes from landfills. *IGBP Newsletter*, **31**: 13-15.

Smith M.L., Bruhn J.N. and Anderson J.B. 1992. The fungus *Armillaria bulbosa* is among the largest and oldest living organisms. *Nature, Lond.* **356**: 428-431.

Smith S.E. and Read D.J. 1997. *Mycorrhizal Symbiosis*. Academic Press, San Diego CA, U.S.A.

Société suisse de pédologie. 1997. *Définition du sol à l'usage du grand public*. Internal document.

Soltner D. 1995. *Les bases de la production végétale. Tome II: le climat*. Coll. Sciences et Techniques agricoles, Ste-Gemmes-sur-Loire, France.

Soltner D. 1996a. *Les bases de la production végétale. Tome I: le sol*. Coll. Sciences et Techniques agricoles, Ste-Gemmes-sur-Loire, France.

Soltner D. 1996b. *Les bases de la production végétale. Tome III: la plante et son amélioration*. Coll. Sciences et Techniques agricoles, Ste-Gemmes-sur-Loire, France.

Southwood T.R.E. 1978. *Ecological Methods*. Chapman and Hall, London.

Spaltenstein H. 1984. *Pédogenèses sur calcaire dur dans les Hautes Alpes calcaires*. Thesis No. 540. EPFL, Lausanne, Switzerland.

Sparks D.L. 1995. *Environmental Soil Chemistry*. Academic Press, San Diego CA, U.S.A.

Speight M.G.D. 1989. *Les invertébrés saproxyliques et leur protection*. European Council, Strasbourg, France.

Speir T.W. and Cowling J.C. 1991. Phosphatase activities of pasture plants and soils: relationship with plant productivity and soil P fertility indices. *Biol. Fertil. Soils*, **12**: 189-194.

SSSA. 1994. *Methods of Soil Analysis. Part 2. Microbiological and Biochemical Properties*. Book Series **5**. Soil Science Society of America, Madison WI, U.S.A.

Stacey G., Burris R.H. and Evans H.J. (eds.). 1992. *Biological Nitrogen Fixation*. Chapman and Hall, New York.

Stal L.J. and Caumette P. (eds.). *Microbial Mats, Structure, Development and Environmental Significance*. Springer Verlag, Berlin.

Stevenson F.J. 1976. Organic matter reactions involving pesticides in soil. In: *Bound and Conjugated Pesticide Residues. Am. chem. Soc. Symp. Series*, **29**.

Stevenson F.J. 1982. *Humus Chemistry: Genesis, Composition, Reactions*. Wiley-Interscience, New York.

Stotzky G. (ed.). 1997. *Soil Biochemistry*. Marcel Dekker Inc., New York.

Stotzky G. and Bollag J.-M. (eds.). 1992. *Soil Biochemistry*, Vol. **7**. Marcel Dekker Inc., New York.

Streefkerk J.G. and Casparie W.A. 1989. *The Hydrology of Bog Ecosystems. Guidelines for Management*. Staatsbosbeheer, Dutch Nat. Forestry Serv., Utrecht, Netherlands.

Strehler C. 1997. *Création et évolution de sols artificiels à base de calcaires et de composts de déchets urbains*. Thesis, University of Neuchâtel, Neuchâtel, Switzerland.

Streit B. and Roser-Hosch S. 1982. Experimental compost cylinder as insular habitats: Colonisation by Microarthropods groups. *Revue suisse Zool.* **89:** 891-902.

Stubbs A. and Chandler P. 1978. *A Dipterist's Handbook*. Amateur Entomologists' Society, Hanworth, U.K.

Succow M. 1988. *Landschaftsökologische Moorkunde*. Gebruder Borntraeger, Berlin.

Sugihara G. and May R.M. 1990. Applications of fractals in ecology. *Trends in Evolution and Ecology*, **5**: 79-86.

Sukachev V.N. 1954. Quelques problèmes théoriques de la phytoécologie. *Essais de botanique*, Acad. Sci. USSR, Moscow-Leningrad, **1**: 310-330.

Summerhayes V.S. and Elton C.S. 1923. Contribution to the ecology of Spitzbergen and Bear Island. *J. Ecol.* **11**: 214-286.

Sumner M.E. (ed.). 2000. *Handbook of Soil Science*. CRC Press, Boca Raton FL, U.S.A.

Suzuki M., Kamekawa K., Sekiya S. and Shiga H. 1990. Effect of continuous application of organic and inorganic fertilizer for sixty years on soil fertility and rice yield in paddy field. In: *Trans. 14th Int. Cong. Soil Sci.* **4**.

Swift M.J., Heal O.W. and Anderson J.M. 1979. *Decomposition in Terrestrial Ecosystems*. Blackwell Scientific Publications, Oxford, U.K.

Tajovsky K., Santruckova H., Hanel L., Balik, V. and Lukesova A. 1992. Decomposition of faecal pellets of the millipede *Glomeris hexasticha* (*Diplopoda*) in forest soil. *Pedobiologia*, **36**: 146-158.

Tansley A.G. 1935. The use and abuse of vegetational concepts and terms. *Ecology*, **16**: 284-307.

Tate R.L. 1987. *Soil Organic Matter. Biological and Ecological Effects*. Wiley-Interscience, New York.

Tester C.F. 1990. Organic amendment effects on physical and chemical properties of a sandy soil. *Soil Sci. Soc. Am. J.* **54**: 827-831.

Thies J.E., Holmes E.M. and Vachot A. 2001. Application of molecular techniques to studies in *Rhizobium* ecology. A review. *Aust. J. Exptl Agric.* **41**: 299-319.

Thorn R.G. and Barron G.L. 1984. Carnivorous mushrooms. *Science, N.Y.* **224**: 76-78.

Thorn R.G. and Barron G.L. 1986. *Nematoctonus* and the tribe *Resupinatae* in Ontario, Canada. *Mycotaxon*, **25**: 231-453.

Thurmann J. 1849. *Essai de phytostatistique appliquée à la chaîne du Jura et aux contrées voisines.* 2 vols. Jent and Gassmann, Berne.

Tjepkema J.D., Schwintzer, R. and Benson D.R. 1986. Physiology of actinorhizal nodules. *Ann Rev. Plant Physiol.* **37**: 209-232.

Toepffer R. 1846. *Voyages en zigzag*. Dubochet, Le Chevalier et compagnie, Paris (republished 1996 by Slatkine, Geneva).

Topp W. 1971. Zur Ökologie der Mülhalden. *Ann. Zool. Fennici* **8:** 194-222.

Toutain F. 1974. *Étude écologique de l'humification dans les hêtraies acidiphiles*. Doctoral thesis, University of Nancy I, Nancy, France.

Toutain F. 1981. Les humus forestiers: structures et modes de fonctionnement. *Revue forestière française*, **XXXIII**(6): 449-477.

Toutain F. 1987. *Les humus forestiers: biodynamique et mode de fonctionnement*. Centre de documentation pédagogique, Rennes, France.

Travé J., André H.M., Taberly G. and Bermini F. 1996. *Les Acariens Oribates. Études en Acarologie* **1.** Agar and SIALF, Wavre, Belgium.

Trochain J.-L. 1980. *Écologie végétale de la zone intertropicale non désertique.* Université Paul Sabatier, Toulouse, France.

Tuittila E.-S., Rita H., Vasander H., Nykänen H., Martikainen P.J. and Laine J. 2000. Methane dynamics of a restored cut-away peatland. *Global Change Biol.* **6**: 569-581.

Tunlid, A., Baird, B.H., Trexler, M.B., Olsson, S., Findlay, R.H., Odham, G. and White D.C. 1985. Determination of phospholipid ester-linked fatty acids and poly-beta-hydroxybutyrate for the estimation of bacterial biomass and activity in the rhizosphere of the rape plant *Brassica napus*. *Can. J. Microbiol.* **31**: 1113-1119.

Tuor U., Winterhalter K. and Fiechter A. 1995. Enzymes of white-rot fungi involved in lignin degradation and ecological determinants for wood decay. *J. Biotechnol.* **41**: 1-17.

Turner M.G. and Gardner R.H. (eds.). 1991. *Quantitative Methods in Landscape Ecology*. Ecological Studies 82. Springer Verlag, New York.

Tutin T.G., Heywood V.H., Burges N.A., Valentine D.H., Walters S.M. and Webb D.A. 1964*ff*. *Flora Europaea.* Cambridge University Press, Cambridge.

UNESCO. 1969. Biologie des sols. *Recherches sur les ressources naturelles,* **IX**. UNESCO, Paris.

USDA. 1999. *Soil Taxonomy: A Basic System of Soil Classification for Making and Interpreting Soil Surveys*. 2nd edition. USDA, Natural Resources Conservation Service, Washington DC.

Vadi G. and Gobat J.-M. 1998a. Un paradoxe apparent: les PODZOSOLS sur calcaire dans le Jura. *Communication* 1247, *World Cong. Soil Sci.*, Montpellier, France.

Vadi G. and Gobat J.-M. 1998b. Le paradoxe de la podzolisation en domaine jurassien. Aspects pédologiques et phytosociologiques. *Bull. Soc. neuchâtel. Sci. nat.* **121:** 79-91.

Vadi G., Le Bayon C. and Gobat J.-M. 2002. Influence of earthworms (Oligochaetes, Lumbricidae) on the soil structuring in recent fluviatile deposition. Poster presentation, 7th International Symposium Earthworm Ecology, Cardiff, Wales, U.K., 1-6 September 2002.

Valloton R. 1983. La lutte biologique contre les nématodes phytoparasites. *Revue suisse Agric.* **5**: 245-316.

Van Breemen N. (ed.). 1998. *Plant-induced Soil Changes: Processes and Feedbacks.* Kluwer Academic Publishers, Dodrecht, Netherlands (Reprinted from *Biogeochemistry,* **42**).

Van Breemen N. and Buurman P. 1998. *Soil Formation*. Kluwer Academic Publishers, Dordrecht, Netherlands.

Van Cleve K., Oechel W.D. and Hom J.L. 1990. Response of black spruce *(Picea mariana)* ecosystems to soil temperature modifications in interior Alaska. *Can. J. For. Res.* **20**: 291-302.

Van der Drift J. 1965. The effects of animal activity in the litter layer. In: *Experimental Pedology. Proceedings of the 11th Easter School in Agricultural Sciences*. University of Nottingham. pp. 227-235.

Van der Drift J. and Witkamp M. 1960. The significance of the breakdown of oak-litter by *Enoicyla pusilla. Burm. Arch. Neerl. Zool.* **XIII**: 486-492.

Van Rhijn P. and Vanderleyden J. 1995. The *Rhizobium*-plant symbiosis. *Microbiol. Rev.* **59**: 124-142.

Van Straalen N.M. and Ernst W.H.O. 1991. Metal biomagnification may endanger species in critical pathways. *Oikos,* **62**: 255-256.

Van Straalen N. M., Butoyvsky R.O., Pozarzhevskii A.D., Zaitsev A.S. and Verhoef S.C. 2001. Metal concentration in soil and invertebrates in the vicinity of a metallurgical factory near Tula (Russia). *Pedobiologia,* **45**: 451-466.

Vance A.E.D., Brookes P.C. and Jenkinson D.S. 1987. Microbial biomass measurement in forest soils: the use of the chloroform fumigation-incubation method in strongly acid soils. *Soil Biol. Biochem.* **19**: 697-702.

Vandel A. 1960, 1962. Isopodes terrestres I and II. *Faune de France* **64** and **66**. Library of the Faculty of Sciences, Paris.

Vanden Berghen C. 1982. *Initiation à l'étude de la végétation*. National Botanical Garden of Belgium, Meise, Belgium.

Vannier G. 1970. *Réactions des Microarthropodes aux variations hydriques du sol.* Éditions CNRS, Paris.

Vasander H. and Starr M. 1992. Carbon cycling in boreal peatlands and climate change. *SUO, Mires and Peat,* **43**: 4-5.

Vaucher-von Balmoos C. 1997. *Étude entomologique de six zones de transition entre tourbières acides et zones agricoles dans le Haut-Jura suisse.* Thesis, University of Neuchâtel, Neuchâtel, Switzerland.

Vaughan D. and Malcolm R. (eds.). 1985. *Soil Organic Matter and Biological Activity.* Nijhoff/Junk, Dordrecht, Netherlands.

Velde, B. (ed.). 1995. *Origin and Mineralogy of Clays*. Springer Verlag, New York.

Vernadsky W. 1929. *La biosphère.* Alcan, Paris.

Viaud M., Pasquier A. and Brygoo Y. 2000. Diversity of soil fungi studied by PCR-RFLP of ITS. *Mycol. Res.* **104**: 1027-1032.

Visser S. 1995. Ectomycorrhizal fungal association in jack pine stands following wildfire. *New Phytol.* **129**: 389-401.

Vittoz P., Theurillat J.-P., Zimmermann K. and Gallandat J.-D. (eds.). 1995. Contribution à la flore et à la végétation des Alpes. *Diss. Botan.* **258** (J.-L. Richard jubilee volume). J. Cramer, Stuttgart, Germany.

Voroney R.P. and Paul E.A. 1984. Determination of Kc and Kn *in situ* for calibration of the chloroform fumigation-incubation method. *Soil Biol. Biochem.* **16**: 9-14.

Wallén B. 1986. Above and below ground dry mass of the three main vascular plants on hummocks on a subarctic peat bog. *Oikos,* **46**: 51-56.

Wallwork J.A. 1970. *Ecology of Soil Animals*. McGraw-Hill, London.

Wallwork J.A. 1976. *The Distribution and Diversity of Soil Fauna.* Academic Press, London.

Wardle D.A. 1995. Impacts of disturbance on detritus food webs in agro-ecosystems of contrasting tillage and weed management practices. *Adv. Ecol. Res.* **26**: 105-185.

Warner B.G. 1990. Testate amoebae. (Protozoa: Rhizopoda). In: *Methods in Quaternary Ecology* (Geoscience Canada Reprint Series **5**). pp. 65-74.

Waterhouse D.F. 1974. The biological control of dung. *Scientific American*, April 1974: 101-109.

Weber J.-M. 1990. La fin de la loutre en Suisse. *Cahier de l'environnement,* **128**. Office fédéral de l'environnement, des forêts et du paysage, Berne.

Werner D. 1997. Die Dipterenfauna verschiedener Mülldeponien und Kompostierungsanlagen in der Umgebung von Berlin unter besonderer Besichtigung ihrer Ökologie und Bionomie. *Studia Dipterologica.* Supplement **1:** 1-176.

Wheatley A. (ed.). 1990. *Anaerobic Digestion: a Waste Treatment Technology*. Elsevier, London and New York.

Wheeler B.D., Shaw S.C., Fojt W.J. and Robertson R.A. (eds.). 1995. *Restoration of Temperate Wetlands.* Wiley and Sons, Chichester, U.K.

Whittaker R.H. 1975. *Communities and Ecosystems.* Macmillan, New York.

Widmer F. and Beffa R. 2000. *Aide-mémoire de biochimie et de biologie moléculaire.* 2nd edition. Coll. Tec & Doc, Lavoisier, Paris; EM Inter, Cachan, France.

Wildi O. and Orloci L. 1996. *Numerical Exploration of Community Patterns.* SPB Academic Publishers, The Hague.

Williams R.T. and Crawford R.L. 1983. Effects of various physicochemical factors on microbial activity in peatlands: aerobic biodegradative processes. *Can. J. Microbiol.* **29**: 1430-1437.

Wilson E.O. 1992. *The Diversity of Life.* Harvard University Press, Cambridge MA, U.S.A.

Wilson W.S. (ed.). 1991. *Advances in Soil Organic Matter Research: the Impact on Agriculture and the Environment.* The Royal Society of Chemistry, Cambridge.

Winogradsky S.N. 1949. *Microbiologie du sol. Problèmes et méthodes. Cinquante ans de recherches.* Masson, Paris.

Wise D. (ed.). 1987. *Bioenvironmental Systems.* Vol. III. CRC Press, Boca Raton FL, U.S.A.

Wood M. 1995. *Environmental Soil Biology.* Blackie Academic and Professional, London.

Woodin S.J. and Lee J.A. 1987. The effects of nitrate, ammonium and temperature on nitrate reductase activity in *Sphagnum* species. *New Phytol.* **105**: 103-115.

Wüthrich M. and Matthey W. 1978. Les Diatomées de la tourbière du Cachot (Jura suisse). II. Associations et distribution des espèces caractéristiques. *Rev. Suisse Hydrol.* **40**: 87-103.

Young J.M., Kuykendall L.D., Martínez-Romero E., Kerr A. and Sawada H. 2001. A revision of *Rhizobium* Frank 1889, with an emended description of the genus, and the inclusion of all species of *Agrobacterium* Conn 1942 and *Allorhizobium undicola* de Lajudie *et al.* 1998 as new combinations: *Rhizobium radiobacter, R. rhizogenes, R. rubi, R. undicola* and *R. vitis. Int. J. Syst. Evol. Microbiol.* **51**: 89-103.

Zachariae G. 1979. *Spuren tierischer Tätigkeit im Boden des Buchenwaldes.* Verlag Paul Varey, Hamburg and Berlin.

Zaitsev A.S. and van Straalen N.M. 2001. Species diversity and metal accumulation in oribatid mites *(Acari, Oribata)* of forests affected by a metallurgic plant. *Pedobiologia,* **45**: 467-479.

Zanella A., Tomasi M., De Siena C., Frizzera L., Jabiol B. and Nicolini G. 2001. *Humus Forestali. Manuale di Ecologia per il riconoscimento e l'Interpretazione. Applicazione alle Faggete*. Edizioni Centro di Ecologia Alpina, Turin, Italy.

Zantua M.I. and Bremner J.M. 1977. Stability of urease in soils. *Soil Biol. Biochem.* **9**: 135-140.

Zeikus J.G. and Ward J.C. 1974. Methane formation in living trees: a microbial origin. *Science, N.Y.* **184**: 1181-1183.

Zettel J. 1999. *Blick in die Unterwelt. Ein illustrierter Bestimmungsschlüssel zur Bodenfauna*. Verlag Agrarökologie, Berne-Hannover.

Zettel J. and von Allmen H. 1982. Jahresverlauf der Kälterresistenz zweier Collembolen-Arten in den Berner Voralpen. *Rev. suisse Zool.* **89**: 927-939.

UNITS OF MEASURE

Physical, chemical and biological units used in the book
(in particular from CRM, CRP, CRC, 1992)

Domain	Abbr. or symbol	Unit (SI or old)
'Acidity' (log $1/[H_3O^+]$)	pH	—
Absorbance (see optical density)	—	—
Acceleration of gravity	*G*	$G = 9.81$ m s^{-2}
Active lime	—	% (of dry soil)
Albedo	a	% (of R_{si})
Base saturation percentage	S/T, V	%
Biomass	—	kg m^{-2}, t ha^{-1}
Bulk density	dA (= ρA)	g cm^{-3}
Cation exchange capacity	CEC, T	cmol (+) kg^{-1} (dry soil) meq/100 g (dry soil) [1 cmol (+) kg^{-1} = 1 meq/100 g]
Diffusion coefficient (of oxygen)	D	cm^2 s^{-1}
Electrical conductivity	γ	dS m^{-1}; µS cm^{-1}
Energy	*W* or *E*	J (joule) = newton metre = kg m^2 s^{-2}
Exchange acidity	EA	cmol (+) kg^{-1} (dry soil) meq/100 g (dry soil) [1 cmol (+) kg^{-1} = 1 meq/100 g]
Extractability (of organic matter)	EXT	% (of total organic C)
Force	*F*	N (newton) = kg m s^{-2}
Gram-equivalent (milli-)	eq (meq)	Atomic mass in g (mg)/ionic valence
Heat , thermal energy	*W* or *E*	J (joule) = newton metre = kg m^2 s^{-2}; cal (calorie) [1 J = 0.24 cal; 1 cal = 4.19 J]
Heat capacity, specific heat	*c*	J kg^{-1} K, cal g^{-1} K (K = degree Kelvin)
Hydraulic conductivity = permeability	K	cm h^{-1}
Molarity (concentration of a solute)	M	gram mole L^{-1}
Molecular or atomic mass	—	Da (dalton)

(*Contd.*)

(*Contd.*)

Domain	Abbr. or symbol	Unit (SI or old)
Optical density (= absorbance)	O.D.	—
Particle density	d	—
Porosity	—	% (of total volume)
Potential, gravitational	ψ_z	MPa
Potential, hydraulic (= pressure)	ψ_P	MPa
Potential, matric (= capillary potential)	ψ_m	MPa
Potential, matric (expressed in log ψ_m)	pF	—
Potential, osmotic	ψ_s	MPa
Potential, total hydric	ψ_t	MPa
Power (energy/time)	*P*	kg m^2 s^{-3}, J s^{-1} (watt = W) [1 W = 0.0014 hp] hp (horse-power) [1 hp = 736 W]
Pressure (= force/area)	*p*	Pa (pascal) = N m^{-2} = kg m^{-3} s^{-2}; atm (atmosphere) = kg cm^{-2} [1 atm = $1.013 \cdot 10^5$ Pa = 1.013 bar]; bar [1 bar = 10^5 Pa = 0.98 atm]; mm Hg [1 mm Hg = 133.3 Pa]
Primary production, gross	GP_1, GPP	g m^{-2} y^{-1}, t ha^{-1} y^{-1}, kcal m^{-2} y^{-1}
Primary production, net	NP_1, NPP	g m^{-2} y^{-1}, t ha^{-1} y^{-1}, kcal m^{-2} y^{-1}
Quantity of information	—	bit, shannon
Radiation (solar irradiance, net radiation, thermal radiation)	R_{si}, R_n, R_t	kW m^{-2}, cal cm^{-2} min^{-1}
Redox potential	Eh	mV (millivolt)
Rhizosphere coefficient	R/S	—
Solar constant	I_o	KW m^{-2}; cal cm^{-2} min^{-1}
Specific gravity (=density)	ρ	g cm^{-3}, kg m^{-3}
Temperature	θ	°C (degree Celsius); K (degree Kelvin) [1°C = 1 K]
Total exchangeable bases (basic cations)	S	cmol (+) kg^{-1} (dry soil) meq/100 g (dry soil) [1 cmol (+) kg^{-1} = 1 meq/100 g]
Viscosity	η	daP (decapoise); kg m^{-1} s^{-1} (pascal second = Pa s) [1 Pa s= 1 daP]
Water content	θ	% (of dry soil)

INDEX

This index contains all the terms defined in the book (the page where each is defined is marked **bold**) as well as others of general interest, even if they have not been defined here. Some very general words often used (e.g. organic matter, mineral constituents) do not appear in the Index; the reader is requested to refer to the Table of Contents. The pages marked '*ff*' (e.g. 105*ff*) indicate that the term also appears on the following page or a few consecutive pages.

Terms separated by a slash (/) are considered synonyms. Between parentheses they indicate the domain or context of use. All occurrences of pedological References (e.g. CALCOSOLS, FLUVIOSOLS) are indexed, with the exception of references to the abstracted Tables 5.2, 5.3 and 5.4 in Chapter 5. Only the higher categories of soil organisms (e.g. Cyanobacteria, Basidiomycetes, Nematoda, Collembola, Formicidae) are listed; specific genera or species are not referred to. In this case, the pages identified in **bold** point to the general description of the group.

About the Authors

Jean-Michel Gobat is since 1987 Professor of General Ecology and. Biological Soil Science at the University of Neuchâtel (Switzerland), where he heads the Department of Plant Ecology and Phytosociology. After a Ph.D. in ecology of the boundary between acid peatlands and alkaline marshes, he completed a course in numerical ecology at CNRS, Montpellier, and one in biological soil science at CNRS, Vandoeuvre-les-Nancy, France. This enabled him direct his researches on understanding the relationships between vegetation and soils, with special emphasis on the evolution of organic matter. The recent publications from his laboratory, which integrate approaches based on phytosociology, soil biology and dynamic modelling of ecosystems, primarily pertain to alluvial zones, marshes and peatlands, and wooded pastures. J.-M. Gobat was president of the Swiss Society of Soil Science from 1996 to 1998.

Michel Aragno is Professor of General Microbiology and Mycology since 1978 at the University of Neuchâtel, where he heads the Laboratory of Microbiology. After a Ph.D. in the physiology of a parasitic fungus of vineyards and courses in soil microbiology at the Pasteur Institute, Paris, and bacterial physiology at the University of Göttingen (Lower Saxony, Germany), he focused his researches on bacterial ecology and physiology, especially of chemolithoautotrophic bacteria and nitrogen-fixing bacteria. The milieus studied are waters, soils and geothermal ecosystems. His applied researches are concerned with optimization of biological processes of organic-waste management. In the area of soils, his laboratory is studying the biodiversity of the rhizosphere and associative nitrogen fixation. Applied research has concentrated on the biological process of composting, microbial phenomena associated with garbage-dump management and the biomethanization of wastes.

Willy Matthey, after study at Neuchâtel (Ph.D. in ecology of insects of peatlands, Oxford and Calgary, was appointed Professor responsible for teaching animal ecology, entomology and soil zoology at the University of Neuchâtel. He headed the Laboratory of Animal Ecology and Entomology from its creation in 1972 till 1994. His researches and those of his collaborators have been concerned primarily with the ecology of arthropods of peatlands and of soil, with a constant desire to base all ecological interpretation on solid systematic knowledge. Willy Matthey is also interested in problems related to protection of nature—he presided over the Scientific Commission of the Swiss National Park, in Graubünden—and in education in ecology.